T0315153

Computer Aided Design and Manufacturing

Wiley-ASME Press Series

Corrosion and Materials in Hydrocarbon Production: A Compendium of Operational and Engineering Aspects
Bijan Kermani and Don Harrop

Design and Analysis of Centrifugal Compressors
Rene Van den Braembussche

Case Studies in Fluid Mechanics with Sensitivities to Governing Variables
M. Kemal Atesmen

The Monte Carlo Ray-Trace Method in Radiation Heat Transfer and Applied Optics
J. Robert Mahan

Dynamics of Particles and Rigid Bodies: A Self-Learning Approach
Mohammed F. Daqaq

Primer on Engineering Standards, Expanded Textbook Edition
Maan H. Jawad and Owen R. Greulich

Engineering Optimization: Applications, Methods and Analysis
R. Russell Rhinehart

Compact Heat Exchangers: Analysis, Design and Optimization using FEM and CFD Approach
C. Ranganayakulu and Kankanhalli N. Seetharamu

Robust Adaptive Control for Fractional-Order Systems with Disturbance and Saturation
Mou Chen, Shuyi Shao, and Peng Shi

Robot Manipulator Redundancy Resolution
Yunong Zhang and Long Jin

Stress in ASME Pressure Vessels, Boilers, and Nuclear Components
Maan H. Jawad

Combined Cooling, Heating, and Power Systems: Modeling, Optimization, and Operation
Yang Shi, Mingxi Liu, and Fang Fang

Applications of Mathematical Heat Transfer and Fluid Flow Models in Engineering and Medicine
Abram S. Dorfman

Bioprocessing Piping and Equipment Design: A Companion Guide for the ASME BPE Standard
William M. (Bill) Huitt

Nonlinear Regression Modeling for Engineering Applications: Modeling, Model Validation, and Enabling Design of Experiments
R. Russell Rhinehart

Geothermal Heat Pump and Heat Engine Systems: Theory and Practice
Andrew D. Chiasson

Fundamentals of Mechanical Vibrations
Liang-Wu Cai

Introduction to Dynamics and Control in Mechanical Engineering Systems
Cho W.S. To

Computer Aided Design and Manufacturing

Zhuming Bi
Professor of Mechanical Engineering
Department of Civil and Mechanical Engineering
Purdue University Fort Wayne
Indiana
USA

Xiaoqin Wang
Associate Professor of Manufacturing Engineering
School of Mechanical Engineering
Nanjing University of Science and Technology
Nanjing
China

This Work is a co-publication between John Wiley & Sons Ltd and ASME Press

This edition first published 2020
© 2020 John Wiley & Sons Ltd

This Work is a co-publication between John Wiley & Sons Ltd and ASME Press

All rights reserved. No part of this publication may be reproduced, stored in a retrieval system, or transmitted, in any form or by any means, electronic, mechanical, photocopying, recording or otherwise, except as permitted by law. Advice on how to obtain permission to reuse material from this title is available at http://www.wiley.com/go/permissions.

The right of Zhuming Bi and Xiaoqin Wang to be identified as the authors of this work has been asserted in accordance with law.

Registered Offices
John Wiley & Sons, Inc., 111 River Street, Hoboken, NJ 07030, USA
John Wiley & Sons Ltd, The Atrium, Southern Gate, Chichester, West Sussex, PO19 8SQ, UK

Editorial Office
The Atrium, Southern Gate, Chichester, West Sussex, PO19 8SQ, UK

For details of our global editorial offices, customer services, and more information about Wiley products visit us at www.wiley.com.

Wiley also publishes its books in a variety of electronic formats and by print-on-demand. Some content that appears in standard print versions of this book may not be available in other formats.

Limit of Liability/Disclaimer of Warranty
MATLAB® is a trademark of The MathWorks, Inc. and is used with permission. The MathWorks does not warrant the accuracy of the text or exercises in this book. This work's use or discussion of MATLAB® software or related products does not constitute endorsement or sponsorship by The MathWorks of a particular pedagogical approach or particular use of the MATLAB® software.

In view of ongoing research, equipment modifications, changes in governmental regulations, and the constant flow of information relating to the use of experimental reagents, equipment, and devices, the reader is urged to review and evaluate the information provided in the package insert or instructions for each chemical, piece of equipment, reagent, or device for, among other things, any changes in the instructions or indication of usage and for added warnings and precautions. While the publisher and authors have used their best efforts in preparing this work, they make no representations or warranties with respect to the accuracy or completeness of the contents of this work and specifically disclaim all warranties, including without limitation any implied warranties of merchantability or fitness for a particular purpose. No warranty may be created or extended by sales representatives, written sales materials or promotional statements for this work. The fact that an organization, website, or product is referred to in this work as a citation and/or potential source of further information does not mean that the publisher and authors endorse the information or services the organization, website, or product may provide or recommendations it may make. This work is sold with the understanding that the publisher is not engaged in rendering professional services. The advice and strategies contained herein may not be suitable for your situation. You should consult with a specialist where appropriate. Further, readers should be aware that websites listed in this work may have changed or disappeared between when this work was written and when it is read. Neither the publisher nor authors shall be liable for any loss of profit or any other commercial damages, including but not limited to special, incidental, consequential, or other damages.

Library of Congress Cataloging-in-Publication Data

Names: Bi, Zhuming, author. | Wang, Xiaoqin (Writer on manufacturing), author.
Title: Computer aided design and manufacturing / Zhuming Bi, Xiaoqin Wang.
Description: First edition. | Hoboken, NJ : John Wiley & Sons, Inc., 2020. | Series: Wiley-ASME press series | Includes bibliographical references and index.
Identifiers: LCCN 2019047705 (print) | LCCN 2019047706 (ebook) | ISBN 9781119534211 (hardback) | ISBN 9781119534235 (adobe pdf) | ISBN 9781119534242 (epub)
Subjects: LCSH: Manufactures–Computer-aided design.
Classification: LCC TS155.6 .B53 2020 (print) | LCC TS155.6 (ebook) | DDC 658/.5–dc23
LC record available at https://lccn.loc.gov/2019047705
LC ebook record available at https://lccn.loc.gov/2019047706

Cover Design: Wiley
Cover Image: © Monty Rakusen/Getty Images

Set in 9.5/12.5pt STIXTwoText by SPi Global, Chennai, India

10 9 8 7 6 5 4 3 2 1

Contents

Series Preface

The Wiley-ASME Press Series in Mechanical Engineering brings together two established leaders in mechanical engineering publishing to deliver high-quality, peer-reviewed books covering topics of current interest to engineers and researchers worldwide.

The series publishes across the breadth of mechanical engineering, comprising research, design and development, and manufacturing. It includes monographs, references and course texts.

Prospective topics include emerging and advanced technologies in Engineering Design; Computer-Aided Design; Energy Conversion & Resources; Heat Transfer; Manufacturing & Processing; Systems & Devices; Renewable Energy; Robotics; and Biotechnology.

Preface

Manufacturing Engineering is to apply mathematics and science in practice to design, manufacture, and operate products or systems. *Manufacturing engineers* are responsible for designing, developing, and running manufacturing systems to produce competitive goods. In most universities' engineering curricula, the courses in a manufacturing program cover the fundamentals such as *mathematics, physics, computing engineering, management*, as well as disciplinary subjects such as *materials science, machine designs, solid mechanics, thermodynamics, fluid mechanics, manufacturing processes*, and *design optimization*. Currently, most engineering curricula cover as many sub-disciplines of mechanical and manufacturing engineering as possible and students take their options to be specialized in one or more sub-disciplines. Therefore, existing curricula are mostly discipline-oriented.

Designs of modern products and processes are very complex, and complex designs rely heavily on computer aided technologies. However, computing and information technologies have been developed in a way that is very different from the manufacturing engineering education. Numerous computer aided tools become commercially available; moreover, the number and capabilities of computer tools keep increasing continuously. These tools are usually application-oriented and they have been developed by integrating design theories in multiple disciplines, rather than the theories and methods in an individual discipline. Due to the strongly decoupling of multiple disciplines, the classification of sub-disciplines in traditional engineering curricula is not aligned with the varieties of computer aided tools.

The misalignment of the discipline-oriented curricula and a variety of functional computeraided tools cause the dilemma in manufacturing engineering education. *On the one hand*, the sub-disciplines in manufacturing engineering are highly diversified. Keeping an increasing number of elective courses in an engineering programme becomes a great challenge since the number of required credit hours for an engineering degree has been continuously reduced in public university systems. *On the other hand*, engineering universities are facing an increasing pressure to train students for the appropriate set of knowledge and skills of using advanced computeraided tools to solve complex design problems. There is an emerging need to re-design engineering curricula so that the CAD/CAM-related courses can be aligned with discipline-oriented curricula.

We deal with this misalignment by proposing a new course framework to integrate *Computer Aided Design* (CAD), *Computer Aided Manufacturing* (CAM), and other Information Technologies (ITs) in one course setting. In the framework, the inclusion of engineering disciplines and computer aided technologies are driven by the needs in designing products and manufacturing processes. While the theories in engineering disciplines are used to clarify design and analysis problems, computer aided tools are presented as the corresponding solutions. In addition, the framework is modularized and the course contents can be customized for some specialties.

The current version is customized to meet education needs of students in *Mechanical Engineering*, *Manufacturing Engineering*, and *Industrial Engineering* at junior or senior levels. It consists of three parts: Part I – *Computer Aided Design* (CAD), Part II – *Computer Aided Manufacturing* (CAM), and Part III – *System Integration*. Part I includes Chapter 2 – *Computer Aided Geometric Modelling*, Chapter 3 – *Knowledge-Based Engineering*, Chapter 4 – *Platform Technologies*, Chapter 5 – *Computer Aided Reverse Engineering*, and Chapter 6 – *Computer Aided Machine Design*. Part II includes Chapter 7 – *Group Technology and Cellular Manufacturing*, Chapter 8 – *Computer Aided Fixture Design*, Chapter 9 – *Computer Aided Manufacturing (CAM)*, Chapter 10 – *Simulation of Manufacturing Processes*, and Chapter 11 – *Computer Aided Design of Tools, Dies, and Moulds (TDMs)*. Part III includes Chapter 12 – *Digital Manufacturing (DM)*, Chapter 13 – *Direct and Additive Manufacturing*, and Chapter 14 – *Design for Sustainability (D4S)*.

The concepts of CAD, CAM, and system integration are not new; a number of textbooks on these subjects have been on the market for a long time. However, CAD, CAM, and system integration technologies have been greatly advanced recently due to the rapid development of IT and computer hardware and software systems. Most of the available textbooks in relevant subjects become obsolete in terms of design theories, methods, and technical coverages. Modern manufacturing systems in the digital era become ever more complicated, and product and system designs demand more advanced computer tools to deal with the complexity, varieties, and uncertainties. In contrast to other textbooks in similar areas, this book is featured (i) to update computeraided design theories and methods in modern manufacturing systems and (ii) to cover mostly advanced computeraided tools used in digital manufacturing. It will be an ideal textbook for undergraduate and graduate students in *Mechanical Engineering*, *Manufacturing Engineering*, and *Industrial Engineering* and can be used as a technical reference for researchers and engineers in mechanical and manufacturing engineering or computer aided technologies.

Zhuming Bi
Professor of Mechanical Engineering
Department of Civil and Mechanical Engineering
Purdue University Fort Wayne, Indiana, USA

Xiaoqin Wang
Associate Professor of Manufacturing Engineering
School of Mechanical Engineering
Nanjing University of Science and Technology, Nanjing, China

About the Companion Website

The companion website for this book is at

www.wiley.com/go/bi/computer-aided-design

The website includes:

- PDFs
- PPTs

Scan this QR code to visit the companion website.

1

Computers in Manufacturing

1.1 Introduction

1.1.1 Importance of Manufacturing

The life quality of human being relies on the availability of products and services from primary industry, secondary industry, and tertiary industry. According to the *three-sector theory* (Fisher 1939), *the primary industry* relates to the economic activities to extract and produce raw materials such as coal, wood, and iron. *The secondary industry* relates to the economic activities to transfer raw or intermediate materials into goods such as cars, computers, and textiles. *The tertiary industry* relates to the economic activities to provide services to customers and businesses. The secondary industry supports both the primary and tertiary industries, since the businesses in the secondary industry take the outputs of the primary industry and manufacture finished goods to meet customers' needs in the tertiary industry. In contrast to the wealth distribution or consumption in the tertiary industry, the secondary industry creates new wealth to human society (Kniivila 2018).

A manufacturing system can be very simple or extremely complex. Figure 1.1a shows an example of blacksmithing where some simple farming tools are made from iron (Source Weekly 2012). Figure 1.1b shows an example of a complex car assembly line, which is capable of making Ford Escape cars (Automobile Newsletter 2012). Despite the difference in complexity, both of them are good examples of a manufacturing system since *manufacturing* refers to the production of merchandise for use or sale using labour and machines, tools, chemical and biological processing, or formulation (Wikipedia 2019a). Manufacturing is one of fundamental constitutions of a nation's economy. Manufacturing businesses dominate the secondary industry. Powerful countries in the world are those who take control of the bulk of the global production of manufacturing technologies. Over the past hundreds of years, advancing manufacturing has been the strategic achievement of the developed counties to sustain their national wealth and global power. The importance of manufacturing to a nation has been discussed by numerous of researchers and organizations. For example, a summary of the importance to the USA economy is given by Flows (2016) and Gold (2016) as follows:

1. Manufacturers contributed $2.2 trillion with ~12% of gross domestic product (GDP) to the USA economy in 2015.

Computer Aided Design and Manufacturing, First Edition. Zhuming Bi and Xiaoqin Wang.
© 2020 John Wiley & Sons Ltd. This Work is a co-publication between John Wiley & Sons Ltd and ASME Press.
Companion website: www.wiley.com/go/bi/computer-aided-design

(a) (b)

Figure 1.1 A manufacturing system can be very simple or complex (a). Blacksmithing (Source weekly 2012), (b). Ford assembly line at Kansas City (Automobile Newsletter 2012).

2. The manufacturing multiplier effect is stronger than in other sectors. For $1.00 spent in manufacturing, $1.81 is added to other sectors of the economy. Manufacturing has the *highest multiplier effect*. Gold (2016) argued that the impact of manufacturing has been greatly underestimated; it is supported by the findings of the Manufacturers Alliance for Productivity and Innovation (MAPI) Foundation that the total impact of manufacturing on the economy should be 32% of GDP and that the full value stream of manufactured goods for final demand was equal to $6.7 trillion in 2016.
3. Manufacturing employs sizeable workforces. The manufacturing sector provides ~17.4 million jobs, or over 12.3 million.
4. Manufacturing pays premium compensation. Manufacturing workers earnt a high average of $81 289 annually in 2015.
5. Manufacturing dominates US exports; the United States is the No. 3 manufacturing exporter.
6. The US attracts more investment than other countries and foreign investment in US manufacturing grows; the foreign direct investment in manufacturing exceeded $1.2 trillion in 2015. New technologies allow manufacturers to alter radically the way they innovate, produce, and sell their products moving forward, improving efficiency and competitiveness.

1.1.2 Scale and Complexity of Manufacturing

From a system perspective, a manufacturing system can be described by the *inputs*, *outputs*, *system components*, and *their relations*, as shown in Figure 1.2. The system is modelled in terms of its *information flow* and *materials flow*, respectively. System inputs and outputs are involved at the boundaries of a manufacturing system in its surrounding business environment. For example, the materials from suppliers are system inputs and the final products delivered to customers are system outputs. System components include all of the manufacturing resources for designing, manufacturing, and assembling of products as well as

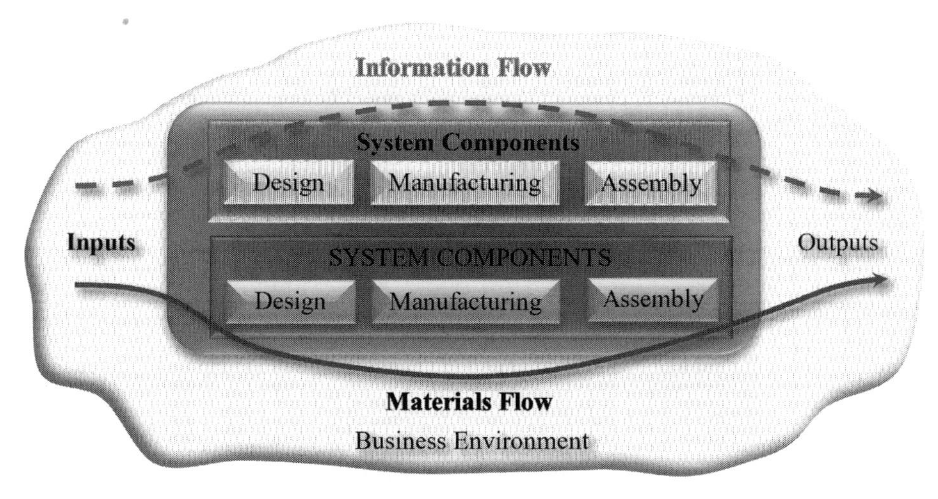

Figure 1.2 Description of a manufacturing system.

other relevant activities such as transportations in the system. In addition, a virtual twin in the information flow is associated with a physical component in the materials flow for decision-making supports of manufacturing businesses.

In the evolution of manufacturing technologies, the scale and complexity of manufacturing systems have been growing constantly. Note that both the scale and complexity of a system relates to the number and types of inputs, outputs, and system components that transform inputs to outputs. Figure 1.3 shows the impact of the evolution of system paradigms on the complexity of manufacturing systems (Bi et al. 2008). The evolution of system paradigms is divided into the phases of craft systems, English systems, American systems, lean production, flexible manufacturing systems (FMSs), computer integrated manufacturing (CIM), and sustainable manufacturing.

Historically, the manufacturing business began with *craft systems* where some crude tools were made from objects found in nature. The system inputs were simple objects and the requirements of the products were basic functions. In the 1770s, James Watt improved Thomas Newcomen's steam engines with separate condensers, which triggered the formation of *English systems*. In an English manufacturing system, machines partially replaced human operators for heavy and repetitive operations, the power supply became an essential part of the manufacturing source, and the production was scaled to make functional products for profit. In the 1800s, Eli Whitney introduced interchangeable parts in manufacturing that allowed all individual pieces of a machine to be produced identically. Thus, mass production became possible, the manufacturing processes began to be distributed, and system inputs in general assembly companies included parts and components. The criteria of system performance were prioritized with productivity and product quality. Mass production in the *American system* paradigm brought the rapid growth of manufacturing capacities that led to the saturation of manufacturing capacities in comparison with global needs. The global market became so competitive that the profit margin was such that without consideration of cost savings in the manufacturing processes profits would be insufficient to sustain manufacturing business. The *lean production* paradigm was conceived in Japan to optimize

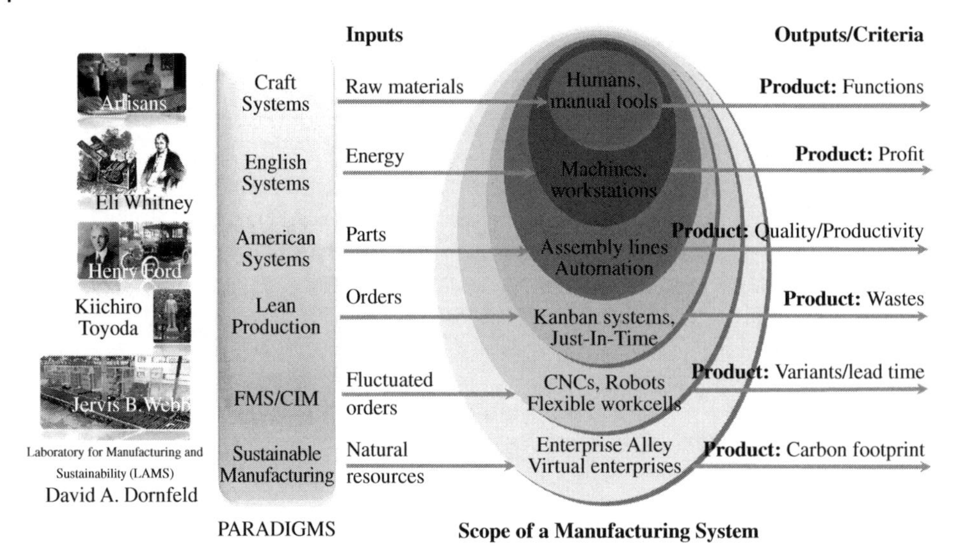

Figure 1.3 The growth of scale and complexity of manufacturing systems (Bi et al. 2014). (*See color plate section for color representation of this figure*).

system operation by identifying and eliminating waste in production, thus reducing product cost to compensate for the squeezed profit margin. Most recently, *sustainable manufacturing* paradigms were developed to optimize manufacturing systems from the perspective of the product life cycle. This was driven by a number of factors, such as global warming, environmental degradation, and scarcity of natural resources. Manufacturing system paradigms are continuously evolving. The trend of the evolution in Figure 1.3 has shown that manufacturing systems are becoming more and more complicated in terms of the *number of system parameters*, the *dependence on system parameters*, and *their dynamic characteristics* with respect to time. The engineering education for human resources must evolve to meet the growth needs of the manufacturing industry.

1.1.3 Human Roles in Manufacturing

Computer aided technologies (CATs) in manufacturing are of the most interest in this book and are widely adopted to replace humans in various manufacturing activities and decision-making supports. To appreciate the applications of CATs, the roles of the human being in manufacturing systems are firstly discussed to explore the possibilities of automated solutions.

As shown in Figure 1.4, the importance of human being in a manufacturing system has been widely discussed. In developing the Purdue system architecture, Li and Williams (1994) classified manufacturing activities into the activities in information/control flow and material flow, respectively. Human resources are needed to accomplish the tasks in both information and material flows. For example, human labourers are commonly seen in an assembly plant to accomplish manual assemblies in the material flow; technicians are needed by small and medium sized companies (SMEs) to generate codes and run computer

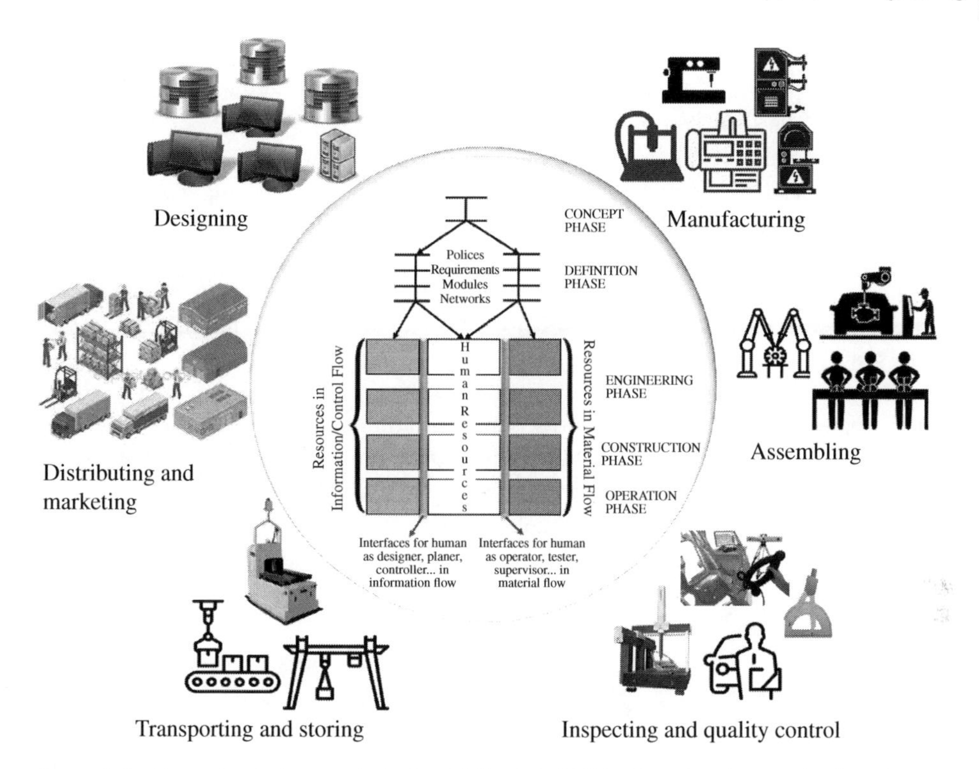

Figure 1.4 Human's role in manufacturing (Ortiz et al. 1999). (*See color plate section for color representation of this figure*).

numerical controls (CNCs) in the information/control flow. From the perspective of a product lifecycle (Ortiz et al. 1999), human resources are needed at every stage from designing to manufacturing, assembling, inspecting, transporting, marketing, and so on.

Human resources will certainly play an essential role in the future of manufacturing where manufacturing technologies and human beings are being integrated more closely and more harmoniously than ever before. While the focus should be shifted to the effective human–machine interactions to synergize both strengths of human beings and machines, manufacturing technologies should be advanced to balance the strengths and limitations of human resources optimally.

With the rapid development of information technologies (IT), CATs are replacing human beings for more and more decision-making support. The design and operation of a manufacturing system involves numerous decision-making undertakings at all levels and domains of manufacturing activities. In any engineering decision-making problem, one can follow the generic procedure with a series of design activities: (i) defining the scope and boundary of a design problem and its objective, (ii) establishing relational models among inputs, outputs, and system parameters, (iii) acquiring and managing data on current system states, and (iv) making decisions according to given design criteria. In the information flow of a manufacturing system, each entity normally has its capabilities to acquire the input data, process data, and make the decision as an output data.

1.1.4 Computers in Advanced Manufacturing

The performance of a manufacturing system can be measured by many criteria. Some commonly used evaluation criteria are *lead-time*, *variants*, and *volume*s of products, as well as *cost* (Bi et al. 2008). Manufacturing technologies have advanced greatly to optimize system performances. Figure 1.5 gives a taxonomy of available enabling technologies in terms of the *strategies*, *domains*, and *product paradigms* of businesses to optimize systems against the aforementioned evaluation criteria. In the implementation of production, the majority of advanced technologies, such as CIM, FMS, Concurrent Engineering (CE), Additive Manufacturing (AM), and Total Quality Management (TQM), are enabled by CATs.

The advancement of CATs can be measured by their capabilities in dealing with the growing scales and complexities of systems and the autonomy of system responsiveness. Figure 1.6 shows the evolution of CATs from the perspective of these three measures (Bi et al. 2014), together with computer applications in manufacturing using numerical control (NC)/CNC workstations, FMSs, CIM, distributed manufacturing (DM), and predictive

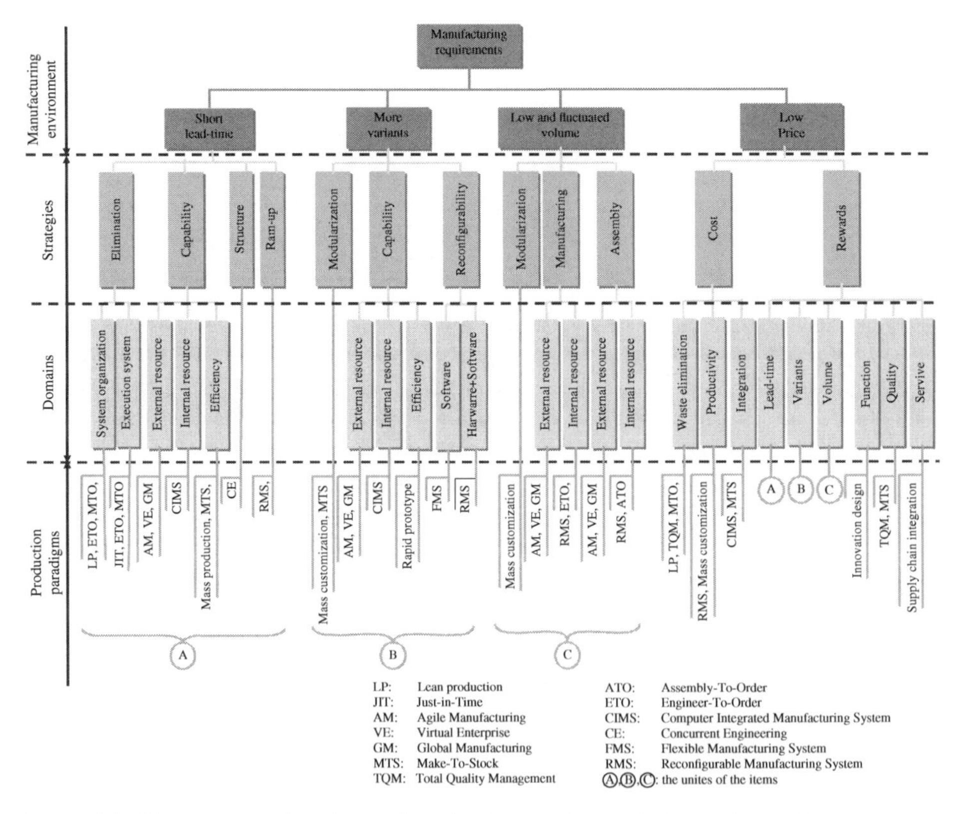

Figure 1.5 The strategies, domains, and production paradigms of advanced manufacturing technologies (Bi et al. 2008). (*See color plate section for color representation of this figure*).

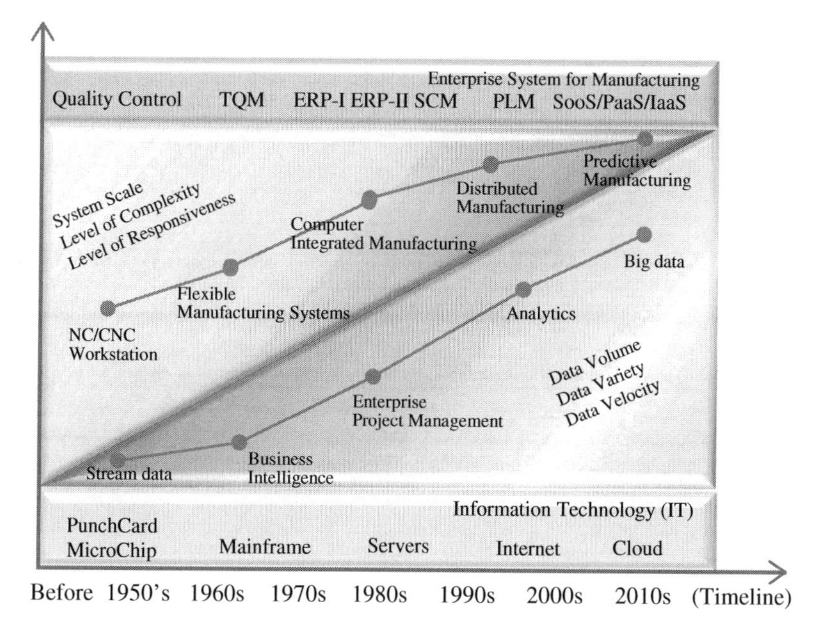

Figure 1.6 Evolution of computer aided technologies in manufacturing (Bi and Cochran 2014). (*See color plate section for color representation of this figure*).

manufacturing (PM). Typical computer aided tools to support these enabling technologies are *Quality Control* (QC), TQM, *Enterprise Requirements Planning* (ERP-I), *Enterprise Resources Planning* (ERP-II), *Product Lifecycle Management* (PLM), *Software as a Service* (SaaS), *Platform as a Service* (PaaS), and *Infrastructure as a Service* (IaaS), respectively. Correspondingly, the capacities of software systems to deal with volume, variety, and velocity of the data have been increased gradually from stream data early in the digital era to big data now. IT hardware systems must be capable of processing data in a timely manner. The computing environments have evolved from Microchip, mainframe, servers, the Internet, to today's Cloud.

1.2 Computer Aided Technologies (CATs)

Computer aided technologies provide the aids for design, analysis, manufacture, and assembly of products and for design, planning, scheduling, controlling, and operations of production systems. Various CATs have been gradually known and adopted by engineers since CATs and tools were developed in the late 1950s. Wikiversity (2019) divides the historical development of computer aided design (CAD) into the stages of two-dimensional, three-dimensional, and parametric designs. Many other researchers discussed the development of CAD technologies by marking some significant theoretical contribution and products as milestones.

Table 1.1 Milestones in the 60 year history of CAD development (Computer History 2019a).

Year	Products, developers, and features
1957	PRONTO was developed by Patrick Hanratty as the first commercial numerical control programming system. It sparked everything that is CAD, known as the building block of everything CAD.
1960	Sketchpad was developed by Ivan Sutherland as the first tool with a graphic user interface. Users wrote with a light pen on an x–y pointer display, let users constrain properties in a drawing, and created the use of objects and instances.
1966	Computer Aided Design and Drawing (CADD) was developed by McDonell-Douglas. It was used to create layouts and geometry of designs and could be customized and improved for specific uses.
1967	The Product Design Graphics System (PDGS) was developed by Ford and used internally at Ford and its partners as CAD/CAM systems.
	Digigraphics was developed by Itek as the first commercial CAD system with a unit price of $500 000; only six copies were sold.
1970	SynthaVision was developed as the first available commercial solid modelling program.
1971	Automated Drafting and Machining (ADAM) was developed by Patrick Hanratty as an interactive graphics design, drafting, and manufacturing system. It was written in Fortran and designed to work on virtually every machine. Today, nearly 80% of CAD programs can be traced back to the roots of ADAM.
1975	ComputerVision by Kenneth Versprille was where the rational B-spline geometry was added to CAD systems.
1977	CADAM was used by Lockheed to pioneer the applications of CAD in aerospace engineering.
1978	Unigraphics was developed by Siemens as a high-end and easy to use software; it was used by many corporations and set a new gold standard for CAD software at that time.
1980	MiniCAD was introduced as the bestselling CAD software on Mac computers.
1981	Geometric Models (GEOMODs) were developed and featured geometric precision and accuracy due to the modelling capability using a non-uniform rational B-spline (NURBS).
1982	AutoCAD was developed by Autodesk as the first CAD software made for Personal Computers (PCs) instead of mainframe computer workstations.
1987	Pro/Engineer was developed as the first mainstream CAD tool incorporating the ideas of Sketchpad. It was based on solid models, history-based features, and the uses of design constraints. It marked a high point in CAD history.
1994	AutoCAD version 13 was released with 3D modelling capabilities. The Standard for the Exchange of Product Model Data (STEP) was initially released as a new format and international standard of 3D models for data exchanges.
1995	eCATALOG was developed by Cadenas as the solution of digital product catalogs with multiple native CAD formats.
	SolidWorks 95 was developed by Dassault Systems as another software that succeeded in ease of use, and allowed more engineers than ever to take advantage of 3D CAD technology.
	Solid Edge was developed by Siemens as a Product Lifecycle Management (PLM) software. It was Window-based and provided solid modelling, assembly modelling, and a 2D orthographic view.

(continued)

Table 1.1 (Continued)

Year	Products, developers, and features
1996	Computer-Aided Three-Dimensional Interactive Application (CATIA) Conference Groupware was developed by Dassault Systems as the first CAD tool allowing users to review and annotate CATIA models with others over the Internet.
1999	Inventor was developed by Autodesk; it aimed to be more intuitive, simple, but allowed complex assemblies to be created in a shortened time.
2012	Autodesk 360 was developed whose computing was moved to the cloud.
2013	The first application (APP) for 3D CAD manufacturers was developed.
2015	Onshape was developed as a completely cloud-based CAD program.
2017	PARTSolutions was provided by Cadenas to help manufacturers with future proof of their catalog by keeping up to date with future native formats, versions, and revisions.

The first system of CATs in manufacturing was developed by Patrick Hanratty in 1957 as a programming system for numerical control. The major innovation in CATs occurred in 1963 when Ivan Sutherland, for his PhD thesis at MIT, created Sketchpad, which was based on a GUI (Graphical User Interface) to generate *x*–*y* plots. The innovation in Sketchpad pioneered the use of object-oriented programming in modern CAD and CAE (Computer Aided Engineering) systems. In the 1960s, the extensive works were developed in the aircraft, automotive, machine control, and electronics industries for three-dimensional modelling and the programming for numerical control. A few significant works were published as the fundamentals of the CAD theory, such as the mathematical representations of polynomial curves and surfaces by Bezier and Casteljau (Citroen Automotive Company) and Coons at MIT (Ford Motor Company).

Due to the rapid growth of computing power and reduction of hardware sources, three-dimensional (3D) modelling techniques are now widely used in video games, robotics, simulation of discrete systems, medical imaging and diagnosis, and in computer-controlled surgeries. As a milestone, the Manufacturing and Consulting Services Inc. (MCS) had developed CAD technologies in their commercially available Automated Drawing and Machining (ADAM) (Cadhistory 2019). The software was used and updated by McDonnell Douglas as *Unigraphics* and by Computer Vision as CADDs. Later, 3D printing technologies were developed and 3D printing technology evolved from 3D polygonal modelling to some objects with highly curvy surfaces, machine objects, and many other objects. 3D printing gave a boost to the development of computer aided manufacturing (CAM) technologies. As a few examples of successful companies in the CAM fields, (i) 3D Systems Corporation specializes in converting 3D solid or scanned models into physical objects; (ii) Stratasys Ltd. adopts additive manufacturing (AM) for direct manufacture of end parts; (iii) Intuitive Surgical Inc. designs, manufactures, and markets da Vinci surgical systems, as well as related instruments and accessories; and (iv) iRobot Corporation designs, develops, and markets robots for consumer, defence, security, telemedicine, and video collaboration. Caudill and Barnhorn (2018) provided a comprehensive summary of the milestones in the 60-year development of CATs, which is given in Table 1.1.

1.3 CATs for Engineering Designs

1.3.1 Engineering Design in a Manufacturing System

Design science uses scientific methods to analyse the structure of technical systems and their relationships with the environment. The aim is to derive rules for the development of these systems from system elements and their relationships. *Design methodology* is a concrete course of action for the design of technical systems that derives its knowledge from design science and cognitive psychology, and from practical experience in different domains.

Designs of products and processes are essential to manufacturing systems. For example, some typical activities to design a product are (i) a *functional design* to determine functional modules and features and their relations, (ii) a *parametric design* to determine geometrics and dimensions of parts, (iii) a *tolerance analysis* of geometric dimensioning and tolerances (GD&Ts) to determine the quality, position, and shape of all parts, and (iv) an *assembly design* to determine the assembly relations of parts and components.

1.3.2 Importance of Engineering Design

A manufacturing system is understood from the technological and economic perspectives. Technologically, a manufacturing system is to transform raw materials into final products via a set of operations. Economically, a manufacturing system is a process to add values to final products via a set of economic transactions associated with manufacturing processes. Making a profit is always a primary goal to entrepreneurs. The profit can be maximized in two ways: (i) to reduce costs on no value-added activities and (ii) to increase the sale price by providing a corresponding value to the customer.

Engineering design plays a significant role in implementing these two strategies. Figure 1.7 shows the impact of the activities of design, manufacturing processes, raw materials, management, and marketing on the overall product costs. The impacts are measured by the percentages of overall product cost affected by the activities in a certain

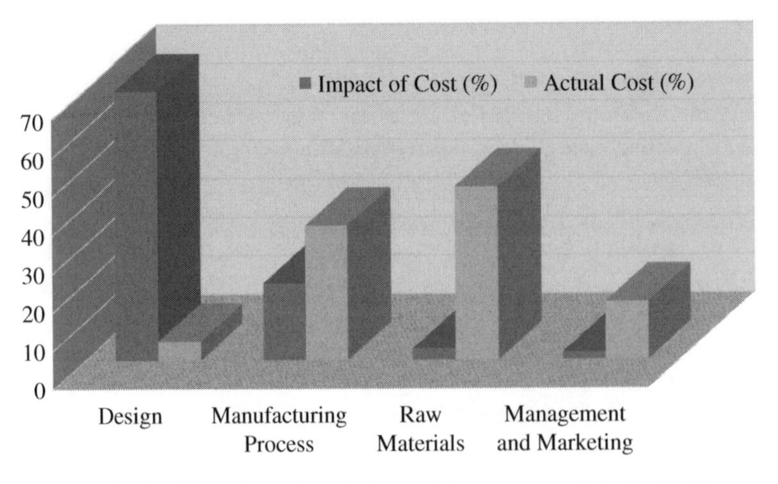

Figure 1.7 Significant impact of design activities on overall product cost.

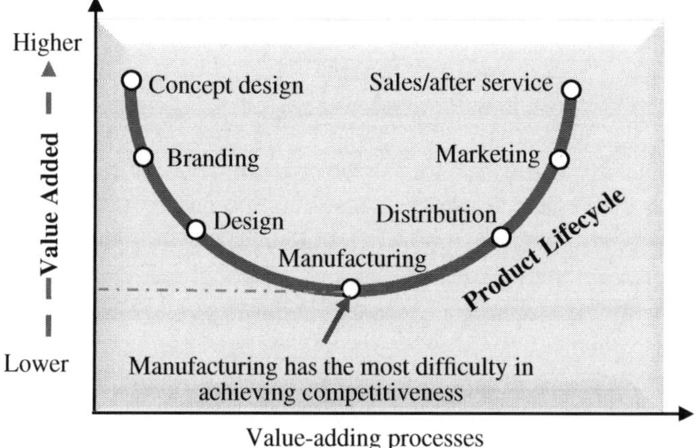

Figure 1.8 Value-added chain smile curve (IEC 2019; ITC 2019).

category. Even though the minimal percentage of the actual cost is related to the design activities, they mainly affect the overall product cost.

In Figure 1.8, the value-added chain smile curve by ITC (2019) shows the importance of design effectiveness on the possibility of added values of products in contrast to competitors. Business activities in a manufacturing system can be classified based on its involvement with hardware systems. The more dependence on hardware systems, the less chances a company gains in competitiveness on adding more values to products. Along with the product lifecycle, the more design activities are involved, the better the chances to gain competitiveness by adding more values of products than competitors.

The importance of engineering design can also be reflected by the additional costs a company may spend to fix some defects and errors occurring to different phases of a product lifecycle. *The first-time correct* is an idea goal to make highly diversified products and can only be achieved when CAD tools are capable of eliminating all of the design defects at the design stage. Figure 1.9 shows that, conventionally, the errors with a 75% fixing cost are

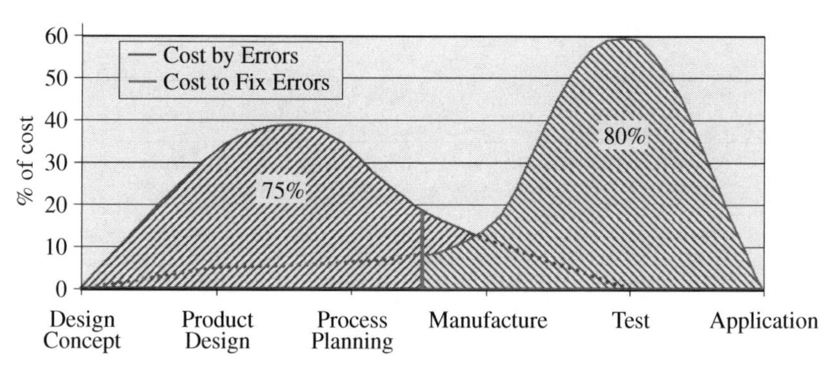

Figure 1.9 Relative costs to fix errors at different phases of product lifecycle.

made at the design stage of products, but the errors fixed at the manufacturing stage or later take around 80% of the fixed cost. It is clear that the earlier an error or defect is identified, the less cost is needed to fix it.

1.3.3 Types of Design Activities

In this section, the role of CATs in the design aspect is analysed, the nature of engineering designs is discussed, and the potential for using CAT technologies for design problems with different creativity levels is explored. Based on the level of creativity, Designwork (2016) classifies the engineering design problems into types: *routine design*, *redesign*, *selection design*, *parametric design*, *integrated design*, and *new design* (Table 1.2).

Corresponding to the needs to involve computers, human–machine interaction, and human designers, a rougher classification based on the level of creativity and innovation may facilitate further discussion. To this end, the designs can be regrouped as:

1. *New design* for a design from scratch to conceptual design, detailed design, and assembly design to the final product to meet specified design requirements.
2. *Incremental design* for a design subjected to given product structures, change individual parts, and functions to meet additional requirements of products.
3. *Routine design* for a design subjected to given functionalities, topological relations, layouts, and parametrize dimensions for the design of product families.

The level of creativity and innovation can be characterized based on the types of *solution space* and *design variables*. Goel (1997) corresponded routine design, innovative design, and new creative design to different combinations of solution space and design variables (Table 1.3).

Table 1.2 Types of design activities.

Type	Description	Example
Routine design	To perform a design by following existing standards and codes that outline the steps and computations for certain products or systems.	Follow the American Society of Mechanical Engineers (ASME) Boiler and Pressure Vessel Codes to design a pressure vessel.
Redesign	To revise an existing design when FRs have been changed in a dynamic environment.	Reprogram a robot when the tag points on the motion trajectory are changed.
Selection design	To find a solution by selecting appropriate components from an existing design inventory.	Select standardized fasteners to join two metal plates.
Parametric design	To determine design variables in a given conceptual structure for an optimized performance.	Minimize the materials usage of a cylindrical container subject to the given volume.
Integrated design	To design and assemble components as an integrated product or system to meet strongly coupled FRs.	Design a robotic configuration for a given task in a modular robot system.
New design	To design a product or system from scratch to meet emerging FRs.	Design a patentable product or system.

Table 1.3 Characteristics of solution spaces and design variables.

Level of creativity		Routine design	Incremental design	New design
Solution space	Structure	Known	Known	Unknown
	Search procedure	Known	Unknown	Unknown
Design variables	Types	Fixed	Fixed	Changed
	Ranges	Fixed	Changed	Changed

The characteristics of the solution space and design variables determines whether or not an engineering design belongs to a 'routine', 'incremental', or 'new' design. Both humans and computers can compete to accomplish some design activities; however, computers and human designers are good at different things, and it is desirable to synergize the strengths of designers and computers to achieve the effectiveness of engineering designs. Many researchers have discussed the differences between humans and computers. Table 1.4 summarizes the role differences of human beings and computers in engineering design processes.

1.3.4 Human Versus Computers

A manufacturing system has been set up to transfer raw materials into final products by performing a series of manufacturing and assistive activities in a system. Section 1.1.3 showed that a conventional manufacturing system relies heavily on human resources in both materials flow and information flow. With the continuous growth of scale and the complexity of systems, human designers and operators approach their limits to make

Table 1.4 Comparison of human designers and computers.

	Human designers	Computers
Strengths	• Identifying design needs • Brainstorming to think solutions 'out of the box' • Engineering intuition and a big knowledge base • Selecting design variations • The flexibility to deal with changes • Qualitative reasoning • Psychologically, human decision is more trusted than artificial intelligence • Predict trends, patterns, or anomalies, and • Learn from experience	• Fast speed, reliable, endurance, and consistent • Capable of exploring a large number of options • Carry out long, complex, and laborious calculations • Store and efficiently search large databases and • Provide information on design methodologies, heuristic data, and stored expertise
Weaknesses	• Easily tired and bored • Cannot do micro manage • Biased and inconsistent • Prone to make errors • Not good at quantified reasoning • Incapable of utilizing the data presented in an awkward manner	• Difficult to synthesize new rules • Limited knowledge base • No common sense

decisions or operate machines in cost-effective ways. To sustain vital manufacturing systems, CATs are expected to fully or partially substitute for a human being in a material or information flow. Table 1.4 gives a brief comparison of human designers and computers in terms of their strengths and weaknesses.

1.3.5 Human and Machine Interactions

It is critical to take advantage of modern computing technologies to improve the effectiveness of engineering designs. The limitations of human designers can be addressed by computer programs. As shown in Figure 1.10, human designers and computer aided tools are good at providing a *new design* and a *routine design*, respectively. Statistics shows that the percentages for providing a new design, increment design, and parametric design are 24%, 56%, and 20%, respectively. The total new design only took 24%. If the requirements of standardization are high, the parametric design is higher, where 50% of design activities are of a routine parametric design.

While the majority of engineering design activities are incremental designs, the computer assisted interactive design can maximize the synergized strengths of both human designers and computer aided tools. This implies that engineers might spend a lot of time to repeat existing work by others. Therefore, it becomes mandatory to utilize CAD/CAM to improve productivity. CAD/CAM is designed to synergize the methodologies, tools, and expertise to solve design problems of CAD/CAM applications.

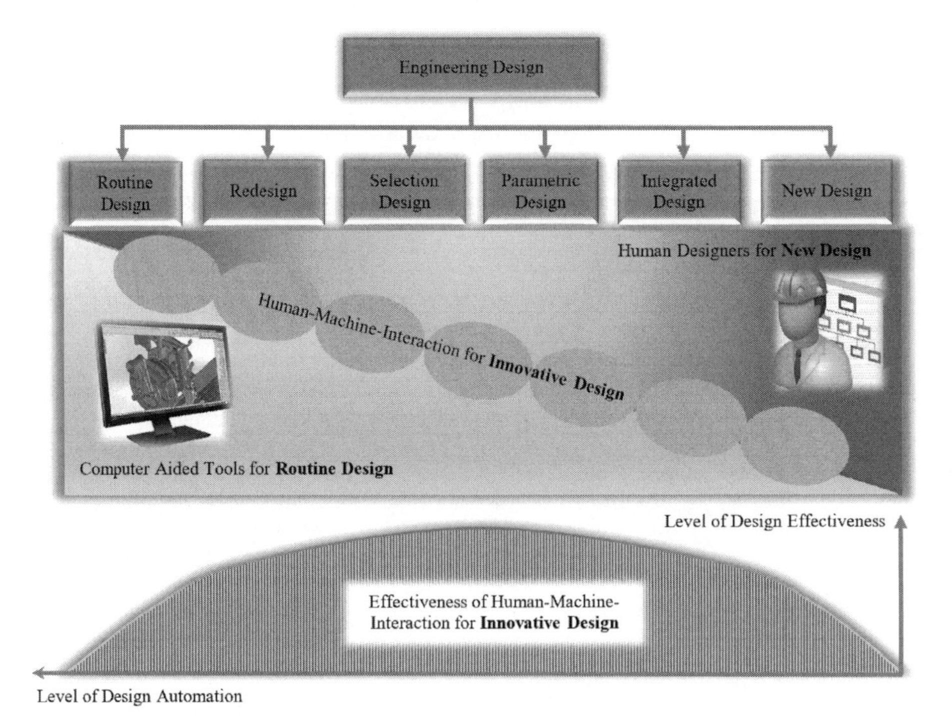

Figure 1.10 Human designers and computers in engineering design.

1.4 Architecture of Computer Aided Systems

System architecture is used to represent system components and their relations. Like any computer application system, the components in a computer aided system can be classified as hardware components or software components, as shown in Figure 1.11. The hardware components are further classified into *computers*, *peripheral equipment*, and *servers and networks* for Manufacturing Resources. The software components are classified into *system software*, *support software*, and *application software*.

1.4.1 Hardware Components

Hardware components of computers are organized based on their functions: computing units focus on data processing and other peripheral devices are for inputs, outputs, storage, and communications (Savage and Vogel 2013).

Computing units differ in *size* for mass and *power* for computing speed, and common categories are *supercomputers*, *mainframes*, and *microcomputers*. Mainframes are usually larger than microcomputers, but modern microelectronics allows very large power systems such as computer workstations to be packed into small spaces. Supercomputers are very advanced and expensive, and are characterized as having the fastest computing speeds for the most complex problems. The most critical measure of computer power is Million Instructions per second (MIPS). Moravec (1998) discussed the evolution of computer power/cost as shown in Figure 1.12, which shows that computers doubled in capacity every two years since 1945. This predicted speed was used as an indicator by computer manufacturers. They had to make new products whose computing speeds exceed the predicted speed, and those who failed to catch up with the increase in computing speed would lose business. In the 1980s, the doubling time contracted to 18 months and the computer performance in the late 1990s doubled every 12 months. Accordingly, the cost for computer power has been greatly reduced. The level of MIPS was predicted to be millions per $1000 in 2020. The core of the computing unit is the processor. Evolution of

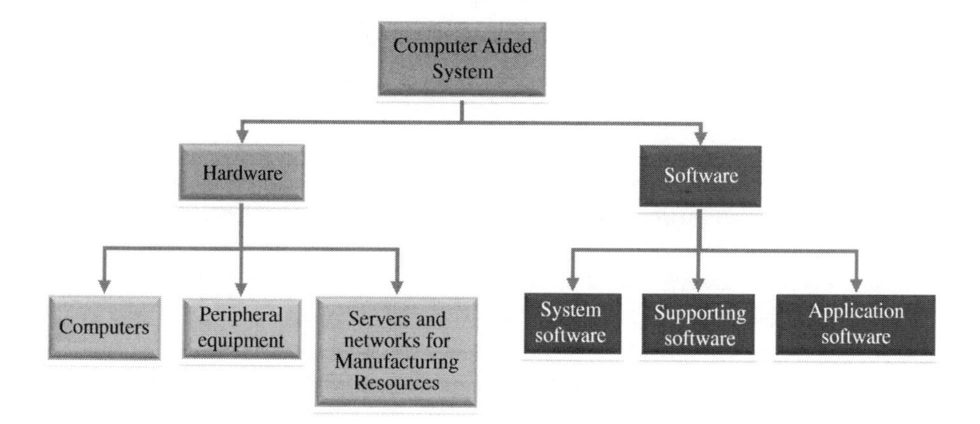

Figure 1.11 Architecture of computer aided systems.

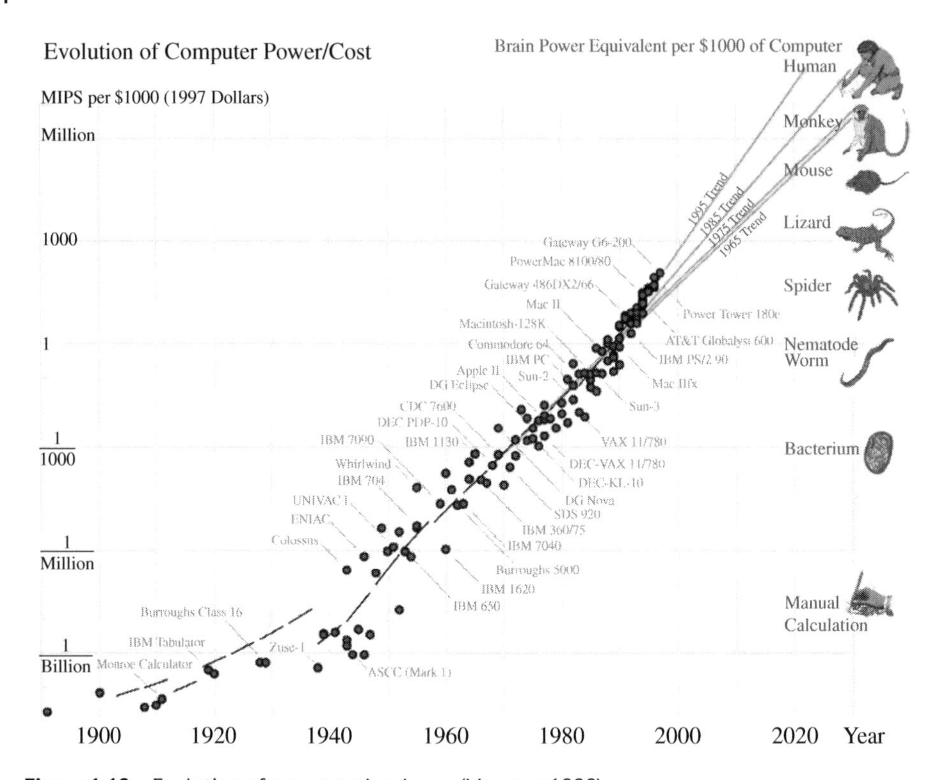

Figure 1.12 Evolution of computer hardware (Moravec 1998).

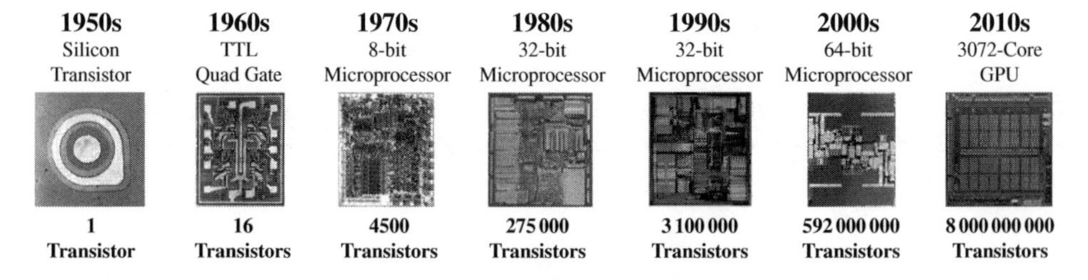

Figure 1.13 Evolution of semiconductors for processors (Computer History 2019a).

the computer processors is shown in Figure 1.13. The growth of processing capabilities matches Moore's Law – *transistor density on integrated circuits doubles about every two years.*

Another important specification of a computing unit is memory, which is a type of integrated circuit used to store data. The memory for the storage of the data for immediate use in a computer is called *primary storage*. Primary storage such as random-access memory (RAM) operates at a high speed. *Secondary storage* such as external drives provides slow-to-access data but offers higher capacities (Wikipedia 2019b). The idea of computer memory came with the usage of punch cards as memory by Charles Babbage in 1837; it was not until 1932 that Gustav Tauschek invented drum memory. Later in 1946, magnetic core memory became popular, which was attributed to the application of Williams-Kilburn

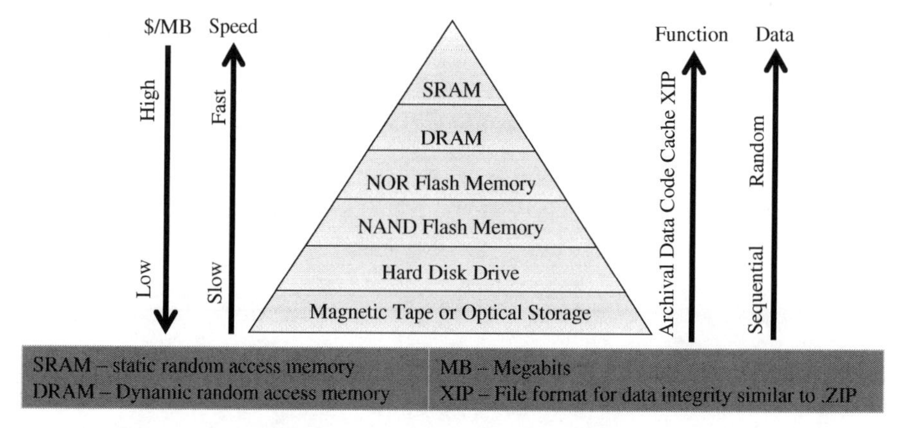

Figure 1.14 Main types of computer memories (Computer History 2019b).

tubes. Magnetic core RAM was introduced by MIT and the patent on pulse transfer controlling device by An Wang in 1955. Thereafter, dynamic random-access memory (DRAM), phase-change random-access memory (PRAM), static random-access memory (SRAM), due drive rate (DDR) RAM, and solid state RAM were gradually developed (Computer Hope 2019). Bhatt and Die (2015) give a summary of existing solutions of computer memories as well as a comparison of the main specifications on cost, speed, function, and data type in Figure 1.14. The more powerful a computer aid system is, the higher the requirements to the computer and memories are.

Computers to an information system are machines to a manufacturing system. Computers serve as the transformers to transfer input data to output data. Peripheral devices serve as the interfaces for computers to input raw data and output processed data in applications. Computers at an early time have limited choices of input and output devices such as punch-cards and printers. Today, many types of devices can be used as inputs and output devices of computer systems. Figure 1.15 shows a classification of computer devices for human machine interfaces. A peripheral device can be a *unidirectional input*, *unidirectional output*, or *bidirectional input and output* device (Wole 2018). It is worth noting that with an increase in the capabilities of computer aided systems, more and more smart devices, such as indoor Global Positioning Systems (GPSs), haptic systems, and 3D printers, can be connected to computer systems directly as input and output devices.

1.4.2 Computer Software Systems

A computer aided system runs on computers and consists of a set of functional components at different layers of information, from the layer for hardware interfaces to the layers of operating systems, networking, database, sophisticated computer aided tools, and finally to the system layer of applications. Figure 1.16 shows the architecture of a computer aided software system. The functional components in the system are generally encapsulated and used independently. These functional components are classified into four groups: (i) *software for hardware* operation interacting with computers, printers, plotters, and so on, (ii) *system software* associated with operating systems and networking, (iii) *support*

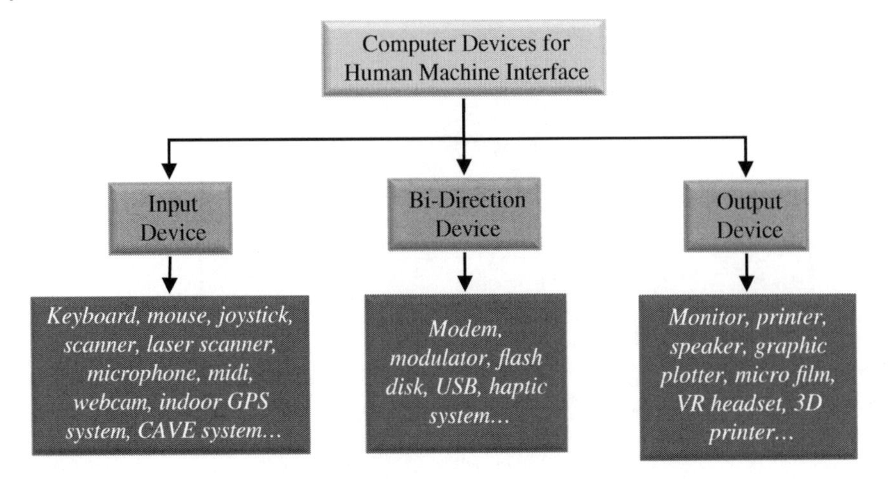

Figure 1.15 Types of peripheral devices for inputs and outputs.

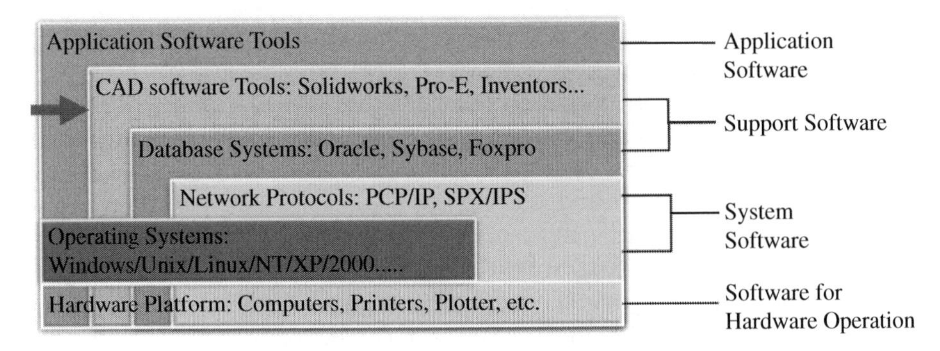

Figure 1.16 Computer aided software system architecture.

software consisting of database systems and CAD software, and (iv) *application software* as comprehensive computer aided systems.

The architecture of a computer aided software system is modularized; it allows the developers to customize the selections of functional components and integration of the selected components to meet the design needs at different domains and levels of manufacturing systems. It seems obvious that the capabilities of computer aided systems will be continuously expended due to the rapid development of information systems in computing systems, networking systems, and databases. In the next section, the impact of networking and cloud technologies on computer aided systems is briefly discussed.

1.4.3 Servers, Networking, and Cloud Technologies

As an information system, a computer aided software system needs the capabilities to deal with the variety, complexity, and changes of design needs. Three basic strategies to deal with the variety, complexity, and changes are to (i) modularize system architecture, (ii) make functional components flexible, and (iii) support system integration (Bi and Zhang 2001a,b; Bi et al. 2008). To implement these strategies, functional modules must be interoperated, collaborated, and integrated. Therefore, servers and networking technology play a significant role to advance CATs. A *server* acts as an agent to accept and perform tasks and to generate the results of a task to requesters. *Networking* allows a number of the servers working collaboratively to solve a holistic problem, whose solution consists of a set of sub-solutions that are used to divide subproblems with the given constraints. For example, networking makes the following technologies possible in computer aided application systems.

Parallel computing is a type of computation in which many calculations or the execution of processes are carried out at the same time. Large-scale design problems should be decomposed into smaller ones, so that the sub-solutions can be obtained by using a number of computing resources concurrently. Parallel computing may refer to parallel processing by a set of central computing units (CPUs) or graphic processing units (GPUs) in one computer or parallel supercomputers containing hundreds or thousands of processors, networks of workstations, multiple-processor workstations, and embedded systems (Foster 1995). Computer aided systems, especially CAE systems, need parallel processing capabilities to achieve scalability in solving complex design problems, which may involve multiple physics, transient dynamics, or large-scale coupled models.

Hardware in the loop (HIL) simulation is a type of real-time simulation for the design of real-time control systems. In HIL, the machine or physical part of the system is networked with the control model through actuators and sensors. The rest of the system is represented by the mathematic model in the simulation. HIL assists in developing and testing complex real-time embedded systems while considering the complexity of the real-world system. HIL simulation shows how a control model responds to realistic virtual stimuli in a real-time manner and can be used to determine whether the mathematic model for a physical system is valid (MathWorks 2018).

Distributed database is a storage solution where a common processor accesses the data in a distributed system. Data may be stored in multiple networked computers in the same location or can be dispersed over a network of interconnected computers. A system server distributes collections of data across network servers or independent computers over intranets, extranets, or the Internet. Computer aided systems benefit from the distributed databases in increasing the scalability, modularity, reliability, and flexibility of systems (Wikipedia 2019c).

Cloud computing is the delivery of computing services such as data collections, computing, data mining, analytics, storage, servers, and more over the Internet for faster innovation, flexible resources, and economies of scale (Microsoft 2019). Computer aided systems greatly benefit from cloud computing over a number of aspects. *Firstly*, cloud computing allows single users, freelance designers, and SMEs to participate in the entire design processes of complex products; large and multinational companies can work with several hundred SMEs and individual designers to seek design solutions. *Secondly*, this helps to reduce an overall cost of the design and development processes. Cloud computing helps to reduce the investment on computer aided systems. Traditionally, commercial software tools are associated with the licensing and training costs as well as the additional cost on maintenance, software and hardware upgrades, IT personnel, power requirements, and rental costs for additional space. Cloud solutions offer an alternative where licensing costs are substituted with a subscription fee and the expenditure is substantially reduced since

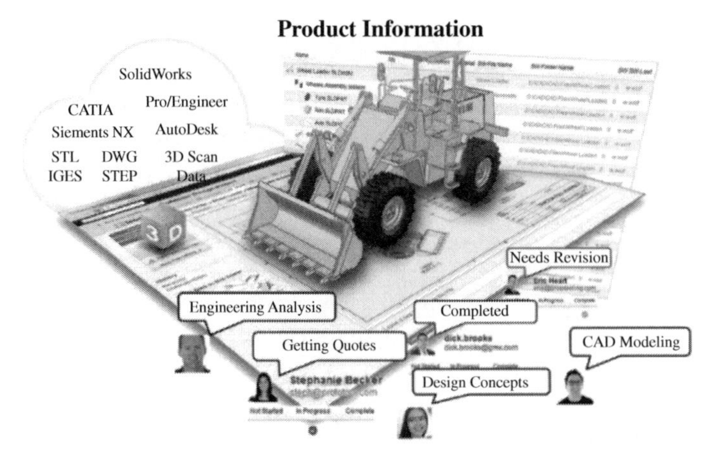

Figure 1.17 Computer aided collaboration in virtual environment (Wu et al. 2015; Wu 2014).

the resources are shared by users in the cloud. *Thirdly*, most of the cloud services offer pay-as-you-go options, where one pays only for the used computational resources. Such services reduce the overall cost for both the provider and the client since the service loads are customized to the needs (Harish 2018).

Manufacturing systems became highly distributed for enhanced flexibility and adaptability to meet the needs of regional markets promptly. The high-speed 3G, 4G, and 5G wireless networks helped to mobilize product information so that the PLM could be accessed by mobile apps. From the perspective of information technology, we have entered the fourth industrial revolution, where nearly every device will be networked, which allows continuous data streams to populate in memory databases. The Internet of Things (IoT) will transform the manufacturing sector in the coming years. More and more cloud-based solutions will be available to manufacturing enterprises to support their operations (Bi et al. 2014; Wang et al. 2014; Morley 2014).

Virtual enterprise (VE) is a temporary alliance of businesses for partners to share core competencies, resources, and skills in order to take advantage of emerging business opportunities. VE is facilitated by computer networks (Wikipedia 2019d). The core competencies, resources, and skills in VE are mainly for CATs. Figure 1.17 shows that using the virtual environment over the Internet, designers across different regions, domains, enterprises, and disciplines can collaborate on all of the CAD activities involved in the lifecycle of product development (Wu et al. 2015; Wu 2014).

1.5 Computer Aided Technologies in Manufacturing

A manufacturing system is involved in numerous decision-making activities and computers outperform human beings at many tasks in both materials and information flows, such as *machine operation, planning and scheduling, engineering, analysis, data acquisition and sharing*, *computing*, *data storage*, *data retrieval*, and *inspection* (Cummings 2014; Sotala 2012). The importance of computer aided technologies can be clearly evidenced

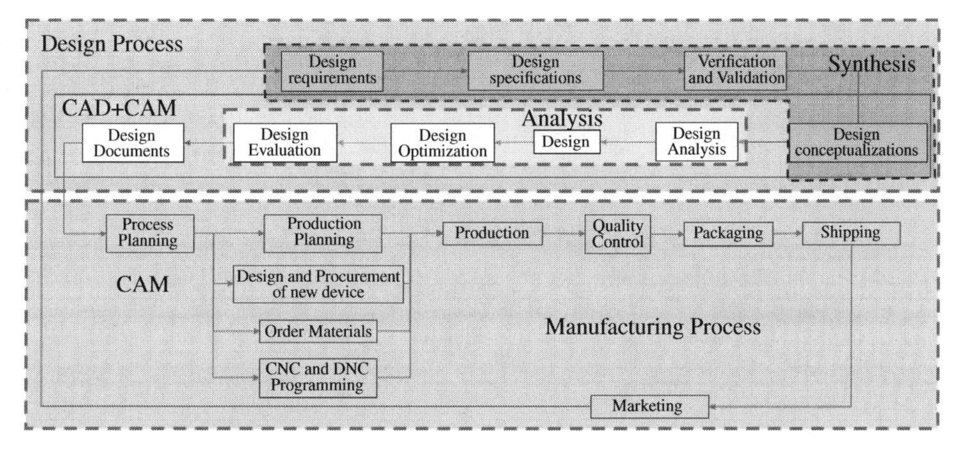

Figure 1.18 CATs in designing, manufacturing, and assembling and system integration.

by the growing number of computer aided tools exemplified in Figure 1.5. The rapidly developing information technologies (IT) make all of these advanced manufacturing technologies practical.

Figure 1.18 shows some typical manufacturing activities in a product lifecycle from the identification of design requirements of products to the delivery of final products to end-users. The fulfilment of these manufacturing activities is mostly assisted by computer programs. For example, CAD tools are used to create, modify, and optimize the design of parts, products, processes, and systems by using computer systems. CAM tools use computer software to control machine tools and related machinery in the manufacture of workpieces. CAD/CAM tools provide an integrated solution to bridge CAD and CAM systems. Figure 1.19 gives some typical computer aided tools under the categories of

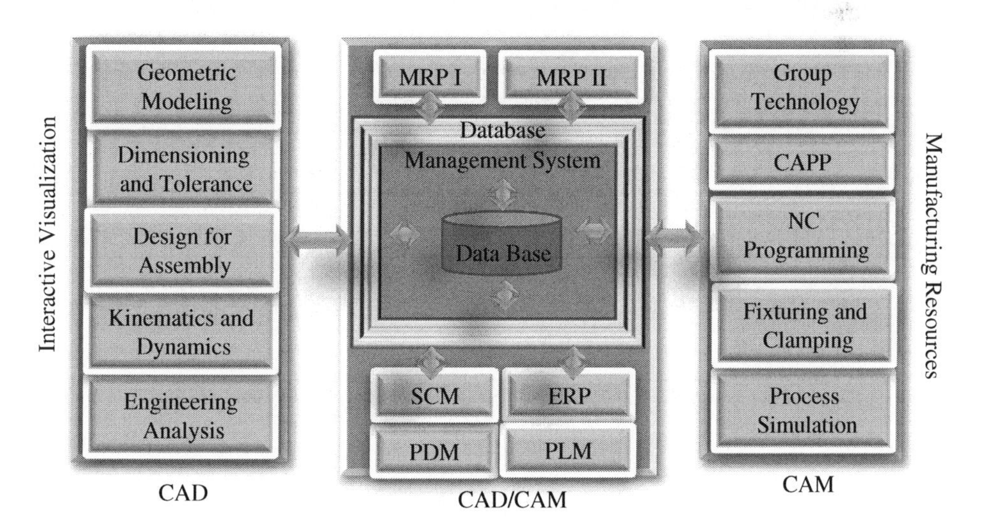

Figure 1.19 Typical computer aided tools in CAD, CAM, and CAD/CAM. (*See color plate section for color representation of this figure*).

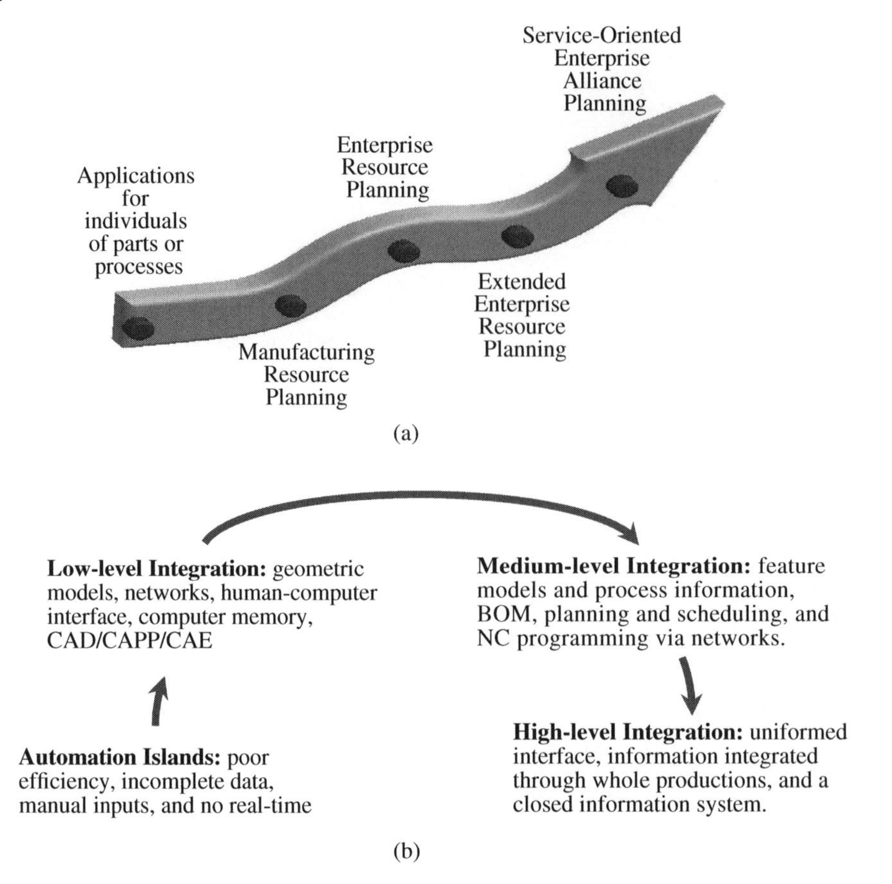

Figure 1.20 The evolution of computer aided technologies in manufacturing, (a) The increasing varieties of system functionalities and (b) the increasing level of system integrations.

CAD, CAM, and CAD/CAM tools. For example, the CAD tools for *geometric modelling, dimensioning and tolerance, design for assembly, kinematic and dynamic simulation*, and *engineering analysis* will be covered in this book.

Figure 1.20 shows that the capabilities of CATs have been continuously expanded in two aspects: (i) the variety of functionalities from isolated applications for individuals of parts or processes at a lower level to the planning for service-oriented enterprise alliances at a higher level and (ii) the level of system integration from isolated system components to holistic integration across enterprises.

1.6 Limitation of the Existing Manufacturing Engineering Curriculum

Manufacturing engineering is to apply mathematics and science in practice to design, manufacture, and operate products. Engineers in the manufacturing sector focus on design, development, and operation of manufacturing systems to make competitive products. The

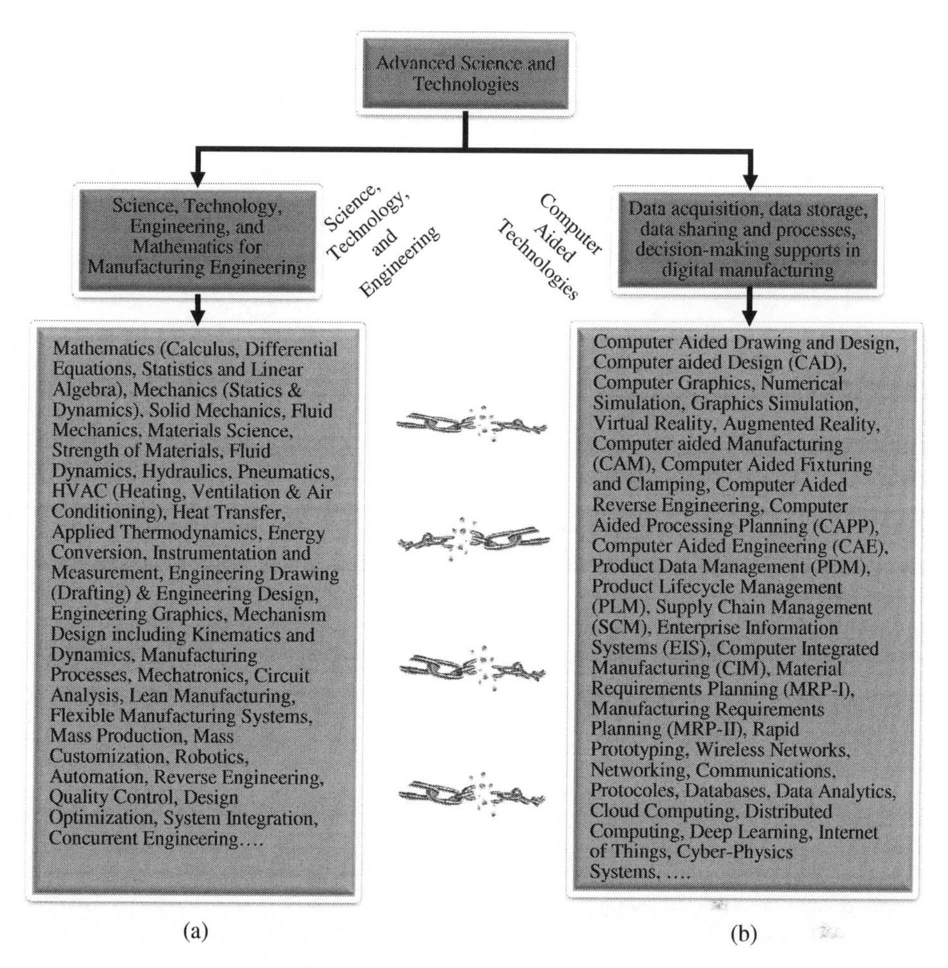

Figure 1.21 Mismatch of subdisciplines and computer aided tools in manufacturing engineering. (a) Subdisciplines in manufacturing engineering and (b) computer aided tools in digital manufacturing. (*See color plate section for color representation of this figure*).

existing engineering curricula usually include some core courses in *mathematics*, *physics*, *computing engineering*, and *management*, as well as some sophisticated courses in mechanical and manufacturing engineering such as *materials science*, *statics and dynamics*, *thermodynamics*, and *fluid mechanics*. Engineering curricula are generally designed to cover as many sub-disciplines of mechanical and manufacturing engineering as possible. Students have options to specialize in one or more sub-disciplines. Some typical courses for the bachelor's degree in design in manufacturing engineering are listed in Figure 1.21a (Wikipedia 2017). From this perspective, existing curricula are mostly discipline-oriented.

From the perspective of computer aided technologies, numerous computer aided tools become commercially available. However, these software tools are application-oriented, and most of the tools are developed based on the theories in multiple disciplines. Figure 1.20b shows a list of commonly used computer aided tools in the manufacturing sector. Due to the strong decoupling of multidisciplinary knowledge in these software tools, the classification of disciplines in manufacturing engineering is not well aligned with the classification of available computer aided tools. Figure 1.20 shows that there is no one-to-one correspondence between sub-disciplines and available computer aided tools.

The misaligned engineering curricula and a broad scope of computer aided tools in manufacturing pose a great challenge in the teaching of manufacturing engineering. *On the one hand*, the sub-disciplines in manufacturing engineering are so diversified that an ever-increasing number of elective technical courses are needed in engineering programmes. Meanwhile, public education systems are facing the pressure to reduce the number of credit hours for college degrees. Taking as an example the mechanical engineering program at Purdue University, Fort Wayne, the number of required credit hours for a bachelor degree has been reduced from 126 in the spring of 2012 to 120 in the spring of 2017 (Bi and Mueller 2016). *On the other hand*, engineering programmes are responsible for preparing students for an appropriate set of knowledge and skills using advanced computer aided tools; however, more and more computer aided tools are becoming commercially available and so their functionalities need to be upgraded and expanded continuously. This proves to be a great challenge to integrate disciplinary theories and computer aided tools in the limited selection of engineering courses.

1.7 Course Framework for Digital Manufacturing

The concern on the discipline-oriented curricula has attracted a great deal of attention in recent years. A number of educational programmes were proposed and implemented to address this issue. For example, the Engage Program sponsored by the National Science Foundation (NSF) aimed to increase the capacity of engineering institutions to retain undergraduate students by facilitating the implementation of three research-based strategies, i.e. (i) improve faculty–student interaction, (ii) improve spatial visualization skill, and (iii) use everyday examples in engineering teaching, to improve educational experiences (Nilsson 2014; Bi and Mueller 2016).

To adapt the rapid advancement of CATs, this books proposes to improve existing discipline-oriented engineering programmes, at least for some upper-level engineering courses. The objective is to develop a new course framework where constitutive elements are not varied with an increase of computer aided tools or the diversification of sub-disciplines.

The design of an engineering course curriculum is similar to the design of any engineering system in the sense that the complexity and dynamic characteristics become two critical factors to deal with when the system is continuously evolving. The modularity concept has proved to be an effective way to deal with system complexity and dynamic characteristics (Bi et al. 2008). In the similar way, the modularity concept is proposed to deal with the misalignment of discipline-oriented curricula and a large variety of computer aided software tools in manufacturing engineering.

Figure 1.22 shows an alternative to the discipline-oriented curriculum. It can be referred to as a 4-P engineering curriculum since the manufacturing fundamental is differentiated for *Product*, *Process*, *Production*, and *Platform*, respectively, for the required system functionalities in a product lifecycle. The objective of the proposed curriculum is to minimize

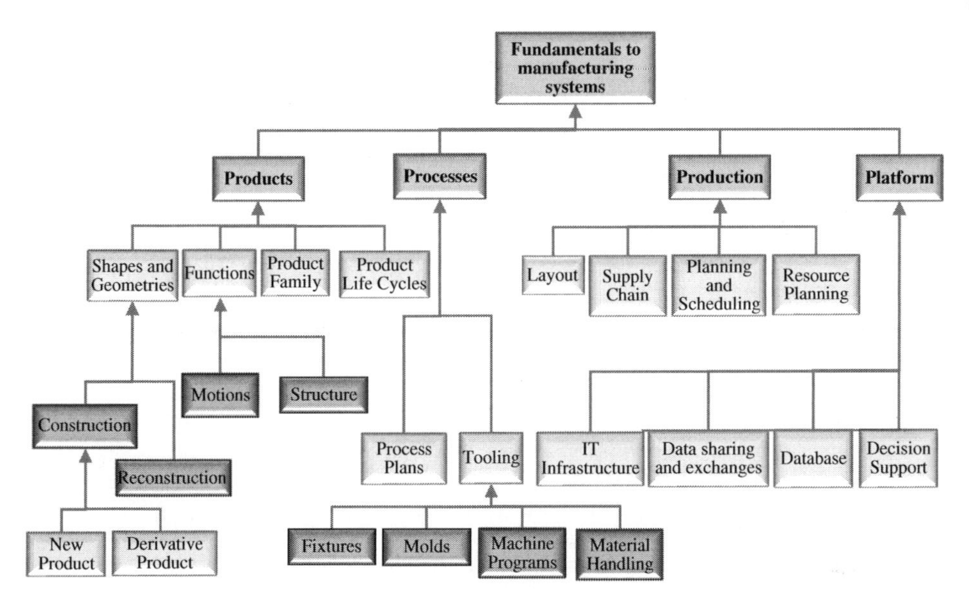

Figure 1.22 Proposed course framework for digital manufacturing.

the impact of the ever-increasing complexity of the system as well as computer aided tools. Since any manufactured product has its own lifecycle, a taxonomy of engineering courses based on the product lifecycle can sustain its consistence, even though the scope of a manufacturing system or computer aided tools may vary with respect to time.

Following the axiomatic theory (Cochran et al. 2016a, 2016b, 2017a, 2017b), the high-level functionalities for the product, process, production, and platform can be further decomposed as a modularized structure. Take an example the functional requirements (FRs) for a product design, FRs have been decomposed further into the designs of geometries, motions, product families, and a sustainable design related to the product life cycle. The granularity of the functionalities can be appropriate to match the functionalities for well-established engineering sub-disciplines as well as available computer aided tools. Due to the modularized structure, the proposed framework has the flexibility to customize the selection of sub-disciplines and corresponding computer aided tools in a specific engineering curriculum.

1.8 Design of the CAD/CAM Course

The modular framework of digital manufacturing provides the flexibility to select course elements to customize the educational needs in a specific engineering programme. This book is written as a CAD/CAM text to achieve two main goals: (i) introduce manufacturing fundamentals, which are not usually covered in depth in traditional mechanical engineering programmes and (ii) expose students to as many computer-aided software tools as possible, so that they can utilize advanced computing tools to deal with the designs related to manufacturing processes.

Table 1.5 Examples of existing CAD/CAM textbooks (Wang and Bi 2018).

Author(s)	Title	Year	Publishers
M. Groover and E. Zimmers	CAD/CAM: Computer-Aided Design and Manufacturing	1984	Pearson
P. Martin, N.E. Larsen, and David D. Hansen	Computer Aided Design in Control and Engineering Systems	1986	Pergamon
M. Bedworth, R. Henderson, and P.M. Wolfe	Computer-Integrated Design and Manufacturing	1991	McGraw-Hill International
M. Groover and E. Zimmers	CAD/CAM: Computer-Aided Design and Manufacturing	1993	CRC Press
Jami J. Shah and Martti Mäntylä	Parametric and Feature-Based CAD/CAM: Concepts, Techniques, and Applications	1995	Wiley
Nanua Singh	Systems Approach to Computer-Integrated Design and Manufacturing	1996	Wiley
Kunwoo Lee	Principles of CAD/CAM/CAE	1999	Pearson, Elsevier
Alberto Paoluzzi	Geometric Programming for Computer Aided Design	2003	Wiley
T.C. Chang, R.S. Wysk, and H.P. Wang	Computer-Aided Manufacturing	2003	Prentice Hall
Ibrahim Zeid	Mastering CAD/CAM (Engineering Series)	2004	McGraw-Hill
André Chaszar	Blurring the Lines: Computer-Aided Design and Manufacturing in Contemporary Architecture	2006	Wiley
Khoi Hoang	Computer-Aided Design and Manufacture	2011	McGraw-Hill Custom Publishing
Kuang-Hua Chang	Product Manufacturing and Cost Estimating using CAD/CAE	2013	Academic Press
Kuang-Hua Chang	Product Design Modeling using CAD/CAE	2014	Academic Press

1.8.1 Existing Design of the CAD/CAM Course

The importance of computer aided technologies in manufacturing engineering has been well recognized. The majority of higher educational institutions offer one or several CAD/CAM courses in their engineering programmes. However, selecting an appropriate textbook proves to be a challenge since all of the textbooks are too sophisticated in certain subjects but lack in coverage on a broad scope of disciplines and CAD/CAM tools. In our primary survey, we found a few common CAD/CAM textbooks that were adopted by different institutions, as shown in Table 1.5.

Since the information technology (IT) has developed so rapidly in recent years, most of the textbooks in Table 1.5 are out of the date, and the last three recent ones cover only the integration of CAD and CAM. No appropriate textbook has been found that has a wide coverage of subdisciplines and computer aided tools.

1.8.2 Customization of the CAD/CAM Course

Undergraduates in mechanical engineering with a minor in manufacturing engineering must understand theoretical fundamentals related to the design of products and manufacturing processes. On the other hand, the theories and methods relevant to high-level planning, scheduling, or computer implementation might not be the first priority for them. To meet our specified teaching needs, the engineering curriculum in Figure 1.23 can be utilized to choose appropriate contents for our students. The selected course elements are highlighted in Figure 1.23. The covered CAD/CAM theory and methods are illustrated in Figure 1.24 and the corresponding computer aided technologies and tools are accordingly specified in Figure 1.25. The customized CAD/CAM course consists both of the theoretical part in Figure 1.24 and the practical training part in Figure 1.25. To sustain the independence for the selection of course elements in a modularized course framework, the axiomatic design theory has been applied to map the theoretical part in Figure 1.24 to the computer-aided tools in Figure 1.25 (Bi and Mueller 2016).

As an introduction to the CAD/CAM course, human designers and computers are compared to clarify the roles of computers for design activities at different design phases of products and systems. The computer applications in engineering are overviewed. The course structure is presented and the main design concepts related to products and manufacturing processes are discussed. Figure 1.26 illustrates the organization of the customized CAD/CAM book. Each of the selected concepts corresponds to a section/chapter in the course. Besides the introduction section, all of the other sections include the laboratories where one functional module of the CAD/CAM software tool is utilized to illustrate the

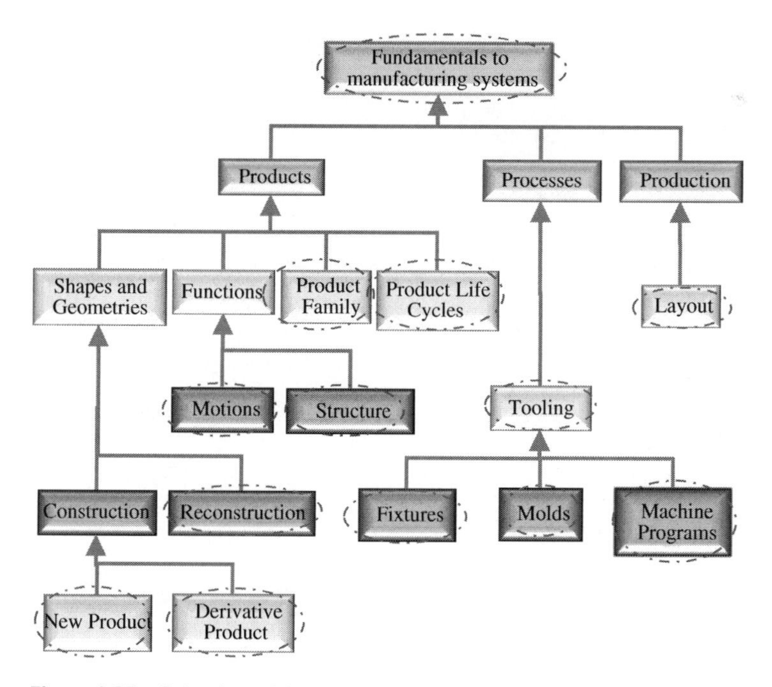

Figure 1.23 Selective subjects in a new CAD/CAM course.

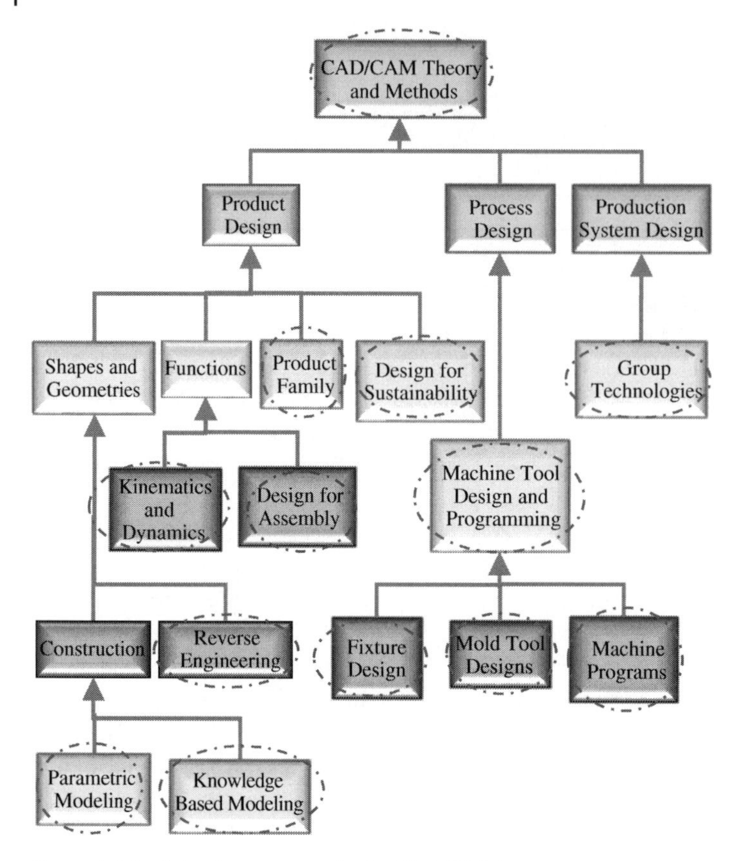

Figure 1.24 Selective concepts in the CAD/CAM theory.

application of computer aided technology to formulate and solve real-world engineering design problems. The unique features of this course framework include (i) coverage of a broad scope of sub-disciplines so that the students can understand better the design challenges over the product lifecycles in a minimal class setting, (ii) selection of course elements that is oriented where individual functional modules are available in commercial computer aided software tools to handle the formulated design problems in corresponding disciplines, and (iii) emphasis on the self-guiding exploration of a comprehensive CAD/CAM software tool, so that students in mechanical or manufacturing engineering are able to formulate multidisciplinary problems and use the correct computer aided tools to solve problems effectively.

1.9 Summary

Through the discussion in this chapter, we find that the majority of existing CAD/CAM course designs are out of date due to the rapid development of CATs. There is a misalignment between the classification of sub-disciplines and the types of emerging computer aided tools. To address these two concerns, a new course framework has been presented

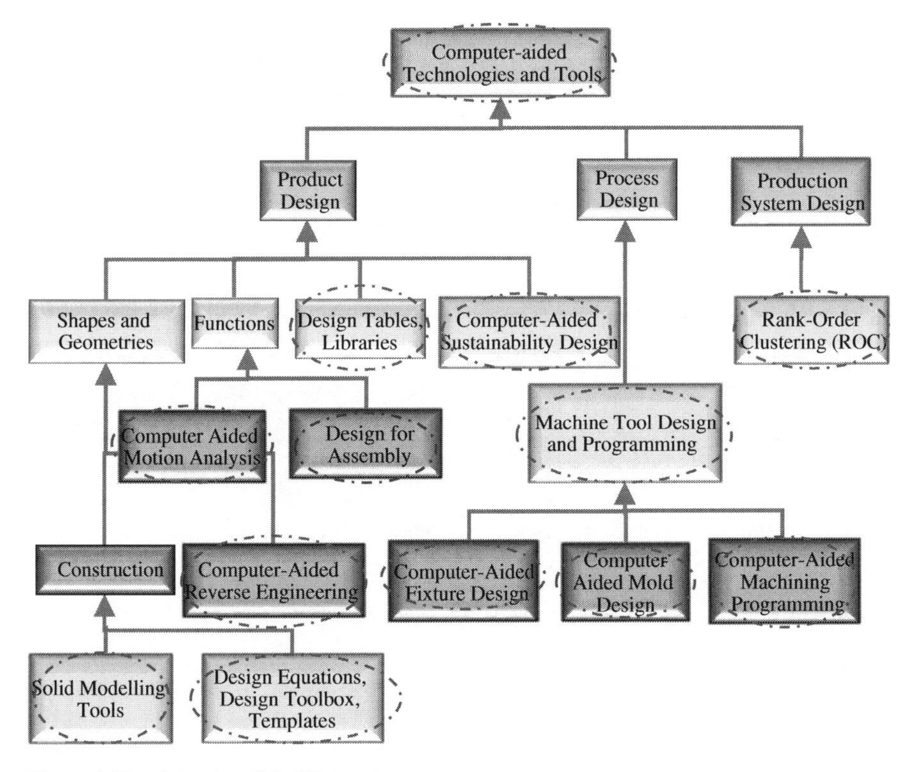

Figure 1.25 Selective CAD/CAM tools.

for Digital Manufacturing. It is unique in the sense that (i) the course topics are defined based on the design needs in product lifecycles, which are likely to be changed over time; (ii) the types of computer aided tools are application-oriented for manufacturing systems, so that the correspondence of course topics can be readily mapped to existing computer aided tools for training; (iii) the framework is modularized so that educators have flexibility in selecting course subjects to tailor the digital manufacturing teaching needs of their degree programmes; and (iv) emphasis is put on the balance of theoretical knowledge and the skills involved in using CAD/CAM tools, where each course topic corresponds to training in the use of computer aided tools to solve real-life design problems.

The proposed framework of a Digital Manufacturing course is modularized and expandable. The following chapters cover a limited number of computer aided tools, but the continuous effect of using multiple aspects is expected to broaden the coverage of contemporary computing aided systems.

1.10 Review Questions

1.1 Discuss how CAD/CAM helps in modern manufacturing? Elaborate on any one aspect.

1.2 What is CAD/CAM?

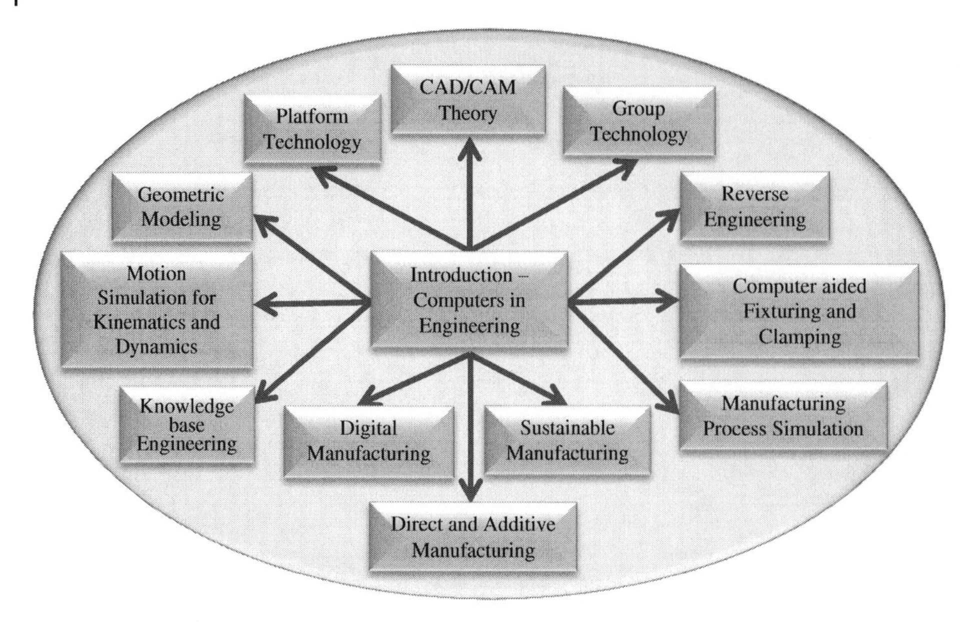

Figure 1.26 Customized outline of the CAD/CAM book.

1.3 What are basic elements of a CAD/CAM system?

1.4 What are the objectives of using CAD/CAM?

1.5 What advantages does the CAD/CAM approach offer in NC programming?

1.6 List a number of enabling technologies related to CAD/CAM and discuss their functions.

References

Automobile Newsletter (2012). The 15 top-producing American car plants. https://www.automobilemag.com/news/the-15-top-producing-american-car-plants-151801.

Bhatt, A., and Die, M. (2015). Solid state drives. http://meseec.ce.rit.edu/551-projects/spring2015/1-4.pdf.

Bi, Z.M. and Cochran, D. (2014). Big data analytics with applications. *Journal of Management Analytics* 1 (4): 249–265.

Bi, Z.M. and Mueller, D. (2016). Integrating everyday examples in mechanical engineering courses for teaching enhancement. *International Journal of Mechanical Engineering Education* 44 (1): 16–28.

Bi, Z.M. and Zhang, W.J. (2001a). Flexible fixture design and automation: review, issues and future directions. *International Journal of Production Research* 39 (13): 2867–2894.

Bi, Z.M. and Zhang, W.J. (2001b). Modularity technology in manufacturing: taxonomy and issues. *International Journal of Advanced Manufacturing Technology* 18 (5): 381–390.

Bi, Z.M., Lang, S.Y., and Shen, W. (2008). Reconfigurable manufacturing systems: the state of the art. *International Journal of Production Research* 46 (4): 967–992.

Bi, Z.M., Xu, L.D., and Wang, C. (2014). Internet of Things for enterprise systems of modern manufacturing. *IEEE Transactions on Industrial Informatics* 10 (2): 1537–1546.

Cadhistory (2019). Chapter 15 Patrick Hanratty and manufacturing & consulting services, http://www.cadhistory.net/15%20Patrick%20Hanratty%20and%20MCS.pdf.

Caudill, L., and Barnhorn, A. (2018). 60 Years of CAD Infographic: the history of CAD since 1957. https://partsolutions.com/60-years-of-cad-infographic-the-history-of-cad-since-1957.

Cochran, D., Umair, M., and Bi, Z.M. (2016a). Incorporating design improvement with effective evaluation using manufacturing system design decomposition (MSDD). *Journal of Industrial Information Integration* 2: 65–74.

Cochran, D.S., Hendricks, S., Barnes, J., and Bi, Z.M. (2016b). Extension of manufacturing system design decomposition to implement manufacturing systems that are sustainable. *ASME Journal of Manufacturing Science and Engineering* 138: 101006-1–101006-10.

Cochran, D., Kim, Y.-S., Foley, J., and Bi, Z.M. (2017a). Use of the manufacturing system design decomposition for comparative analysis and effective design of production systems. *International Journal of Production Research* 55 (3): 870–890.

Cochran, D., Arinez, J.F., Collins, M.T., and Bi, Z.M. (2017b). Modeling of human-machine interaction in agile production systems. *Enterprise Information Systems* 11 (7): 969–987.

Computer History (2019a). The silicon engine: a timeline of semiconductors in computers. https://www.computerhistory.org/siliconengine.

Computer History (2019b) Timeline of computer history. https://www.computerhistory.org/timeline/memory-storage.

Computer Hope (2019). Computer hard drive history. https://www.computerhope.com/history/memory.htm.

Cummings, D. (2014). Man versus machine or man + machine? *IEEE Intelligent Systems* 29: 1–9.

Designworks (2016). Types of design. http://www-eng.lbl.gov/~dw/services/TypesOfDesign/TypesOfDesign.htm.

Fisher, A.G.B. (1939). Production, primary, secondary and tertiary. *Economic Record* 15 (1): 24–38.

Flows, C. (2016). A strong manufacturing sector fuels economic growth. https://www.forbes.com/sites/realspin/2016/11/21/a-strong-manufacturing-sector-fuels-economic-growth/#4185e2d07f3e.

Foster, I. (1995). Parallelism and computing. https://www.mcs.anl.gov/~itf/dbpp/text/node7.html.

Goel, A.K. (1997). Design, analogy, and creativity. https://pdfs.semanticscholar.org/29ed/10edb3e1e65ded62c58577d3e2fd13a8f2cf.pdf.

Gold, S. (2016). Manufacturing's economic impact: so much bigger than we think. https://www.mapi.net/blog/2016/02/manufacturings-economic-impact-so-much-bigger-we-think.

Harish, A. (2018). Cloud solution in the world of CAD and CAE. https://www.simscale.com/blog/2017/05/cloud-solutions-secure.

IEC (2019). Factory of the future. https://www.iec.ch/whitepaper/pdf/iecWP-futurefactory-LR-en.pdf.

International Trade Centre (ITC) (2019). *Global Value Chains in Services: A Case Study on Costa Rica*. Geneva: ITC, 2014. vi, 17 pages. Doc. No. DMD-14-257.E. http://www.intracen.org/uploadedFiles/intracenorg/Content/Publications/AssetPDF/cover%20Global%20Value%20Chains%20in%20Services%20-%20a%20Case%20Study%20of%20Costa%20Rica.pdf.

Kniivila, M. (2018). Industrial development and economic growth: implications for poverty reduction and income inequality. http://www.un.org/esa/sustdev/publications/industrial_development/3_1.pdf (last accessed on 24 April 2018).

Li, H., and Williams, T.J. (1994). A formalization and extension of the Purdue Enterprise Reference Architecture and the Purdue Methodology. http://www.ict.griffith.edu.au/~bernus/taskforce/archive/pera/hong/THESIS-HONG-LI.pdf.

MathWorks (2018) What is hardware-in-the-loop simulation? https://www.mathworks.com/help/physmod/simscape/ug/what-is-hardware-in-the-loop-simulation.html.

Microsoft (2019). What is cloud computing? https://azure.microsoft.com/en-us/overview/what-is-cloud-computing.

Moravec, H. (1998). When will computer hardware match the human brain. *Journal of Evolution and Technology* 1: 1–12. https://jetpress.org/volume1/moravec.pdf.

Morley, M. (2014). The evolution of the digital manufacturing business. https://blogs.opentext.com/the-evolution-of-the-digital-manufacturing-business-2.

Nilsson, T. L. (2014). Why am I learning this? Using everyday examples in engineering to engage female (and male) students in the classroom. *The 121 ASEE Annual Conference and Exposition*, Indianapolis (15–18 June 2014), paper ID #10398.

Ortiz, A., Lario, F., and Ros, L. (1999). Enterprise integration – business processes integrated management: a proposal for a methodology to develop enterprise integration programs. *Computers in Industry* 40 (1999): 155–171.

Savage, T.M. and Vogel, K.E. (2013). *An Introduction to Digital Multimedia*, 2e. Jones & Bartlett Learning, ISBN-10: 144968839X.

Sotala, K. (2012). Advantages of artificial intelligences, uploads, and digital minds. *International Journal of Machine Consciousness* 4 (1): 275–291. https://doi.org/10.1142/S1793843012400161.

Source Weekly (2012). Traditional heavy metal: blacksmithing and Rock 'N Roll are both alive and well at Orion Forge. https://www.bendsource.com/bend/traditional-heavy-metal-blacksmithing-and-rock-andaposn-roll-are-both-alive-and-well-at-orion-forge/Content?oid=2140249.

Wang, X. and Bi, Z.M. (2018, 2018). New CAD/CAM Course Framework in Digital Manufacturing. *Computer Applications in Engineering Education*: 1–17. https://doi.org/10.1002/cae.22063.

Wang, C., Bi, Z.M., and Xu, L.D. (2014). IoT and cloud computing in automation of assembly modeling systems. *IEEE Transactions on Industrial Informatics* 10 (2): 1426–1434.

Wikipedia (2017). Three-sector theory. https://en.wikipedia.org/wiki/Three-sector_theory.

Wikipedia (2019a). Manufacturing. https://en.wikipedia.org/wiki/Manufacturing.

Wikipedia (2019b). Computer memory. https://en.wikipedia.org/wiki/Computer_memory.

Wikipedia (2019c) Distributed database. https://en.wikipedia.org/wiki/Distributed_database.

Wikipedia (2019d). Virtual enterprise. https://en.wikipedia.org/wiki/Virtual_enterprise.

Wikiversity (2019). Computer-aided design/history, present and future. https://en.wikiversity.org/wiki/Computer-aided_design/History,_Present_and_Future.

Wole, A. (2018). The evolution: computer input and output. https://shopinverse.com/blogs/technology/the-evolution-computer-input-and-output.

Wu, D. (2014). Cloud-based design and manufacturing: a network perspective. PhD thesis. Georgia Institute of Technology. https://smartech.gatech.edu/bitstream/handle/1853/53029/WU-DISSERTATION-2014.pdf.

Wu, D., Rosen, D.W., Wang, L., and Schaefer, D. (2015). Cloud-based design and manufacturing: a new paradigm in digital manufacturing and design innovation. *Computer Aided Design* 59: 1–14.

Part I

Computer Aided Design (CAD)

2

Computer Aided Geometric Modelling

2.1 Introduction

Any *matter* for an object in the world exists in one of four distinct states, i.e. *solid*, *liquid*, *gas*, and *plasma*. The solid state is a common state for the majority of products around us. To investigate the behaviours of solids, it is necessary to know its geometry and shape. Similarly, to investigate the behaviours of fluid, gas, or plasma in a certain volume, the geometry and shape of boundary surfaces of a fine volume is also needed to be defined.

For centuries, geometry has played its crucial role in the development of many scientific and engineering disciplines such as astronomy, geodesy, mechanics, ballistics, civil and mechanical engineering, ship building, and architecture. The importance of the study on geometry has been shown in this century in automobile and aircraft manufacturing. Since geometry is primarily visual, geometry becomes a unique and particularly exciting branch of mathematics. Geometry became a branch of mathematics at the end of the nineteenth century; however, great designs in the history were always inspired by observation and intuition on geometric shapes (Gallier 2008). *Geometric modelling* is a branch of *applied mathematics* and *computational geometry*; it studies the methods and the algorithms for the mathematical representation of geometries and shapes. Geometric modelling serves for the visualization of objects and lays the foundation for *computer graphics*, which is the construction of models of scenes from the physical world and their visualization as images.

Geometric modelling is as important to computer aided technologies as the governing equilibrium equations are to classical engineering fields such as mechanics and thermal fluids. As shown in Figure 2.1, the design of a part, product, or system usually begins with geometric modelling, so that the physical objects to be designed can be represented virtually in computers. Geometric modelling techniques and algorithms are used to model objects, and the dimensions and spatial constraints of objects are inputs via graphic user interfaces (GUI) of modelling tools. Once the objects are modelled, any information associated with the objects can be utilized in the design processes or decision-making supports involved in a product lifecycle. For example, virtual geometric models can be used by *computer graphics* to visualize the designs before physical products are made. *Engineering drawings* needed in manufacturing processes can be directly generated from solid models of products, and all of the annotations relating to dimensioning and tolerance can be included in the drawings.

Computer Aided Design and Manufacturing, First Edition. Zhuming Bi and Xiaoqin Wang.
© 2020 John Wiley & Sons Ltd. This Work is a co-publication between John Wiley & Sons Ltd and ASME Press.
Companion website: www.wiley.com/go/bi/computer-aided-design

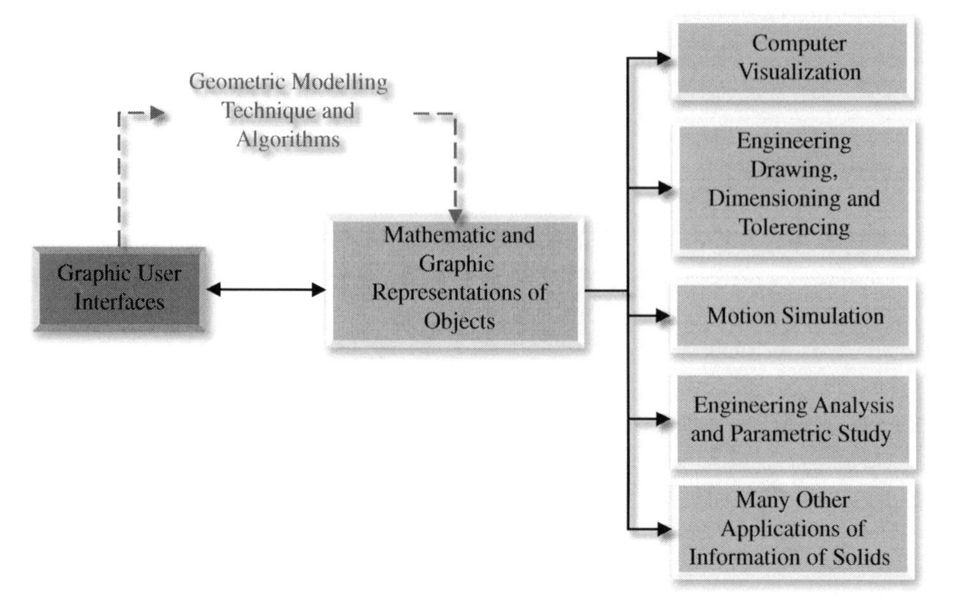

Figure 2.1 Role of geometric modelling in computer aided systems (CAD).

If a product is a machine with relative motions among system components, a *motion study* can be defined upon the computer aided design (CAD) model to investigate the relations of driving forces and motions. In addition, *engineering analysis* can be performed at any stage of the product design process. In the following chapters, you will learn how virtual models of products can be utilized to solve design problems in mould designs, engineering analysis, simulation of manufacturing processes, evaluation of system sustainability, and so on.

2.2 Basic Elements in Geometry

Geometry relates to the properties and relations of basic geometric elements such as *points*, *lines*, *surfaces*, *solids*, and higher dimensional analogues.

Figure 2.2 shows an example of basic geometric elements of an object. An object is represented by its geometric elements such as *points*, *edges*, *surfaces*, and *solids*. In addition, basic geometric elements are at different levels of the topological tree of an object. As shown in Figure 2.2, *points* and *nodes* are at the lowest level, *edges* formed by points or nodes are at the second level, *surfaces* formed by the boundary edges are at the third level, and solids with a given volume formed by a set of watertight boundary surfaces are at the upper level.

Geometry includes the relations of these geometric elements. These relations can be *spatial* or *topological*. To represent spatial relations of geometric elements, some references have to be established, and common references can be *points*, *axes*, *planes*, and *coordinate systems*. To represent topological relations, one has to be familiar with some common logical operations such as *union*, *subtraction*, and *intersection*.

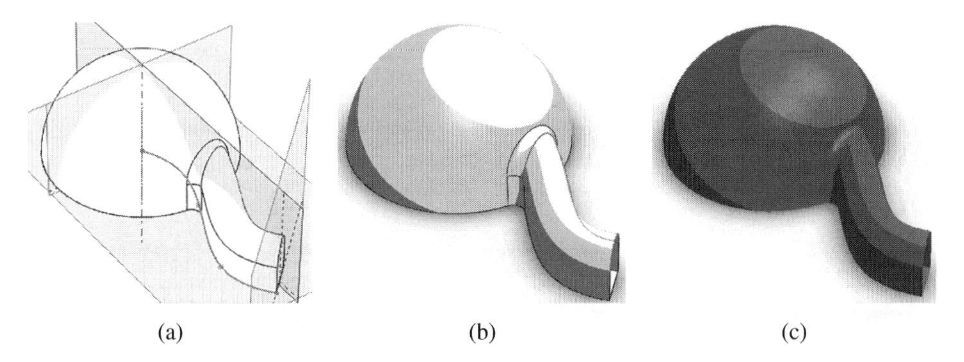

Figure 2.2 Example of basic geometric elements. (a) Points, nodes, lines, edges, axes, and planes. (b) Patches, surfaces, and operations. (c) Volumes, features, solids, and operations.

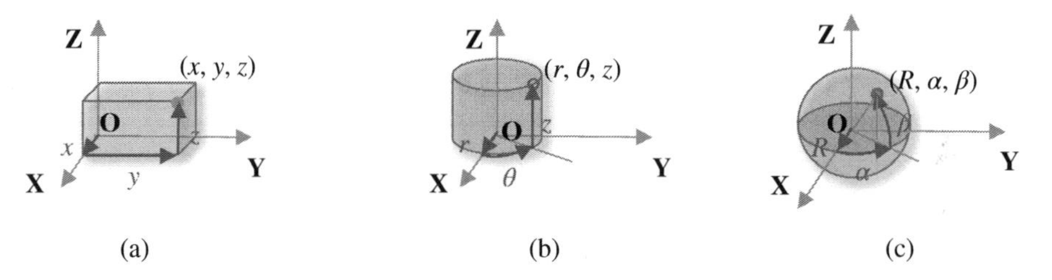

Figure 2.3 Three commonly used coordinate systems. (a) Cartesian coordinate system. (b) Cylindrical coordinate system. (c) Spherical coordinate system.

2.2.1 Coordinate Systems

Any geometric element in a three-dimensional space has six *degrees of freedom* (DOF). To describe the spatial relation between two geometric elements in solids, a reference coordinate system must be established. Figure 2.3 shows three most commonly used coordinate systems (CSs), i.e. a *Cartesian coordinate system*, *cylindrical coordinate system*, and *spherical coordinate system*.

A Cartesian CS consists of three axes that are mutually perpendicular to each other. A position in the Cartesian CS is defined by its distances to the origin (x, y, z) projected on three axes (X, Y, Z). A cylindrical CS consists of two linear axes (X and Z) and one rotational axis. Correspondingly, a position in the cylindrical CS is defined by two scalar variables and one angular variable, i.e. (r, θ, z). A spherical CS consists of two rotational axes and one translational axis. The position of a point in the spherical CS is specified by three variables: the radial distance (R) of that point from a fixed origin, its polar angle (α) measured from a fixed zenith direction, and the azimuth angle (β) of its orthogonal projection on a reference plane that passes through the origin and is orthogonal to the zenith, measured from a fixed reference direction on that plane, (R, α, β).

A user of a CAD system should always be aware that a reference CS is essential for geometric modelling of objects in computers. As shown in Figure 2.4, any commercial CAD software comes with a default coordinate system in modelling templates. The default coordinate system is usually a Cartesian CS. In Figure 2.4, the default Cartesian CS is $\boldsymbol{O-XYZ}$ and the planes of $\boldsymbol{O-XY}$, $\boldsymbol{O-XZ}$, $\boldsymbol{O-YZ}$ are defined as a default *front plane*, *top plane*, and *right plane*, respectively.

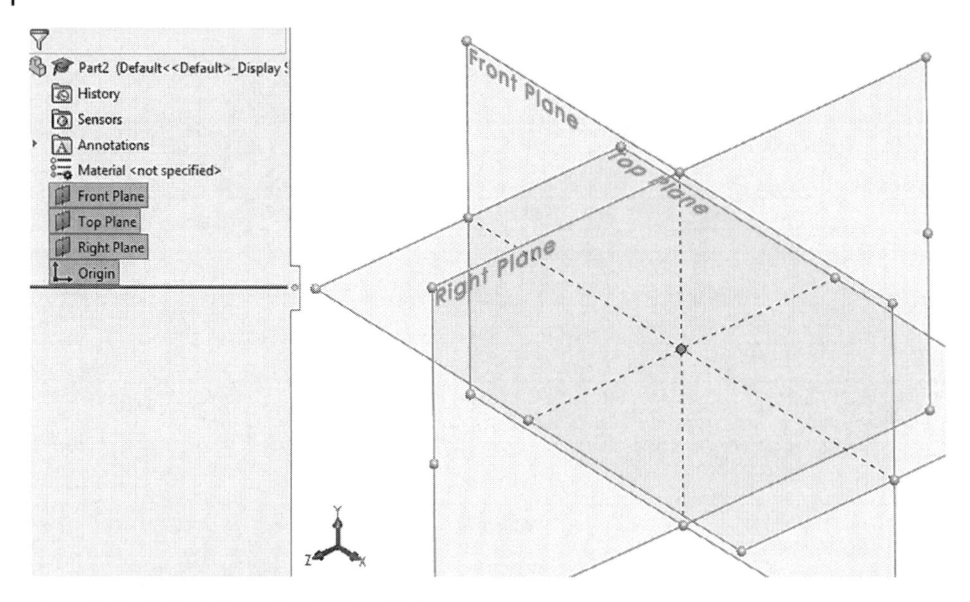

Figure 2.4 Default Cartesian CS in a CAD system.

2.2.2 Reference Points, Lines, and Planes

Other than the references of coordinate systems, other elements such as points, lines, and planes can also be used as the references in geometric modelling. Moreover, geometric elements at a high level can be represented by a number of points at the lowest level. Figure 2.5 shows a point in the reference CS that is completely defined by giving its three coordinate values (x, y, z) along x-, y-, and z-axes. For a given geometric element, the same amount of information can be defined in many different ways. For example, a point P is relevant to the high-level elements such as lines or surfaces in the sense that a point can be an end-point of these elements. Therefore, a reference point can be defined directly based on its relations to existing geometric elements.

Figure 2.5 shows that a point can be defined in many different ways: Figure 2.5a to f shows that the point is defined as the centre of an arc, the centre of a face, the interaction of two lines that interact, the projected points of a vertical line on a plane, the intersection of a line and a plane, and the intersection of three planes, respectively.

Mathematically, a point $\boldsymbol{P}$ in the Cartesian coordinate system can be represented as

$$\boldsymbol{P} = x \cdot \boldsymbol{i} + y \cdot \boldsymbol{j} + z \cdot \boldsymbol{k} = \begin{Bmatrix} x \\ y \\ z \end{Bmatrix} \tag{2.1}$$

where x, y, and z are the coordinates and $\boldsymbol{i}$, $\boldsymbol{j}$, and $\boldsymbol{k}$ are the unit vectors along the axes of $\boldsymbol{X}$, $\boldsymbol{Y}$, and $\boldsymbol{Z}$, respectively.

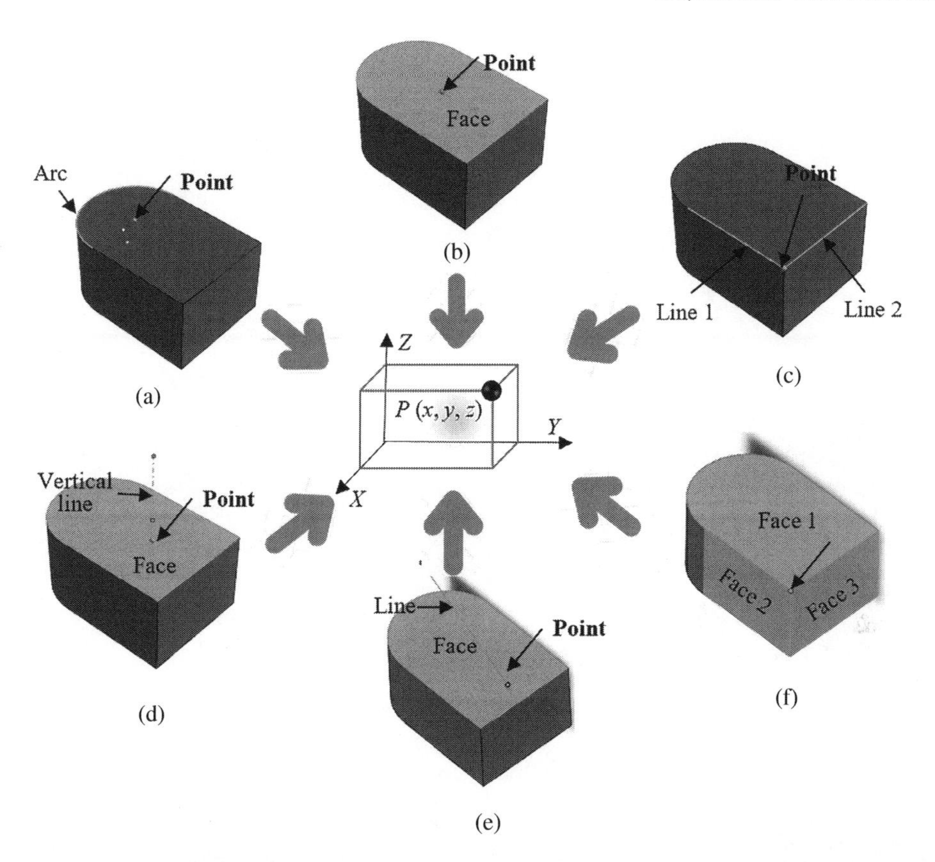

Figure 2.5 The definition of a point from existing reference geometries. (a) Arc centre. (b) Centre of face. (c) Intersection of two lines. (d) Projection of a vertical line on a plane. (e) Interaction of a line and plane. (f) Interaction of three planes.

As shown in Figure 2.6b, a line $\boldsymbol{L}$ in the Cartesian coordinate system can be represented as the line segment connected by two points $\boldsymbol{P}_1$ (x_1, y_1, z_1) and $\boldsymbol{P}_2$ (x_2, y_2, z_2) as

$$\boldsymbol{P} = \begin{Bmatrix} x \\ y \\ z \end{Bmatrix} = \begin{Bmatrix} x_1 \\ y_1 \\ z_1 \end{Bmatrix} + t\boldsymbol{v} = \begin{Bmatrix} x_1 \\ y_1 \\ z_1 \end{Bmatrix} + t \begin{Bmatrix} x_2 - x_1 \\ y_2 - y_1 \\ z_2 - z_1 \end{Bmatrix} \tag{2.2}$$

where $\boldsymbol{v}$ is the vector along the line determined by $\boldsymbol{P}_1$ and $\boldsymbol{P}_2$, t is an independent variable, and $\boldsymbol{P}(x, y, z)$ is an arbitrary point on the line.

Similar to the definitions of a reference point in Figure 2.5, a reference line can be defined directly based on its relationship to existing geometric elements. Figure 2.7a to d shows that the line is defined as the connection between two end-points, the intersecting lines of two faces, the rotational axis of a cylindrical surface, and the line passing through a point and perpendicular to a plane, respectively.

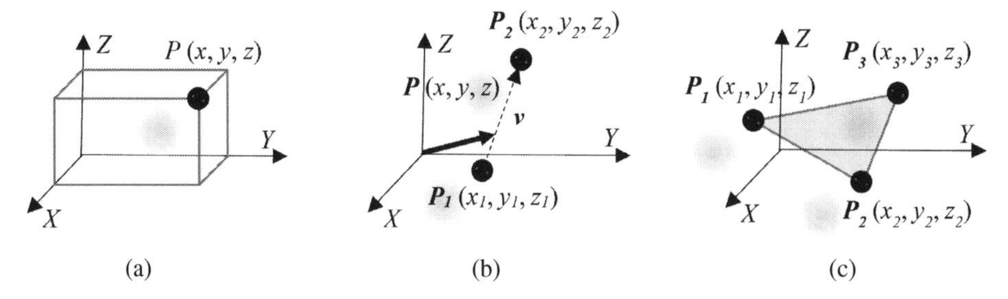

Figure 2.6 Using points to represent lines and planes. (a) Point. (b) Line. (c). Plane.

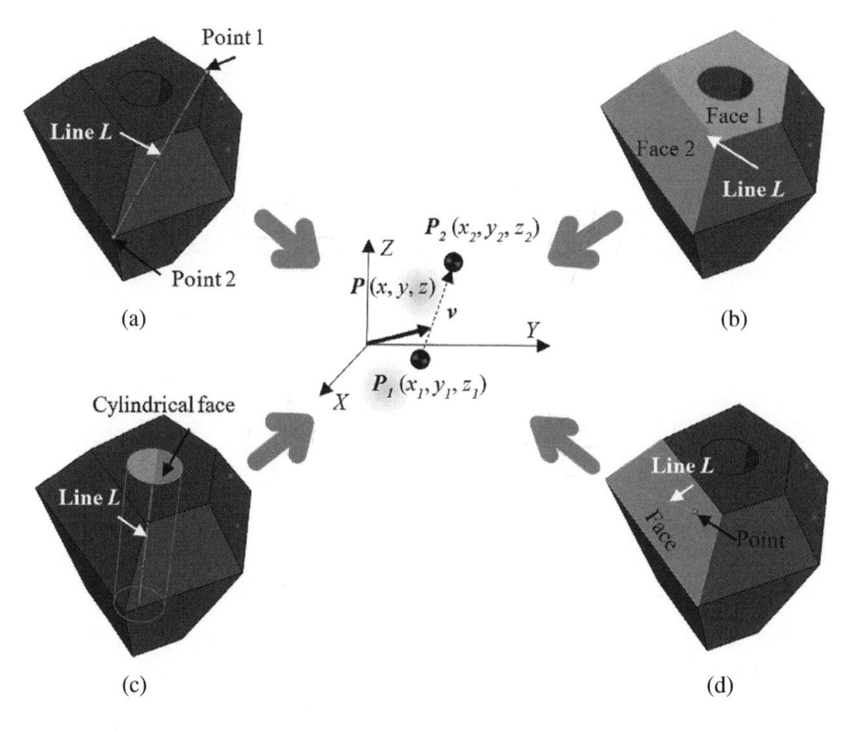

Figure 2.7 The definition of a reference line from existing geometric elements. (a) Line by two points. (b) Line by two interacting planes. (c) Line by a cylindrical face. (d) Line by passing a given point perpendicular to a plane.

As shown in Figure 2.6c, a plane ***PL*** in an arbitrary direction can be formed by three known points $\boldsymbol{P}_1$, $\boldsymbol{P}_2$, and $\boldsymbol{P}_3$. More generically, it can also be described as

$$a_1x + b_1y + c_1z + d_1 = 0 \tag{2.3}$$

where a_1, b_1, c_1, and d_1 are constant coefficients and (x, y, z) is an arbitrary point on ***PL***.

Similar to the definitions of a reference line in Figure 2.7, a reference plane be defined directly based on its relationship to existing geometric elements. Figure 2.8a to d shows that

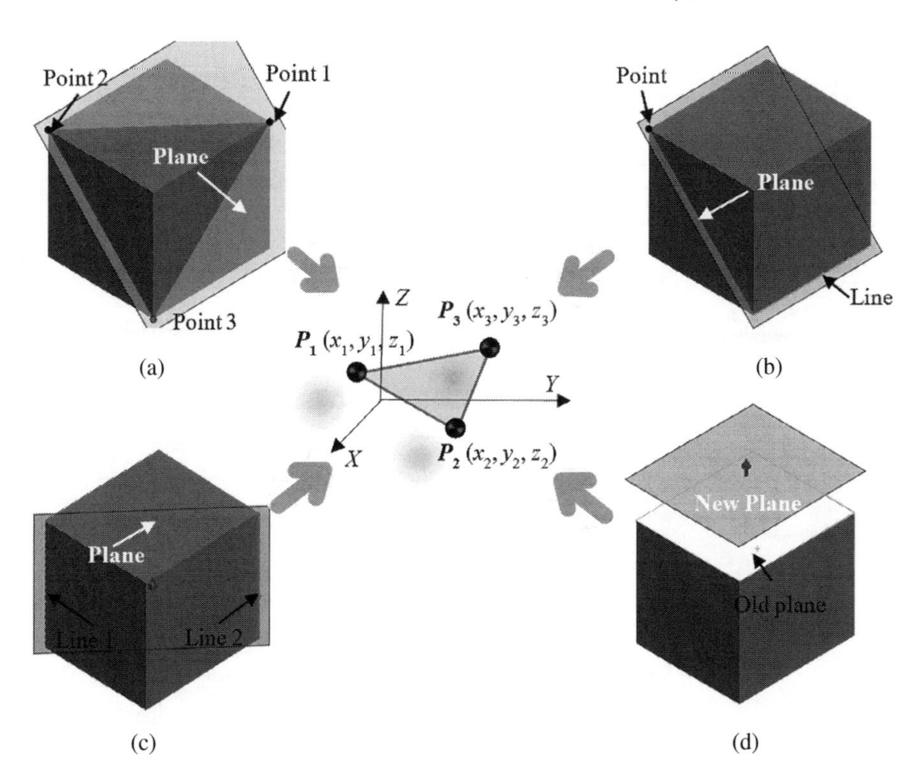

Figure 2.8 The definition of a reference plane from existing geometric elements. (a) Plane by three points. (b) Plane by a through point and line. (c) Plane by two parallel or interacting lines. (d) Plane by an offset of an existing plane.

the reference plane can be defined by specifying three points on the plane, one point and a line, two lines, and a distance to an existing paralleled plane, respectively.

2.2.3 Coordinate Transformation of Points

When an object consists of multiple geometric elements, the position and orientation of each element affect the geometry of the object. To place an element at the correct position and orientation, the coordinate transformation is often required. Table 2.1 shows the common types of coordinate transformation for a point. The coordinate transformation is performed *point by point* and the common coordinate transformations include *translation*, *scaling*, *rotation*, *mirroring*, and *projection*. In Table 2.1, the second column gives the explanations of these coordination transformations, and the third column gives the mathematical representation and graphic illustration of each type of transformation.

2.2.4 Coordinate Transformation of Objects

A point only includes positional data while an object includes both positional and orientational data in space. In addition, it may be convenient to represent the position and

Table 2.1 Coordinate transformation of a point.

Transformation	Features	Illustration
Translation	A translation is the simplest transformation and is the translation when the point $P\,(x, y, z)$ is moved by the vector $\boldsymbol{d}(d_x, d_y, d_z)$ to a new point $P'(x', y', z')$.	$\begin{Bmatrix} x' \\ y' \\ z' \end{Bmatrix} = \begin{Bmatrix} x \\ y \\ z \end{Bmatrix} + \begin{Bmatrix} d_x \\ d_y \\ d_z \end{Bmatrix}$
Scale	In case of scaling, every coordinate value of $\boldsymbol{P}\,(x, y, z)$ is multiplied by a constant. If the constants are the same along three axes, this corresponds to a *uniform scaling* (i.e. $C_x = C_y = C_z$). Otherwise, it is a *non-uniform scaling*.	$\begin{Bmatrix} x' \\ y' \\ z' \end{Bmatrix} = \begin{bmatrix} C_x & 0 & 0 \\ 0 & C_y & 0 \\ 0 & 0 & C_z \end{bmatrix} \begin{Bmatrix} x \\ y \\ z \end{Bmatrix}$
Rotation	A rotation refers to the rotation around a specified axis with an angle (i.e. θ_x, θ_y, or θ_z along the x, y, and z axes, respectively). A generic rotation along a specific axis can be decomposed as a series of aforementioned rotations.	$\begin{Bmatrix} x' \\ y' \\ z' \end{Bmatrix} = \begin{bmatrix} 1 & 0 & 0 \\ 0 & \cos\theta_x & -\sin\theta_x \\ 0 & \sin\theta_x & \cos\theta_x \end{bmatrix} \begin{Bmatrix} x \\ y \\ z \end{Bmatrix}$ $\begin{Bmatrix} x' \\ y' \\ z' \end{Bmatrix} = \begin{bmatrix} \cos\theta_y & 0 & \sin\theta_y \\ 0 & 0 & 0 \\ -\sin\theta_y & 0 & \cos\theta_y \end{bmatrix} \begin{Bmatrix} x \\ y \\ z \end{Bmatrix}$ $\begin{Bmatrix} x' \\ y' \\ z' \end{Bmatrix} = \begin{bmatrix} \cos\theta_z & -\sin\theta_z & 0 \\ \sin\theta_z & \cos\theta_z & 0 \\ 0 & 0 & 0 \end{bmatrix} \begin{Bmatrix} x \\ y \\ z \end{Bmatrix}$

Mirror	The mirror of an object is defined with respect to a reference plane, i.e. O-YZ, O-XZ, and O-XY planes, respectively.	$\begin{Bmatrix} x' \\ y' \\ z' \end{Bmatrix} = \begin{bmatrix} -1 & 0 & 0 \\ 0 & 1 & 0 \\ 0 & 0 & 1 \end{bmatrix} \begin{Bmatrix} x \\ y \\ z \end{Bmatrix}$ $\begin{Bmatrix} x' \\ y' \\ z' \end{Bmatrix} = \begin{bmatrix} 1 & 0 & 0 \\ 0 & -1 & 0 \\ 0 & 0 & 1 \end{bmatrix} \begin{Bmatrix} x \\ y \\ z \end{Bmatrix}$ $\begin{Bmatrix} x' \\ y' \\ z' \end{Bmatrix} = \begin{bmatrix} 1 & 0 & 0 \\ 0 & 1 & 0 \\ 0 & 0 & -1 \end{bmatrix} \begin{Bmatrix} x \\ y \\ z \end{Bmatrix}$ 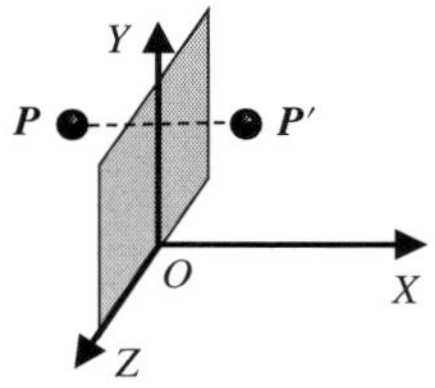
Projection	The transaction for projection computes the coordinates $\boldsymbol{P}'(x', y', z')$ of a point $\boldsymbol{P}(x, y, z)$ projected on a plane with a distance d to the observer.	$\begin{Bmatrix} x' \\ y' \\ z' \end{Bmatrix} = \begin{bmatrix} \frac{f}{z} & 0 & 0 \\ 0 & \frac{f}{z} & 0 \\ 0 & 0 & \frac{f}{z} \end{bmatrix} \begin{Bmatrix} x \\ y \\ z \end{Bmatrix}$ 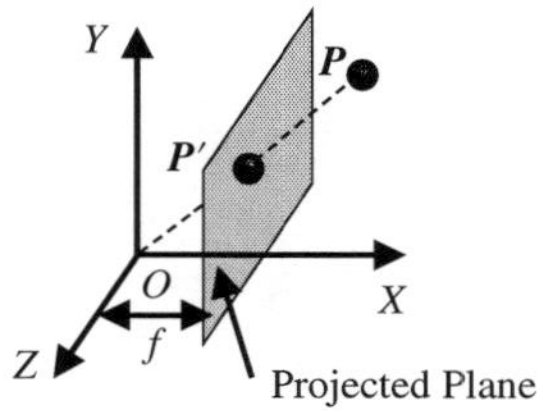

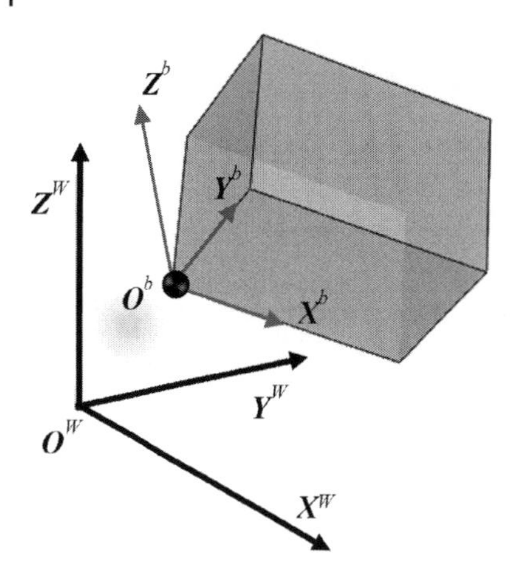

Figure 2.9 Object coordinate system in a world coordinate system.

orientation by a local coordinate system (LCS) attached to an object. Figure 2.9 shows such a CS, which is called an object coordinate system ($\boldsymbol{O^b} - \boldsymbol{X^b Y^b Z^b}$).

In Figure 2.9, a reference origin $\boldsymbol{O^b}(\boldsymbol{o^b_x}, \boldsymbol{o^b_y}, \boldsymbol{o^b_z})$ is used to represent the position of the object with respect to the world coordinate system ($\boldsymbol{O^w} - \boldsymbol{X^w Y^w Z^w}$) and the orientation of the object is represented by three unit vectors ($\boldsymbol{X^b}$, $\boldsymbol{Y^b}$, $\boldsymbol{Z^b}$). Furthermore, a *homogeneous coordinate* (x, y, z, h) and the corresponding 4 by 4 *homogenous matrix* $[\boldsymbol{T}]_{4x4}$ in Eq. (2.4) are widely used to facilitate the matrix manipulations for the coordinate transformation of objects:

$$[\boldsymbol{T}]_{4\times4} = \begin{bmatrix} \boldsymbol{X^b} & \boldsymbol{Y^b} & \boldsymbol{Z^b} & \boldsymbol{O^b} \\ \boldsymbol{0} & \boldsymbol{0} & \boldsymbol{0} & \boldsymbol{1} \end{bmatrix} = \begin{bmatrix} X^b_x & Y^b_x & Z^b_x & O^b_x \\ X^b_y & Y^b_y & Z^b_y & O^b_y \\ X^b_z & Y^b_z & Z^b_z & O^b_z \\ 0 & 0 & 0 & 1 \end{bmatrix} \tag{2.4}$$

where $[\boldsymbol{T}]_{4\times4}$ is the homogenous matrix for the representation of an object CS $\{\boldsymbol{O^b} - \boldsymbol{X^b Y^b Z^b}\}$ in the world coordinate system ($\boldsymbol{O^w} - \boldsymbol{X^w Y^w Z^w}$).

Note that the homogenous matrix in Eq. (2.4) includes six independent variables: three for the position and the rest for the orientation.

A homogenous transformation matrix can be generalized to deal with any transformation of objects in a coordinate system. Figure 2.10 shows the generalized matrix of homogeneous

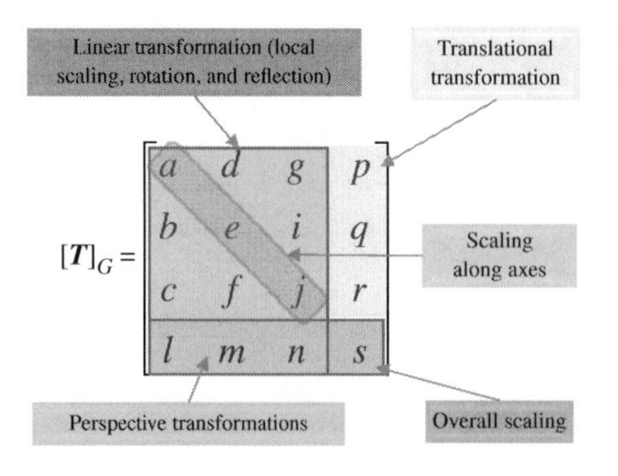

Figure 2.10 Impacts of elements in a generalized homogeneous matrix.

transformation. The elements of the matrix in different fields affect the transformation of an object in different ways. For examples, the elements (a, e, j) scale the coordinates of x, y, and z, the elements (p, q, r) translate the coordinates of x, y, and z, and the element s gives an overall inverse scale of the object.

The generic homogeneous matrix $[\boldsymbol{T}]_G$ can be customized to implement some common coordinate transformations of an object. For example, the translation matrix $[\boldsymbol{T}]_{TR}$ of an object can be simplified as

$$[\boldsymbol{T}]_{TR} = \begin{bmatrix} 1 & 0 & 0 & p \\ 0 & 1 & 0 & q \\ 0 & 0 & 1 & r \\ 0 & 0 & 0 & 1 \end{bmatrix} \tag{2.5}$$

where $[\boldsymbol{T}]_{TR}$ is the 3D translation matrix and p, q, r are the translational distances of a point from its original position along the $\boldsymbol{x}$, $\boldsymbol{y}$, and $\boldsymbol{z}$ axes, respectively.

The scaling matrix $[\boldsymbol{T}]_{SC}$ of a 3D object can be defined as

$$[\boldsymbol{T}]_{SC} = \begin{bmatrix} a & 0 & 0 & 0 \\ 0 & e & 0 & 0 \\ 0 & 0 & j & 0 \\ 0 & 0 & 0 & 1 \end{bmatrix} \tag{2.6}$$

where $[\boldsymbol{T}]_{SC}$ is the 3D scaling matrix and a, e, j are the scaling factors along the $\boldsymbol{x}$, $\boldsymbol{y}$, and $\boldsymbol{z}$ axes, respectively.

Example 2.1 Given a cube with the size of $1 \times 1 \times 1$ at its origin in Figure 2.11, determine the vertices of the object after the lengths along the $\boldsymbol{x}, \boldsymbol{y}, \boldsymbol{z}$ axes are scaled up by 2, 3, and 1, respectively.

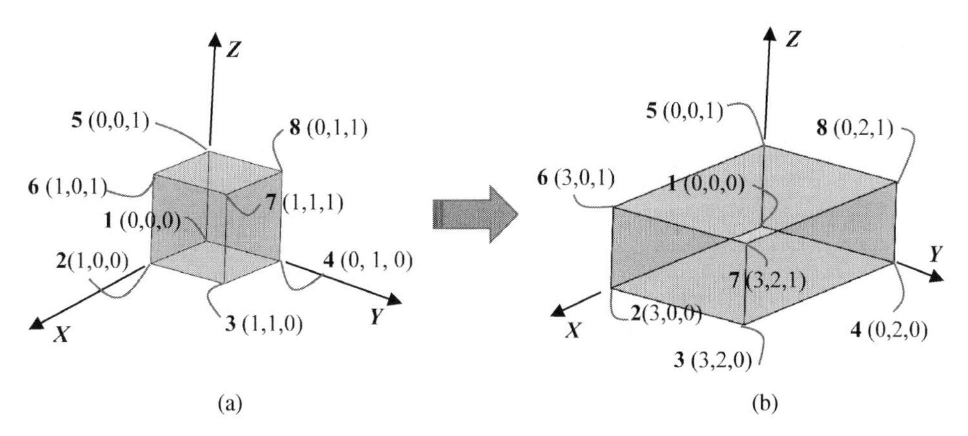

Figure 2.11 Example of a scaling matrix of a 3D object. (a) Before transformation. (b) After transformation.

Solution

Before the transformation, the set of eight vertices (1, 2, …, 8) in their homogeneous coordinates is written as

$$[\boldsymbol{P}]_{org} = \begin{bmatrix} 0 & 1 & 1 & 0 & 0 & 1 & 1 & 0 \\ 0 & 0 & 1 & 1 & 0 & 0 & 1 & 1 \\ 0 & 0 & 0 & 0 & 1 & 1 & 1 & 1 \\ 1 & 1 & 1 & 1 & 1 & 1 & 1 & 1 \end{bmatrix}$$

Using the given scales along the x, y, z axes, the scaling matrix Eq. (2.6) is given as

$$[\boldsymbol{T}]_{SC} = \begin{bmatrix} 2 & 0 & 0 & 0 \\ 0 & 3 & 0 & 0 \\ 0 & 0 & 1 & 0 \\ 0 & 0 & 0 & 1 \end{bmatrix}$$

Therefore, the set of vertices of the object after scaling transformation can be found as

$$[\boldsymbol{P}]_{new} = [\boldsymbol{T}]_{SC} \cdot [\boldsymbol{P}]_{org} = \begin{bmatrix} 0 & 2 & 2 & 0 & 0 & 2 & 2 & 0 \\ 0 & 0 & 3 & 3 & 0 & 0 & 3 & 3 \\ 0 & 0 & 0 & 0 & 1 & 1 & 1 & 1 \\ 1 & 1 & 1 & 1 & 1 & 1 & 1 & 1 \end{bmatrix}$$

A *3D reflection* refers to a transformation of mirroring an object about a specified plane. A reflection transformation matrix can be treated as a special case of a scaling matrix, where the scale of the corresponding axis of the reflected coordinate is set as a negative value of −1. Note that the axis for the reflection is aligned with the normal reflecting or mirroring plane. Therefore, the reflection coordination transformations about ***YZ***, ***XZ***, and ***XY*** planes are expressed by $[\boldsymbol{T}]_{RE\_YZ}$, $[\boldsymbol{T}]_{RE\_XZ}$, and $[\boldsymbol{T}]_{RE\_XZ}$, respectively, as follows:

$$\left.\begin{aligned} [\boldsymbol{T}]_{RE\_YZ} &= \begin{bmatrix} -1 & 0 & 0 & 0 \\ 0 & 1 & 0 & 0 \\ 0 & 0 & 1 & 0 \\ 0 & 0 & 0 & 1 \end{bmatrix} \\ [\boldsymbol{T}]_{RE\_XZ} &= \begin{bmatrix} 1 & 0 & 0 & 0 \\ 0 & -1 & 0 & 0 \\ 0 & 0 & 1 & 0 \\ 0 & 0 & 0 & 1 \end{bmatrix} \\ [\boldsymbol{T}]_{RE\_XY} &= \begin{bmatrix} 1 & 0 & 0 & 0 \\ 0 & 1 & 0 & 0 \\ 0 & 0 & -1 & 0 \\ 0 & 0 & 0 & 1 \end{bmatrix} \end{aligned}\right\} \tag{2.7}$$

Example 2.2 Figure 2.12a shows an object in the coordinate system (***O-XYZ***) with the specified set of vertices as follows:

$$[\boldsymbol{P}]_{org} = \begin{bmatrix} 1 & 2 & 2 & 1 & 1 & 2 & 2 & 1 & 1 & 2 & 2 & 1 \\ 1 & 1 & 4 & 4 & 1 & 1 & 2 & 2 & 2 & 2 & 4 & 4 \\ 1 & 1 & 1 & 1 & 3 & 3 & 3 & 3 & 2 & 2 & 2 & 2 \\ 1 & 1 & 1 & 1 & 1 & 1 & 1 & 1 & 1 & 1 & 1 & 1 \end{bmatrix}$$

Determine the reflected objects about the ***XY***, ***YZ***, ***XZ*** planes, respectively.

Solution

Using Eq. (2.7), the vertices of the reflected objects about the ***XY***, ***YZ***, ***XZ*** planes are determined and illustrated in Figure 2.12b to d, respectively. Taking as an example determination of the reflected object about the *XY* plane,

$$[\boldsymbol{P}]_{new} = [\boldsymbol{T}]_{RE-XY} \cdot [\boldsymbol{P}]_{org}$$

$$= \begin{bmatrix} 1 & 2 & 2 & 1 & 1 & 2 & 2 & 1 & 1 & 2 & 2 & 1 \\ 1 & 1 & 4 & 4 & 1 & 1 & 2 & 2 & 2 & 2 & 4 & 4 \\ -1 & -1 & -1 & -1 & -3 & -3 & -3 & -3 & -2 & -2 & -2 & -2 \\ 1 & 1 & 1 & 1 & 1 & 1 & 1 & 1 & 1 & 1 & 1 & 1 \end{bmatrix}$$

where $[\boldsymbol{T}]_{RE-XY}$ is given in Eq. (2.7) for the reflected object about the ***XY*** plane.

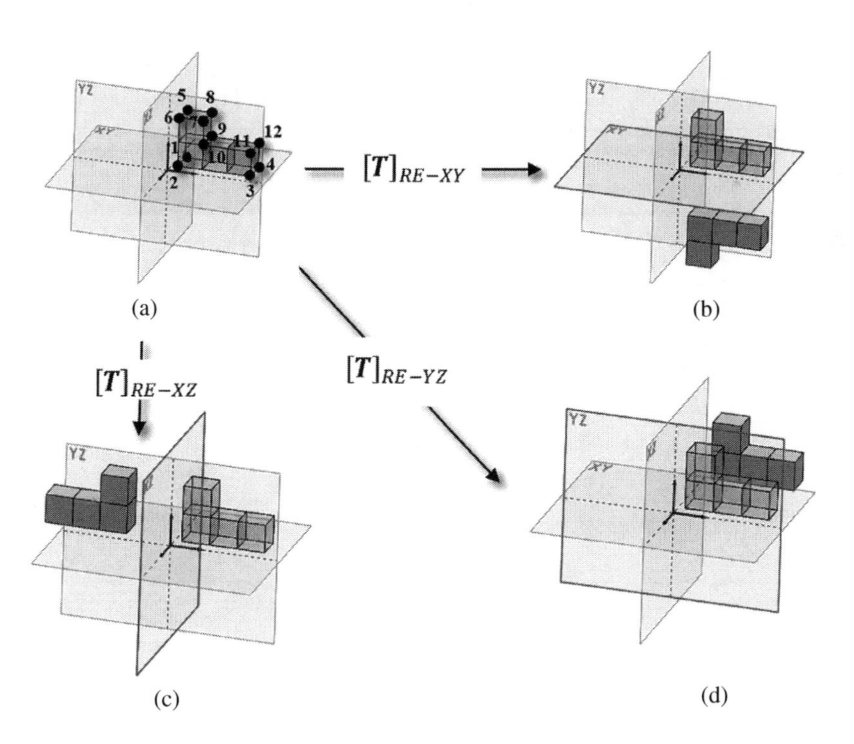

Figure 2.12 Reflection coordinate transformation matrices of an object. (a) Object at the original position. (b) Reflected object about the **XY** plane. (c) Reflected object about the **YZ** plane. (d) Reflected object about the **YZ** plane.

The coordinate transformation matrix for an object with *a shearing transformation* is represented as (TutorialsPoint 2018)

$$[\boldsymbol{T}]_{SH} = \begin{bmatrix} 1 & d & g & 0 \\ b & 1 & i & 0 \\ c & f & 1 & 0 \\ 0 & 0 & 0 & 1 \end{bmatrix} \tag{2.8}$$

where $[\boldsymbol{T}]_{SH}$ is the shearing transformation matrix, d, g, i, b, c, f are the shearing factors of $\boldsymbol{y}$ along $\boldsymbol{x}$, $\boldsymbol{z}$ along $\boldsymbol{x}$, $\boldsymbol{z}$ along $\boldsymbol{y}$, $\boldsymbol{x}$ along $\boldsymbol{y}$, $\boldsymbol{x}$ along $\boldsymbol{z}$, and $\boldsymbol{y}$ along $\boldsymbol{z}$, respectively. A shearing is a relative angle deformation between two axes. Therefore, $[\boldsymbol{T}]_{SH}$ has to be symmetric, i.e. $b = d$, $g = c$, and $i = f$. In other words, $[\boldsymbol{T}]_{SH}$ only has three independent variables.

For a point $\boldsymbol{P}$ in its homogeneous coordinate $(x, y, z, 1)$, Eq. (2.8) is applied to define its new position P' $(x', y', z', 1)$ as

$$\left. \begin{array}{l} \begin{Bmatrix} x' \\ y' \\ z' \\ 1 \end{Bmatrix} = [\boldsymbol{T}]_{SH} \begin{Bmatrix} x \\ y \\ z \\ 1 \end{Bmatrix} = \begin{Bmatrix} x + d \cdot y + g \cdot z \\ b \cdot x + y + i \cdot z \\ c \cdot x + f \cdot y + z \\ 1 \end{Bmatrix} \\ b = d, g = c, i = f \end{array} \right\} \tag{2.9}$$

Example 2.3 Given a cube with the size of $1 \times 1 \times 1$ at its origin in Figure 2.13, assume that node 1 is fixed and node 5 is sheared from (1, 1, 1) to (2, 2, 2), and determine the matrix for 3D shearing and the new positions of all vertices after shearing.

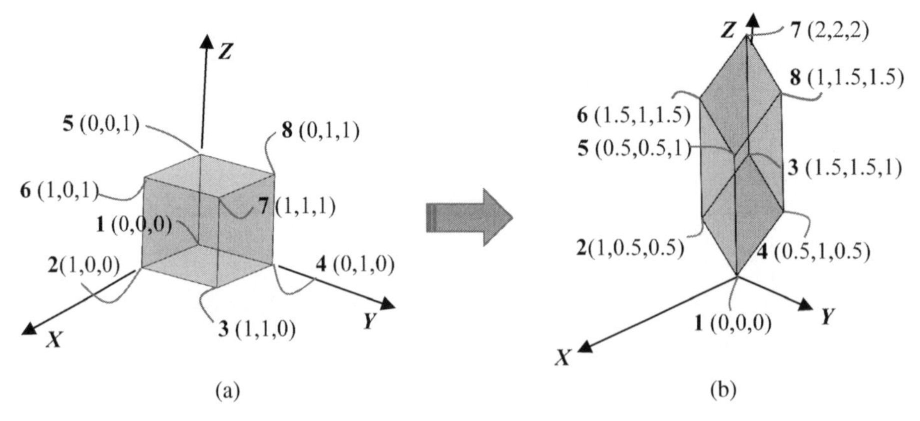

Figure 2.13 Example of the 3D shearing transformation of an object. (a) Before transformation. (b) After transformation.

Solution

To determine the shearing matrix, six independent variables must be determined in Eq. (2.9). Since the positions of nodes 1 and 5 are known before and after shearing, Eq. (2.9) is applied to determine the conditions that these variables must satisfy:

$$\left.\begin{array}{l}\text{For Node 5}: \begin{Bmatrix} 2 \\ 2 \\ 2 \\ 1 \end{Bmatrix} = \begin{Bmatrix} 1+d+g \\ b+1+i \\ c+f+1 \\ 1 \end{Bmatrix} \\ b=d, g=c, i=f \end{array}\right\} \tag{2.10}$$

Solving Eq. (2.10) gives $b = d = c = g = f = i = 0.5$.

Figure 2.13b shows the sheared object whose vertices after shearing are determined as

$$[\boldsymbol{P}]_{new} = [\boldsymbol{T}]_{SH} \cdot [\boldsymbol{P}]_{org} = \begin{bmatrix} 1 & 0.5 & 0.5 & 0 \\ 0.5 & 1 & 0.5 & 0 \\ 0.5 & 0.5 & 1 & 0 \\ 0 & 0 & 0 & 1 \end{bmatrix} \begin{bmatrix} 0 & 1 & 1 & 0 & 0 & 1 & 1 & 0 \\ 0 & 0 & 1 & 1 & 0 & 0 & 1 & 1 \\ 0 & 0 & 0 & 0 & 1 & 1 & 1 & 1 \\ 1 & 1 & 1 & 1 & 1 & 1 & 1 & 1 \end{bmatrix}$$

$$= \begin{bmatrix} 0 & 1 & 1.5 & 0.5 & 0.5 & 1.5 & 2 & 1 \\ 0 & 0.5 & 1.5 & 1 & 0.5 & 1 & 2 & 1.5 \\ 0 & 0.5 & 1 & 0.5 & 1 & 1.5 & 2 & 1.5 \\ 1 & 1 & 1 & 1 & 1 & 1 & 1 & 1 \end{bmatrix}$$

The elements a to j in the reflection transformation matrix of Eq. (2.7) can also be determined for a rotation transformation matrix along an axis in a 3D space. The corresponding matrix can be derived from the rotational matrices for the points in Table 2.1. Figure 2.14 shows the rotational transformation matrices when $\boldsymbol{x}$, $\boldsymbol{y}$, and $\boldsymbol{z}$ are chosen as a rotational axis for a rotational angle of θ_x, θ_y, and θ_z, respectively.

A series of the coordinate transformations, such as translation, scaling, shearing, rotation, and reflection, can be combined as a comprehensive transformation of a point or object. In such a case, the combined transformation matrix can be determined as

$$[\boldsymbol{T}]_{COM} = [\boldsymbol{T}]_N [\boldsymbol{T}]_{N-1} \cdots [\boldsymbol{T}]_i \cdots [\boldsymbol{T}]_1 \tag{2.11}$$

where $[\boldsymbol{T}]_{COM}$ is the combined transformation matrix, $i = 1, 2, \ldots, N$ is the order of N transformation operations, and $[T]_i$ is the transformation matrix for the ith operation.

Note that the order of the multiplications of the matrices is reversed with that of the operations.

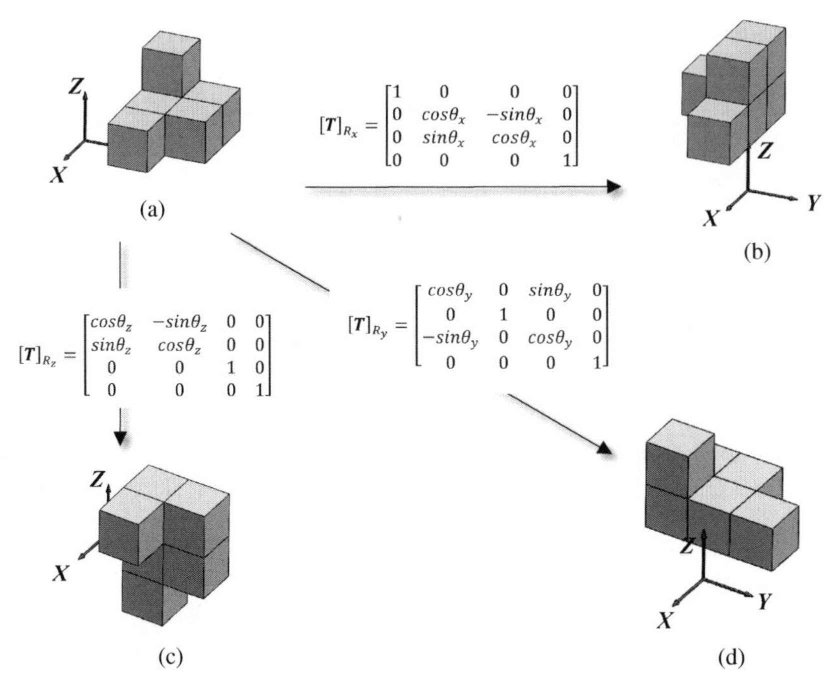

Figure 2.14 Rational coordinate transformation of an object. (a) Object in the original position. (b) The x-axis rotation (θ_x). (c) The z-axis rotation (θ_z). (d) The y-axis rotation (θ_y).

Example 2.4 Given a point $P(1, 3, 2)$ in a reference coordinate system (***O-XYZ***), determine its new position $\boldsymbol{P'}$ after the following operations: (i) rotate 45° with respect to the ***X*** axis, (ii) translate it with a vector of (0.5, 2, 1), (iii) rotate −30° with respect to the ***Z*** axis, and (iv) mirror the position about the ***XY*** plane.

Solution

Using the coordinate transformations in Table 2.1, the corresponding transformation matrix of each operation is determined as follows:

$$[\boldsymbol{T}]_1 = [\boldsymbol{T}]_{\boldsymbol{Rx}} = \begin{bmatrix} 1 & 0 & 0 & 0 \\ 0 & \cos 45^\circ & -\sin 45^\circ & 0 \\ 0 & \sin 45^\circ & \cos 45^\circ & 0 \\ 0 & 0 & 0 & 1 \end{bmatrix} \qquad [\boldsymbol{T}]_2 = [\boldsymbol{T}]_{\boldsymbol{TR}} = \begin{bmatrix} 1 & 0 & 0 & 0.5 \\ 0 & 1 & 0 & 2 \\ 0 & 0 & 1 & 1 \\ 0 & 0 & 0 & 1 \end{bmatrix}$$

$$[\boldsymbol{T}]_3 = [\boldsymbol{T}]_{\boldsymbol{Rz}} = \begin{bmatrix} \cos 30^\circ & -\sin 30^\circ & 0 & 0 \\ \sin 30^\circ & \cos 30^\circ & 0 & 0 \\ 0 & 0 & 1 & 0 \\ 0 & 0 & 0 & 1 \end{bmatrix} \qquad [\boldsymbol{T}]_4 = [\boldsymbol{T}]_{\boldsymbol{RE-XY}} = \begin{bmatrix} 1 & 0 & 0 & 0 \\ 0 & 1 & 0 & 0 \\ 0 & 0 & -1 & 0 \\ 0 & 0 & 0 & 1 \end{bmatrix}$$

Therefore, using Eq. (2.11),

$$[T]_{COM} = [T]_4[T]_3[T]_2[T]_1 = \begin{bmatrix} 0.866 & -0.3535 & 0.3535 & -0.567 \\ 0.5 & 0.612262 & -0.61226 & 1.982 \\ 0 & -0.707 & -0.707 & -1 \\ 0 & 0 & 0 & 0 \end{bmatrix}$$

$$P' = [T]_{COM} \cdot P = [T]_{COM} \cdot \begin{Bmatrix} 0.5 \\ 2 \\ 1 \\ 1 \end{Bmatrix} = \begin{Bmatrix} -0.4875 \\ 2.844262 \\ -3.121 \\ 1 \end{Bmatrix}$$

2.3 Representation of Shapes

A solid object has a geometric shape. *A solid model* is a digital representation of the geometry of an existing or envisioned physical object. A solid can be modelled in two basic methods: (i) designers may specify points, curves, and surfaces to define the mathematical representation of boundary surfaces and knit them together as a finite volume of the object; (ii) designers may select the models of simple shapes, such as blocks or cylinders, with specified dimensions, positions, and orientations, and combine them using logic operations such as union, intersection, and difference. The resulting representation is an unambiguous, complete, and detailed digital approximation of the object geometry or an assembly of objects (such as a car engine or an entire aeroplane).

Figure 2.15 shows that a solid object must be a watertight fine volume (*B*), which is defined by a number of boundary faces (*F*). In turn, a face *F* consists of a number of edges

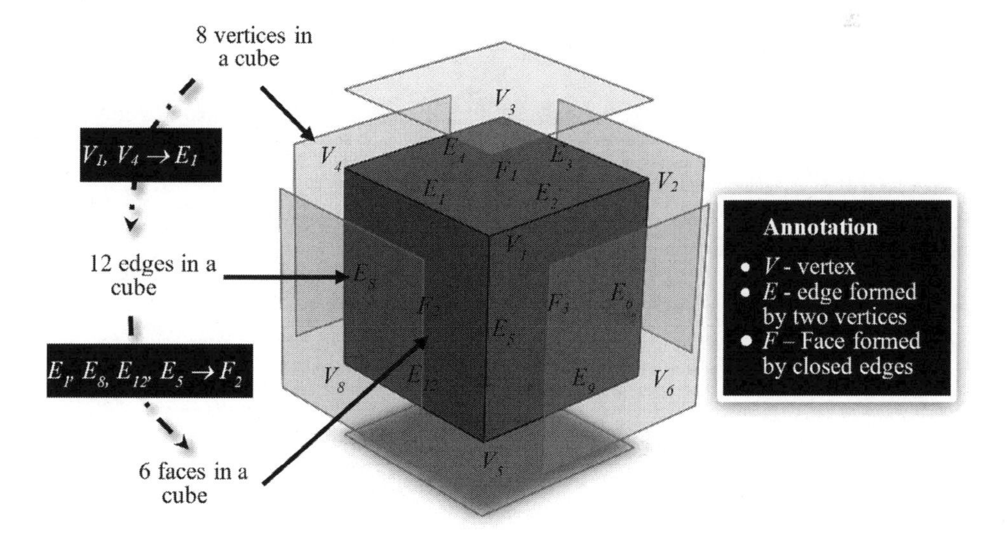

Figure 2.15 Vertices, edges, and faces of a solid.

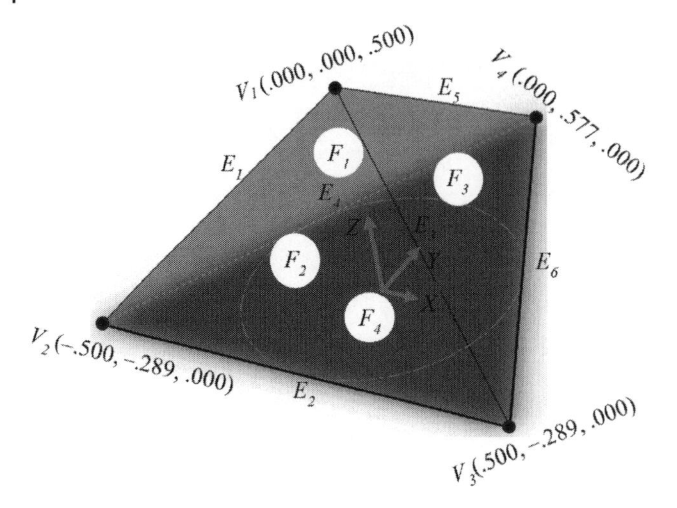

Figure 2.16 Relational structure of a pyramid object.

(E) and an edge E consists of a number of vertices (V). The information about vertices, edges, faces, and volumes is at different levels. In order to work with the solid models in the computer systems, they should be stored in the computer memory, which allows processing, converting, and ultimately displaying them. Different data structures can be used to represent solids.

2.3.1 Basic Data Structure

The information of solids must be stored in computers. Three popular methods to represent the data structure of solids are *relational structure, hierarchical structure*, and *network structure.*

A relational structure defines a set of lists for vertices, edges, and faces, and these lists are stored in term of arrays in a relational database. Taking an example of the pyramid in Figure 2.16, its rational structure includes the lists of vertices, edges, and faces shown in Table 2.2.

Table 2.2 Relational structure of pyramid object.

List of vertices	List of edges	List of faces
V_1 (0.000, 0.000, 0.500)	$E_1\ (V_1, V_2)$	$F_1\ (E_1, E_4, E_5)$
V_2 (−0.500, −0.289, 0.000)	$E_2\ (V_2, V_3)$	$F_2\ (E_1, E_2, E_3)$
V_3 (0.500, −0.289, 0.000)	$E_3\ (V_1, V_3)$	$F_3\ (E_3, E_6, E_5)$
V_4 (−0.000, −0.577, 0.000)	$E_4\ (V_2, V_4)$	$F_4\ (E_2, E_6, E_4)$
	$E_5\ (V_1, V_4)$	
	$E_6\ (V_3, V_4)$	

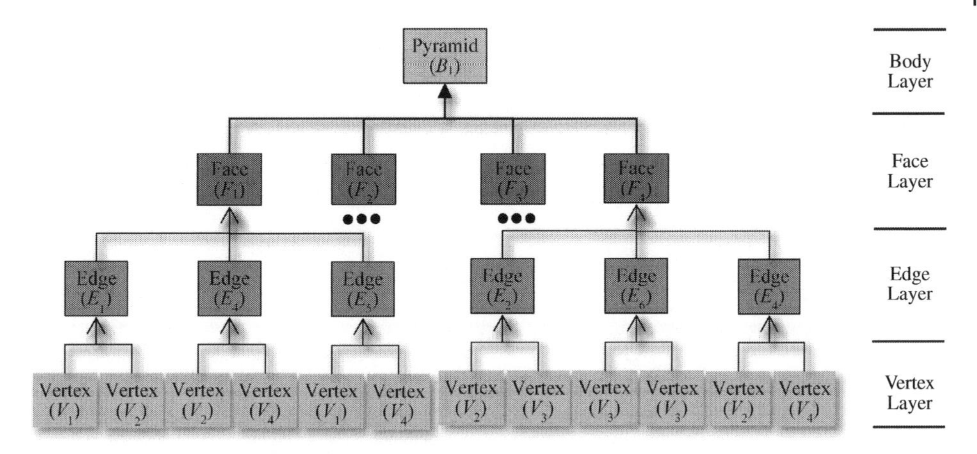

Figure 2.17 Hierarchical structure of a pyramid object.

A solid can also be represented by a hierarchical structure where geometric elements are organized by a tree-like layer structure. For a pyramid object, the highest layer is the solid body, which is bounded by four faces, each face being bounded by three edges and each edge being defined by two vertices. The complete hierarchical structure of a pyramid object is shown in Figure 2.17.

A hierarchical structure includes redundant information since the same geometric elements may have relations with multiple elements at high levels. Such redundancies can be eliminated by using a network data structure. As shown in Figure 2.18, a network data structure uses data pointers to represent the topological connections of geometric elements. The number of data pointers for a geometric element type can be varied based on the number of connections the element has with others.

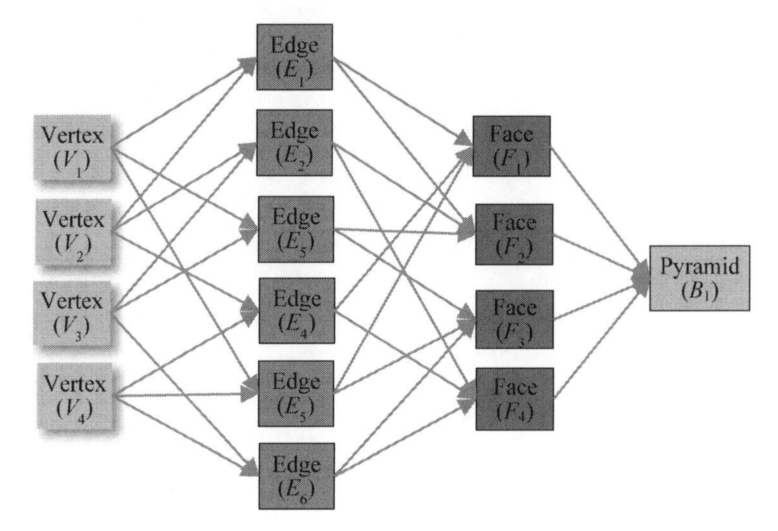

Figure 2.18 Network structure of a pyramid object.

Table 2.3 Representation of 2D and 3D curves.

Curvy features	Representation		Example
2D curve	Explicit	$\left.\begin{array}{r} x = x(t) \\ y = y(t) \\ t \in [0,1] \end{array}\right\}$	$\left.\begin{array}{c} x = x_0 + R \cdot \cos(2\pi t) \\ y = y + R \cdot \sin(2\pi t) \\ t \in [0,1] \end{array}\right\}$
	Implicit	$f(x, y) = 0$	$(x - x_0)^2 + (y - y_0)^2 = R^2$
3D curve	Explicit	$\left.\begin{array}{r} x = x(t) \\ y = y(t) \\ z = z(t) \\ t \in [0,1] \end{array}\right\}$	$\left.\begin{array}{c} x = x_1 \cdot t + x_2 \cdot (1 - t) \\ y = y_1 \cdot t + y_2 \cdot (1 - t) \\ z = z_1 \cdot t + z_2 \cdot (1 - t) \\ t \in [0,1] \end{array}\right\}$
	Implicit	$\left.\begin{array}{r} f_1(x, y, z) = 0 \\ f_2(x, y, z) = 0 \end{array}\right\}$	$\left.\begin{array}{c} \dfrac{x - x_1}{y - y_1} = \dfrac{x_2 - x_1}{y_2 - y_1} \\ \dfrac{z - z_1}{y - y_1} = \dfrac{z_2 - z_1}{y_2 - y_1} \end{array}\right\}$

2.3.2 Curvy Geometric Elements

Most objects have curvy boundary edges and surfaces. A curvy edge has one independent variable. As shown in Table 2.3, a curvy edge in 2D and 3D can be represented explicitly or implicitly in terms of a normalized length variable t from the starting point to the ending point.

The complexity of a 3D curve can be measured by the order of polynomial terms in its mathematic model for piecewise interpolation. Given a number of control points on the curve, different interpolation methods lead to different results for 3D curves.

The mathematic models for 2D or 3D curves in Tables 2.3 and 2.4 can be readily expanded to represent 3D surfaces as follows:

In an *explicit form*:

$$\left.\begin{array}{l} x = x(u, v) \\ y = y(tu, v) \\ z = z(tu, v) \\ u, v \in [0, 1] \end{array}\right\} \tag{2.12}$$

where u, v are normalized independent variables of surface.

Table 2.4 High-order curves.

Interpolation	Representation	Illustration
Lagrange	$\left.\begin{array}{l} P(t) = \sum_{i=0}^{n-1} L_i(t) \cdot P_i \\ L_i(t) = \dfrac{\prod_{j\neq i}(t - t_i)}{\prod_{j\neq i}(t_j - t_i)} \end{array}\right\}$	
Bezier	$\left.\begin{array}{l} P(t) = \sum_{i=0}^{n} b_{i,n}(t) \cdot P_i \\ b_{i,n}(t) = \begin{pmatrix} n \\ i \end{pmatrix} t^i (1-t)^{n-i} \\ t \in [0,1]; \begin{pmatrix} n \\ i \end{pmatrix} = \dfrac{n!}{i!(n-i)!} \end{array}\right\}$	
Cubic spline	$P(t) = a_1 + ta_2 + t^2 a_3 + t^3 a_4$ where $t = [0, 1]$ and the coefficient vectors a_1, a_2, a_3, and a_4 are selected to satisfy $\left.\begin{array}{l} P(0) = P_i; P(1) = P_{i+1} \\ P'(1) = P_i(0); P^{\varepsilon}(1) = P^{\varepsilon}(0) \end{array}\right\}$	

In an *implicit form*:

$$f(x, y, z) = 0 \tag{2.13}$$

An example of a spherical surface in Figure 2.19 can be represented mathematically as
In an *explicit form*:

$$\left.\begin{array}{l} x = x_0 + R \cdot \cos(2\pi u) \sin(\pi v) \\ y = y_0 + R \cdot \sin(2\pi u) \sin(\pi v) \\ z = z_0 + R \cdot \cos(\pi v) \\ \quad u, v \in [0, 1] \end{array}\right\} \tag{2.14}$$

In an *implicit form*:

$$(x - x_0)^2 + (y - y_0)^2 + (z - z_0)^2 = R^2 \tag{2.15}$$

In computer aided geometric modelling, 2D and 3D curves are commonly used to generate 3D surfaces. Depending on surface features, 3D surfaces can be classified into *swept*, *ruled-generated*, or *free-formed* surfaces, as shown in Table 2.5.

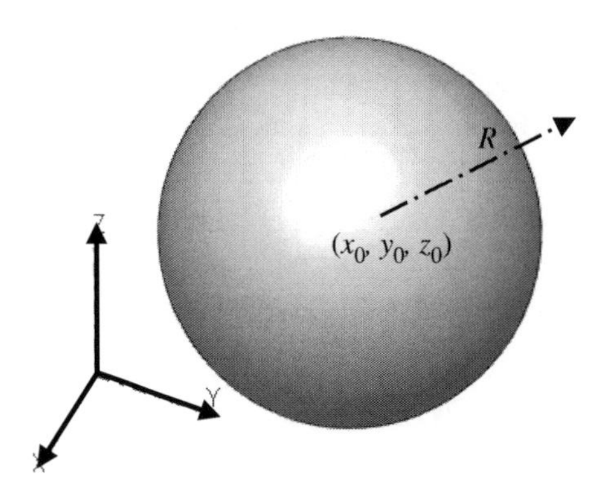

Figure 2.19 Representation of a spherical surface.

Table 2.5 Types and features of surfaces.

Surface type	Feature	Illustration
Swept	A swept surface is defined by two elements, i.e. driving curve (*D*) and guide curve (*G*) or trajectory. The driving curve (*D*) can be open or closed. The guide curve (*G*) will run along *D* with the constant contact point.	
Ruled	A ruled surface is defined by three *3D* curves. The *G* curve drives along the *D*1 curve and leans in *D*2. In the first case, *D*1 and *D*2 are divided into equal segments and the end points of these segments are connected by *G*. In the second case, the *G* curve just leans to *D*1 and *G* will be parallel in every position. Other variations can be generated from a ruled surface by application of a non-constant *G* curve.	
Freeform generated	If a surface cannot be described by analytic or moving curves, they are called free-form or sculpture surfaces. Control points are used to determine the surface. The mathematic presentation of these surfaces is similar to the spline curves. The parametric surface description uses two independent variables (u, v).	

Table 2.6 Manifold and non-manifold examples in 1D, 2D, and 3D.

N-dimension	Manifold example	Non-manifold example	
1D (line)	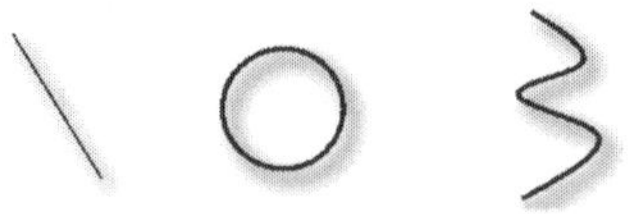		The intersecting point is not locally homogeneous.
2D (surface)			The intersecting line is not locally homogeneous.
3D (solid)			The geometry mixes 2D and 3D entities.

Geometric modelling is used to create a virtual geometric representation of a real or imagined object, which includes information of the shape, dimensions, and materials of an object. Many methods have been developed to model geometrics of products. A better understanding of the theoretical basics of geometric modelling helps in (i) improving modelling efficiency and (ii) shortening the learning curves of various CAD systems. While every method has its limitations, no universal solution is available that satisfies all demands for geometric models in itself. To select a modelling method, one must ensure the validity of the geometries.

A *manifold* is a topological space that locally resembles a Euclidean space near each point. In an n-dimensional manifold, any neighbourhood of a point is homoeomorphic to the Euclidean space of dimension n. Accordingly, computer geometric models can be classified based on *manifolds* or *non-manifolds*. Table 2.6 shows a few examples of manifolds or non-manifolds; a non-manifold usually includes some entities of different dimensions (1D, 2D, or 3D).

In a computer representation, the information about physical objects is digitized. In other words, a free-form curve or surface is represented by a set of straight-line segments or flat patches. The *geometric topology* concerns the connectivity of geometric elements. As shown in Figure 2.20, not all geometric elements can be connected together for a valid geometric topology.

A valid polyhedral in a 3D space should be homomorphic to a sphere and the validity of the geometry can be evaluated using *the Euler–Poincare Law* as

$$\left.\begin{array}{ll} F - E + V - L = 2(B - G) & \text{general geometries} \\ F - E + V = 2 & \text{simple solids} \\ F - E + V - L = B - G & \text{open objects} \end{array}\right\} \quad (2.16)$$

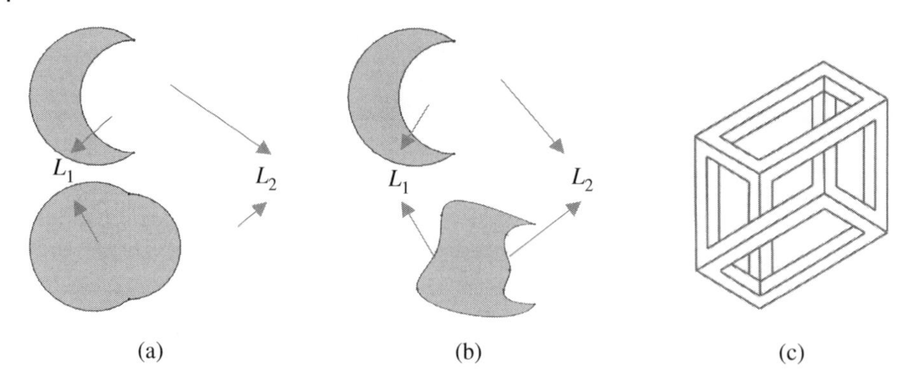

Figure 2.20 Examples of valid and invalid geometries. (a) Same geometries with different topologies. (b) Different geometries with the same topologies. (c) Invalid geometry.

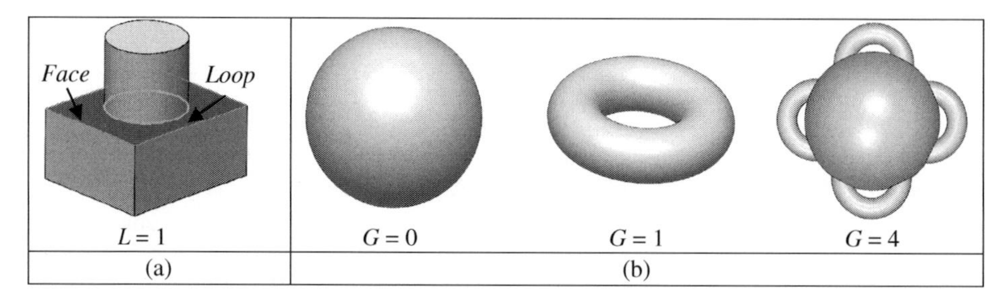

Figure 2.21 Inner loop and genus examples. (a) Inner loop example. (b) Genus examples.

where F, E, V, B, L, and G are the numbers of faces, edges, vertices, bodies, inner loops on faces, and genuses in a geometry, respectively.

The meanings of inner loops and genuses are illustrated by the examples in Figure 2.21.

Example 2.5 To use the Euler–Poincare Law to justify the validity of simple objects in the first column of Table 2.7.

Solution

As shown in the middle three columns of Table 2.7, the numbers of faces (F), edges (E), and vertices (V) are counted and the valid condition for a simple solid using the Euler–Poincare Law is applied. It shows that all objects satisfy the conditions as simple objects.

Example 2.6 To use the Euler–Poincare Law to justify the validity of open objects in the first column of Table 2.8.

Solution

As shown in the middle six columns of Table 2.8, the number of faces (F), edges (E), vertices (V), bodies (B), inner loops on faces (L), and genuses (G) in a geometry is counted and the valid condition for a simple solid by the Euler–Poincare Law is applied. It shows that all of the objects satisfy the conditions as open objects.

Table 2.7 Examples of simple objects.

Example	F	E	V	$F-E+V=2$
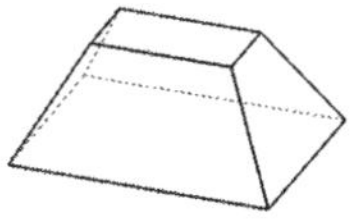	6	12	8	$6-12+8\equiv 2$
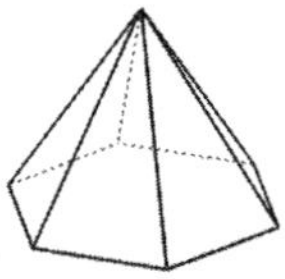	7	12	7	$7-12+7\equiv 2$
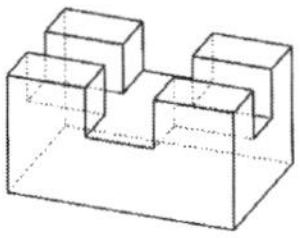	18	48	32	$18-48+32\equiv 2$
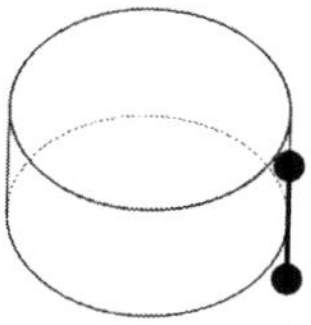	3	3	2	$3-3+2\equiv 2$
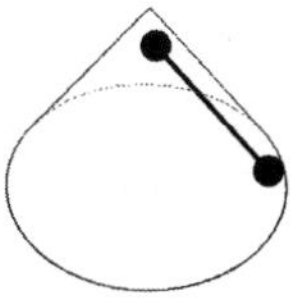	2	2	2	$2-2+2\equiv 2$
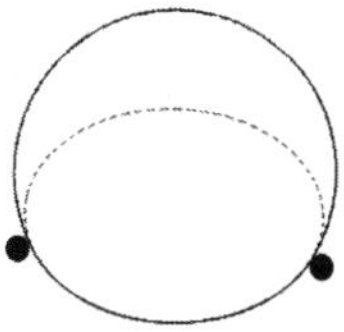	2	2	2	$2-2+2\equiv 2$

Example 2.7 To use *the Euler–Poincare law* to justify the validity of generic objects in the first column of Table 2.9.

Solution

As shown in the middle six columns of Table 2.9, the numbers of faces (F), edges (E), vertices (V), bodies (B), inner loops on faces (L), and genuses (G) in a geometry are counted and the valid condition for a simple solid using the Euler–Poincare Law is applied. It shows that all generic objects satisfy the conditions as generic objects.

Table 2.8 Examples of open objects.

Example	*F*	*E*	*V*	*L*	*B*	*G*	$F-E+V-L=B-G$
	0	1	2	0	1	0	$0-1+2–0 \equiv 1-0$
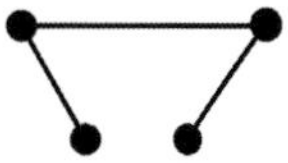	0	3	4	0	1	0	$0-3+4-0 \equiv 1-0$
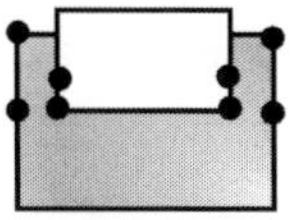	2	8	8	1	1	0	$2-8+8-1 \equiv 1-0$
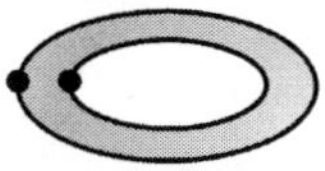	2	2	2	1	1	0	$2-2+2-1 \equiv 1-0$
	1	4	4	0	1	0	$1-4+4-0 \equiv 1-0$
	5	12	8	0	1	0	$5-12+8-0 \equiv 1-0$

Table 2.9 Examples of generic objects.

Example	*F*	*E*	*V*	*L*	*B*	*G*	$F-E+V-L=2(B-G)$
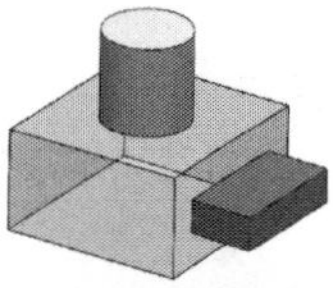	13	27	18	2	1	0	$13-27+18-2 \equiv 2(1-0)$
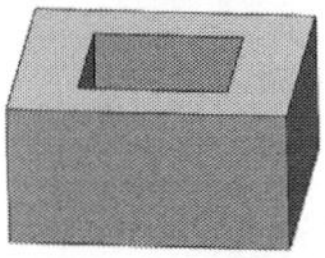	10	24	16	2	1	1	$10-24+16-2 \equiv 2(1-1)$
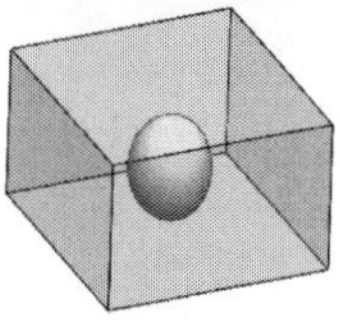	7	13	10	0	2	0	$7-13+10-0 \equiv 2(2-0)$
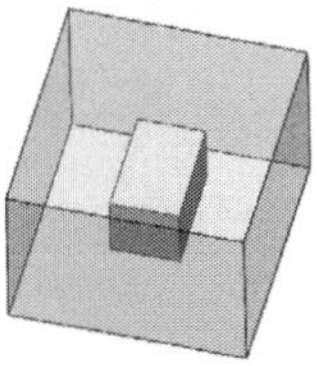	12	24	16	0	2	0	$12-24+16-0 \equiv 2(2-0)$

2.3.3 Euler–Poincare Law for Solids

Objects in the physical world are solid. A solid object consists of solid elements and the process of adding, merging, uniting, or deleting solid elements in a solid object is commonly called *Euler operation*. To sustain a valid solid object, the second condition in the Euler–Poincare Law, which is usually called the Euler formula ($F - E + V = 2$), must be satisfied. This leads to the following features of a solid object:

1. Each vertex must have at least three edges to meet together
2. Each edge must be shared by two and only two faces.
3. Any face must be homomorphic to a disk with no holes; it is simply connected and bounded by a single ring of edges.
4. A solid must be simply connected with no through hole.

2.4 Basic Modelling Methods

The methods of geometric modelling differ from one to another based on the level of information completeness of the features in objects. Figure 2.22 shows the evolution of modelling methods. *Wireframe modelling* is historically developed for computer aided drawing but only for the representation of key vertices and boundary edges. *Solid modelling* is the most advanced modelling technique since it is capable of representing the complete information of solids and beyond. Other methods such as *surface modelling*, *boundary surface modelling* (also called *B-Rep* or *Mantle modelling*), and *space decomposition* are between the extremes of wireframe modelling and solid modelling in terms of the completeness of information about solids.

2.4.1 Wireframe Modelling

A wireframe model represents the boundary edges of an object; these edges can be of lines, arcs, and curves. A wireframe model does not include the upper-level information such as

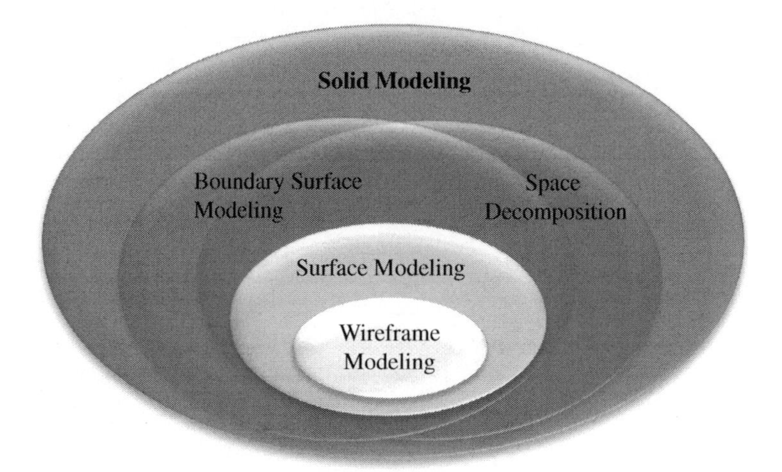

Figure 2.22 Variety of geometric modelling methods.

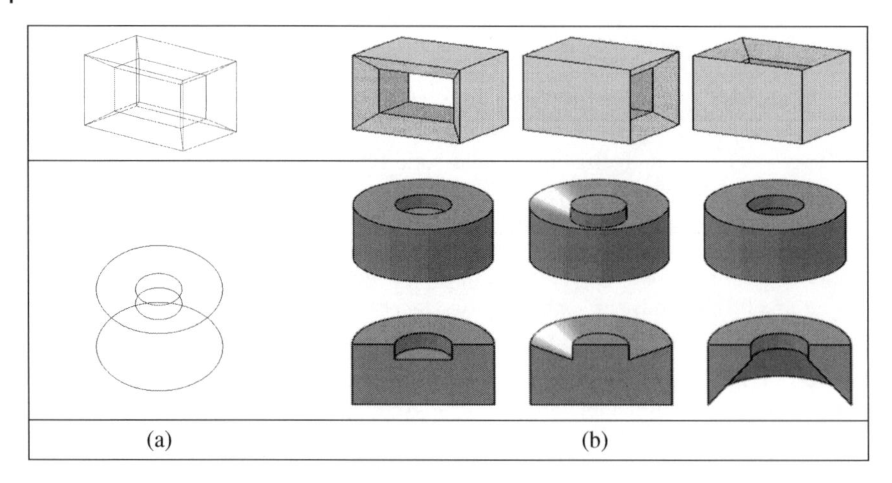

Figure 2.23 Ambiguity examples of wireframe models. (a) Wireframe model. (b) Possible solid geometries.

boundary surfaces or volumes. In addition, the results from wireframe modelling have the following limitations:

- All of the edges are displayed as elements in an image and the visibility caused by overlapping is not identifiable.
- No high-level information related to solids and masses such as surface areas or masses is available.
- The primary data from wireframe modelling is the coordinates of vertices; therefore, the preparation, importing, and processing of modelling data are very time-consuming and error-prone.
- The wireframe modelling method is incapable of designing shapes and specifying more complex forms due to the need for a large number of data points.

In addition, due to incompleteness of solid information, a wireframe model may cause the *ambiguity* of a represented geometry. Figure 2.23 shows two examples whose wireframe models in the left column are not able to distinguish the actual geometrics of objects since their corresponding wireframe models are the same.

On the other hand, a wireframe model is very concise and fundamental. It can be used as the supporting technique for other modelling methods.

2.4.2 Surface Modelling

Surface modelling aims to create the models for finite, non-open surface patches of free forms. Note that the boundary surfaces of an object are formed by positioning and joining surface patches with the specified continuity restrictions. A surface modelling method does not represent the topological information of geometric elements such as knitting of boundary surfaces into a watertight volume. Figure 2.24 shows an example of a modelled surface from the surface modelling method. A surface model differs from a solid model since there is no information of the thickness.

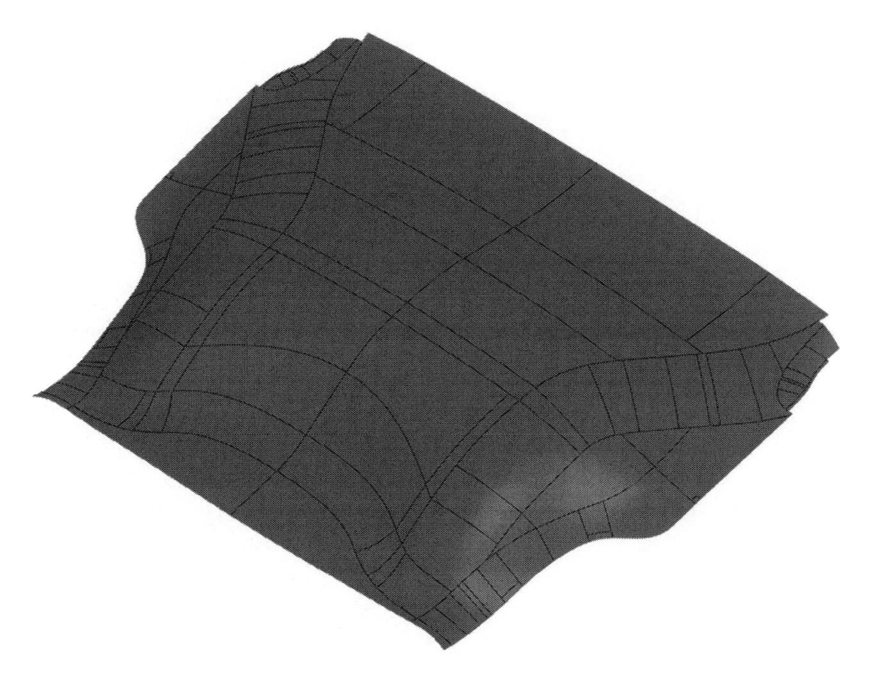

Figure 2.24 Surface model example.

Differing from a wireframe model, a surface model has sufficient information to determine hide and shade displays when multiple surfaces are involved; however, a surface model does not include information of volumes or masses. Therefore, it is usually unsuitable to be used in simulations for engineering analysis unless the object is a sheet part with uniform thickness perpendicular to surfaces.

2.4.3 Boundary Surface Modelling (B-Rep)

Boundary surface modelling (B-Rep) is used to define the finite and closed cover of an object (the mantle) upon a surface model. In B-Rep modelling, it is assumed that each physical object has an unambiguously determinable boundary surface, which is a continuous closing set of surface patches. Since a finite volume is defined, the B-Rep method provides a comprehensive topological characterization of solids.

In defining the volume of solids, B-Rep modelling utilizes a surface model with all of its determined boundary surface patches, and the normal vector of each surface patch is then determined. A watertight volume can be finally determined by collecting all spatial points at the internal sides of all the surrounding boundary surfaces. This is implemented by the half-space concept.

Mathematically, a *half space* can be expressed as

$$\boldsymbol{H} = \{\boldsymbol{P} : \boldsymbol{P} \in E^3 \, and f(\boldsymbol{P}) < 0\} \tag{2.17}$$

which shows that point P belongs to a half space E^3 if the condition $f(P) < 0$ can be satisfied where $f(P) = 0$ is the equation of the surface in an implicit form. The examples of surface equations for some common geometric elements are given in Table 2.10.

Table 2.10 Equations of common surfaces.

Name	Illustration	Implicit equation $f(P) = 0$
Flat *XY* plane		$\{(x, y, z): z = d\}$
Cylinder		$\{(x, y, z): x^2 + y^2 = R^2\}$
Cone		$\{(x, y, z): x^2 + y^2 = kz^2\}$
Sphere		$\{(x, y, z): x^2 + y^2 + z^2 = R^2\}$
Torus		$\{(x, y, z): (x^2 + y^2 + z^2 - R^2{}_2 - R^2{}_1)^2 = 4R^2{}_2(R^2{}_1 - z^2)\}$

The half space on one side of the surface is empty while the other one is filled with a material. The B-Rep method assumes that the volume occupied by an object is bounded by surfaces of infinite extension. Each surface divides space into two regions of infinite extension. The volume of the solid S is the intersection (common portion) of half spaces H_i $(i = 1, 2, \ldots, N)$ where

$$S = \cap\left(\sum_{i=0}^{N} H_i\right) \tag{2.18}$$

Figure 2.25 shows an example where the B-Rep method is applied to define a rectangle body (Figure 2.25b) as an intersected volume of seven surfaces of infinite extension (*left*, *right*, *rear*, *front*, *top*, *bottom*, and *central cylinder*) (Figure 2.25a).

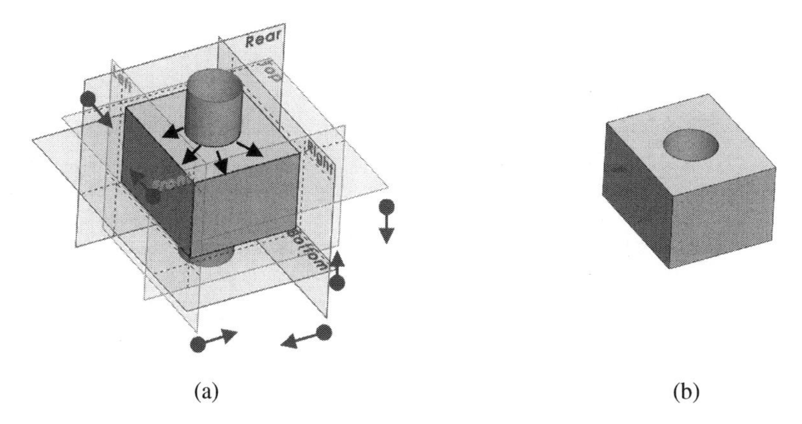

Figure 2.25 Example of half-spaces in B-Rep method. (a) Seven surfaces with an infinite extension. (b) Enclosed volume by seven half spaces.

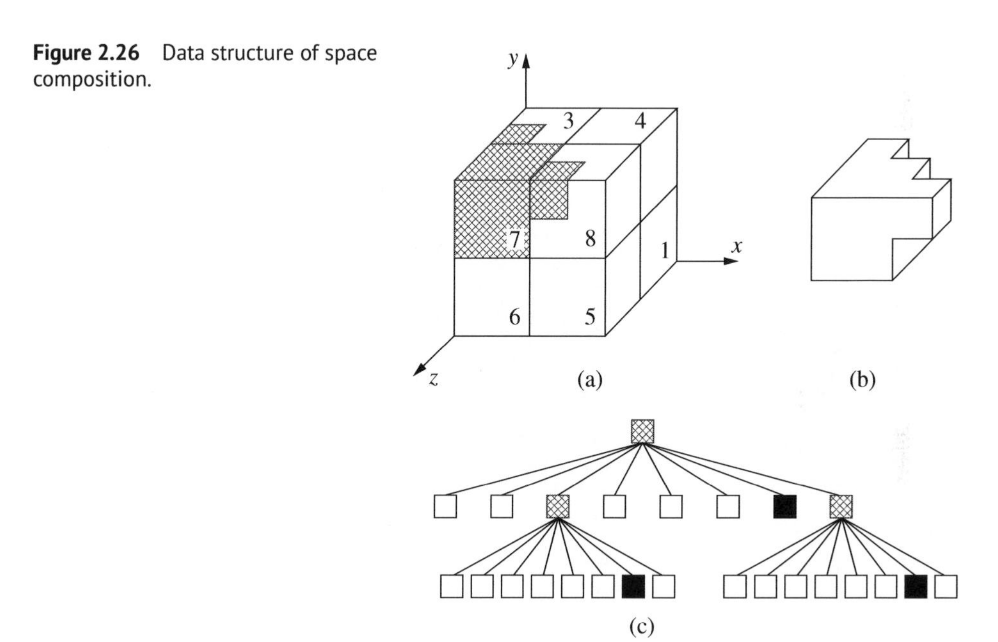

Figure 2.26 Data structure of space composition.

2.4.4 Space Decomposition

In a space decomposition, an object is represented by a collection of isomorphic cells. The sizes of isomorphic cells are very small, being several orders of the magnitude smaller than the dimensions of an object. The space decomposition method is very popular in numerical simulations such as *finite element analysis* and *boundary element analysis*.

Figure 2.26 illustrate how the space decomposition is processed as well as the corresponding data structure. The space decomposition is performed as the following procedure:

1. It divides a finite space into eight parts (producing octants).
2. It then examines each space region as to whether they are fully or partly filled.

3. Partial regions that are totally filled up or are not filled at all can be excluded from further investigations.
4. The partially filled octants are continuously refined until the required accuracy is achieved.

Figure 2.27 shows examples of using the space decomposition method to represent objects (Bi and Kang 2014). In the left column, the datasets of point clouds are obtained by 3D scanning, and in the right column, the point clouds have been transformed into solid geometries using the space decomposition method.

In the numerical simulations, the finite volume of solid is decomposed into small cells, so-called *isomorphic cells*; these cells are usually smaller by several orders of magnitude than the dimensions of the solid itself. The space decomposition in a numerical simulation is also called a meshing process and Figure 2.28 shows two example models where the solids are decomposed into a set of small cells called elements.

2.4.5 Solid Modelling

In solid modelling, the geometry of an object is modelled by a set of solid primitives, which are assembled into an object using composition operations. The modelling procedure

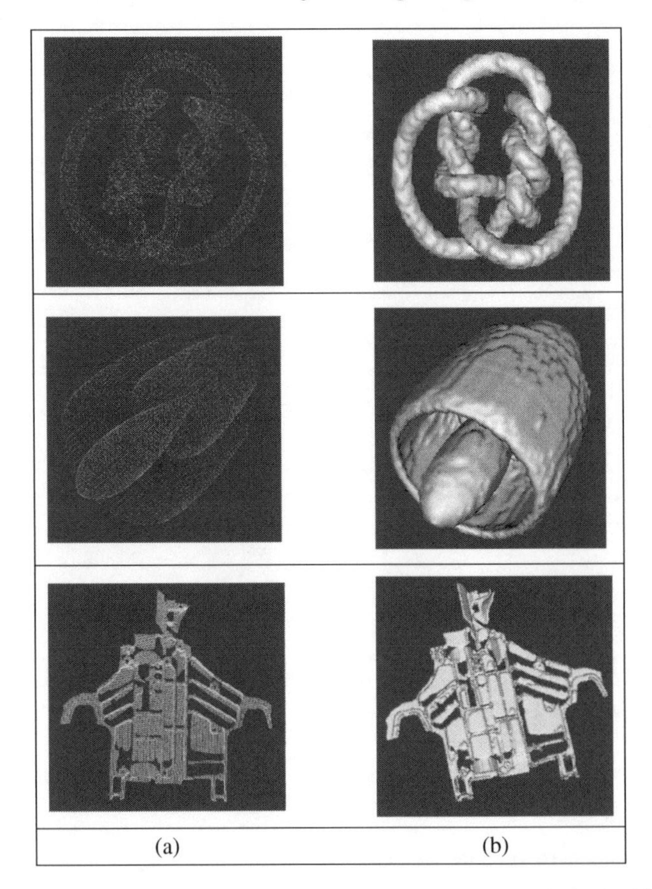

Figure 2.27 Examples of solid objects using a space decomposition method (Bi and Kang 2014). (a) Point cloud datasets of objects. (b) Solids and surfaces from the space decomposition method.

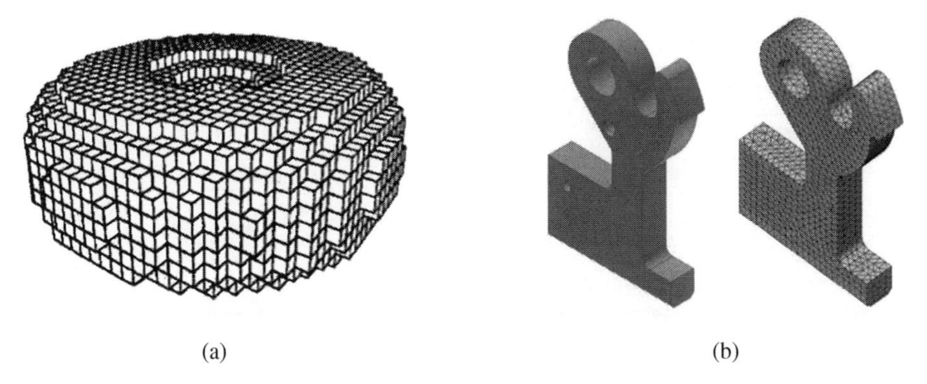

Figure 2.28 Examples of space decomposition in numerical simulation. (a) Solid A. (b) Solid B.

to combine elemental solids using composition operations is commonly known as *Constructive Solid Geometry* (CSG) modelling. The geometric representation from solid modelling gives complete information about physical objects. CSG modelling is based on the following assumptions:

1. A constitutive object is a rigid solid; the object has a concrete and invariant shape not affected by spatial location or position.
2. An object fills the space occupied by it homogeneously; the inside positions of the object are always connected to the complementary of the model through boundary surfaces.
3. The extension of an object is finite; the model can be mapped to a 2D plan for computer visualization.
4. An object can be generated as a composition of a finite number of solid primitives and the modelled object model can be stored in a computer.
5. An object can be modelled as a closed set in terms of rigid solid motions.

2.4.5.1 Solid Primitives

Solid primitives are the simplest solids used in CSG. As shown in Figure 2.29, most commonly used primitives include *cuboids*, *cylinders*, *prisms*, *spheres*, *cones*, and *tori*. Note that different solid modelling tools may use different solid primitives in CSG.

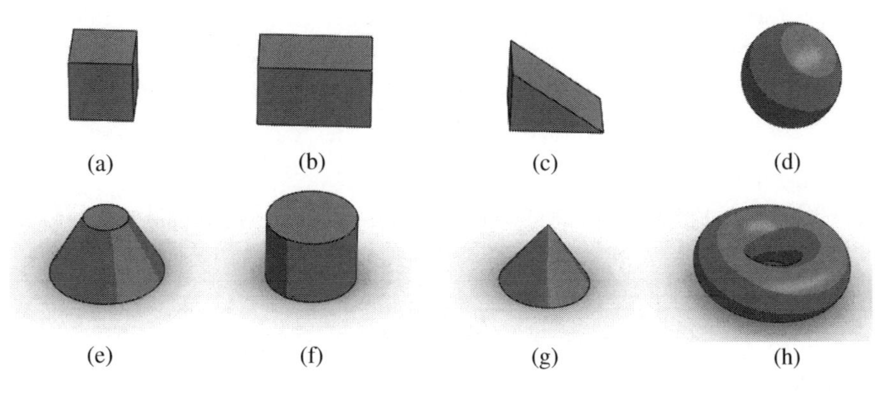

Figure 2.29 Examples of solid primitives. (a) Cuboid. (b) Rectangle cuboid. (c) Prism. (d) Sphere. (e) Tapped cylinder. (f) Cylinder. (g) Cone. (h) Torus.

Table 2.11 Modelling tools for customized solid primitives.

Modelling tool	Explanation	Example
Extruding	Extruding creates a solid by moving a two-dimensional profile along a straight path.	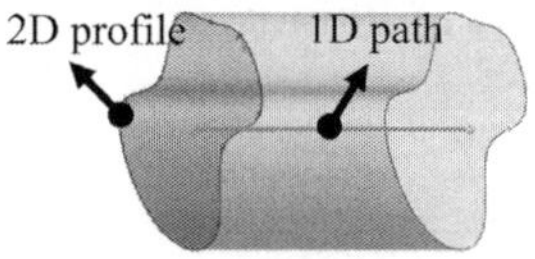 2D profile; 1D path
Revolving	Revolving creates a solid by revolving a two-dimensional profile along an axis.	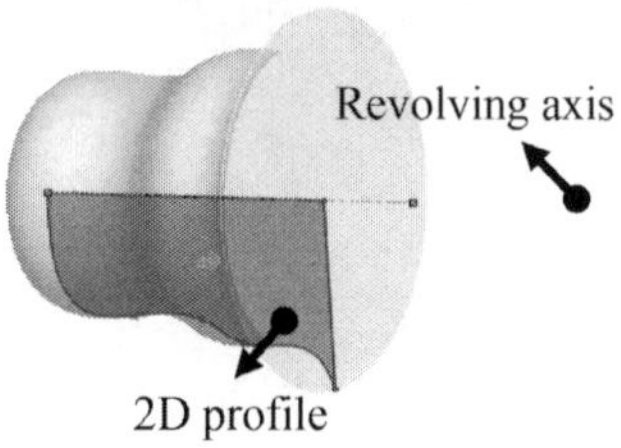 Revolving axis; 2D profile
Sweeping	Sweeping creates a solid by translating a 2D profile along a 3D path with or without one or a few of 3D guide curves.	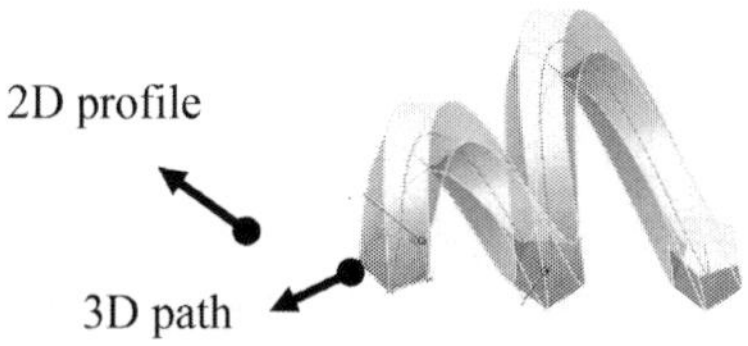 2D profile; 3D path
Lofting	Lofting creates a solid by specifying and connecting vertices on a series of 2D profiles.	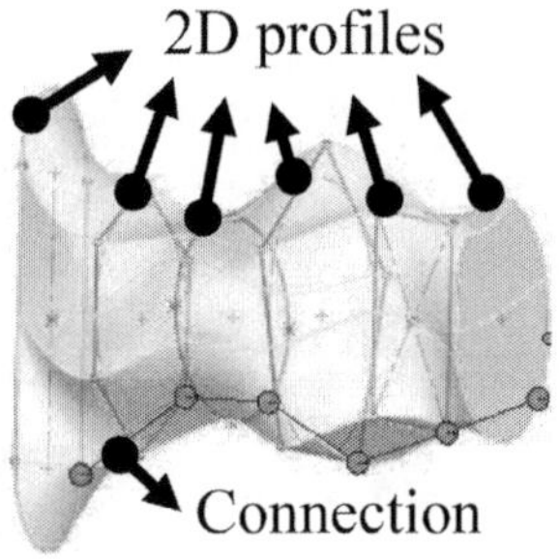 2D profiles; Connection

For an object with a complex geometry, solid primitives can be customized. Table 2.11 lists some common methods used to create solid primitives (i.e. design features) in a solid modelling tool. These modelling tools include *extruding*, *revolving*, *sweeping*, *lofting*, and many others.

2.4.5.2 Composition Operations

In CSG, an object is constructed from solid primitives by composition operations. Eligible composition operations are the Boolean operations on sets, i.e. *union* (∪), *intersection* (∩), and *difference* (\), as well as associated geometric transformations of those sets.

For a pair of solid primitives with given postures, different composition operations lead to different outcomes of assembled objects. Figure 2.30 shows three different objects from the compositions of a rectangular cuboid and a cylinder primitive.

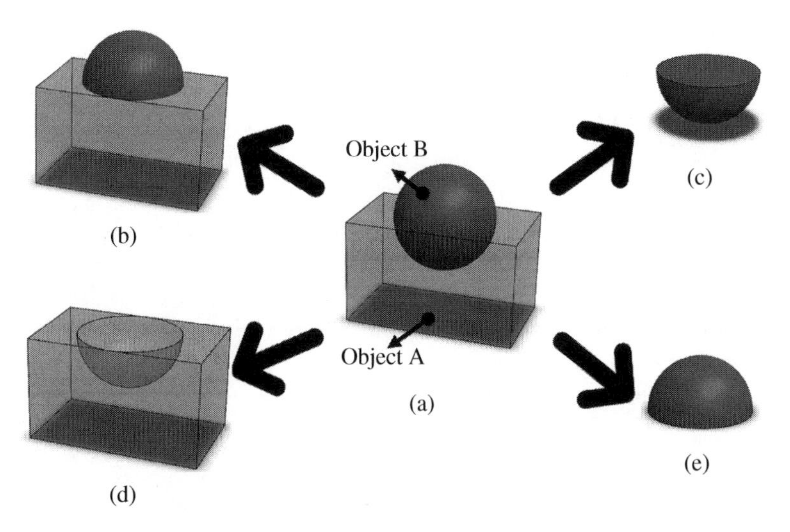

Figure 2.30 Differences of Boolean operations. (a) Two primitives at given positions. (b) A∪B. (c) A∩B. (d) A\B. (e) B\A.

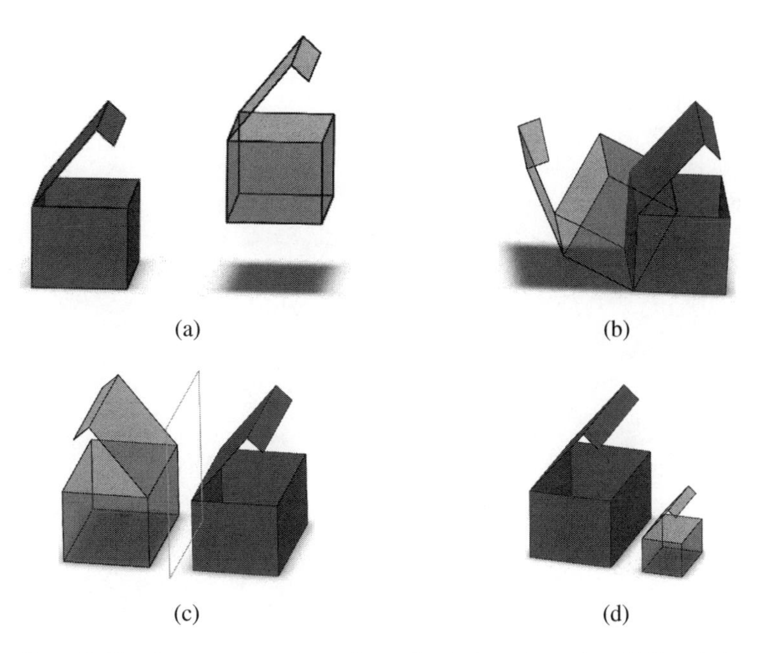

Figure 2.31 Common operations of coordinate transformation. (a) Translating. (b) Rotating. (c) Mirroring. (d) Scaling.

Other than the Boolean operations among solid primitives, a coordinate transformation can be applied to individual primitives before a Boolean operation takes place. Common operations of coordinate transformation include *translating*, *copying*, *rotating*, *mirroring*, and *scaling*. Figure 2.31 shows some examples of these operations.

2.4.5.3 CSG Modelling

Two basic elements of a solid model are (i) elemental geometric solids T_i ($i = 1, 2, 3, \ldots, N$), where N is the number of available solid primitives of the object, and (ii) the set of composition operations $\otimes$ for a Boolean operation ('∪', '∩', or '\') of geometric solids.

A series of composition operations can be expressed as

$$T_C = ((T_1 \otimes T_2) \otimes T_{3,} \ldots) \tag{2.19}$$

where T_C stands for the assembled object, T_i ($i = 1, 2, 3, \ldots, N$) are the primitives in the object, and $\otimes$ is one of the Boolean operations ('∪', '∩', and '\').

2.4.5.4 Modelling Procedure

The procedure of CSG modelling is as follows:

1. Define a set of solid primitives.
2. Create dimensional variables and constraints of solid primitives.
3. Transform each solid primitive into the appropriate position, applying the generalized set operations.
4. Finally, combine elemental solids to generate a unique and legal solid model.

The above procedure shows that the resulting solid model is affected by (i) the set of solid primitives and dimensions, (ii) locations and orientations of solid primitives, and (iii) the types and orders of the Boolean operations of solid primitives. Figure 2.32 shows different solid models from the same set of solid primitives.

2.4.5.5 Data Structure of CSG Models

Figure 2.32 shows that a CSG model consists of geometric and topological data, which is similar to a model from the B-Rep modelling process. A *data structure* of a CSG model does not give the model shape directly; instead, it consists of a number of solid primitives

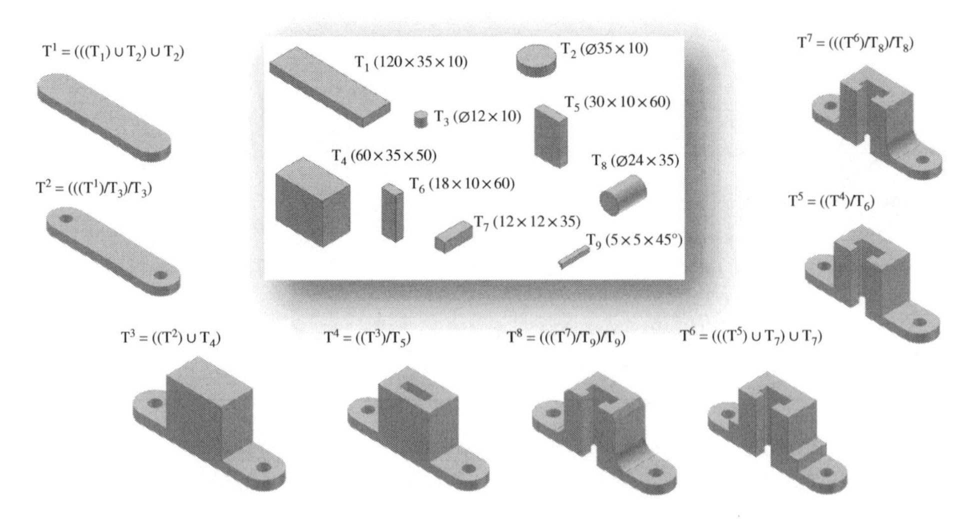

Figure 2.32 Different solid models from the same set of solid primitives.

and combines them through a procedural description. More specifically, the data structure corresponds to a *graph* and *tree* as the history of applying the Boolean operations on solid primitives, in such a way that:

1. The topological information is stored in a binary tree format.
2. The outer leaf nodes correspond to solid primitives.
3. The interior nodes are the Boolean operations over solid primitives or components.

Figure 2.33 shows the data structure of an example CSG model. It consists of two data types: (i) three solid primitives, i.e. two blocks with different dimensions and one cylinder, and (ii) a topological tree, i.e. one union operation of two blocks and then one difference operation of the united component with the cylinder.

Since CSG modelling begins with solid primitives, CSG models are always valid and they have complete and unambiguous information of solids. In addition, the primitives are defined directly at the volume level and low-level entities such as vertices, edges, and faces are defined implicitly. However, an entity at any level of solid can be utilized and accessed readily when it is needed. The advantages of CSG modelling are:

1. It constructs a solid model with the minimized steps.
2. It leads to a concise database with less storage since the entities at the low level are represented implicitly.
3. It provides a complete history of the model, which is retained but can be altered at any phase of product design.
4. A CSG model can be easily converted to the corresponding boundary representation.

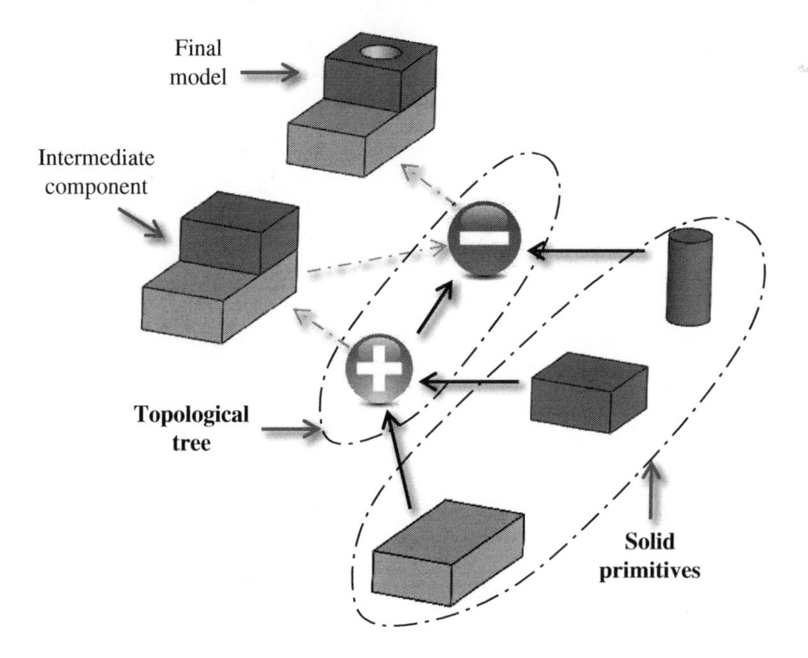

Figure 2.33 Data structure example of a CSG model.

The tools for CSG modelling were introduced in the 1980s. However, early CSG modelling tools suffered from a number of limitations as follows:

1. They can only provide far fewer types of basic solid primitives in modelling than the vast number of varieties that would be required in engineering practice.
2. CSG modelling does not adequately support engineering thinking. It implies that a final solid model is created from a theoretical sketch through continued modifications. In other words, traditional CSG modelling is rather used to reconstruct solid models than to actually design.
3. CSG modelling does not provide a comprehensive description of solids to be modelled. Early CSG models do not cover the information on microgeometry, materials, and physical characteristics. All of the above are important for operation, manufacturing, and control of products.

2.5 Feature-Based Modelling with Design Intents

There are numerous ways to create computer models using a given geometry. However, geometry and shape of an object have their purpose in a product and it is very helpful for a designer to take design intent into account when creating computer representations of solids.

Design intent is a term used to describe how the model should be created and how it should behave when it is changed. Design intent should be built into the model according to how dimensions and relations are established, since changes to a model will yield a different result for each different design intent.

Design intent is not just about the size and shape of features, but it can be extended to cover *tolerances*, *manufacturing processes*, *design constraints*, and *relationships of features* and *dimensions*. The use of design intents is an effective approach to build a parametric model of a part that is fully constrained and easy for modification. For example, sketches can be dimensioned to reflect design intents in parametric modelling. If the design intents for the part have not been adequately considered, the model might be useless from a practical viewpoint (Rynne 2006).

With regards to design changes, geometric features created at the design phase are closely related to manufacturing activities at other phases of a product cycle. It is ideal that designers are able to model solids directly with expected features. The advancement of computer aided methods in Figure 2.34 shows the growing expectation of the functional requirements (FRs) on computer aided technologies, which is from two-dimensional drafting and drawing to high-level solid modelling with consideration of design intents.

Feature-based modelling refers to the construction of object geometry as a combination of form features with design intents. The designer specifies the features in engineering terms, such as holes, slots, or bosses, rather than in geometric terms, such as circles or boxes. The concepts of features can also be extended to include non-graphic information. This information can be used in activities such as drafting, numerical control (NC), finite-element analysis, and kinematic analysis. Furthermore, feature-based packages frequently record the geometric construction and modification sequences used in building the model.

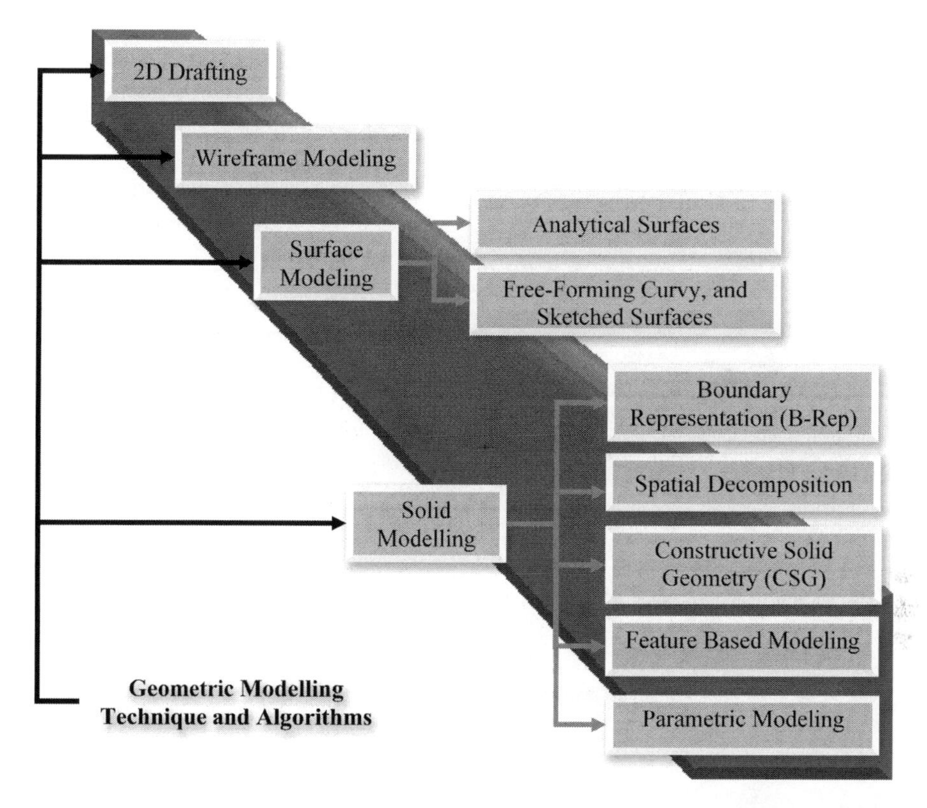

Figure 2.34 From 2D drafting and drawing to interactive solid modelling with design intents.

Bunge (1983) gave the principles of feature-based modelling as:

1. The physical world consists of things that are considered to be objects regardless of their contents. Objects can be characterized by their *features*, known or to be detected by scientific instruments. Features are quality and quantity characteristics, together with the correlations between them.
2. In terms of design, products and their various parts can be interpreted as *objects*, while features are characteristics associated with them. *Relations* between characteristics are described and regulated by correlations and restrictions.
3. As regards mechanical products, the geometric form is of primary importance in respect of material realization; therefore, it seems to be natural that the geometry of objects is derived from given features and their relations.

Features in an object can be defined by three aspects: (i) the characteristics of geometric shape, (ii) the characteristics of processes, and (iii) the ontological interpretation for the meanings of features. Figure 2.35 shows an example of a machined part with some geometric features with logical associations of points, edges, and surfaces of the part. Figure 2.36 shows an example of classified manufacturing features for prismatic parts by Šibalija et al. (2013), while Figure 2.37 shows a few exemplified parts in a chair model with their ontological features.

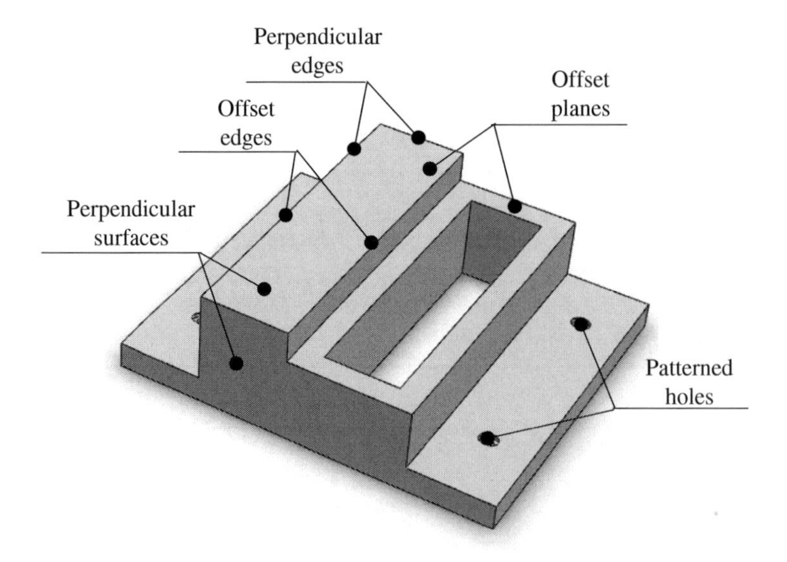

Figure 2.35 Example of geometric features.

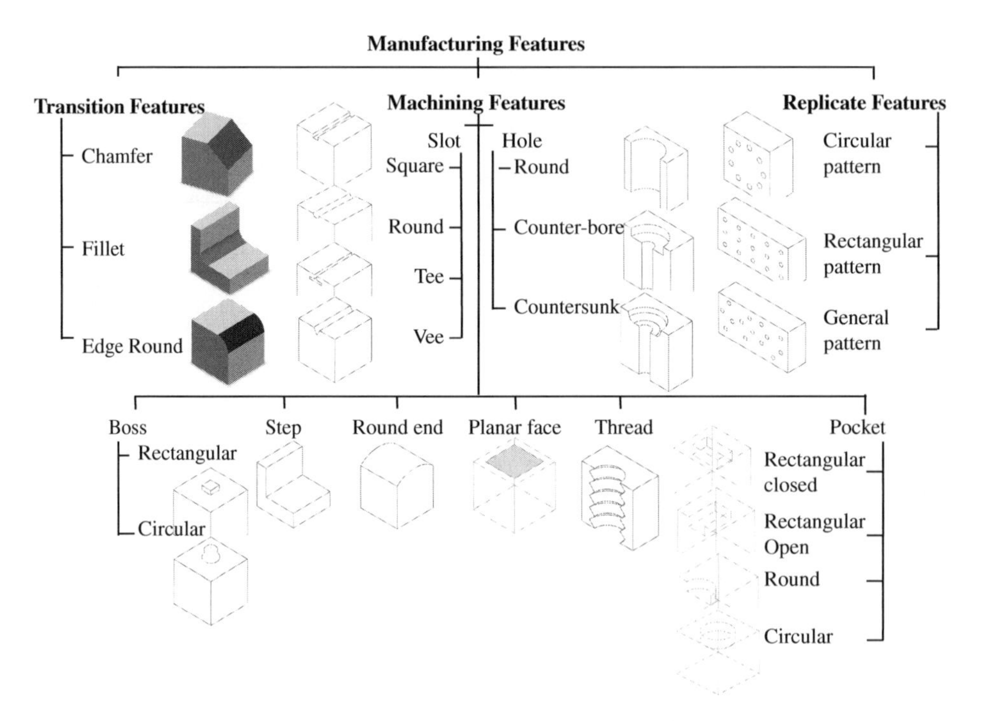

Figure 2.36 Classified manufacturing features for prismatic parts (Šibalija et al. 2013).

ITEM NO.	ONTOLOGICAL FEATURES
1	Seat and back supports
2	Rod for adjustment
3	Connector of seat and support
4	Main cylinder
5	Base frame
6	Caster wheels

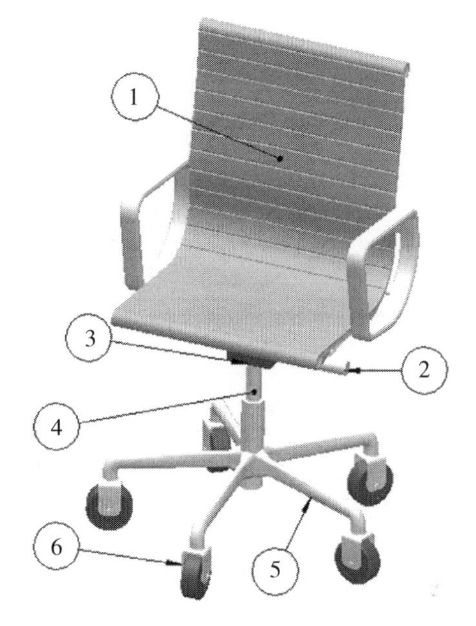

Figure 2.37 Examples of ontological features in a chair model.

2.6 Interactive Feature-Based Modelling Using CAD Tools

A SolidWorks modelling environment is used to illustrate the application of interactive feature-based modelling. As shown in Figure 2.38, a part usually consists of a number of features. The corresponding solid of a feature can be added or removed from the volume of the part model by an *Operator*, which is associated with the feature type. The geometry and shape of a feature can be created by built-in tools (e.g. *Fillet* or *Chamber*) or sketch tools (e.g. *Extruded Boss* or *Cut*). The parameters for dimensions and relations are defined when a feature is created. Each built-in tool is equipped with a *wizard* to guide user inputs. The sketch tools provide the flexibility for the user to define the dimensions of a feature by hand. Finally, all sketches, paths, dimensions, and constraints are defined upon certain references such as points, axes, planes, and coordinate systems.

When an original CAD model of a part is available, any entities (edges, axes, surfaces, planes, or features) in the model can be identified and reused as the references to modify or create new features. Note that the information of any entity in the model is included and can be accessed through the model tree interfaces. Figure 2.39 shows some major feature-based modelling tools in SolidWorks. These tools are associated with the functional requirements to create and modify different types of features, as shown in Figure 2.38. Designers should be aware of these tools and know how to access them when they are needed.

Figure 2.40 shows the procedure of modifying or creating a feature in a solid model. It begins with determination of *design intent*. Design intent represents the strategies to create a certain feature. For example, a cylindrical shape can be created by *Extruding*, *Revolving*, *Lofting*, or *Sweeping*, but the amount of information in creating a feature and the effort in modifying the feature are different from one another. The design intent is your selection of

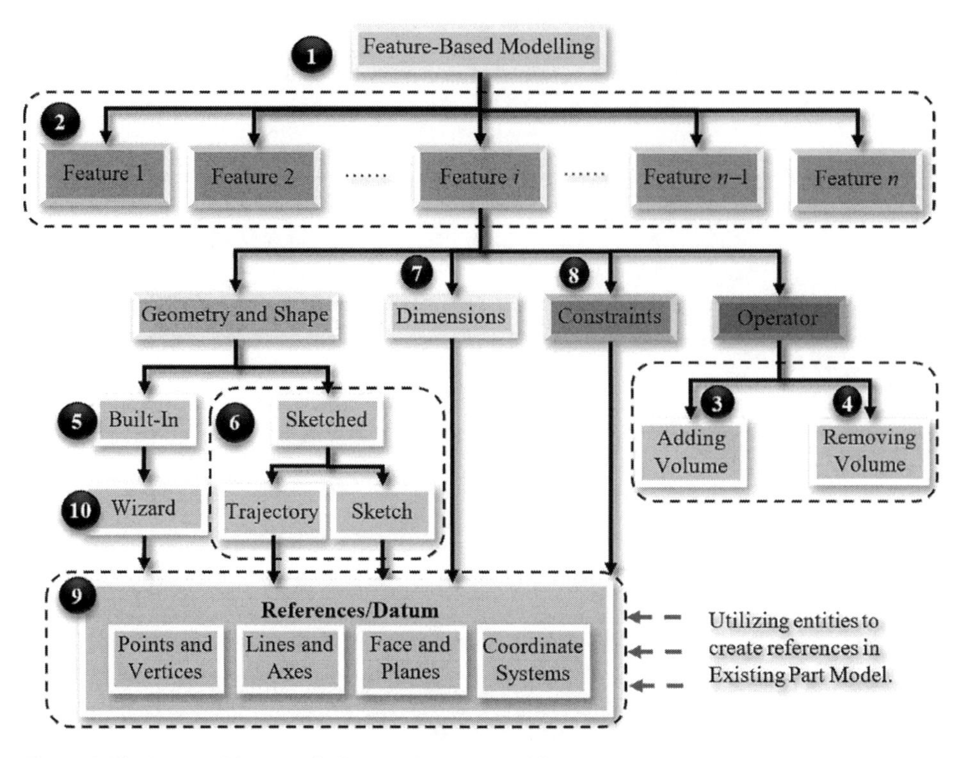

Figure 2.38 Types of features in feature-based modelling.

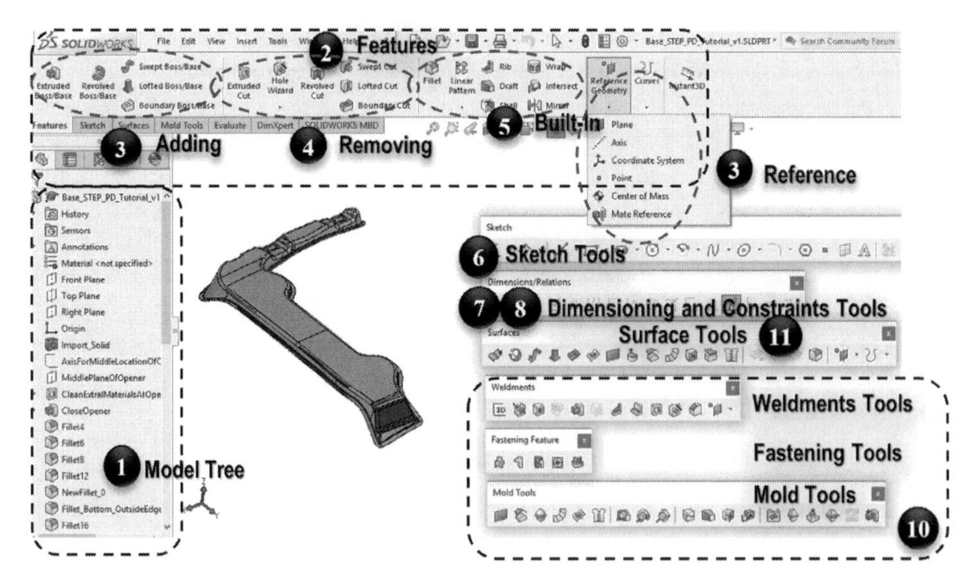

Figure 2.39 Feature-based modelling tools in SolidWorks. (*See color plate section for color representation of this figure*).

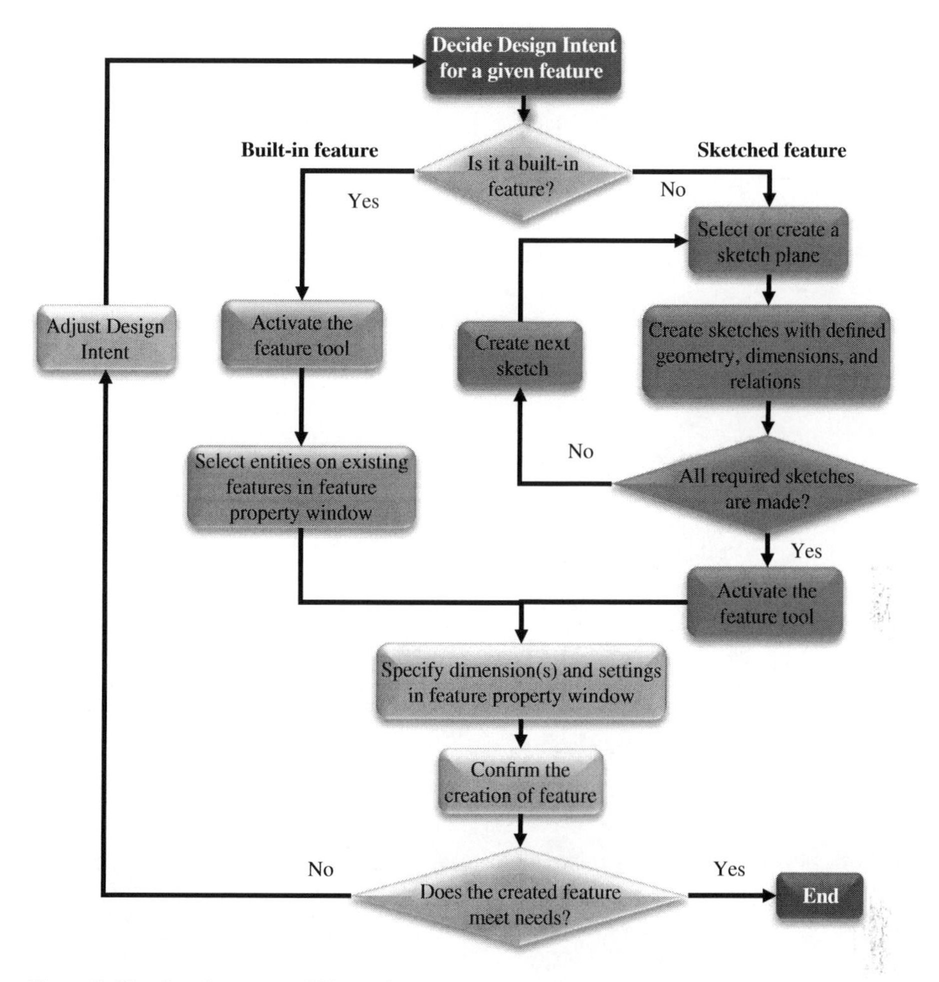

Figure 2.40 Creating or modifying a feature in feature-based modelling. (*See color plate section for color representation of this figure*).

modelling technique when a number of modelling options are available. Multiple design intents might be involved for the same feature. Generally, a SolidWorks (SW) feature can either be a *built-in feature* or a *sketched feature*. A built-in feature, such as *Fillet* or *Standard Hole*, is created upon an existing feature, and all the required inputs are specified in the feature property window. A *sketched feature*, such as *Extrusion*, *Sweep*, or *Revolve*, requires the user to create one or more sketches to define the geometry and/or paths, and the rest of the inputs are defined in the feature property window. Revising or adding a feature in a complex part model with many free-form boundary surfaces is not a trivial task. Designers are usually required to try different modelling strategies (for example, selecting and creating a sketch plane) to create a feature successfully.

Figures 2.41 and 2.42 show two examples where a built-in feature and a sketched feature are created, respectively.

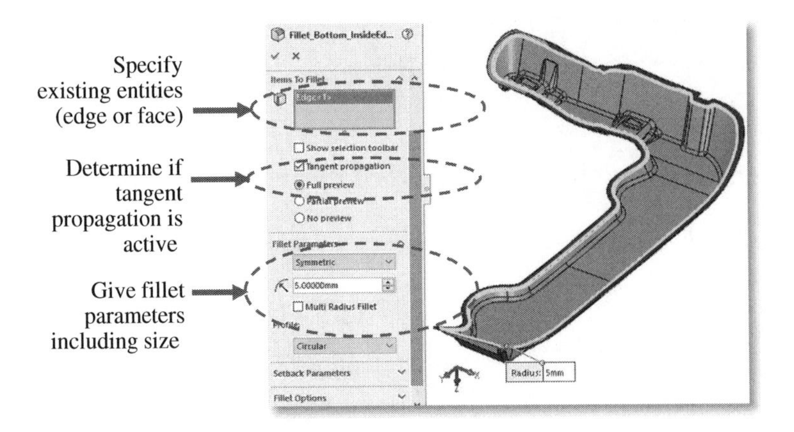

Figure 2.41 Example of creating a built-in feature.

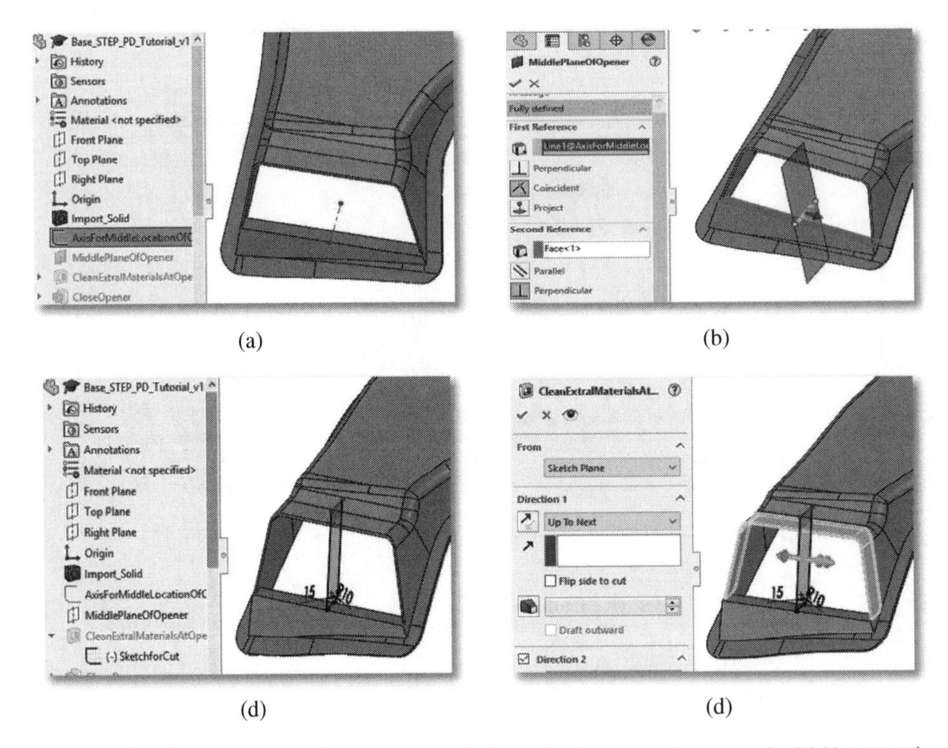

Figure 2.42 Example of creating a sketched feature. (a) Create a reference axis. (b) Use an axis and a flat face to create a reference axis. (c) Create a sketch on a new plane. (d) Create an extrude cut at the outlet.

2.7 Summary

A solid object consists of (i) geometric elements at different levels, from vertices to edges, faces, volumes, and features, and (ii) the topological relations of geometric elements. Geometric modelling is used to create computer representations of solid objects, while geometric modelling lays the foundation for using computer aided techniques. The commonly used modelling methodologies include wireframe modelling, surface modelling, space decomposition, constructive solid geometric modelling, and feature-based modelling.

The information of geometric representation of an object from different modelling methods may be different. A wireframe model consists entirely of points, lines, and curves; it does not have the information for face or volume, and no topological data are needed in modelling. A surface model stores the topological information in corresponding objects, but a surface model can still be ambiguous in some cases. Both a surface model and a solid model can be utilized to identify a shading area and a solid model has complete, valid, and unambiguous spatial addressability. In addition, a lower-level model (i.e. a wireframe) can be extracted from a high-level surface or solid model.

The same geometry can be modelled in different ways. However, it is desirable to take into consideration the design intents in geometric modelling. Design intent is beyond the size and shape of features and can be extended to manufacturing features such as tolerances, manufacturing processes, and design constraints. The use of design intents is an effective approach to build a parametric model of a part that is fully constrained and easy to modify. By far, feature-based modelling is the most advanced method, where a part model is modelled as a set of features as well as their topological relations. Modern computer aided software tools support feature-based modelling methods.

2.8 Modelling Problems

2.1 Create solid models for drawings with standard views (front, right, top, and isometric views) and annotated masses for the parts in Figure 2.43 (unit: inch per pound; thickness: 0.25 inch; materials: grey iron).

2.2 Create solid models for drawings with standard views (front, right, top, and isometric views) and annotated masses for the parts in Figure 2.44 (unit: mm per gram; materials: grey iron).

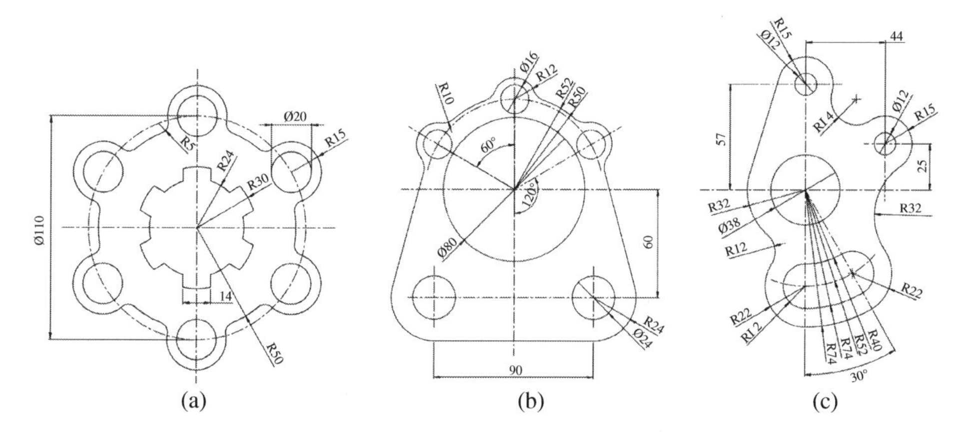

Figure 2.43 Drawings for modelling Problem 2.1.

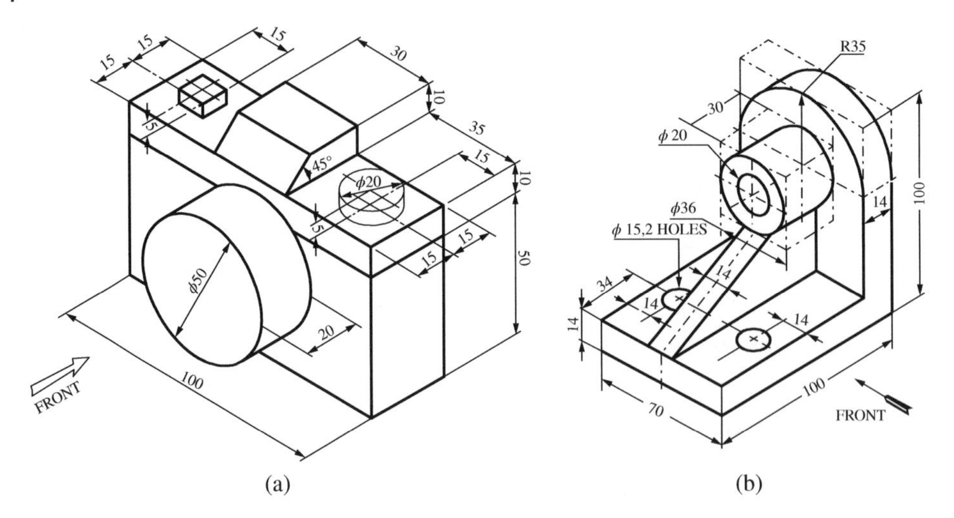

Figure 2.44 Drawings for modelling Problem 2.2.

2.3 Create a pipe model with the dimensions specified in Figure 2.45. Have a few examples of design intents you incorporate in the modelling process and create a similar drawing to that in Figure 2.45 with the answers about design intents and total mass in grams.

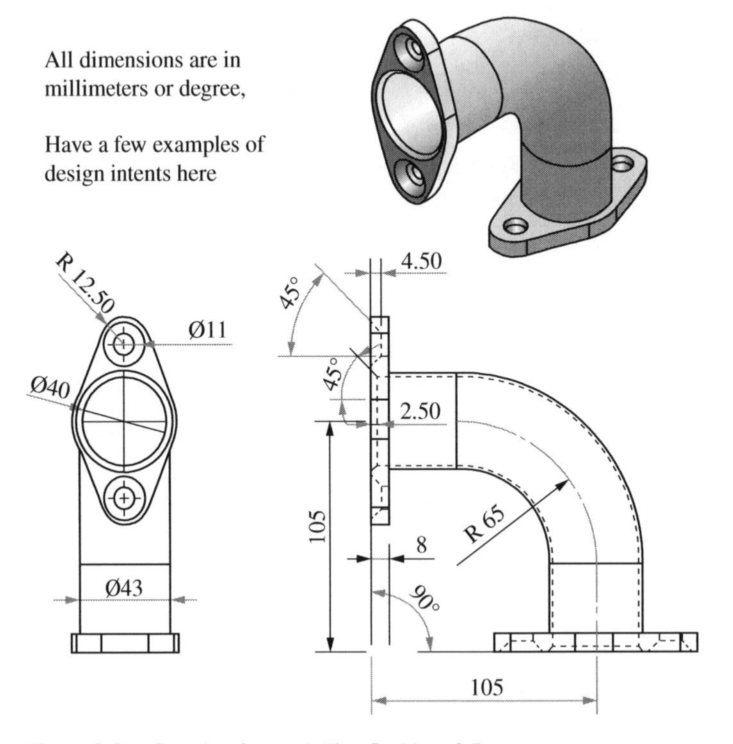

Figure 2.45 Drawing for modelling Problem 2.3.

References

Bi, Z.M. and Kang, B. (2014). Sensing and responding to the changes of geometric surfaces in flexible manufacturing and assembly. *Enterprise Information Systems* 8 (2): 225–245.

Bunge, M. (1983). *Treatise on Basic Philosophy, Epistemology and Methodology I: Exploring the World*, vol. 5. Dordrecht: D. Reidel Publishing Company.

Gallier, J. (2008). Curves and surfaces in geometric modelling: theory and algorithms. https://www.cis.upenn.edu/~jean/tabcont.pdf.

Rynne, A. (2006). Design intent. https://www3.ul.ie/~rynnet/designintent-solidworks.php.

Šibalija, T.V., Majstorović, V.D., Erčević, B.M., and Erčević, M.M. (2013). Process planning for prismatic parts in digital manufacturing. *The 7th International Working Conference on Total Quality Management – Advanced and Intelligent Approaches*, Belgrade, Serbia (4–7 June 2013). https://pdfs.semanticscholar.org/8ed9/a00902e634e5e9a4dec39763d383119a12d0.pdf.

TutorialsPoint (2018). 3D transformation. https://www.tutorialspoint.com/computer_graphics/3d_transformation.htm.

3

Knowledge-Based Engineering

3.1 Generative Model in Engineering Design

An *engineering design* consists of (i) a *design analysis* where a design candidate is analysed against the expected design criteria and (ii) a *design synthesis* where a group of potential design candidates are explored, analysed, and compared to find the optimized solution to an engineering problem. An engineering design process begins with the formulation of a design problem. In formulating a design problem, the design needs are interpreted and represented by a set of *functional requirements* (*FRs*), and design synthesis is performed to define a set of *design parameters* (*DPs*) and a design space for these DP_S. The relational models of DP*s* and FRs are established in design analysis, and these models are further utilized in design synthesis, so that the design candidates can be compared to optimize the solution to a specified design problem (Bi and Zhang 2001a,b; Bi et al. 2014).

Figure 3.1 shows that an engineering design can be viewed as *an information system* to transfer the given set of FRs into an engineering solution with the specified optimum DPs. Other than functionalities of products, common FRs of a product include size and weights, predicted loads, accuracy, durability, and service supports, while DPs of the engineering design include functional models, associated manufacturing and assembly processes, and so on. To derive DPs from the given FRs, the *generative models* of DPs and FRs are developed and utilized to evaluate engineering designs with a set of DPs are given.

In developing generative models, knowledge reuse is critical to optimize design solutions, improve design efficiency and productivity, and shorten design cycles. Engineering designs are classified into *creative*, *innovative*, and *routine* designs (Bi 2018). The level of automation and productivity of a design process greatly relies on the level of the knowledge reuse at all of the phases of a product lifecycle, such as designs, manufacturing, and assemblies of products.

3.2 Knowledge-Based Engineering

A computer aided design (CAD) system can be viewed as a toolbox with a wide collection of design tools for the functionalities, from the creations of basic geometric elements to sustainability analysis of the product. Such a system allows engineers to focus on a higher level of creative designs. Design tools automatically perform routine and non-creative design

Computer Aided Design and Manufacturing, First Edition. Zhuming Bi and Xiaoqin Wang.
© 2020 John Wiley & Sons Ltd. This Work is a co-publication between John Wiley & Sons Ltd and ASME Press.
Companion website: www.wiley.com/go/bi/computer-aided-design

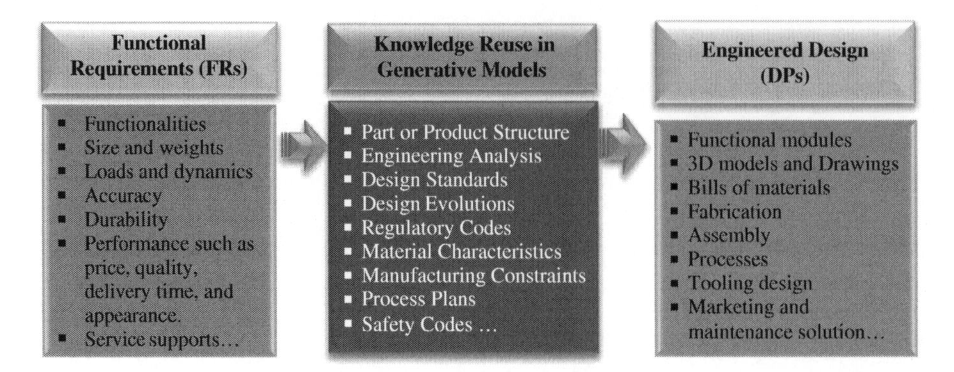

Figure 3.1 Engineering design process. (*See color plate section for color representation of this figure*).

activities such as reuses of sketches, features, parts, and components at detail design levels. Similarly, a computer aided engineering system can reduce manual intervention greatly in meshing and defining boundary conditions and loads.

A computer aided system aims to maximize knowledge reuse in engineering design; the possibilities of knowledge reuse should be explored at all stages of a design process. Figure 3.2 shows the procedure of product designs, which consists of two phases of *top-down modelling* and *bottom-up modelling*. With regard to geometric modelling of an object, a CAD tool can be either *morphological knowledge reuse* for geometries or *topological knowledge reuse* for relations. The level of knowledge reuse depends on the commonality of

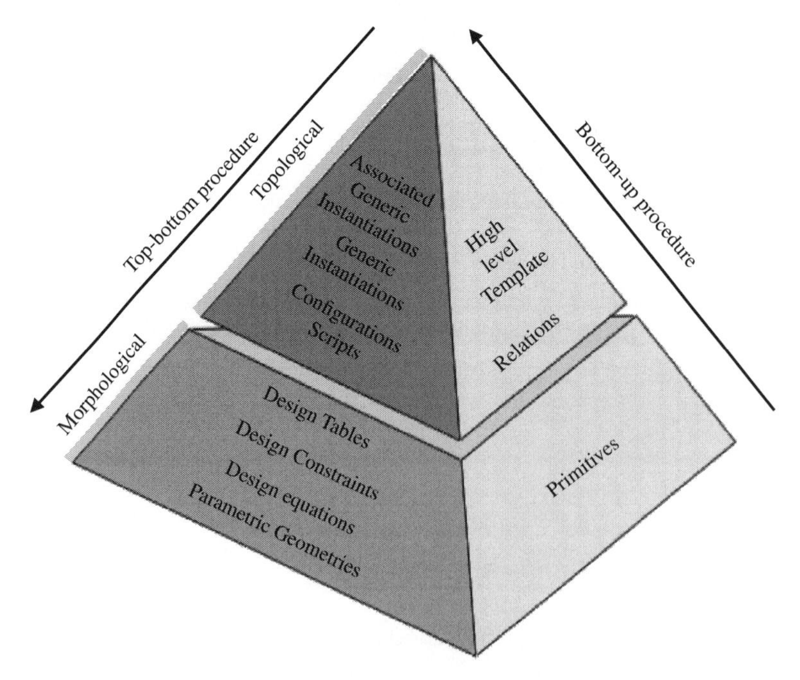

Figure 3.2 Morphological and topological level of geometric automation. (*See color plate section for color representation of this figure*).

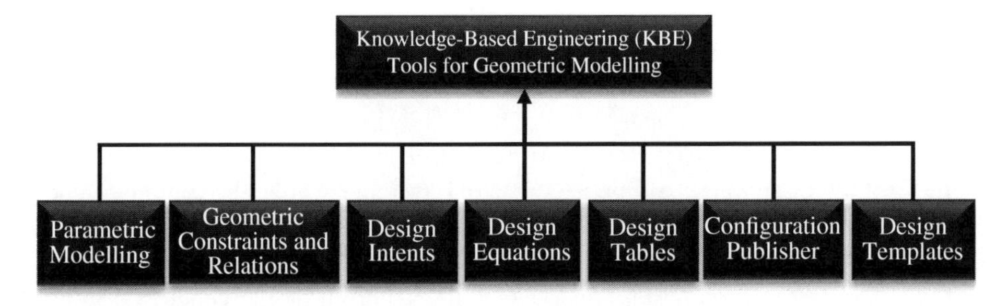

Figure 3.3 Widely used knowledge-based engineering (KBE) tools.

one product with existing products. The higher level a geometric component or element is, the less content it shares its commonality with others. Therefore, a geometric design tool with the maximized reusability is usually developed at the lowest bottom level; on the other hand, the commonalities of high-level components are the operations of relations rather than geometry, where the design tools for high-level knowledge reuse are for topological modelling activities (Tarkian 2012).

Knowledge-based engineering (KBE) aims to reuse knowledge and experience, which includes proprietary design and manufacturing practices. KBE is used to develop a formalized process for capturing and retaining methods, techniques, and practices used for design and product development. The skills and expertise that an engineer acquires to become productive are knowledge based. In other words, these types of skills and expertise might not be directly from formal education or training in classrooms.

KBE and CAD are closely related to each other. On the one hand, KBE is a type of *rules-based engineering*. It incorporates the CAD modelling techniques with *engineering knowledge*, *design automation (DA)*, and *best practices*, where products are designed, manufactured, assembled, and used. The fundamentals to KBE are the theories on CADs, design analysis and optimization, knowledge shares, and system integration. KBE addresses the engineering rules, governmental regulations, planning, and market research. KBE is aligned with concurrent engineering philosophy where all of the design factors must be taken into consideration simultaneously in creating, manufacturing, marketing, and using a product. On the other hand, CAD is an essential part of KBE even though they are different. CAD addresses mechanical designs to find design solutions of dimensions, geometries, materials, and manufacturing of products (Mohammed et al. 2008). Figure 3.3 shows some widely used KBE-based tools for geometric modelling, including *parametric modelling*, *geometric constraints and relations*, *design intents*, *design equations*, *design tables*, *configurations*, and *configuration publishers*. These tools will be discussed in detail in this chapter.

3.3 Parametric Modelling

Legacy computer aided systems are very limited in terms of their capabilities to support knowledge reuse. A geometric modelling process in these systems is mostly *dimensions driven*, which implies that creating a geometric element requires knowing all of its actual dimensions. The modelled geometry is rigid since any change has to be made from dimensions. In addition, no association can be defined even though two models are closely relevant from a certain perspective. Taking an example of automated drafting and drawing tools, three standard views of the same object are created, respectively, and, of course, a change made on one view cannot be reflected automatically on another view.

Modern computer aided systems support *parametric modelling* where the information for any attribute of an object can be parametrized. In other words, the computer aided system provides the mechanisms to preserve a design parameter for an attribute. Such an attribute can be associated with other attributes or objects as long as relationships exist. Designers have their freedom to set and change the values of design parameters at any time in any model where design parameters can be accessed. A design parameter links to its unique storage for the corresponding attribute. If the value of a design parameter is changed in a model, such a change occurs to all of the associated objects or attributes. When knowledge about a product design is defined in terms of design parameters, such knowledge can be easily reused.

The parametric modelling technique was introduced in 1980s (Ault 1999). In parametric modelling, *design parameters* can be defined for geometric dimensions and their relations, and *design features* can be defined for predefined geometric elements and construction tools. Figure 3.4 shows a hierarchical representation of system model from parametric

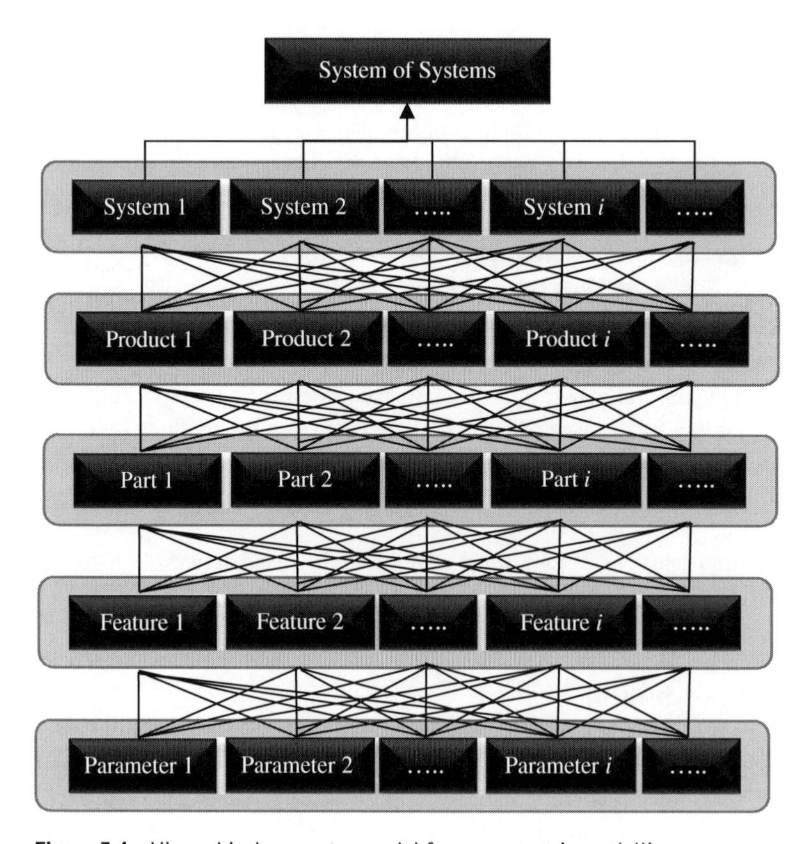

Figure 3.4 Hierarchical computer model from parametric modelling.

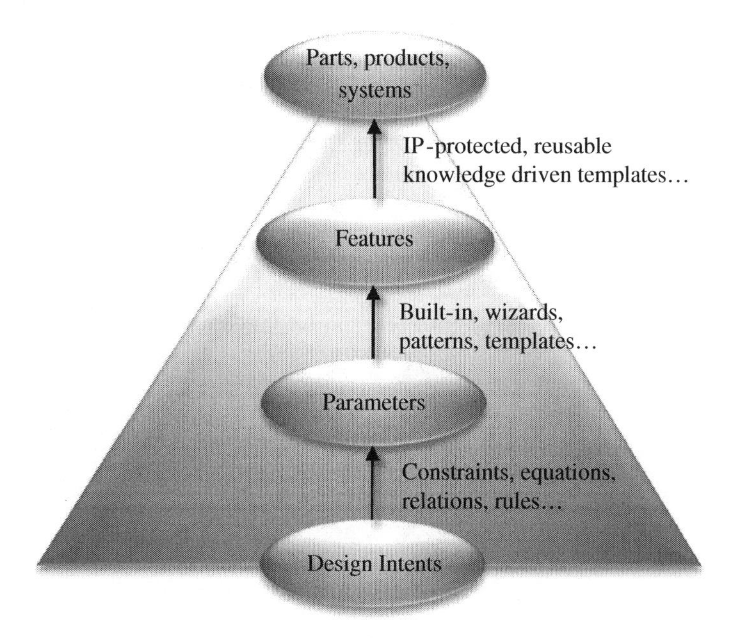

Figure 3.5 Knowledge-based engineering (KBE) for parametric modelling.

modelling. At an abstract level, a set of features can be added, intersected, or subtracted as a *part*, a set of parts can be assembled into a *product*, a set of products and machine tools can be configured as a *system*, and a set of systems can interact as a *system of systems*, sequentially. The parameters and features in a model provide numerous ways of reusing knowledge for the design of parts, products, and systems. Any attributes in a model can be represented by design parameters and these parameters can be changed unambitiously any time during the modelling process.

KBE provides some effective tools for designers to define parameters, features, and their relations in geometric modelling. In the modelling process shown in Figure 3.5, design intents are reflected by design parameters, constraints, equations, and tables. High-level geometric elements can be defined by knowledge reuses via *built-in features*, *patterns*, *configurations*, and *templates* (Dassault Systems 2007).

3.3.1 Define Basic Geometric Elements

Solid objects are modelled in a three-dimensional space. To create a solid model, low-dimensional geometric elements such as *points*, *lines*, *planes*, *curves*, and *surfaces* are needed for knowledge reuse.

3.3.1.1 Parametrized Points

Points are the lowest level of geometric elements. Points are often reused to define other geometric element types. For example, points can be used as the vertices of lines or curves or as the references for dimensioning and relations. Figure 3.6 shows that a point can be defined by its coordinates in a given coordinate system. Any change on the point can be

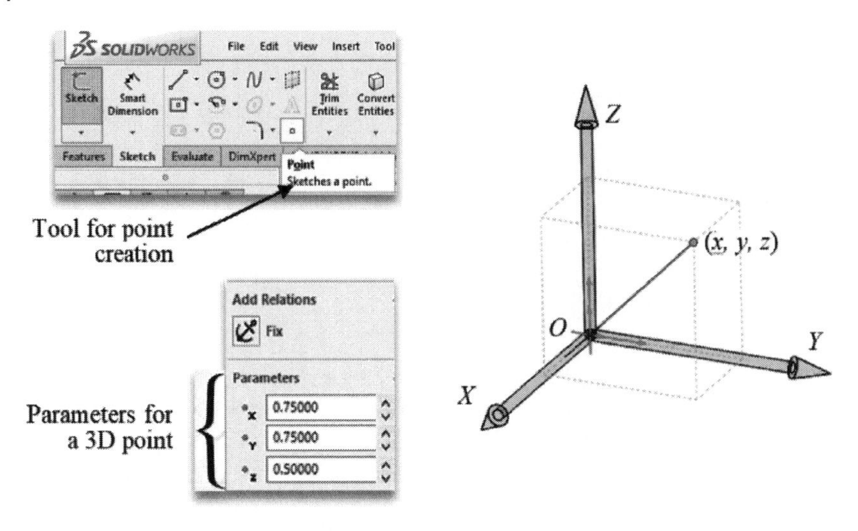

Figure 3.6 Point and its coordinates in three-dimensional space.

Table 3.1 Common approaches to define points.

Approach	Illustration	Parameters
Point on a curve	s	A curve and a distance of a new point from a reference point
Point on a plane		A plane and a direction of projection
Point on a surface		A surface and a direction of projection
Point at a circle/sphere centre		A circle/sphere and its centre
Point tangent on a curve		A curve and a specified point for the tangent direction
Point between two points	s_1 s_2	Two points and a ratio of the distances to these points
Points spaced on a curve		A curve and number of points

made by updating three coordinates associated with its position relative to the origin in the given coordinate system.

Due to numerous possible associations of a point with other geometric elements, a point may be created directly by reusing knowledge of available geometric elements. Table 3.1 shows some common ways to define new points.

3.3.1.2 Parametrized Lines

As shown in Figure 3.7, a line has two basic constitutive elements, i.e. two vertices at its two ends. *Lines* are often needed as *guide lines*, *references*, *axes*, or *directions* to create other geometric elements. For examples, lines can be used as the tangent directions for splines, the guide curves for sweeping features, or the generative curves or paths for surfaces. A line can be created or edited by specifying its two end-points.

Table 3.2 shows some common ways to define a new line or axis from existing geometric elements.

3.3.1.3 Parametrized Surfaces

In a CAD tool, a template of part models includes three default planes (***XY***, ***YZ***, and ***XZ*** planes as the front, top, and right planes of {***O-XYZ***}). While a geometric element of object can be at an arbitrary orientation, a plane aligned with such an orientation is very useful to define the dimensions of geometric element. Therefore, flat planes are customized as references to define dimensions and constraints on sketches. Mathematically, a plane in a coordinate system {***O-XYZ***} is represented as

$$Ax + By + Cz + D = 0 \tag{3.1}$$

where A, B, C, and D are four constants about the plane and $\mathbf{P}(x, y, z)$ is an arbitrary point on the plane.

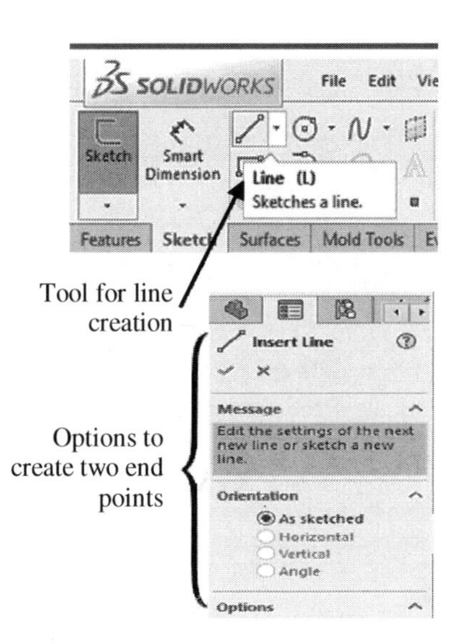

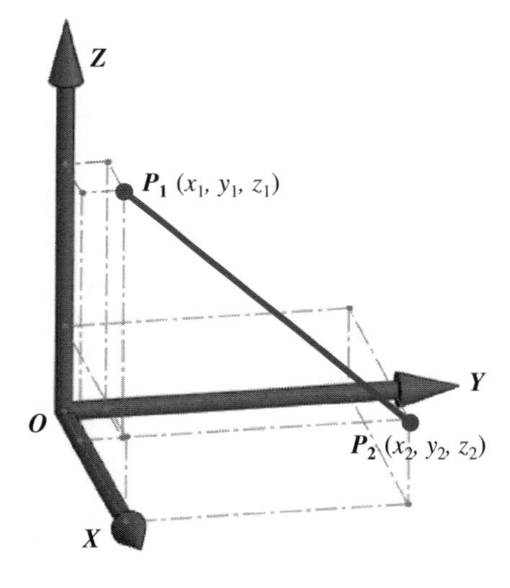

Figure 3.7 Define a line and its parameters.

Table 3.2 Common approaches to define lines.

Approach	Illustration	Parameters
Line by a point and direction		A reference point and a direction
Line by a point and an angle with curve		A reference point, a curve, and an angle
Line by a point and tangent curve		A reference point and a curve
Line by a point and normal to plane		A reference point and a plane
Line by bisecting		Two existing lines

Table 3.3 Equations for default front, top, and right plane.

Default plane	*A*	*B*	*C*	*D*	Plane equation
Front (***XY***)	0	0	1	0	$z = 0$
Top (***YZ***)	1	0	0	0	$x = 0$
Right (***XZ***)	0	1	0	0	$y = 0$

Note that Eq. (3.1) includes only three independent parameters, since one of A, B, C, and D can be selected freely without affecting the condition given by Eq. (3.1). Some software tools allow a plane to be defined by specifying the values of A, B, C, and D directly. Taking the examples of default ***XY***, ***YZ***, and ***XZ*** planes, the corresponding coefficients of A, B, C, and D are given in Table 3.3.

Alternatively, the information for three independent parameters is obtained by other means. Figure 3.8 shows an example that a plane is be defined by specifying three non-collinear points $\boldsymbol{P}_i(x_i, y_i, z_i)$, where $i = 1, 2, 3$. Since these three points are on the plane, the constants for the plane are determined from the following conditions:

$$\left.\begin{aligned} Ax_1 + By_1 + Cz_1 + D = 0 \\ Ax_2 + By_2 + Cz_2 + D = 0 \\ Ax_3 + By_3 + Cz_3 + D = 0 \end{aligned}\right\} \tag{3.2}$$

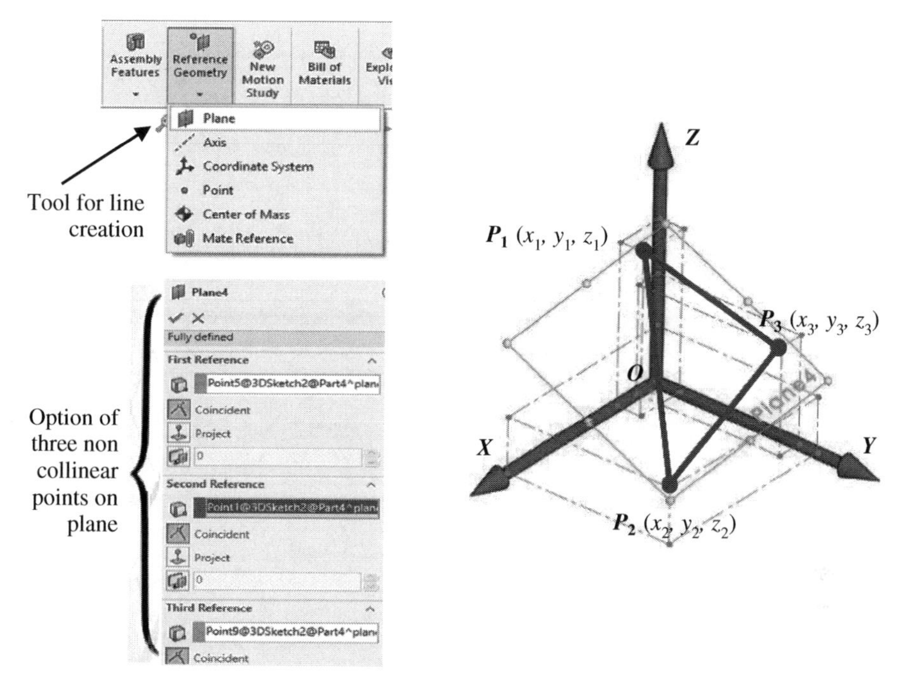

Figure 3.8 Define a plane and its parameters.

Equation (3.2) leads to the constants of the plane as

$$\left.\begin{aligned} A = -D\frac{\begin{vmatrix} x_1 & y_1 & z_1 \\ x_2 & y_2 & z_2 \\ x_3 & y_3 & z_3 \end{vmatrix}}{M} \quad B = -D\frac{\begin{vmatrix} x_1 & 1 & z_1 \\ x_2 & 1 & z_2 \\ x_3 & 1 & z_3 \end{vmatrix}}{M} \quad C = -D\frac{\begin{vmatrix} x_1 & y_1 & 1 \\ x_2 & y_2 & 1 \\ x_3 & y_3 & 1 \end{vmatrix}}{M} \\ M = \begin{vmatrix} x_1 & y_1 & z_1 \\ x_2 & y_2 & z_2 \\ x_3 & y_3 & z_3 \end{vmatrix} \end{aligned}\right\} \tag{3.3}$$

where A, B, and C are determined by Eq. (3.3) from D, and D can be freely selected.

Figure 3.8 shows the interface to create a new plane in SolidWorks. The information for the constants of the plane equation is from up to three reference geometric elements. When three points are used as references, these points are specified as the first, second, and third references in defining a plane passing through these points. The order of points makes no change on the defined plane.

Similarly, a plane can be defined based on the information of other geometric elements. Table 3.4 shows some common ways to define new plane.

3.3.1.4 Parametrized Curves

Curves are defined as guide-lines or references to create other geometric elements or as the boundaries of surfaces. A curve can be two-dimensional or three-dimensional, and either curve type needs the information for a number of vertices in corresponding 2D or

Table 3.4 Common approaches to define planes.

Approach	Illustration	Parameters
Plane by a distance from offset plane	d	A reference plane and a distance
Plane by a parallel reference and a point		A reference plane and point
Plane by a line, angle with a reference plane	θ	A reference line, plane, and an angle
Plane by two lines		Two co-plane lines
Plane by a line and a point		Two existing lines
Plane by a planar curve		A reference planar curve
Plane by a point and normal to a curve		A curve and a point
Plane in the middle of two others		Two parallel planes

3D space. A curve only has one independent variable. Usually, the mathematic model of a curve takes the length to a starting point as the independent variable to calculate coordinates of points on the curve. The roles of vertices on the curve depend on the selected underlying mathematic models of the curve. Some vertices are confined to be located on the curve; others are used as control points to determine the first, second, or ever high-order derivatives of the curve with respect to its length. A parametrized curve can be defined directly if one knows its analytical representation. Figures 3.9 and 3.10 show the graphic user interfaces (GUIs) where different approaches can be applied to define a 2D or 3D curve in SolidWorks, respectively.

A curve can also be defined based on the information of existing geometric elements. Table 3.5 shows some common ways used to define a new curve in CAD systems.

3.3.1.5 Parametrized Surfaces

An object may have free-form boundary surfaces. Modelling of an object with complex geometry requires the boundary surfaces to be defined. Different from a 2D or 3D curve, a free-form surface is controlled by two independent variables, each corresponding to a set of curves. These two sets of curves are formed by a *generator* and a *path* in a parametrized surface. Table 3.6 shows some common approaches where a parametrized surface is defined. In addition, a surface can also be defined as a mesh of the vertices with known coordinates.

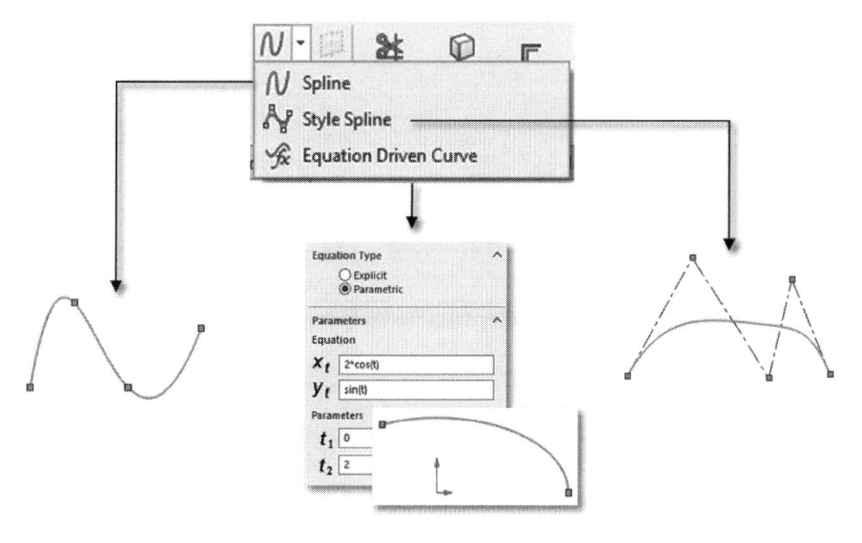

Figure 3.9 Define a 2D curve and its parameters in SolidWorks.

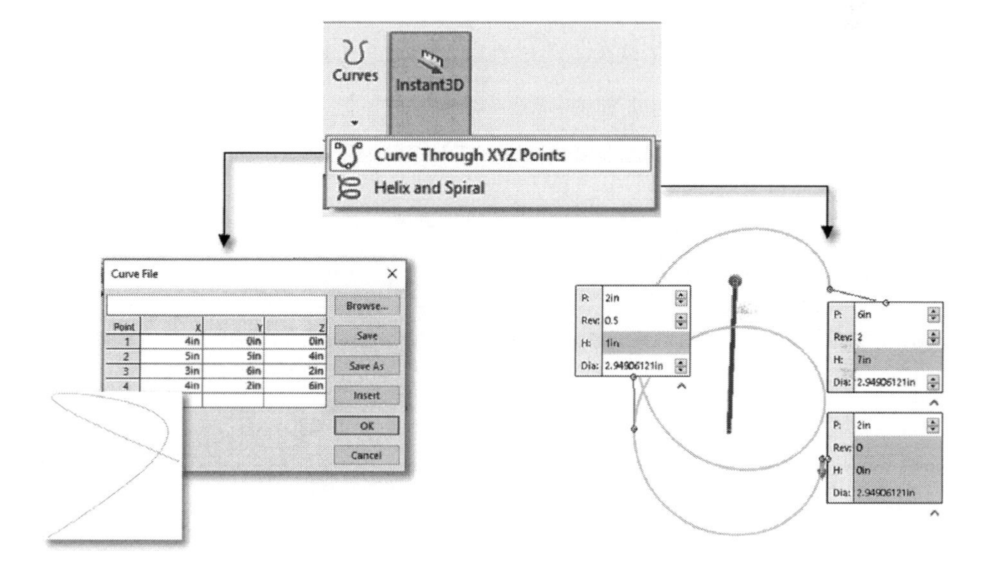

Figure 3.10 Define a 3D curve and its parameters in SolidWorks.

If a solid is to be created, a set of surfaces are selected as boundary surfaces, which are knitted to form a watertight fine volume for the solid. If the modelling process does not require creating a solid from surfaces, they are kept as they are as two-dimensional geometric elements.

3.3.2 Types of Parameters

Parameters in a model are critical since the parameters can be used to (i) access directly to attributes of features, relations, geometric elements, and parts for changes, (ii) sustain the

Table 3.5 Common approaches to define curves.

Approach	Illustration	Parameters
Curve by projecting a 2D curve on a surface		A reference 2D curve and a surface
Curve by intersecting two surfaces	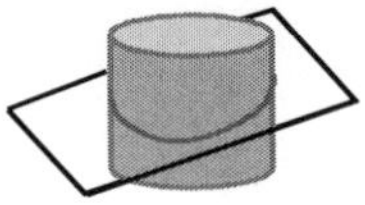	Two reference surfaces
Curve by offsetting a curve	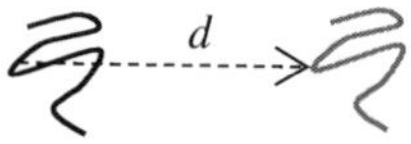 d	A reference curve and the direction and distance of offsetting
Curve by mirroring a curve about a plane		A reference and mirroring plane
Hexic curve from a circle by specifying varying pitches, revolutions, and diameters	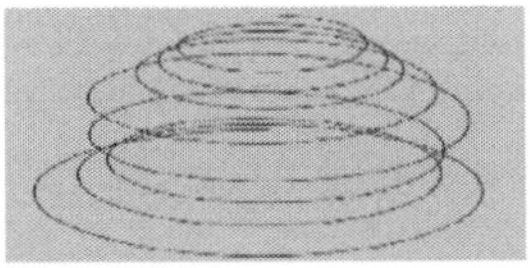	A reference circle and a table for the varying pitches, revolutions, and diameters

Table 3.6 Common approaches to define surfaces.

Approach	Illustration	Parameters
A surface by extruding a 2D profile		A 2D profile and extruding direction
A surface by revolving a 2D profile		A 2D profile and revolving axis
A surface by sweeping a 2D profile along a 3D path		A reference curve and the direction and distance of offsetting
A surface by lofting a number of 2D profiles		A number of 2D sketches and the order and location of connections

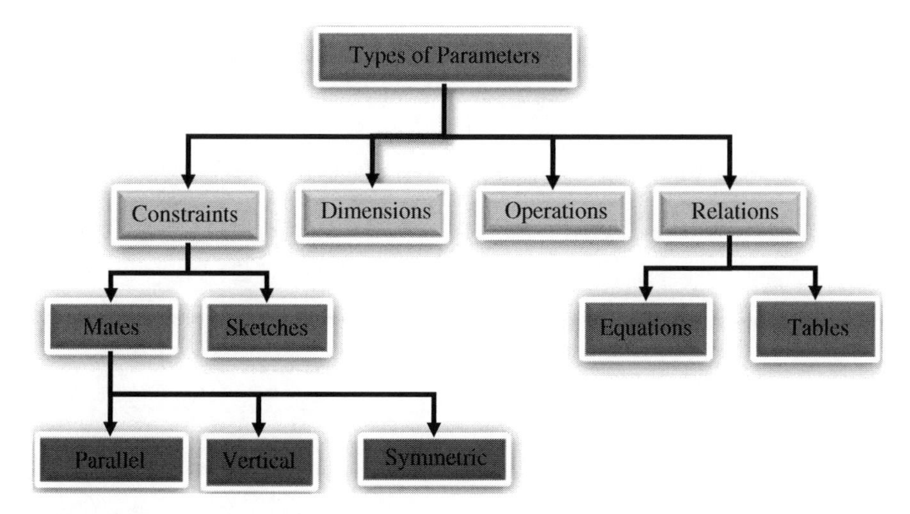

Figure 3.11 Types of parameters.

consistence of information since the attributes of solids are sustained by same parameters, where the change at one place will update the attribute in all relevant models automatically, and (iii) support parametric modelling where geometric models are driven by user-defined design parameters.

Parametric modelling uses various parameters to represent dimensions and geometric attributes. As shown in Figure 3.11, a parameter can represent different attributes, but it is usually at the lowest level of part geometry or an assembly of part geometries. A parameter can relate to *a constraint* such as a relation of two entities in a sketch or an assembly made of the geometric features from two parts. A parameter can be a *dimension* such as a length, width, or angle. A parameter can be an *operation* such as the spacing dimension or the number of instances in a linear or circular pattern. A parameter can be a *relation* of multiple parameters such as the relations bonded by a design equation or design table. Figure 3.12 shows some examples of the types of parameters. Figure 3.12a shows the parameters for linear and angular dimensions in a sketch; the parameters can be created by using the '*smart dimension*' tool in SolidWorks. Figure 3.12b shows examples of logic parameters for the constraints of geometric entities, as, for example, the tangent relation of two arcs or a line and an arc. In creating a new geometric entity, SolidWorks has the capability to detect all types of constraints applicable to certain entity; one can simply confirm the constraint if it applies. Alternatively, a constraint can be defined from scratch by selecting relevant entities. Figure 3.12c shows the parameters associated with the construction pattern tool where the parameters for spacing and the number of instances are embedded in the construction tool. Figure 3.12d shows a design table where the design parameters are associated with configurations. Note that the uses of design equations and design tables will be discussed extensively in sequential sections. As a summary, any piece of information in a parametric model is rooted in the parameters associated with the model.

In a solid modelling tool, every dimension, feature, configuration, part, or assembly has *a parameter identity* (*ID*). Using an ID allows the information of associated entities, such as size, shape, visibility, and state of sketches, features, parts, and assemblies, to be traced back.

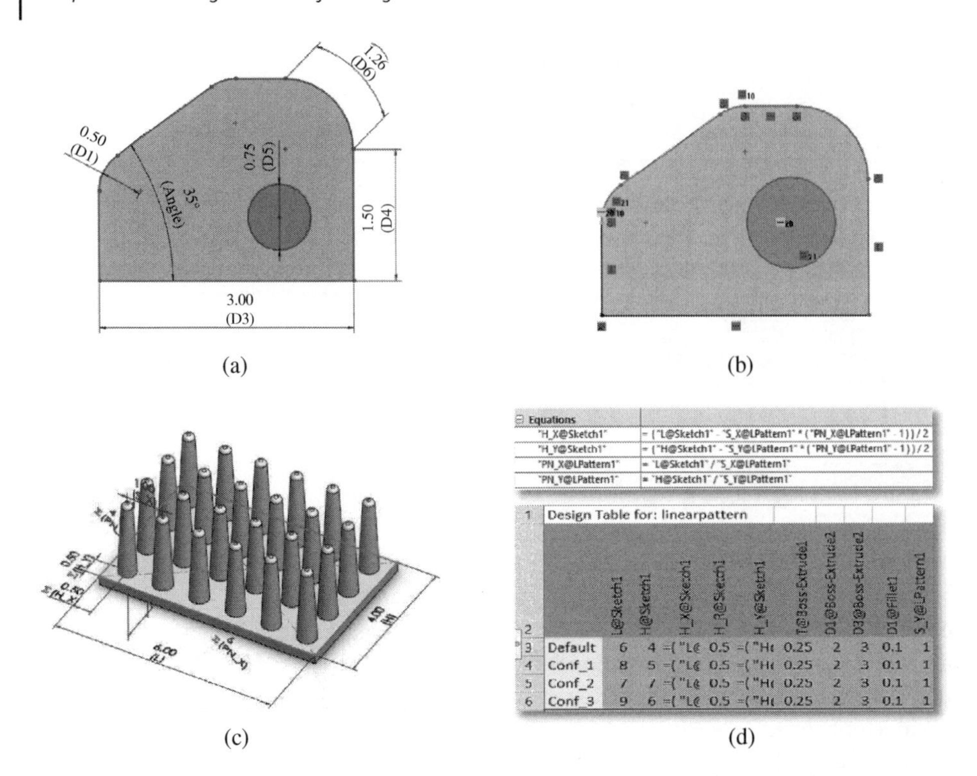

Figure 3.12 Examples of different types of parameters. (a) Parameters for dimensions. (b). Parameters for constraints. (c) Parameters for operations. (d) Parameters for relations.

Figure 3.12 also shows that aside from the classification based on the types of associate attributes in Figure 3.11, a parameter can be one of different types, such as *real*, *integer*, *string*, or *Boolean operation*. In addition, not all of the parameters have to be defined by modellers.

A computer aided system will create a default parameter for an attribute automatically when it is needed. From this perspective, a parameter in a model can be a *user parameter*, which is created by a modeller. A parameter can be an *intrinsic parameter*, which is created for any geometry or features automatically by the software tool. The difference between a user parameter and an intrinsic parameter is that the constraint for a user parameter is '*hard*' and that for an intrinsic parameter is '*soft*'.

Hard and soft constraints show the difference when the parameters need to be changed. A 'soft' constraint means that the parameter is *dependent* and its value can be changed automatically if the current value conflicts with other constraints. A 'hard' constraint means that the parameter is *independent* and the model does not allow the software to change its value automatically, even if the current value conflicts with other constraints. Instead, the software gives an alert to the modeller either to change other constraints or to update the parameter to eliminate conflicts.

A parameter can also be *continuous* or *discrete*. A continuous parameter, such as the dimensions for length or width of a solid, can take any value within the given range.

A discrete parameter, such as the number of instances in a pattern feature, can only be one of pre-specified integrals.

In addition, a parameter can be defined at different levels of model such as at a sketch level, part level, assembly level, product level, or system level.

3.3.3 Geometric Constraints and Relations

In parametric modelling, a software tool is capable of detecting various attributes and defining them as intrinsic parameters in defaults. However, if the modeller decides to take full control of a certain attribute, he or she needs to define such an attribute as a user parameter. Depending on the type of attribute, the corresponding user parameters can be defined in different ways. In the following, the definitions of dimensional and geometric parameters are discussed.

3.3.3.1 Dimensional Parameters

Figure 3.13 shows some examples of intrinsic parameters and user parameters. A user parameter is defined and assigned by the modeller. Figure 3.13a shows how some dimensional parameters were created for the sketch and features. An intrinsic parameter is automatically created by the software based on the dependence of this parameter on other attributes. Figure 3.13b shows that the software tool defines a list of the properties of objects as intrinsic parameters whose values depend on the given geometry and density of objects, and can be changed automatically when the geometry or materials are changed.

In defining a user parameter, Figure 3.14 shows that a dimensional parameter on a sketch is usually created by '*smart dimensions*' in SolidWorks. It provides a flexible means to create an appropriate type of dimensional parameter based on selected geometric entities.

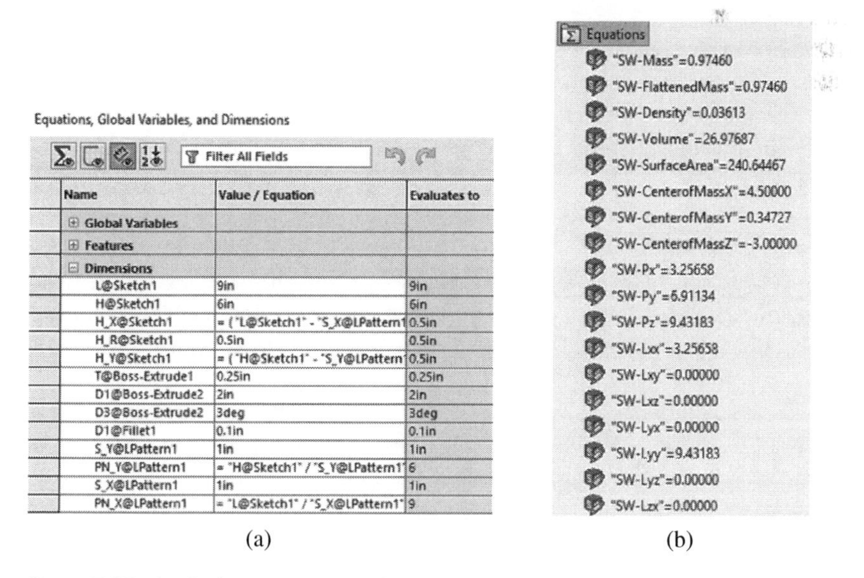

Figure 3.13 Intrinsic parameters and user parameters. (a) User parameters. (b) Intrinsic parameters.

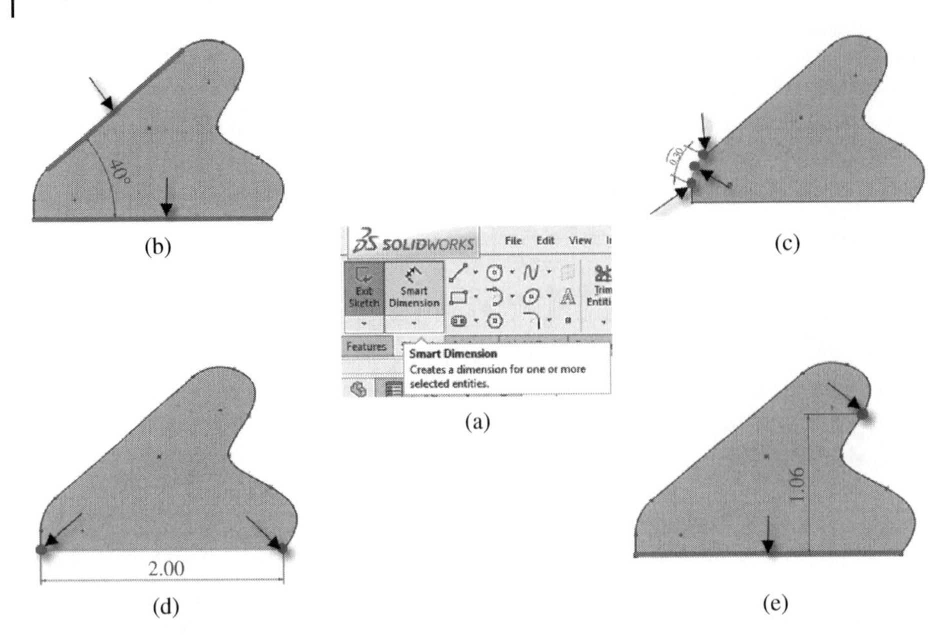

Figure 3.14 Defining a dimensional parameter by 'smart dimension' in SolidWorks. (a) Activating 'smart dimensions'. (b) Angle by two lines. (c) Arc length by three vertices. (d) Distance by two vertices. (e) Distance by a vertex and line.

Figure 3.14a shows that the interface activates the 'smart dimensions' tool. Figure 3.14b defines *an angle* when two lines are selected. Figure 3.14c defines a *curve length* when two vertices and an intermediate point on a curve are selected. Figure 3.14d and e define a *distance* when two vertices are selected or a vertex and a line are selected, respectively.

In parametric modelling, parameters are used to (i) allow direct access to dimensional and geometric attributes and make easy changes when they are needed, (ii) eliminate redundancy since the attributes are determined by their parameters and access to the parameters are through their storage in computer memory, and (iii) sustain the consistence since the same parameters in multiple models can be simultaneously updated when the changes are made in one model.

3.3.3.2 Geometric Constraints

A *geometric constraint* is a non-numerical geometric relationship between two or more entities. Most geometric constraints are used to represent design constraints of geometric entities in two-dimensional sketches. Figure 3.15 shows a list of commonly used geometric constraints in commercial CAD software tools.

In creating a sketch, SolidWorks is capable of detecting applicable geometric constraints of one entity with others. When a geometric constraint is detected, the designer receives the feedback and then makes the justification if such a constraint should be applied to the entity to be modelled. The designer can simply continue the next action if the identified constraint is accepted; otherwise, the designer has to take the next action when the constraint becomes inactive. To create or modify the relations in a sketch, SolidWorks provides a '*Display/Delete Relations*' tool for users to review, create, modify, and delete geometric

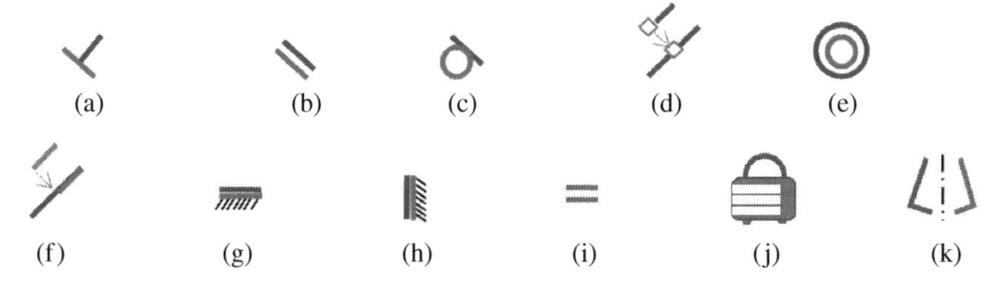

Figure 3.15 Types of geometric constraints in sketches. (a) Perpendicular. (b) Parallel. (c) Tangent. (d) Endpoint coincident. (e) Central coincident. (f) Collinear. (g) Horizontal. (h) Vertical. (i) Equal. (j) Fix. (k) Symmetric.

constraints at any time. Figure 3.16a shows the interface where the 'Display/Delete Relations' is activated. Figure 3.16b shows when a point and a line are selected and applicable constraints are that the point is a *Midpoint* of the line segment, the point is *Coincident* with the line segment, or selected entities are *Fix*. Figure 3.16c shows when two line segments are selected and applicable constraints are that two lines are *Horizontal*, *Vertical*, *Collinear*, *Perpendicular*, *Parallel* with each other, two lines have *Equal* length, or the selected lines are *Fix*. Figure 3.16d shows when two points are selected and applicable constraints are the connections of two points are *Horizontal*, *Vertical*, two points are *Merge*, or the selected two points are *Fix*. Depending on the selected entities, applicable constraints vary.

Geometric constraints at the assembly level are called assembly mates, which will be discussed later.

3.4 Design Intents

A computer aided system is able to capture design intents and represent them as parameters, constraints, relations, and modelled behaviours. For example, a designer can first create an approximate sketch and later the dimensions of the sketch can be refined exactly. Designers can change the dimensions of sketches and features at any time. In general, *design intents* show how to (i) proceed modelling activities efficiently for a number of design features and (ii) take into consideration possible changes and facilitate the changes occurring to geometric entities. Design intents are used to simplify the modelling process and accommodate the predicted changes in engineering design. In addition, some common senses in parametric modelling can be interpreted as design intents.

3.4.1 Default Location and Orientation of a Part

In computer modelling, the location and orientation of a part are defined with respect to a given coordinate system. The location, orientation, or the origin of a part are often taken as a reference in order to create some features on the part. For example, (i) if a part is symmetric about a plane, it becomes convenient to set a default plane of the reference coordinate system as the symmetric plane and let the origin of the part pass through the selected plane. (ii) When the properties of an object are evaluated, the alignment of an inertial coordinate system with the reference coordinate system can simplify the calculation. Based on the above

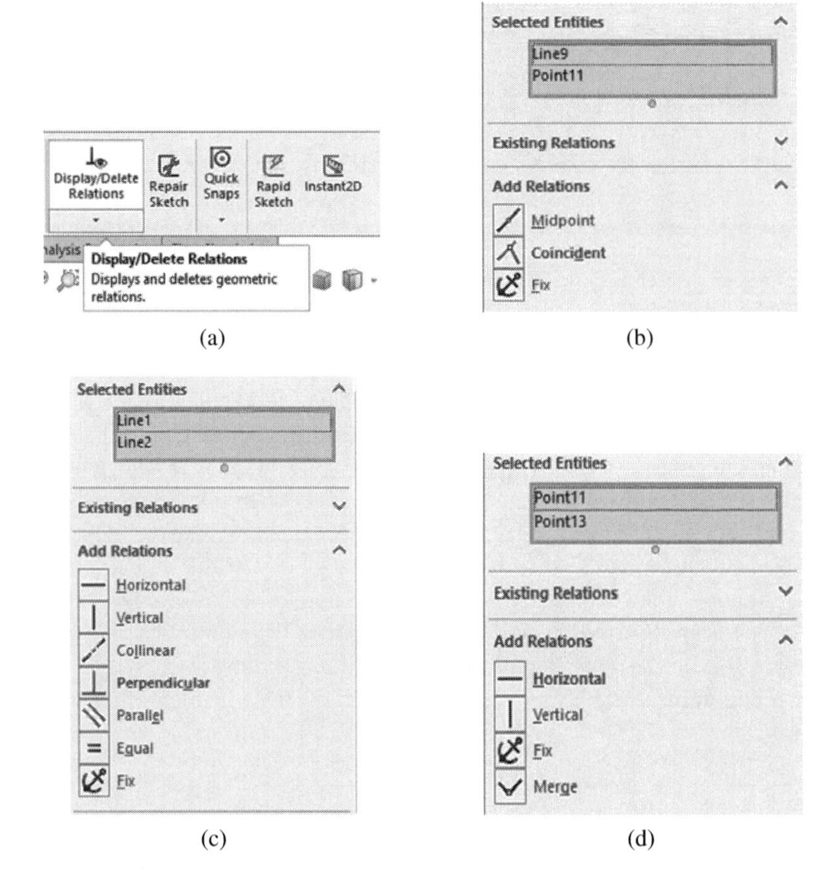

Figure 3.16 Define the relations in a sketch. (a) Activating 'Display/Delete Relations'. (b) Relations of a line and point. (c) Relations of two lines. (d) Relations of two points.

discussion, a design intent can be taken to align the origin and posture of the part with the default coordinate system shown in Figure 3.17. Doing so will allow all of the entities, such as the origin and default of the front, right, and top planes, to be used as the reference in creating new parameters and features for the part.

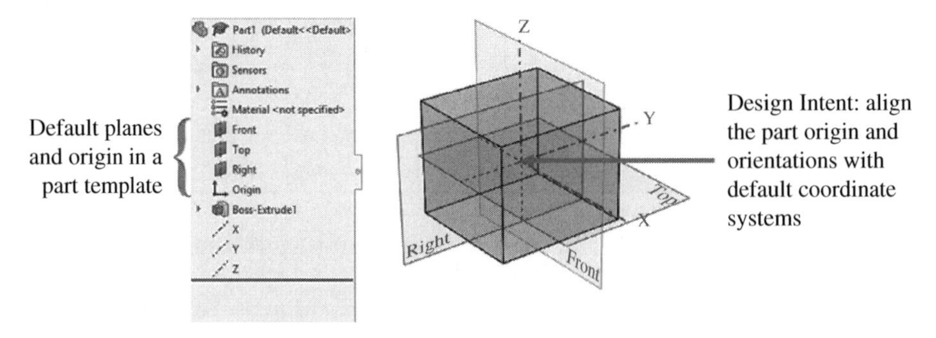

Figure 3.17 Selection of a default location of part as a design intent.

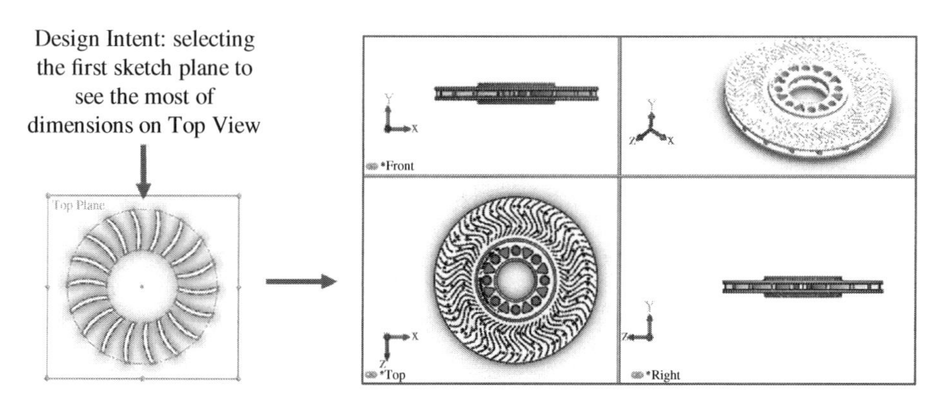

Figure 3.18 Selection of first sketch plane as a design intent.

3.4.2 First Sketch Plane

Engineering drawings are often needed when making parts in manufacturing processes. In general, engineers expect to review most of the dimensions either in top or front view of a part. The contents projected from one orthogonal direction depend on the sketch plane of the first feature of the part. Therefore, a design intent can be set for the selection of the first sketch plane. This sketch plane should be selected to ensure that the majority of the dimensions are visible. Figure 3.18 shows a disk part with all of the dimensional details on the O-XZ plane. If the designer tends to view the majority of dimensions on the top view, the top plane should be selected as the first sketch to take consideration of this design intent.

3.5 Design Equations

Design equations can be used to represent the design intents where geometric dimensions adhere to a specified set of rules. A *design equation* represents the dependent relation of one parameter on other parameters. Design rules are an effective way to standardize parts. Note that standardization affects the cost of a product. Cost is one of the primary goals of product design and the standardization of parts or components allows product costs to be reduced via mass production. Due to standardization, independent parameters may become dependent on each other. Figure 3.19 shows an example of a set of design variables of threads, which become dependent due to standardization. Machine elements such as screws, nuts, gears, and bearings are all standardized in certain ways and their design variables become dependent when design equations are applied.

Defining a design equation follows the syntax of programming in other languages. To create a design equation in modelling, a variable on the left side of the equation is a dependent one and the variables cited on the right side of the equation are either *global variables* or *independent variables*. In addition, the values of *global variables* are specified in a separated section. Independent variables are defined in the model using a dimensioning tool such as the *smart dimension* in SolidWorks. Moreover, a design equation may include other parameters such as file properties, dimension measurements, or mathematical functions. For examples, SolidWorks allows the operators or mathematical functions given in Table 3.7 in the definition of a design equation.

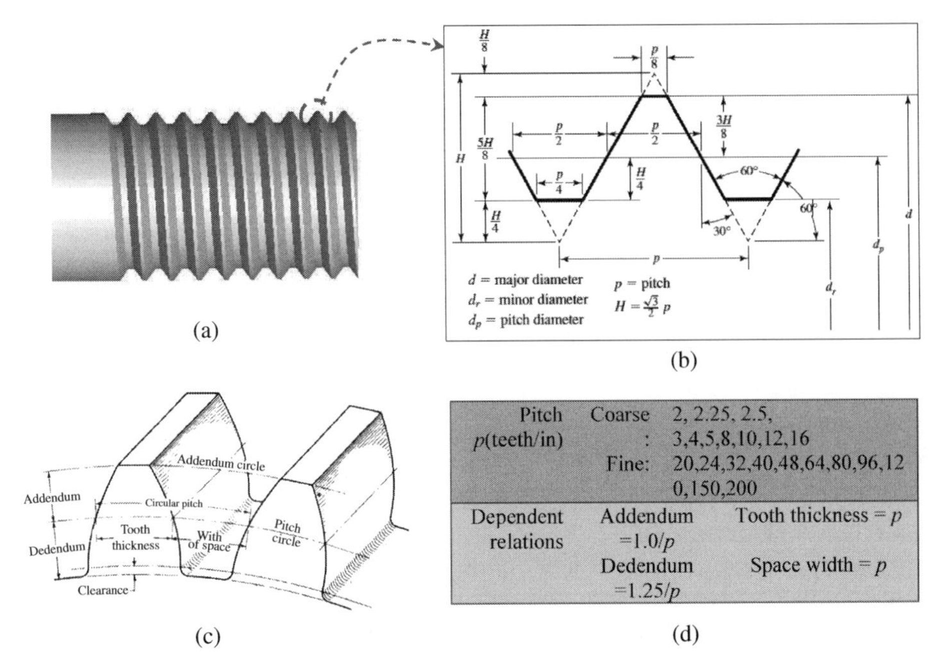

Figure 3.19 Dependent variables in threads. (a) Thread. (b) Detailed view of dependent variables on a thread. (c) Gear teeth. (d) Dependent variables on a gear tooth.

Table 3.7 The operators and functions in the definition of design equations.

Operator or function	Meanings	Operator or function	Meanings
+	Addition	*arccos(x)*	Find the angle of the cosine ratio *x*
–	Subtraction	*atn(x)*	Find the angle of the tangent ratio *x*
*	Multiplication	*arcsec(x)*	Find the angle of the secant ratio *x*
/	Division	*arccosec(x)*	Find the angle of the cosecant ratio *x*
^	Exponentiation	*arccotan(x)*	Find the angle of the cotangent ratio *x*
sin(x)	Fine the sine ratio of angle *x*	*abs(x)*	Find the absolute value of *x*
cos(x)	Fine the cosine ratio of angle *x*	*exp(x)*	Find *e* raised to the power of *x*
tan(x)	Fine the tangent ratio of angle *x*	*log(x)*	Find the natural log of *x* to the base of *e*
sec(x)	Fine the secant ratio of angle *x*	*sqr(x)*	Find the square root of *x*
cosec(x)	Fine the cosecant ratio of angle *x*	*int(x)*	Find the integer of *x*
cotan(x)	Find the cotangent ratio of angle *x*	*sng(x)*	Find the sign of *x* as '–1' or '1'.
arcsin(x)	Find the angle of the sine ratio *x*	*pi*	Refers to the ratio of the circumference

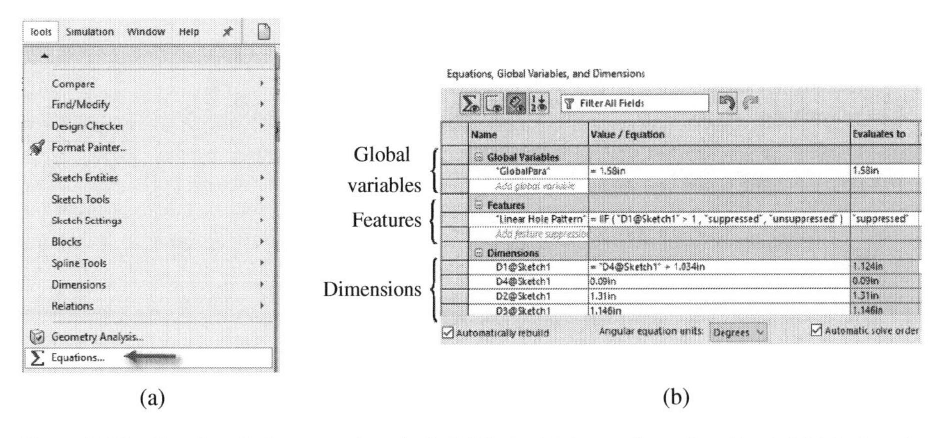

Figure 3.20 Creating design equations in SolidWorks. (a) Access 'equations' tool. (b) Interface to define equations for 'global variables', 'features', and 'dimensions'.

Figure 3.20a shows the interface used to activate the '*equations*' tool in SolidWorks. It is within the '*tool*' menu. Figure 3.20b shows three sections of the 'equations' tool. It consists of the sections for *global variables*, *feature variables*, and *dimensional variables*. A variable on the left side is determined by the expression on the right side of the equation and its initialized value is given in the last column.

Example 3.1 Create a model of a Lego piece for the requirements of (i) a rectangular shape with four pegs on the male face and one peg on the female side, (ii) the wall thickness is given as 2 mm, and (iii) the male and female faces must be matched seamlessly.

Solution

A Lego piece with the outside dimensions of 50 (mm) × 50 (mm) and an extruded thickness of 10 (mm) is modelled to create all the variables of dependent variables. Figure 3.21 shows that these variables include $Width_I$, $Width_O$, $Height_I$, $Height_I$, $PegO_D$, $PegI_D$, and Ref_D. All of the required relations are defined as a set of equations in Figure 3.21d.

3.6 Design Tables

A useful design intent is to create a single CAD model for the whole part family. A design table is used to create a variety of configurations for the part family. A CAD model for a part family consists of (i) a solid part model where all of the design variables and features are defined and (ii) a *design table* where the values of the design variables or the states of the design features are specified for all parts in the family. A part in the family is called a *configuration*, and the values or states of the controlled design variables are specified in a single row.

There is a difference in dealing with the dependence of design variables by a design table and design equation. A design table controls the dimensions and feature states of parts in the family. A design table deals with the dependence of *discrete variables* or features in a part family. It differs from a design equation, which deals with the *dependence of continuous*

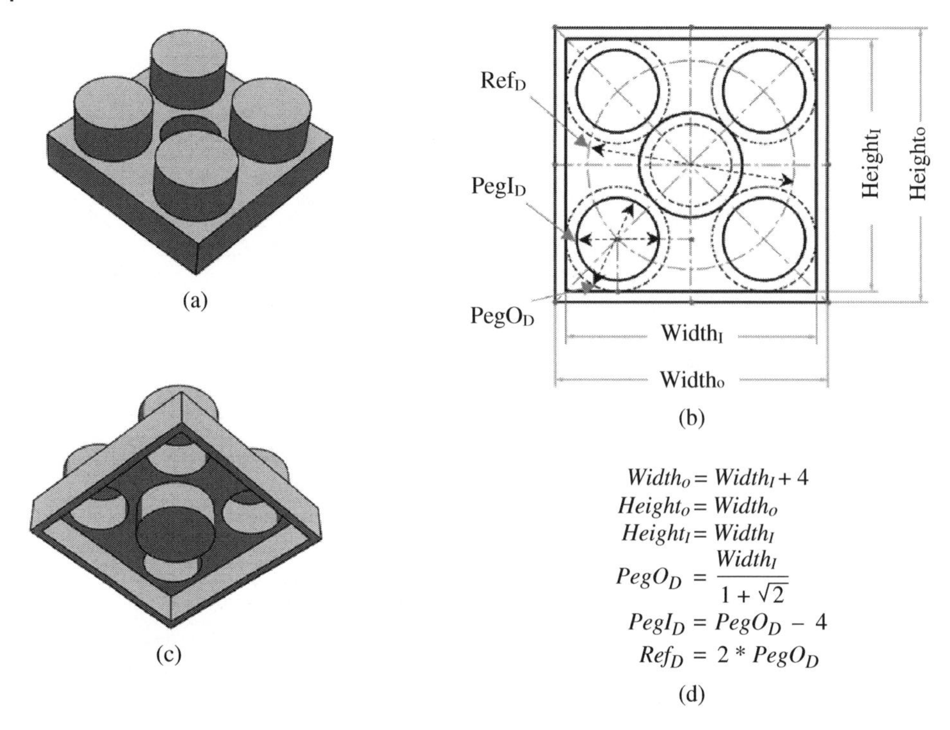

Figure 3.21 Design equations in a Lego piece model (unit: mm). (a) Male surface. (b) Main dimensions. (c) Female surface. (d) Equations for dependent relations.

variables in a part model. Design tables are effective when representing design variations from the combination of design variables.

Designers and companies are always seeking ways to innovate and create new products to meet customers' needs. Figure 3.22 shows three basic ways for innovations and creations: (i) '*copying*' to apply existing ideas in new applications, (ii) '*transforming*' to improve existing ideas for new functionalities and better performance, and (iii) '*combining*' to synergize existing ideas as a comprehensive solution to complex applications with more functions. It is clear that not all of the innovations are made from scratch. One has to utilize as many existing technologies, methodologies, theories, tools, and models as possible to reduce the design cycle and make the innovations prolific. On the other hand, modern products are highly diversified to meet personalized needs, and it is uncommon that the same design model can be used for a long time. Instead, designers can benefit greatly by taking an existing model, stretching and pulling it, and revising it as a new model. Parametric modelling techniques including design tables support knowledge reuse in designing processes.

In a solid model, every constraint, dimension, feature state, part state, mate state, or configuration has an associated parameter ID; the values of these parameters give detailed information such as *sizes*, *shapes*, *visibilities*, and the *states* of features, parts, mates, or assemblies. Moreover, parameters can be either '*dependent*' or '*independent*'. The parameters included in a design table are associated with the solid model, and the values of these parameters are specified for each configuration. From the perspective of KBE, the design table tools allow (i) modification of sketches, features, parts, and assemblies in a single

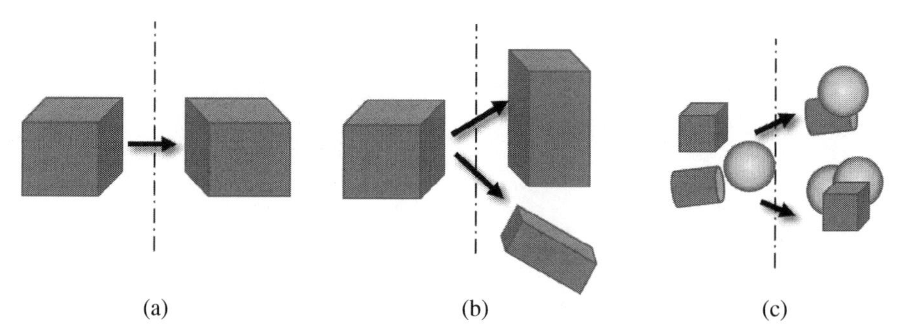

Figure 3.22 Three common ways of innovations and creations. (a) Copying. (b) Transforming. (c) Combining.

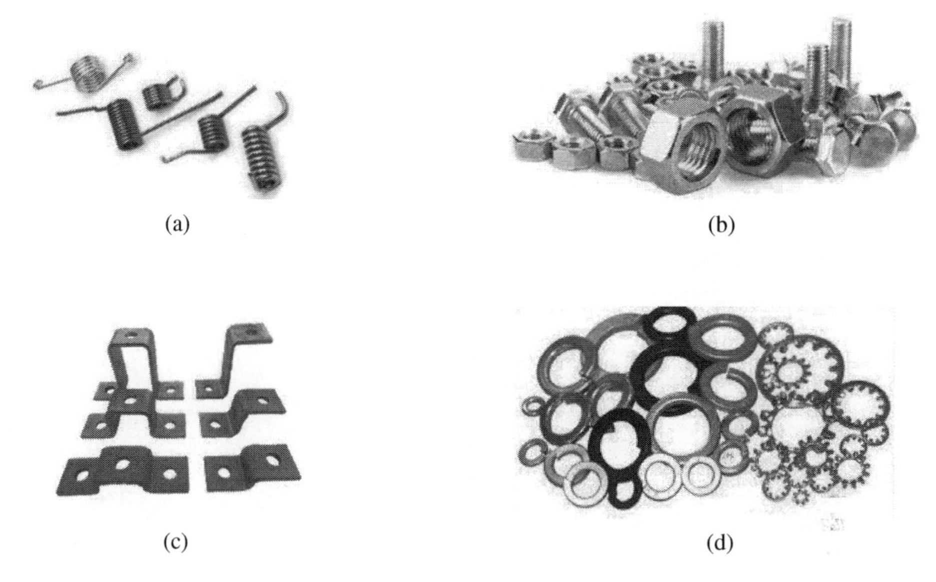

Figure 3.23 Examples of using a design table for part families. (a) Springs. (b) Threads, screws, and nuts. (c) Brackets. (d) Washers.

source, (ii) derivation of unique components from existing components, and (iii) building intelligence such as parametrical constraints and design equations in a design table. However, one CAD model can only work on one design table at a time.

In parametric modelling, design tables are applicable at both *part level* and *assembly level*. Figure 3.23 shows some common machine elements, such as springs, bolts, brackets, and washers, where design tables are used to represent the part families. Figure 3.24 shows a few of the product families where design tables are used to represent the variations of part types and configurations for valves, cylinders, and actuators.

A design table consists of the configurations for a part or product family. The configurations can be created in two different ways. Figure 3.25a shows an example where a number of configurations are generated manually before a design table is created. Figure 3.25b shows that new configurations are directly derived from a design table where each row

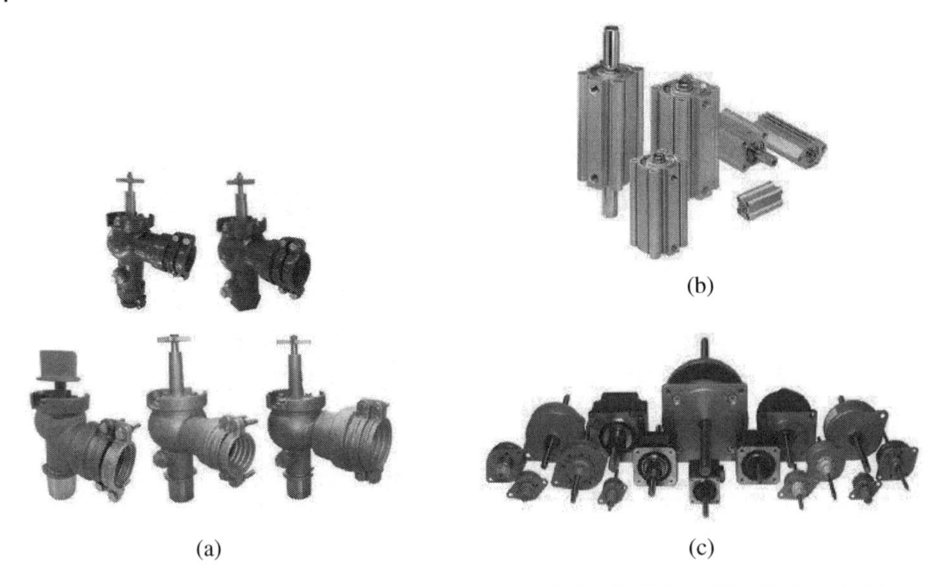

(a) (b) (c)

Figure 3.24 Examples of using a design table for assemblies. (a) Valves. (b) Cylinders. (c) Motors.

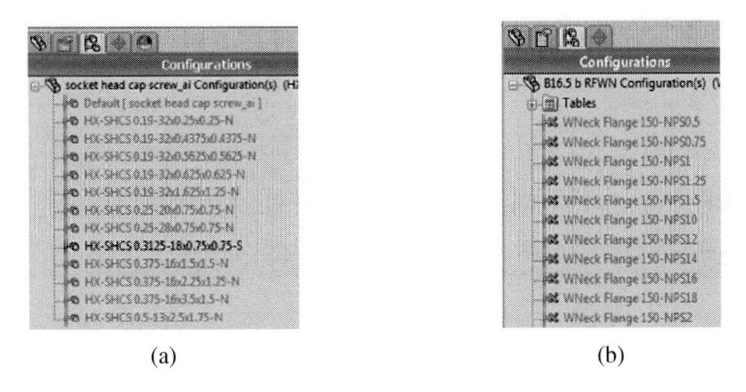

(a) (b)

Figure 3.25 Creating configurations in a design table. (a) Manually created configurations. (b) Configurations generated by a design table.

in the table defines the properties of design variables for the corresponding configuration. Each part or assembly model allows at most one design table. In defining a design table, the syntax of the table must be followed in order to associate the design table with the original model. In SolidWorks, a design table is created using the Microsoft Excel program, and all of the design variables and the feature or part statuses in the table have their links to the corresponding attributes in solid models.

Figure 3.26 shows the procedure to creating a design table for a part family. It begins with the creation of the solid model with all of the controllable variables, features, and properties. It also helps to create a number of configurations manually by assigning different values to design variables that will be included in the design table. Design tables are supported by common parametric modelling software tools. In SolidWorks, the *design*

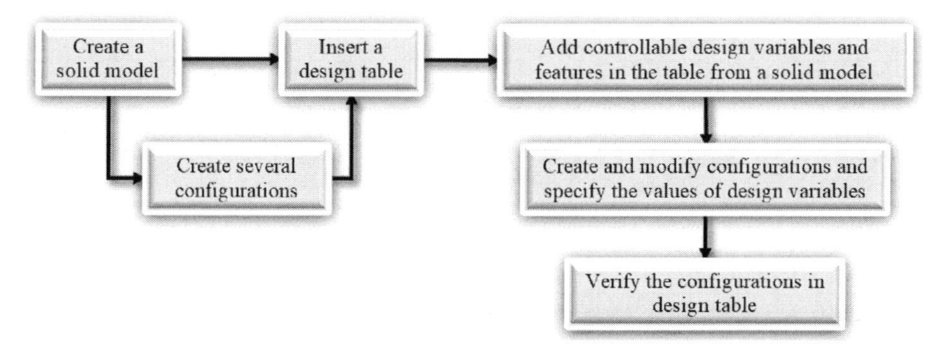

Figure 3.26 Procedure for creating a part model with a design table.

Table 3.8 Types of design variables in a design table (SolidWorks 2019).

	In either a part model or an assembly model	**Only in a part model**	**Only in an assembly model**
Design variables	• Tolerance type • Configuration specific properties • Model colour • Linear and radial pattern • Spacing and instances in derived configurations • Lighting state • Equation state • Sketch relationship state • Mass properties • Centre of gravity	• Feature state • Configuration of base or split part • Dimension values	• Component state • Mate state • Referenced configuration • Display state • Assembly feature state (cuts) • Dimension and mate values • Bill of materials (BOM) part number • Expand in BOM

table tool in included in the '*tool*' menu. Once the design table tool is activated, one can select controllable design variables and specify their values in individual configurations. Note that any design variable included in the table must be linked to its attributed part in the solid model. By double clicking the attribute in the solid model, SolidWorks can automatically add the corresponding variable in the row of design variables with the correct link to the original attribute. Table 3.8 shows the types of design variables in a part or assembly model.

In SolidWorks, the design table tool is included in the folder of '*Tables*' under the '*Insert*' menu. Figure 3.27a shows the path to access the '*Design Table*' tool as 'Insert' → 'Tables' → 'Design Table'. As shown in Figure 3.27b, a design table can be initialized in three ways: '*blank*' for an empty design table from template, '*auto-create*' for a design table with detailed guides in selecting design variables, features, and configurations, and '*from file*' for importing a pre-defined design table. A design table from 'auto-create' provides a detailed guidance on how to complete the design table. Figure 3.27c shows

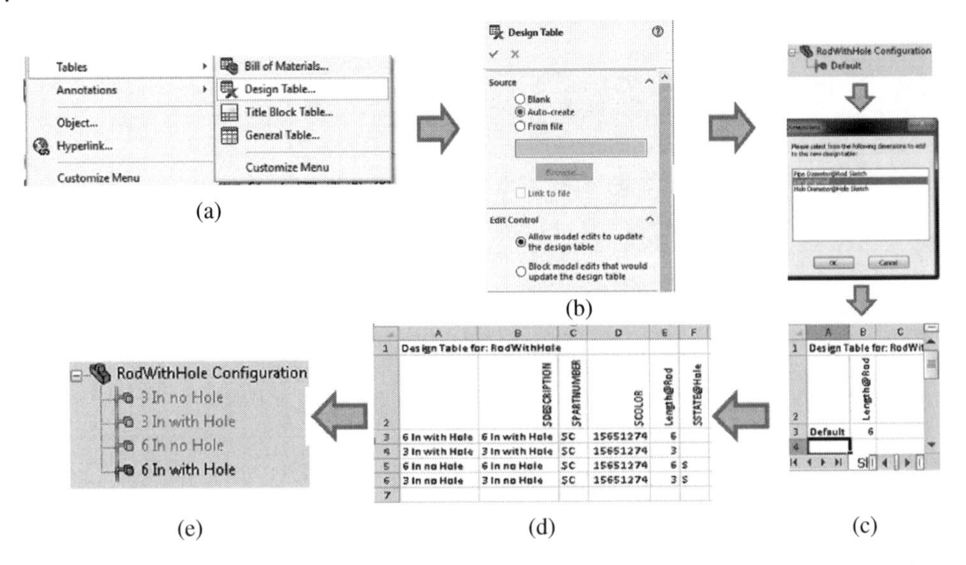

Figure 3.27 Defining a design table in SolidWorks. (a) Insert a design table. (b). Ways to initialize a table. (c) Add parameters in a table. (d) Completed design table. (e) Configurations in a design table.

a pop-up window with a list of all the dimensional variables and attributes of the solid model; users can decide if certain variables or attributes should be included in the table. Once the selections are confirmed, they are added into the design table automatically. In addition, users have their options to add additional design variables or attributes into the design tables if they are not included in the list generated by the software. The next step is to create new rows to include new configurations in the table. As shown in Figure 3.27d, each row corresponds to one configuration. The values of design variables and attributes are specified in the row. After the design table is completed, new configurations are listed under '*Configuration Manager*', as shown in Figure 3.27e.

Example 3.2 Create a part model for a set of six sockets as shown in Figure 3.28d.

Solution

A parametric model is designed with four variables '*InA_D*', '*In_D*', '*Out_D*', and '*Out_Depth*'. Figure 3.28a shows the model. The procedure in Figure 3.27 is followed to create a design table and Figure 3.28c shows the content with the configurations for all six variants.

To evaluate the value of a parameter in a design table, a design equation can be incorporated. As shown in Figure 3.29, if a design equation is used in the table, the corresponding field must be preceded by a single apostrophe '′' and an equal sign '=', i.e. '′='. The single apostrophe '′' implies that the design equation is exported to the Excel file as an expression for a design equation rather than a text string. In Figure 3.29, design equations are used to evaluate the parameters '*Width*' and '*Height*' in each configuration. For the block with the length, width, and height, only the length is independent as the width is determined by the length and the height is determined by the width based on the golden ratio (0.618). Note that at most one design table is allowed in a part or assembly model.

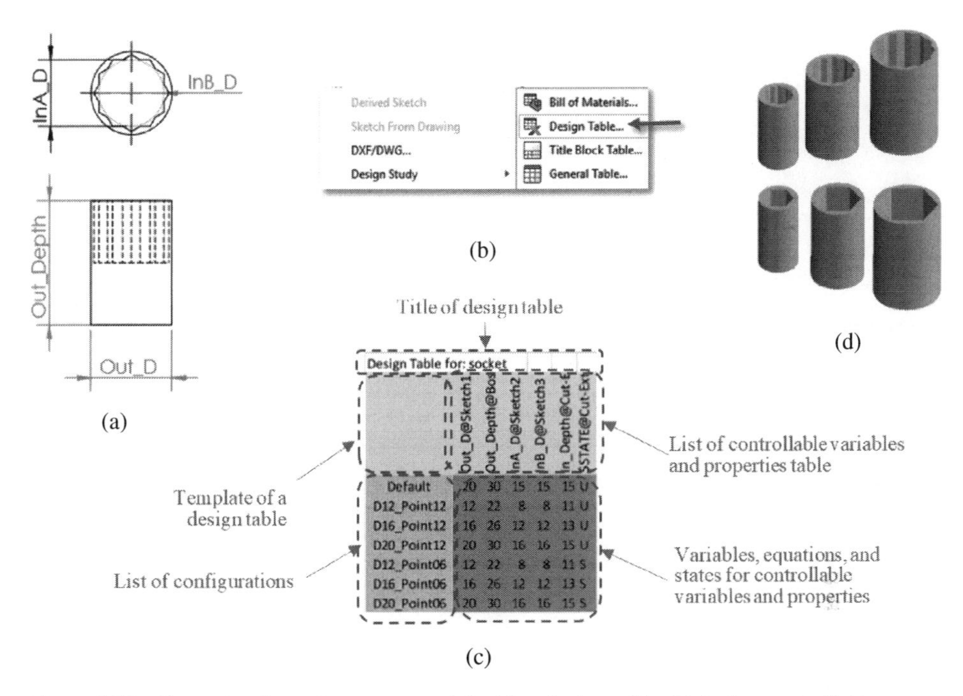

Figure 3.28 Example of creating a part model with a design table. (a) Define controllable variables and feature in a model. (b) Create a 'design table' in the 'tools' menu. (c) Create a design table for a part family. (d) Six configurations in model.

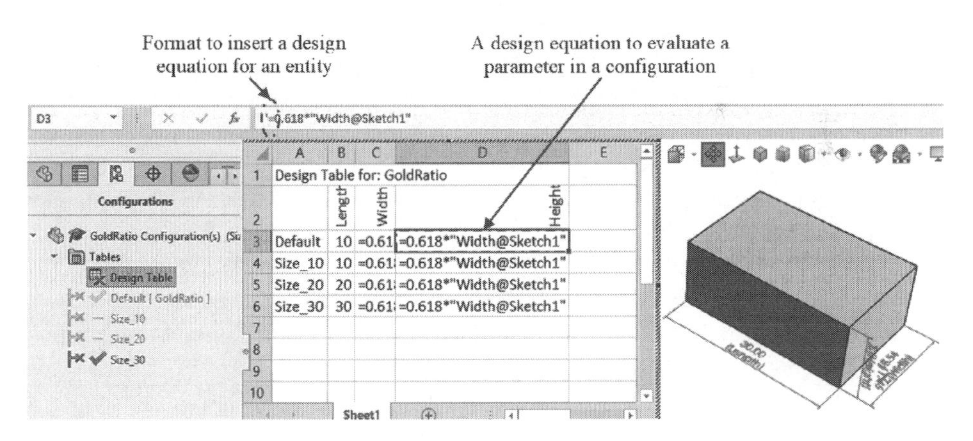

Figure 3.29 Example of creating a part model with a design table.

3.7 Configurations as Part Properties

A product is usually assembled from a set of parts. More variants can be generated by using different configurations of parts in an assembly. The *configuration publisher* tool allows users to change design variables of parts in an assembly model through the *Property Manager*. Once a configuration is published, the design variables are treated as the

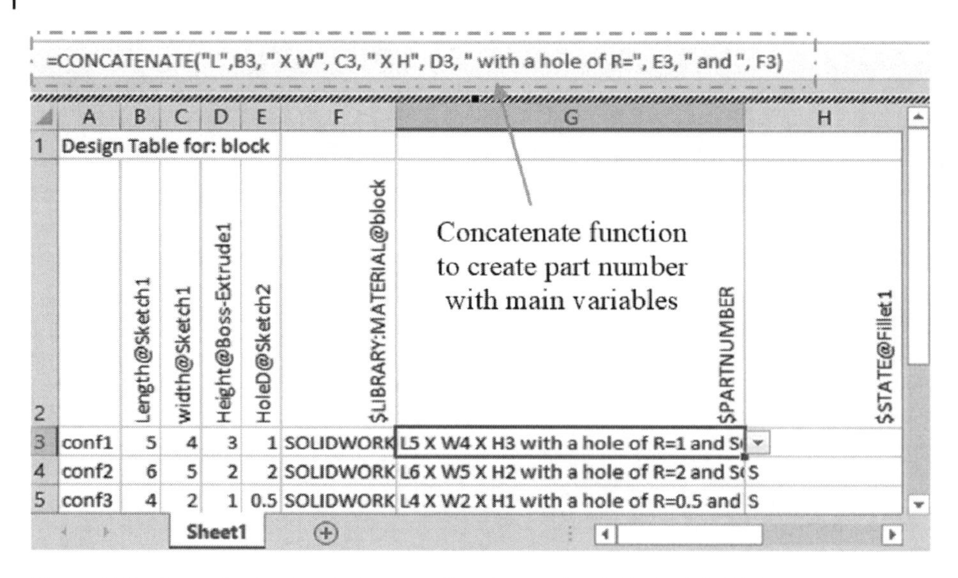

Figure 3.30 Example of using design equations in a design table in SolidWorks.

properties of the configuration, which can be accessed and changed in product modelling. Once a part is inserted in a model, the configuration publisher tool in SolidWorks prompts users to specify the values for the design variables in order to determine which configuration is selected. Such a tool is especially useful for standardized parts and assemblies, which are used frequently but with many configurations. The descriptive information from configuration names is insufficient to tell which part is actually used in assembly.

A configuration publisher tool is available when a part or assembly model includes one or more configurations. The tool applies to a design table, which includes all of the parameters that might be changed from one configuration to another. In addition, the parameter '*$partnumber*' can be used to control the name of the new configuration. As shown in Figure 3.30, *$partnumber* can be concatenated from the texts or values of main control parameters.

Figure 3.31 shows the interface used to activate the '*Configuration Publisher*' tool in the Configuration Manager. After all of the desired configurations are generated, right click on the part or component name (block configurations' in the example) in the Configuration Manager and choose 'Configuration Publisher' to activate the tool.

The configuration publisher tool obtains all of the configurations from a design table. If a design table is not ready, the software tool gives the user the prompt to create a design table before he or she proceeds to the next steps. As shown in Figure 3.32, the configuration publisher then shows a list of the properties (controls) used to filter through the configurations. These properties can be dimensions, materials, and suppression states. The user needs to drag these controls from the left panel into the centre pane when the *edit* tab is selected.

Selecting a control in the centre pane allows editing of its properties on the right. Figure 3.33 shows that the appropriate attribute names can be assigned to control variables in the design table.

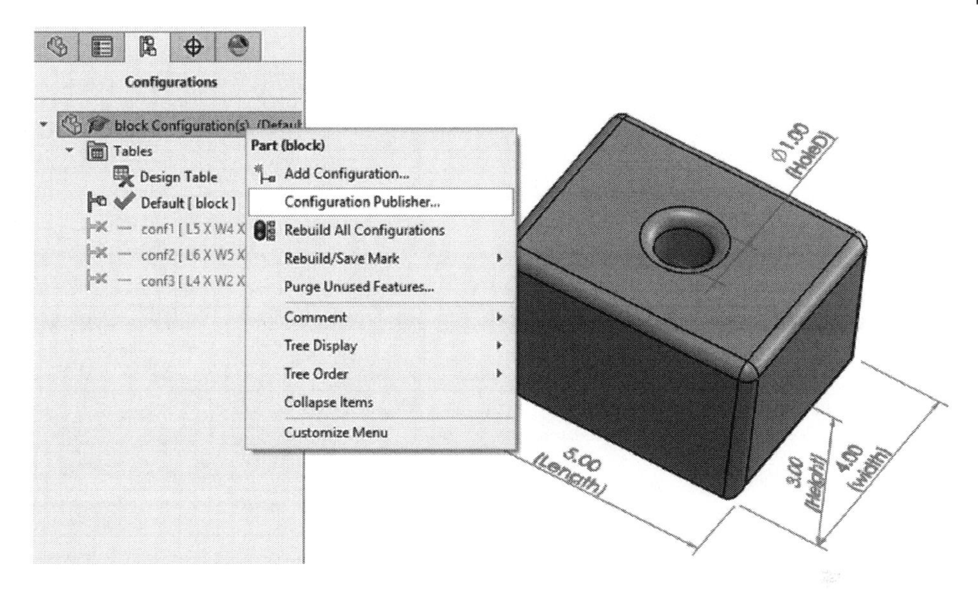

Figure 3.31 Example of using the concatenation function to create part numbers.

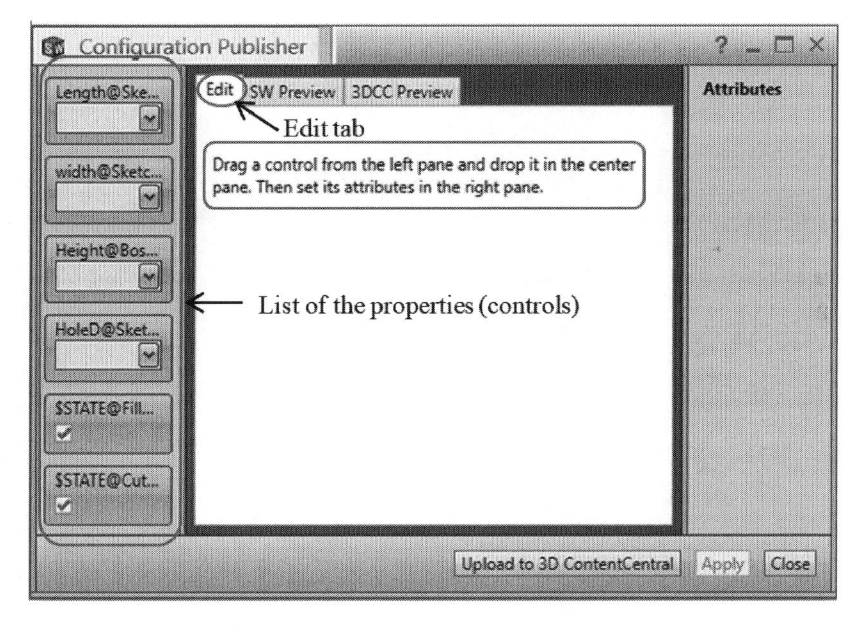

Figure 3.32 Activate 'Configuration Publisher' tool in SolidWorks.

After editing is completed, the *SW Preview* tab allows the user to preview how the Configuration Publisher looks when it becomes active. If the user is satisfied with the results, it is important to click the '*apply*' button to confirm acceptance of the changes. When a part with the defined configuration publisher is inserted in an assembly model, SolidWorks prompts the user to choose the set of control parameters to specify the

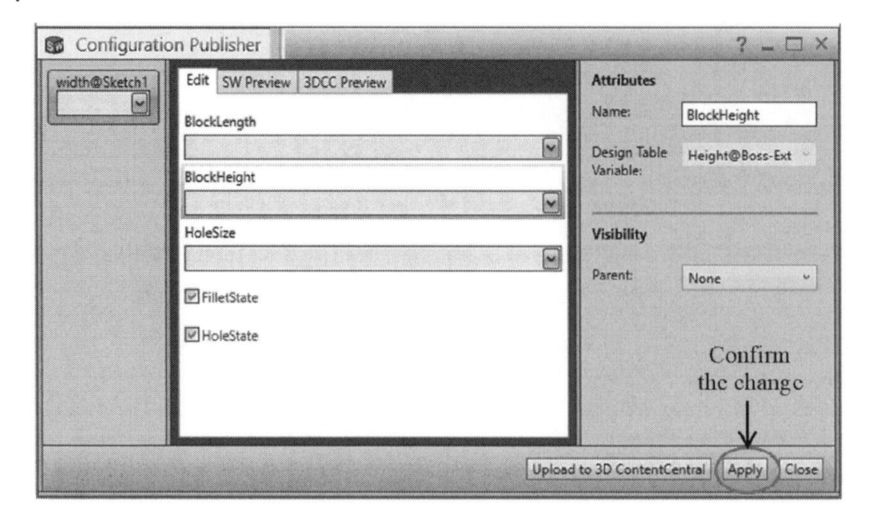

Figure 3.33 Showing a list of the properties for filtering.

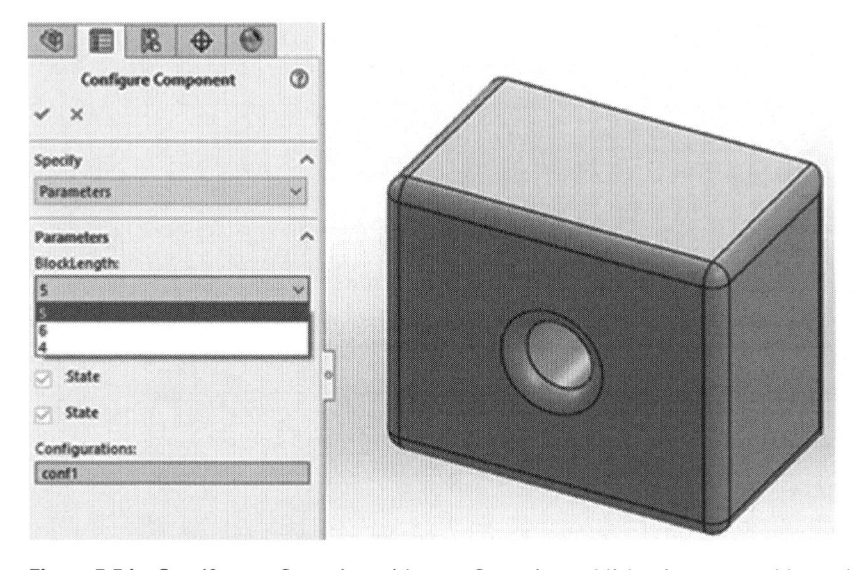

Figure 3.34 Specify a configuration with a configuration publisher in an assembly model.

configuration. SolidWorks then filters through the remaining parameters until a unique configuration is identified (Figure 3.34).

3.8 Design Tables in Assembly Models

A design table in an assembly model works in the similar way to that in a part model. It may include dimensions, custom properties, and the states of parts as controllable parameters. A dimensional parameter can be a *mate dimension* or *reference geometry dimension*. The state

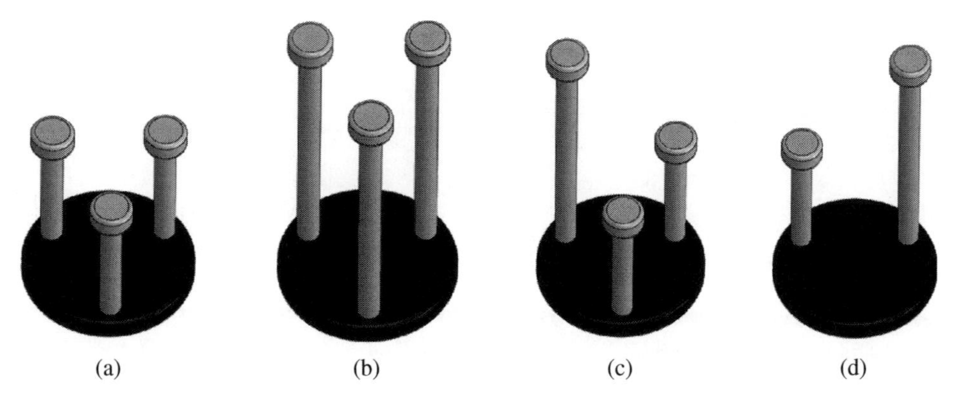

Figure 3.35 Example of an assembly model with the variants from a part level. (a) Configuration A. (b) Configuration B. (c) Configuration C. (d) Configuration D.

Figure 3.36 Design table at an assembly model.

Design Table for: DesignTableinAssembly

	$Configuration@Rod<1>	$Configuration@Rod<5>	$Configuration@Rod<6>	$State@Rod<6>
Default	conf4	conf4	conf4	R
A	conf1	conf1	conf1	R
B	conf3	conf4	conf3	R
C	conf1	conf3	conf1	R
D	conf4	conf2	conf3	S

of a part can be '*suppressed*' or '*resolved*', which is similar to the states of feature '*suppressed*' or '*unsuppressed*' in a design table of a part model.

However, a design table in an assembly model is not able to deal with the variants at a part level directly. One may not suppress or unsuppress a part feature or change a part dimension. If an assembly model has to define the variants at both part and assembly levels, one has to define the part variants as the configurations in a part model and use the part configuration as a controllable parameter in the design table of the assembly file.

Figure 3.35 shows an example of an assembly model with four variants related to part configurations. The model is assembled from a *base* part and three instances of a *rod* part. The corresponding design tables in the assembly model and the rod part model are given in Figures 3.36 and 3.37, respectively.

Note that the names of part configurations and part states are input manually into the row of controllable parameters following the syntax of the design table in SolidWorks.

1	Design Table for: Rod					
2		$partnumber	BaseR@Sketch2	BaseH@Boss-Extrude1	CapH@Boss-Extrude2	$STATE@Boss-Extrude2
3	Default	=CONCATENATE("R-",C3,"H-",D3,"Cap",E3)	0.5	3	0.5	U
4	conf1	=CONCATENATE("R-",C3,"H-",D3,"Cap",E3)	0.5	3	0.5	U
5	conf2	=CONCATENATE("R-",C3,"H-",D3,"Cap",E3)	0.5	3	0.5	U
6	conf3	=CONCATENATE("R-",C3,"H-",D3,"Cap",E3)	0.5	6	0.5	U
7	conf4	=CONCATENATE("R-",C3,"H-",D3,"Cap",E3)	0.5	6	0.5	U

Figure 3.37 Design table at a part model.

1. For the configuration of a part or component:
   $Configuration@component_name<instance number>, e.g. $Configuration@Rod<1>
2. For the state of a part or component:
   $State@component_name<instance number>, e.g. $State@Rod<6>

3.9 Design Tables in Applications

Over 98% of manufacturing enterprises are small or medium sized enterprises (SMEs). An SME has limited design expertise but makes a variety of products to meet market needs. KBE becomes extremely importance since (i) product designs need the experiences and skills from designers, but human resources at an SME are dynamic. Experienced designers are likely to explore new opportunities at other enterprises and new designers get the shortened period of training to take over designs. (ii) Modern products are mostly complicated, involving multiple disciplines, while an individual designer only possesses a limited set of knowledge and skills in the specialties they have studied. Design knowledge and skills are utilized in terms of design intents in product designs and the KBE tools make it feasible for SMEs to embed design intents in product models.

A KBE tool such as design tables has demonstrated its effectiveness in engineering practice. For example, Beach (2012) introduced a successful case from the Rotomation Inc. (Rotomation 2019). The company used design tables to model their product families for rotary actuators and indexers (Figure 3.38).

The products varied by five main design variables:

1. Shaft orientation: *pneumatic*, *clockwise* (CW), *counterclockwise* (CCW), or *top centred* (TC).
2. Stroke of angular motion: <2000°.
3. Indexing: a continuous rotation <360°.
4. Seal variations by location and size of *ports* and *wash-down*.

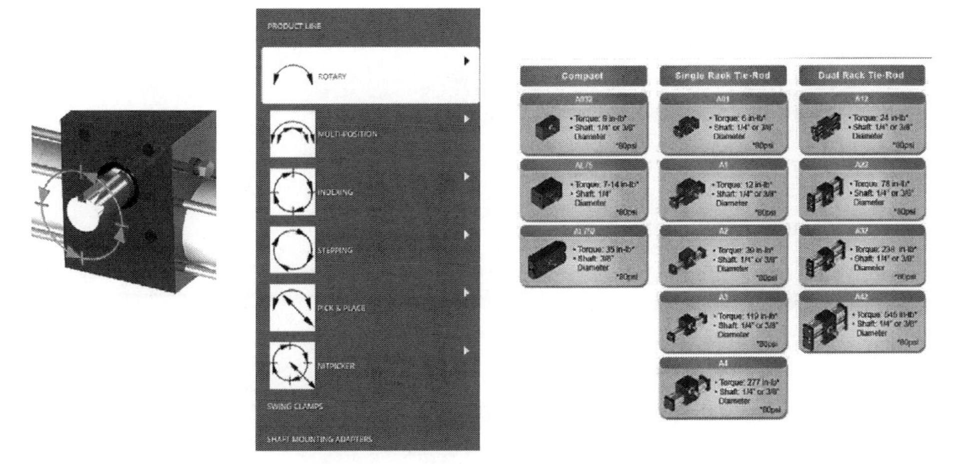

Figure 3.38 The product variants at the Rotomation Inc. (Rotomation 2019). (a) Actuator. (b) Shaft mounting adapters. (c) Variants with rotary adapters.

The company was able to create 1.5 million parts on the most complex model. Design tables allow multilayers of dependency for all possible options and locations of features. The design table tool has been proven to be an ideal tool to model product families.

3.10 Design Templates

The importance of using CAD templates was firstly recognized by software developers, where a CAD template was developed for user defined functions (UDFs) (Arnout 2004). Using CAD templates facilitated the reuses of design intents from one project to another. UDFs were later utilized for the purpose of design automation (DA), where a configurator tool can identify the required template from a library automatically and update the model with user-defined parameters interactively. Ma et al. (2003) presented a framework with object-oriented features consisting of a standard component library for mould design. Users were able to choose the templates with customized user inputs. Halfawy and Froese (2005) presented the paradigm of 'smart objects', which were parametric entities configured and modified by the framework. Context-dependent instantiation of templates was proposed as an effective way to enhance the design space. Similar methods were investigated by Siddique and Boddu (2005) and Ledermann et al. (2005).

The framework by Lederman et al. (2005) used the predefined CAD templates called 'Dynamic Objects', which were automatically inserted and placed in context to generate various repetitive aircraft geometries. Cederfeldt and Sunnersjö (2003) proposed a similar approach with building blocks parametrically retrieved from a library. The main difference was that the proposed building blocks were assembled through the Boolean operations and thus were not fully associative. Other research groups were influenced by Lederman et al. (2005) or independently developed their own dynamic topology frameworks. Danjou et al. (2008) presented an approach based on UDFs as knowledge carriers, where the model was set up by instantiating the UDFs with a few input parameters. They agreed that since the

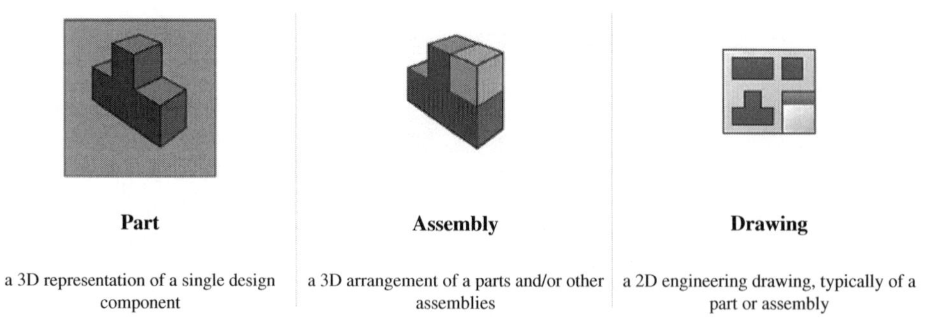

Figure 3.39 Templates for part, assembly, and drawing models in SolidWorks.

UDFs were stored in a library, a fast geometric modification was possible to redefine the design space effectively. Thus, even with the proposed highly automated framework, the designer was not subjected to severe limitations.

To eliminate the identified non-creative work, the CAD templates can be used to create models and automate the modelling processes. The basics can be compared to parametric LEGO® blocks containing a set of design and analysis parameters. These are produced and stored in libraries, giving engineers or a computer agent the possibility to select the templates topologically and then modify the shape of each template parametrically to finally evaluate the generated system with the analysis parameters (Tarkian 2012).

Figure 3.39 shows the templates for part, assembly, and drawing models, which include (i) essential information of models such as the coordinate systems for all three model types and a drawing template for a drawing file, (ii) access to applicable modelling tools, and (iii) model properties and attributes, such as *material properties* and *annotations* in a part model and *scale* and *views* in a drawing model.

The importance of using model templates cannot be overemphasized when a simulation model is defined for numerical simulation in computer aided engineering (CAE). Figure 3.40 shows a module template for a static analysis of a solid object subjected to

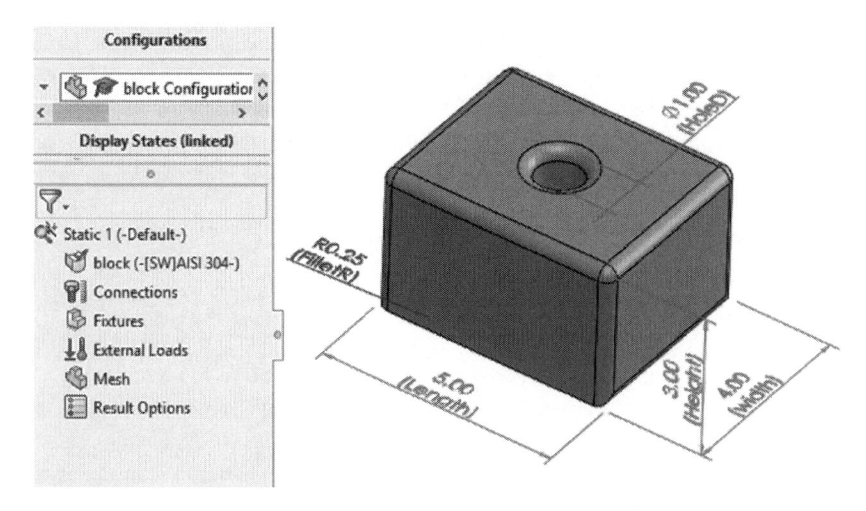

Figure 3.40 Template for static analysis in SolidWorks.

external load and boundary conditions. The template tells the users what inputs are expected to run the simulation and how outputs can be obtained, visualized, and retrieved after simulation.

3.11 Summary

Three basic ways to produce innovative designs are copying, transforming, and combining. These ways towards innovations share commonality on the utilization of existing theories, methodologies, technologies, skills, and tools. Designers should master the theory and skills for KBE to represent, explore, and reuse knowledge for product designs.

Computer aided technologies for KBE need the tools to catch, represent, and use knowledge as design intents in computer models. The representations of geometric elements are reviewed and their modelling methods based on existing geometric entities are introduced. Parametric modelling, design equations, and design tables are discussed; these techniques have proved to be effective in supporting KBE practice.

3.12 Design Problems

3.1 Use design tables in part modelling to create drawing families from the geometries in Figures 3.41 and 3.42.

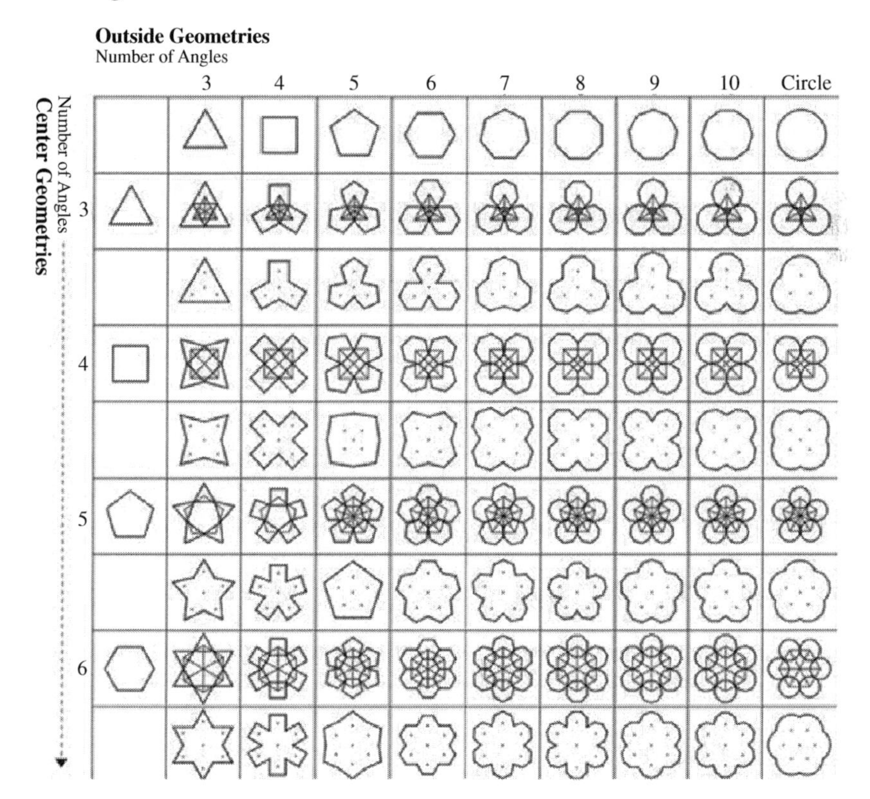

Figure 3.41 Drawing family A.

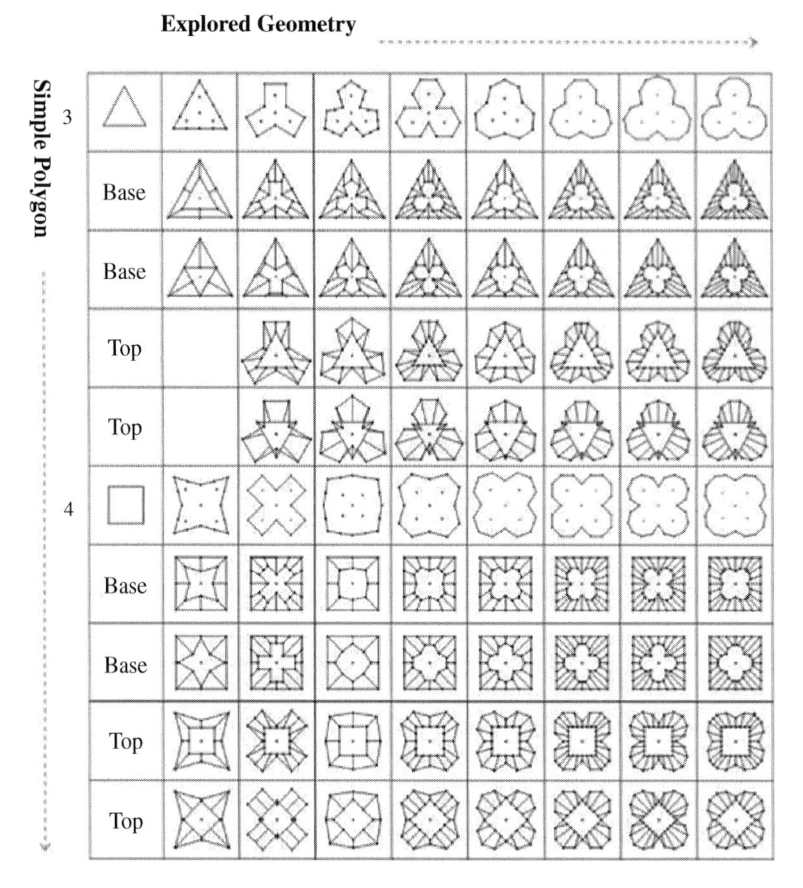

Figure 3.42 Drawing family B.

3.2 Find the masses of part configurations by creating parametric models for the parts in Figures 3.43 and 3.44 (Materials: plain carbon steel; unit: millimetre). The displayed models are for part families where all of the dimensions are scaled proportionally. The highlighted dimensions (lengths) have a set of range with (80, **100**, 120, 160, 200) (Hint: design equations and multiple configurations).

3.3 Using design tables/design equations as major KBE tools, develop a product family model for cell-phone covers for a broad scope of applications, as shown in Figure 3.45.

3.4 Using design table/design equation as major KBE tools, develop a product family model for car snow scrapers for a broad scope of applications as shown in Figure 3.46 (difference of users' heights and strengths, car sizes, colours, additional functions such as brush or motored).

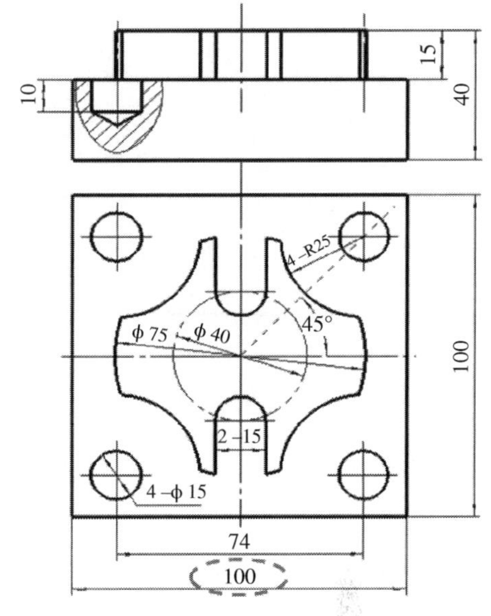

Figure 3.43 Part A.

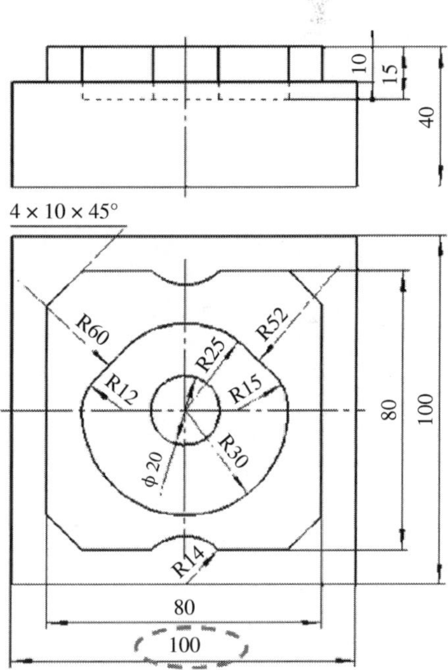

Figure 3.44 Part B.

1. Introduction (backgrounds, FRs for product variants)
2. Conceptual design (physical solutions, design intents to generate variants)
3. Geometric modelling (design tables/design equations) for product family
4. Summary and discussions (materials, cost, manufacturing, marketing, etc.)

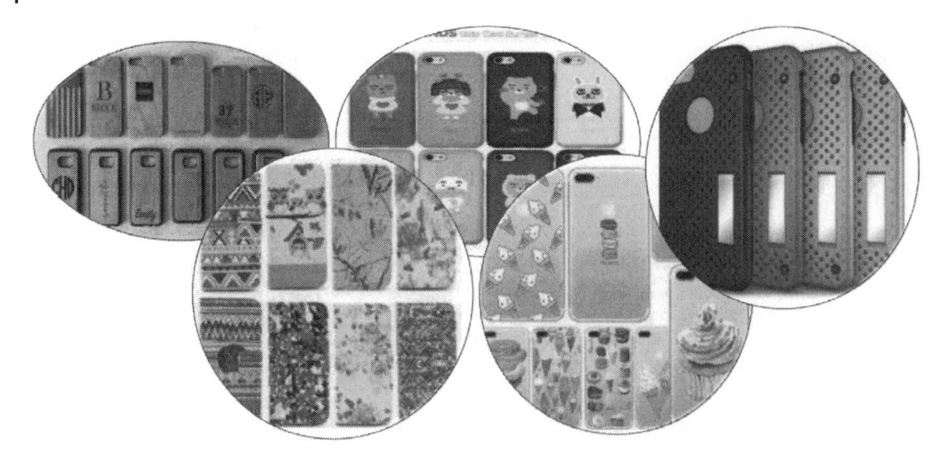

Figure 3.45 Example cell phone cover family.

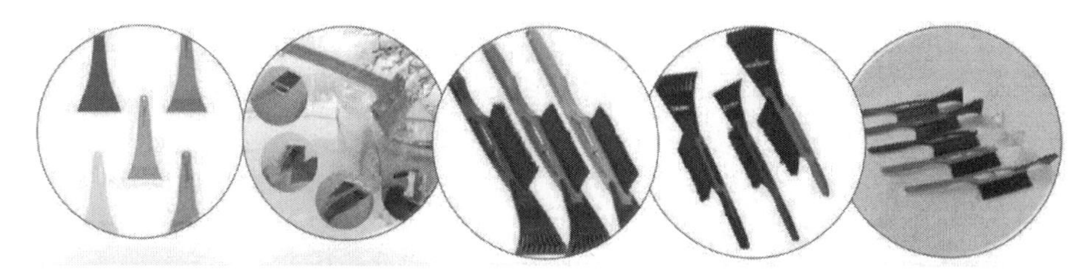

Figure 3.46 Example car snow scraper family.

References

Arnout, K. (2004). From patterns to components, Swiss Federal Institute of Technology, Zurich, Dissertation ETH No. 15500. http://se.inf.ethz.ch/people/arnout/patterns/download/karine_arnout_phd_thesis.pdf.

Ault, H.K. (1999). 3-D geometric modelling for the 21st century. *Engineering Design Graphics Journal* 63 (2): 33–42.

Beach, D. (2012) Configurations and 3DCC Publishers. http://www.cohodesignsllc.com/wp-content/uploads/2012/06/Configurations-PPT_PDF.pdf.

Bi, Z.M. (2018). *Finite Element Analysis Applications: A Systematic and Practical Approach*, 1e. Academic Press ISBN-13: 978-0128099520, ISBN-10:0128099526.

Bi, Z.M. and Zhang, W.J. (2001a). Flexible fixture design and automation: review, issues and future directions. *International Journal of Production Research* 39 (13): 2867–2894.

Bi, Z.M. and Zhang, W.J. (2001b). Modularity technology in manufacturing: taxonomy and issues. *International Journal of Advanced Manufacturing Technology* 18 (5): 381–390.

Bi, Z.M., Xu, L.D., and Wang, C. (2014). Internet of Things for enterprise systems of modern manufacturing. *IEEE Transactions on Industrial Informatics* 10 (2): 1537–1546.

Cederfeldt, M. and Sunnersjö, S. (2003). Solid modeling with dimensional and topological variability. In: *Proceedings of ICED03: International Conference on Engineering Design (19–21 August 2003)*. Stockholm, Sweden: ICED.

Danjou, S., Lupa, N., and Koehler, P. (2008). Approach for automated product modeling using knowledge-based design features. *Computer-Aided Design and Applications* 5 (5): 622–629.

Dassault Systems (2007). Knowledge based engineering. https://d2t1xqejof9utc.cloudfront.net/files/24023/EDU_CAT_EN_KBE_FF_V5R18_toprint.pdf?1374070986 (last accessed 15 May 2018).

Halfawy, M. and Froese, T. (2005). Building integrated architecture/engineering/construction systems using smart objects: methodology and implementation. *Journal of Computing in Civil Engineering* 19: 172–181.

Ledermann, C., Hanske, C., Wenzel, J. et al. (2005). Associative parametric CAE methods in the aircraft pre-design. *Aerospace Science and Technology* 9 (7): 641–651.

Ma, Y.S., Britton, G.A., Tor, S.B., et al. (2003). Standard component library design and implementation for plastic injection mold design with a CAD tool. https://sites.ualberta.ca/~yongshen/index_files/%5BMa%20Y%20S%202003%20ICCA%bD.pdf.

Mohammed, J., May, J., and Alavi, A. (2008). Application of computer aided design (CAD) in knowledge based engineering. http://ijme.us/cd_08/PDF/83%20ENG%20103.pdf.

Rotomation (2019). What's unique about us? https://rotomation.com/index.php/literature/featured-articles?start=4.

Siddique, Z. and Boddu, K. (2005). A CAD template approach to support Web-based customer centric product design. *Journal of Computing and Information Science in Engineering* 5 (4): 381–387.

SolidWorks (2019). Design table configurations. http://help.solidworks.com/2019/english/SolidWorks/sldworks/c_Design_Table_Configurations.htm.

Tarkian, M. (2012). Design automation for multidisciplinary optimization – a high level CAD template approach. Linköping Studies in Science and Technology. Dissertation No. 1479, ISBN 978-91-7519-790-6.

4

Platform Technologies

4.1 Concurrent Engineering (CE)

Concurrent engineering (CE) is also called simultaneous engineering. CE is a method of designing and developing products where the designs at different stages are performed simultaneously rather than consecutively and where the functions of design engineering, manufacturing engineering, and other functions are integrated to reduce the time required to decrease product development time, improve productivity, and reduce manufacturing costs. CE is in contrast to conventional sequential engineering design where design activities are performed in a sequential way.

4.1.1 Brief History

The idea of CE is simple but it has proven to be powerful with many claims of significant product development successes (Smith 1997). CE as a concept has been applied since the 1980s. At the beginning, a product design and its manufacturing process were developed simultaneously, cross-functional groups were used to accomplish integration, and the voice of the customer was included in the product development process. The evolution of CE has been divided into three phases, which are summarized in Table 4.1 (MIT 2019).

4.1.2 Needs of CE

CE has two basic premises: (i) all elements of a product's life-cycle – from functionality, production, assembly, testing, maintenance, environmental impact, and finally disposal and recycling – should be taken fully into consideration at early design phases and (ii) the design activities should all be occurring concurrently. In such a way, errors and redesigns can be discovered early in the design process when the project is still flexible. By locating and fixing these issues early, the design team can avoid what often become costly errors as the project moves to more complicated computational models and eventually into the actual manufacturing of hardware (Wikipedia 2019).

The need of CE has been well illustrated by Mason (2009) and Nicolai and Carichner (2010) with the example of aircraft design shown in Figure 4.1. For the design of a complex product such as an aircraft, if one group in its discipline dominates the design without consideration of the design constraints in other disciplines, there will not be a final product that can meet the functional requirements (FRs) of all design aspects.

Computer Aided Design and Manufacturing, First Edition. Zhuming Bi and Xiaoqin Wang.
© 2020 John Wiley & Sons Ltd. This Work is a co-publication between John Wiley & Sons Ltd and ASME Press.
Companion website: www.wiley.com/go/bi/computer-aided-design

Table 4.1 Brief history of CE (MIT 2019).

1980s
1. Japanese overtook the automotive industry.
2. Researchers and companies experienced an intense period of scrutiny/(critical examination).
3. The tools and methods were developed for the implementation of '*House of Quality & Voice of the Customer*'.
4. *Design for Product Assembly* was proposed and used in 1988.
1990s
1. The concept of lean production was coined in '*The Machine that Changed the World*' by Womack in 1991 (Womack and Jones 2007).
2. New tools and methods, such as Design for Manufacturing (DFM), Design for Manufacturing and Assembly (DFMA), Failure Modes and Effects Analysis (FMEA), Design for Manufacturability/Testability/Assembly/Logistics (Dfx), were developed quickly for the implementation of lean production.
3. CE became the new way for enterprises to run their manufacturing businesses.
4. Cross-functional teams were refined and transformed into matrix organizations, integrated product teams, and so on.
5. Original concept of lean production was enhanced as *Lean Engineering*, *Lean Product Development System*, *Product Lifecycle Management*, and many others.
After 2000s
CE becomes influential; many computer aided tools support CE practice.

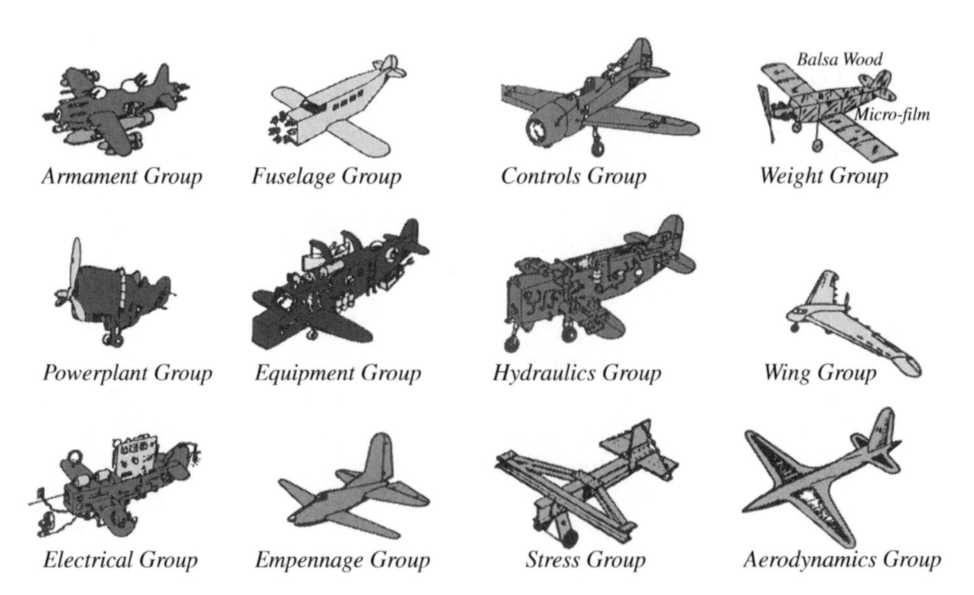

Figure 4.1 Aircraft design when one group is dominant (Mason 2009; Nicolai and Carichner 2010).

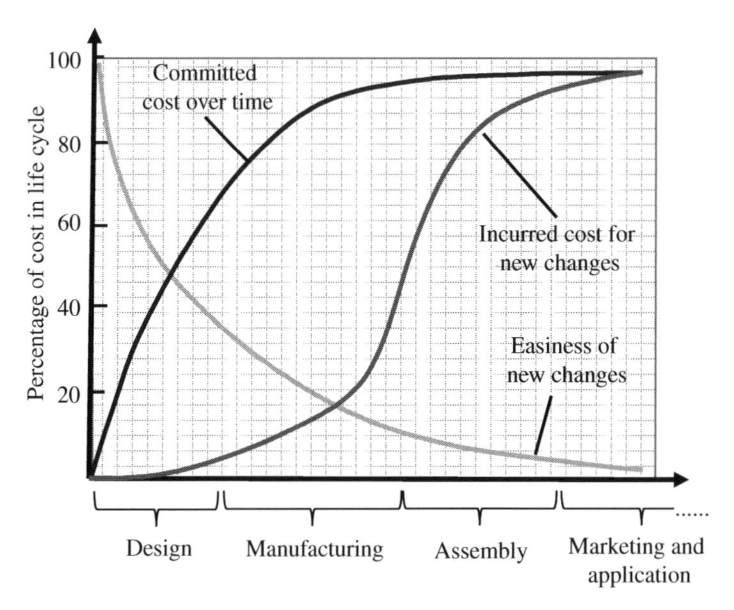

Figure 4.2 Committed cost of product over its lifecycle.

As shown in Figure 4.2, an engineering process is an iterative process where the design errors and defects are gradually eliminated until the product is good enough to meet all design constraints and expectations. A certain cost is associated with an error, and the earlier an error is identified, the less cost is committed to fix it. CE becomes important since design errors and defects can be identified as early as possible due to the simultaneous considerations of design constraints of all design aspects.

CE was proven as an effective tool to reduce the cost and development time for complex products like aircrafts. It is effective for simple products that are involved in the activities of design, manufacturing, and assembly. Heizer and Render (2008) introduced an example of bracket products where the product cost was reduced to $0.80 by CE from $3.50 by traditional sequential design (Figure 4.3). Without affecting the functionalities, the number of parts was reduced from 5 to 3, the number of assembling processes was reduced, and the quality of the new product was improved due to the reductions of parts and joints.

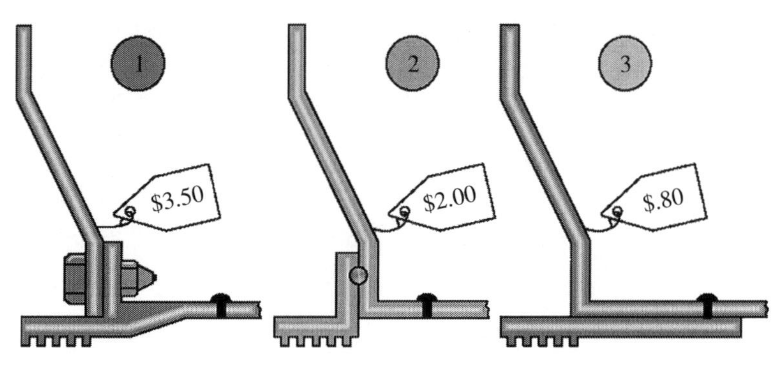

Figure 4.3 Example of cost reduction by CE (Heizer and Render 2008).

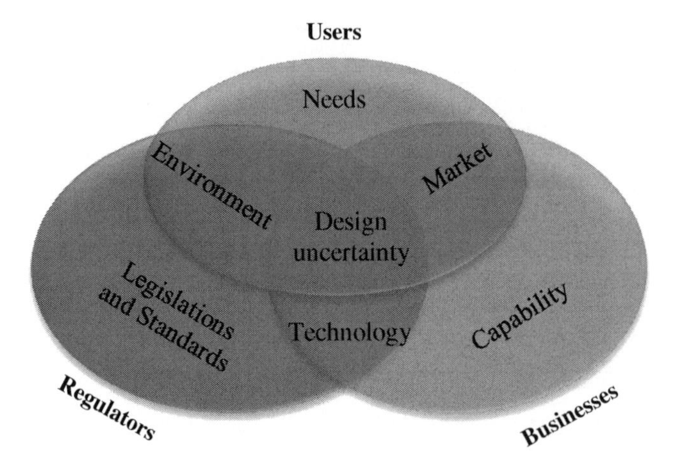

Figure 4.4 Challenges in CE practice (Nadadur et al. 2012). (*See color plate section for color representation of this figure*).

4.1.3 Challenges of CE Practice

Since CE aims to deal with the design constraints at all aspects of a product lifecycle, it is usually a complex system engineering project that involves technical, cultural, and environmental challenges. Nadadur et al. (2012) and Mason (2009) discussed some challenges of CE projects related to users' needs, regulators, and businesses, shown in Figure 4.4. To meet users' needs in a globalized market, one design challenge for enterprises was for the development of their product platforms to meet global and regional markets, make products varied in functionalities, regulations, and businesses, and satisfy high-end and low-end customers at affordable costs. A CE project must also address organizational challenges raised from, for example, distributed teams, cultural differences, dynamics and uncertainties, technological differences, localized designs, and market differences (Mason 2009).

4.1.4 Concurrent Engineering (CE) and Continuous Improvement (CI)

CE emphasizes that the design activities for all aspects of manufacturing have to be taken into consideration simultaneously. This raises a concern since not all of the information about certain activity is available when it is taken into account in the design. This implies that the designs have been refined continuously when the information of the activities becomes more detailed. As shown in Figure 4.5, continuous improvement (CI) is an ongoing improvement of products, services, or processes through incremental and breakthrough improvements. CI seeks 'incremental' improvement over time instead of a 'breakthrough' improvement at once. From the perspective of the product lifecycle, the information about products is accumulated gradually from conceptual design, detailed design, manufacturing, and assemblies to operation, and this process is repetitive as long as CI adds new values to products. Therefore, CE is usually integrated with CI.

As shown in Figure 4.5, CE or CI requires information integration across disciplines and time domains. Information integration is used to blend data from different sources, making it more useful and valuable than it was before. Information integration combines technical

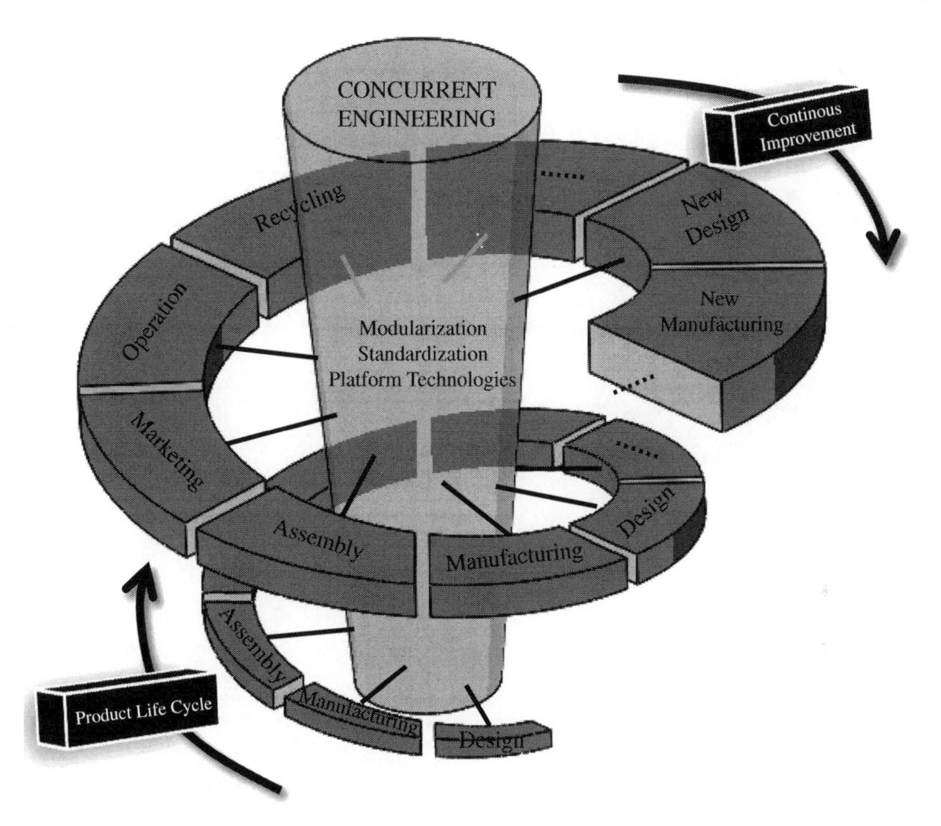

Figure 4.5 CE and Continuous Improvement (CI) in the product design cycle. (*See color plate section for color representation of this figure*).

and business processes used to combine data from disparate sources into meaningful and valuable information. Information integration aims to make the data comprehensive and more usable for reviews, consolidation, propagation, virtualization, federation, and warehousing (Globalscape 2017).

In CE of product or system designs, platform technologies help to migrate and extract value using the capabilities of a complete data integration, data quality, and data governance solutions. Platform technologies have been evolved with CE. The importance of the platform technologies has been increased as companies enter global markets. Figure 4.6 shows product platforms for robotic tool changers, industrial robots, smart watches, and electronic products. A vital product platform becomes essential for the sustainable product development in business domains in which enterprises are specialized.

Instead of forming individual cross-functional teams for different products, CE tends to develop a product platform from where the product can be upgraded or a new product can be spanned from existing ones with minimal design effort. The benefits of using platform technologies include:

1. Leverage resources across multiple development projects.
2. Reduce time-to-market, shorten lead-time, and lower costs.
3. Maximize commonality, reuse, and standardization.
4. Increase efficiency and improve responsiveness.

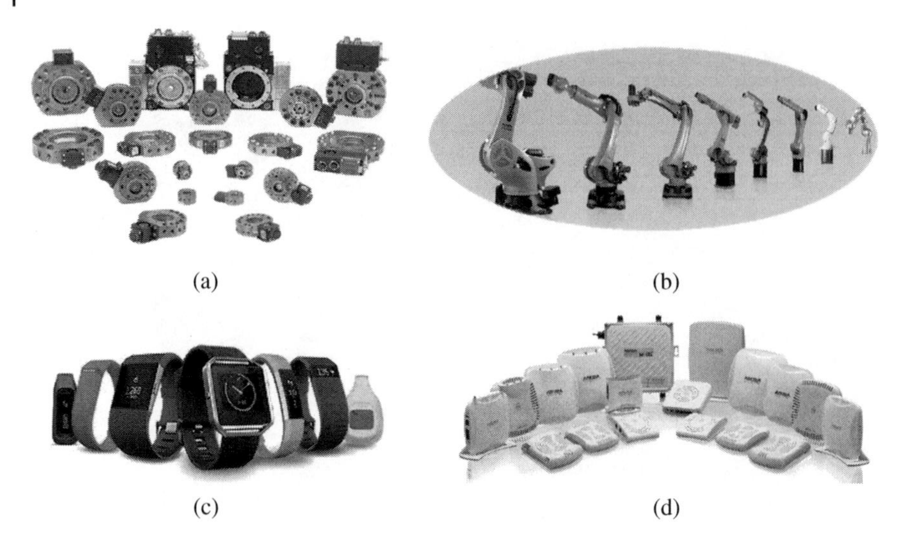

Figure 4.6 Examples of product platforms. (a) Robotic tool changers. (b) Industrial robots. (c) Consumer products (smart watches). (d) Electronic products (routers).

4.2 Platform Technologies

As shown in Figure 4.5, platform technologies are listed together with standardization and modularization to cope with the complexity of technical products and systems. Standardization, modularization, and platforms (S/M/P) had their origins in manufacturing, but CE brings S/M/P together, so that the maximum values of enterprises can be carried over in continuous improvement of product developments (Gepp et al. 2016). S/M/P aims at finding the commodities of design, manufacturing, and assemblies to maximize knowledge reuse. Due to its importance, many relevant terminologies have been proposed by researchers in different fields. Figure 4.7 shows some commonly used terminologies relevant to platform technologies.

Among the concepts in Figure 4.7, standardization and modularization are the two most significant ones. Standardization is used to implement and develop technical standards based on the consensus of different parties that include companies, users, interest groups, standards organizations, and governments. Standardization helps to maximize compatibility, interoperability, safety, repeatability, or quality, and facilitates commoditization of former custom processes. Modularization refers to the design practice of using functional modules to create, assemble, develop, and upgrade products or systems. Modularization aims at cost reduction, a shortened delivery time, and more diversified products.

4.3 Modularization

The importance of modularization was first recognized in software engineering, where modularization was widely applied for code reusability. In general, modularization allows designers to divide a problem into a set of subproblems and the solution to the problem can then be obtained by uniting the solutions to divided subproblems.

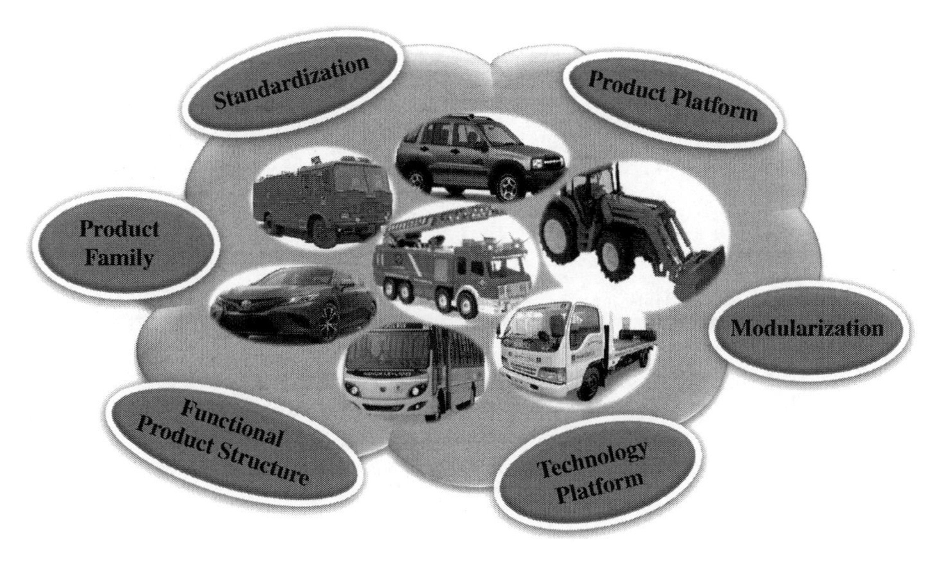

Figure 4.7 Relevant terminologies of platform technologies.

In modularization, a complex or big problem is first divided into parts at the first level. Every part at this level will be checked to see if a solution exists; if not, such a part will be further divided until the part is small and simple enough to have an existing solution. From the perspective of the problem-solving process, modularization corresponds to the idea of breaking a problem into smaller parts and obtaining the solution to each part separately. To enable modularization, abstraction is needed to encapsulate the details of the implementation of the parts. As a whole, the concepts of abstraction and modularization lay the foundation of object-oriented programming in software engineering. Nowadays, software programs are mostly embedded systems and programmers rely greatly on existing libraries and frameworks by others. The novelty of a program relates to the customization of logic of reusing and assembling existing infrastructure, libraries, frameworks, and functional modules (Sun Microsystem 2007).

Modularization tends to be an ideal paradigm to deal with the distributed nature since modern manufacturing environments are distributed, turbulent, and dynamic. Different from a monolithic integral system with a tightly coupled nature, a modularized system consists of a set of encapsulated modules that can be maintained and upgraded independently with less effect on the system. Modularization is so important that it was suggested that modularization or platform technologies should be the mindset of a designer, especially a product designer or an engineer. Modularization should be self-evident and fundamental for cross-functional team practice (MIT 2019).

Modularization is highly stimulated by a number of factors, such as the trend of distributed manufacturing and the ever-growing complexity of modern products. For example, a car has an estimated 30 000 parts, an airplane such as the Boeing 737 includes 367 000 parts, and an Airbus A380, shown in Figure 4.8, has around 4 million parts. Modularization is involved in 1500 manufacturing companies in 30 countries to make 2.5 million different

Facts and figures (Airbus 380)

- max 853 seating capacity
- max 575-ton take-off weight
- ~4 million individual components
- ~2.5 million part numbers
- 1500 manufacturing companies in 30 countries
- 19,000 bolts in fuselage, and additional 4,000 to attached both wings
- 220 windows and 16 doors

Figure 4.8 Needs of modularization by example of a complex product (Airbus 2019).

parts (Airbus 2019; Lennane 2017). It is impractical to design all of the parts of a complex product by CE from scratch without using existing functional modules.

Besides product variants, Stake (2000) identified 12 generic drivers for modularization and platform technologies, which are summarized in Table 4.2.

4.4 Product Platforms

Standardization, modularization, and platform technologies lead to modularized product platforms. In this section, three product platform examples are introduced to gain insights into (i) the relations of product platforms of S/M/P; (ii) the methods of developing product platforms; (iii) the architecture of product platforms at the levels for parts, modules, components, subsystems, to systems; (iv) platform-related design factors such as the industrial sector, company size, product types, cost, technology lifecycles, timescales, customer behaviour, and risks; and (v) the importance of using product platforms.

Huang (2000) provided valuable managerial guidelines in implementing overall platform-based product family development. There are generally two approaches for product family design. One is the top-down method that adopts platform-based product family design (Simpson 2019). The other is the bottom-up approach, which implements a family-based product design through redesign or modification of constituent components of the product.

Figure 4.9 shows the differences of the Black & Decker Corporation (B&D) motors before and after modularization and standardization (Simpson 2019). Universal motors are the most commonly used components in power tools and the B&D is one of the leading manufacturers for power tools and accessories in the world. B&D faced a challenge to introduce more and more motor components for 122 basic tools with hundreds of variants. The solution in Figure 4.8 was a product platform that was able to derive all product variants of universal motors. The product platform was developed using the bottom-up method. In a bottom-up method, the structures of legacy products were analysed and decomposed and the common base and components for geometry and axial profiles were identified to determine the product platform. The resulting platform in Figure 4.9 shows that it could accommodate the variants of two main parameters at their interfaces:

- a static length of motion ranging from 0.80 to 1.75 in and
- an output power ranging from 60 to 650 W.

Table 4.2 Twelve drivers for modularization and platform technologies (Stake 2000).

Product development and design	*Carryover*	A part or component that is not likely to be changed during the life of the product platform and enables heavy investment in production technology.
	Technology evolution	Parts are likely to undergo changes when customers demand or technology changes. This needs to accommodate the interfaces so new technology can be adopted.
	Planned product changes	Parts that the company intends to develop and change.
Variance	*Difference specification*	Designers should allocate all the variations to as few product parts as possible in customizing products.
	Styling	Styling modules contain visible parts of the product that can be altered to create different variants.
Production	*Common unit*	Common unit is similar to the shared functions across products.
	Process and/or organization	Parts requiring the same production process are clustered together.
Quality	*Separate testing*	The possibility of separately testing each module before delivery to find assembly.
Purchase	*Supplier availability*	Purchase standard modules for external vendors.
After sales	*Service and maintenance*	Parts exposed to service and maintenance may be clustered for a service module to be able to be done quickly.
	Upgrading	Give customers the possibility of changing the product in future.
	Recycling	The number of materials in each module should be minimized, and easily recyclable materials should be kept in a separate recycling module.

Using the developed product platform made it possible to fully automate the assembly process, the labour cost was cut from $0.14 to $0.02, and the total unit cost of materials, labour, and overheads was cut from $0.51 to $0.31.

Figure 4.10 shows three platforms of Walkman products at Sony Inc., which was recognized as one of most consistent innovators of consumer and industrial products; one of its successful products was the Walkman in the 1980s (Sanderson and Uzumeri 1998). Sony dominated the portable stereo market with three basic platforms: Walkman II (WM2), WMDD, and WM20. This innovation led Sony to take 45–50% of market share while other leading companies including General Electric (10–12%), Panasonic (6–8%), Sanyo (4–6%), and Aiwa (4–6%) shared the rest.

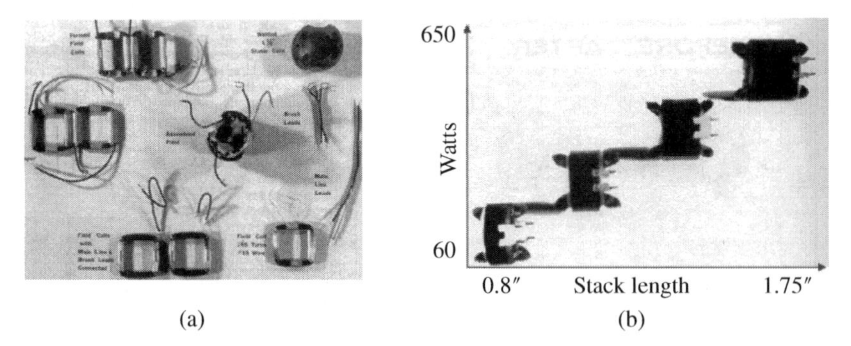

Figure 4.9 Universal motors by Black & Decker using a bottom-up method (Simpson 2019). (a) Electric motor field components prior to standardization. (b) Variants from a product platform.

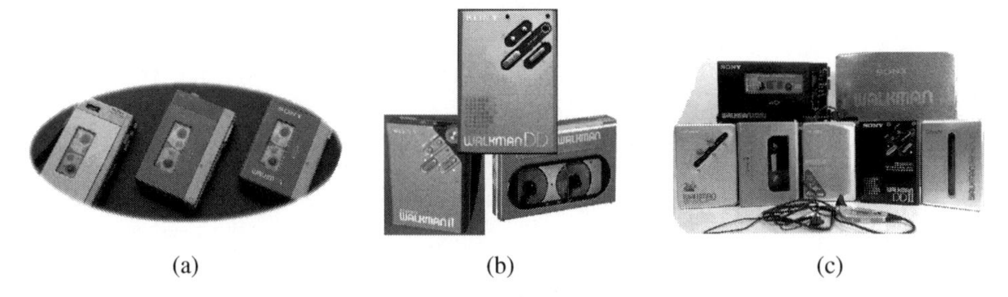

Figure 4.10 Platforms of Walkman products at Sony Inc. (Sanderson and Uzumeri 1995; Simpson 2019). (a) Various products such as TPS L2 before 1980. (b) Product platform for WM-II, WM-DD, and WM-20 in 1982–1984. (c) Over 85% of new products driven from the defined platform after 1984.

The remaining 85% of Sony's models were produced from minor rearrangements of existing features and cosmetic redesigns of external cases. Incremental changes accounted for only 20–30 of the 250+ models introduced in the US (Sanderson and Uzumeri 1995).

For continuous improvement of complex products, product platforms have evolved and been perfected continuously with new product developments. Figure 4.11 shows an example of the evolution of product platforms at Volkswagen (VW). VW had started their journey by including an internal form of modularity as an enhancement to their traditional automotive platform approach. The A-platform in the 1990s was a very successful base for 19 different vehicles, spanning the VW, Skoda, Seat, and Audi brands. VW planned 19 vehicles based on their A-platform. VW estimated development and investment cost savings of $1.5 billion per year using platforms. This program introduced the Modular Strategy into Volkswagen, and the success of this experience led to the more comprehensive and aggressive Modular Toolkit Strategy (Johnson 2013).

The aforementioned examples show that (i) the development of product platforms relies on standardization, modularization, and platform technology. (ii) Either the bottom-up method or the top-down method can be used to develop a product platform. (iii) Similar to a product structure, the product platform applies to all levels from parts, modules,

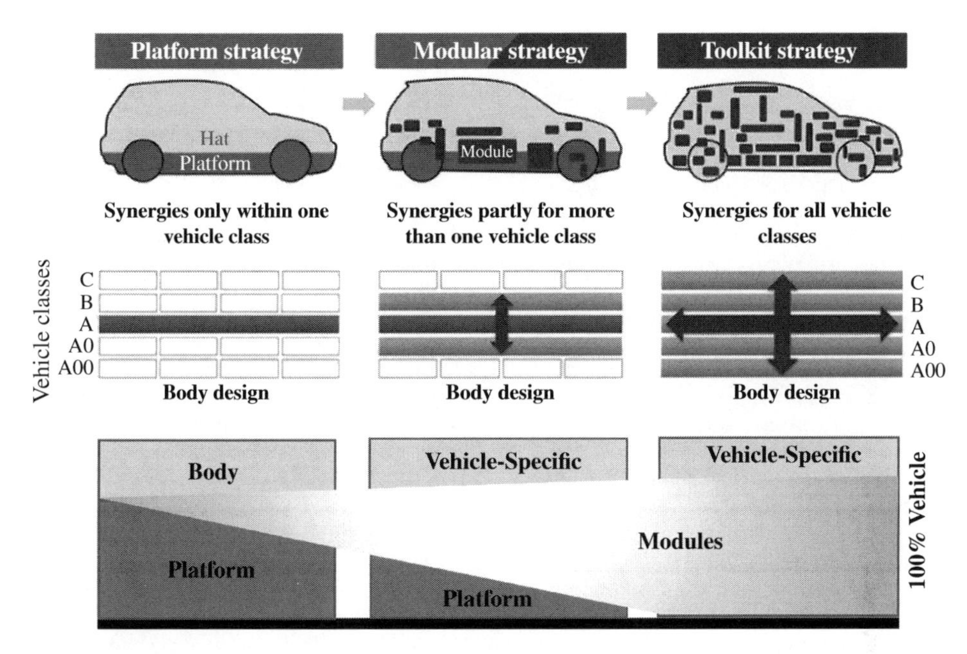

Figure 4.11 Evolution of product platforms at Volkswagen (Johnson 2013; Kreindler 2014).

components, and subsystems to systems and products. (iv) The product platform is affected by all design factors related to products, such as the industrial sector, company size, product types, cost, technology life cycles, timescales, customer behaviour, and risks. (v) The success of a product platform can bring significant economic benefits to manufacturing enterprises. Moreover, product platforms offer the opportunities of new product developments, as discussed in the following section.

4.5 Product Variants and Platform Technologies

With market globalization and saturation of manufacturing and service capabilities, customer markets have been highly diversified and segmented. To sustain or increase market share of certain products, companies have been effortless, except for providing more personalized products for customers. Customizing products and services have been among the most critical means to deliver true value of products for customers and to achieve competitive advantages over business opponents. However, the challenge is not to customize products and services itself; it is how to customize products in the most profitable way.

Either integral products or modular products can be made to provide customers with diversified functional requirements on products (Bi and Zhang 2001). A brief comparison of integral and modular products is given in Table 4.3. An integral product consists of blocks, with each block having one or more function elements optimized for given functionalities.

Table 4.3 Structure comparison of integral and modular products (Huang 2000).

Integral products	Modular products
• A collection of components that implement some functions of a product is called a *block*. • The functional elements of a product are implemented using more than one block. • A single block implements many functional elements. • The interactions between blocks are ill-defined and may be incidental to the primary function of the product. • Product performance can be enhanced through an integrated architecture. • Changing a block in an integrated product may influence many functional elements and require changes to several related blocks.	• A collection of components that implement some functions of a product is called a *module*. • A module implements one or a few functional elements in their entirety, and the functions of modules are the same as those of blocks in an integral product. • The interactions between modules are well defined and are the generally fundamental to the primary function of the product. • Product performance may not be enhanced by a modular architecture. • Changing a few isolated functional elements of a product may not affect the design of other modules.

Table 4.4 Advantage of modular product design over integral product design (Mikkola and Gassmann 2003).

Integral product design	Modular product designs
• Interactive learning • High levels of performance through proprietary technologies • Systemic innovations • Superior access to information • Protection of innovation from imitation • High entry barriers for component suppliers • Craftsmanship	• Task specialization • Platform flexibility • Increased number of product variants • Economies of scale in component commonality • Cost savings in inventory and logistics • Reduced costs by easy maintenance • Shorter product life cycles through incremental improvements (upgrading and adaptations) • Flexibility in component reuse • Independent product development • Feasibility of outsourcing • System reliability due to high production volume and experience curve

The product has to be completely redesigned when any of the functional requirements are changed. A modular product consists of a number of modules with an encapsulated structure that fulfills given functional requirements independently. The interfaces are well designed so that the selected modules can be assembled easily as a product variant. Changing an individual module has no effect on interfaces.

A further comparison of integral and modular products is given in Table 4.4, which shows the advantages of modular products (Mikkola and Gassmann 2003). In summary, modular

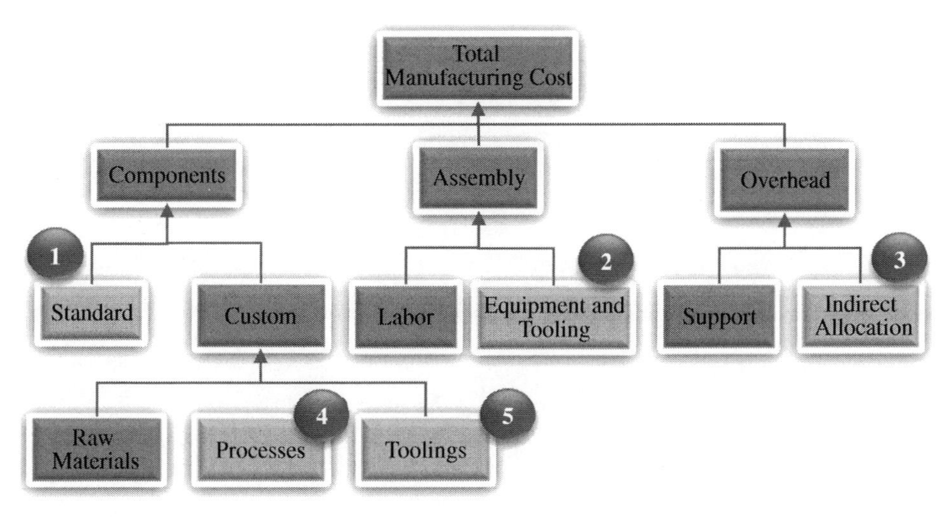

Figure 4.12 Example of using the platform technologies for the cost reduction of various manufacturing operations.

products based on platform technologies allow companies to make product variants in profitable and sustainable ways.

Platform technologies may bring significant cost savings. *Product cost* refers to the sum of the expenditures of all activities in the product life cycle from acquisition of raw materials, design, manufacturing, assembly, marketing, and operations, to the disposal of the retired product.

From the perspective of manufacturing, the manufacturing-related cost can generally be decomposed into a *fixed cost* and a *variable cost*. A fixed cost is incurred in a predetermined amount, regardless of the number of units produced, i.e. setting up a factory work area or cost of an injection mould. A variable cost is incurred in direct proportion to the number of product units (i.e. material cost). Taking an example of the manufacturing cost in Figure 4.12, the standardization and modularization from platform technologies would significantly reduce the unit cost of a product in five highlighted activities.

Platform technologies bring radical changes to the activities at all domains and levels of business in enterprises. Figure 4.13 shows how platform technologies are used in enterprises to create numerous opportunities for engineers to reduce product costs by modularizing product structures.

The implementation of platform technologies is among the most powerful ways of achieving this in practice, offering a reduction of the lead time for products and quotations. It also achieves faster and more qualified responses to customer inquiries, fewer transfers of responsibility and fewer specification mistakes, a reduction of the resources spent for the specification of customized products, and the possibility of optimizing the products according to customer demands. Platform technologies add new values to products by directly facilitating matches between customers' needs and manufacturers' requirements. Platform technologies save cost by facilitating the collaboration of designers across the business domains and stages (https://www.applicoinc.com/blog/what-is-a-platform-business-model).

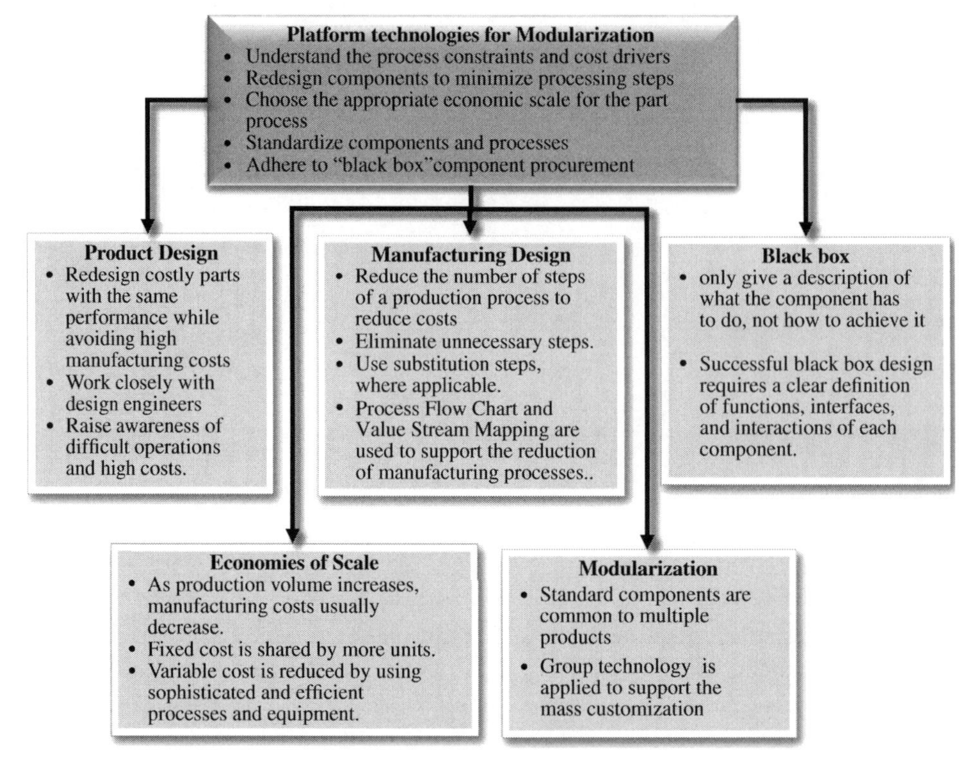

Figure 4.13 Platform technologies for cost savings.

4.6 Fundamentals to Platform Technologies

Modularized products from platform technologies promise the benefits of high-volume production (which arises from producing standard modules) and, at the same time, the ability to produce a wide variety of products that are customized for individual customers. Modularization of products is treated as the goal of good engineering design practice (Huang 2000).

Intensive researches have been conducted to develop systematic methodologies that support CE teams to adopt platform-based product developments (Wang 2010; Simpson 2019). Zha and Sriram (2006) gave an overview of research progress on the methodologies and tools of S/M/P for system-level goals, shown in Figure 4.14. The S/M/P is driven by an enterprise needs to make and deliver products with required functionalities and volume at a shortened lead-time and an affordable cost. S/M/P tools have been developed to fulfil all of the activities in a product lifecycle, such as product developments, materials selections, designs, manufacturing, assemblies, marketing and sales, and system operations. However, no universally applicable tool has been available to cover the designs of all aspects, as existing S/M/P tools have their emphases on modular product architecture, product platform, platform representation, variety design and synthesis, and product family architecture.

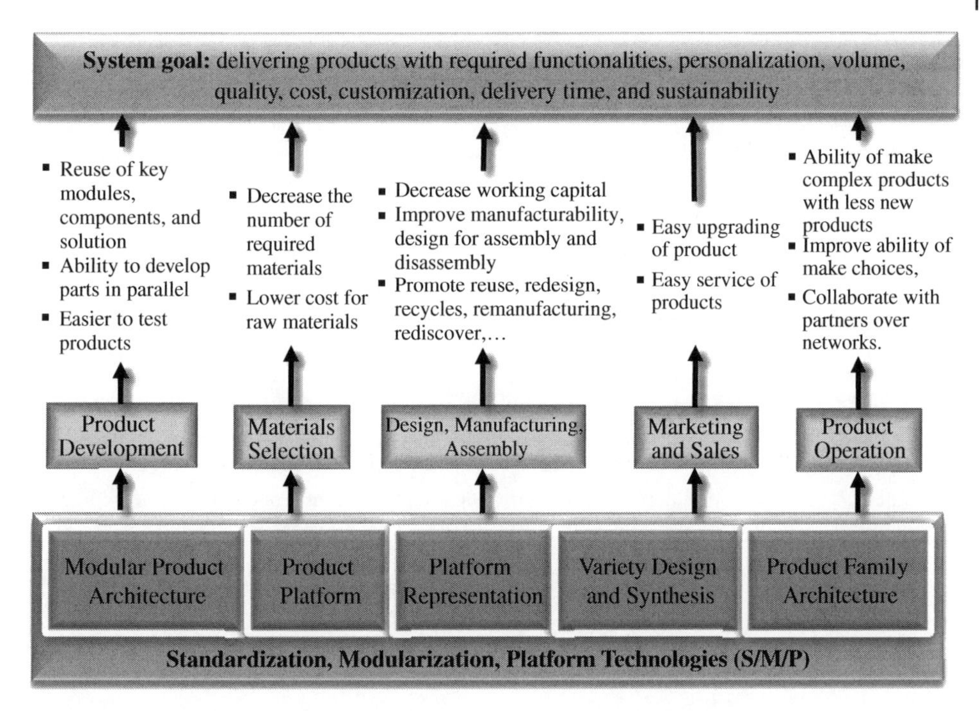

Figure 4.14 Overview of platform-based product family design methods (Zha and Sriram 2006).

Figure 4.14 includes some relevant but different terminologies for modular products and platforming technologies. Table 4.5 gives a clarification of some commonly used terminologies related to modularization.

A product platform consists of a number of modules, and the modules in the platform are classified in terms of their commonality of usage in a product family. Figure 4.15 shows that three types of modules are common modules, variant modules, and unique modules. A *common module* is required to configure any product variant, a *variant module* is used to configure some product variants, and a *unique module* is only needed to configure a specific product variant. Note that a product variant is configured from a set of selected modules, where the configuration of each product variant represents the topology of assembly relationships.

As shown in Figure 4.16, the goal of developing and improving a product platform is to maximize the commonalities across product variants by (i) maximizing common modules, (ii) managing variant modules in either a *mass production* way or a *mass customization* way, (iii) minimizing unique modules, and (iv) utilizing a product platform to develop innovations and new products.

Products are varied in two ways, i.e. *firstly*, by selecting the types and numbers of modules and, *secondly*, by assembling the selected modules in a specified topological structure. Figure 4.17 shows three variants configured from the selected common module (A), variant module (B), and unique modules such as I, J, and M. Note that assembling topology contributes to product variants. Even the same set of constitutive modules are used, although the products may be different if the modules are assembled differently.

Table 4.5 Key concepts relevant to modularization.

Concepts	Explanation
Product Platform:	A product platform is the collection of common elements, especially the underlying core technology, implemented across a range of products (McGrath 1995). A product platform consists of (i) a set of platform elements and (ii) architectural rules which are able to generate anticipated product variants from existing platform elements. A product platform allows an increased leverage and re-use manufacturing resources across multiple product lines.
Platform Element:	A platform element is a building block which can be used in product variant; elements in a product can be varied based on the constraints that the platform applies on product.
Product Family:	A product family is a group of related products that share common features, parts, and subsystems; it is developed to satisfy a variety of markets.
Variant or Derivative:	A product variant or derivative is an individual product which can be built from the product platform. Two ways to build a product variant are (i) adding, removing, and/or substituting one or more modules in a base product from a product family; (ii) scaling or 'stretching' the platform in one or more dimensions over a base product in a product family.

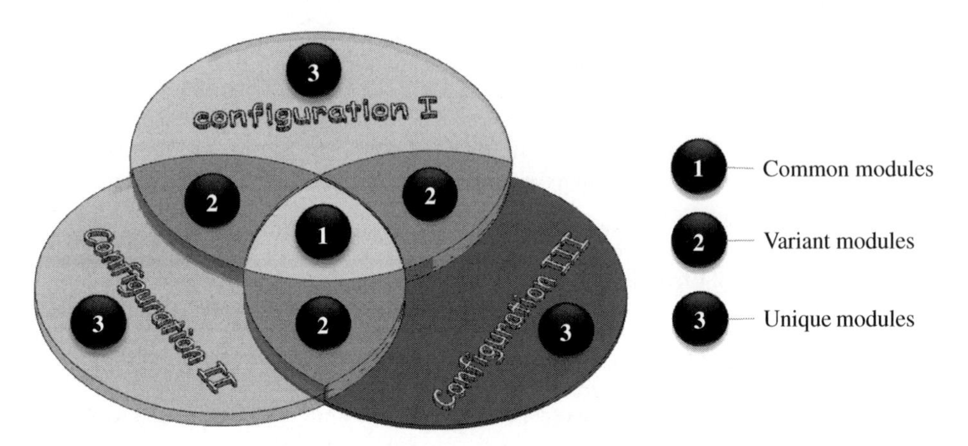

Figure 4.15 Types of modules in a product platform.

Simpson (2019) provides the automobile platform at Ford, shown in Figure 4.18. At Ford, a product platform represents a common architecture for assembly sequences, jointing configuration, and system interfaces. A platform specifies the set of subsystems and their interfaces, from which a family of derivative products can be produced in the way of mass customization. Table 4.4 shows that the automobile platforms are developed using the bottom-up method, where the common components are determined by analysing the commonalities of legacy products.

Ford announced a reduction in its 16 distinct vehicle platforms to nine platforms in 2014. It was expected to cut costs by 20% (Wikipedia 2018). As shown in Table 4.6, Ford implemented such a swift move by using modular platforms that shared the components as many

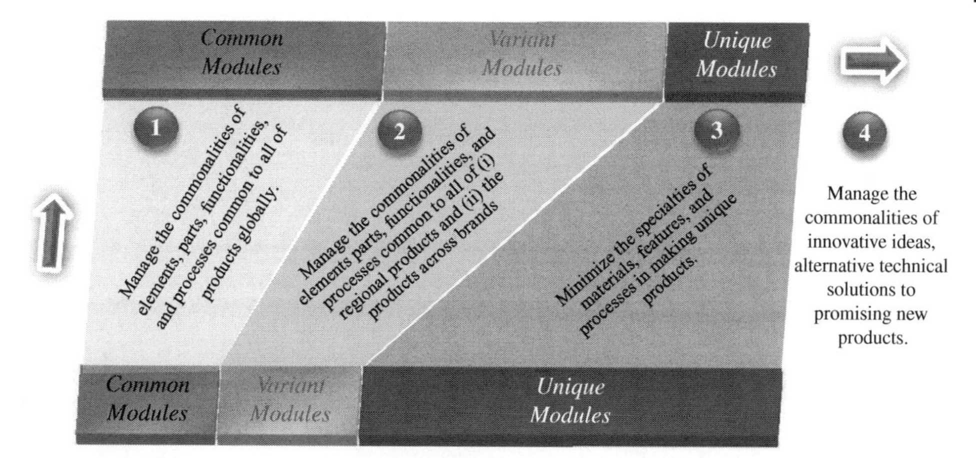

Figure 4.16 Prioritizing commonalities for different module types.

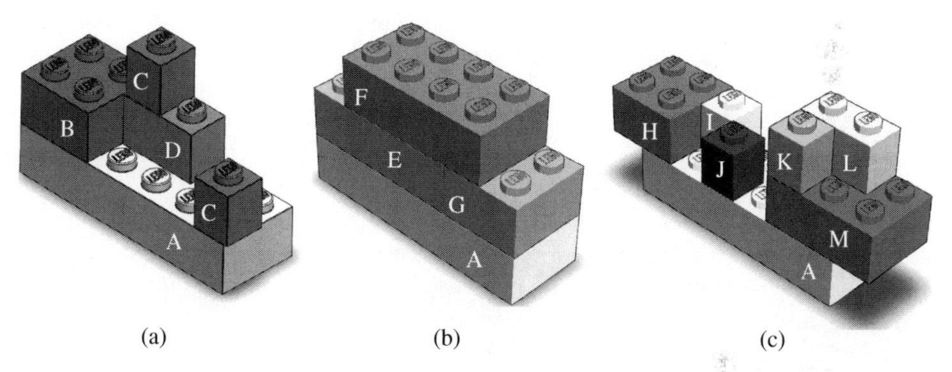

Figure 4.17 Configure product variants from a product platform. (a) A + B + C(2) + D. (b) A + E + F + G. (c) A + B + H + I + J + K + L + M.

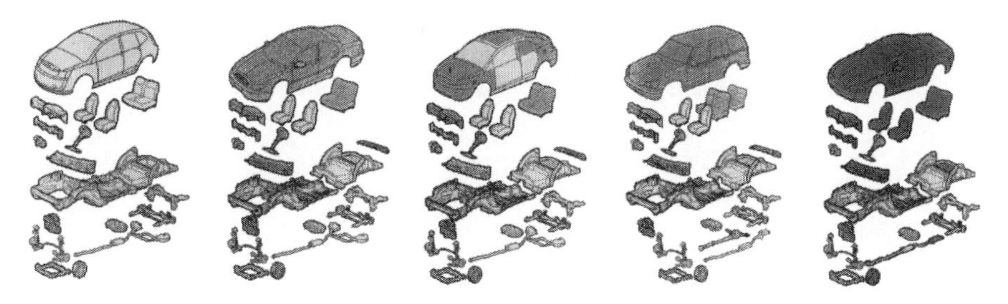

Figure 4.18 Ford auto platform with subsystems and interfaces (Simpson 2019).

times as possible among the vehicles with different brands. The number of product platforms in one company site would be less. For example, the Fort factory in Detroit only used five platforms for all of its vehicles, i.e. a front-wheel-drive unibody (i.e. small cars and crossovers), a rear-wheel-drive unibody, a commercial-vehicle unibody (i.e. a van), a body-on-frame platform (i.e. trucks) and an all-electric car platform (Holems 2018).

Table 4.6 Connections of automobile platforms with old models at Ford (Wikipedia 2018).

	Old models	Ford platforms
Subcompact cars	B-Mazda DA	B3
Compact cars	CE14, Mazda, C170	C1, C2
Mid-size cars	Fox, D186, CDW27, Mazda G	CD3, CD4, DEW, EUCD
Full-size cars	1949 Ford, 1952 Ford, 1957 Ford, 1960 Ford, 1965 Ford, 1969 Ford, Panther, Full-Size Lincoln, EA26, EA169, E8	Ford D3
Other cars	FN10/MN12, SN-95, D2C/S197	S550
Trucks/SUVs/ Crossovers/Vans	CD2, E265, U2, V, VN1, VN, Mazda S.	PC/P3, T1, D4, V363N, V227

If a company designs new products one at a time, the focus on individual customers and products often results in a failure to embrace commonality, compatibility, standardization, or modularization among different products or product lines. The end result would be the diversification of products and components with proliferating variety and costs, with unwanted costs eroding profit margins. Therefore, it is important to put the development of product platforms as the priority, so that a steady stream of new products can be derived from the platform continuously in order to meet market changes.

The assembly options in a product platform produce a large member of product variants. At the assembly level, product variants are generated by different strategies of interfaces. Huang (2000) classifies the interfaces into *component-swapping modularity*, *component-sharing modularity*, and *bus modularity*. Table 4.7 gives the explanation, definition, and explanation of these interface types.

4.7 Design Procedure of Product Platforms

The design procedure for a product platform is similar to design of a product; except that in designing a product platform, a number of product variants are covered, and the design is emphasized on the identification of commonalities of functional modules. Table 4.8 shows the procedure of product platform deigns. The design begins with the definition of customer needs on product variants, and it follows by interpreting customer needs as abstract-level FRs. FRs are then divided into sub-FRs until the corresponding technical solutions have been obtained. The feasible solutions for sub-FRs are evaluated, compared until module types and assembling options are finalized as a modular product platform.

4.8 Modularization of Products

Steps 2 to 4 in the design procedure of product platforms (Table 4.8) show the modularization of products. In this section, the methods and tools used to fulfil tasks in these steps are discussed.

Table 4.7 Three interface types in a product platform.

Interface strategy	Principle	Example
Component-swapping modularity	It occurs when two or more types of basic components are paired with one module, thus creating different product variants in a product family.	Screw driver: swappable parts in the same family
Component-sharing modularity	It is complementary to component-swapping modularity. Various modules share same basic components and create product variants belonging to different product families.	Linear actuators, basic components to many product family such as machine tools and robots
Bus modularity	It occurs when a module can be matched to any number of basic components. Bus modularity allows varying the number and locations of basic components in a product while component-swapping and component-sharing modularity allow only varying the types of basic components.	I/O modules in numerous part families with communications

4.8.1 Classification of Functional Requirements (FRs)

The FRs of products in a family must be decomposed to group these FRs in terms of their commonalities. According to the classification of module types in Figure 4.15, FRs of products are decomposed and classified into *common FRs*, *variant FRs*, and *unique FRs*. Figure 4.19 gives an example of the classification of Lego products. FRs of all products in the family are decomposed, analysed, and grouped into common, variant, and unique FRs. Once the solutions to these FRs are implemented, the number and types of modules can be customized for product variants.

4.8.2 Module-Based Product Platforms

The axiomatic design theory (ADT) can be used to guide the decomposition of FRs of products. As shown in Figure 4.20, a design process in ADT refers to the mappings of four domains, *Customer Requirements* (CRs), FRs, *Design Parameters* (DPs), and *Physical*

Table 4.8 The design procedure of product platforms.

Steps	Tasks	Methods
1	Define and interpret customer needs	Data collection, interviewing, and interpreting
2	Transform customer needs in to abstract-level functional requirements (FRs)	Determination of product strategies and application of Quality Function Deployment (QFD) for the requirements of modularity.
3	Decomposition of abstract-level FRs into detailed sub-FRs.	The generation of black box models and function chains such as serial or parallel chains.
4	a. Create a model from which modules can be identified	The aggregation of function chains as a functional model.
	b. Create a technical solution to the functional models of identified functional modules.	The determination of technical solutions for the specified functions.
5	a. Define product architecture	Application of heuristic methods to identify modules for the functional models (dominant flow, branching flow, and conversion transmission).
	b. Design technical solution for components and parts	Identification of possible modules from technical solutions using the module indication matrix with module drivers.
6.	Generate modular concepts	Creation of rough geometric layouts and searching existing components.
7	Evaluate concepts	Evaluation, comparing, and selection of the concepts by testing the interfaces between the modules.

Solutions (PSs). ADT uses matrix methods to analyse the transformation from one domain to another systematically.

In a mapping from FRs to DPs, ADT provides two axioms to guide the mapping process, i.e. (i) the *Independence Axiom* states that when there are two or more FRs, an ideal solution is that each FR is satisfied by its corresponding DP and (ii) the *Information Axiom* states that a specific DP will deliver the functional performance required to satisfy the FR. DPs must logically satisfy the set of given FRs. Axiomatic design involves interplay between (i) what we want to achieve (FRs) and (ii) how we choose DPs to achieve FRs.

At a certain level of a design hierarchy, FRs constitute the $\{\boldsymbol{FR}\}$ vector of these FRs in the functional domain. Similarly, DPs form the $\{\boldsymbol{DP}\}$ vector in the physical domain. The relationship between $\{\boldsymbol{FR}\}$ and $\{\boldsymbol{DP}\}$ can be represented by a design matrix $[\boldsymbol{A}]$:

$$\begin{Bmatrix} FR_1 \\ FR_2 \end{Bmatrix} = [\boldsymbol{A}] * \begin{Bmatrix} DP_1 \\ DP_2 \end{Bmatrix} = \begin{bmatrix} A_{11} & A_{12} \\ A_{21} & A_{22} \end{bmatrix} * \begin{Bmatrix} DP_1 \\ DP_2 \end{Bmatrix} \tag{4.1}$$

where $[\boldsymbol{A}]$ is called the design matrix for the relations of $\{\boldsymbol{FR}\}$ and $\{\boldsymbol{DP}\}$.

Equation (4.1) can also be written in a differential form, where the elements of the design matrix are given by

$$A_{ij} = \frac{\partial FR_i}{\partial DP_j} \tag{4.2}$$

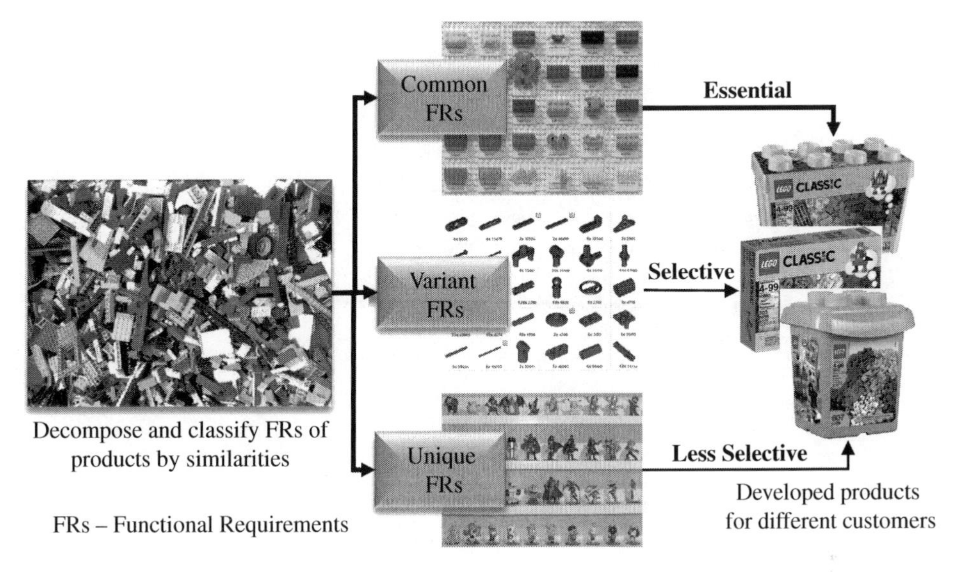

Figure 4.19 Modularization of a product family. (*See color plate section for color representation of this figure*).

If the derivative in Eq. (4.2) can be quantified, the result reflects the significance of DP_j in satisfying the functional requirement FR_i.

The zigzagging decomposition leads to a hieratical structure of a product platform, where DPs are determined for all FRs, respectively. The corresponding PSs are unnecessary at the same level. As shown in Figure 4.21, the modules at the lowest level can be grouped into new modules at the component level to reduce the complexity of the modular products.

Figure 4.22 shows an example where ADT was used to define a helmet product platform (Pandremenos and Chryssolouris 2009). The helmet shows four main components, which are treated as DPs. These parameters were obtained by the mapping matrix from FRs to DPs in Figure 4.22b. The design matrix $[\boldsymbol{A}]$ in Figure 4.22 is diagonal, which means that design parameters are fully decoupled to fulfil their FRs. The design solutions satisfy the two design axioms.

Simpson (2019) shows the platform of the Braun coffee makers in Figure 4.23. After the zigzagging mapping of FRs and DPs, a DP will be identified for each FR. Ideally, it is one-to-one correspondence between FRs and DPs. The physical modules are defined by grouping DPs, and these modules are further classified based on the commonalities in product variants into common modules, variant modules, and unique modules.

4.8.3 Scale-Based Product Family

A product platform can be modularized by *scaling* or *stretching* one or more dimensions to satisfy the variety of market niches. An example of scale-based product families is the Boeing 737 product platform (Wikiwand 2019), shown in Figure 4.24. It consists of products of initial 100 and 200 models, classic 300, 400, and 500 models, and next generation 600, 700, 800, and 900 models. The new product Boeing 777 was also stretched from existing models.

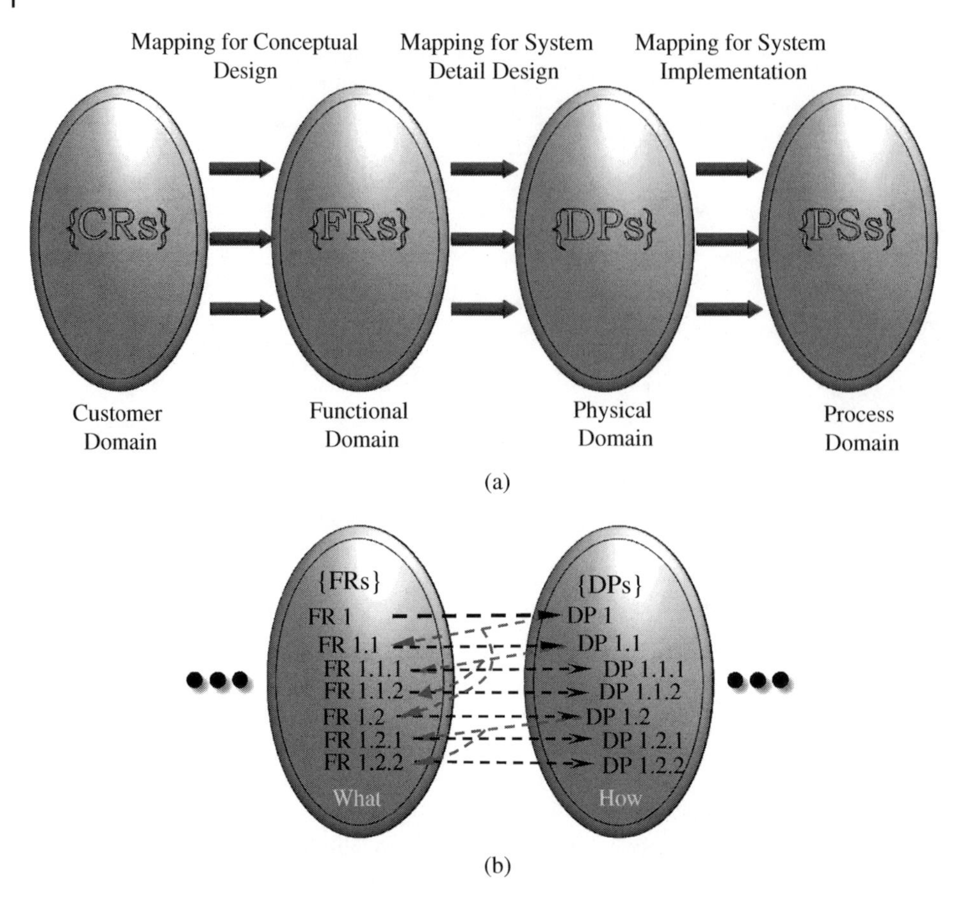

Figure 4.20 Zigzagging decomposition in axiomatic design theory (ADT). (a) Mapping in four domains of axiomatic design. (b) Zigzagging decomposition. (*See color plate section for color representation of this figure*).

Rothwell and Gardiner (1990) reported that Rolls Royce scaled aircraft engines to satisfy a variety of new requirements and expedite corresponding testing and certification. For example, the incremental improvements and variations were made to increase thrust and reduce fuel consumption. The common model Rolls Royce Turbomeca (RTM322) was for turboshaft, turboprop, and turbofan engines. When it is scaled up by a factor of 1.8, it becomes the Rolls Royce RB550 series.

4.8.4 Top-Down and Bottom-Up Approaches

The need to use a product platform brings new requirements for product development. As shown in Figure 4.25, each product is optimized for its specified functions without the need to consider reusability. On the other hand, new products are mostly defined from scratch since the components in legacy products are not modularized and ready for reuse.

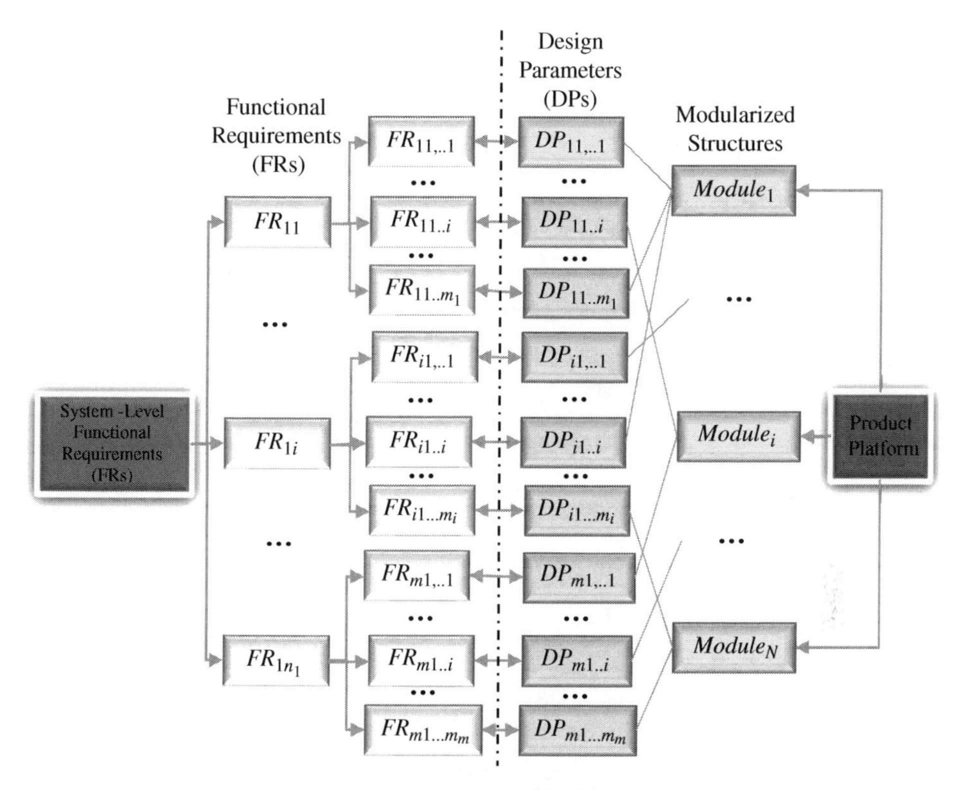

Figure 4.21 Structure of module-based product platform (Golfmann and Lammers 2014). (a) Zigzagging decomposition. (b) FRs and DPs. (c) Mapping of FRs and DPs.

Therefore, the FR of a product platform must be taken into consideration in a product design lifecycle.

Figure 4.5 has shown that CE and CI are integrated in product lifecycles. Therefore, the development of a product platform may be involved at any stage of the product lifecycle. At an earlier stage when an enterprise begins to punch into the markets, it does not have existing products whose components can be reused; a product platform has to be developed using *a top-down approach* from an abstract level to a detail level. Figure 4.26 show the idea of the top-down approach.

In a top-down approach, a company strategically designs a product platform to manage and develop product families using module-based and/or scale-based derivatives. Examples of companies using the top-down approach for their product platforms are Volkswagen and Sony. The developed platform enables technology leverage, research and development, economy of scale and scope, and organic growth (MIT 2019).

At the stage of continuous improvement when an enterprise begins to expand its market share, a number of product variants exist for reuse and a product platform can be developed using *a bottom-up approach* from the detail level to the abstract level. Figure 4.27 shows the idea of bottom-up approaches. In the bottom-up approach, a company redesigns or consolidates a group of distinct products by standardizing components to improve economies of

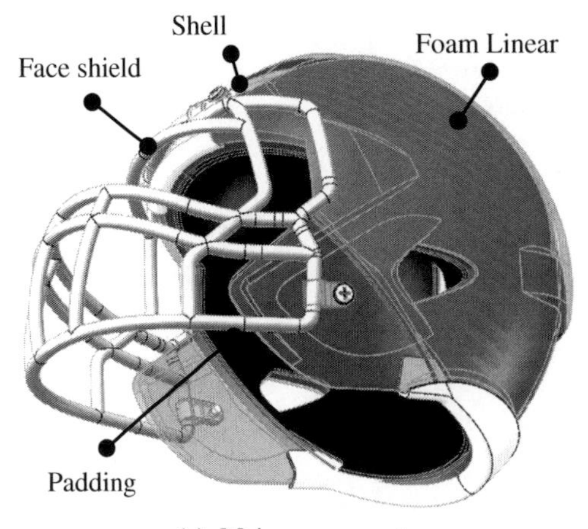

(a). Main components

Functional Requirements (*FRs*)		Design Parameters (DPs)	
FR_1	Prevent penetration	DP_1	Shell
FR_2	Absorb energy	DP_2	Foam liner
FR_3	Provide comfortable	DP_3	Padding
FR_4	Protect fact and ensure visibility	DP_4	Face shield

(b). FRs and DPs

$$\begin{bmatrix} DP_1 \\ DP_2 \\ DP_3 \\ DP_4 \end{bmatrix} = \begin{bmatrix} A_{11} & 0 & 0 & 0 \\ 0 & A_{22} & 0 & 0 \\ 0 & 0 & A_{33} & 0 \\ 0 & 0 & 0 & A_{44} \end{bmatrix} \begin{bmatrix} FR_1 \\ FR_2 \\ FR_3 \\ FR_4 \end{bmatrix}$$

(c). Mapping of FRs and DPs

Figure 4.22 Example of decomposition of Functional Requirements (FRs).

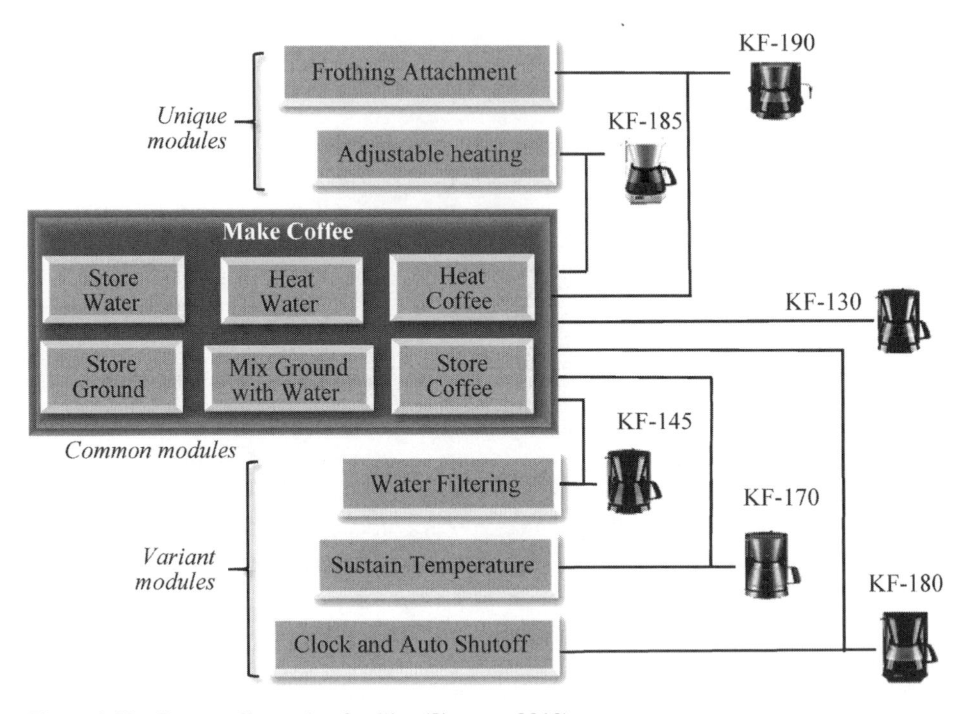

Figure 4.23 Braun coffee maker families (Simpson 2019).

scale and reduce inventory. Examples of companies using a bottom-up approach for their product platforms are Lutron and Black & Decker. Lutron makes customizable lighting control systems for commercial and residential applications, including hotel lobbies, ballrooms, conference rooms, and exec offices. Lutron is able to customize lighting systems using its product platform from existing projects with up to 20 standardized components and over 100 models (Pessina and Renner 1998).

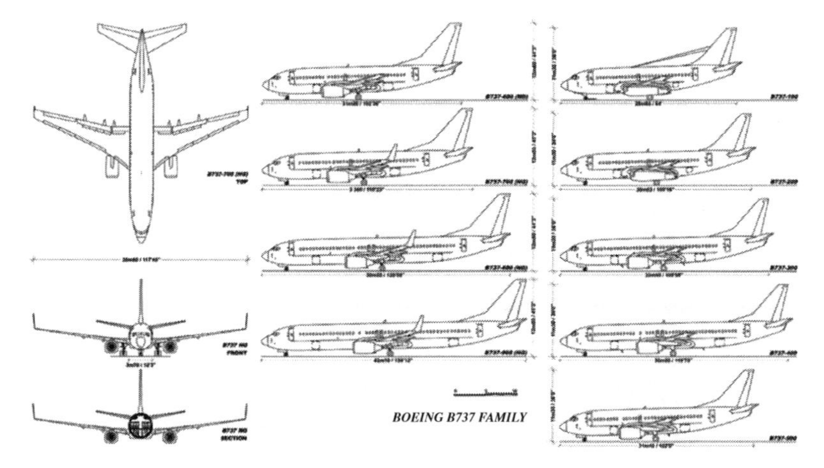

Figure 4.24 The Boeing 737 family (Wikiwand 2019).

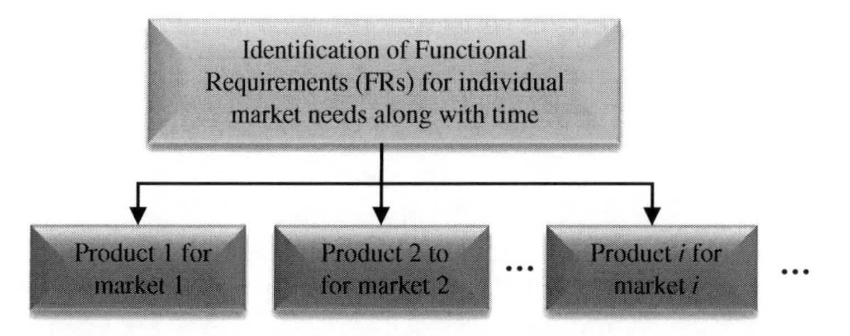

Figure 4.25 Traditional product design method without consideration of product platform.

4.9 Platform Leveraging in CI

In earlier times, manufacturing companies faced a sellers' market where customers' demands exceeded manufacturing supplies, product structures were static, and products were niche-specific. As shown in Figure 4.28, products were diversified in terms of segments and targeted customers. However, some limitations of static product structures were (i) the product designs might be easily copied by competitors, (ii) a higher cost of manufacturing and capital investments was involved without the consideration of commonalities of products, and (iii) the opportunities for synergy in marketing development were confined. These factors led to the myriad of products, higher costs, and low profit margins when products became more and more diversified.

Nowadays, manufacturing companies are in the buyers' market where global manufacturing capabilities greatly exceed customer demands. Products are continuously upgraded at a fast pace and the companies are highly pressured to make new products to meet customer personalized needs. Since new products are derived from a product platform, the product platform must be continuously evolved and upgraded in the product lifecycle (Figure 4.5). Meyer (1997) suggested to use the market segmentation grid to identify and

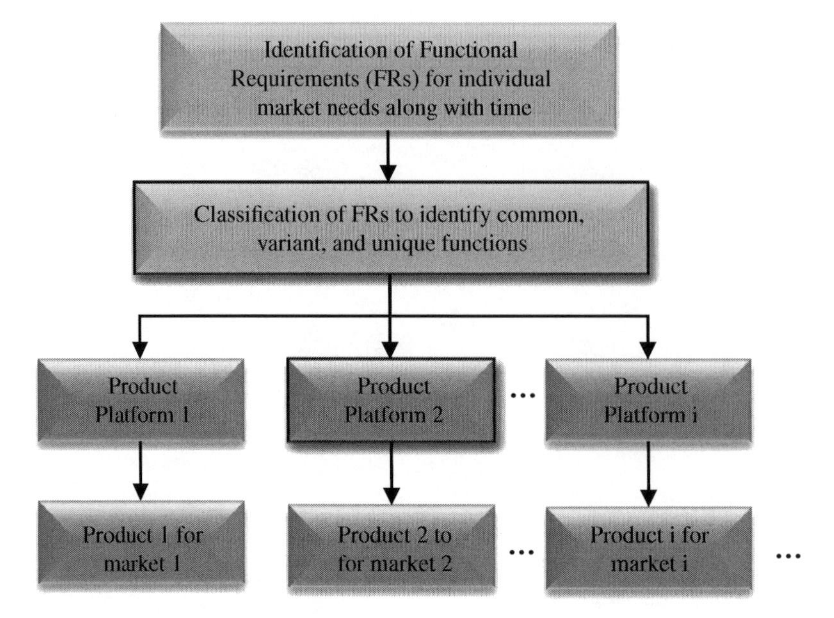

Figure 4.26 Top-down approach.

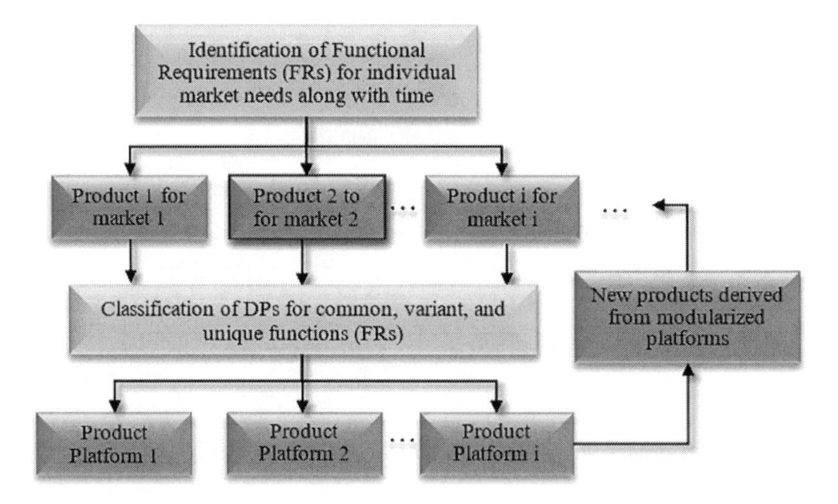

Figure 4.27 Bottom-up approach.

map platform leveraging strategies. Accordingly, product platforms can be leveraged by *horizontal leveraging*, *vertical leveraging*, and *beachhead leveraging*.

Figure 4.29 shows the scenario of horizontal leveraging. Commonalities of the products across the segments are identified to determine the modules, and multiple product platforms are developed to target customers at the low end, middle end, and high end, respectively. Companies can benefit from horizontally leveraged platforms to (i) introduce a series of relevant products for different customer groups without having to 'reinvent the

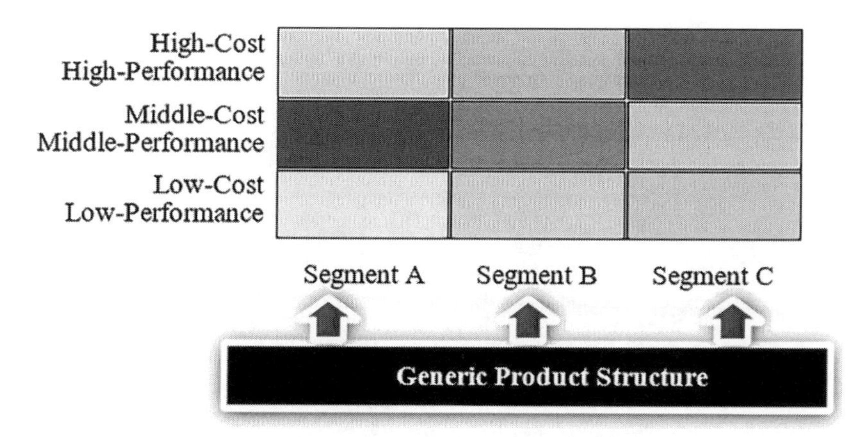

Figure 4.28 Traditional static product structure with no leveraging.

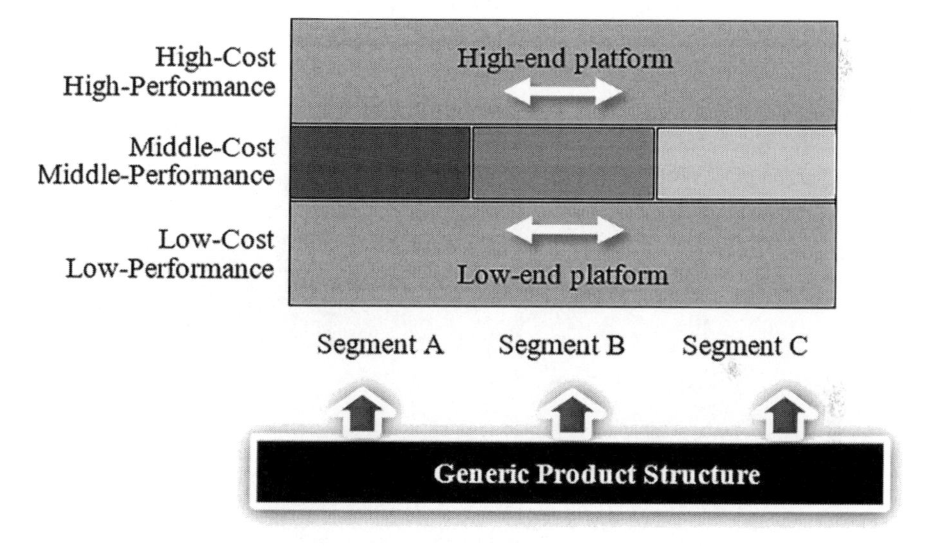

Figure 4.29 Horizontal leveraging product platforms.

wheel', (ii) develop new products in a shortened time and with less risk since the applied technologies are proven in other market segments, and (iii) minimize manufacturing procurement and retooling costs. Figure 4.30 shows the horizontally leveraged platforms of B&D power suppliers across the segments of saws, drills/drives, and lighting products (MIT 2019).

Figure 4.31 shows the scenario of vertical leveraging. Commonalities of the products across targeted customers are identified to determine the modules and multiple product platforms are developed for different segments, respectively. Companies can benefit from vertically leveraged platforms to (i) leverage the knowledge for customer needs within a given market segment and (ii) reduce the design and manufacturing costs of products similar to the case of horizontal leveraging. Figure 4.32 shows the vertically leveraged

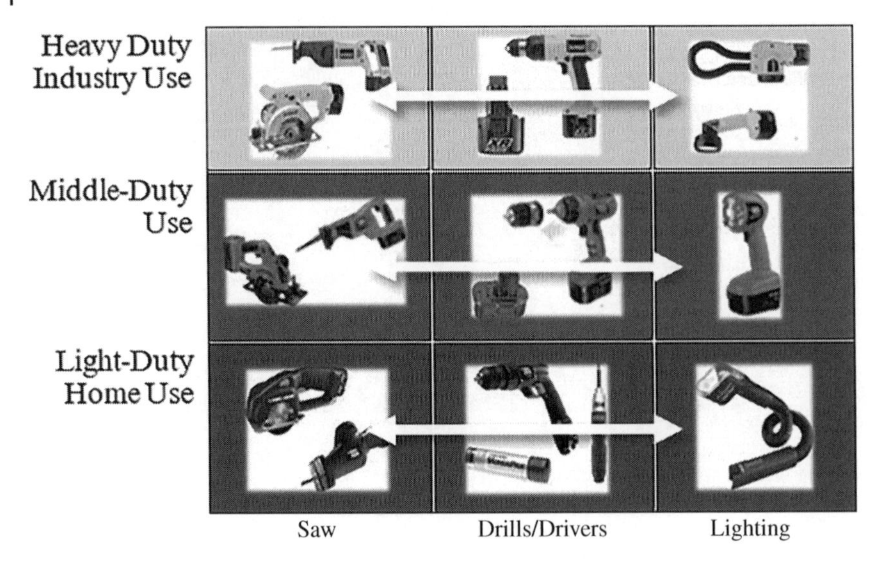

Figure 4.30 Horizontally leveraged B&D power platforms (MIT 2019).

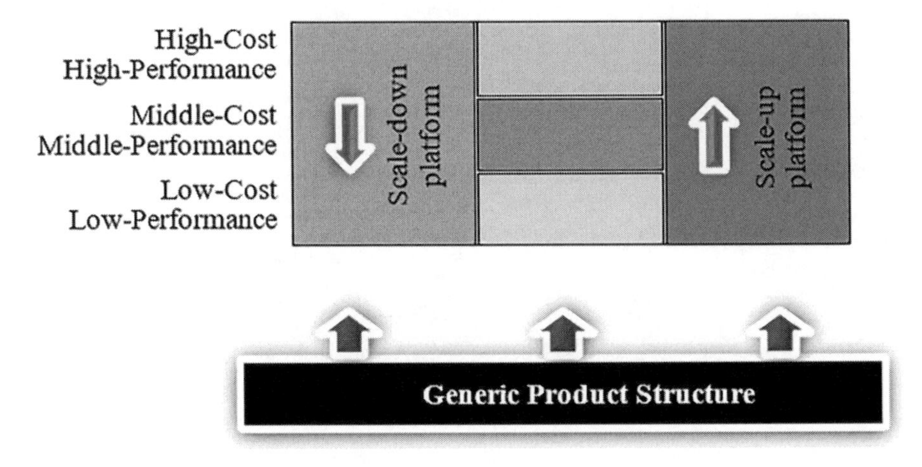

Figure 4.31 Vertical leveraging product platforms.

platforms of Gillette shavers across targeted customers of high end, middle end, and low end (MIT 2019).

Figure 4.33 shows the scenario of beachhead leveraging. The commonalities of the products across targeted customers and segments are analysed and identified to determine the modules, and multiple product platforms are developed based on other meaningful classifications, respectively. Companies can benefit from beachhead leveraging to (i) develop a low-cost platform, an effective platform and efficient processes, (ii) scale up performance characteristics of the low-cost platform to appeal to the needs of mid- and high-end users, (iii) extend the platform for customers in different market segments, and (iv) combine the extensions and scaling to provide step-up functions required by mid- and high-end users

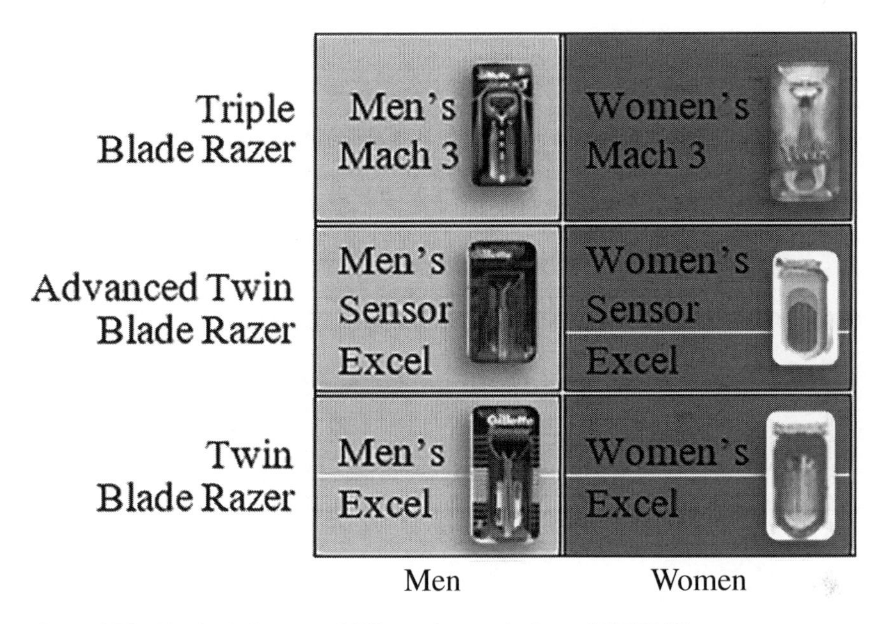

Figure 4.32 Vertically leveraged Gillette shaver platforms (MIT 2019).

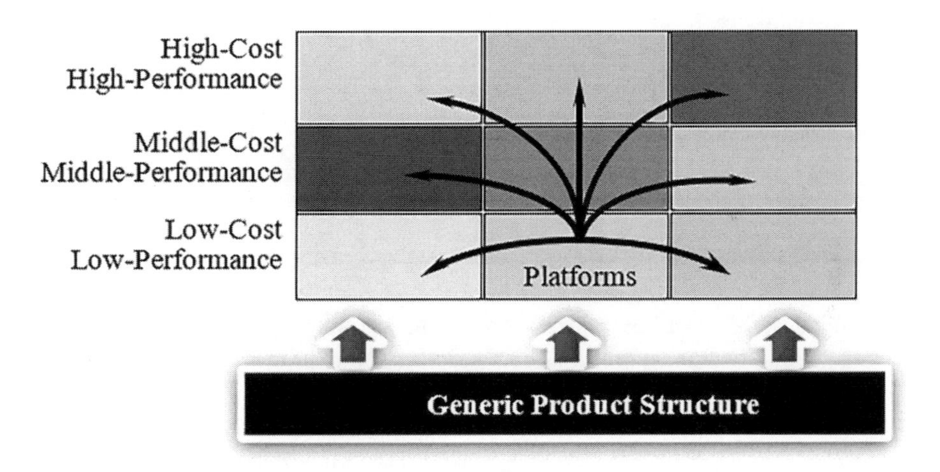

Figure 4.33 Beachhead leveraging product platforms.

in other segments. Figure 4.34 shows the beachhead leveraged platforms of FlexiBlade ice scrapers with small, medium, and large sizes (MIT 2019).

4.10 Evaluation of Product Platforms

In Section 4.9, it was shown that different leveraging strategies lead to different product platforms. To evaluate and select product platforms, *market-based* or *product-based* evaluations can be used (Blecker et al. 2003). Each evaluation method consists of a

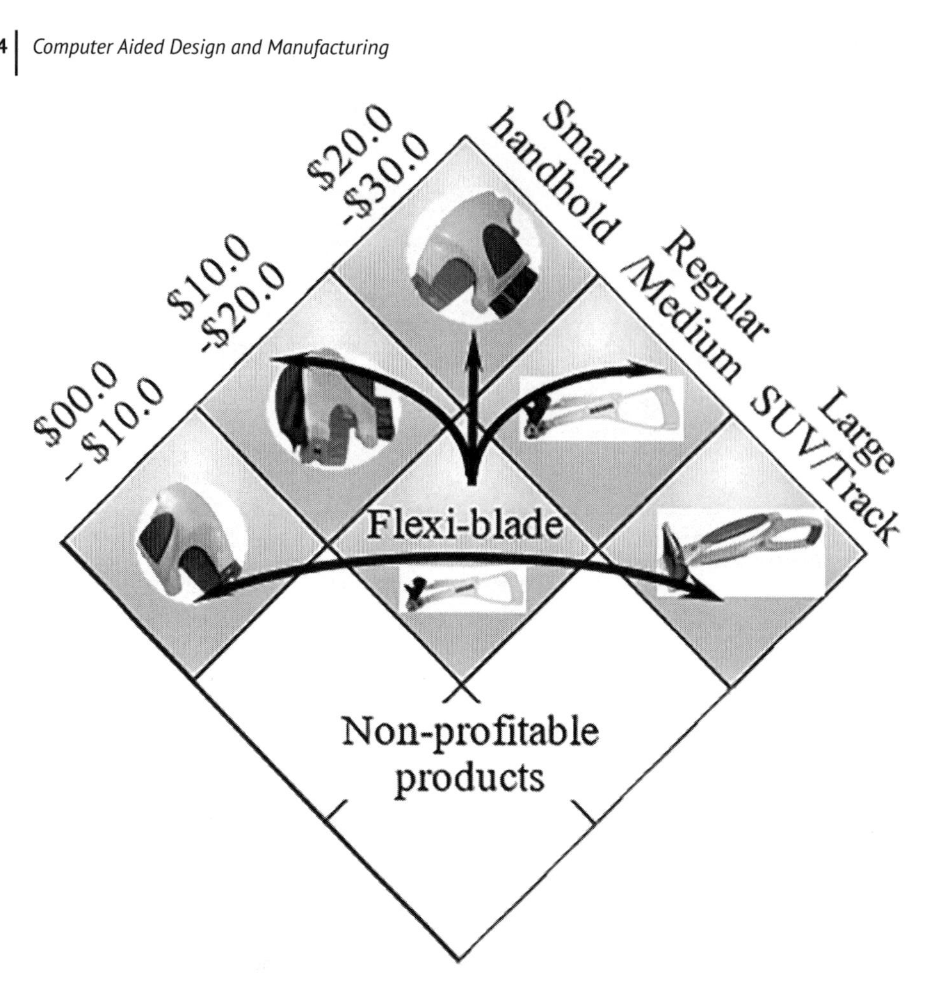

Figure 4.34 Beachhead leveraged ice scraper platforms (MIT 2019).

set of quantified design metrics. Tables 4.9 and 4.10 provide lists of main metrics and corresponding quantification approaches in market-based and product-based evaluations, respectively (MIT 2019).

Most of the metrics in Tables 4.9 or 4.10 are mostly related to each other. It is impractical to incorporate many design criteria or develop a universal criterion that can be applied to the design of any product platform. Therefore, practical methods for the evaluation of product platforms vary from one case to another. Shararah et al. (2018) emphasized the reusability in the evaluation of product platforms, which evaluate the reusability at three levels: (i) the *system-level*, where the solution of a whole system can be reused either by adopting it directly or as a derived solution in one application family, (ii) the *component-level*, where the components of a system solution can be reused to fulfil sub-FRs of another application, and (iii) the *function-level*, where the design concept of a well-defined function can be reused.

Bi et al. (2003) adopted a similar method to evaluate robot platforms, shown in Figure 4.35. Different from traditional robots with an integral structure, a modular robot system consists of a set of modules that can be selected and assembled in a variety of configurations. Each configuration can fulfil a specific task optimally. Therefore, the

Table 4.9 Metrics in product-based evaluation of product platforms.

Metrics	Quantification
Commonality of product processes	The number of common processes divided by the total processes of productions
Commonality of purchasing processes	The number of standardized purchasing processes divided by the total purchasing processes
Setup durations	The average production time divided by the lead time of product
Utilization of manufacturing resources	The total used time divided by the total time (including idle and maintenance time) of a manufacturing resource
Work-in-process turnover	The total market revenue divided by value of work-in-progress inventory
Delivery time	The expected delivery time divided by the actual delivery time

Table 4.10 Metrics in market-based evaluation of product platforms.

Metrics	Quantification
Time to market	The total time in production from raw material to final product
Multiple use	The number of product variants divided by the total number of possible configurations from available modules
Used variety	The number of perceived variants divided by the number of all possible variants
Platform efficiency	The total of research and development time for derivative products divided by the development time of the base platform.
Interface complexity	The assembly time for one interface divided by the ideal assembly operation time
Customizable attributes	The number of new customizable attributes and the number of eliminated customizable attributes.

variety of configurations from the given set of modules can be used to evaluate system configurability. Modular robot platforms are evaluated using the following critical steps.

4.10.1 Step 1. Representation of a Modularized Platform

ADT is applied to represent FRs of modular platforms. Three FRs at the system level are defined as follows:

1. *FR1* – module-level varieties for a wide selection of modular functions in terms of dimensions, power outputs, and precision
2. *FR2* – assembly-level varieties for flexibility in assembling two or more modules in different modes and orientations
3. *FR3* – topology-level varieties where common topologies of configurations such as serial, parallel, or hybrid configurations can be formed from pre-defined modules.

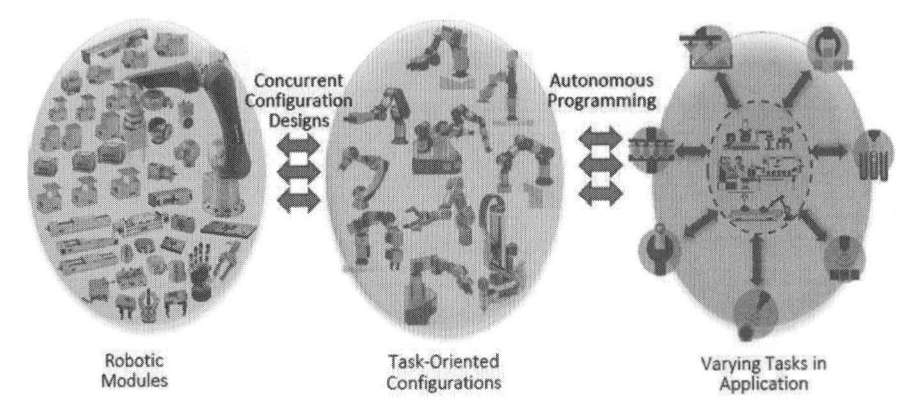

Figure 4.35 Modular robot systems.

Accordingly, the following three DPs are identified to meet these FPs:

1. *DP1* – a variety of modules including joint modules, link modules, and specialized modules such as base modules and end-effector modules, gripping modules, and wrist modules.
2. *DP2* – a variety of the interface ports of modules where the connections take place to connect one module with the other. The types of possible interface ports are for the connections of two joint modules, two link modules, or between a joint module and a link module. Note that an interface port has its associate dimensions and only two modules with matched dimensions can be connected satisfactorily.
3. *DP3* – a number of platform or hybrid modules with multiple assembly ports for connections of multiple modules. A platform module can be either a base platform or an end-effector platform in a parallel configuration and a hybrid module is a connector of multiple joint modules to form a hybrid configuration.

4.10.2 Step 2. Mapping a Modular Architecture for Robot Configurations

Robot configuration variants are derived from a system architecture. The mapping of feasible robot configurations from a system architecture must be established to look into the impact of design parameters on product variants. Figure 4.36 shows such a mapping model for the modular platform where a number of design variables of a robot configuration are identified.

4.10.3 Step 3. Determine Evaluation Criteria of a Product Platform

The system configurability is evaluated by the number of robot configurations against the design space of the robots formed by continuous design variables. However, it raises an issue of evaluating the platform purely based on the number of robot configurations. For example, if a modular platform can produce 100 different configurations, are these configurations sufficient for certain applications? This issue can be addressed by (i) eliminating *isomorphic configurations* (i.e. different configurations but with the same functions) and (ii) comparing the set of feasible configurations with the number of configurations formed by discretized design variables in the design space of robot structures.

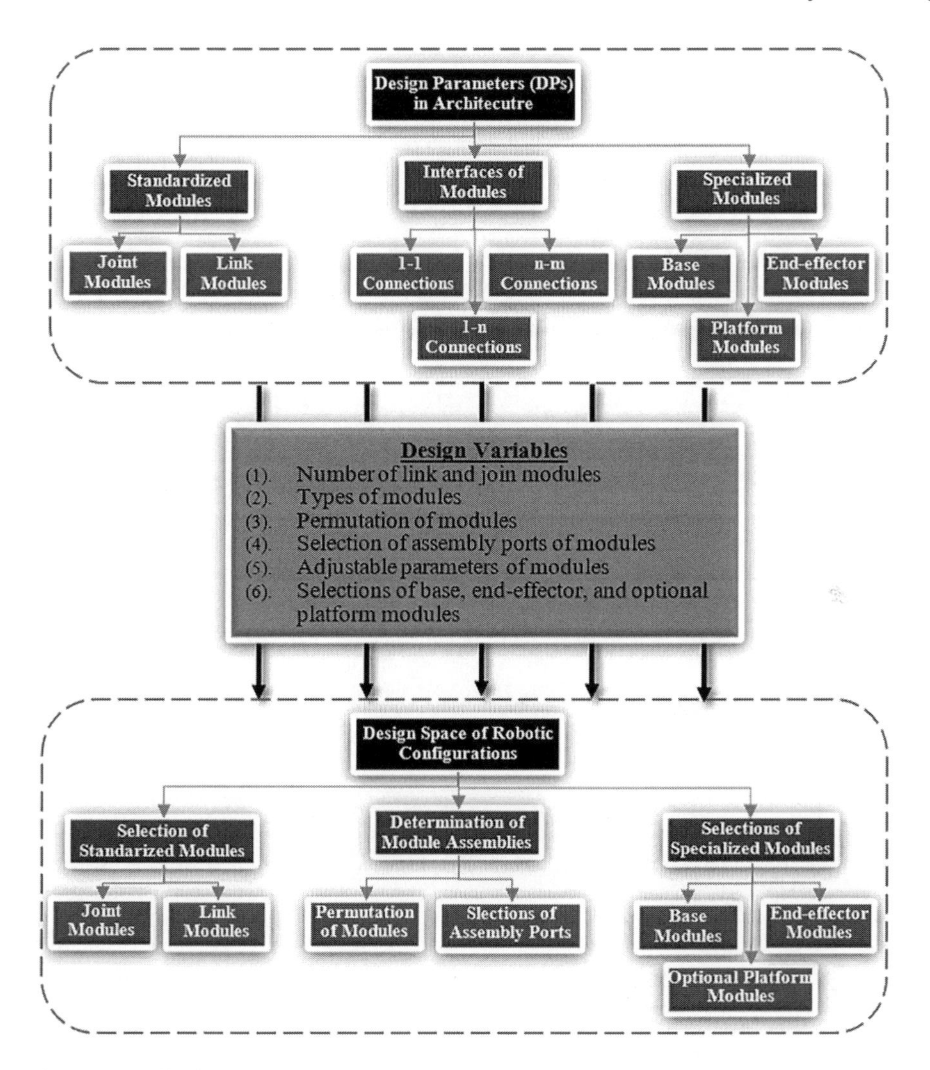

Figure 4.36 Design variables in modular robot platform based on ADT.

Any robot consists of a number of motion axes and the Denavit–Hartenberg (DH) notations are often used to describe spatial relations of motion axes. If an open-loop robot is of interest, the variants of the DH notation can be used to determine the required number of configurations in a discretized design space of the robot structure. As shown in Figure 4.37, a robotic structure can be mathematically represented by a set of DH parameters (a_i, d_i, α_i, θ_i), where the meanings of these variables are explained in the figure, where $i = 1, 2, \ldots, n$, and n is the number of degrees of freedom (DoF). The difference between the DH parameters of two robots gives the difference between FRs on the robots.

The FRs of a robot are determined by the spatial relationships of its motion axes. In return, the DH parameters for spatial relations of motion axes determine the FRs of robot structures. The variants of DPs for the given FRs can also be represented by selecting DH parameters and assembling the motion axes of DF parameters correspondingly. Taking an example of a serial robot structure with 3 DoF, the design space of the robot configurations is

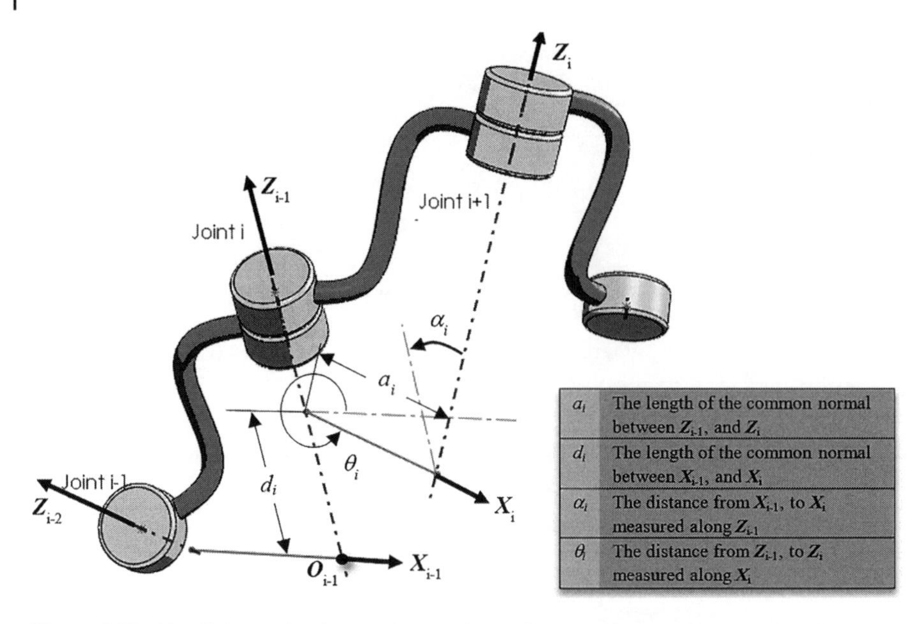

Figure 4.37 The DH notation for spatial relations of two motion axes.

Table 4.11 The variants of a set of DH parameters.

Motion type	a_i	d_i	α_i	θ_i
(i) Rotational (θ_i)	(i) Zero	(i) Zero	(i) 0	(i) Variable
(ii) Translational (α_i)	(ii) Discretized values	(ii) Discretized values	(ii) $\pm\pi/2$	(ii) 0
		(iii) Variable	(iii) π	(iii) $\pm\pi/2$
				(iv) π

determined by discretizing the set of DH parameters for each pair of motion axes. Table 4.11 gives an example of the variants of a set of D-H parameters in terms of motion types, the orientation of the motion axis with precedent axis, and the resolutions of dimensional changes.

In Table 4.11, the resolution for the discretization of a length parameter (a_i or d_i) depends on (i) the ranges of these design variables under consideration and (ii) how significantly it affects the robotic behaviour. Assuming that the variants of a 3DoF serial robot are considered and the ranges of a_i or d_i are specified as (0.0 m, 1.0 m), the design space of a robot configuration can be defined as

$$
\begin{aligned}
&a_i \in (0.0, 0.1, \dots, 0.9)\\
&d_i \in (0.0, 0.1, \dots, 0.9, variable)\\
&\alpha_i \in \left(-\frac{\pi}{2}, 0, \frac{\pi}{2}, \pi\right)\\
&\theta_i \in \left(-\frac{\pi}{2}, 0, \frac{\pi}{2}, \pi, variable\right)
\end{aligned}
\tag{4.3}
$$

where $i = (1, 2, 3)$ is the index of the motion axis.

Table 4.12 Number of robot configurations based on discretized DH parameters (DH-CS).

	The 1st axis	The 2nd axis	The 3rd axis	Total
Variants	40 + 40	400 + 160	2	89 600

Since the variants of the base or the end-effector module do not yield robot configurations, for the first motion axis, one can simplify $d_1 = 0$ or $\theta_1 = 0$ when the first joint is rotational or translational. For the third motion axis, one can simplify $d_3 = 0$ and $\alpha_3 = 0$ for a rotational joint and $a_3 = 0$ and $\theta_3 = 0$ for a translational joint.

Consequently, the variations of the first set of DH parameters are determined as:

The first joint is rotational	or	The first joint is translational
$a_1 \in (0.0, 0.1, \ldots, 0.9)$		$a_1 \in (0.0, 0.1, \ldots, 0.9)$
$d_1 \in (0.0)$		$d_1 \in (variable)$
$\alpha_1 \in \left(-\frac{\pi}{2}, 0, \frac{\pi}{2}, \pi\right)$		$\alpha_1 \in \left(-\frac{\pi}{2}, 0, \frac{\pi}{2}, \pi\right)$
$\theta_1 \in (variable)$		$\theta_1 \in (0)$

The variations of the second set of DH parameters are determined as:

The second joint is rotational	or	The second joint is translational
$a_2 \in (0.0, 0.1, \ldots, 0.9)$		$a_2 \in (0.0, 0.1, \ldots, 0.9)$
$d_2 \in (0.0, 0.1, \ldots, 0.9)$		$d_2 \in (variable)$
$\alpha_2 \in \left(-\frac{\pi}{2}, 0, \frac{\pi}{2}, \pi\right)$		$\alpha_2 \in \left(-\frac{\pi}{2}, 0, \frac{\pi}{2}, \pi\right)$
$\theta_2 \in (variable)$		$\theta_2 \in \left(-\frac{\pi}{2}, 0, \frac{\pi}{2}, \pi\right)$

The third set of DH parameters varies only when different motion types are chosen, i.e. rotational or translational. Therefore, the FRs of a modular robot platform for system configurability are evaluated by the total number of robot configurations. For the topology of a 3 DoF serial robot, the total number of variations is based on Table 4.12.

4.10.4 Step 4. Evaluate Platform Solutions

Three robot platforms are compared to illustrate the evaluation of robot platforms (Bi et al. 2003). The first platform in Figure 4.38 consists of five link modules, a rotary joint with seven assembly patterns, and a translational joint with five assembly patterns. The link and joint modules can be connected directly, but no direct connection is allowed for two joint modules.

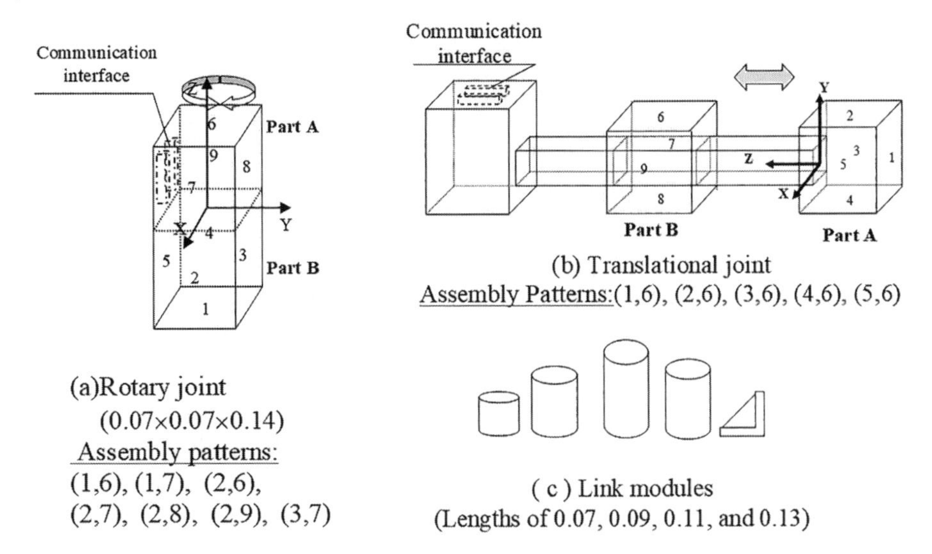

Figure 4.38 Robot platform design I. (a) Rotary joint (0.07 × 0.07 × 0.14) assembly patterns: (1,6), (1,7), (2,6), (2,7), (2,8), (2,9), (3,7). (b) Translational joint assembly patterns: (1,6), (2,6), (3,6), (4,6), (5,6). (c) Link modules (lengths of 0.07, 0.09, 0.11, and 0.13).

The second robot platform in Figure 4.39 consists of five link modules, a rotary joint module, and a translational joint module. These modules must be connected through a connector. The link module has multiple choices of assembly patterns. Note that the assembly ports on a module are dimensioned so that a link module in Figure 4.39d provides 10 assembly ports and the link module in Figure 4.38a provides only six assembly ports.

The third robot platform in Figure 4.40 consists of a link module, a rotary joint module, and a translational joint module. One connector is used to connect the joint and link modules. A joint module or a link module has multiple assembly patterns. More than one link module can be used between two joint modules. However, at most two link modules are permitted between two joint modules in order to achieve the required precision.

Table 4.13 gives the evaluation results of three robot platforms. The comparison is made by system configurability, which is measured by the ratio of the number of configurations from a platform (PS-CS) and the discretized DH parameter (DH-CS). All three platforms can generate a limited number of robot configurations. System III has the most robot configurations (based on DH-CS) but it has a relatively large number of modules; moreover, multiple assembly ports on the joint module are more complicated than the parts on a link module. Platform II is the best choice due to its efficiency in mapping PS-CS to DH-CS. Overall, using more link modules can assist in improving system configurability.

4.11 Computer Aided Tools (CAD) for Platform Technologies

Computer aided tools (CAD) become prevalent in designing, verifying, and validating new products in virtual environments. Virtual design allows design faults to be detected and fixed before the products are manufactured, assembled, and tested. For the design and

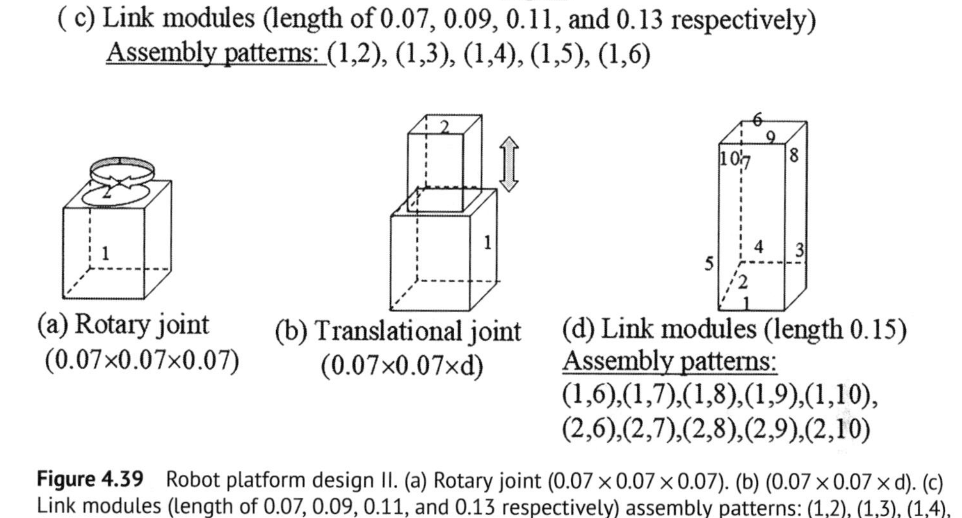

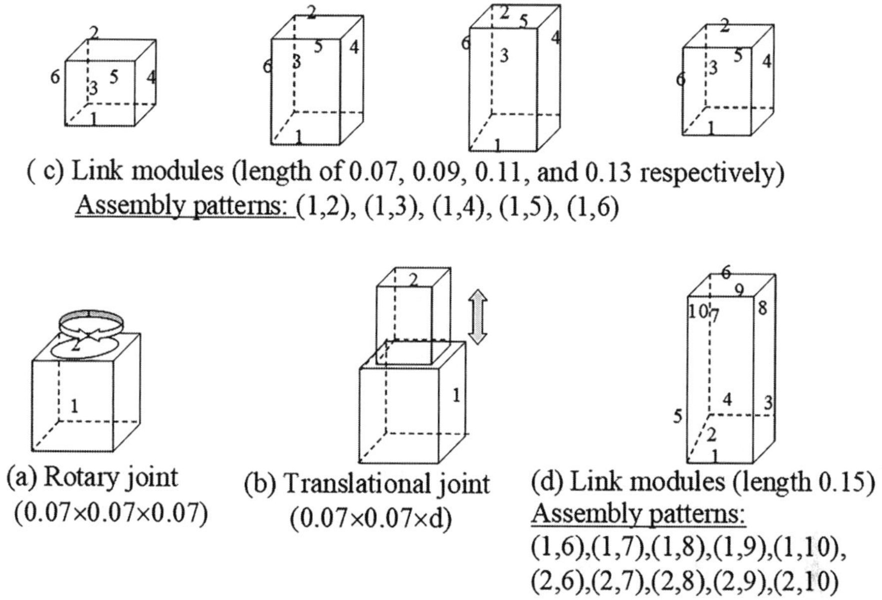

Figure 4.39 Robot platform design II. (a) Rotary joint (0.07 × 0.07 × 0.07). (b) (0.07 × 0.07 × d). (c) Link modules (length of 0.07, 0.09, 0.11, and 0.13 respectively) assembly patterns: (1,2), (1,3), (1,4), (1,5), (1,6). (d) Link modules (length 0.15) assembly patterns: (1,6), (1,7), (1,8), (1,9), (1,10), (2,6), (2,7), (2,8), (2,9), (2,10).

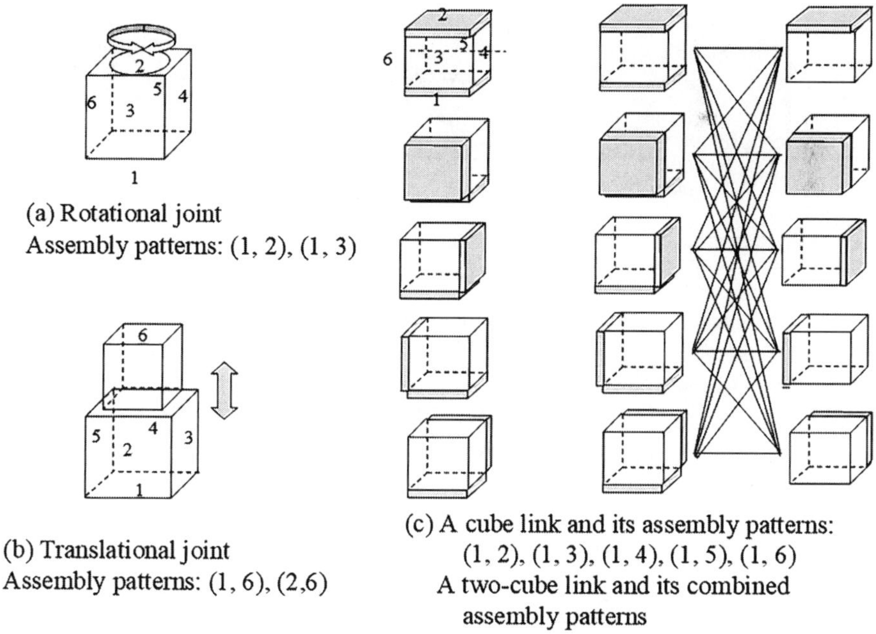

(a) Rotational joint
Assembly patterns: (1, 2), (1, 3)

(b) Translational joint
Assembly patterns: (1, 6), (2,6)

(c) A cube link and its assembly patterns:
(1, 2), (1, 3), (1, 4), (1, 5), (1, 6)
A two-cube link and its combined
assembly patterns

Figure 4.40 Robot platform design III. (a) Rotational joint assembly patterns: (1,2), (1,3). (b) Translational joint assembly patterns: (1,6), (2,6). (c) A cube link and its assembly patterns: (1,2), (1,3), (1,4), (1,5), (1,6), a two-cube link and its combined assembly patterns.

Table 4.13 Evaluation and comparison of three robot platforms.

	PS-CS	DH-CS/expected DH-CS		
I	43 200	518	0.00578	0.0120
II	7200	663	0.00740	0.0920
III	57 600	2648	0.0295	0.0458

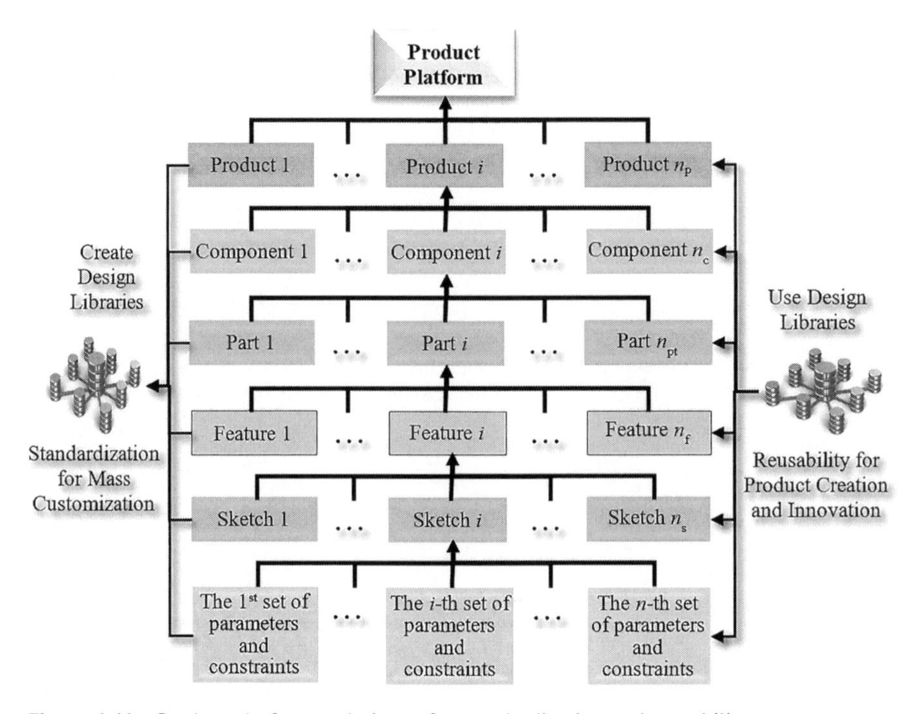

Figure 4.41 Product platform techniques for standardization and reusability.

implementation of product platform technologies, modern computer aided tools have their capabilities in creating and upgrading design toolboxes and libraries at the sketch level, module level, component level, and system level.

Product cost relates to the degrees of standardization of the product design at all levels. The more standardized features, parts, and components that are used, the less design effort is required, and the lower the cost of manufacturing and assemblies will be. For the sake of product evolution, it is beneficial to store the solutions and models to specified FRs. On the other hand, when existing products need to be improved, maximizing the utilization of reusable features or other graphical elements helps to innovate a new product with aspects of (i) reusing existing features of product design in the current solution to a new application, (ii) using new innovations by combining and assembling existing functional modules, and (iii) using the more standardized features so that less cost occurs to add to the manufacturing costs (Figure 4.41).

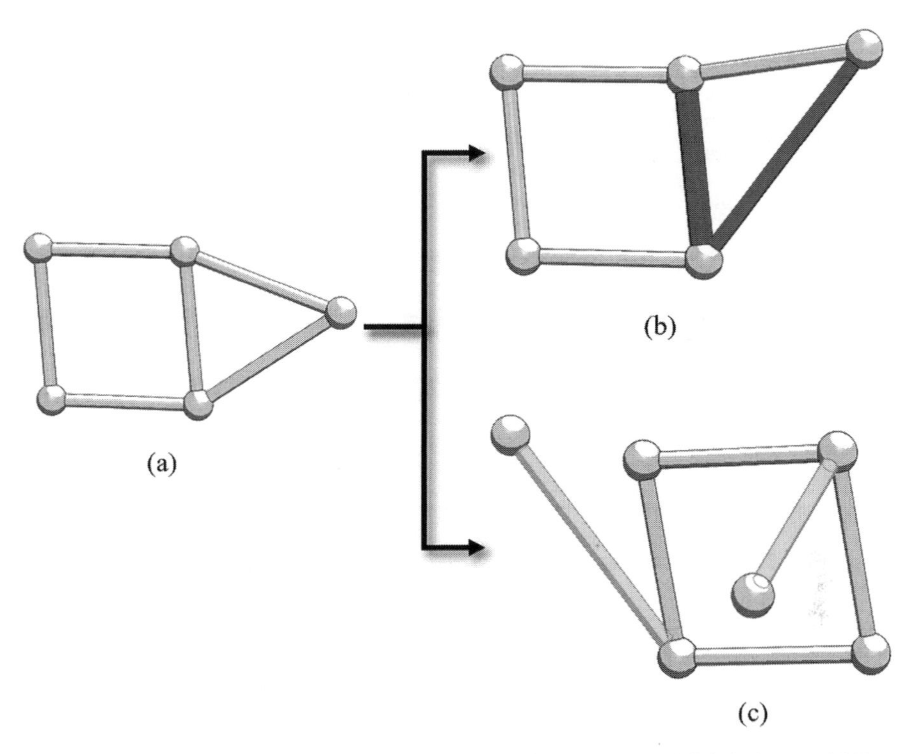

Figure 4.42 Difference of morphological and topological changes. (a) Original truss. (b) Truss after morphological change. (c) Truss after topological change.

4.11.1 Modelling Techniques of Product Variants

Products are varied in dimensions and topologies. Accordingly, two types of computer aided techniques for product variants are *morphology* or *topology*. Figure 4.42 shows the difference between the two types of technique. A morphological change is made on a dimensional variable such as thickness, width, length, and orientation of objects. A morphological change is made by continuous operations such as shrinking, stretching, and transforming of objects. While a topological change corresponds to an increase, removal, or restructure of objects, a morphological change is made by a continuous operation such as tearing and gluing.

In knowledge-based engineering (KBE), the topological changes can be implemented by adding or modifying features at the topological level, while morphological changes can be implemented by modifying an existing entity at the same level of entity or below. The corresponding tools in computer aided modelling can be referred to as *design toolboxes* and *design libraries*, respectively.

4.11.2 Design Toolboxes

Design refers to a set of actions to develop new or better solutions to customer needs based on existing ones (Jordan 2013). Figure 4.43 shows an example of a rotary actuator design, which is built from many standardized parts such as fasteners, gears, and screws. Models for standardized parts and processes should be stored as knowledge in design toolboxes.

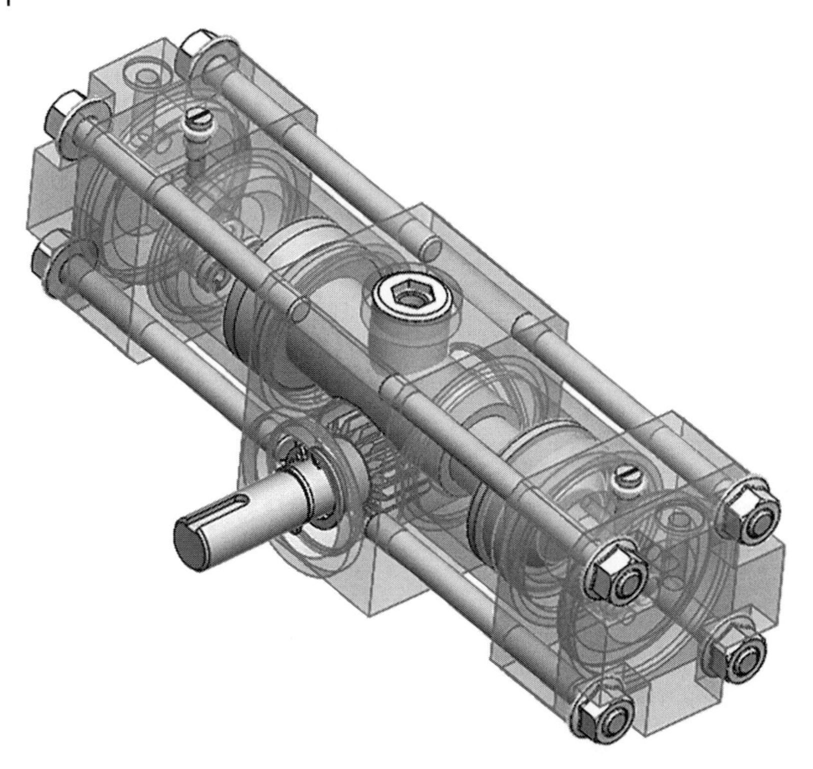

Figure 4.43 Example of rotary actuator with many standardized parts or features.

A toolbox is a suite of powerful tools to maintain features, parts, components, configurations, on product platforms. Toolboxes support the reuses of existing features, parts, components, and configurations in new designs. The toolbox allows us to avoid the reinvention of wheels, for example, in a product design from standardized or existing parts or components. A toolbox usually contains a vast library of standard hardware components that can easily be dropped into a design when needed. For example, standardized components can easily be configured to populate *Bills of Material* (BOM) with part number descriptions or any other custom information that needs to be easily displayed to the user (GoEngineer 2019). The modern CAD tools, such as SolidWorks, are equipped with some design toolboxes and libraries for part reuses. For example, an assembled product usually consists of many machined screws, bolts, and nuts. These parts are standardized and can be stored as design libraries to be reused when they are needed. SolidWorks has an *add-ins* of *smart fasteners*. Table 4.14 shows the advantages of using a smart fastener design toolbox in computer aided modelling (Ideal-parts 2019).

SolidWorks includes other design toolboxes for standard bearings and the channels, beams, and weldments with the main features specified in Table 4.15 (Ideal-parts 2019).

4.11.3 Custom Design Libraries

Standardization, modularization, and platforms (S/M/P) are critical enablers to CE and CI. Design toolboxes and libraries are effective techniques for KBE and for implementation of

Table 4.14 Advantages of using design toolboxes.

Modelling without *smart fasteners*	**Modelling with *smart fasteners***

• Each fastener has to be individually created
• Tricky to set up (too many choices)
• US-centric
• Unstructured naming convention (DIY)
• Unstructured file repository
• No PhotoView360 materials assigned
• No mass-properties assigned
• No Real View properties assigned
• It's expensive!
• Upgrade issues
	• Fast, easy, and inexpensive upgrade for users of SolidWorks Standard edition. Get all the parts you need with Ideal Parts!
	• Hugh selection of Nuts and Bolts
	  • 45 metric bolts, screws, nuts, washers, pins, and clips in more than 2000 commercially available sizes
	  • 4200 bolts, screws, nuts, washers, pins, clips, etc.
	  • Drag and drop from the Design Library
	  • SmartMates snap fastener to feature
	  • PhotoView360 materials assigned
	  • RealView properties assigned
	  • Correct SG assigned
	  • Consistent naming convention
	  • Stay whole in section views
	• Metal Sheet Fasteners
	  • More than 400 PSM® parts
	  • Library feature files SLDLFP
	  • Drag and drop placement
	  • Sketch driven pattern placement
	  • Correct callouts in drawings

mass customization. In computer aided modelling, creating design libraries for common features, parts, components, and configurations could minimize repetitive designs, shorten product design cycles, and reduce product development costs.

The procedure to create and store an entry in a design library of SolidWorks is shown in Table 4.16 (3Dvision 2015). The part and its features, shown in Figure 4.44, are used as an example.

Using design libraries supports the development of product platforms and improves the efficiency and productivity of modelling processes in the sense that sketches, blocks, features, parts, annotations, and assemblies can be reused easily. The idea of design libraries is similar to an object-oriented design approach in Software Engineering. An entity, whether or not it is simple or complex, can be defined as an *object class* with a black or grey box, and can be stored in a design library for reuses. When a new model has similar entities, the models of these entities can be instantiated directly from the stored object class.

4.12 Summary

Design of a complex product or system involves design constraints and goals in multiple disciplines. These design factors have to be taken into consideration simultaneously to (i) maximize the design space of feasible solutions for given FRs and (ii) simplify the design process by decomposing FRs at system level into sub-FRs at the levels of subsystems, components, parts, features, and sketches. Standardization, modularization, and platforms (S/M/P) tend to be critical enablers to support CE in a product lifecycle.

Table 4.15 Main features of design toolboxes of bearings and weldments in SolidWorks.

Toolbox of bearings	Toolbox of channels, beams, and weldments
• More than 7000 parts	• 600+ OneSteel™ and Tubemakers™ profiles
• SmartMates snap to features and creates mates	• Also included as standalone parts (SLDPRT)
• Correct PhotoWorks material	• Drag, drop, mate, and modify
• Correct RealView material	• Write-protected master files to avoid errors
• Correct weight	• Mid and end planes included for easier mating
• Includes engineering data (abutment, max. speed and load)	• Embedded spreadsheets inc. engineering data
• Proper section display in technical drawings	• Channels beams fully integrates with SolidWorks Weldments feature
• Includes seals, rod-ends, spherical bearings, etc.	• Original profiles also still supplied as SLDPRT files!
• Fully customizable in Microsoft Excel	
• Bearings section correctly in drawings	
• Pop rivet material/shank fit visually indicated	
• Visual indicator to bolt thread engagement	
• Cir-clip groove data embedded in configuration properties	
• Bearing data embedded	

A product platform supports knowledge reuse for CIs in a product lifecycle. The ADT is an ideal tool to use to modularize product structures. Three methods can be used to develop a product platform: the bottom-up, top-bottom, and combined beachhead methods. To maximize the knowledge reuse in developing or continuously improving a product platform, engineers should be able to use some advanced tools in modern CAD packages, such as design toolboxes, design libraries, and other add-ins in SolidWorks.

4.13 Design Projects

4.1 Design a gearbox with a 4.0 gear ratio (see a reference model in Figure 4.45), create its CAD model, and maximize the usages of standard components in SolidWorks toolboxes such as fasteners, bearings, and keys.

4.2 Create a design library with a number of icons such as those for a product family, or logos for department, university, or company and create several parts where multiple features from the design library can be reused (see a reference model in Figure 4.46):

1. Discuss the needs of knowledge reuse for product platforms.
2. Document your procedure to create a design library.
3. Document your procedure to apply the features from the design library.
4. Create drawings for your part(s) with features you have added in the design library.

Table 4.16 Steps to create and store an entry in design library.

Step	Task	Explanation
1	Create the model	One has to build or import the model one time. It is desirable to include some design intents of reuse. For example, fixing the geometric dimensions of the features in the part and freeing the positions of features. If an additional dimension is added to locate a feature, a reference is needed to locate the feature when it is reused.
2	Rename important dimensions to logical names	It is optional but makes the entry more user friendly. The names of dimensions can be displayed by (i) turning on *Show Feature Dimensions* by right clicking on *the Annotations* in *the Feature Tree* and (ii) activate *Display Dimension Names* in the *view* menu. You can rename a dimension when it is selected. Renaming a feature or part can be directly done in the Feature tree.
3	Change the appearances of entry	It is optional and can be done in the Feature Tree. By right-clicking the feature, one can activate the *appearance* icon for appearance changes. In addition, one may orient and position the model since the current view generates the thumbnail graphic in the design library.
4	Add entry to library	The design library can be accessed in the *Task Pane*. The user needs to specify the folder where the new entry has to be stored in, click the *Add To Library* button at the top in the *Property Manager* that comes up on the left under *Items To Add*, and be sure to select the feature(s) which would be added as a *library feature*. Alternatively, users can pre-select the feature(s) in the feature tree, hit the *Add to Library* button, and hit *OK* to add the feature in the library.
5	Organize library features	The file format in a design library is *.sldlfp* (library feature part). In the *Dimensions* folder, drag locking dimensions into the *Locating Dimensions* folder. In default, any of other dimensions in the dimensions folder can be modified when the feature is reused. To confirm the changes, save the .sldlfp file again and close all open files.
6	Reuse items in design library	The items in a design library can be reused anytime by dragging and dropping it into any model. Any dimensions other than locking dimensions ones can be specified when the feature is reused.

Figure 4.44 Creating a feature, part, component, or configuration in a design library.

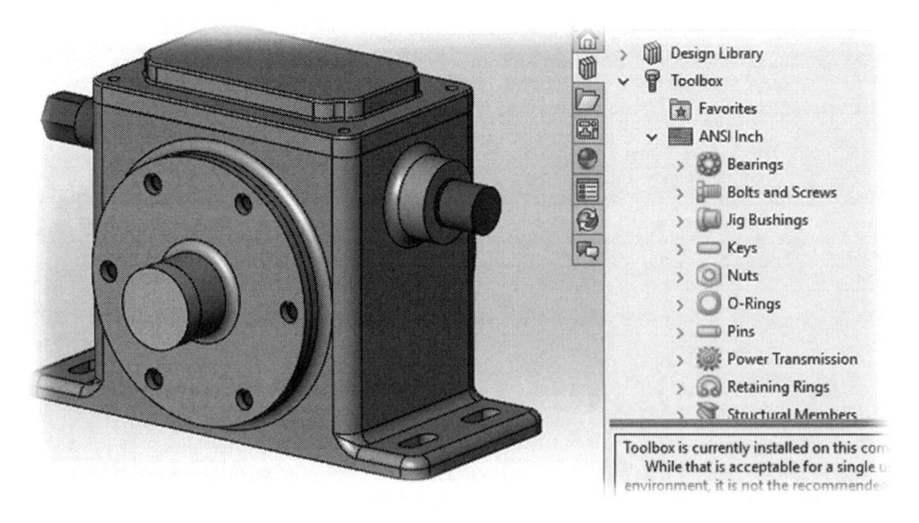

Figure 4.45 Example of gearbox models for design project 1.

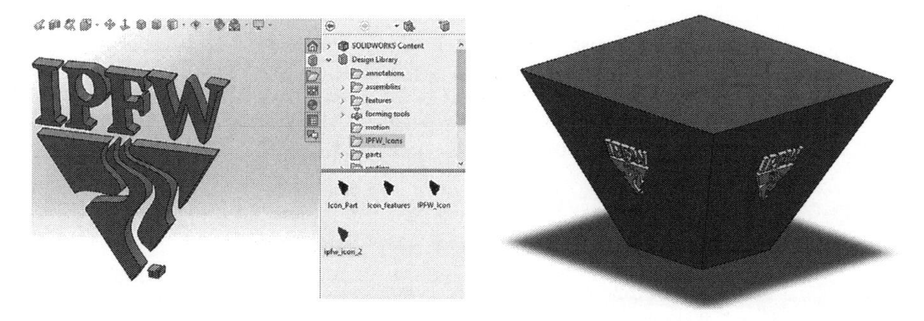

Figure 4.46 Example of creating a custom design library for knowledge reuse for design project 4.2.

References

3Dvision (2015). Creating SolidWorks library features from scratch. https://blogs.solidworks.com/tech/2015/10/creating-solidworks-library-features-scratch.html.

Airbus (2019). Facts and figures. https://www.airbus.com/content/dam/corporate-topics/publications/backgrounders/Backgrounder-Airbus-Commercial-Aircraft-A380-Facts-and-Figures-EN.pdf.

Bi, Z.M. and Zhang, W.J. (2001). Modularity technology in manufacturing: taxonomy and issues. *International Journal of Advanced Manufacturing Technology* 18 (5): 381–390.

Bi, Z.M., Gruver, W.A., and Zhang, W.J. (2003). *Adaptability of reconfigurable robot systems. Proceedings of the 1003 IEEE international Conference on Robotics and Automation*, 2317–2322. Taipei, Taiwan (14–19 September 2003): IEEE.

Blecker, T., Abdelkafi, N., Kaluza, B., and Friedrich, G. (2003) Key metrics system for variety steering in mass customization. *2nd World Congress on Mass Customization and Personalization*, Munich, Germany (6–8 October 2003).

Gepp, M., Foehr, M., and Vollmar, J (2016). Standardization, modularization, and platform approaches, in the engineer-to-order business – review and outlook. *2016 Annual IEEE Systems Conference (SysCon)*, Orlando, Florida, USA (18–21 April 2016). IEEE. DOI: 10.1109/SYSCON.2016.7490549.

Globalscape (2017). Five types of data integration you need to know. https://www.globalscape.com/blog/5-types-data-integration.

GoEngineer (2019). SolidWorks toolbox library setup and utilization. http://www.goengineer.com/knowledge-base/solidworks-toolbox-library-setup-utilization.

Golfmann, J. and Lammers, T. (2014). Modular product design: reducing complexity, increasing efficacy. *Performance* 7 (1): 58–63.

Heizer, J. and Render, B. (2008). *Principles of Operations Management*, 7e. Pearson/Prentice Hall Inc. ISBN-10: 0132343282.

Holems, J. (2018). Ford to save money by using just five platforms for all its models. https://www.cnet.com/roadshow/news/ford-platform-sharing-plan (accessed 29 September 2018).

Huang, C.-C. (2000). Overview of modular product development. *Proceedings-National Science Council Republic of China Part a Physical Science and Engineering* 24 (3): 149–165.

Ideal-parts (2019). The essential SolidWorks libraries. http://www.ideal-parts.com/downloads/IDEAL_PARTS_PRESENTATION.pdf.

Johnson, L. (2013). Modularity: a growing management tool because it delivers real value. modularmanagment.com.

Jordan, M. (2013). Design toolbox. https://www.slideshare.net/MartinJordan/design-toolbox-27383401.

Kreindler, D. (2014). Heads rolling at Volkswagen over MQB fumbles. https://www.thetruthaboutcars.com/2014/08/heads-rolling-volkswagen-mqb-fumbles.

Lennane, A. (2017). Aerospace – the longstar, 5(2017): 1–5. https://theloadstar.co.uk/wp-content/uploads/Aerospace.pdf.

Mason, W. H. (2009). Why airplanes look like they do. http://www.dept.aoe.vt.edu/~mason/Mason_f/SD1L3.pdf.

McGrath, M.E. (1995). *Product Strategy for High-Technology Companies*. New York: Irwin Professional Publishing.

Meyer, M.H. (1997). Revitalize your product lines through continuous platform renewal. *Research Technology Management* 40 (2): 17–28.

Mikkola, J.H. and Gassmann, O. (2003). Managing modularity of product architecture: toward an integrated theory. *IEEE Transactions on Engineering Management* 50 (2): 204–218.

MIT (2019). Fundamental concepts of platforming examples, definitions, approaches, and metrics. http://web.mit.edu/deweck/Public/ESD39/Lectures/ESD39s.02.Platform-Def-Examples.pdf.

Nadadur, G., Kim, W., Thomson, A.R. et al. (2012). Strategic product design for multiple global markets. In: *ASME 2012 International Design Engineering Technical Conferences and Computers and Information in Engineering Conference*, vol. 7, 837–848. IDETC/CIE https://doi.org/10.1115/DETC2012-70723.

Nicolai, L.M., and Carichner, G.E. (2010). Fundamentals of aircraft and airship design: Volume I – Aircraft design. http://buildandfly.shop/wp-content/uploads/2018/05/Fundamentals-of-Aircraft-and-Airship-Design-AIAA-ISBN9781600867514-Leland-Nicolai.pdf.

Pandremenos, J., and Chryssolouris, G. (2009) Modular product design and customization. *Proceedings of the 19th CIRP Design Conference – Competitive Design*, Cranfield University (30–31 March 2009), 200 pp. CIRP.

Pessina, M.W. and Renner, J.R. (1998). Mass customization at Lutron Electronics – a total company process. *Agility and Global Competition* 2 (2): 50–57.

Rothwell, R. and Gardiner, P. (1990). Robustness and product design families. In: *Design Management: A Handbook of Issues and Methods* (ed. M. Oakley), 279–292. Cambridge, MA: Basil Blackwell Inc.

Sanderson, S. and Uzumeri, M. (1995). Managing product families: the case of the Sony Walkman. *Research Policy* 24 (1995): 761–782.

Shararah, M.A., El-Kilany, K.S., and El-Sayed, A.E. (2018). Component based modeling and simulation of value stream mapping for lean production systems. https://www.extendsim.com/images/downloads/academic/grants/shararah-paper.pdf.

Simpson, T.W. (2019). ME/IE 546: designing product families. https://www.mne.psu.edu/simpson/courses/me546/lectures/me546.01.overview.pdf.

Smith, R.P. (1997). The historical roots of concurrent engineering fundamentals. *IEEE Transactions on Engineering Management* 44 (1): 67–78.

Stake, R.B. (2000). On conceptual development of modular products. Doctoral thesis. Department of Production Engineering, Royal Institute of Technology, Stockholm.

Sun Microsoftsystem (2007). The benefit of modular programming. https://netbeans.org/project_downloads/usersguide/rcp-book-ch2.pdf.

Wang, Q. (2010). *Understanding Platform-Based Product Development: A Competency-Based Perspective*. Eindhoven: Technische Universiteit Eindhoven. https://doi.org/10.6100/IR673035.

Wikipedia (2018). List of Ford platforms. https://en.wikipedia.org/wiki/List_of_Ford_platforms. Wikipedia (2019). Concurrent engineering. https://en.wikipedia.org/wiki/Concurrent_engineering.

Wikiwand (2019). Boeing 737. http://www.wikiwand.com/en/Boeing_737.

Womack, J.P. and Jones, D.T. (2007). *The Machine that Changed the World: The Story of Lean Production-- Toyota's Secret Weapon in the Global Car Wars that Is Now Revolutionizing World Industry*. Free Press ISBN-10: 0743299795.

Zha, X.F. and Sriram, R.D. (2006). Platform-based product design and development: a knowledge intensive support approach. *Knowledge-Based Systems* 19 (7): 524–543.

5

Computer Aided Reverse Engineering

5.1 Introduction

Engineering involves designing, manufacturing, assembling, and system operations. At its highest abstract level, engineering can be classified into *forward engineering* (FE) and *reverse engineering* (RE). On the one hand, forward engineering is rather traditional and the engineering design process begins with a high-level conceptual design, gradually reduces to low-level detail designs, and finally to physical implementation. On the other hand, RE begins with examination of physical objects at a low level and gradually moves up to a high-level virtual representation of a product model with embedded knowledge. RE aims to develop a virtual computer model for an existing physical part, component, subassembly, or product when the computer models of artefacts are not available to engineers. In general, RE is a system analysis methodology to (i) create computer models for existing physical objects, (ii) explore knowledge from reversely engineered models, and (iii) identify the components and their relations to existing products or systems. RE is an effective method to re-use knowledge by others. Since today's manufacturing enterprises are highly pressured into making new products in a shortened development time, RE has gained a great deal of popularity in recent years.

FE or RE is involved at the development stage of products in order for designers to gain more understanding of products in terms of wisdom, data, information, and knowledge. As shown in Figure 5.1, FE begins with design ideas and concepts with abstract information and terminates with implemented products with the detailed information of products. In contrast, RE begins with physical products with little information of its designs and may terminate at any stage when the expected knowledge of product designs is obtained. RE is to recover the authentic knowledge of product design through an RE process. Note that the information accumulated in an RE process may be distorted, incomplete, or extended in comparison to the knowledge embedded in physical objects.

Due to the rapid development of information technologies (IT), RE has been widely used in many fields such as *medical applications*, *militaries*, *software engineering*, *entertainments*, *automotive industries*, *consumer products*, and *mechanical systems*. RE is used as an effective tool to enable competitors to shorten technological gaps with opponents. When a new product becomes accessible, competitors tend to use RE in order to (i) explore the know-hows of the product, (ii) duplicate and improve the design, and (iii) develop the upgraded version of the existing product.

Computer Aided Design and Manufacturing, First Edition. Zhuming Bi and Xiaoqin Wang.
© 2020 John Wiley & Sons Ltd. This Work is a co-publication between John Wiley & Sons Ltd and ASME Press.
Companion website: www.wiley.com/go/bi/computer-aided-design

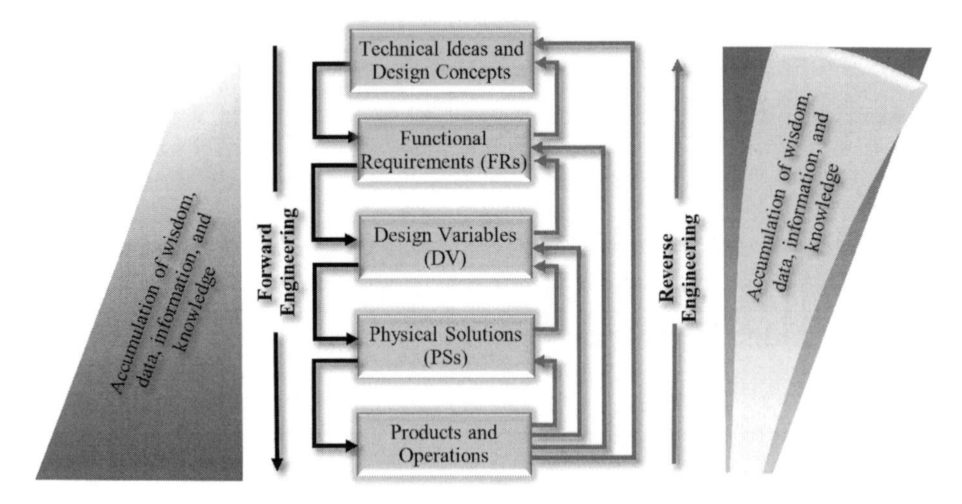

Figure 5.1 Forward engineering (FE) and reverse engineering (RE). (*See color plate section for color representation of this figure*).

As shown in Figure 5.2, RE consists of *a part-to-CAD process* and *a CAD-to-part process*. The application fields in medical fields, construction, archaeology, and consumer products usually have products with free-form surfaces. Taking as an example the process of designing a consumer product (Acadtechnologies 2019), an original design of a part geometry may be created from clay, rubber, plaster, or wood. External surfaces are scanned

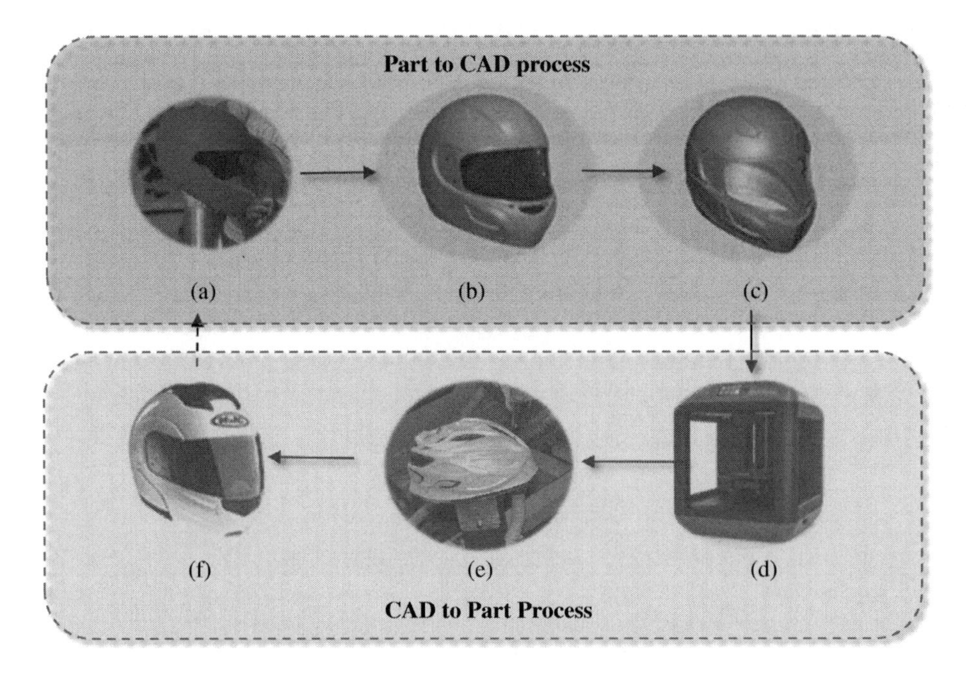

Figure 5.2 Part-to-CAD and CAD-to-part processes in RE. (a) Clay model. (b) Scanned points cloud. (c) Parametric model. (d) Rapid prototyping. (e) Tooling and manufacturing. (f) Physical product. (*See color plate section for color representation of this figure*).

to obtain points-cloud models and these models are further processed to create solid models. Following the part-to-CAD process, rapid prototyping can be used to prototype the part. In addition, the computer aided design (CAD) model can be utilized to design manufacturing tools such as a mould for injection moulding or a fixture for holding a part. The second part of RE is the CAD-to-part process.

We can greatly reduce the time of product development. In the intensely competitive global market, manufacturers are constantly seeking new ways to shorten the times to market of new products. Recent techniques such as so-called *rapid product development* (RPD) take the advantages of RE to assist designers in developing new products in a shortened lead-time. For example, injection-moulding companies are forced to drastically reduce development times for dies and tools. By using RE, the model of a part to be fabricated can be quickly captured in a digital form, and this model can then be used to generate tooling models, which can be used to prototype manufacturing tools sequentially.

With regards to the applications of RE in manufacturing, NPD (2019) gives a summary of exemplifying scenarios where RE is often adopted to meet companies' design needs as follows.

1. A product is discontinued by original manufacturers, but the company intends to reuse its design.
2. The documentation of the original product is incomplete for required duplication or upgrading.
3. The vendor of an original product no longer exists, but the product still has its commercial market demand.
4. The technical documents of products are lost or do not exist at all.
5. A product needs to be improved in order to eliminate some faulty features, such as parts that exceed stress concentration factors or wear in a machine.
6. The features of a product have to be identified, strengthened, and extended to product families.
7. A product made by a competitor is analysed to identify the weaknesses and strengths of the product and gain a competitive advantage of the company's own product.
8. A legacy product is investigated to inspire new avenues for innovations.
9. The supplier of original parts are unwilling or are incapable of providing parts continuously but the parts are essential for the company to make its own products
10. The suppliers of the original machine tools are unwilling to provide new machines or services continuously, but these manufacturing resources are essential to run manufacturing processes of host companies.
11. A company needs to update obsolete materials or antiquated manufacturing processes with more current, less-expensive technologies.

5.2 RE as Design Methodology

The relationship between FE and RE is that of chickens and eggs. The evolution of human being's civilization relies greatly on the accumulation of scientific knowledge, discoveries, inspiration, wisdom, experiences, and technologies, generation by generation. *Standing on the shoulders of giants* is one of Newton's famous quotations. It tells us to discover the truths by building on previous discoveries made by others (Wikipedia 2019a). RE is a design

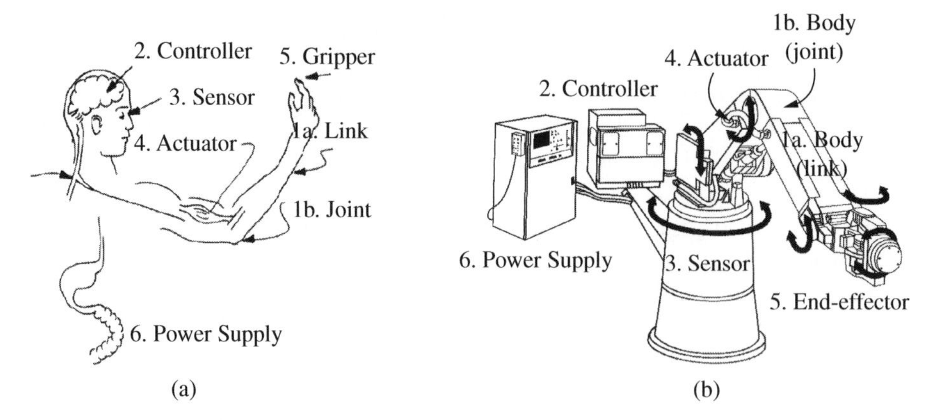

Figure 5.3 Correspondence of manipulator and human body. (a) Human being. (b) Manipulator.

methodology where a physical object is taken as the reference to explore how it works, how it can be improved, and how similar concepts could be adopted for other applications.

The things in nature are always the best source of inspiration. Numerous great inventions have been directly inspired by things in nature. Taking one example, the Wright brothers observed how birds fly and mimicked their wings, shapes, and flying modes in the first ever aircraft in 1903 (Spenser 2011). By reversing engineering practice, the body structures, and moving mechanisms of birds, men were able to build aeroplanes. The shapes of birds helped to address the conflicted design requirements such as aerodynamics, agility, strengths, stability, and lifting weight. Figure 5.3 shows a comparison of the main components of a human body and a manipulator as another example of RE. All of the main components in a manipulator, such as links, joints, controller, sensors, actuators, end-effector, and power supplies, are mimicked from counter-components in a human body.

More recent great inventions inspired by things in nature are illustrated by Hennighausen and Roston (2015) in Figure 5.4. The included examples are (i) the exterior design of a bullet train similar to a bird's long beak, (ii) a device to collect water like a desert fog, (iii) the exterior design of a car looking like a fish head, and (iv) a construction with a water-cube structure.

Historically, RE as a design methodology greatly progressed during World War II (WWII) and the Cold War. A few well-known examples of RE in history are (i) Jerry cans reversely engineered by British and American forces from gasoline cans made by the Germans, (ii) the Tupolev Tu-4 reversely engineered by the Soviets based on American B-29 bombers in 1944, (iii) the PGM-11 Redstone rocket reversely engineered by the American and Western Allies based on the V2-rocket by the Germans, and (iv) the K-13/R-3S missile reversely engineered by the Soviets based on the AIM-9 Sidewinder by the USA in 1958 (Srivastava and Mishra 2015). In these successful cases, RE was used to analyse the design of an existing device or system. The objective was to gain an understanding of physical objects, since this was a prerequisite for duplication, upgradation, or improvement.

Other than a part-CAD process and a CAD-part process, another relevant concept of RE is *digital modelling*. In general, digital modelling specifically refers to the procedure of generating a digital representation of an object by computer technologies such as

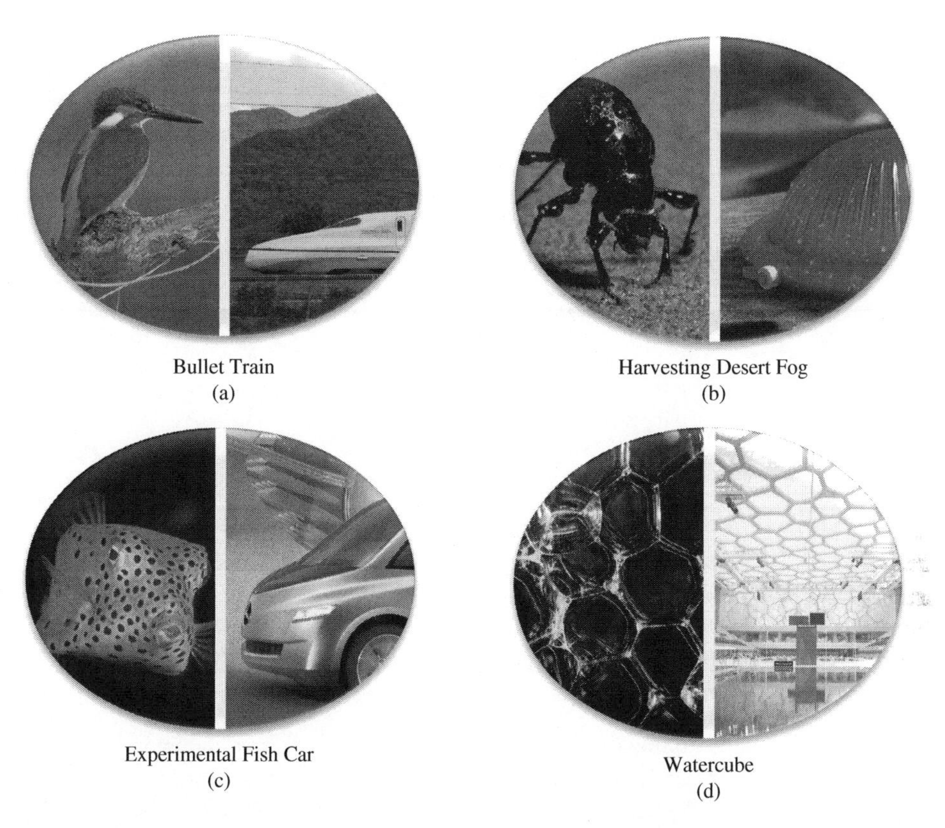

Figure 5.4 Example of inventions inspired by nature (Hennighausen and Roston 2015). (a) Bullet train. (b) Harvesting desert fog. (c) Experimental fish car. (d) Watercube.

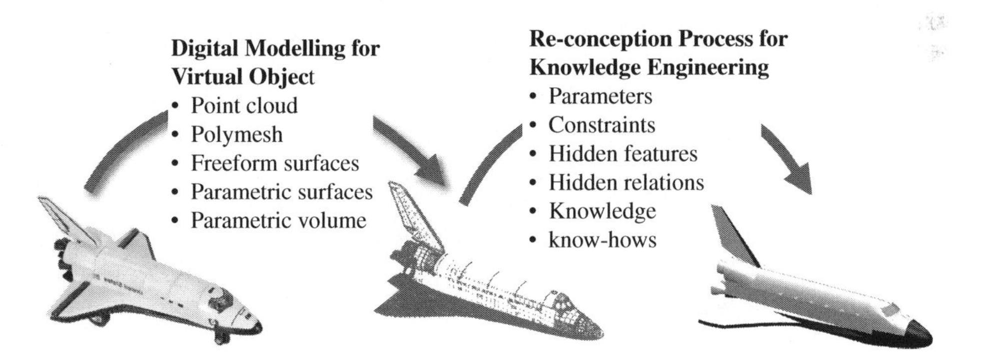

Figure 5.5 Digital modelling and re-conception process in RE (Bernard et al. 2010).

three-dimensional scanning. As shown in Figure 5.5, digital modelling is often an essential part of RE used to obtain virtual object models in the formats of point clouds or parametric surfaces (Bernard et al. 2010). The tasks involved in an RE project go far beyond digital modelling. For example, RE requires application of knowledge-based engineering methods to reason and explore hidden features, relations, and design know-hows of products.

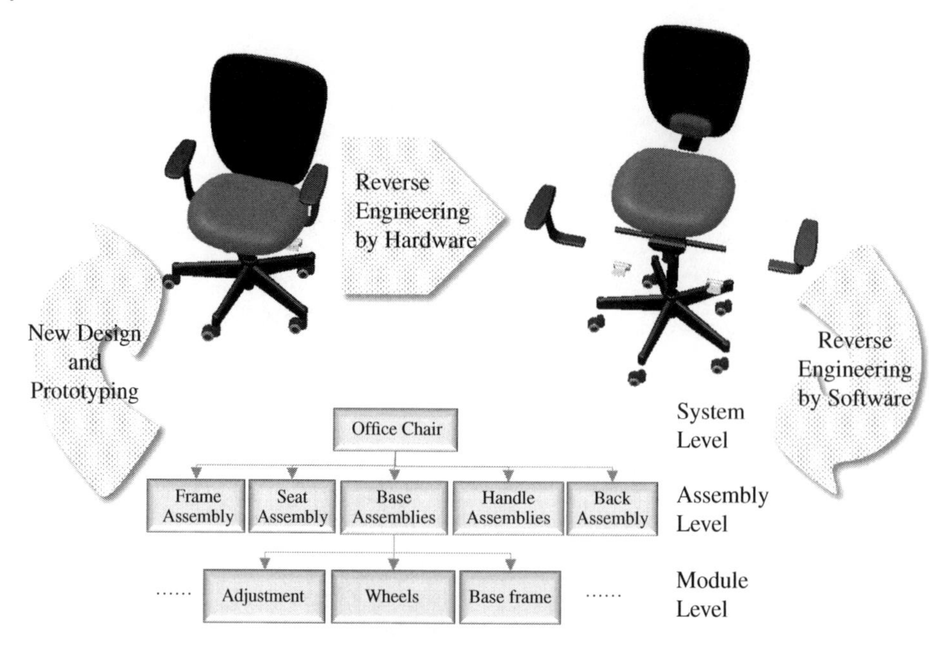

Figure 5.6 Example of reverse engineering at different levels.

The role of digital modelling in RE is illustrated by the RE example shown in Figure 5.6. In reversely engineering an office chair, RE begins with digital modelling by (i) disassembling the product into parts and acquiring point clouds using the hardware system and (ii) processing the point clouds for low-level geometric models of parts. After digital modelling, the reconception process of RE follows to generate the high-level parametric models of parts and explore the product structure and assembly relations of parts.

RE is capable of capturing and reusing knowledge from physical objects. However, RE can be cost-effective only when a product to be engineered shows its potential return on investments (ROI). Moreover, ROI can generally be reflected by demanded quantities of reproduced products on the market. Therefore, cost analysis and life-cycle analysis are needed to justify the necessity for an RE project.

In the rest of the chapter, RE of mechanical parts is of the most interest. Especially, we focus on *computer aided reverse engineering* (CARE) techniques and tools for (i) the generation of virtual solid models from existing physical objects and (ii) the identification of parameters, constraints, and features for virtual solid models.

5.3 RE Procedure

An RE project begins to acquire point cloud data from the boundary surfaces of an object. Three-dimensional sensors, such as the coordinate measuring machine (CMM), laser scanners, and computed tomography (CT) or cameras, are typical hardware devices to perform

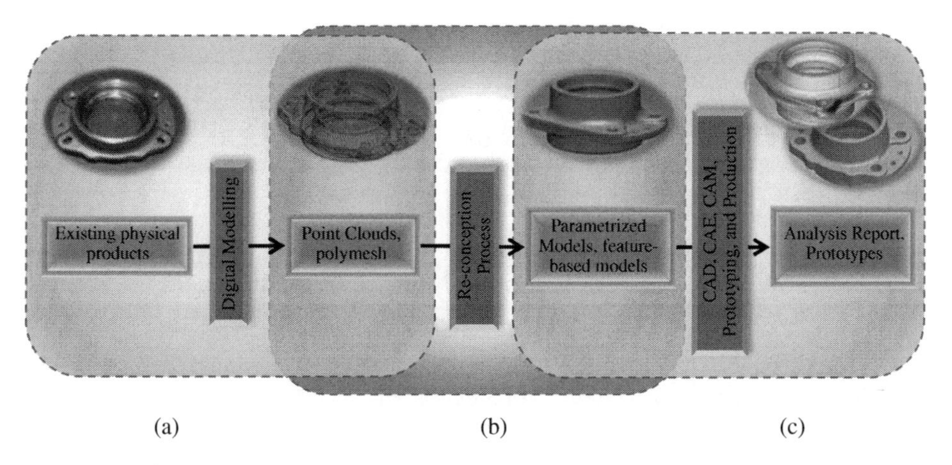

Figure 5.7 Three phases of RE. (a) *The first phase* to create point cloud or polymesh model from existing objects. (b) *The second phase* to create parametrized models for point cloud or polymesh. (c) *The third phase* to utilized an RE model for CAD, CAE, CAM, and prototypes.

such tasks. With the point clouds, the next step is to create a polymesh, which may consist of a number of parametric surface patches. A polymesh is then cleaned and further exported to a CAD package for parametric modelling.

As shown in Figure 5.7, an RE project can be divided into three phases, i.e. *digital modelling, re-conception process*, and *CAD/CAE/CAM and prototyping*. Digital modelling is at the first phase. It is the process of creating a computer model for the physical object that exactly replicates the form of the object. A data acquisition hardware system such as a 3D scanner can be used to capture the 3D data of the object, and the data is then transferred to computing resources where the data sets from different sources are cleaned, filtered, fused, and processed to create polymesh models. A re-conception process is at the second phase. It uses knowledge-based engineering approaches to identify the features and develop parametrized models of solids. The third phase relates to the application of RE. The solid models from the re-conception process can be fully utilized to support computer aided engineering (CAE), computer aided manufacturing (CAM), rapid prototyping, and even for new product development based on reverse engineered models.

5.4 Digital Modelling

Representing an object in its computer model is the first step to take advantage of the computer. However, a parametric model is not always available to represent objects in many cases. For example, (i) archaeological objects, such as pottery vessels and ancient weapons, have existed for hundreds even thousands of years before computers came into our lives; (ii) creatures and natural objects grow up naturally without digital models; (iii) although an object is an artefact that is based on a computer model, undocumented changes occur in production or application, which leads to the difference between an as-designed model and an as-built object. In any of the aforementioned cases, digital modelling can be applied to capture data and represent physical objects virtually. Digital modelling allows designers to gain an understanding of physical objects through an analysis of structures, governing principles, functions, and operations.

5.4.1 Types of Digital Models

Depending on the stage of RE, reconstructed models of objects can be of different types, i.e. *point clouds*, *polygonal models*, and *meshes*.

Point clouds. The cleaned data from a data acquisition is usually represented in the form of point clouds. A *point cloud* is the set of points with measured coordinates, which are collected from visible surfaces of an object. A point is described by its X, Y, and Z coordinates on a surface of the object with respect to a reference coordinate system. A point cloud can be organized or unorganized depending on the type of data acquisition system. Figure 5.8 shows the three types of point clouds (Chen et al. 2018). The characteristic of point clouds could greatly affect the number of computations for a high-level of RE. For example, an organized point cloud can easily be converted into a mesh. Generally, it is desirable to convert an unorganized point cloud into an organized one by dividing the volume/surface into cells for object and mapping points into nodes or elements of cells (Bi and Kang 2014).

Polygonal models. These occur if an object is scanned to acquire data from visible surfaces. The object is firstly represented by unstructured three-dimensional data in the form of point clouds. The next step of RE is to convert point clouds into *polygonal models*. A polygonal model is also called a *mesh*, which is a faceted (or tessellated) model consisting of many triangles. In a mesh, the facets are formed by connecting points within a point cloud. A mesh model is of limited usage except for the purposes of visualization, verification, and initial inspections.

NURBS models. As shown in Figure 5.9, the mesh model from a point cloud can be converted into a parametric model called a *non-uniform rational basis spline* (NURBS) surface. NURBS surfaces are wrapped over the polygonal wire frames. The wrapped surfaces in the model are smoother than those of polygonal models. If a set of boundary surfaces in an

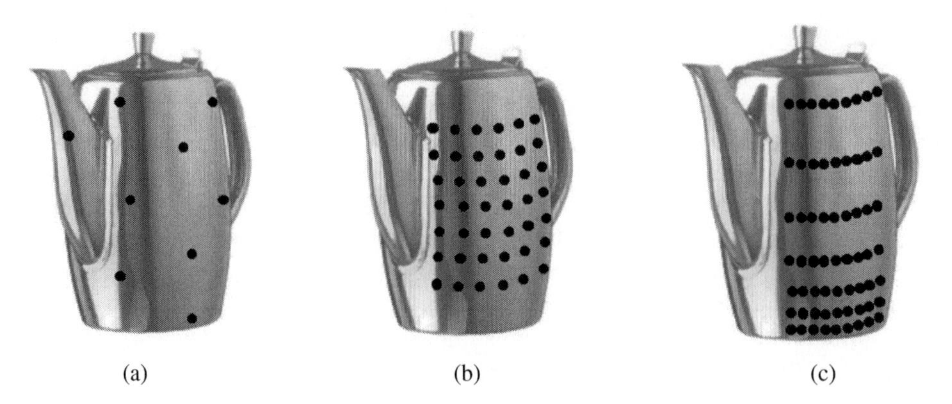

Figure 5.8 Three types of point clouds (Chen et al. 2018). (a) Unorganized. (b) Organized – grid type. (c) Organized – line type.

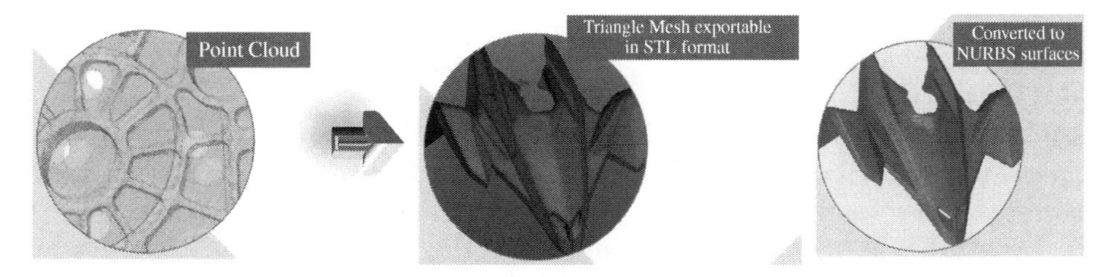

Figure 5.9 Steps in creating a parametric surface from a point cloud (Creaform3D 2019).

NURBS can be used to form a watertight volume, these surfaces can be knitted together to form a solid called a 'dumb' solid. The word '*dumb*' implies that the corresponding model has no history of parameterization.

Hybrid solid models. NURBS surface models can be fully exploited to identify the features and parametrize the dimensions and constraints of the corresponding solids. The resulting models with identified features and parameters have all the characteristics of a solid model from FE. However, such models sustain the information of all NURBS surfaces from which the features and parameters are derived. The models with both information of surfaces and solids are called hybrid solid models.

5.4.2 Surface Reconstruction

If the geometric representation of an object is based on the data acquired from boundary surfaces of the object, such an RE process for a surface model is called *surface reconstruction*. Hoppe (1994) formulated a surface reconstruction problem as: *given the partial information of an unknown surface, construct, to the extent possible, a compact representation of the surface*.

The surface reconstruction technique was traditionally applied in medical imaging, cartography, computer arts, and visualization of cultural and scientific heritages. Mijajlovic et al. (2004) developed a Matlab toolbox to analyse 3D images using unorganized point clouds. Rocchini et al. (2001) presented a suite of tools specially designed to reconstruct heritage images. Rebolj et al. (2008) introduced a system capable of monitoring a construction process through comparison of scheduled and as-build construction models. A vast number of works in manufacturing were for RE, where surface models were generated from physical objects (Várady et al. 1997; Thompson et al. 1999; Pernkopf 2005; Fisher 2002). Bi and Lang (2007), Biegebauer et al. (2002), and Vincze et al. (2002) introduced a sensor-based robotic painting system and part geometries were acquired by a vision system with the robotic programs generated by computers to perform painting operations.

5.4.3 Algorithms for Surface Reconstruction

Surface reconstruction from point clouds is fundamental to RE. Some comprehensive literatures on the algorithms of surface reconstruction were given by Gois et al. (2004), Schall and Samozino (2005), and Azernikov et al. (2003). For example, Azernikov et al. (2003) classified the methods of surface reconstruction into *computational geometry approaches* and *computer graphics approaches*.

A computational geometry approach focuses on the *piecewise-linear interpolation* of unorganized points. It defines a surface as a carefully chosen subset of the *Delaunay triangulation* in a Cartesian coordinate system. Cazals et al. (2004) discussed the approach of using the Delaunay triangulation for surface reconstruction in detail. Amenta et al. (2001) used the medial axis transform to approximate a surface with points. In the medial axis transform, an object was represented as an infinite union of balls, and an approximated surface was called a *power crust*. Bernardini et al. (1999) proposed the *Ball-Pivoting Algorithm* to reconstruct surfaces from point clouds. The basic idea behind the ball-pivoting algorithm was that three points form a triangle if a ball of a user-specified radius touched these points without an inclusion of other point(s). In execution, the ball-pivoting algorithm began with a seed triangle and progressed when the ball pivoted around an edge until it touched another point to form the next triangle. Abdel-Wahab et al. (2005) compared some computer geometry-based approaches including *the crust*, *the power crust*, *the tight cocone*, and *the ball pivoting algorithm* in terms of reconstruction quality, memory usages, and computing times. They concluded that the crust and power crust algorithms showed a balanced trade-off between the execution time and memory usage. On the other hand, the ball-pivoting algorithm was able to minimize the execution time and memory usage in contrast to the tight Cocone algorithm. In general, the computational cost of these algorithms depended on how efficiently the Delaunay triangulation was generated.

A computer graphic approach aims at the visualization quality of surface models from point clouds. A computer graphic approach does not emphasize the closeness of the interpolated surface with raw point samples. Hoppe et al. (1992), Hoppe (1994), and Neugebauer and Klein (1997) pioneered the development of computer graphic approaches using signed distance functions. In their approaches, an approximated surface was treated as a zero set of an estimated signed distance function. It was able to infer surface topology with boundary curves automatically. Freedman (2004) proposed an incremental technique to reconstruct surfaces. It did not require an embedding space of the surface. Thus, sizes of the embedding space might vary arbitrarily without substantially affecting the complexity of the algorithm. Zhao et al. (2001) used the variational and partial differential methods in surface reconstruction.

5.4.4 Limitations of Existing Algorithms

In comparison with applications in other sectors, manufacturing applications require a higher precision of CAD models than RE. Moreover, if the CAD models from RE are utilized for planning, scheduling, and controlling in production, the surface reconstruction process requires high efficiency and automation, so that the production system can respond to detected changes promptly. The aforementioned algorithms face a number of obstacles in meeting the requirements of high precision and automation.

Dataset complexity. The complexity of a dataset is measured by a number of the factors, such as sizes of point samples, the variety of embedded features, and uncertainties and accuracy of points. The existing algorithms have limitations in dealing with the complexity of datasets robustly. The size of a dataset is a primary concern, and affects the success of surface reconstruction. It is particularly true for cloud-based applications where a large amount of data has to be transferred and processed remotely. Engel et al. (1999) provided a trade-off solution to control the number of vertices that have to be reconstructed, transmitted, and rendered. If an RE project is too complex, it should be decomposed into a number of subprojects with manageable sizes of datasets.

Processing time. Different from RE applications in other sectors where the modelling time is not very critical, the operations in a production line are confined by the cycle time, since the production system might not be affected by the delay of a single operation. When RE is used in a manufacturing system, for example acquiring the part surface for surface treatment, the time for surface reconstruction must be less than the cycle time of the production line. The available algorithms are not effective enough to meet the requirements of processing times.

Automation. Machine intelligence relates to the capabilities of the machine to deal with uncertainties and changes in application. Sensors are essential components to detect part or geometric changes, so control programs can be adjusted to accommodate changes. If a change to be detected is part of the geometry, the computer aided system must be able to translate the point clouds into an acceptable CAD model without human intervention. In all of the discussed algorithms, considerable manual interventions are needed to reconstruct surfaces.

5.4.5 Data Flow in Surface Reconstruction

To reconstruct a surface, stereo cameras or laser sensors are often used to capture 3D range data from visible surfaces of an object. After a point cloud is acquired, it is processed to identify geometric changes for corresponding tasks in applications. The conversion from an acquired point cloud to a geometric model of a part is a complicated process. Figure 5.10 shows some typical tasks in the conversion.

- *Data acquisition.* A physical object is scanned to capture point clouds from visible surfaces. When an object is large or complex, the data from multiple views and segments must be acquired to get sufficient information over surfaces.

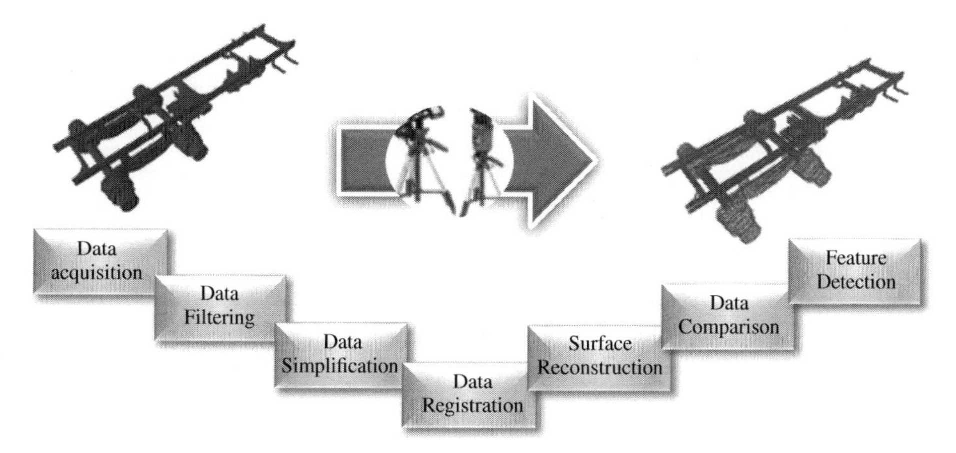

Figure 5.10 Critical tasks from data acquisition to reconstructed surface/model (Bi and Kang 2014).

- *Data filtering*. Raw data includes noise, distortion, and invalid data caused by hardware systems and/or the environment. The raw data must be filtered to remove unwanted and noise data.
- *Data registration and integration*. A data acquisition system such as a laser scanner is capable of collecting points from a surface visible to the scanner. Multiple views or continuous paths are needed to acquire the data on the entire surface. Therefore, the datasets from different views must be integrated to model whole surfaces. The datasets are obtained with respect to different references. The registration is used to determine the transformation of the data from two different views so that the data can be integrated under the same coordinate system. Integration is the process of creating a single surface representation from two or more range images.
- *Surface reconstruction*. By using a raw point cloud or volumetric data from an unknown surface, surface reconstruction generates a surface model that approximates the actual surface on an object.
- *Data simplification and smoothing*. Raw data from a scanning process includes a large amount of data. Data simplification and smoothing reduce the amount of data without affecting the quality of a reconstructed surface. Data simplification can also accelerate engineering analysis, where geometry and shape play a major role, such as finite element analysis, collision detection, visibility testing, shape recognition, and visualization. Simplification aims at high efficiency for storage, transmission, computation, and visualization.
- *Data comparison*. RE is occasionally used to detect the discrepancy between a virtual model and a physical object. A reference virtual model can be available in a data comparison. A data comparison is used to identify the difference between an actual physical object and a reference virtual model. Data comparisons can be applied in alignment, inspection, surface control, or CAD compare. For example, (i) in feature detecting, point clouds can be used to measure geometric elements like a plane, cylinder, circle, sphere, and boundaries. (ii) In monitoring and controlling, as-designed and as-built models are compared so that the deviation (average error), tolerance, and distribution can be evaluated.
- *Feature detection*. Features are basic elements of CAD models, especially of products from machining processes. Examples of the features are edges, holes, bosses, dimensions, locations, and orientations. Feature detection is used to identify certain features or validate if a certain dataset includes a specified feature.

5.4.6 Surface Reconstruction Algorithm

In this section, an algorithm for surface reconstruction is introduced as an example (Bi and Kang 2014). *Surface reconstruction* is defined as a surface U from a set of unorganized data points $P = \{p_1, p_2, \ldots, p_n\}$, which are assumed to be on or close to this surface. Hoppe's algorithm is used as the benchmark algorithm. The key task is to estimate the scalar distance of an arbitrary point $p \in R^3$ to the approximated surface U. The distance is signed and defined as $d_u(p) = s(p) \cdot d(p, U)$, where $s(p)$ denotes the side of the surface where p lies.

Knowing the signed distance function d_u is equivalent to knowing the surface U. Therefore, the algorithm is to estimate d_u from the data points and then to extract an approximation of its *zero set*. The estimated scalar distance $\hat{d}_U$ is associated with an oriented plane with each of the data points. At each data point p_i, its *oriented plane* is the plane $T_p(x_i)$ tangent to the linearly approximated surface at this point. The tangent plane is represented as a centre point o_i together with a unit normal vector $\hat{n}_i$. The scalar $\hat{d}_U$ of an arbitrary point $p \in R^3$ to $T_p(x_i)$ is defined by $d_i(p) = (p - o_i) \cdot \hat{n}_i$.

The centre and normal of $T_p(x_i)$ are determined by gathering together the group of points of P within the distance $\rho + \delta$ of p_i (where are ρ and δ are the parameters estimating the sampling density and noise). This set is denoted by $N_{bhd}(x_i)$ and called the neighbourhood of p_i. The centre o_i is taken to be the centroid of $N_{bhd}(x_i)$, and the normal $\hat{n}_i$ is determined using principal component analysis. After $\hat{d}_U$ is defined for all points in the area of interest, a *contouring algorithm* is applied to extract an approximation to $Z(\hat{d}_U)$ in the form of a mesh.

The Hoppe algorithm illustrated the limitations of the following aspects:

1. The underlying surface must be a manifold. However, the acquired surfaces in many applications may fail to meet this assumption.
2. The underlying surface should satisfy the linearization, which depends on the parameters of ρ and δ. However, ρ and δ are specified based on the density and noise of the given samples, but the density and noise vary greatly in data acquisition. It is difficult or even impossible to specify one value of ρ or δ that is appropriate to the whole point cloud.
3. Computing is highly demanded. For each data point, the tangent plane has to been determined from a set of neighbouring points. This requires determining the centroid of the neighbouring set and the normal using principal component analysis.

Note that the key idea of the surface reconstruction is to define a scalar function to represent the distance of an arbitrary point to the underlying surface. To overcome the aforementioned limitations, the Hoppe algorithm is modified to reconstruct the surface from a point cloud. Figure 5.11 shows the procedure of surface reconstruction, Figure 5.12 shows the impact of a data point on neighbouring nodes, and Table 5.1 gives an explanation of critical tasks involved in surface reconstruction using a lookup table (Table 5.2).

The algorithm proceeds through the calculation of the scalar field, taking each cell with eight nodes at a time, and then calculates the polygon(s) to represent a part of the iso-surface that passes through this cube. Individual polygons are then fused into an approximated surface.

In applying a contouring algorithm, an index to a pre-calculated array of 256 possible polygon configurations ($2^8 = 256$) within the cube is pre-defined. The value after all eight scalars are checked is the actual index to the polygon configuration array. If the index belongs to one of 16 cases in Figure 5.13, each vertex of the polygons is determined on the cube's edge by linearly interpolating the two scalar values that are connected by that edge.

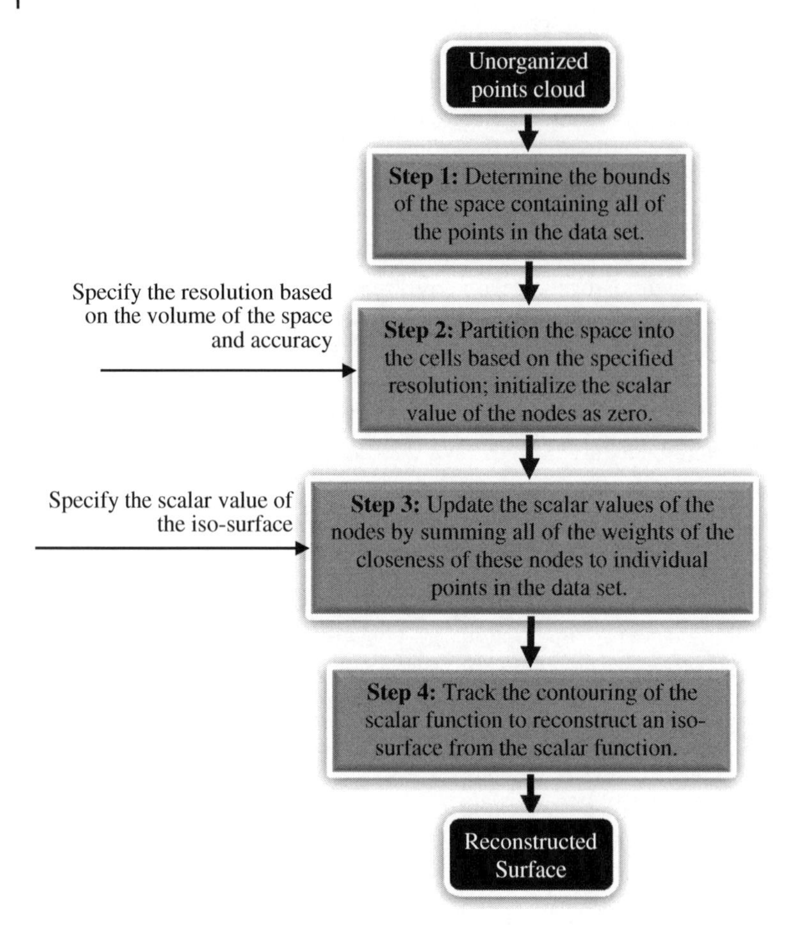

Figure 5.11 The procedure of surface reconstruction.

By directly using the space partition and scalar values obtained in step 1, the calculation for contouring tracking can be reduced significantly.

5.4.7 Implementation Examples

The proposed algorithm in Section 5.4.6 was implemented with an open-source Insight Segmentation and Registration Toolkit (ITK) and Visualization ToolKit (VTK) package (MeVisLab 2019). The Hoppe algorithm and power crust algorithm were also implemented as benchmarking algorithms. Note that the power crust algorithm for 3D surface reconstruction was based on the medial axis transform and was guaranteed to produce a geometrically and topologically correct approximation to the surface when the data is sufficiently dense. However, sufficient density is very hard to achieve in practice. The tested datasets except the last one were downloaded from the Hoppe website (Hoppe 2009). Table 5.3 shows the reconstructed surfaces from three algorithms.

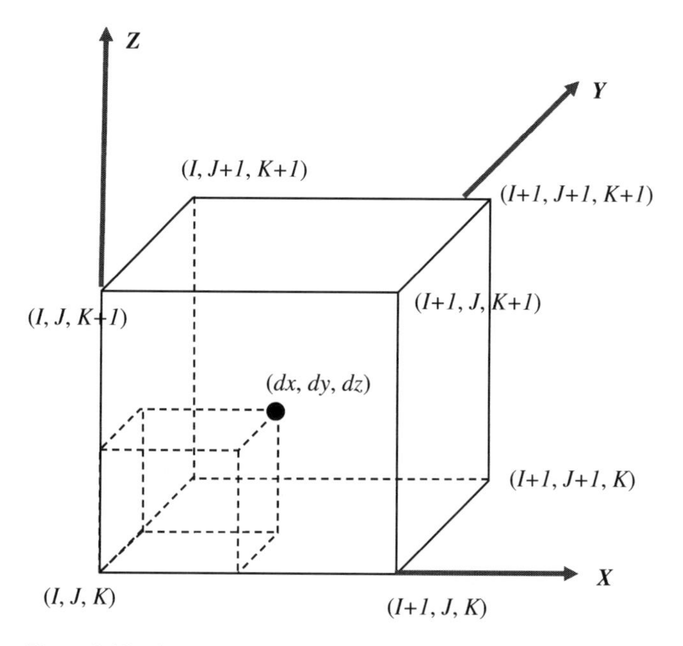

Figure 5.12 A data point and its weights on the neighbouring nodes.

Table 5.1 Critical activities in a surface reconstruction algorithm (Bi and Kang 2014).

Step	Critical activities
1	A dataset of unorganized points is obtained to determine the finite spatial volume to include all of the points. This volume is bounded along three axes of the Cartesian coordinate system.
2	It is impractical to calculate the scalar distance of all points to the boundary surface U to be determined in the defined space. To reduce computing, the space partition in finite element analysis (FEA) is adopted where the space is decomposed as a set of the finite cubes. As a result, only the scalar distance between these nodes to the surface U needs to be calculated.
3	The scalar distance is defined by mapping the points of the data set to the node based on the distance of each point to the concerned node. As shown in Figure 5.12, to minimize the calculation, the weighted function is applied where the weights are determined by the lookup Table 5.2. One point has its mapped weights only on the nodes of the cell where this point is located. The weights to the nodes of any other cells are zero. The scalar value corresponding to each point is actually a measure of the probability of the node belonging to the underlining surface. The higher the value, the higher is the possibility that this node is on or near the surface. From this point of view, this value is always larger or at least equal to zero. Different from the distance function in the Hoppe algorithm, no sign is required for the scalar value based on probability.
4	Once the scalar functions over the divided space are determined, the contouring algorithm is applied to generate an iso-surface. The problem of extracting an iso-surface is well studied. By applying the modified approach, the space has been partitioned and the scalar functions at the nodes have been evaluated. These results can be directly applied in the contouring tracking.

Table 5.2 Lookup table for determination of the weights.

Nodes	Weights			
	x-axis	*y*-axis	*z*-axis	Total
(I, J, K)	$(1-dx)/12$	$(1-dy)/12$	$(1-dz)/12$	$(3-dx-dy-dz)/12$
$(I, J, K+1)$			$dz/12$	$(2-dx-dy+dz)/12$
$(I, J+1, K)$		$dy/12$	$(1-dz)/12$	$(2-dx+dy-dz)/12$
$(I, J+1, K+1)$			$dz/12$	$(1-dx+dy+dz)/12$
$(I+1, J, K)$	$dx/12$	$(1-dy)/12$	$(1-dz)/12$	$(2+dx-dy-dz)/12$
$(I+1, J, K+1)$			$dz/12$	$(1+dx-dy+dz)/12$
$(I+1, J+1, K)$		$dy/12$	$(1-dz)/12$	$(1+dx+dy-dz)/12$
$(I+1, J+1, K+1)$			$dz/12$	$(dx+dy+dz)/12$

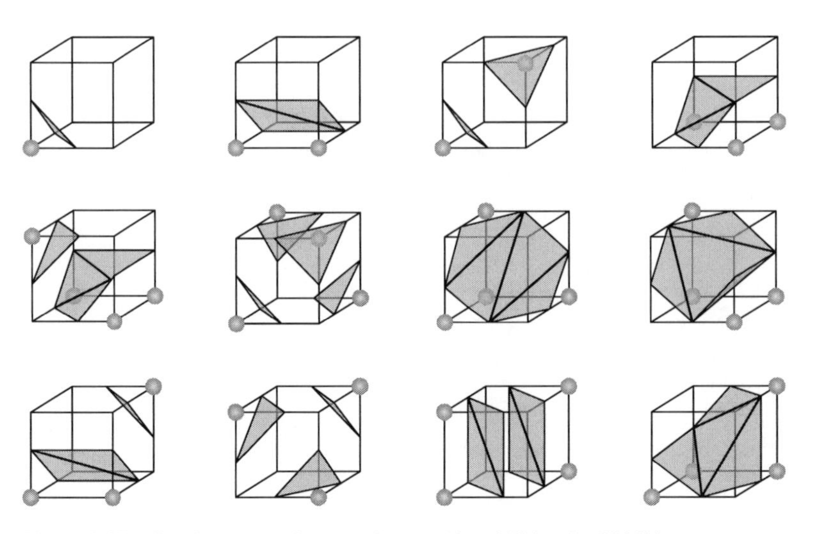

Figure 5.13 Special cases of contouring tracking (Wikipedia 2019b).

5.5 Hardware Systems for Data Acquisition

The implementation of an RE project requires both of hardware and software systems. *A hardware system* acquires point clouds or volumetric data from existing objects. Different working principles or physical phenomena are used for hardware systems to interact with the surface or volume of objects. *A software system* processes point clouds and transfers the dataset into computer models of objects.

Over decades, the extensive use of digital modelling techniques has yielded enormous economic and quality benefits to manufacturing companies. However, the pace of leveraging digital technologies has yet to catch up with the high demands for reconstructing computer models for complex objects with a close tolerance. Some desirable characteristics of hardware systems are high speed, high accuracy, large working volume, the ability

Table 5.3 Comparison of three algorithms on different point clouds (Bi and Kang 2014).

Point cloud data	Surface reconstruction algorithm	Power crust surface reconstruction	New algorithm

Table 5.3 (Continued)

Point cloud data	Surface reconstruction algorithm	Power crust surface reconstruction	New algorithm
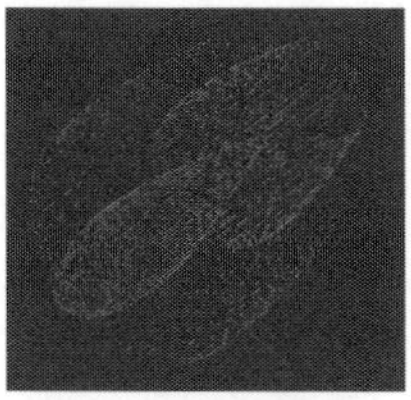	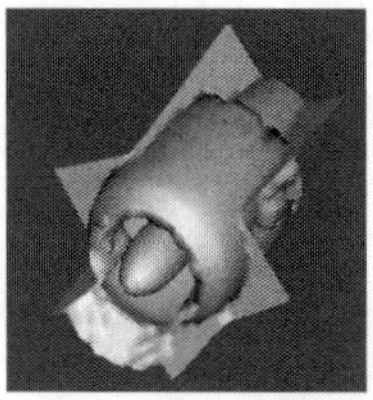		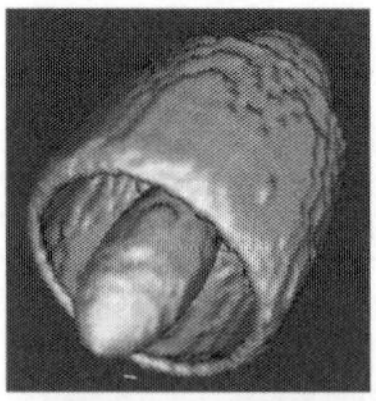
	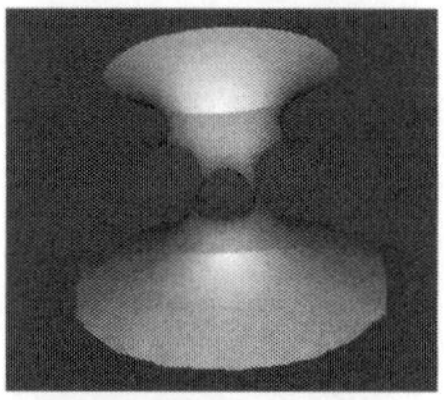	No result in a reasonable time	
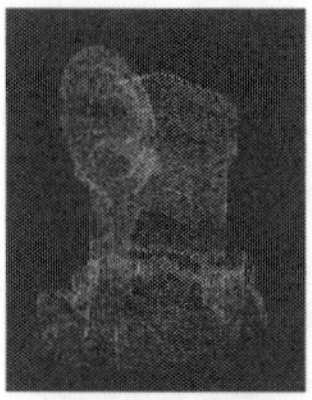	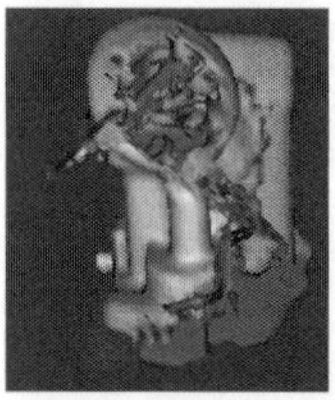	No result in a reasonable time	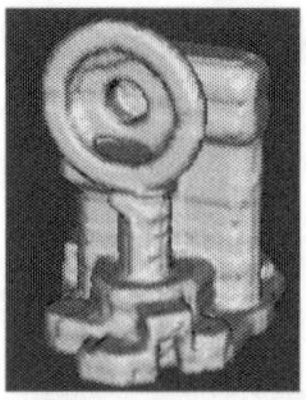
	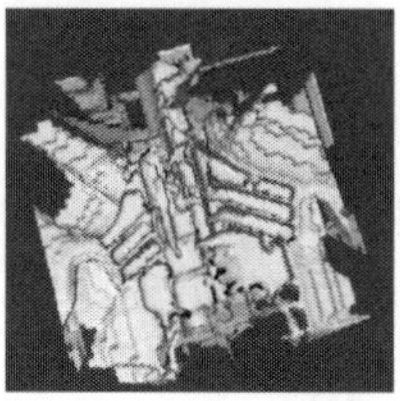	No result in a reasonable time	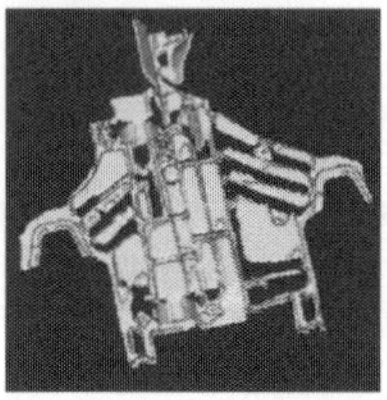

to measure features on simple or complex parts, the ability to measure parts with a wide scope of surface finishes from freshly machined to dark paints, the operability in ambient factory conditions, and the support to real-time data analysis (Johnston 2006).

Many types of data acquisition systems are commercially available. While there exists no universal hardware system that is capable of handling all of the RE projects in different applications, the selection of a hardware system needs to take into consideration many factors including accuracy, speed, working volume, reliability, and cost. In this section, the hardware systems used to collect point clouds for boundary surfaces of an object are of interest. Different physical phenomena are utilized to create the point clouds for surfaces and faces of parts.

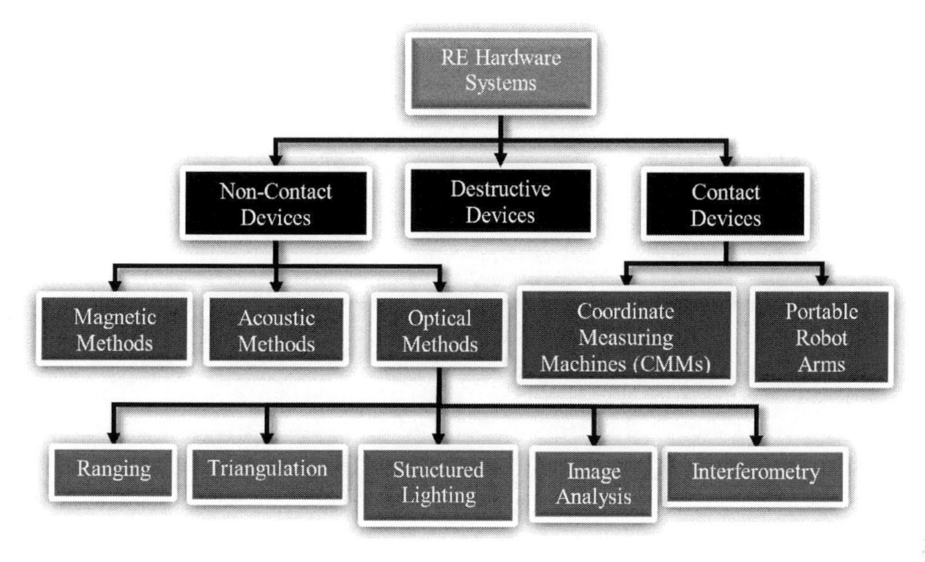

Figure 5.14 Classification of hardware systems for data acquisition.

5.5.1 Classification of Hardware Systems

Peng and Sanchez (2017) classified data acquisition devices in terms of the technical principle a device applies. As shown in Figure 5.14, a device can be non-contact, destructive, and a contact one. A *contact device* feels an object by a mechanical probe on the surface. A *non-contact device* uses a medium such as light, ultrasound, or magnetics to collect the geometric data. A *destructive device* acquires the volumetric data slide by slide. Therefore, it destructs the physical object for data acquisition. In any case, a mathematic model is needed to determine the coordinates of points on the surfaces of an object. In manufacturing, non-contact devices are more prevalent than contact devices due to efficiency and flexibility.

A destructive or contact device makes physical contact with an object to collect information of positions point by point. Figure 5.15 shows the structures of destructive or contact devices. Table 5.4 lists a few commercial products as well as main specifications.

A destructive or contact device usually possesses high accuracy of geometric shapes of objects. These devices are typically used in engineering projects and first article inspections, especially when a target object has some important features such as lines, planes, or curves. Figure 5.16 shows a scenario where a contact device is used to digitize a physical object. It only acquires a finite number of the points collected on the grid surface and the dataset are organized, which can be directly converted to a mesh with a high efficiency. A contact device is a non-destructive technology that captures the shape of physical objects using a surface contact roller. The idea was conceived decades ago but was not commercialized quickly due to a slow transmission speed and small storage capacity. The first handheld 3D scanner was developed in the 1980s, used an inaccurate contact probe, and took weeks to complete a simple scan. It was not until the 1990s that the optical technology had caught up, allowing the scanning of fragile objects and colour scans.

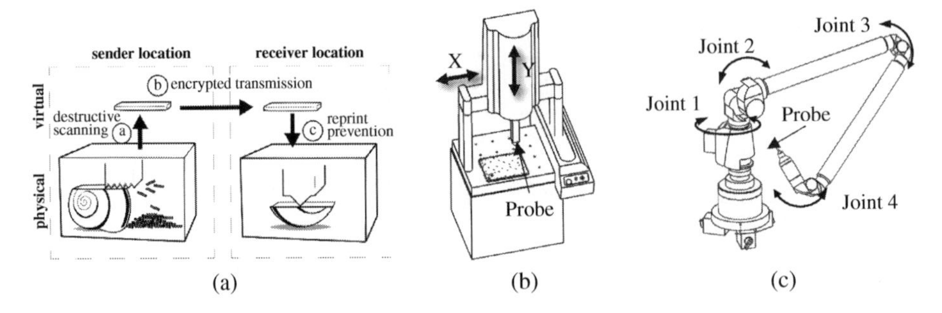

Figure 5.15 Examples of destructive and contact devices. (a) Destructive scanning (slicing) to keep uniqueness (Mueller et al. 2015). (b) Conceptual view of a CMM. (c) Faro arm–platinum articulated arm with mechanical probe.

Table 5.4 Typical contact devices for reverse engineering (Pham and Hieu 2008).

Methods	Supplier	Products	Range (mm)	Precision/ resolution	Automation
Mechanical arm with touch-probe	Faro Technology	FaroArm Advantage	1200–3700	±0.090 ~ 0.431 mm	Manual operation
		FaroArm Platinum		±0.018 ~ 0.086 mm	
	Immersion Corp.	MicroScribe MX	<1270	~ 0.1016 mm	
		MicroScribe MLX	<1670	~ 0.1279 mm	
CNC with a scanning probe	Roland DGA Corp	Picza PIX-30	305 × 203 × 60	Pitch resolution of 0.050 mm along X and Y axes and 0.025 mm along Z axis	Programmable
		MDX-15	150 × 100 × 60		
		MDX-20	200 × 150 × 60		
CMM with touch probe	Mitutoyo	Euro-C-121210	1205 × 1205 × 1205	Resolution of 0.001 mm along X, Y, and Z axes	
CNC or CMM with scanning probe	Renishaw Inc.	Renscan 200	Depend on CNC or CMM	Resolution of 0.001 mm along X, Y, and Z axes	

A non-contact data acquisition system can be *passive* and *active*. *A passive system* works on naturally reflected light from physical objects. Such a system does not use an additional power source in sensing an object. On the other hand, an *active system* emits signals in a certain form, such as structured lights or sonar pulses on an object. The coordinates of points on the surface are calculated using the method of time-of-flight (ToF).

In general, a passive system uses *shape-from-shading*, *shape-from-motion*, or *passive-stereo-vision* to acquire 3D coordinates of points. A shape-from-shading method uses a

Figure 5.16 Meshed points from a contact device. (a) Physical object. (b) Meshed points. (c) Contact RE device.

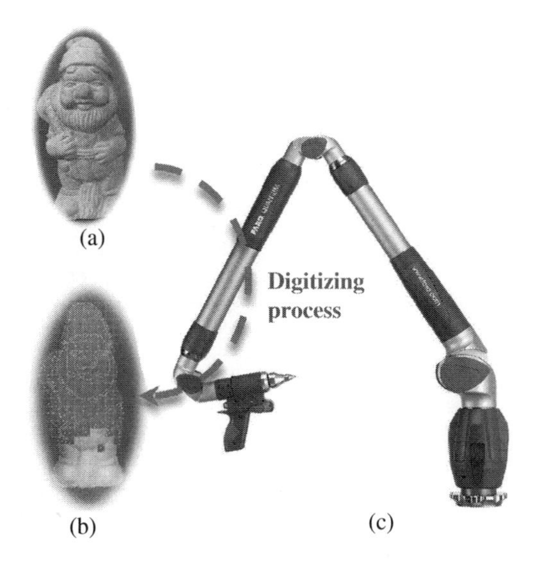

single camera on an object under different lighting conditions. By studying the changes in brightness over a surface and deploying the constraints in the orientation of surfaces, the depth information of an object can be determined. A shape-from-motion method moves the relative position of the object and camera to create a set of images in a sequence of motions. A *passive-stereo-vision* method uses two or more cameras to observe an object and determines the coordinates of points by triangulation.

A passive system does not require external energy sources. It is less sensitive to environment and is applicable for a portable vision platform. However, shape-from-shading and shape-from-motion methods have their limitations in acquiring 3D data due to three reasons: (i) the sensitivity to the illumination and surface reflectance properties of an object, (ii) the incapability to cope with non-uniform surface textures, and (iii) the difficulty of inferring an absolute depth. A passive-stereo-vision method faces the challenge of finding the correlations of pixels in two images. The correlation is made by extracting and matching features in two images. However, both the feature extraction and matching are complex and computationally intensive, and a depth map may not be generated in an acceptable period of time.

An active system needs an additional power source to emit signals and a receiver to get responding signals from objects. Incoming signals can be transmissive and reflective. A *transmissive signal* is generated when an incoming signal on one side passes through the solid of an object on the other side. As shown in Figure 5.17a, an industrial CT is a typical transmissive data acquisition device. *A reflective signal* is generated when an incoming signal is reflected back by an exterior surface of the object. In addition, an incoming signal can be *optical* or non-optical, such as *microwave*, *radar*, or *ultrasonic* (Peng and Sanchez 2017). Figure 5.17b and c shows examples of optical and non-optical devices, respectively.

For an active system, commonly used methods for calculating the coordinates of points are *triangulation*, *time-of-flight* or *laser pulse*, *interferometry*, and *photogrammetry*.

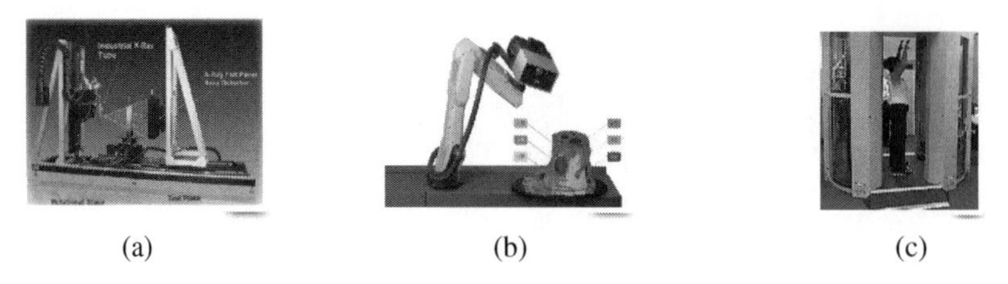

Figure 5.17 Examples of active data acquisition systems. (a) Transmissive sensing. (b) Reflective-optical sensing. (c) Reflective–non-optical sensing.

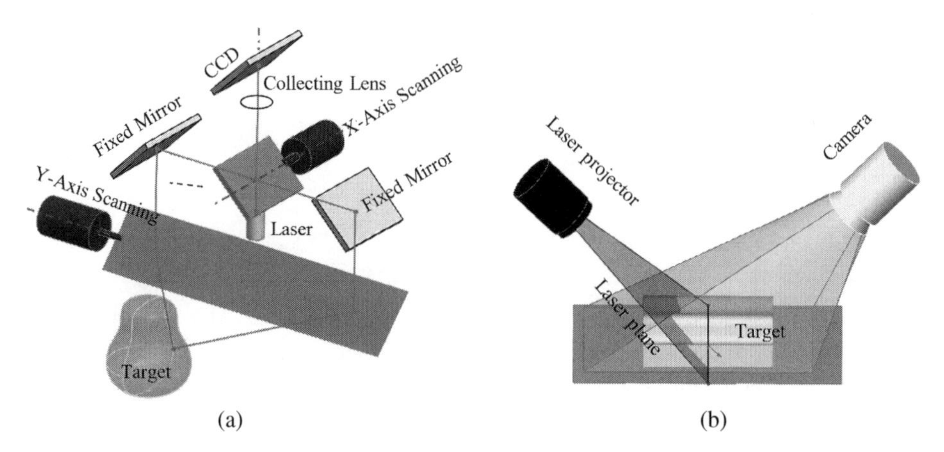

Figure 5.18 Single-point and line scanning methods (Boehler et al. 2002). (a) Single-point scanning. (b) Line scanning.

Triangulation is based on the spatial relations of three objects, i.e. *a target point* on object to be sensed, a *light source*, and a *detector*. The coordinates of the point are calculated based on the measured distances of the light source to the object, the object to the detector, and the light source to the detector. As shown in Figure 5.18, a triangulation-based sensor can be single-point scanning or line scanning. A *single-point sensor* acquires the range information point by point, whereas a *line scanner* projects a laser line to the object at once and the detection of a complete line profile of the object simultaneously. A line scanner must compromise between the field of view and depth resolution. In addition, a line scanner has relatively poor immunity to ambient light.

A *time-of-flight or laser pulse* scanner works on the ToF principle. The surface of object reflects the light back towards a receiver, the flight time or phase difference between the emitter and receiver is measured to detect the distance from the scanner to a point on object. Time-of-flight (ToF) refers to the travelling time from emission to reception.

As shown in Figure 5.19, the coordinates of a point with respect to a reference coordinate system is calculated from (i) the distance of the scanner to the point, (ii) the bearing (horizontal angle from a known line) to the point, and (iii) the vertical angle (angle from gravity) to the point. Since the speed of light (laser beam) is known and the time of travel of the laser beam can be measured, one can find the distance of the point from the scanner.

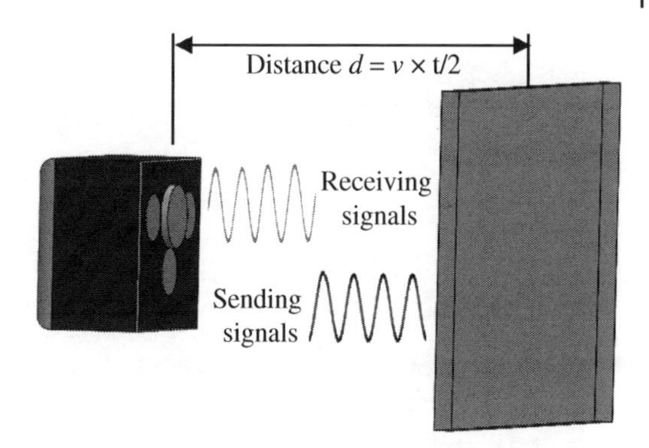

Figure 5.19 Using the time-of-flight concept to calculate distance.

Figure 5.20 Example of the time-of-flight laser sensor (Sick 2019).

A time-of-flight system requires very accurate timing; the precision of long-range applications is typically at millimetre or centimetre levels. Therefore, a time-of-flight system is mostly suited for the distance measurement in a medium to long range. For example, the Sick sensor in Figure 5.20 has a repeatability of 0.5–5 mm, a working range of 10–30 m, and a sampling rate of 3000 s^{-1} (Sick 2019).

An *interferometry* scanner uses lights with multiple stripes or patterns, which are projected on to an object simultaneously. Using two precisely matched pairs of gratings, the amplitude of the projected light is spatially modulated and the camera grating demodulates the viewed pattern to create interference fringes. The phases of interference fringes are proportional to the distance. Interferometry methods are mostly used for objects with relatively large flat surfaces and small depth variations. Although the accuracy for an equivalent depth of view is comparatively smaller than that of a line-scanning or single-point laser scanner, an incoherent light can be used to remove speckle noises associated with the laser. This results in a smoothness of data and the possibility of acquiring colour texture. Figure 5.21 shows an example of an image acquired by an interferometry scanner.

Image analysis reconstructs 3D points of an object and the coordinates of 3D points are determined by image analysis on 2D images taken on the object. The technique of reconstructing 3D points from 2D images is known as *photogrammetry*. Photogrammetry is the

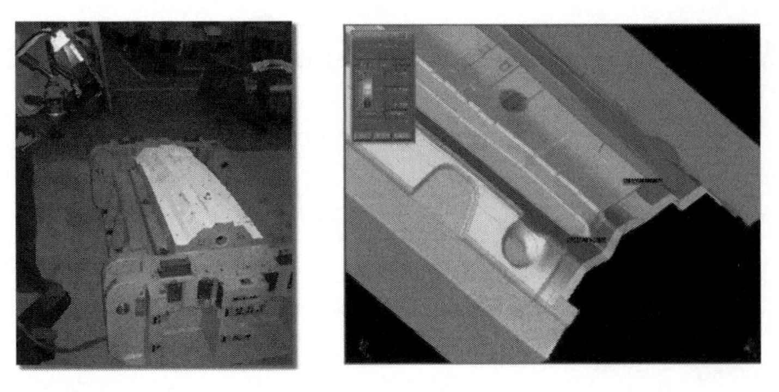

Figure 5.21 An image example acquired by an interferometry sensor (www.capture3d.com).

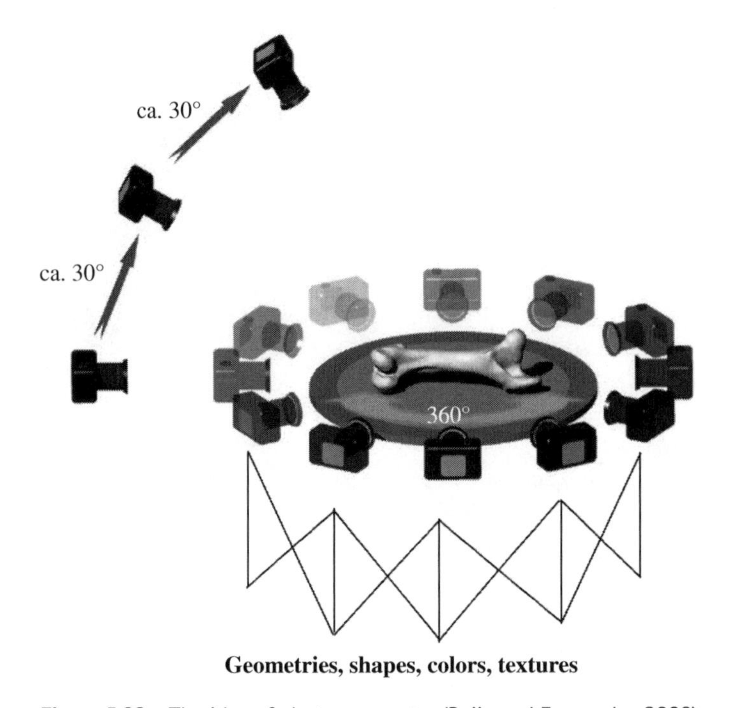

Figure 5.22 The idea of photogrammetry (Rajia and Fernandes 2008).

science of making the measurements from photographs to recover the exact positions of surface points. Photogrammetry is as old as modern photography (Rajia and Fernandes 2008).

Figure 5.22 shows the idea of photogrammetry (Rajia and Fernandes 2008). In contrast to 3D scanning, photogrammetry can capture not only geometric data, but also other data types such as colours, appearances, and textures. To calculate the coordinates of 3D points, the photography needs additional steps to compare and correspond to image pixels. Figure 5.23 shows the difference of data-acquisition processes between 3D scanning and photogrammetry. An additional step is needed by photogrammetry to derive point clouds from a set of 2D images.

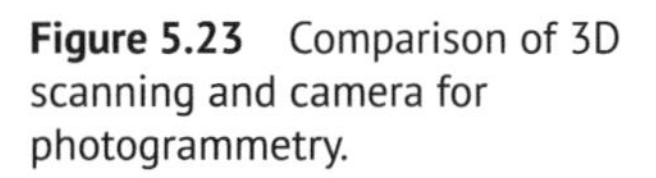

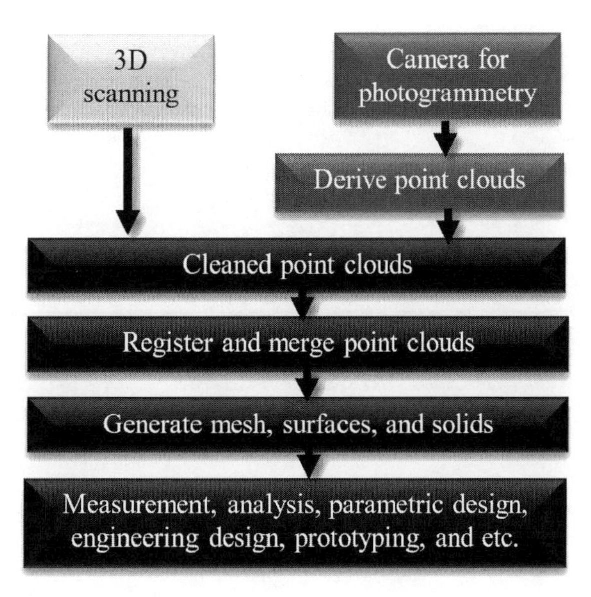

Figure 5.23 Comparison of 3D scanning and camera for photogrammetry.

As a summary, in choosing a hardware system for a specific RE application, critical design factors in selection are *precision*, *speeds*, *resolutions*, *measurement ranges*, *the influence of interfering lights*, and *the field of view*.

5.5.2 Positioning of Data Acquisition Devices

When a part size is large or the points on all visible surfaces in different orientations need to be detected, a positioning system is used to move the sensor or the object so that the data can be acquired from different views.

A positioning system include new variables, which affect relative positions of the scanner and the object. Therefore, the precision of the positioning system directly affects the accuracy of the point cloud. It is important to take a look at how an object is positioned relative to a scanner. Figure 5.24 show available positioning methods: *portable CMM scanning*, *tracked scanning*, *structured lighting*, and *portable scanning*.

Portable CMM scanning. A CMM or a measuring arm is equipped with a fixed probe or touch trigger probe. It is also possible to mount a 3D scanning head. A portable scanner is usually positioned by the encoders embedded in the driving motors. A probing tool can be mounted on the portable CMM, which makes it possible to integrate the scanning and probing processes for high efficiency. Note that a portable CMM is fixed but a physical arm moves the sensor around the object; the scanner can only work at a low speed to ensure good accuracy of the measurement.

Tracked scanning. An optical tracking device can be integrated to track a measurement device including a 3D scanner. An optical tracking device uses the markers (either passive or active ones) that are bonded with a scanner optically for positioning. A tracked 3D scanner is able to provide very good accuracy throughout the working volume. This avoids physical contact with the object to be measured. However, the tracker must have a clear and

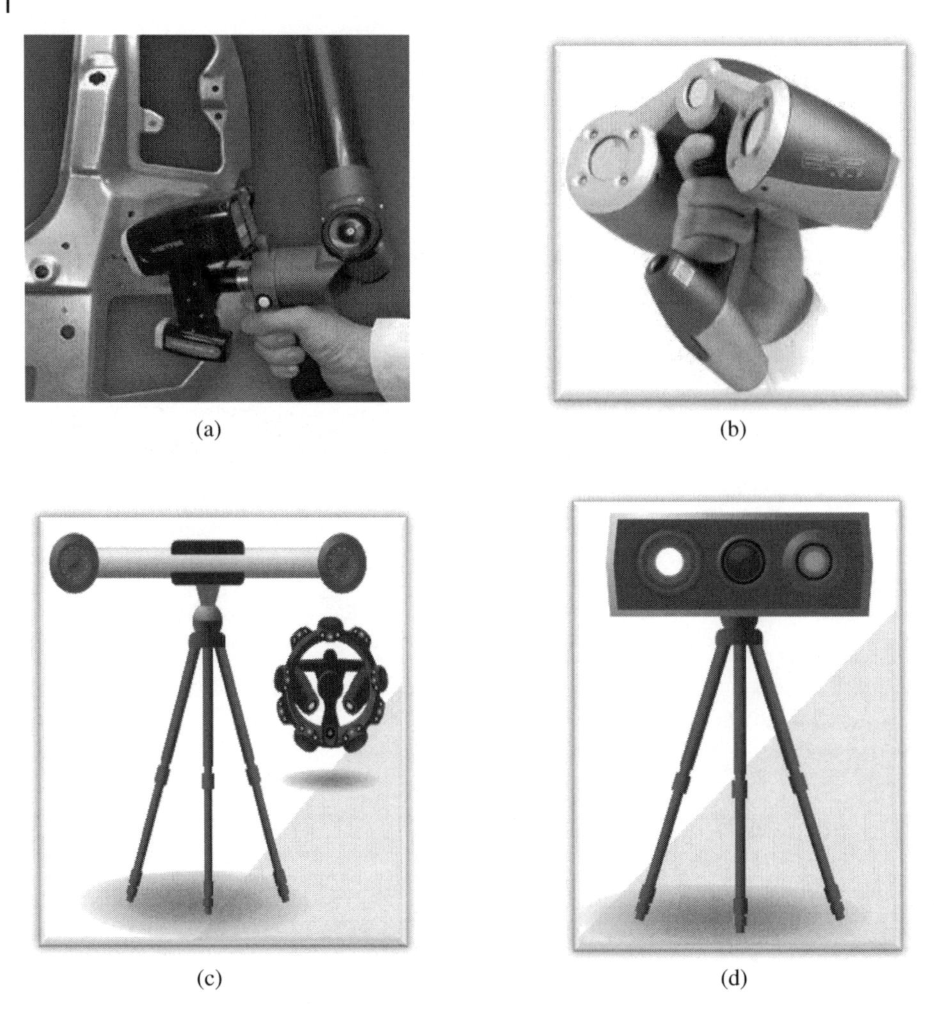

Figure 5.24 Positioning methods in scanning. (a) Mechanical encoding. (b) Tracked scanning. (c) Structured lighting. (d) Portable scanning.

direct line of sight to the 3D scanner. Tracked scanning increases the complexity of system operation since more scanning parameters are introduced. Tracked 3D scanning is usually more expensive than portable 3D scanning.

Structured lighting. In structured lighting, a pattern of lights is projected on to a part and the pattern of returned lights is analysed to determine how the pattern is distorted when lights hit the object. Either a liquid crystal display (LCD) projector or a scanned or diffracted laser beam can be used to generate the light pattern, and one or more sensors are used to record the projected pattern. If only one camera is used, the relative position of the projector to the camera must be determined in advance. If two cameras are used, the stereoscopic pair must be calibrated in advance. High-end structure light scanners generate very high-quality data. They deliver an excellent resolution, which allows for the smallest features on the object to be captured. While white-light scanners can acquire large quantities of data in one scan, multiple scans are required in most cases to cover all angles on more complex parts, which is very time consuming.

Portable scanning. The majority of portable scanners are able to track their own position by self-tracking. In the operation, one or more laser lines are projected on to the object, while white-light devices project a light and shade pattern. The feedback of the light and the shaded pattern is collected and merged to analyse deformed projections and position targets. Portable scanning can be transported or combined with multiple positioning devices. Portable scanning provides positional accuracy as well as flexibility of object features and texture positioning. Portable scanning has a high sampling rate; for example, advanced portable scanning can acquire more than half a million points per second and rebuild the 3D triangle mesh lie during the scanning process.

However, self-positioning is made based on a local reference. This means that the errors in a world coordinate system can stack up as the scanning volume grows. It is possible to combine other technologies such as photogrammetry and positioning targets to minimize system errors. Note that these additional steps might increase the setup time and limit the size of the objects.

5.5.3 Control of Scanning Processes

No control is needed for a fixed scanner since it has a pre-defined window and resolution. However, a sophisticated sensor is integrated with a position device to offer a range of feasible resolutions and the flexibility of selecting one or more individual windows for scanning.

In such a case, the scanning process must be controlled to register and fuse point clouds from different views. Scott et al. (2003) conducted the survey and concluded that there is no general-purpose solution for view planning of a high-quality object reconstruction/inspection. The challenge is how to accommodate the needs of surface reconstruction in the scanning control. Whether or not the target object is simple or complex, one has to consider efficiency, accuracy, and robustness of data acquisition. The efficiency relates to the computational complexity of the view planning algorithm in terms of both time and memory. In addition, timeliness is probably the more important factor. Note that performance and efficiency are conflicting requirements. View planning algorithms and imaging environments restricted to a one- or two-dimensional viewpoint space can at best provide limited-quality, limited-coverage models for a restricted class of objects. There is a need to measure and model object reflectance to handle shiny surfaces and compensate for a limited camera dynamic range. View planning techniques are also required to avoid or compensate for the effects of geometric step edges, reflectance step edges, and multiple reflections.

Either a single-point or a line scanner has a limited scope of views at a particular time. Therefore, a motion mechanism must be integrated with a sensor to acquire the data from different perspectives. The scanning path is determined by the sensor type. As shown in Figure 5.25, the larger the scope of view at one position, the shorter is the path and the more spacing there is between the two path segments.

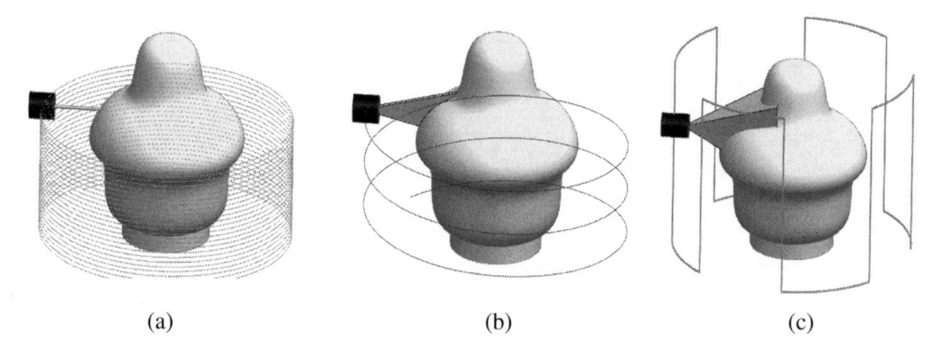

(a) (b) (c)

Figure 5.25 Scanning paths and sensor types. (a) *Single-point scanning:* acquires one point at a time; it takes a long time and travel to complete the scanning process over an object. (b) *Line scanning:* acquires a line at a time; it takes a short time and travel to complete the scanning process over an object. (c) *Area scanning:* acquires one area at a time; it takes the shortest scanning time, but it was very difficult to achieve good accuracy and fuse data.

The camera for photogrammetry does not infer the distance information between the camera and the object. The *close-range photogrammetry* (CRP) generates a 3D model of the object from a series of 2D images using the *scale-invariant feature transform (SIFT) algorithm*. SIFT automates the workflow of feature matching among multiple neighbouring images to calculate the coordinates of points. No sophisticated control is needed to move a camera, as one could take images for any position and orientation. However, a successful comparison of two contiguous images must ensure that there is over 80% overlapping. The image quality is critical to define the correspondences of images.

Figure 5.26 shows the general procedure of running an automated CRP as well as some major factors that affect the performance of CRP (Barnes et al. 2018). To ensure the success of photogrammetry, an image capturing process should avoid some unfavourable situations, such as dark, reflective, transparent surfaces or surfaces with uniform textures or solid colours. In addition, shadows are treated as noise in image data; the date acquiring process should avoid moving light sources or shadows by the objects beyond the scope of interest.

5.5.4 Available Hardware Systems

A data acquisition hardware system is usually tailored to the specific needs of an RE project. For example, Biegebauer et al. (2002) used a laser triangulation sensor to obtain a range of images of different parts of components such as gearboxes for motors, compressor tanks, steering columns for trucks, and rear-view mirrors for a conveyor. The sampling rate was up to 600 scans per second and the resolution for a scan line width of 2 m was 1.2 mm. Teutsch et al. (2005) developed an optical laser scanner for digital modelling of industrial work pieces. The scanner measures an object by integrating the methods of structured lighting with triangulation. For applications where the time factor is not critical, conventional data acquisition methods such as CMMs, shape from shading, and CT can be adopted. For other applications, the time factor is considered with other criteria, such as cost, quality, precision, and the flexibility to be integrated with other applications. Ingensand (2016) and

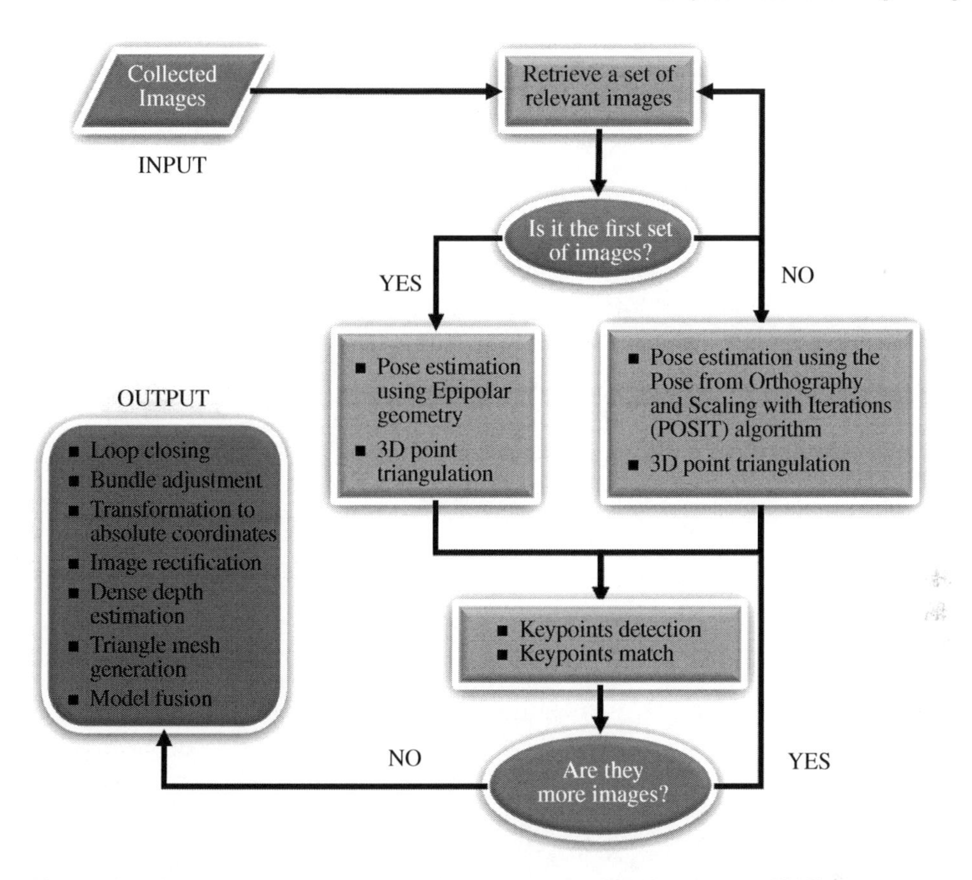

Figure 5.26 Procedure and design factors in an automated CRP (Barnes et al. 2018).

Boehler and Marbs (2002) provide basic characteristics of popular 3D scanners and their manufacturers in Table 5.5.

5.6 Software Systems for Data Processing

After a point cloud is required, the data must be processed before it can be utilized to reconstruct surfaces and extract features of computer models. The main tasks in data processing are *data filtering*, *data registration and integration*, *feature detection*, *simplification*, *segmentation*, and *3D reconstruction*.

5.6.1 Data Filtering

The acquired data must be filtered to eliminate invalid data. A part of the point data can be invalid for many reasons: (i) the points collected from the background of the object, (ii) the points from obstacles between the scanner and the object (such as trees and other objects in the foreground, moving persons or traffic, or atmospheric objects such as dust or rain); (iii) the points from the laser spot at edges; (iv) the points due to multiple reflections

Table 5.5 Some manufacturers and data acquisition devices (Ingensand 2016; Boehler and Marbs 2002).

Manufacturer	Type	Range (m)	Software	Website
Optech	ILRIS-3D	0.8–1.5, 3–1500	Polyworks ILRIS-3D software	www.optech.ca www.geo-konzept.de
ShapeGrabber	Scan heads, portable classic, automated system	0.33–1.75	ShapeGrabber SGCentral	www.shapegrabber.com
INO	3D laser profiling sensor	0.3	INO software	www.ino.ca
3D Scanners Metres	ModelMake D50, 100, 200 ModelMake Z35, 70, 140 (handler scanner)	0.05–0.2	KUBE CAS Corrosion Analysis Software	www.3dscanners.com
Tyzx	DeepSea camera 3 cm, 6 cm, 22 cm, DeepSea G2	0.2–2.7	SEER Software	www.tyzx.com
Point Grey Research	Bumblebee 2 Bumblebee XB2	n/a	FlyCapture SDK Triclops SDK, Censys3D SDK, Multiclops	www.ptgrey.com
Konica Minolta	VIVID Series – 910 VIVID Series – VIVID 9i	0.5–2.5	Polygon Editing Tool Software Photogrammetry PSC-1	www.se.konicaminolta.us
Genex Technologies	Rainbow 3D Camera	0.2–0.37	3D Mosaic 3D Surgeon	www.genextech.com/pages/601/Rainbow_3D_Camera.htm
3rdTech	DeltaSphere-3000 3D Digitizer	0.5–15	SceneVision-3D	www.3rdtech.com
CALLIDUS Precision Systems	CALLIDUS CT 180, 900, 3200	0.18	3D-Extractor	www.callidus.de
Leica Geosystems	Scan Station	0.3	Cyclone, CloudWorx	www.leica-geosystems.com
FARO	LS 420, LS 840, LS 880	20, 40, 80	FARO Scene	www.faro.com
I-SiTE Pty Ltd	I-SiTE 4400 LR	0.15–0.7	I-SiTE Studio	www.isite3d.com
MetricVision	MV224, MV260 XC50 Cross Scanner XC50-LS Cross Scanner LC50, LC15	0.07–0.195	CMM Software	www.metris.com
Riegl Laser Measurement Systems	LMS-Zxxx series	0.35–1.0	RiSCAN PRO	www.riegl.com

Zoller+Froehlich GmbH	IMAGER 5006 IMAGER 5003	53.5	Z+ laser control Light Form Modeller (LFM) Visual Sensor Fusion (VSF) A + F ProjectView 3D Reconstructor	www.zf-laser.com
RSI GmbH	DigiScan2000	0.4	RapidForm2000 PhotoModeler Pro 5	www.rsi.gmbh.de/maine.htm
Roland Corp	LPX-60/600 3D	0.3–0.4	Rapid 2007 Webinar	www.rolanddga.com
Inus Technology Inc.				www.rapidform.com
Bytewise Measurement Systems	CTWIST	0.032	Profile360™ Profile Measurement System	www.bytewise.com
Micro-Epsilon	LLT2800-25, 100100	0.025–0.1	scanControl 2800	www.me-us.com
NextEngine, Inc.	Desktop 3D Scanner	n/a	ScanStudio™ CORE	www.nextengine.com
Steinbichler	COMET 5	0.42–1.7	T-Scan Software system	www.steinbichler.de
ARIUS3D	ARIUS3D	6	Pointstream 3d	www.arius3d.com
Breuckmann	OptoScan Smartscan	0.36 0.72	OPTOCAT for Windows 3D-Alignment	www.breuckmann.com
MicroScribe	MicroScribe MX MicroScribe MLX	0.63 0.84	MicroScribe Utility Software	www.emicroscribe.com
GOM mbH	ATOS 3D Digitizer	1.6	ATOS software	www.gom.com/EN
MEL Mikroelektronik GmbH	M2D, M2DW, M20D-XF	1.2	MEL Software	www.melsensor.de
Cyberware	Model shop Whole body 4 and X	1	Cyberware software	www.cyberware.com
Laser Design	DS, RE, PS	1	Surveyor Scan Control RealScan	www.laserdesign.com
Vitronic	Vitus	1	Vitus	www.vitronic.de
Polhemus	FastScan	0.8	FastScan software	www.fastscan3d.com
TC2	Body scanner	0.8	3D Body Measurement software	www.tc2.com
Nextec	Hawk	0.3	Hawk software	www.nextec-wiz.com
Kreon	ZEPHYR KZ 25, 50, 100	0.2	Polygonia	www.kreon3d.com
Perceptron	Contour Probe Sensor	0.1	ScanWorks	www.perceptron.com

of laser beams; (v) the points due to the range differences caused by different reflectivity over surfaces; and (vi) erroneous points caused by bright objects. Unfortunately, eliminating erroneous points from a point cloud have to be assisted interactively by a human being (Boehler et al. 2002).

The acquired data may also be filtered to reduce the noise level. If the object is known to be smooth, the application of a low-pass or median filter can improve the data considerably. It should be noted, however, that filtering will affect the valid data from the object in the same way. If an object consists of smooth sections and edges, filtering may not be advisable at the early stages of processing. The data may also be filtered to reduce the number of points in a dense area. Point thinning can have a similar effect as filtering if those points having large deviations from an intermediate surface are preferably deleted. If several scans are taken from different observation stations, it may be advisable to delay point thinning until all measurements are combined in the registration process.

Most of the available software tools have basic filtering functions. For example, Matlab toolboxes include filtering programs to reduce the number of points, remove isolated points, and reduce blurring. The filters to remove isolated points are based on various criteria. For example, if a neighbourhood is given, the software can determine the number of neighbouring points for each point of the scanned image. If this number is under a specified threshold, the point is considered isolated and is removed from the contour matrix. The group of filters to reduce blurring consists of assorted functions for smoothing data. For each point, a neighbouring block is determined and a new value is generated based on the chosen criteria.

5.6.2 Data Registration and Integration

Data registration serves two purposes: (i) to merge multiple point clouds that are taken from different observation points and (ii) to transform point clouds from the coordinate system where the data is taken to the global coordinate system where it is merged. The most critical step to perform the registration is to define a 'distance' between two images. The zero distance corresponds to a perfect overlap of two images. The most successful methodology for the data registration and integration is the *iterative closest point* (ICP). In ICP, the distance between two objects is defined as the closest Cartesian distance between two objects. Since the introduction of ICP by Chen and Medioni (1991) and Besl and McKay (1992), many variants have been introduced to the original ICP concept.

The approaches of registration can be pair-wise and global (Mahadevan et al. 2001). A *pair-wise registration* is simple since it works only with two views at a time and it is continuously processed until all views are registered. The advantage of a pair-wise approach is its simplicity. However, the pair-wise registration can accumulate errors that occurred when minimizing the error for the pair. A *global registration* deals with all points simultaneously to approximate an unknown surface.

The difficulty in performing registration is to deal with outliers and local minima for high efficiency. While the iterative closest point algorithm is efficient, it is unable to deal with outliers or avoid local minima. On the other hand, an optimization algorithm such as simulated annealing is effective in tackling local minima. Luck et al. (2000) suggested integrating simulated annealing with the iterative closest point method for surface reconstruction.

ICP is successful when a close initial transformation is provided between two images. However, determination of an initial transformation is not an easy task. Murino et al. (2001) made the pre-alignment based on matching between the branches of 3D skeletons extracted from the two images. They demonstrated that 3D skeletons were successfully aligned to ensure the convergence of ICP. Gelfand et al. (2005) developed an algorithm to automate registrations without any assumptions about initial positions. A descriptor was defined and computed for each data point based on local geometry and a few feature points were automatically picked from the data shape according to the uniqueness of the descriptor value at that point. A branch-and-bound algorithm was used to define the initial correspondence of the two views.

5.6.3 Feature Detection

Feature detection is used to recover a high-level geometric description from the lower-level geometric representation of a part. Several methods have been proposed to extract a high-level, feature-based description from lower-level part geometry (Fischer and Shpitalni 1992; Vandenbrande and Requicha 1993; Regli et al. 1994). Thompson et al. (1999) used some geometric primitives for RE. These primitives could be directly imported into a CAD model without loss of the semantics and topological information. Schindler and Bauer (2003) applied model-based methods for building reconstructions. The 3D points were segmented into a coarse polyhedral model with a robust regression algorithm. The geometry of the model was refined with the pre-defined shape templates to recover a CAD model of construction.

5.6.4 Surface Reconstruction

Surface reconstruction from point clouds is fundamental in many applications. Gois et al. (2004), Schall and Samozino (2005), and Azernikov et al. (2003) investigated works surface reconstruction. The algorithms for surface reconstruction have been thoroughly discussed in Section 5.4.3.

5.6.5 Surface Simplification

While most surveys ignored the issue of surface simplification, Schroeder et al. (2004) introduced an algorithm using local operations on geometry and topology to reduce the number of triangles in a triangle mesh. The implementation is for the triangle mesh and is applicable for other types of polygon meshes. Hoppe et al. (1992) developed an energy minimization approach where the energy function consists of three terms: a *distance energy* that measures the closeness of fit, a *representation energy* that penalizes meshes with a large number of vertices, and a *regularizing term* that conceptually places springs of rest length zero on the edges of the mesh. The minimization algorithm divided the simplification problem into two nested subproblems: an inner continuous minimization and an outer discrete minimization. The search space consisted of all mesh homeomorphisms to the starting mesh. Pauly et al. (2002) analysed and compared a number of surface simplification methods for point-sampled geometry. They implemented incremental and hierarchical clustering, iterative simplification, and particle simulation algorithms to create approximations of point-based models with a lower sampling density. To compare the quality of the simplified surfaces, they also designed a new method for computing numerical and visual error estimates for point-sampled surfaces.

5.6.6 Segmentation

Segmentation is to partition an image into a number of homogeneous segments. Many methods have been developed for image segmentation. Li et al. (2003) classified the methods into *histogram-based techniques*, *edge-based techniques*, *region-based techniques*, *Markov random field-based techniques*, *hybrid methods* that combine edge and region methods, the *level set method*, and the *morphological watershed transform*. Zhang (2001) and Chan and Zhu (2003) reviewed the most significant works.

If the data in a segment is treated as a feature, segmentation relates closely to the feature extraction or fitting in some way. Some research studies focused on the fitting issue. Many of the fitting methods (Fisher 2002; Werghi et al. 2000; Ahn et al. 2002) have focused on recovering patches of simple geometric surfaces, which are then connected together resulting in a B-rep (Boundary Representation) model. Rabbani et al. (2004) presented some methods for fitting CAD models to point clouds. Constructive Solid Geometry (CSG) was used to represent the models due to its flexibility and compactness. A CSG tree was converted to a B-rep (or a triangular mesh or a point cloud) for approximating the orthogonal distance of a given point from the model surface. The notion of Internal Constraints was introduced to represent the geometric relationships among constituent components of a CSG tree.

5.6.7 Available Software Tools

Software tools for data processing include an extensive collection of modules for different purposes, ranging from scanner controlling to 3D modelling. For example, Rocchini et al. (2001) introduced a 3D scanning software suite for *range map alignment*, *range map merging*, *mesh editing*, and *mesh simplification*.

All of the venders of 3D scanners listed in Table 5.6 provide software tools to support data processing. However, most of them are rudimentary. The high demand for greater capability has led to many stand-alone software products. Table 5.6 lists some of the available stand-alone software tools for 3D data processing (Boehler et al. 2002, http://scanning.fh-mainz.de/scanninglist.php). While the market is growing, it is likely that consolidation will take place due to a growing number of competitors and the trend towards large and comprehensive computer aided tools.

5.7 Typical Manufacturing Applications

There is an increasing demand for accurate, as-built, 3D models of existing industrial sites in many sectors. For example, Rabbani et al. (2004) recommended application areas including (i) planning (clash detection, decommissioning, design changes), (ii) revamping and retrofitting of old sites, (iii) implementing services based on virtual and augmented reality,

Table 5.6 Independent 3D scanning software tools.

Producer	Software tool	Hyper link
Braintech Canada Inc.	eVisionFactory 4.0, VOLTS-IW	www.braintech.com
InnovMetric	Polyworks	www.innovmetric.com/Manufacturing/home.aspx
3D Veritas	3D-Veritas	www.3dveritas.com
Metrologic Group	Metrolog II	www.metrologic.fr
Octocom AG	OctoCAD	www.zf-laser.com/e_octocad.html
SDRC	Imageware Surfacer	www.mayametrix.com/surfacer
Z + F UK Ltd	Light Form Modeller	praxis.zf-uk.com/index.html
Inn.Tec s.r.l.	Reconstructor	www.reconstructor.it
INUS Technology	RapidForm	www.rapidform.com
Kubit GmbH	PointCloud	www.kubit.de
Phocad GmbH	Phidias	www.phocad.de
UGS	Imageware	www.ugsplm.de
Raindrop Geomagic	Geomagic Studio	www.geomagic.com
Pointools	Pointools View	www.pointools.com
Free Open Source	MeshLab	www.meshlab.sourceforge.net
3D3 Solutions	FlexScan3D	www.3d3solutions.com

Sources: Boehler et al. 2002; http://scanning.fh-mainz.de/scanninglist.php.

(iv) off-site training, (v) safety analysis, and (vi) detecting changes and defects. In fact, surface reconstruction and digital manufacturing have been successfully applied in RE, inspection and measurement, digital mockup and simulation, medical applications, multimedia, and arts and museums. Typical applications in manufacturing are as follows.

RE and rapid prototyping. Products with one or small volumes are required to be manufactured for special purposes, such as for a new product demonstration before the product is finalized. It is a common situation where the concept of the product comes from an object without a computer model; therefore, a vision system is applied to capture the data from a physical object and generate its parametric model.

Part positioning and alignment. A high-precision machining operation needs to know the exact location of a part. A vision system can be applied to detect the position when a part is fixed and the detection result can be used to modify its corresponding machining program or alignment of the part on a machine tool for an ideal position. The tolerance of part positioning can be accommodated in such a way. In some situations, such as the milling operation of a cast part, the margin of operation can be changed from one part to another. A vision system can be applied to capture the real dimension of an actual part, so that an optimized fixture position can be determined for the required quality.

Inspection. Inspection is a critical step to ensure product quality. CMMs are widely used to perform inspections. However, there are some limitations of CMMs: (i) a part has to be placed on the CMM to proceed with the inspection – in many cases, out of the production line; (ii) it takes a long time to do the inspection and a 100% inspection is impossible for most products; and (iii) the contact inspection may damage the part surface. A vision-based inspection system is expected to address all of the aforementioned issues.

Virtual assembly. In developing a new product, prototyped parts have to be assembled together to eliminate potential mistakes of assembly modelling. Those parts are usually fabricated individually without or with fewer considerations of assemblies. Trails and errors are needed for reordering the assembly sequence, relocating and reorientation of parts in an assembly operation, and changing the physical parts to fit them in the assembly. A vision-based system can acquire and generate CAD models of parts, identify critical problems, and accelerate the assembly process in a virtual environment.

Robot automation in assembly, welding, and surface treatments. Automation relies on industrial robots. Robots have to be programmed to execute tasks. A robot program is usually dedicated to one task. However, it is desirable that a vision-based system can be applied to capture the changes of a task and that a software tool can respond to the changes automatically by generating new programs for the task.

5.8 Computer Aided Reverse Engineering (CARE)

This section overviews a number of commercial computer aided tools for reverse engineering (CARE). As discussed in Section 5.6, many computer aided packages have been developed to support CARE. Figure 5.27 shows some examples of popular commercial and open-source software tools. In general, commercial software packages have more advanced functionalities while open-source tools have better flexibility to accommodate customized programs for the second development. Some software packages, such as Sketchup, Inventor, and Blender, have both commercial and open-source versions on the market.

Two most critical capabilities of CARE tools are to (i) convert sensed data into point clouds and polygonal models and (ii) create parametric models from polygonal models.

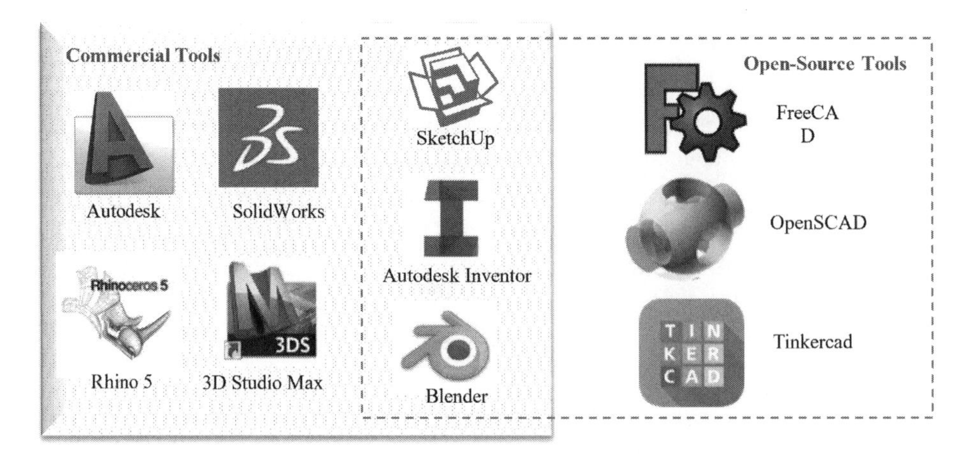

Figure 5.27 Commercial and open-source software tools for RE.

The Autodesk *Recap* by Autodesk and the *ScanTo3D* by SolidWorks Inc. are used as examples to illustrate how commercial CARE tools can be used to perform tasks (i) and (ii), respectively.

5.8.1 Recap to Convert Sensed Data into Polygonal Models

Among the data acquisition methods in generating raw data for RE, the methods of triangulation, time-of-flight or laser pulse, and interferometry show the models in the form of point clouds directly. The hardware venders usually provide the software tools to convert point clouds to polygonal models, so the integrated system can be optimized for better performance. However, the raw data from the photogrammetry is a large number of 2D images. The sophisticated third-party software tools and high-performance computing are required to process 2D images and convert them into polygonal models.

The Autodesk Recap Pro can be used to show how a set of 2D images are converted into a polygonal model. As shown in Figure 5.28a, the software tool takes point cloud, data from a mobile device, or 2D photos as inputs in an RE project. The generation of mesh or polygonal models are fully supported by cloud computing. The services can be subscribed and accessed from any operating system. The underlining algorithms for this software tool come from an image-based modelling technique called *camera for photogrammetry* (Autodesk 2019). Generated models can be outputted as point clouds (.rcs) or other mesh formats such as .obj, .rcm, .fbx, and .ipm.

Figure 5.29 shows three main steps used by Autodesk Recap Pro for an RE project.

The result of surface construction can be downloaded in a *.rcm* format. As shown in Figure 5.30a, the model in *.rcm* format can be further edited, re-topologized, analysed, and the final model can be exported in different formats such as *.obj*, *.stl*, and *.fbx*, as shown in Figure 5.30b.

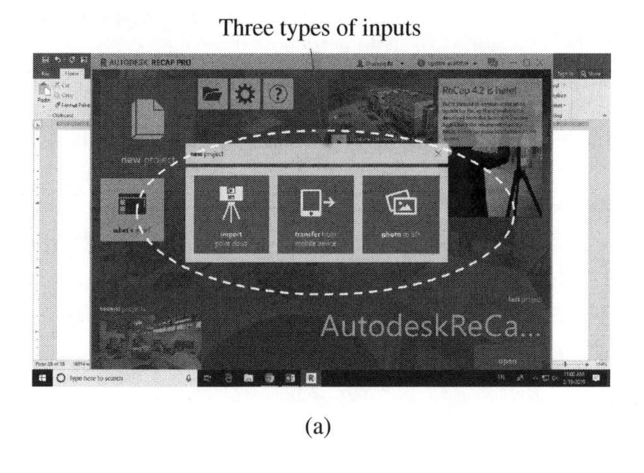

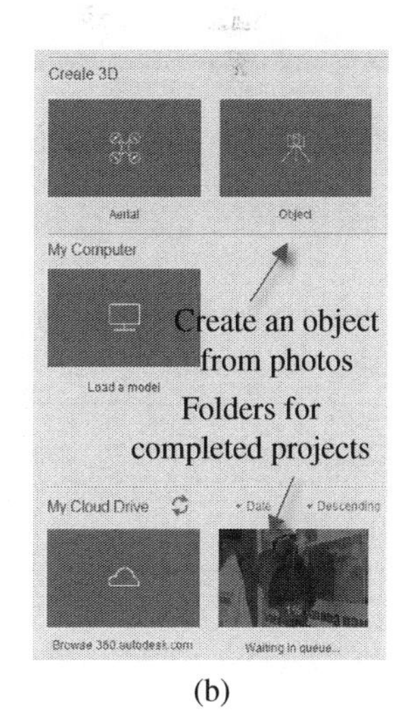

Figure 5.28 Cloud-service for an RE project by Autodesk Recap Pro. (a) Launching an RE in Autodesk Recap Pro. (b) Cloud-computing for an RE.

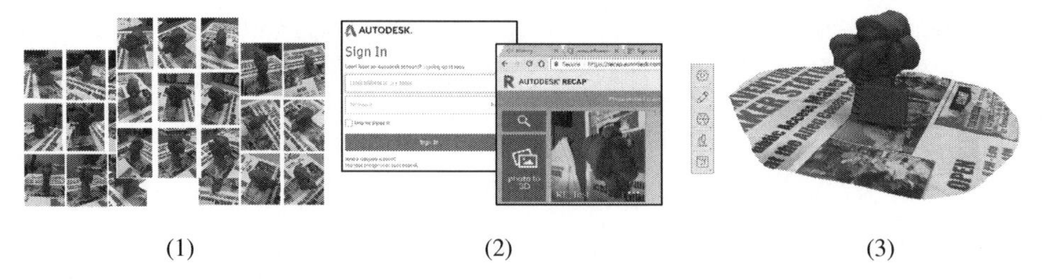

Figure 5.29 The steps of using Recap Pro for an RE project. (1) Capture sufficient photos of objects: take photos from different locations, directions, and altitudes of objects. (2) Create an RE project. Create or access a user account at https://accounts.autodesk.com/users, launch ReCap Pro, create a new project, select the mesh quality, the export file types, upload the photos, and confirm the submission. (3) Wait for the completion of surface reconstruction and download the polygonal model in the *.rcm* format. The uploaded images will be processed via the online service by cloud computing. After processing, one can review the model online or download the model for use in other software platforms. (*See color plate section for color representation of this figure*).

Figure 5.30 Processing and exporting a polygonal model in Recap Pro. (a) Editing a polygonal model in Recap Pro. (b) Exporting a polygonal file. (*See color plate section for color representation of this figure*).

In reconstructing a 3D model, the photogrammetry calculates the locations, orientations, and distortions of cameras by comparing and matching pixels in a series of photos. The quality and sufficiency of photos are two fundamental factors needed to reconstruct a 3D model accurately. The data should be sufficient so that each photo in the collected set should have at least 80% overlapping with another photo. The following guides can be used to produce good photos for photogrammetry (Autodesk 2019):

1. *Lighting conditions*. Avoid any place where the light creates a strong contrast shadow over the object of interest. In a closed space, avoid using a flash and place an object with diffuse lights without shadows. In an open space, place an object under direct and strong lights, and try to take photos in the morning or at dawn.
2. *Cameras and setting up*. Try to use a lens other than plastic lenses to achieve better sharpness. Take photos with a fixed lens to avoid focal changes. To ensure sharp photos, use a tripod and a clicker to stabilize the camera. Use a wide depth of field to reduce the amount of light coming into the camera.

3. *Shooting strategies.* Ensure the object of interest is always within the photo and fill the view window as large as possible. Use a rich background rather than a monochromatic background. If a monochromatic background cannot be avoided, place some small objects around the point of interest so that the algorithms can use these objects in matching. Try to make the scene as still as possible and avoid moving any objects during shooting. Take photos all around the object to cover each detail with multiple overlapping photos. Take a photo at each 5–15 degrees at the same height as the object.

5.8.2 ScanTo3D for Generation of Parametric Models

A computer model with point clouds or polygonal patches only includes a portion of the geometric information. It is not the destination for the majority of RE projects. To further explore design intents, knowledge, and wisdom beyond geometry and shapes, the model must be cleaned, processed, and parametrized for inspection, feature recognition, and redesigns.

In Figure 5.31, the *ScanTo3D* by SolidWorks is used as an example to show how a point cloud or polygonal model can be cleaned, processed, and parametrized. *Firstly*, input the acquired data in the form of a point cloud or polygonal model; *secondly*, run a mesh

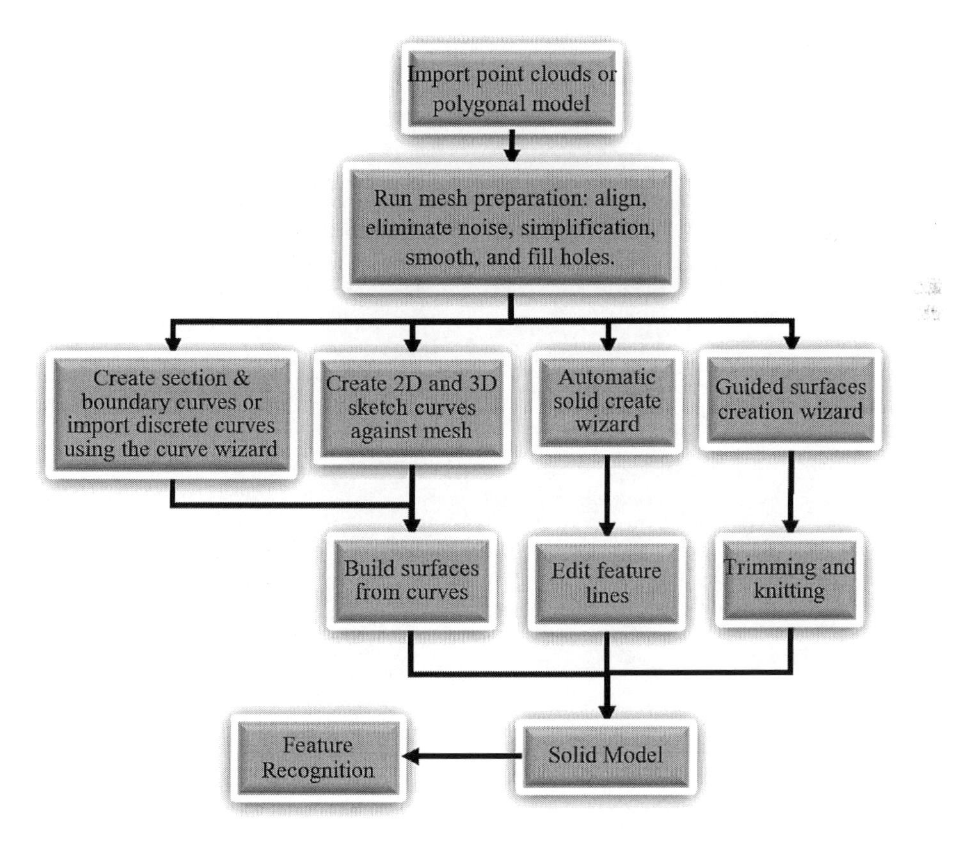

Figure 5.31 Procedure of generating a parametrized solid model by ScanTo3D by SolidWorks.

preparation to align, simplify, smooth data, eliminate noise, and fill holes if no watertight volume can be defined; *thirdly*, a number of tools for parametric modelling can be applied to define section and boundary curves, create lines, surface, and finally the solids of the object; and *fourthly*, run *feature recognition* to find if the object includes some features such as cylinders, extrusions, and fillets.

5.8.3 RE of Assembled Products

The algorithms and software tools presented in Sections 5.4 and 5.6 are for a single solid. If an assembled product has to be reversely engineered, the *Divide and Conquer* technique can be applied, i.e. the product is decomposed into a number of parts or components and the RE process is performed on individual parts or components. RE on an assembly gains an understanding of individual parts as well as their assembly relations. Figure 5.32 shows an example of the RE project on an assembled product. The models of critical parts and components can be further investigated in details to explore embedded design knowledge and wisdom.

The reconstructed models can be fully utilized to enhance the product development. Figure 5.33 shows a few examples where further modifications are made on reversely engineered models. The model in Figure 5.33a is scaled for an enlarged and reduced model from the same data. 3D data can be used to replicate or restore damaged objects. The model in Figure 5.33b is utilized to replicate or restore damaged objects. Using 3D data, it becomes

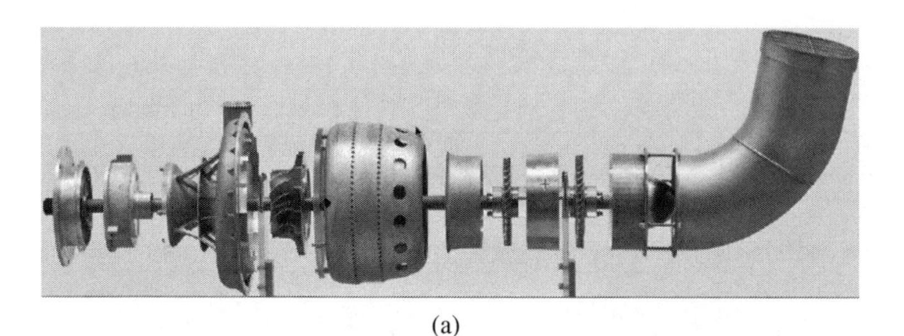

(a)

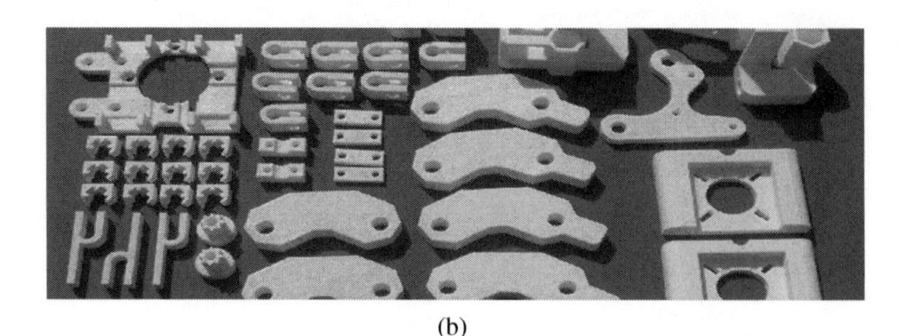

(b)

Figure 5.32 Divide and conquer an assembled product in reverse engineering (Gonuke 2017). (a) Physical product to be reversely engineered. (b) Individually reverse engineered parts.

(a) (b) (c)

Figure 5.33 Examples of sequential operations after digital modelling in RE (Gonuke 2017). (a) Scaling. (b) Restoring. (c) 3D prototyping.

Table 5.7 Trend of reverse engineering (RE) techniques (Gonuke 2017).

Year	Milestones
2015	Submillimetre scanning is perfected, bringing with it endless possibilities.
2017	The US Army launches the Expeditionary Warrior program, providing 3D scanned and printed drones and equipment to deployed soldiers.
2022	3D scanning market hits peak, with more than 60% of all companies using the technology, creating a $15 billion industry.
2029	The first mass-produced 3D captured and printed vehicles are released for sale.

possible to recover an object that has been damaged by weather, neglect, natural disasters, etc. The model in Figure 5.33c is a prototype of the new version of parts based on the reconstructed model.

5.9 RE – Trend of Development

RE technologies have been widely adopted in today's manufacturing. Due to its great potential, the studies on RE have attracted many researchers and organizations. Gonuke (2017) predicted that the researchers in the field of RE would make the following breakthroughs, with time being as shown in Table 5.7.

5.10 Summary

Knowledge-based engineering on physical products by others begins with RE. In RE, the data is acquired from objects, it is then processed and converted to a polygonal model, and the polygonal model is further analysed to build a solid model, recognize features, and recover and reuse design intents of original objects. RE plays an important role in generating technological innovations and new products at an accelerated pace. Numerous hardware and software tools are available to practice CARE.

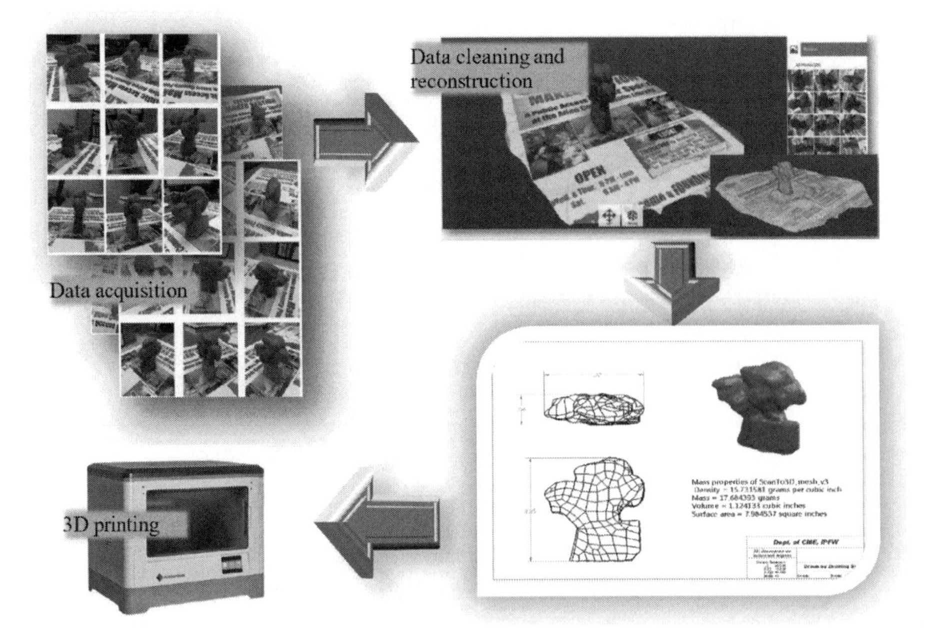

Figure 5.34 Example of a CARE design project.

5.11 Design Project

5.1 Reverse engineer a physical object (such as a part in Figure 5.34) to create its 3D Model and determine its physical properties such as mass, volume, and inertial properties using a cellphone or digital camera, Autodesk Recap Pro, and SolidWorks ScanTo3D.

1. Select a physical object to be modelled in your RE project. Place an object on a flat surface. Make sure that photos can be taken of the object from all sides and angles.
2. Use a cellphone or digital camera to take over 50 photos from different positions and angles.
3. Create a user account at www.autodesk.com, download and install ReCap Pro, create an RE project in ReCap Pro, upload all photos, convert them into a mesh file, download the .rcm file, and use ReCap to export .rcm file as an .obj file or .stl file for its use in SolidWorks.
4. Use SolidWorks – ScanTo3D to clean the imported mesh file, fix the mesh, create a surface model, fix the surface model, and generate a solid model from the watertight surface model.
5. For the solid model, assign the materials and evaluate mass and volume properties.
6. Document your project idea, procedure, the challenges and solution, and result of reverse engineering.

References

Abdel-Wahab, M.S., Hussein, A.S., Taha, I., and Gaber, M.S. (2005). An enhanced algorithm for surface reconstruction from a cloud of points. In: *GVIP 05 Conference* (19–21 December 2005). Cairo, Egypt: CICC www.icgst.com.

Acadtechnologies (2019). Reverse engineering. https://www.acadtechnologies.com/services/reverse-engineering.

Ahn, S.J., Rauh, W., Cho, H.S., and Warnecke, H.J. (2002). Orthogonal distance fitting of implicit curves and surfaces. *IEEE Transactions on Pattern Analysis and Machine Intelligence* 24 (5): 620–638.

Amenta, N.C., Sunghee, K., and Ravi, K. (2001). The power crust, unions of balls and the medial axis transform. *Computational Geometry: Theory and Applications* 19: 127–153.

Autodesk (2019). Recap: reality capture and 3D scanning software for intelligent model creation. https://www.autodesk.com/products/recap/overview.

Azernikov, S., Miropolsky, A., and Fischer, A. (2003). Surface reconstruction of freeform objects based on multiresolution volumetric method. *Transactions of the ASME, Journal of Computing and Information Science in Engineering* 3: 334–337.

Barnes, A., Simon, K., and Wiewel, A. (2018). From photos to models: strategies for using digital photogrammetry in your project. https://sparc.cast.uark.edu/assets/webinar/SPARC_Photogrammetry_Draft.pdf.

Bernard, A., Laroche, F., Sebastien, R., and Durupt. A. (2010). New trends in reverse engineering: augmented-semantic models for redesign of existing objects. *Ouvrage Collect of Reverse-Engineering*, 2010, 27.

Bernardini, F., Mittleman, J., Rushmeier, H. et al. (1999). The ball-pivoting algorithm for surface reconstruction. *IEEE Transactions on Visualization and Computer Graphics* 5 (4): 349–359.

Besl, P.J. and McKay, N.D. (1992). A method for registration of 3-D shapes. *IEEE Transactions on Pattern Analysis and Machine Intelligence* 14 (2): 239–256.

Bi, Z.M. and Kang, B. (2014). Sensing and responding to the changes of geometric surfaces in flexible manufacturing and assembly. *Enterprise Information Systems* 8 (2): 225–245.

Bi, Z.M. and Lang, S.Y.T. (2007). A framework for CAD and sensor based robotic coating automation. *IEEE Transactions on Industrial Informatics* 3 (1): 84–91.

Biegebauer, G., Pichler, A., and Vincze, M. (2002). *Detection of Geometric Features in Ranges Images for Automated Robotic Spray Painting of Lot Size One*. IROS.

Boehler, W. and Marbs, A. (2002). 3D scanning instruments. In: *Proceedings of the International Workshop on Scanning for Cultural Heritage Recording*, Corfu, Greece (1–2 September 2002), 9–12.

Boehler, W., Heinz, G., Marbs, A., and Siebold, M. (2002). 3D scanning software: an introduction. In: *International Workshop on Scanning for Cultural Heritage Recording*, Corfu, Greece (1–2 September 2002), 47–51.

Cazals, F., Giesen, J., and Yvinec, M. (2004). Delaunay triangulation based surface reconstruction: a short survey. http://hal.inria.fr/inria-00070609/en.

Chan, T., and Zhu, W. (2003). Level set based shape prior segmentation. Technical Report, UCLA, California.

Chen, Y., and Medioni, G. (1991). Object modeling by registration of multiple range images. *Proceedings of the IEEE Conference on Robotics and Automation* (1991).

Chen, S., Tian, D., Vetro, A., and Kovacevic, J. (2018). Fast resampling of three-dimensional point clouds via graphs. *IEEE Transactions on Signal Processing* 66 (3): 666–681.

Creaform3D (2019). An introduction to 3D scanning https://www.creaform3d.com/sites/default/files/assets/technological-fundamentals/ebook1_an_introduction_to_3d_scanning_en_26082014.pdf.

Engel, K., Westermann, B., and Ertl, T. (1999). Isosurface extraction techniques for web-based volume visualization. In: *Proceedings of the Conference on Visualization '99*, San Francisco, California, United States, 139–146.

Fischer, A. and Shpitalni, M. (1992). Encoding and recognition of features by applying curvatures and torsion criteria to boundary representation. In: *Proceedings of ASME 1992 Winter Annual Meeting-Symposium on Concurrent Engineering*, 69–84.

Fisher, R.B. (2002). Applying knowledge to reverse engineering problems. In: *Proceedings of Geometric Modeling and Processing*, Riken, Japan, 149–155.

Freedman, D. (2004). Surface reconstruction, one triangle at a time. In: *CCCG 2004*, Montreal, Quebec (9–11August 2004), 15–19.

Gelfand, N., Mitra, N.J., Guibas, L.J., and Pottmann, H. (2005). Robust global registration. *Proceedings of the Third Eurographics Symposium on Geometry Processing* (4–6 July 2005), Vienna, Austria.

Gois, J.P., Filho, A.C., Nonato, L.G., and Biscaro, H.H. (2004). Surface reconstruction: classification comparisons, and applications. In: *Proceedings of CILAMCE'2004, XXV Iberian Latin American Congress on Computational Methods – CILAMCE, Fortaleze*, 1–15.

Gonuke (2017). Nuclear 3D scanning – ATE Central. https://atecentral.net/downloads/962.

Hennighausen, A., and Roston, E. (2015). 14 Smart inventions inspired by nature: biomimicry. https://www.bloomberg.com/news/photo-essays/2018-10-25/bloomberg-s-week-in-pictures.

Hoppe, H. (1994). Surface reconstruction from unorganized points. PhD thesis, University of Washington. http://research.microsoft.com/en-us/um/people/hoppe/thesis.pdf and http://hhoppe.com/.

Hoppe (2009). Demos and publication, http://hhoppe.com/

Hoppe, H., DeRose, T., Duchamp, T., et al. (1992). Surface reconstruction from unorganized points. *ACM SIGGRAPH*, pp. 71–78.

Ingensand, H. (2016). Metrological aspects in terrestrial laser-scanning technology. *Proceedings of the 3rd IAG/12th FIG Symposium*, Baden (22–24 May 2006).

Johnston, K. (2006). Automotive applications of 3D laser scanning. www.metronsys.com.

Li, H., Elmoataz, A., Fadili, J. and Ruan, S. (2003). An improved image segmentation approach based on level set and mathematical morphology. *Proceedings of SPIE* 5286: 851.

Luck, J., Little, C., and Hoff, W. (2000). Registration of range data using a hybrid simulated annealing and iterative closest point algorithm. In: *Proceedings of the IEEE International Conference on Robotics and Automation*, San Francisco (24–28 April 2000), 3739–3744.

Mahadevan, S., Pandzo, H., Bennamoun, M., and Williams, L. (2001). A 3D acquisition and modelling system. In: *Proceedings of the Acoustics, Speech, and Signal Processing, 2001 on IEEE International Conference (7–11 May)*, 1941–1944.

MeVisLab (2019). ITK and VTK Integration. https://www.mevislab.de/mevislab/features/itk-and-vtk-integration.

Mijajlovic, Z., Jovanovic, B., Maric, F, and Maric, M. (2004). MATLAB Toolbox for analysis of 3D images. https://www.researchgate.net/publication/28808984_Matlab_toolbox_for_analysis_of_3D_images.

Mueller, S., Fritzsche, M., Kossmann J., and Schneider, M. (2015). Scotty: relocating physical objects across distances using destructive scanning, encryption, and 3D printing, DOI: 10.1145/2677199.2680547. https://www.researchgate.net/publication/271524512.

Murino, V., Fusiello, A., Castellani, U., and Ronchetti, L. (2001). Pre-aligned ICP for the reconstruction of complex object. In: *Proceedings of the Italy–Canada 2001 Workshop on 3D Digital Imaging and Modeling Applications of Heritage, Industry, Medicine & Land*, Padova, Italy (3–4 April 2001).

Neugebauer, P.J., and Klein, K. (1997). Adaptive triangulation of objects reconstructed from multiple range images. *IEEE Visualization '97 – Late-Breaking Hot Topics Session* (October 1997). IEEE.

NPD Solution (2019). What is reverse engineering? http://www.npd-solutions.com/reverse-engineering.html.

Pauly, M., Gross, M., and Kobbelt, L.P. (2002). Efficient simplification of point-sampled surfaces. *Proceedings of the Conference on Visualization '02*, Boston, Massachusetts (27 October to 1 November 2002).

Peng, Q., and Sanchez, H. (2017). 3D digitizing technology in product reverse design. http://citeseerx.ist.psu.edu/viewdoc/download?doi=10.1.1.458.6322&rep=rep1&type=pdf.

Pernkopf, F. (2005). 3D surface acquisition and reconstruction for inspection or raw steel products. *Computers in Industry* 56: 876–885.

Pham, D. and Hieu, L. (2008). Reverse engineering – hardware and software. In: *Reverse Engineering (eds. V. Raja and K. Fernandes), Springer Series in Advanced Manufacturing*. London: Springer https://doi.org/10.1007/978-1-84628-856-2_3.

Rabbani, T., and Heuvel, van den (2004). Method for fitting CSG models to point clouds and their comparison. *Computer Graphics and Imaging*, Kauai, Hawaii (17–19 August 2004).

Rajia, V. and Fernandes, K.J. (2008). *Reverse Engineering – an Industrial Perspective*. London: Springer Series in Advanced Manufacturing, Springer-Verlag. ISBN: 13: 781846288555.

Rebolj, D., Babič, N.C., Magdič, A. et al. (2008). Automated construction activity monitoring system. *Advanced Engineering Informatics* 22 (4): 493–503.

Regli, W., Gupta, S., and Nau, D. (1994). Feature recognition for manufacturability analysis. Technical Report ISR TR94-10, University of Maryland.

Rocchini, C., Cignoni, P., Montani, C., et al. (2001) A suite of tools for the management of 3D scanned data. http://vcg.isti.cnr.it/publications/papers/3d_dima.pdf.

Schall, O. and Samozino, M. (2005). Surface from scattered points: a brief survey of recent developments. In: *1st International Workshop towards Semantic Virtual Environments*, Villars, Switzerland (16–18 March 2005), 138–147.

Schindler, K. and Bauer, J. (2003). A model-based method for building reconstruction. In: *Proceedings of the First IEEE International Workshop on Higher-Level Knowledge in 3D Modeling and Motion Analysis (17 October 2003)*, 74.

Schroeder, W., Martin, K., and Lorensen, B. (2004). *Visualization Toolkit*, 3e. Kitware Inc. ISBN: 1-930934-12-2.

Scott, W.R., Roth, G., and Rivest, J.-F. (2003). View planning for automated three dimensional object reconstruction and inspection. *ACM Computing Surveys* 35 (1): 64–96.

Sick Sensor Intelligence (2019). Mid range distance sensor. https://www.sick.com/us/en/distance-sensors/mid-range-distance-sensors/c/g176373.

Spenser, J. (2011). How the airplane got its shape. https://vintagespace.wordpress.com/2011/08/31/how-the-airplane-got-its-shape.

Srivastava, M. and Mishra, M. (2015). A view of reverse engineering. *International Journal of Advanced Engineering Science and Technological Research* 1 (4): 40–43.

Teutsch, C., Isenberg, T., and Trostmann, W. (2005). Evaluation and correction of laser-scanned point clouds. In: *Proceedings of SPIE, 5665* (eds. J.-A. Beraldin, S.F. El-Hakim, A. Gruen and J.S. Walton), 172–183. Videometrics VIII.

Thompson, W.B., Owen, J.C., Germain, H.J.S. et al. (1999). Feature-based reverse engineering of mechanical parts. *IEEE Transactions on Robotics and Automation* 15 (1): 57–66.

Vandenbrande, J.H. and Requicha, A.A.G. (1993). Spatial reasoning for the automatic recognition of machinable features in solid models. *IEEE Transactions on Pattern Analysis and Machine Intelligence* 15 (12): 1269–1285.

Várady, T., Martin, R., and Cox, J. (1997). Reverse engineering of geometric models – an introduction. *Computer-Aided Design* 29 (4): 255–269.

Vincze, M., Pichler, A., Biegelbauer, G., et al. (2002). Automatic robotic spray painting of low volume high variant parts. *Proceedings of the 33rd ISR International Symposium on Robotics* (7–11 October 2002).

Werghi, N., Fisher, R.B., Ashbrook, A., and Robertson, C. (2000, 2000). Shape reconstruction incorporating multiple non-linear geometric constraints. *Computer-Aided Design* 31 (6): 363–399.

Wikipedia (2019a). Standing on the shoulders of giants. https://en.wikipedia.org/wiki/Standing_on_the_shoulders_of_giants.

Wikipedia (2019b). Marching cubes. http://en.wikipedia.org/wiki/Marching_cubes.

Zhang, Y. (2001). A review of recent evaluation methods for image segmentation. In: *Proceedings of the International Symposium on Signal Processing and Its Applications*, Kuala Lumpur, Malaysia (13–16 August 2001), 148–151.

Zhao, H.K., Osher, S., and Fedkiw, R. (2001). Fast surface reconstruction using the level set method. *IEEE Workshop on Variational and Level Set Methods in Computer Vision* (*VLSM 2001*) (July 2001).

6

Computer Aided Machine Design

6.1 Introduction

Any processes in a manufacturing system are performed by machines in one form or another. A machine is a mechanical structure that transfers power or movement in an intended action. A machine consists of both fixed parts and moving parts. Fixed parts keep still relative to a reference base and moving parts change their positions with time while being in machine operation. Moving parts or components are mostly built upon machine elements such as flywheels, pulleys, belts, linkages, conjugate cams, spindles, chains, cranks, and gears.

Machine design is one of the most important branches of engineering design. Machine design can be formulated as an optimization problem with multiple constraints and criteria. As shown in Figure 6.1, the design space of an optimization problem is formed by selected design variables (DVs) and corresponding ranges. The design constraints define the boundaries of the design space and the design criteria guide the searching process to obtain an optimum for the given functional requirements (FRs). Many optimization algorithms, such as the steepest descent, conjugate gradient and quasi-Newton algorithms, genetic algorithms, and other heuristic algorithms, have been developed to search the optimization solution effectively. In addition, multiple design criteria may be in conflict with each other and the trade-off of multiple criteria can be made by techniques such as the weighted method or the Parato set method.

The complexity of an optimization problem depends on the number of design variables, design constraints, and design criteria as well as their coupling relations. Figure 6.2 shows the design variables in three important domains that affect the complexity of a machine design, i.e. the *motion domain*, the *discipline domain*, and the *time domain*. In the motion domain, the design complexity grows with the numbers of independent motions and coupled dependent motions. In the discipline domain, the design complexity grows with the dimension of objects, the number and types of objects, and multidisciplines. In the time domain, the design complexity grows with the increasing impact of the time factor on machine behaviours.

Traditional simple machines, such as levers, wheels and axles, incline planes, wedges, pulleys, and gear transmissions, can be designed manually. However, machines in a modern manufacturing system become more and more complicated and require computer aided tools to deal with the design challenges in their product cycles. Figure 6.3 illustrates

Computer Aided Design and Manufacturing, First Edition. Zhuming Bi and Xiaoqin Wang.
© 2020 John Wiley & Sons Ltd. This Work is a co-publication between John Wiley & Sons Ltd and ASME Press.
Companion website: www.wiley.com/go/bi/computer-aided-design

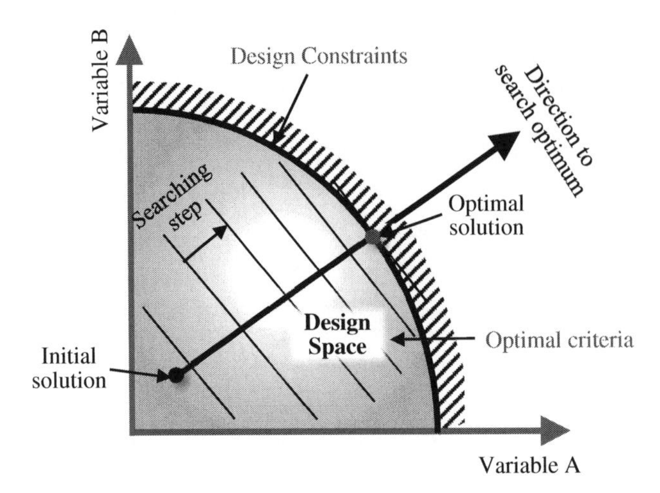

Figure 6.1 Multiple design constraints and criteria in optimization.

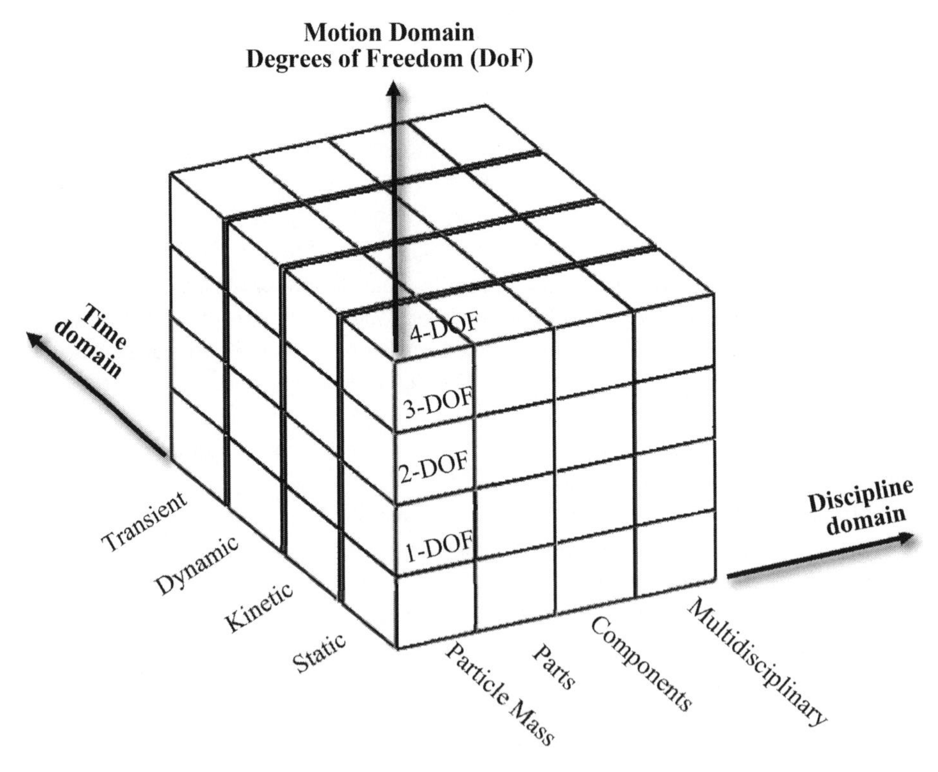

Figure 6.2 Growing complexity of design of modern machines.

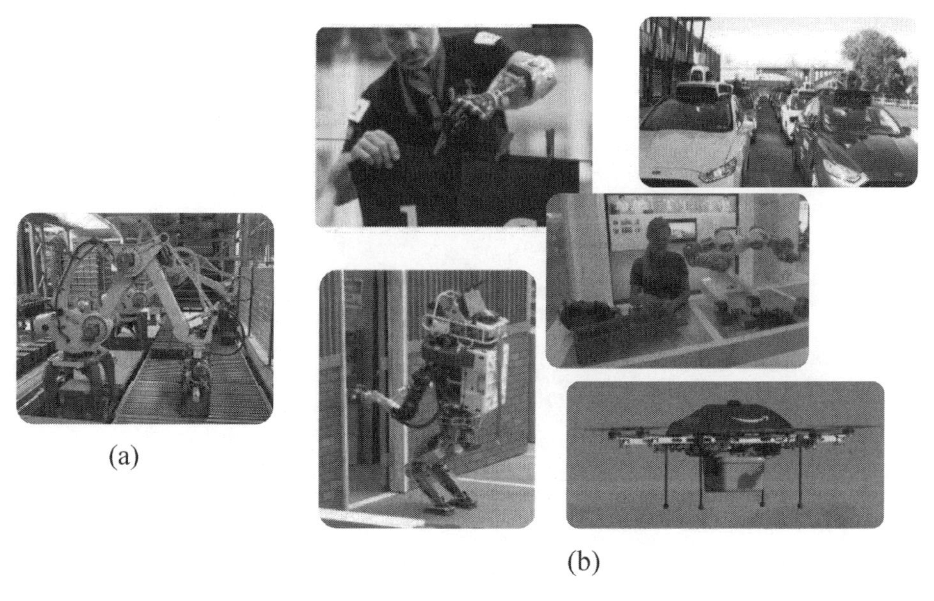

(a)

(b)

Figure 6.3 Example of complexity growth of machine design. (a) Traditional machines in well-structured environment with less challenges. (b) Complex machines in open and dynamic environments with more and more challenges.

the trend of robotic systems as an example of complexity growth of machine designs. Traditionally, industrial manipulators (Figure 6.3a) have a fixed base and operate in a well-structured environment; these manipulators are not designed to deal with unpredicted uncertainties and changes in environments. With the emerging needs of the capabilities for the collaboration and interaction with human beings, robots must be designed to be able to sense changes, detect events, and respond to objects in an open ill-structured environment safely. The complexity of robotic design has been increased greatly in terms of the degrees of motions, the number of sensors and actuators, and time-dependent characteristics.

Machine design is an essential step to build a machine since it determines how a machine should be structured, assembled, and functioned to meet the design requirements by users. Similar to the design of other products, basic elements in a machine design are *conceptions*, *inventions*, *visualizations*, *calculations*, *refinements*, and *specifications of details* that determine the machine structure. The design process is iterative and Figure 6.4 shows some critical steps. The axiomatic design theory (ADT) can be used as a guide for the design process. *Firstly*, the product needs in the customers' domain are analysed. *Secondly*, the identified needs are technically interpreted as the functional requirements (FRs) of products. *Thirdly*, design variables (DVs) for specified FRs are explored to define a design space of technical solutions. *Fourthly*, design analysis and synthesis are iteratively performed to evaluate and compare design alternatives until an optimized solution is obtained. *Finally*, the finalized design is prototyped, tested, and validated before it can actually be deployed in production. If a design fault or defect is identified, the above steps are repeated until all FRs can be satisfied optimally. As shown in Figure 6.4, designers can heavily rely on computer aided tools for design synthesis, design analysis, parametric studies, prototyping, and production.

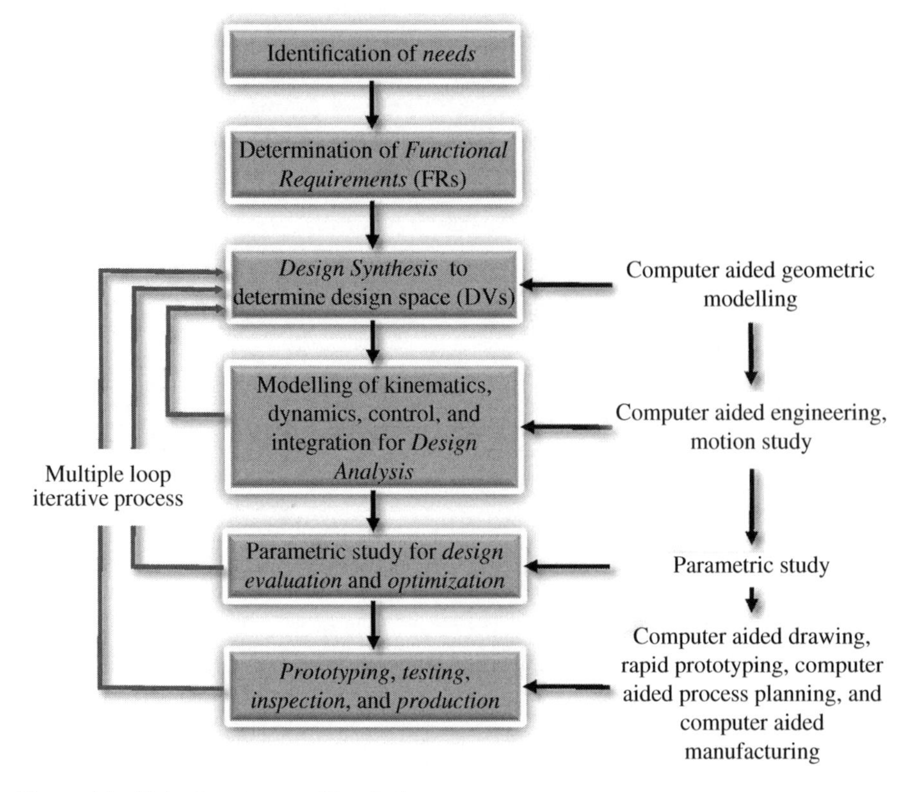

Figure 6.4 Main steps in a machine design process.

6.2 General Functional Requirements (FRs) of Machines

A design is the functional soul of any man-made creation. A design ends up expressing itself in successive outer layers of the product or service. Steve Jobs stated that '*design is not just what it looks like and feels like. Design is how it works*' (BrainyQuote 2019). To evaluate whether or not a design works, a set of functional requirements (FRs) should be firstly defined and the relationships between design variables (DVs) and system outputs are modelled to justify whether or not a specific design meets FRs.

For a machine design, engineers may consider and define FRs in multiple aspects such as functions, mechanical structure, environmental constraints, and economic, social, and political factors. Note that an FR can be *qualitative*, such as the appearance and attractiveness of a consumer product, or *quantifiable*, such as the maximum power of a machine. In design synthesis, the mathematical models should be developed to correlate DVs and FRs, so that the design alternatives of DVs can be evaluated against the given set of FRs. Table 6.1 shows the main aspects for quantifiable FRs of new machines.

It is interesting to note that FRs are interpreted as *design constraints* and *design criteria* or objectives in Figure 6.1. Therefore, FRs can be divided into *hard FRs* for design constraints and *soft FRs* for design criteria. The hard FRs give the strict constraints that a feasible solution must satisfy. The soft FRs define optimized goals of design solutions: the

Table 6.1 Categories of FRs in machine design.

Category of FRs	Typical design variables for FRs
Overall geometry	Structure, topology, dimensions, and footprints
Motion of parts	Type, direction, velocities, acceleration, and kinematics
External forces	Load direction, magnitude, load, and impact
Required energy	Heating, cooling, conversion, and pressure
Materials	Properties, transportation, and enhanced treatment
Control system	Electrical, hydraulic, mechanical, and pneumatic
Information flow	Inputs, outputs, communication, data changes, and visualization

better performance of a design solution in terms of soft FRs, the closer this solution is to an optimum one. Taking an example of a gear transmission design, the gear ratio, materials, and centre distance may be the hard FRs (or design constraints), and cost, hardness, and thickness may be the soft FRs, which are used as optimum criteria.

6.3 Fundamentals of Machine Design

Modern machines or products are mostly mechatronic systems, but they are built upon mechanical systems. Mechanics of mechanical systems is a field of study within applied mechanics that investigates the behaviour of a mechanical system subject to mechanical loads, such as the driven motions of objects and the deflections of machine elements.

A machine consists of a set of *links*, which are assembled together by *joints* to perform its functions. Therefore, the methods to represent links and joints are critical to machine designs. In this section, the representations of links and joints are discussed and the motions of mechanical systems are investigated after.

6.3.1 Link Types

In conceptual design of a mechanical system, any rigid body in a machine is referred to as a link. A mechanical system is an assembly of the links with the pre-determined topology for the connections. Links in a mechanical system can be classified based on the number of positions that can be used as ports to connect other links. As shown in Figure 6.5, a link is called (a) a binary link, (b) a ternary link, (c) a quaternary link, (d) a pentagon link, and (e) a hexagon link if the link has the corresponding number of ports that can be connected to other links, respectively. Note that a connection of one link to another implies that additional motion constraints are made for the relative motion of two links.

6.3.2 Joint Types and Degrees of Freedom (DoFs)

The motion of an object or a system is represented by the degrees of freedom of motion. The *degrees of freedom* (DoFs) is the number of minimal and independent parameters

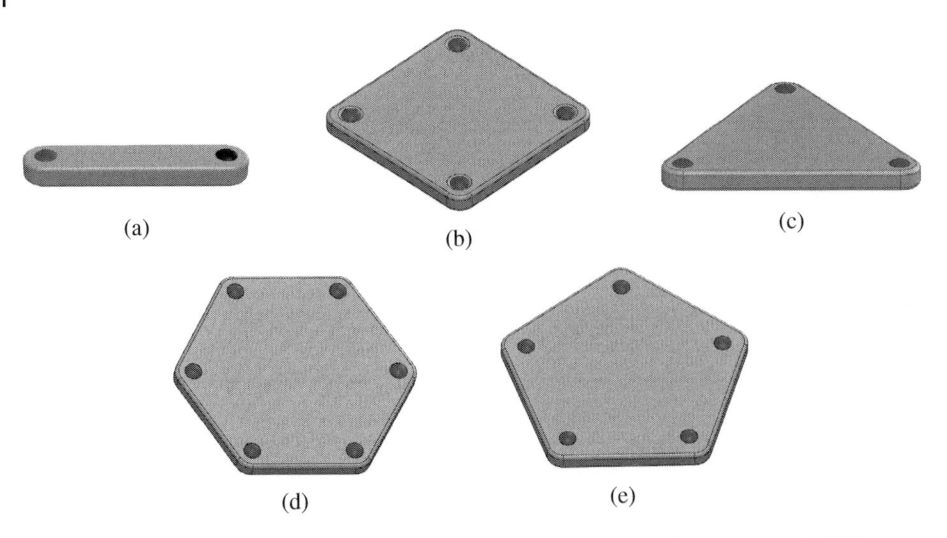

Figure 6.5 Classification of links. (a) Binary link. (b) Quaternary link. (c) Ternary link. (d) Hexagon link. (e) Pentagon link.

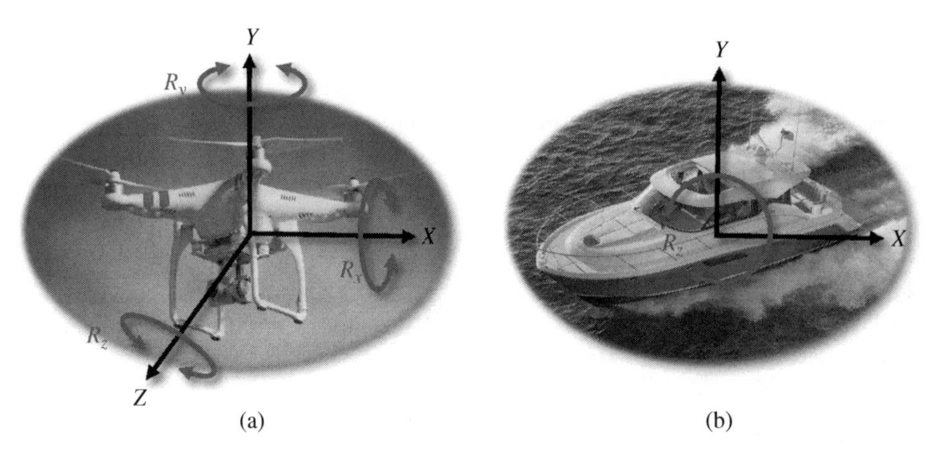

Figure 6.6 Free object and its degrees of freedom (DoF) of motion. (a) Six DoFs for an object in 3D space. (b) Three DoFs for an object in 2D space. (*See color plate section for color representation of this figure*).

(measurements) required to define the position and orientation of an object or system at any instant in time.

As shown in Figure 6.6a, a free body in 3D space possesses 6 DoFs. It can be translated along X, Y, Z axes and it can be rotated along X, Y, Z axes, respectively. A free body in a 2D space possesses 3 DoFs. As shown in Figure 6.6b, it can be translated along X and Y axes and it can be rotated along the Z axis, which is perpendicular to the plane of motion O-XY.

When two links are joined, certain constraint(s) apply to the jointed links, and the joint type determines the motion constraints. Figure 6.7 shows six joint types where the direction(s) of unconstrained motion(s) are illustrated.

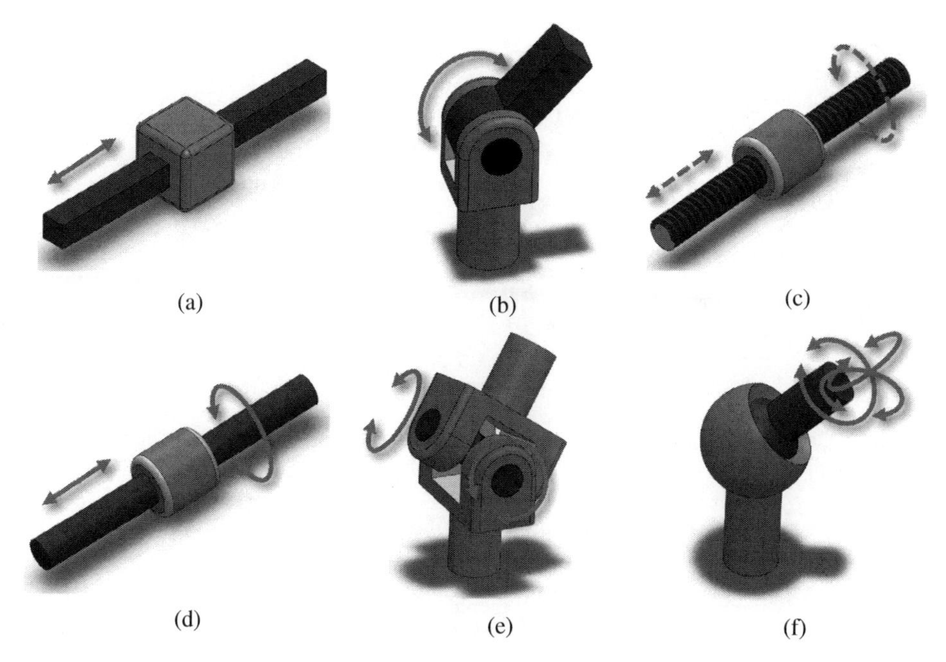

Figure 6.7 Classification of joints and the degrees of freedom of motion. (a) Prismatic joint. (b) Revolute joint. (c) Screw joint. (d) Cylindrical joint. (e) Universal joint. (f) Spherical joint. (*See color plate section for color representation of this figure*).

A *prismatic joint*, shown in Figure 6.7a, has one translational motion and a translation or rotation along any one of five other directions is constrained. A *revolute joint*, shown in Figure 6.7b, has one rotational motion. A *screw joint*, shown in Figure 6.7c, allows the translation and rotation along the same axis simultaneously, but these two motions are coupled and the joint only has one DoF. Different from a screw joint, a *cylindrical joint*, shown in Figure 6.7d, has one transition and one rotation along the same axis while these two motions are independent; the joint has two DoFs. A *universal joint* in Figure 6.7e and a *spherical joint* in Figure 6.7f have two and three rotations without any translations, respectively.

6.3.3 Kinematic Chains

A mechanical system without the consideration of its energy source and ground component is called a kinematic chain. A *kinematic chain* specifically refers to the topology of the assembly of rigid bodies (or links) by joints.

Figure 6.8a shows an *open-loop kinematic chain* where links carry one upon another in a serial. Due to a lower number of constrained motions occurring to links, an open-loop kinematic chain usually has a large range of motion but with limited strength to carry external loads. Any link or joint brings new sources of error; the errors from these links and joints are stacked up linearly. Theoretically, an open-loop chain has a relatively low accuracy.

A *closed-loop kinematic chain*, shown in Figure 6.8b, has two or several special links with more ports; these links are connected to a group of other links simultaneously. The load

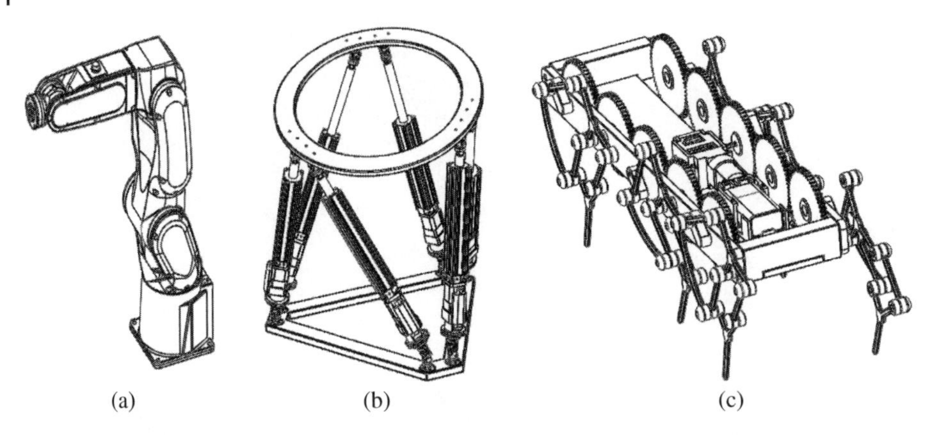

Figure 6.8 Classification of kinematic chains. (a) Open-loop. (b) Closed loop. (c) Combined chain.

on such a link will be shared by other connected links. Therefore, a closed-loop kinematic chain has a better capability to carry external loads. Since the same link is connected with others, the errors from other joints or links are averaged instead of stacked; therefore, a closed-loop kinematic chain is expected to have a better motion accuracy. On the other hand, it has a relatively small range of motion due to more constrained motion occurring in the links.

Both open-loop or closed-loop chains have their advantages and disadvantages. A *combined chain* in Figure 6.9 is introduced to make the trade-off between the loading capability and the range of motion; closed-loop sub-chains are used where the loads are large in comparison to the links, and open-loop sub-chains are used where the links need a large accessible space for the given tasks.

6.3.4 Mobility of Mechanical Systems

When the links, the joints, and the assembly topology are given, the mobility of a mechanical system can be determined. The *mobility* of a mechanical system (M) is quantified as the number of degrees of freedom (DoFs) of the system.

The DoF of a mechanical system is defined with respect to a selected reference frame called a ground reference frame:

$$M = \lambda(l - j - 1) + \sum_{i=1}^{j} f_i \tag{6.1}$$

where

M is degrees of freedom (DoFs) of the system
l is the total number of links, including the fixed link
n is the total number of joints
f_I is the degree of freedom of relative motion between the element pairs of the ith joint
λ is an integer $\lambda = 3$ for a plane mechanism and $\lambda = 6$ for a spatial mechanism

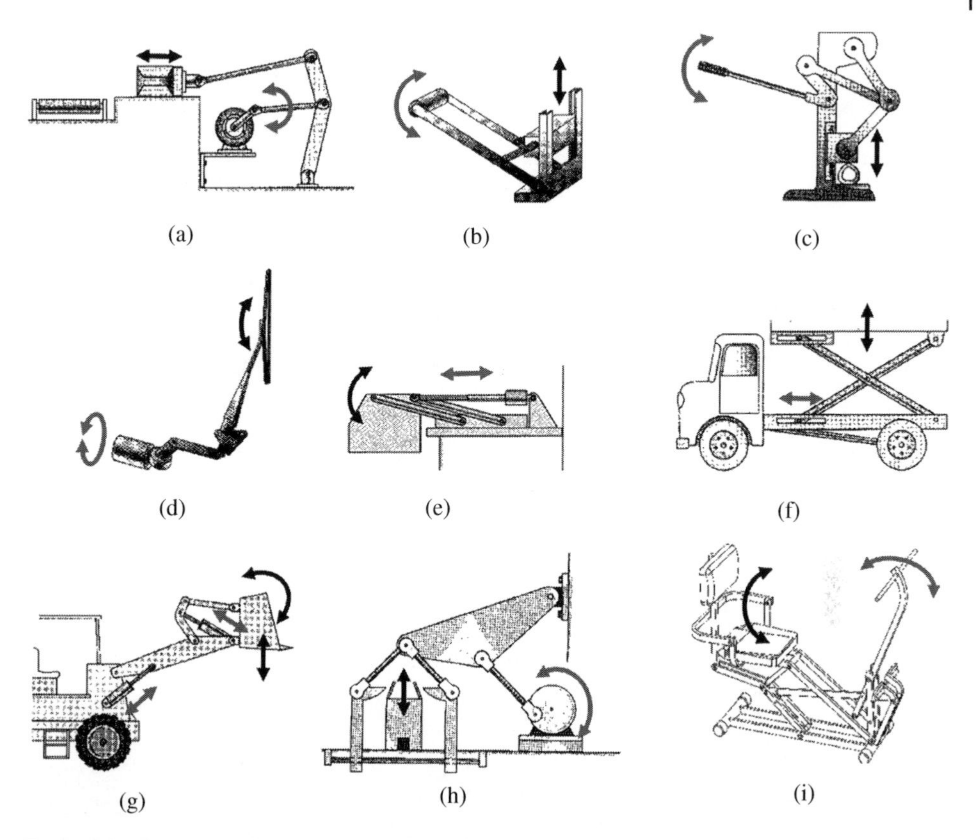

Figure 6.9 Examples of simple machines. (a) Package mover on an assembly bench. (b) Can crusher. (c) Simple press. (d) Car window wiper. (e) Microwave carrier to assist people in a wheelchair. (f) Lift platform in truck. (g) Front loader. (h) Box closer. (i) Mechanism on an exercise machine.

If a mechanism is planar and all of the joints are low-pair joints (prismatic joint or revolute joint), Eq. (6.1) can be simplified as the Gruebler equation as

$$M = 3(l - 1) - 2j \tag{6.2}$$

where

M is the degrees of freedom (DoFs) of a planar mechanism
l is the total number of links, including a fixed link
j is total number of low-pair joints

In a machine, the independent motions occur to active joints and active joints are driven by motors. The motions of active joints are transferred to the end-effector link where the task is performed. Figure 6.9 shows some simple machines with the driving motion (in red) and the driven motion at the end (in black). All of the example machines except Figure 6.9g have one DoF input and output, which implies that the same FR of motions can be satisfied by numerous solutions of the machine design.

At the conceptual design stage, the designer should be able to analyse the DoF when the structure of the machine is given.

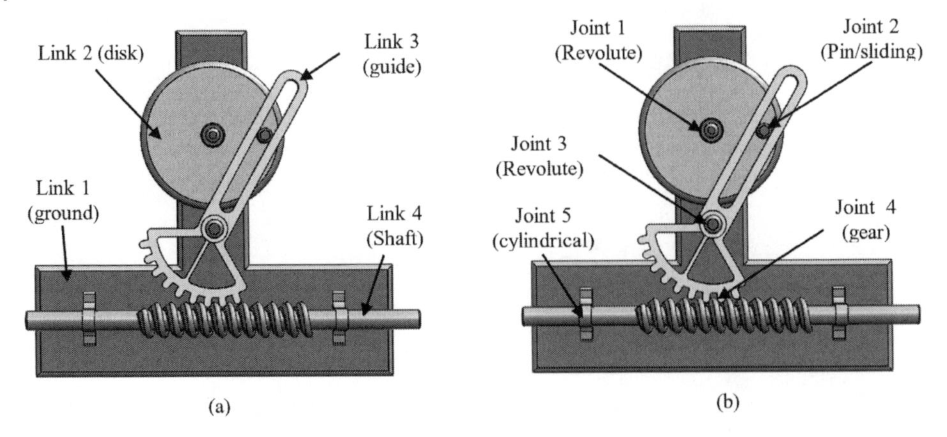

Figure 6.10 Mobility of a quick-return mechanism. (a) Indexed links. (b) Indexed joints.

Example 6.1 Evaluate the DoF of the quick-return mechanism shown in Figure 6.10.

Solution

As shown in Figure 6.10, the quick-return mechanism is a planar mechanism. Let the DoF of a link in a plane be $\lambda = 3$. The links and joints of the quick-return mechanism are denoted in Figure 6.10a and b, respectively. This shows that the total number of links including the ground link is $l = 4$. The total number of joints is $j = 5$. Except for the second and fifth joints $f_i = 2$ $(i = 2, 5)$, other joints have one DoF, i.e. $f_i = 1$ for $(i = 1, 3, \text{and } 4)$. Therefore, Eq. (6.1) is used to calculate the mobility of the quick-return mechanism as

$$M = \lambda(l - j - 1) + \sum_{i=1}^{j} f_i = 3(4 - 5 - 1) + (1 + 2 + 1 + 1 + 2) = 1 \tag{6.3}$$

Example 6.2 Calculate the DoF of the pumping mechanism in Figure 6.11.

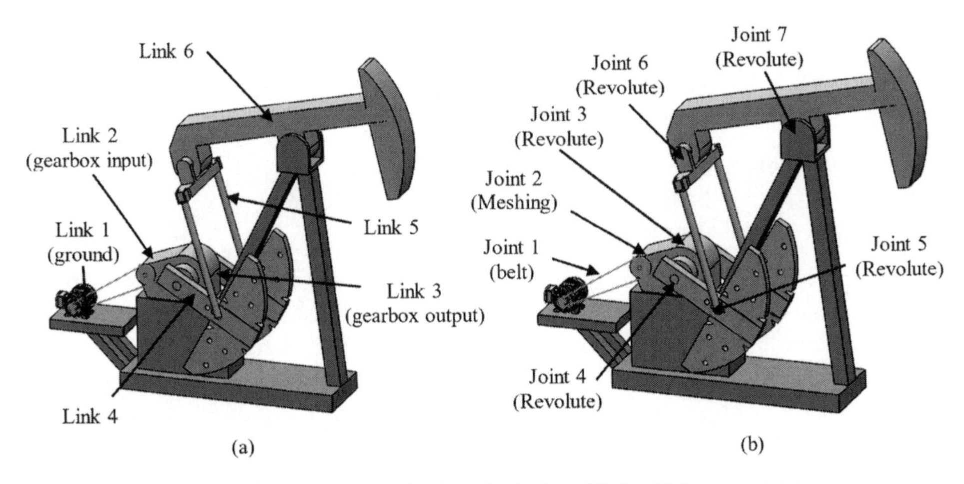

Figure 6.11 Mobility of a pumping mechanism. (a) Indexed links. (b) Indexed joints.

Solution

As shown in Figure 6.11, the pumping mechanism is a planar mechanism. Let the DoF of a link in a plane be $\lambda = 3$. The links and joints of the pumping mechanism are denoted in Figure 6.11a and b, respectively. It shows that the total number of links including the ground link is $l = 6$ and the total number of joints is $j = 7$. All of the joints have one DoF, i.e. $f_i = 1$ (for $i = 1, 2, 3, 4, 5, 6, 7$). Applying Eq. (6.1) yields the mobility of a pumping mechanism as

$$M = \lambda(l - j - 1) + \sum_{i=1}^{j} f_i = 3(6 - 7 - 1) + (1 + 1 + 1 + 1 + 1 + 1 + 1) = 1 \tag{6.4}$$

Example 6.3 Determine the mobilities of the two industrial robots shown in Figure 6.12.

Solution

The articulated robot in Figure 6.12a is a mechanism in 3D Cartesian space. Let the DoF of each link in the 3D space be $\lambda = 6$. The links and joints of the articulated robot are denoted in Figure 6.12a. The mechanism has a total of seven links and six revolute joints. Applying Eq. (6.1) gives the mobility of the given articulated robot as

$$M = \lambda(l - j - 1) + \sum_{i=1}^{j} f_i = 6(7 - 6 - 1) + (1 + 1 + 1 + 1 + 1 + 1) = 6 \tag{6.5}$$

The tripod parallel robot in Figure 6.12b is a mechanism in the 3D Cartesian space. Let the DOF of each link in the 3D space be $\lambda = 6$. The links and joints of the parallel robot are denoted in Figure 6.12b. It is a parallel structure with three symmetrical branches. Each branch consists of two links and three joints with its ends attached to the ground link and the end-effector link, respectively. Therefore, it has a total of eight links including the ground link and nine joints, i.e. three prismatic joints, three revolute joints, and three spherical joints. Applying Eq. (6.1) gives the mobility of the tripod parallel robot as

$$M = \lambda(l - j - 1) + \sum_{i=1}^{j} f_i = 6(8 - 9 - 1) + 3(1 + 1 + 3) = 3 \tag{6.6}$$

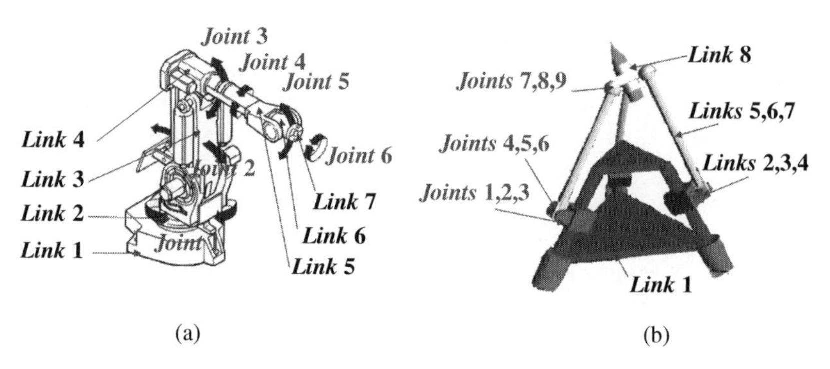

Figure 6.12 Mobility evaluation for industrial robots. (a) Articulated robot. (b) Tripod parallel robot.

6.4 Kinematic Synthesis

Kinematic synthesis is used to design a mechanism to obtain a specified motion or force at its output. Kinematic synthesis involves three main steps: (i) type synthesis, (ii) number synthesis, and (iii) dimensional synthesis (Hayes 2018; Reuleaux 1876). *Type synthesis* is used to define a design space with all suitable types and topologies of mechanisms. A mechanism of a machine consists of machine elements and the exemplifying types of machine elements are linkages, gears, cams and followers, belts and pulleys, and chains and sprockets. *Number synthesis* is used to determine the connections and topologies of the selected links for the given DoF of a mechanism. *Dimensional synthesis* is used to determine the dimensions of all selected machine elements for a mechanism.

6.4.1 Type Synthesis

Different from a mechanism, a machine requires generating a force to affect objects. Whatever the complexity of a machine is, its mechanical structure consists of an assembly of a set of basic mechanisms to transfer forces. Those basic mechanisms were identified by the German kinematician Franz Reuleaux (1876).

The selection of the type of mechanism needed to perform a specified motion depends to a great extent on design factors such as manufacturing processes, materials, safety, reliability, space, and economics, which are arguably outside the field of kinematics. Figure 6.13 shows some basic mechanisms that are often used as functional modules to build complex systems.

Even if the decision is made to use a planar four-bar mechanism, the type issue is still incomplete since it is followed with questions such as 'What is the type of a four-bar mechanism?' Note that a four-bar mechanism consists of four joints to connect four bars in a loop. Each joint has its choices of 'revolute' (R) or 'prismatic' (P) and there are 16 possible combinations of R and P pairs for the design of a four-bar mechanism. Figure 6.14a and b shows two extreme cases where the types of joints are 'revolute' or 'prismatic', respectively. In addition, any one of the four links can be the fixed link, and this leads to a different motion.

6.4.2 Number Synthesis

Number synthesis is the second step in the process of mechanism design. It determines the number of *DoF* and the number of required *links* and *joints*. As shown in Figure 6.15 any low-pair joint in a kinematic chain has one independent variable to describe the relative motion of two connected links. However, such a motion is not necessarily active. If a kinematic chain as a whole has 2 DoF, only two joints need to be active in order to avoid them being under-actuated or over-actuated.

If less actuators are used in a mechanism, the mechanism is referred to as an *under-actuated mechanism*; if more actuators are used in a mechanism, it is referred as an *over-actuated mechanism*. This reveals an issue in the understanding of a DoF concept. For a kinematic mechanism, DoF can be defined for the number of independent joints to be activated in *the joint space*. On the other hand, DoF can also be defined for the number of required independent motions of the tool on the mechanism in the *task space*. If the DoF

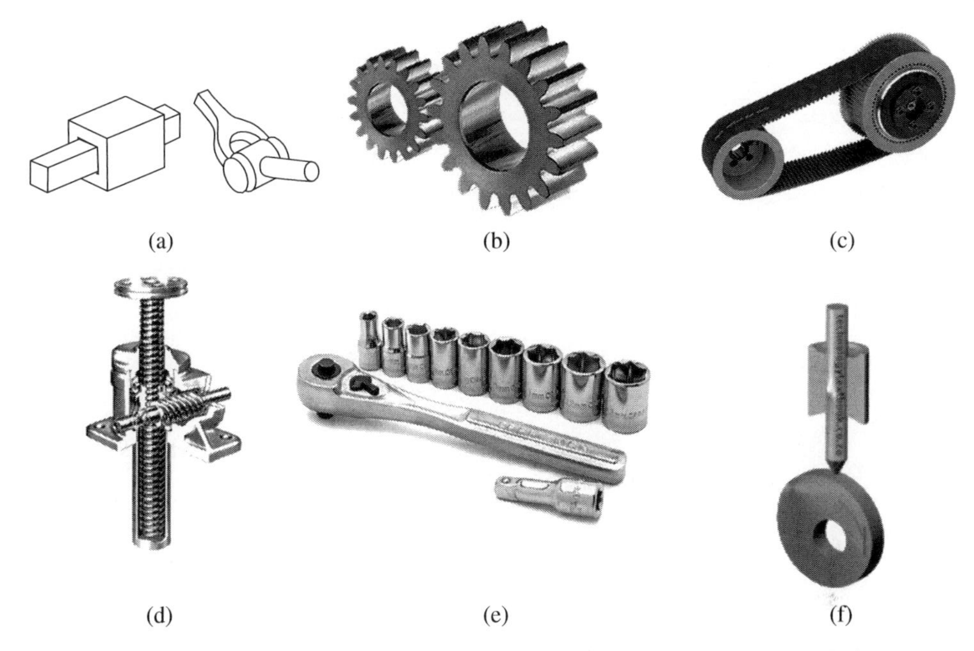

Figure 6.13 Basic mechanisms of machine elements. (a) Eye-bar type link (the lower pairs). (b) Wheel and axle including gears. (c) Tension–compression parts with one-way rigidity such as belts, chains, and hydraulic lines. (d) The screw, which transmits force and motion. (e) Intermittent-motion devices, such as rachets. (f) Cams in their many.

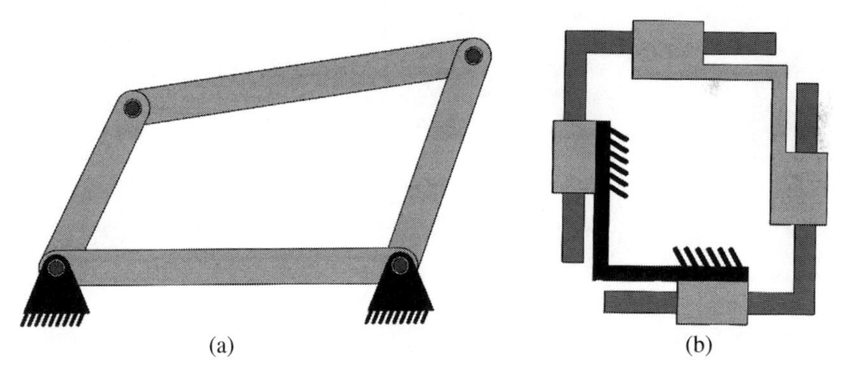

Figure 6.14 Examples of 16 different four-bar mechanisms in type synthesis. (a) RRRR. (b) PPPP.

in the joint space is larger than the DoF in the task space, the kinematic mechanism is an over-actuated mechanism. Figure 6.16 shows an example of an open-chain mechanism consisting of five links and four joints. The required DoF for a body in a plane is $M_{Task} = 3$ and the mobility of the kinematic mechanism is

$$M_{Joint} = \lambda(l - j - 1) + \sum_{i=1}^{j} f_i = 3(5 - 4 - 1) + (1 + 1 + 1 + 1) = 4 \tag{6.7}$$

where M_{Joint} is the mobility of the kinematic mechanism and $\lambda = 3$.

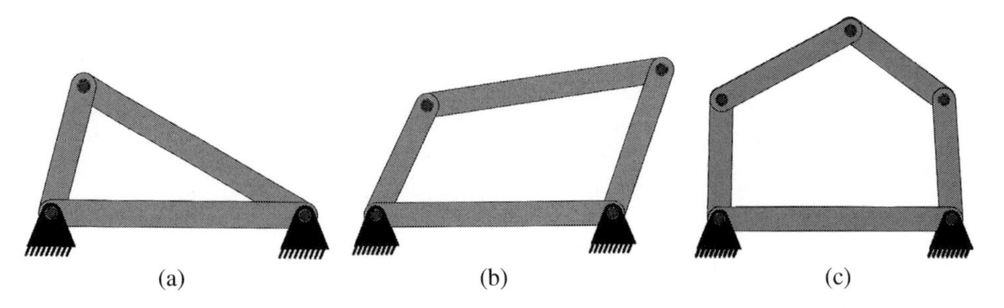

Figure 6.15 Kinematic chains with 0, 1, and 2 DoFs, requiring 0, 1, and 2 inputs. (a) $l = 3, j = 3$, DoF = 0, and no control is required. (b) $l = 4, j = 4$, DoF = 1, and one control is required. (c) $l = 5$, $j = 5$, DoF = 2, and two controls are required.

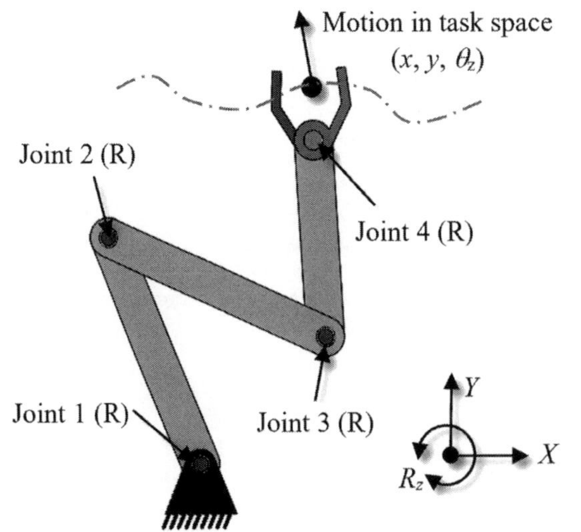

Figure 6.16 Over-actuated kinematic example.

Since $M_{Joint} = 4 > M_{Task} = 3$, the kinematic mechanism in Figure 6.16 is an over-actuated mechanism.

6.4.3 Dimensional Synthesis

The third step in a mechanism design is dimensional synthesis. *Dimensional synthesis* determines kinematic dimensions of links of the mechanism to satisfy the required motion characteristics. For the design of a traditional mechanical structure, a graphical method and analytical method can be selected based on the type of functional requirement (FR) of the mechanism. FRs of traditional mechanisms are classified into *motion generation*, *path generation*, *function generation*, and *dead-centre problems*.

6.5 Kinematics

6.5.1 Positions of Particles, Links, and Bodies in 2D and 3D Space

Kinematics is the study of positions, velocities, accelerations of particles and bodies, and their relations. The positions of a particle in a 2D or 3D space are described in Figure 6.17a and b, respectively. With respect to the coordinate system (CS) (*O-XY*), a position in a 2D space is determined as (x, y) and the position in a 3D space is determined as (x, y, z).

Other than a Cartesian CS, a polar CS, a cylindrical CS, or a spherical CS can also be used to represent a position P in a 2D or 3D space. The coordinates of the position with respect to a different CS can be determined by the coordinate transformation. Taking the examples of Figure 6.1a and b, the coordinates of *P* in a two-dimensional polar CS (r, θ) and three-dimensional cylindrical CS (r, θ, z) are transferred as below:

$$\left.\begin{array}{l} \begin{Bmatrix} x \\ y \end{Bmatrix} \textit{in Cartesian CS} \rightarrow \begin{Bmatrix} r \\ \theta \end{Bmatrix} = \begin{Bmatrix} \sqrt{x^2+y^2} \\ \theta = tan^{-1}\left(\frac{y}{x}\right) \end{Bmatrix} \textit{in polar CS} \\ \begin{Bmatrix} x \\ y \\ z \end{Bmatrix} \textit{in Cartesian CS} \rightarrow \begin{Bmatrix} r \\ \theta \\ z \end{Bmatrix} = \begin{Bmatrix} \sqrt{x^2+y^2} \\ \theta = tan^{-1}\left(\frac{y}{x}\right) \\ z \end{Bmatrix} \textit{in cylindrical CS} \end{array}\right\} \tag{6.8}$$

The positions of link *AB* in a 2D or 3D space are described in Figure 6.18a and b, respectively. With respect to the coordinate system (*O-XY*), the position of link AB in a 2D space is represented by the coordinates of two nodes (x_A, y_A) and (x_B, y_B), and the position in a 3D space is represented by the coordinates of two nodes (x_A, y_A, z_A) and (x_B, y_B, z_B). If

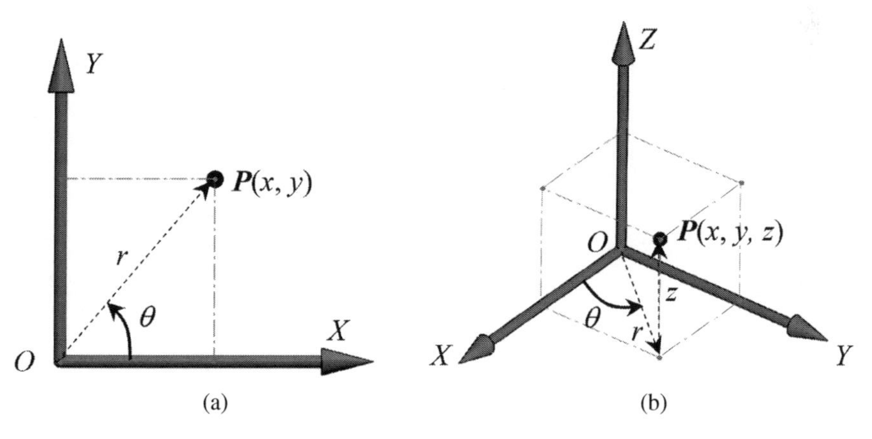

Figure 6.17 A position of particle in 2D and 3D space. (a) Particle *P* in the *O-XY* plane. (b) Particle *P* in the *O-XYZ* coordinate system.

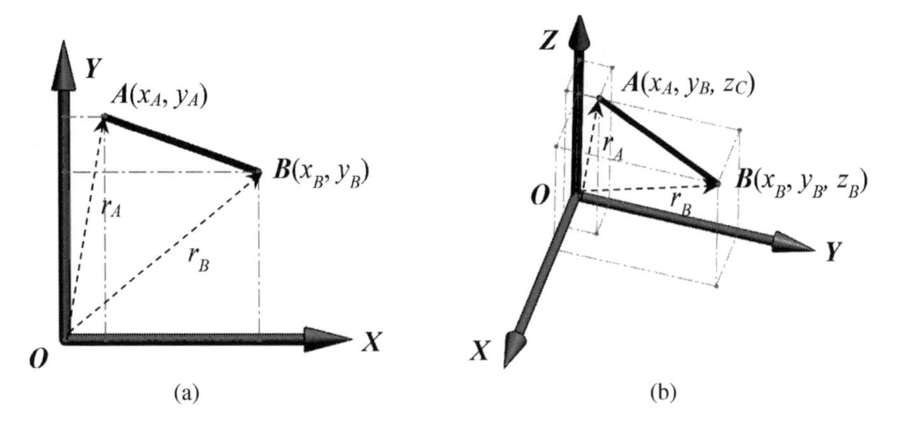

Figure 6.18 A line in 2D and 3D space. (a) Line A–B in the *O-XY* plane. (b) Line A–B in the *O-XYZ* coordinate system.

a CS other than the Cartesian CS is used, the coordinate transformations of the link are performed node by node individually.

Figure 6.19 shows that the presentation of the position and orientation of a body ABC in a 3D space is the Cartesian CS. The coordinates of three nodes on the object are sufficient to determine the position and orientation completely. To facilitate the kinematic and dynamic modelling, a local CS (O'-$X'Y'Z'$) is attached to the body locally, and the position of the body relative to the global CS is determined as

$$[\boldsymbol{T}] = \begin{bmatrix} \boldsymbol{R} & \boldsymbol{O}' \\ 0\ 0\ 0 & 1 \end{bmatrix} = \begin{bmatrix} (\boldsymbol{X}' \cdot \boldsymbol{X}) & (\boldsymbol{Y}' \cdot \boldsymbol{X}) & (\boldsymbol{Z}' \cdot \boldsymbol{X}) & O'_x \\ (\boldsymbol{X}' \cdot \boldsymbol{Y}) & (\boldsymbol{Y}' \cdot \boldsymbol{Y}) & (\boldsymbol{Z}' \cdot \boldsymbol{Y}) & O'_y \\ (\boldsymbol{X}' \cdot \boldsymbol{Z}) & (\boldsymbol{Y}' \cdot \boldsymbol{Z}) & (\boldsymbol{Z}' \cdot \boldsymbol{Z}) & O'_z \\ 0 & 0 & 0 & 1 \end{bmatrix} \tag{6.9}$$

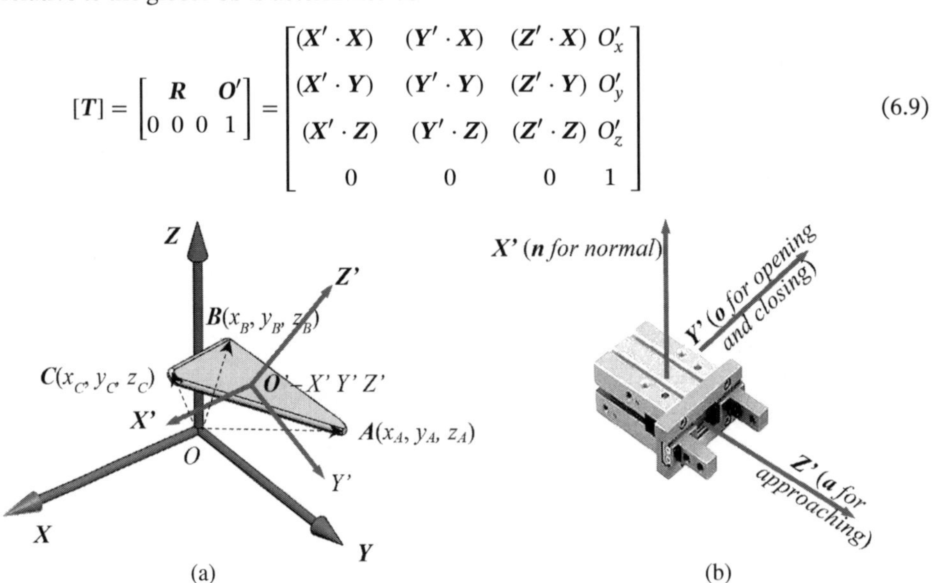

Figure 6.19 An object in 3D space. (a) Object in 3D space. (b) Correspondence of ***x***, ***y***, **z** and ***n***, ***o***, ***a*** in a tool frame.

where $[\boldsymbol{T}]$ is the homogenous representation of the local coordinate system (LCS) of the body with respect to the global CS (*O-XYZ*), $[\boldsymbol{R}]$ is a 3×3 matrix for the orientation of LCS, and $\boldsymbol{O}'$ is a 1×3 vector for the origin of LCS in the global CS. Note that the last row (0, 0, 0, 1) is added to make a square $[\boldsymbol{T}]$ so that the matrix manipulations such as an inverse operation can be valid directly.

As shown in Figure 6.19b, if the body is an end-effector tool in 3D space, the axes (X', Y', and Z') in an LCS are commonly denoted as the axes for $\boldsymbol{n}$ (a normal formed by $\boldsymbol{o}$ and $\boldsymbol{a}$ following a right -hand rule), $\boldsymbol{o}$ (the direction to open or close the tool), and $\boldsymbol{a}$ (the direction to approach an object to be manipulated), respectively. The vectors $\boldsymbol{n}$, $\boldsymbol{o}$, and $\boldsymbol{a}$ can be determined from Eq. (6.9) as

$$[\boldsymbol{T}] = \begin{bmatrix} \boldsymbol{n} & \boldsymbol{o} & \boldsymbol{a} & \boldsymbol{O}' \\ 0 & 0 & 0 & 1 \end{bmatrix} = \begin{bmatrix} (\boldsymbol{X}' \cdot \boldsymbol{X}) & (\boldsymbol{Y}' \cdot \boldsymbol{X}) & (\boldsymbol{Z}' \cdot \boldsymbol{X}) & O'_x \\ (\boldsymbol{X}' \cdot \boldsymbol{Y}) & (\boldsymbol{Y}' \cdot \boldsymbol{Y}) & (\boldsymbol{Z}' \cdot \boldsymbol{Y}) & O'_y \\ (\boldsymbol{X}' \cdot \boldsymbol{Z}) & (\boldsymbol{Y}' \cdot \boldsymbol{Z}) & (\boldsymbol{Z}' \cdot \boldsymbol{Z}) & O'_z \\ 0 & 0 & 0 & 1 \end{bmatrix} \tag{6.10}$$

6.5.2 Motions of Particles, Links, and Bodies

If a particle, link or a body moves, its position changes with respect to time. The movement can be classified into *translational* or *rotational* while the motion of a particle is always translational.

Figure 6.20 shows the description of the particle movement in a 2D or 3D space. If a particle moves in a 2D space, it possesses 2 DoFs; if it moves in a 3D space, it possesses 3 DoFs. The new position of the particle $\boldsymbol{P}'$ with a translation of displacement $\Delta\boldsymbol{P}$ from $\boldsymbol{P}$ can

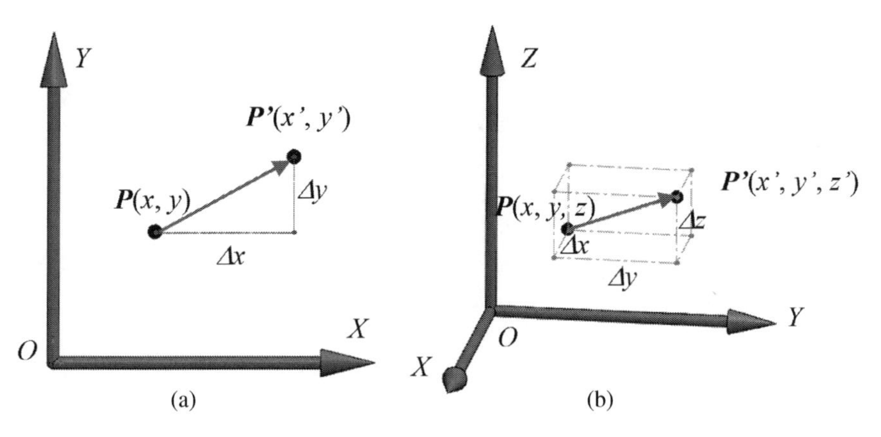

Figure 6.20 A motion of a particle in 2D and 3D space. (a) In the *O-XY* plane. (b) In the *O-XYZ* coordinate system.

be found as

$$\left.\begin{array}{l} \{\boldsymbol{P}\} = \begin{Bmatrix} x \\ y \end{Bmatrix} \rightarrow \{\boldsymbol{P}'\} = \begin{Bmatrix} x' \\ y' \end{Bmatrix} = \{\boldsymbol{P}\} + \{\Delta\boldsymbol{P}\} = \begin{Bmatrix} x+\Delta x \\ y+\Delta y \end{Bmatrix} \ \textit{in 2D space} \\ \{\boldsymbol{P}\} = \begin{Bmatrix} x \\ y \\ z \end{Bmatrix} \rightarrow \{\boldsymbol{P}'\} = \begin{Bmatrix} x' \\ y' \\ z' \end{Bmatrix} = \{\boldsymbol{P}\} + \{\Delta\boldsymbol{P}\} = \begin{Bmatrix} x+\Delta x \\ y+\Delta y \\ z+\Delta z \end{Bmatrix} \ \textit{in 3D space} \end{array}\right\} \tag{6.11}$$

where $\{\Delta\boldsymbol{P}\}$ is the displacement of movement and $\{\boldsymbol{P}'\}$ is the new position after the movement in a 2D or 3D space.

Figure 6.21 shows the description of the movement of link AB in a 2D space. With consideration of the orientation change, the link has three DoFs in the 2D space. It is convenient to describe such a motion with respect to a local CS attached to the link. Without losing generality, node A is selected as the origin and the axis is specified as the direction from node A to B. The movement of link AB can be represented by $(\Delta x, \Delta y, \Delta\theta)$, where $(\Delta x, \Delta y)$ are the translational displacements along axes of X and Y, respectively, and $\Delta\theta$ is the rotational displacement along axis Z. The new position of the particle $\{\boldsymbol{P}'\}$ with a translation of displacement $\{\Delta\boldsymbol{P}\}$ from $\{\boldsymbol{P}\}$ can be found as

$$\left.\begin{array}{l} \{\boldsymbol{A}\} = \begin{Bmatrix} x_A \\ y_A \end{Bmatrix} \rightarrow \{\boldsymbol{A}''\} = \begin{Bmatrix} x''_A \\ y''_A \end{Bmatrix} = \{\boldsymbol{A}\} + \{\Delta\boldsymbol{P}_T\} = \begin{Bmatrix} x_A+\Delta x \\ y_A+\Delta y \end{Bmatrix} \ \textit{for A (origin)} \\ \{\boldsymbol{B}\} = \begin{Bmatrix} x_B \\ y_B \end{Bmatrix} \rightarrow \left\{\begin{array}{l} \{\boldsymbol{B}''\} = \begin{Bmatrix} x''_B \\ y''_B \end{Bmatrix} = \{\boldsymbol{B}\} + \{\Delta\boldsymbol{P}_T\} + \{\Delta\boldsymbol{P}_R\} \\ \quad = \begin{Bmatrix} x_B+\Delta x \\ y_B+\Delta y \end{Bmatrix} + L\begin{Bmatrix} \cos(\theta+\Delta\theta)-\cos\theta \\ \sin(\theta+\Delta\theta)-\sin\theta \end{Bmatrix} \end{array}\right. \textit{for B}\begin{pmatrix}\textit{any point} \\ \textit{on link}\end{pmatrix} \end{array}\right\} \tag{6.12}$$

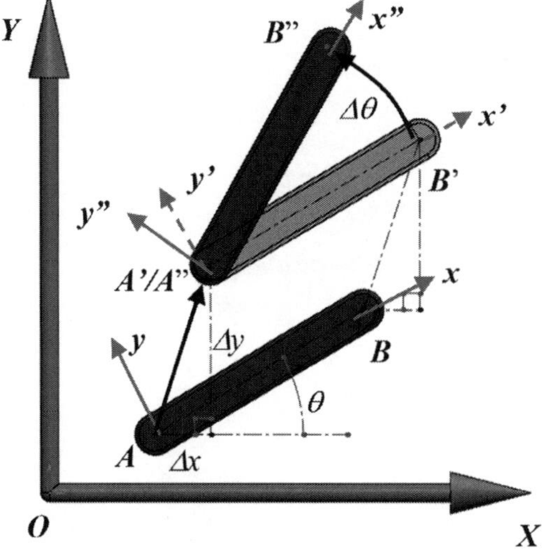

Figure 6.21 General motion of a link in 2D space.

where $\{\Delta \boldsymbol{P}_T\}$ and $\{\Delta \boldsymbol{P}_R\}$ are the displacements by translation and rotation, respectively, L is the length of link AB, θ and $(\theta + \Delta\theta)$ are the directions link AB before and after the motion, and

$$L = \sqrt{(x_B - x_A)^2 + (y_B - y_A)^2}, \theta = tan^{-1}\frac{y_B - y_A}{x_B - x_A}$$

Figure 6.22 shows the description of the generic movement of an object in 3D space. The motion of an object can be described by the coordinate transformation of an LCS from its original position $\{\boldsymbol{o}\text{-}\boldsymbol{xyz}\}$ to the new position $\{\boldsymbol{o}'\text{-}\boldsymbol{x}'\boldsymbol{y}'\boldsymbol{z}'\}$. The homogenous matrix can be used to represent the object at its new position as

$$[\boldsymbol{T}'] = \begin{bmatrix} [\Delta \boldsymbol{R}] & \{\Delta \boldsymbol{P}\} \\ 0\ 0\ 0 & 1 \end{bmatrix} = \begin{bmatrix} \boldsymbol{x}' \cdot \boldsymbol{x} & \boldsymbol{y}' \cdot \boldsymbol{x} & \boldsymbol{z}' \cdot \boldsymbol{x} & \Delta P_x \\ \boldsymbol{x}' \cdot \boldsymbol{y} & \boldsymbol{y}'' \cdot \boldsymbol{y} & \boldsymbol{z}' \cdot \boldsymbol{y} & \Delta P_y \\ \boldsymbol{x}' \cdot \boldsymbol{z} & \boldsymbol{y}' \cdot \boldsymbol{z} & \boldsymbol{z}' \cdot \boldsymbol{z} & \Delta P_z \\ 0 & 0 & 0 & 1 \end{bmatrix} \tag{6.13}$$

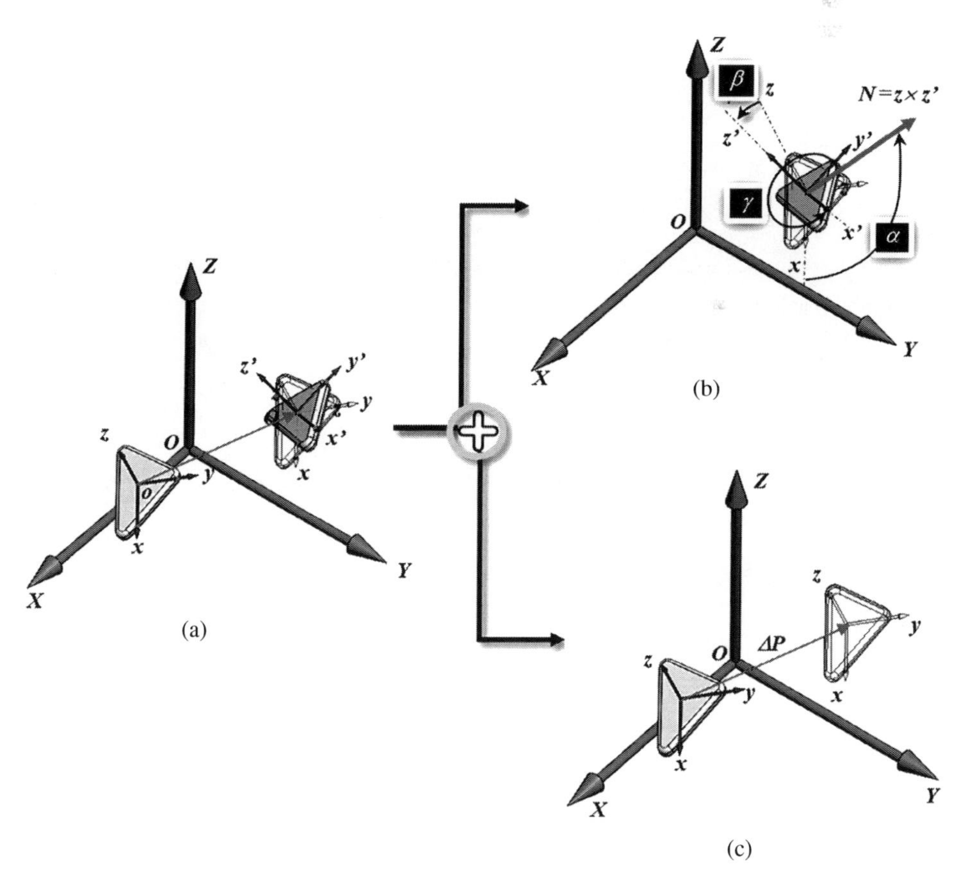

Figure 6.22 General motion of a body in 3D space. (a) Translation and rotation of object in 3D. (b) Rotation object in 3D ($\boldsymbol{\Delta R}$). (c) Translation of an object in 3D ($\boldsymbol{\Delta P}$).

The rotation of an object along an axis of the reference coordinate system is called a *pure rotation*. As shown in Figure 6.22, assume that the axes ($\boldsymbol{x}, \boldsymbol{y}, \boldsymbol{z}$) of the original LCS are aligned with the axes ($\boldsymbol{X}, \boldsymbol{Y}, \boldsymbol{Z}$) of the reference CS except for its origin at (p_x, p_y, p_z), i.e. the position of the object before a pure rotation can be represented as a homogeneous matrix as

$$
{}_2^1[\boldsymbol{T}] = \begin{bmatrix} \boldsymbol{x}\cdot\boldsymbol{X} & \boldsymbol{y}\cdot\boldsymbol{X} & \boldsymbol{z}\cdot\boldsymbol{X} & p_x \\ \boldsymbol{x}\cdot\boldsymbol{Y} & \boldsymbol{y}\cdot\boldsymbol{Y} & \boldsymbol{z}\cdot\boldsymbol{Y} & p_y \\ \boldsymbol{x}\cdot\boldsymbol{Z} & \boldsymbol{y}\cdot\boldsymbol{Z} & \boldsymbol{z}\cdot\boldsymbol{Z} & p_z \\ 0 & 0 & 0 & 1 \end{bmatrix} = \begin{bmatrix} 1 & 0 & 0 & p_x \\ 0 & 1 & 0 & p_y \\ 0 & 0 & 1 & p_z \\ 0 & 0 & 0 & 1 \end{bmatrix} \tag{6.14}
$$

where ${}_1^0[\boldsymbol{T}]$ is the coordinate transformation for the object from LCS {**o**-***xyz***} to the reference CS {***O-XYZ***}.

If the object has a pure rotation (θ_x) along the $\boldsymbol{x}$-axis, as shown in Figure 6.23a, the unit vector along the $\boldsymbol{x}$-axis remains the sameand the unit vectors of along the $\boldsymbol{y}$- and $\boldsymbol{z}$-axes are rotated along the $\boldsymbol{x}$-axis with θ_x. The coordinate transformation of object from original position to the new position after a pure rotation (θ_x) along the $\boldsymbol{x}$-axis is given as

$$
{}_1^0[\boldsymbol{T}] = \begin{bmatrix} \boldsymbol{x}'\cdot\boldsymbol{x} & \boldsymbol{y}'\cdot\boldsymbol{x} & \boldsymbol{z}'\cdot\boldsymbol{x} & 0 \\ \boldsymbol{x}'\cdot\boldsymbol{y} & \boldsymbol{y}'\cdot\boldsymbol{y} & \boldsymbol{z}'\cdot\boldsymbol{y} & 0 \\ \boldsymbol{x}'\cdot\boldsymbol{z} & \boldsymbol{y}'\cdot\boldsymbol{z} & \boldsymbol{z}'\cdot\boldsymbol{z} & 0 \\ 0 & 0 & 0 & 1 \end{bmatrix} = \begin{bmatrix} & & & 0 \\ & [\boldsymbol{R}_x]_{3\times3} & & 0 \\ & & & 0 \\ 0 & 0 & 0 & 1 \end{bmatrix} = \begin{bmatrix} 1 & 0 & 0 & 0 \\ 0 & c\theta_x & -s\theta_x & 0 \\ 0 & s\theta_x & c\theta_x & 0 \\ 0 & 0 & 0 & 1 \end{bmatrix} \tag{6.15}
$$

where '*c*' and '*s*' stand for the *cosine* and *sine* functions, respectively, and ${}_2^1[\boldsymbol{T}]$ is the coordinate transformation for the object from the rotated LCS {**o′**-***x′y′z′***} to the original LCS {**o**-***xyz***}. After a pure rotation of the object along the $\boldsymbol{x}$-axis, the position of the object in the

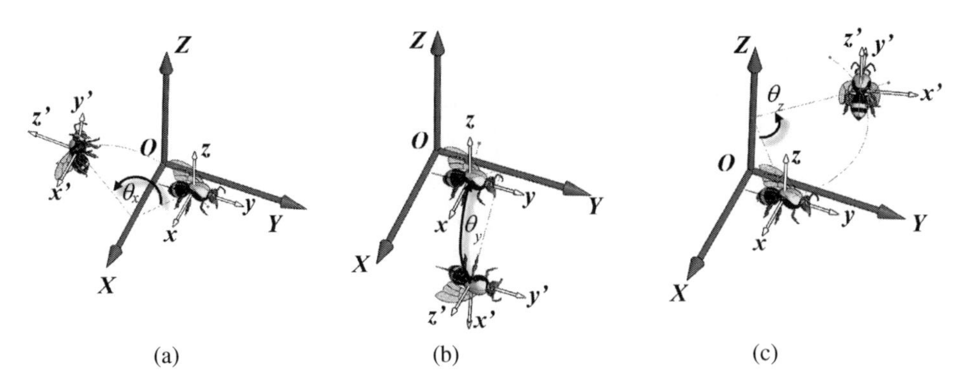

Figure 6.23 Rotation along the axes of CS. (a) Rotation along the *X*-axis. (b) Rotation along the *Y*-axis. (c) Rotation along the *Z*-axis.

reference CS becomes

$$[\boldsymbol{T}]_{\theta_x} = {}^0_1[\boldsymbol{T}] \cdot {}^1_2[\boldsymbol{T}] = \begin{bmatrix} [\boldsymbol{R}_x]_{3\times3} & [\boldsymbol{R}_x]_{3\times3}\{\boldsymbol{P}\}_{1\times3} \\ 0\ 0\ 0 & 1 \end{bmatrix} = \begin{bmatrix} 1 & 0 & 0 & p_x \\ 0 & c\theta_x & -s\theta_x & p_y c\theta_x - p_z s\theta_x \\ 0 & s\theta_x & c\theta_x & p_y s\theta_x + p_z c\theta_x \\ 0 & 0 & 0 & 1 \end{bmatrix} \tag{6.16}$$

where $[\boldsymbol{T}]_{\theta_x}$ is the representation of the object with a pure rotation θ_x along the $\boldsymbol{x}$-axis in the reference CS {***O-XYZ***}.

Due to the symmetry for the selection of an $\boldsymbol{x}$-axis, the representation of the object with a pure rotation θ_y along the $\boldsymbol{y}$-axis or θ_z along the $\boldsymbol{z}$-axis (shown in Figure 6.23b and c) can also be determined readily as

$$[\boldsymbol{T}]_{\theta_y} = \begin{bmatrix} c\theta_y & 0 & -s\theta_y & p_x c\theta_y - p_z s\theta_y \\ 0 & 1 & 0 & p_y \\ s\theta_y & 0 & c\theta_y & p_x s\theta_y + p_z c\theta_y \\ 0 & 0 & 0 & 1 \end{bmatrix} \tag{6.17}$$

$$[\boldsymbol{T}]_{\theta_z} = \begin{bmatrix} c\theta_z & -s\theta_z & 0 & p_x c\theta_z - p_y s\theta_z \\ s\theta_z & c\theta_z & 0 & p_x s\theta_z + p_y c\theta_z \\ 0 & 0 & 1 & p_z \\ 0 & 0 & 0 & 1 \end{bmatrix} \tag{6.18}$$

The rotation of the object can also be along an arbitrary axis other than any of the existing axes in a reference CS. Note that a free object in 3D space has six DoFs. Three DOFs represent the translation $\Delta\boldsymbol{P}$ in three axes (ΔP_x, ΔP_y, ΔP_z), and the other three DoFs are included implicitly in $\Delta\boldsymbol{R}_{3\times3}$ to represent the rotations along the three axes. To represent the rotational motion of an object with three explicit variables, the Euler angles (α, β, γ) can be used. *The Euler angles* are three angles introduced by Leonhard Euler to describe the orientation of a rigid body with respect to a reference coordinate system. The Euler angles are widely used to represent the orientation of a moving object. Given a reference CS (***o-xyz***), any orientation can be achieved by composing three elemental rotations, i.e. the rotations about the axes defined in an LCS attached on to an object.

As shown in Figure 6.23c, assume the starting and finishing positions of the object in {***O-XYZ***} are {***o-xyz***} and {***o′-x′y′z′***}, respectively, and $\Delta\boldsymbol{R}_{3\times3}$ is formed by a series of rotations.

Using the information in Table 6.2, $\Delta\boldsymbol{R}_{3\times3}$ can be expressed in terms of the Euler angles (α, β, γ) as

$$\Delta\boldsymbol{R}_{3\times3} = [\boldsymbol{R}_{z,\alpha}] \cdot [\boldsymbol{R}_{x,\beta}] \cdot [\boldsymbol{R}_{z,\gamma}] = \begin{bmatrix} c\alpha c\gamma - s\alpha c\beta s\gamma & -c\alpha s\gamma - s\alpha c\beta c\gamma & s\alpha s\beta \\ s\alpha c\gamma + c\alpha c\beta s\gamma & -s\alpha s\gamma + c\alpha c\beta c\gamma & -c\alpha s\beta \\ s\beta s\gamma & s\beta c\gamma & c\beta \end{bmatrix} \tag{6.19}$$

Table 6.2 Steps of orientation transformation by Euler angles.

Step	Action
1	Define a tentative rotational axis $\boldsymbol{N}$ as $\boldsymbol{N} = \boldsymbol{z} \times \boldsymbol{z}'$.
2	Take $\boldsymbol{z}$ as the rotational axis for the rotation of α to align $\boldsymbol{x}$ with $\boldsymbol{N}$.
3	Take $\boldsymbol{N}$ (or new $\boldsymbol{x}$ ($\boldsymbol{x}'$)) as the rotational axis for the rotation of β to align $\boldsymbol{z}$ with $\boldsymbol{z}'$.
4	Take $\boldsymbol{z}'$ as the rotational axis for the rotation of γ to align $\boldsymbol{N}$ with $\boldsymbol{x}'$.
5	Use the right-hand rule to determine $\boldsymbol{y}'$ from $\boldsymbol{x}'$ and $\boldsymbol{z}'$.

where $[\boldsymbol{R}_{z,\alpha}]$, $[\boldsymbol{R}_{z,\beta}]$, $[\boldsymbol{R}_{z,\gamma}]$ are the 3×3 matrices of rotation of (α, β, γ) along $\boldsymbol{z}$, $\boldsymbol{x}$, and $\boldsymbol{z}$ in the sequence.

6.5.3 Vector-Loop Method for Motion Analysis of a Plane Mechanism

The motion of a planar mechanism can be tackled by the vector-loop method. In a vector-loop method, a link is represented by its nodes, each node is represented by its coordinates, and the constraints on the kinematic chains of the mechanism are expressed as the loops of vectors. The four-bar mechanism in Figure 6.24 is used to illustrate how the vector-loop method can be used to solve kinematic problems for a planar mechanism.

A mechanism has a number of driving motions as inputs and the motions of its end-effector or tooling tip as outputs. A kinematic model is the mathematic model for the relations of inputs and outputs. A *forward kinematic solver* is used to obtain the output motions when the input motions are given, while an *inverse kinematic solver* is used to calculate the input motions when the required output motions are given. Here, the vector-loop method is applied to solve both forward kinematic and inverse kinematic problems of a four-bar mechanism.

As shown in Figure 6.24, a four-bar mechanism consists of four links and four revolute joints, which are used to connected the links in a loop. Thus, the mobility of the four-bar

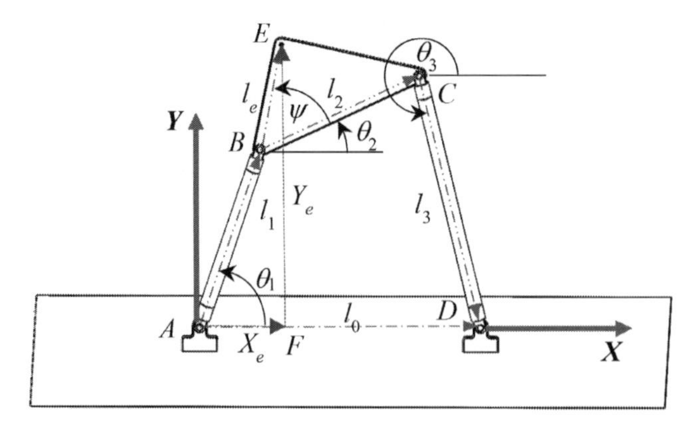

Figure 6.24 Vector-loop method for a four-bar mechanism.

mechanism is calculated as

$$M = \lambda(l - j - 1) + \sum_{i=1}^{j} f_i = 3(4 - 4 - 1) + 4 \times 1 = 1 \tag{6.20}$$

The mechanism has one DoF. Assume link AB is the driving link and the output path is defined at E on the coupler BCE, the input motion is θ_1, and the output is the position of E (X_e, Y_e).

To solve the forward kinematic problem, i.e. calculating the outputs of (X_e, Y_e), $(\dot{X}_e, \dot{Y}_e)$, $(\ddot{X}_e, \ddot{Y}_e)$ based on the given inputs of $\theta_1, \dot{\theta}_1, \ddot{\theta}_1$, the dependent motions of passive joints θ_2 and θ_3 have to be determined by using the vector-loop method. The mechanism includes only one loop as follows:

$$\overrightarrow{\boldsymbol{AB}} + \overrightarrow{\boldsymbol{BC}} + \overrightarrow{\boldsymbol{CD}} - \overrightarrow{\boldsymbol{AD}} = \boldsymbol{0} \tag{6.21}$$

Letting $\theta_4 = 0$ and using the exponential expression of the vectors in Eq. (6.21) gives

$$l_1 e^{j\theta_1} + l_2 e^{j\theta_2} + l_3 e^{j\theta_3} - l_0 e^{j0} = \boldsymbol{0} \tag{6.22}$$

Decomposing the exponential expression of Eq. (6.22) into the constraints of the kinematic chain projected on the X-axis and Y-axis yields

$$\left.\begin{aligned} l_1 c\theta_1 + l_2 c\theta_2 + l_3 c\theta_3 - l_0 = 0 \\ l_1 s\theta_1 + l_2 s\theta_2 + l_3 s\theta_3 = 0 \end{aligned}\right\} \tag{6.23}$$

Equation (6.23) leads to the solutions of θ_2 and θ_3 in terms of θ_1 as

$$\left.\begin{aligned} \theta_3 &= \tan^{-1}\left(\frac{-K_2 \pm \sqrt{K_2^2 - K_3^2 + K_1^2}}{K_3 - K_1}\right) \\ \theta_2 &= \tan^{-1}\left(\frac{l_1 s\theta_1 + l_3 s\theta_3}{l_1 c\theta_1 + l_3 c\theta_3 - l_0}\right) \end{aligned}\right\} \tag{6.24}$$

where K_1, K_2, and K_3 are dependent intermediate parameters as

$$\left.\begin{aligned} K_1 &= 2l_3(l_0 - l_1 c\theta_1) \\ K_2 &= 2l_3 l_1 s\theta_1 \\ K_3 &= l_2^2 - (l_0^2 + l_1^2 + l_3^2 - 2l_0 l_1 c\theta_1) \end{aligned}\right\} \tag{6.25}$$

Note from Eq. (6.24) that the multiple solutions exist for a mechanism with a closed loop. Once θ_2 and θ_3 are solved, another vector loop can be developed to calculate the tooling position $\boldsymbol{E}$ as

$$\overrightarrow{\boldsymbol{AB}} + \overrightarrow{\boldsymbol{BE}} - \overrightarrow{\boldsymbol{AF}} - \overrightarrow{\boldsymbol{FE}} = \boldsymbol{0} \tag{6.26}$$

Using the exponential expression of the vectors in Eq. (6.26) gives

$$l_1 e^{j\theta_1} + l_2 e^{j(\theta_2 + \psi)} - X_e e^{j0^\circ} - Y_e e^{j90^\circ} = \boldsymbol{0} \tag{6.27}$$

Equation (6.27) yields the solution of the forward kinematic problem as

$$\left.\begin{aligned} X_e &= l_1 c\theta_1 + l_2 c(\theta_2 + \psi) \\ Y_e &= l_1 s\theta_1 + l_2 s(\theta_2 + \psi) \end{aligned}\right\} \tag{6.28}$$

where (X_e, Y_e) is the tooling position and θ_2 is calculated from Eq. (6.24) when θ_1 is given.

To model the velocity relation of inputs and outputs, taking the derivation of Eq. (6.28) with respect to time gives

$$\left.\begin{aligned} &-l_1 s\theta_1 \cdot \dot{\theta}_1 - l_2 s\theta_2 \cdot \dot{\theta}_2 - l_3 s\theta_3 \cdot \dot{\theta}_3 = 0 \\ &l_1 c\theta_1 \cdot \dot{\theta}_1 + l_2 c\theta_2 \cdot \dot{\theta}_2 + l_3 c\theta_3 \cdot \dot{\theta}_3 = 0 \end{aligned}\right\} \tag{6.29}$$

Equation (6.29) gives the dependence of $\dot{\theta}_2$ and $\dot{\theta}_3$ on $\dot{\theta}_1$ as

$$\left.\begin{aligned} \dot{\theta}_2 &= \frac{l_1 s(\theta_3 - \theta_1)}{l_2 s(\theta_2 - \theta_3)}\dot{\theta}_1 \\ \dot{\theta}_3 &= \frac{l_1 s(\theta_2 - \theta_1)}{l_3 s(\theta_3 - \theta_2)}\dot{\theta}_1 \end{aligned}\right\} \tag{6.30}$$

Taking the first derivative of Eq. (6.28) with respect to time and using Eq. (6.30) give the velocity relation of inputs and outputs as

$$\left.\begin{aligned} \dot{X}_e &= -l_1\left(s\theta_1 + \frac{s(\theta_3 - \theta_1)s(\theta_2 + \psi)}{s(\theta_2 - \theta_3)}\right)\dot{\theta}_1 \\ \dot{Y}_e &= l_1\left(c\theta_1 + \frac{s(\theta_3 - \theta_1)c(\theta_2 + \psi)}{s(\theta_2 - \theta_3)}\right)\dot{\theta}_1 \end{aligned}\right\} \tag{6.31}$$

Taking the other derivative of Eq. (6.31) with respect to time using Eq. (6.30) gives the acceleration relation of inputs and outputs as

$$\left.\begin{aligned} \ddot{X}_e &= -l_1\left(s\theta_1 + \frac{s(\theta_3 - \theta_1)s(\theta_2 + \psi)}{s(\theta_2 - \theta_3)}\right)\ddot{\theta}_1 \\ &\quad -l_1\left(\begin{aligned} &c\theta_1\dot{\theta}_1 + \frac{c(\theta_2 - \theta_3)s(\theta_3 - \theta_1)s(\theta_2 + \psi)(\dot{\theta}_2 - \dot{\theta}_3)}{s^2(\theta_2 - \theta_3)} \\ &+ \frac{c(\theta_3 - \theta_1)s(\theta_2 + \psi)(\dot{\theta}_3 - \dot{\theta}_1)}{s(\theta_2 - \theta_3)} \\ &+ \frac{s(\theta_3 - \theta_1)c(\theta_2 + \psi)\dot{\theta}_2}{s(\theta_2 - \theta_3)}\dot{\theta}_1 \end{aligned}\right) \\ \ddot{Y}_e &= l_1\left(c\theta_1 + \frac{s(\theta_3 - \theta_1)c(\theta_2 + \psi)}{s(\theta_2 - \theta_3)}\right)\ddot{\theta}_1 \\ &\quad +l_1\left(\begin{aligned} &- s\theta_1\dot{\theta}_1 + \frac{c(\theta_2 - \theta_3)s(\theta_3 - \theta_1)c(\theta_2 + \psi)(\dot{\theta}_2 - \dot{\theta}_3)}{s^2(\theta_2 - \theta_3)} \\ &+ \frac{c(\theta_3 - \theta_1)c(\theta_2 + \psi)(\dot{\theta}_3 - \dot{\theta}_1)}{s(\theta_2 - \theta_3)} \\ &- \frac{s(\theta_3 - \theta_1)s(\theta_2 + \psi)\dot{\theta}_2}{s(\theta_2 - \theta_3)} \end{aligned}\right)\dot{\theta}_1 \end{aligned}\right\} \tag{6.32}$$

To solve the inverse kinematic problem, i.e. calculating the inputs of $\theta_1, \dot{\theta}_1, \ddot{\theta}_1$ based on the given outputs of (X_e, Y_e), $(\dot{X}_e, \dot{Y}_e)$, $(\ddot{X}_e, \ddot{Y}_e)$, Eq. (6.28) can be converted into

$$K_1' c\theta_1 + K_2' s\theta_1 + K_3' = 0 \tag{6.33}$$

where K_1', K_2', K_3' are the intermediate parameters determined by (X_e, Y_e) as

$$\left.\begin{aligned} K_1' &= 2l_1 X_e \\ K_2' &= 2l_1 Y_e \\ K_3' &= -(l_1^2 + X_e^2 + Y_e^2) \end{aligned}\right\} \tag{6.34}$$

Therefore, Eq. (6.33) yields the solution of the inverse kinematic problem as

$$\theta_1 = \tan^{-1}\left(\frac{-K_2' \pm \sqrt{(K_2')^2 - (K_3')^2 + (K_1')^2}}{K_3' - K_1'}\right) \tag{6.35}$$

Equation (6.33) obtains the input velocity of $\dot{\theta}_1$at θ_1 from that of the outputs $(\dot{X}_e, \dot{Y}_e)$as

$$\dot{\theta}_1 = -\frac{\dot{X}_e}{\left[l_1\left(s\theta_1 + \dfrac{s(\theta_3-\theta_1)s(\theta_2+\psi)}{s(\theta_2-\theta_3)}\right)\right]} \text{ or } \dot{\theta}_1 = \frac{\dot{Y}_e}{\left[l_1\left(c\theta_1 + \dfrac{s(\theta_3-\theta_1)c(\theta_2+\psi)}{s(\theta_2-\theta_3)}\right)\right]} \tag{6.36}$$

Note that $\dot{X}_e$ and $\dot{Y}_e$ are dependent and either of them can be used to calculate $\dot{\theta}_1$.

Similarly, the input $\ddot{\theta}_1$ at the given θ_1 and $\dot{\theta}_1$can be obtained from that of the outputs $(\ddot{X}_e, \ddot{Y}_e)$ from the forward kinematic solution in Eq. (6.32). Note that $\ddot{X}_e, \ddot{Y}_e$ are dependent and either of them can be used to calculate $\ddot{\theta}_1$.

The vector-loop method can be extended to take into consideration the geometric constraints in a 3D space. Here, a tripod-based kinematic machine is used as an example to illustrate how the constraints of a kinematic chain can be modelled by the vector-loop method.

The system can generate various tripod-based machine tools. A tripod machine has a base platform and an end-effector platform. These two platforms are connected by three symmetrical legs, and each leg is driven by a linear actuator. A tripod machine can include a passive leg in the middle to restrict the translational motion of the end-effector platform along the same direction. A typical structure is shown in Figure 6.25a and its structural parameters are illustrated in Figure 6.25b.

In Figure 6.25b, the base and end-effector platforms are denoted as $B_1B_2B_3$ and $E_1E_2E_3$, respectively. The base platform is fixed on the ground. The end-effector platform is used to mount a tool or gripper. A passive link is installed between the middle platform $M_1M_2M_3$ and the end-effector platform. The active leg D_iE_i is connected to the end-effector platform by a spherical joint at E_i and to the slide of the active prismatic joint by a universal joint at D_i. The passive leg is fixed on the middle platform at one end and is connected to the end-effector platform by a universal joint at the other end.

A tripod machine tool is usually symmetrical and can be described with the following parameters: the angle α_i $(i = 1, 2, 3)$ between O_bB_i and $\boldsymbol{x}_b$, the angle β_i $(i = 1, 2, 3)$ between O_eE_i and $\boldsymbol{x}_e$, the radius of the base platform l_b, the radius of the end-effector platform l_e, the direction of a guide-way γ, and the length of an active leg l_i.

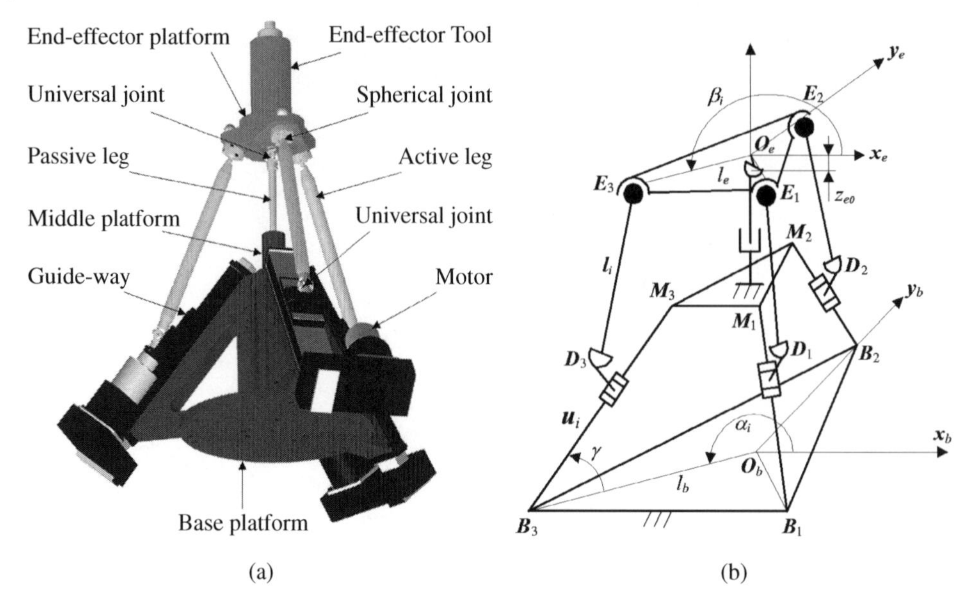

Figure 6.25 A parametric model of a tripod system configuration. (a) System components. (b) Parametric model.

6.5.3.1 Kinematic Parameters and Variables

Tripod-based machine tools vary from each other slightly in terms of kinematic and dynamic models. In this section, the tripod machine tool of Figure 6.25 is used as an example for the derivations of kinematic and dynamic models.

The complete description of the motion of a rigid body requires three rotational parameters $(\theta_x, \theta_y, \theta_z)$ and three translational parameters (x_0, y_0, z_0). However, a tripod machine tool has three degrees of freedom (DoF) motions. For the sample system, the end-effector motion can be denoted by $(\theta_x, \theta_y, \theta_z)$, where θ_x and θ_y are the rotations along $\boldsymbol{x}_e$ and $\boldsymbol{y}_e$, and z_e is the translation along $\boldsymbol{z}_e$, respectively. The x and y translations and z rotation are eliminated ($x_e = y_e = 0$, $\theta_z = 0$) due to the adoption of a passive leg. The posture of the end-effector is represented by

$$T_e^b = \begin{bmatrix} \boldsymbol{R}_e & \boldsymbol{P}_e \\ \boldsymbol{0} & 1 \end{bmatrix} = \begin{bmatrix} c\theta_y & 0 & s\theta_y & 0 \\ s\theta_x s\theta_y & c\theta_x & -s\theta_x c\theta_y & 0 \\ -c\theta_x s\theta_y & s\theta_x & c\theta_x c\theta_y & z_e \\ 0 & 0 & 0 & 1 \end{bmatrix} \tag{6.37}$$

where

c, s denote the cosine and sine functions, respectively
T_e^b is the posture of the end-effector with respect to $\{\boldsymbol{O}_b - \boldsymbol{x}_b\boldsymbol{y}_b\boldsymbol{z}_b\}$
$\boldsymbol{R}_e$ is the 3×3 orientational matrix of the end-effector
$\boldsymbol{P}_e$ is the central position $\boldsymbol{O}_e$ of the end-effector

Figure 6.26 Kinematic constraints.

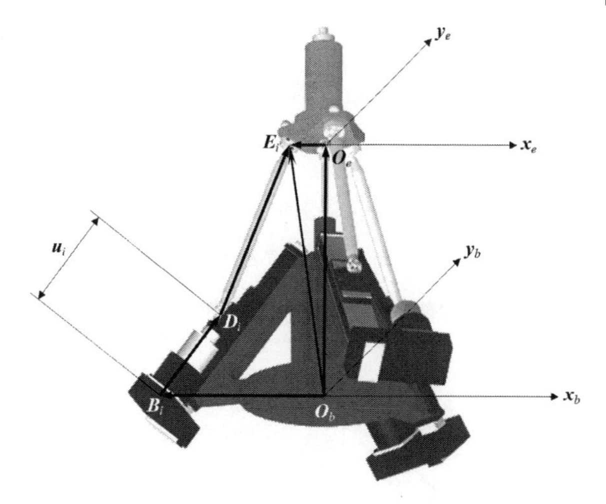

6.5.3.2 Inverse Kinematics

Inverse kinematics is used to find the joint motions when the end-effector motion (θ_x, θ_y, z_e) is known. The motion of the active prismatic joint i is denoted by u_i. As shown in Figure 6.26, a tripod machine tool includes three independent kinematic loops. The kinematic equation in each loop can be used to find one joint parameter. The solution to the inverse kinematic problem can be derived from the condition that the length of the support bar i is fixed.

Firstly, the coordinates of $\boldsymbol{E}_i$ on the end-effector platform are determined as

$$\boldsymbol{P}_{e_i}^b = \begin{bmatrix} x_{e_i}^b \\ y_{e_i}^b \\ z_{e_i}^b \end{bmatrix} = \begin{bmatrix} l_e c\beta_i c\theta_y \\ l_e c\beta_i s\theta_x s\theta_y + l_e s\beta_i c\theta_x \\ -l_e c\beta_i c\theta_x s\theta_y + l_e s\beta_i s\theta_x + z_e \end{bmatrix} \tag{6.38}$$

where $\boldsymbol{P}_{e_i}^b$ are the coordinates of $\boldsymbol{E}_i$ with respect to $\{\boldsymbol{O}_b - \boldsymbol{x}_b\boldsymbol{y}_b\boldsymbol{z}_b\}$.

Secondly, the condition that the support bar i has a fixed length is applied to give

$$|\mathrm{O}_b\mathrm{E}_i - \mathrm{O}_b\mathrm{B}_i - \mathrm{B}_i\mathrm{D}_i| = |\mathrm{D}_i\mathrm{E}_i| \quad (i = 1, 2, 3) \tag{6.39}$$

Finally, the motion of prismatic joint i can be derived from Eq. (6.39) as

$$u_i = \frac{-k_b \pm \sqrt{{k_b}^2 - 4k_c}}{2} \tag{6.40}$$

where

$$k_b = 2(c\gamma(x_{e_i}^b c\alpha_i + y_{e_i}^b s\alpha_i - l_b) - z_{e_i}^b s\gamma)$$
$$k_c = (x_{e_i}^b)^2 + (x_{e_i}^b)^2 + (x_{e_i}^b)^2 + l_b^2 - l_i^2 - 2l_b(x_{e_i}^b c\alpha_i + y_{e_i}^b s\alpha_i)$$

Note that a solution of u_i from Eq. (6.40) should be verified to ensure none of the active or passive joints exceeds its motion limit.

6.5.3.3 Direct Kinematics

Direct kinematics is used to find the end-effector motion (θ_x, θ_y, θ_z) when active joint motions u_i ($i = 1, 2, 3$) are known. The solution to the direct kinematic problem is also derived from Eq. (6.39). To solve the direct kinematic problem, Eq. (6.39) is rewritten in order that z_e and θ_y are expressed in terms of θ_x:

$$z_e^2 + (A_i s\theta_y + B_i)\ z_e + (C_i c\theta_y + D_i s\theta_y + E_i) = 0 \quad (i = 1,\ 2,\ 3) \tag{6.41}$$

where coefficients $A_i \sim E_i$ relate to θ_x. Equation (6.41) gives

$$c\theta_y = -\frac{Fz_e^2 + Gz_e + H}{Kz_e + L} \tag{6.42}$$

$$s\theta_y = -\frac{Iz_e + J}{Kz_e + L} \tag{6.43}$$

where coefficients $F \sim L$ relate to θ_x.

In addition, $c^2\theta_y + s^2\theta_y = 1$. Substituting Eqs. (6.42) and (6.43) into this expression gives

$$M_4 z_e^4 + M_3 z_e^3 + M_2 z_e^2 + M_1 z_e + M_0 = 0 \tag{6.44}$$

where $M_0 \sim M_4$ relates to coefficients $A_i \sim E_i$, $F \sim L$, and θ_x.

Note that Eq. (6.41) has three independent equations and that Eqs. (6.42) and (6.43) have used two of them. The remainder can be derived by substituting Eqs. (6.42) and (6.43) into one of the equations in Eq. (6.41). Giving $i = 1$ in Eq. (6.41),

$$N_3 z_e^3 + N_2 z_e^2 + N_1 z_e + N_0 = 0 \tag{6.45}$$

where $N_0 \sim N_3$ relates to coefficients $A_i \sim E_i$, $F \sim L$, and θ_x.

To have a common solution of z_e for Eqs. (6.44) and (6.45), one must have (Gosselin and Merlet 1994)

$$\begin{vmatrix} M_4 & M_3 & M_2 & M_1 & M_0 & 0 & 0 \\ 0 & M_4 & M_3 & M_2 & M_1 & M_0 & 0 \\ 0 & 0 & M_4 & M_3 & M_2 & M_1 & M_0 \\ N_3 & N_2 & N_1 & N_0 & 0 & 0 & 0 \\ 0 & N_3 & N_2 & N_1 & N_0 & 0 & 0 \\ 0 & 0 & N_3 & N_2 & N_1 & N_0 & 0 \\ 0 & 0 & 0 & N_3 & N_2 & N_1 & N_0 \end{vmatrix} = 0 \tag{6.46}$$

The strategy to solving the forward kinematic model is straightforward: (i) to calculate θ_x from Eq. (6.46), (ii) to calculate z_e from Eq. (6.45), and (iii) to calculate θ_y from Eqs. (6.42) and (6.43).

6.5.4 Kinematic Modelling Based on Denavit–Hartenberg (D-H) Parameters

The vector-loop method is effective to develop kinematic models for a planar mechanism or spatial mechanisms with multiple kinematic loops. However, it is not efficient for a spatial mechanism with a serial or hybrid configuration for flexible rotations, since the vector-loop method represents all of the kinematic constraints in terms of the relations of vector magnitudes projected on to axes.

An alternative method for kinematic modelling of a spatial mechanism is based on the Denavit–Hartenberg (D-H) convention. The D-H convention was proposed by Denavit and Hartenberg (1955) to represent the spatial relations of motion axes in a mechanism. As shown in Figure 6.27, the relations of one motion axis – i with other axes are described by four D-H parameters (θ_i, d_i, $\alpha_{i\text{-}1}$, $a_{i\text{-}1}$), where each motion axis i is associated with a frame i that is attached on link i–1 (Craig 2018).

In the D-H convention, the frames are established and attached to joints where the links are connected. Taking an example of link i in Figure 6.27, it is connected to link $i-1$ by joint $i-1$ and to link $i+1$ by joint i, and link i is associated with frame $i-1$ and frame i. To define frame i, the motion axes with link $i-1$ and link $i+1$ are defined as $\boldsymbol{Z}_{i-1}$ and $\boldsymbol{Z}_i$, respectively, and $\boldsymbol{X}_i$ is defined as the common normal of $\boldsymbol{Z}_{i-1}$ and $\boldsymbol{Z}_i$, $\boldsymbol{O}_i$ is at the intersection of $\boldsymbol{X}_i$ and $\boldsymbol{Z}_i$.

The D-H parameters describe the spatial relation of frame $i-1$ and frame i, and the physical meanings of these parameters are given in Figure 6.47. Once the set of D-H parameters (θ_i, d_i, α_i, a_i) of motion axis i is found, frame i can be represented with respect to frame $i-1$

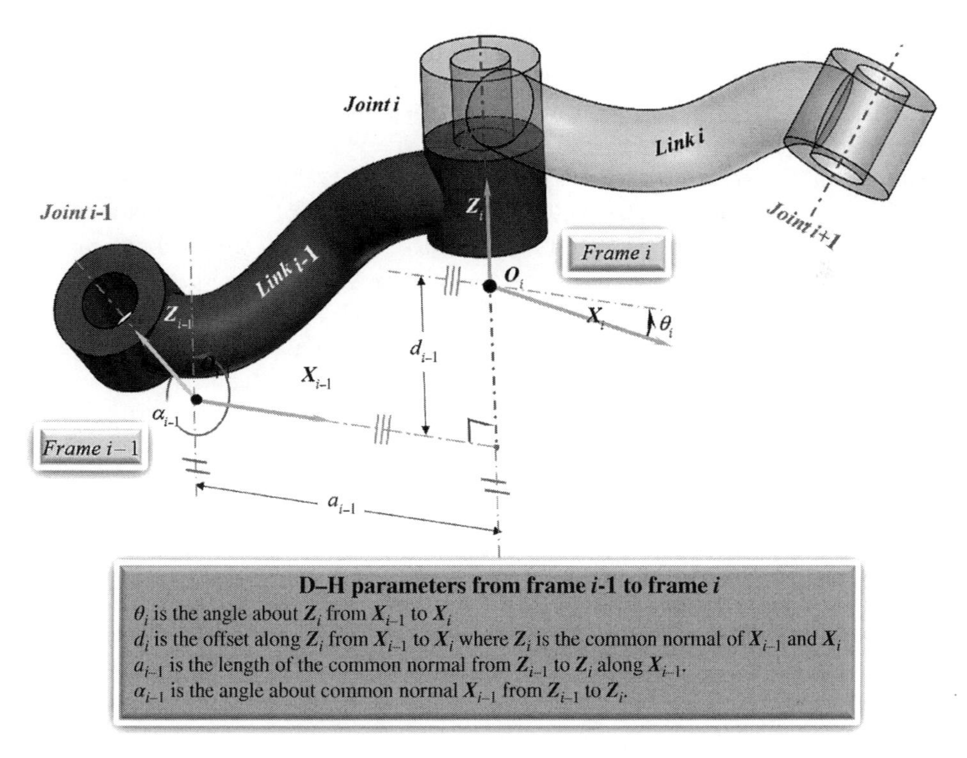

Figure 6.27 D-H convention from frame I – 1 to frame i.

as

$$
\begin{aligned}
{}^{i-1}_{i}\boldsymbol{A} &= \boldsymbol{T}_x(\alpha_{i-1}) \cdot \boldsymbol{T}_x(a_{i-1}) \cdot \boldsymbol{T}_z(d_i) \cdot \boldsymbol{R}_z(\theta_i) \\
&= \begin{bmatrix} 1 & 0 & 0 & 0 \\ 0 & c\alpha_{i-1} & -s\alpha_{i-1} & 0 \\ 0 & s\alpha_{i-1} & c\alpha_{i-1} & 0 \\ 0 & 0 & 0 & 1 \end{bmatrix} \cdot \begin{bmatrix} 1 & 0 & 0 & a_{i-1} \\ 0 & 1 & 0 & 0 \\ 0 & 0 & 1 & 0 \\ 0 & 0 & 0 & 1 \end{bmatrix} \cdot \begin{bmatrix} 1 & 0 & 0 & 0 \\ 0 & 1 & 0 & 0 \\ 0 & 0 & 1 & d_i \\ 0 & 0 & 0 & 1 \end{bmatrix} \cdot \begin{bmatrix} c\theta_i & -s\theta_i & 0 & 0 \\ s\theta_i & c\theta_i & 0 & 0 \\ 0 & 0 & 1 & 0 \\ 0 & 0 & 0 & 1 \end{bmatrix} \\
&= \begin{bmatrix} c\theta_i & -s\theta_i & 0 & a_i \\ s\theta_i c\alpha_{i-1} & c\theta_i c\alpha_{i-1} & -s\alpha_i & -d_i s\alpha_{i-1} \\ s\theta_i s\alpha_{i-1} & c\theta_i s\alpha_{i-1} & c\alpha_i & d_i c\alpha_{i-1} \\ 0 & 0 & 0 & 1 \end{bmatrix}
\end{aligned}
\tag{6.47}
$$

For frame 1 of the first motion axis ($\boldsymbol{Z}_1$) of a mechanism, its D-H parameters are affected by frame 0, i.e. the base CS attached on the ground link. We can simply assume that $\boldsymbol{Z}_0$ is aligned with $\boldsymbol{Z}_1$. If the first motion axis is rotational, $d_1 = 0$ and if the first joint is translational, $\theta_1 = 0$.

For frame *n* of the last motion axis ($\boldsymbol{Z}_\text{n}$) of a mechanism, its D-H parameters are affected by frame $n + 1$, i.e. the tool CS attached on the end-effector link. We set $\boldsymbol{Z}_\text{Tool}$ aligned with $\boldsymbol{Z}_\text{n}$; therefore, one has $a_\text{n} = 0$ and $\alpha_\text{n} = 0$.

The D-H convention has been widely used to establish the kinematic model of robots. As shown in Figure 6.28, the motions involved in a robot are classified into two types: internal motions in the *joint space* and external motions of its end-effector tool in the *task space*. Internal motions are described by joint variables and their derivatives with respect to time ($q_i, \dot{q}_i, \ddot{q}_i$ $(i = 1, 2, \dots, n)$), and external motions are described by the variables for the tooling position and orientation ($\boldsymbol{n}, \boldsymbol{o}, \boldsymbol{a},$ and $\boldsymbol{p}$) as well as their derivatives with respect to time. Note that a forward kinematic problem is to find the position and orientation of tooling when all of the joint variables are given. Using the matrix expression of Eq. (6.47) for frame *i* in frame $i - 1$, the solution to the forward kinematic problem can be readily defined as

$$
\begin{aligned}
{}^B_0\boldsymbol{T} = \begin{bmatrix} \boldsymbol{n} & \boldsymbol{o} & \boldsymbol{a} & \boldsymbol{p} \\ 0 & 0 & 0 & 1 \end{bmatrix} &= {}^B_1\boldsymbol{T} \cdot {}^1_n\boldsymbol{T} \cdot {}^n_T\boldsymbol{T} \\
&= {}^B_0\boldsymbol{T} \cdot [{}^0_1\boldsymbol{A}(q_1) \cdot {}^1_2\boldsymbol{A}(q_2) \cdots {}^{n-1}_{n}\boldsymbol{A}(q_n)] \cdot {}^n_T\boldsymbol{T}
\end{aligned}
\tag{6.48}
$$

where ${}^B_0\boldsymbol{T}$ and ${}^n_T\boldsymbol{T}$ are the matrix representations of frame 0 in a base frame and the tool frame in frame *n*, respectively. ${}^{i-1}_{i}\boldsymbol{A}(i = 2, 3, \dots, n)$ is the matrix representation of frame *i* in frame $i - 1$.

6.5.5 Jacobian Matrix for Velocity Relations

Let the position and orientation of the end-effector tool be represented by a vector of design variable $\{\boldsymbol{x}\}$ and the set of design variables in the joint space as $\{\boldsymbol{q}\}$, the solution to the

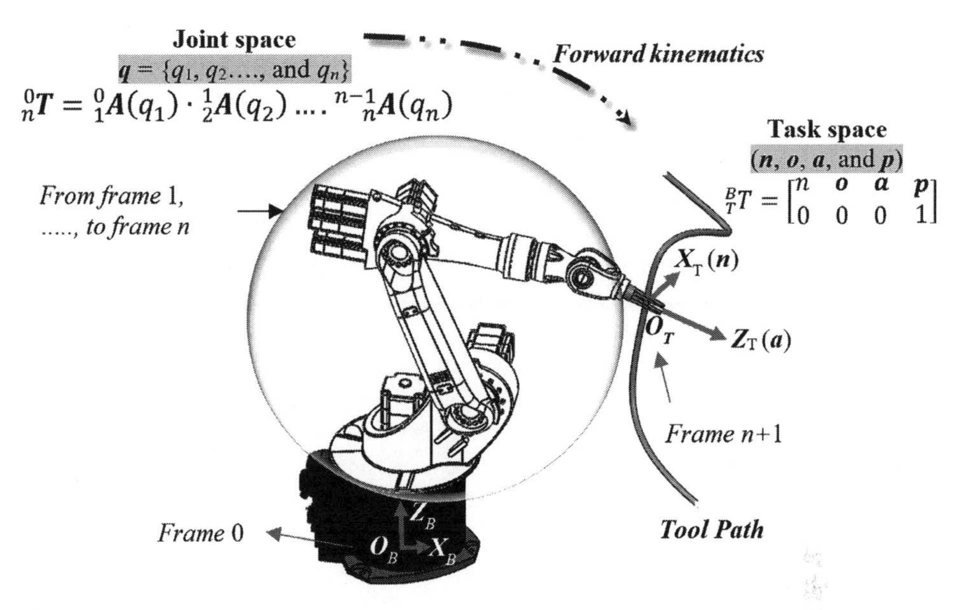

Figure 6.28 D-H convention for a forward kinematic problem.

forward kinematic problem in Eq. (6.48) can be transformed as a mapping function of

$$\{\boldsymbol{x}\} = \begin{Bmatrix} x_1 \\ x_2 \\ \vdots \\ x_m \end{Bmatrix} = \begin{Bmatrix} f_1(\boldsymbol{q}) \\ f_2(\boldsymbol{q}) \\ \vdots \\ f_n(\boldsymbol{q}) \end{Bmatrix} = \{\mathbf{f}(\boldsymbol{q})\} \tag{6.49}$$

where m and n are the number of variables to describe the motions in the task space and joint space, respectively.

Taking the derivative of Eq. (6.49) with respect to time yields

$$\{\delta \boldsymbol{x}\} = [\boldsymbol{J}]\{\delta \boldsymbol{q}\} = \begin{bmatrix} \frac{\partial f_1(\boldsymbol{q})}{\partial q_1} & \cdots & \frac{\partial f_1(\boldsymbol{q})}{\partial q_n} \\ \vdots & \ddots & \vdots \\ \frac{\partial f_m(\boldsymbol{q})}{\partial q_1} & \cdots & \frac{\partial f_m(\boldsymbol{q})}{\partial q_n} \end{bmatrix} \{\delta \boldsymbol{q}\} \text{ or } \{\dot{\boldsymbol{x}}\} = [\boldsymbol{J}]\{\dot{\boldsymbol{q}}\} \tag{6.50}$$

where $[\boldsymbol{J}]$ is the Jacobian matrix used to map the velocities in a task space to the velocities in a joint space.

The Jacobian matrix is not unique since it depends on the representation of the position and orientation of the end-effector tool in the task space. For example, if the direction cosines are used for the orientation and the coordinates in the Cartesian space are used for the position of tolling, the Jacobian matrix has the size of 12×6 for a robot with 6 DoFs. If the orientations are represented by Euler angles, the Jacobian matrix has the size of 6×6 for the same robot. Note that the representation of the position and orientation does not affect the kinematic behaviour of a mechanism. It is desirable to have a kinematic model that is independent of the representation; correspondingly, the Jacobian matrix based on such a kinematic model is unique. It is also called a basic Jacobian matrix (Corke 2011).

The basic Jacobian matrix describes the mapping of joint velocities and is uniquely defined by linear and angular velocities at a given point on the end-effector as

$$\begin{Bmatrix} \boldsymbol{v} \\ \omega \end{Bmatrix}_{6\times1} = [\boldsymbol{J}_0(\boldsymbol{q})]_{6\times n}\dot{\boldsymbol{q}}_{n\times1} \tag{6.51}$$

where $\{\boldsymbol{v}\}$ is the vector of linear velocities, whose elements are the time derivatives of the Cartesian coordinates of the end-effector position vector and $\{\boldsymbol{\omega}\}$ is the vector of an instantaneous quantity related to angular velocities.

It is worthwhile to note that for the orientation representation, there is no physical axis associated with the variables such as Euler angles (α, β, γ). Therefore, their derivatives are not angular velocities along axes. As a matter of fact, the quantities in $\{\boldsymbol{\omega}\}$ do not have a primitive function but instantaneous quantities. However, the time derivative of any representation of the orientation is related to the angular velocity. One has to map $\{\boldsymbol{v}\}$ and $\{\boldsymbol{\omega}\}$ into the position and orientation velocities as

$$\left.\begin{aligned} \dot{\boldsymbol{x}}_p &= \boldsymbol{E}_p(\boldsymbol{x}_p)\boldsymbol{v} \\ \dot{\boldsymbol{x}}_r &= \boldsymbol{E}_r(\boldsymbol{x}_r)\boldsymbol{\omega} \end{aligned}\right\} \tag{6.52}$$

where $\dot{\boldsymbol{x}}_p$ and $\dot{\boldsymbol{x}}_r$ are the time derivatives of the position and orientation parts of the end-effector, respectively. The matrics $\boldsymbol{E}_p$ and $\boldsymbol{E}_r$ are dependent on the particular representation of the position or orientation of the end-effector. Using $\boldsymbol{E}_p$ and $\boldsymbol{E}_r$ allow the basic Jacobian matrix $[\boldsymbol{J}_0]$ to be converted into a Jacobian matrix $[\boldsymbol{J}]$, which is associated with a particular position and orientation representation.

Figure 6.29 shows a kinematic chain with a number of serially connected joints. The velocity of a motion axis with respect to the local frame of the proceeding link depends on the joint type at the connection as

$$\left.\begin{aligned} \boldsymbol{v}_i &= \boldsymbol{z}_i\dot{q}_i \quad \textit{joint j is prismatic} \\ \boldsymbol{\omega}_i &= \boldsymbol{z}_i\dot{q}_i \quad \textit{joint i is revolute} \end{aligned}\right\} \tag{6.53}$$

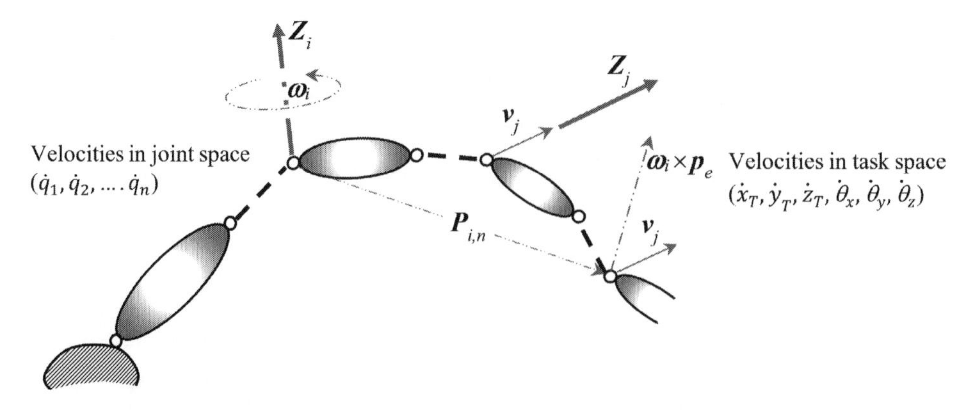

Figure 6.29 Jacobian matrix for relation of velocities.

As shown in Figure 6.29, each motion axis contributes to the motion of the end-effector. A prismatic joint i only generates a linear velocity $\boldsymbol{v}_i$, which is transferred down to the end-effector. A revolute joint adds both linear and angular velocities to the end-effector. Moreover, the linear velocity depends on the spatial relation of the end-effector and the motion axis, which is calculated by a cross-product of $\boldsymbol{\omega}_i$ with the vector towards the reference point of the end-effector. The angular velocity $\boldsymbol{\omega}_i$ is directly transferred down the chain to the end-effector. Therefore, the linear and angular velocities of the end-effector tool are summed as

$$\left.\begin{aligned} \{\boldsymbol{v}_e\} &= \sum_{i=1}^{n}[\epsilon_i \boldsymbol{v}_i + \overline{\epsilon}_i(\boldsymbol{\omega}_i \times \boldsymbol{p}_{i,n})] \\ \{\boldsymbol{\omega}_e\} &= \sum_{i=1}^{n} \overline{\epsilon}_i \boldsymbol{\omega}_i \end{aligned}\right\} \tag{6.54}$$

Substituting Eq. (6.53) into (6.54) gives

$$\begin{Bmatrix} \boldsymbol{v}_e \\ \boldsymbol{\omega}_e \end{Bmatrix} = \begin{Bmatrix} \boldsymbol{J}_v \\ \boldsymbol{J}_\omega \end{Bmatrix} \{\dot{\boldsymbol{q}}\} = \begin{bmatrix} \epsilon_1 \boldsymbol{v}_1 + \overline{\epsilon}_1(\boldsymbol{\omega}_1 \times \boldsymbol{p}_{1,n}) & \cdots & \epsilon_n \boldsymbol{v}_n + \overline{\epsilon}_n(\boldsymbol{\omega}_n \times \boldsymbol{p}_{n,n}) \\ \overline{\epsilon}_1 \boldsymbol{Z}_1 & \cdots & \overline{\epsilon}_n \boldsymbol{Z}_n \end{bmatrix} \begin{Bmatrix} q_1 \\ q_2 \\ \vdots \\ q_n \end{Bmatrix} \tag{6.55}$$

where $\{\boldsymbol{J}_v\}$ and $\{\boldsymbol{J}_\omega\}$ are the components of the Jacobian matrix for linear and angular velocities.

Note that the way to calculate $\{\boldsymbol{J}_v\}$ is not straightforward since all of the relative vectors $\boldsymbol{p}_{i,n}$ $(i = 0, 1, 2, \ldots, n)$ have to be determined. An alternative to calculating $\{\boldsymbol{J}_v\}$ is to take a direct differentiation of the Cartesian coordinates of the reference point of the end-effector as

$$\boldsymbol{v}_e = \begin{Bmatrix} \dot{x}_e \\ \dot{y}_e \\ \dot{z}_e \end{Bmatrix} = \dot{\boldsymbol{x}}_e = \frac{\partial \boldsymbol{p}_{0,n}}{\partial q_1}\dot{q}_1 + \frac{\partial \boldsymbol{p}_{0,n}}{\partial q_2}\dot{q}_2 + \cdots + \frac{\partial \boldsymbol{p}_{0,n}}{\partial q_n}\dot{q}_n \tag{6.56}$$

where $\boldsymbol{p}_{0,\,n}$ is the representation of the reference point on the end-effector tool in frame 0.

Therefore, we have

$$[\boldsymbol{J}] = \begin{bmatrix} \boldsymbol{J}_v \\ \boldsymbol{J}_\omega \end{bmatrix} = \begin{bmatrix} \dfrac{\partial \boldsymbol{p}_{0,n}}{\partial q_1} & \dfrac{\partial \boldsymbol{p}_{0,n}}{\partial q_2} & \cdots & \dfrac{\partial \boldsymbol{p}_{0,n}}{\partial q_n} \\ \overline{\epsilon}_1 {}^{\boldsymbol{0}}_{1}\boldsymbol{Z}_1 & \overline{\epsilon}_2 {}^{\boldsymbol{0}}_{2}\boldsymbol{Z}_2 & \cdots & \overline{\epsilon}_n {}^{\boldsymbol{0}}_{n}\boldsymbol{Z}_n \end{bmatrix} \tag{6.57}$$

where $\overline{\epsilon}_i$ = '0' or '1' $(i = 1, 2, \ldots, n)$, the motion joint i is prismatic or revolute, respectively, ${}^{\boldsymbol{0}}_{i}\boldsymbol{Z}_i$ $(i = 1, 2, \ldots, n)$ is the direction of motion axis i expressed in frame 0, and $[\boldsymbol{J}]$ is the basic Jacobian matrix.

Example 6.4 Find the forward kinematic solution for the 2-DoF robot in Figure 6.30. The robot consists of the ground link, two other links in the body, and two joints to connect these links sequentially. In addition, the tool tip is attached on the last link directly.

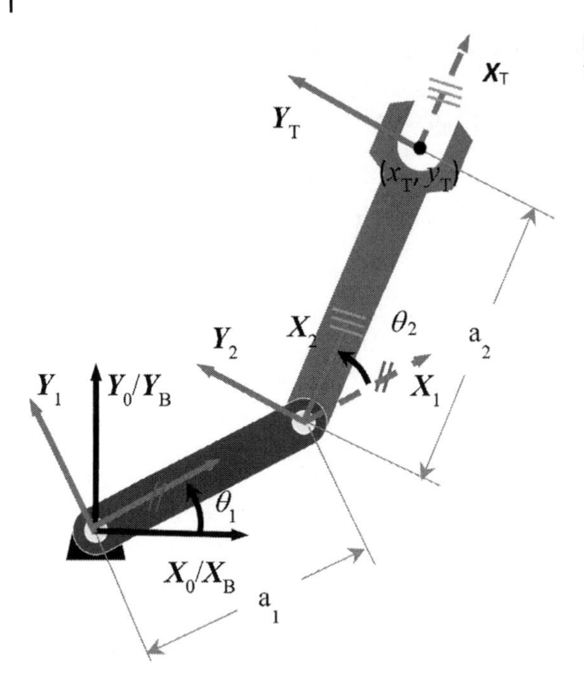

Figure 6.30 D-H convention for a forward kinematic problem.

The D-H parameters for internal two links are determined as follows:

Link	θ_i	d_i	α_{i-1}	a_{i-1}
1	θ_1	0	0°	0
2	θ_2	0	0°	a_1
Tool	0	0	0°	a_2

Accordingly, the D-H matrices for these two links are

$$ {}^{0}_{1}\boldsymbol{A}(\theta_1) = \begin{bmatrix} c\theta_1 & -s\theta_1 & 0 & 0 \\ s\theta_1 & c\theta_1 & 0 & 0 \\ 0 & 0 & 1 & 0 \\ 0 & 0 & 0 & 1 \end{bmatrix} \tag{6.58} $$

$$ {}^{1}_{2}\boldsymbol{A}(\theta_1) = \begin{bmatrix} c\theta_2 & -s\theta_2 & 0 & a_1 \\ s\theta_2 & c\theta_2 & 0 & 0 \\ 0 & 0 & 1 & 0 \\ 0 & 0 & 0 & 1 \end{bmatrix} \tag{6.59} $$

$$ {}^{2}_{T}\boldsymbol{A}(\theta_1) = \begin{bmatrix} 1 & 0 & 0 & a_2 \\ 0 & 1 & 0 & 0 \\ 0 & 0 & 1 & 0 \\ 0 & 0 & 0 & 1 \end{bmatrix} \tag{6.60} $$

In addition, we assume that the base frame is aligned with frame 0, which gives

$$ {}^{B}_{0}\boldsymbol{T} = I_{4\times4} \tag{6.61}$$

where $I_{4\times4}$ is a unit matrix with the size of 4 by 4.

Further, using Eq. (6.48) yields the solution to the forward kinematic problem as

$$ {}^{B}_{T}\boldsymbol{T} = \begin{bmatrix} \boldsymbol{n} & \boldsymbol{o} & \boldsymbol{a} & \boldsymbol{p} \\ 0 & 0 & 0 & 1 \end{bmatrix} = {}^{B}_{0}\boldsymbol{T}\cdot[{}^{0}_{1}\boldsymbol{A}(q_1)\cdot{}^{1}_{2}\boldsymbol{A}(q_2)\cdot{}^{2}_{T}\boldsymbol{A}(0)]$$

$$ = \begin{bmatrix} c\theta_1 & -s\theta_1 & 0 & 0 \\ s\theta_1 & c\theta_1 & 0 & 0 \\ 0 & 0 & 1 & 0 \\ 0 & 0 & 0 & 1 \end{bmatrix} \cdot \begin{bmatrix} c\theta_2 & -s\theta_2 & 0 & a_1 \\ s\theta_2 & c\theta_2 & 0 & 0 \\ 0 & 0 & 1 & 0 \\ 0 & 0 & 0 & 1 \end{bmatrix} \cdot \begin{bmatrix} 1 & 0 & 0 & a_2 \\ 0 & 1 & 0 & 0 \\ 0 & 0 & 1 & 0 \\ 0 & 0 & 0 & 1 \end{bmatrix}$$

$$ = \begin{bmatrix} c(\theta_1+\theta_2) & -s(\theta_1+\theta_2) & 0 & a_1c\theta_1 + a_2c(\theta_1+\theta_2) \\ s(\theta_1+\theta_2) & c(\theta_1+\theta_2) & 0 & a_1s\theta_1 + a_2s(\theta_1+\theta_2) \\ 0 & 0 & 1 & 0 \\ 0 & 0 & 0 & 1 \end{bmatrix} \tag{6.62}$$

Equation (6.62) can be written specifically as the position and orientation of the tool as

$$ \boldsymbol{n} = \begin{Bmatrix} c(\theta_1+\theta_2) \\ s(\theta_1+\theta_2) \\ 0 \end{Bmatrix}, \quad \boldsymbol{o} = \begin{Bmatrix} -s(\theta_1+\theta_2) \\ c(\theta_1+\theta_2) \\ 0 \end{Bmatrix}, \boldsymbol{a} = \begin{Bmatrix} 0 \\ 0 \\ 0 \end{Bmatrix}, \ \boldsymbol{p} = \begin{Bmatrix} a_1c\theta_1 + a_2c(\theta_1+\theta_2) \\ a_1s\theta_1 + a_2s(\theta_1+\theta_2) \\ 0 \end{Bmatrix} \tag{6.63}$$

After the displacement of $\boldsymbol{p}$ is calculated from the angular displacements of two joints in Eq. (6.63), the velocity and acceleration of $\boldsymbol{p}$ can be obtained by taking the derivatives of Eq. (6.63) with respect to time as

$$ \dot{\boldsymbol{p}} = \begin{Bmatrix} \dot{x}_T \\ \dot{y}_T \end{Bmatrix} = [\boldsymbol{J}_1] \begin{Bmatrix} \dot{\theta}_1 \\ \dot{\theta}_2 \end{Bmatrix}$$

$$ \ddot{\boldsymbol{p}} = \begin{Bmatrix} \ddot{x}_T \\ \ddot{y}_T \end{Bmatrix} = [\boldsymbol{J}_1] \begin{Bmatrix} \ddot{\theta}_1 \\ \ddot{\theta}_2 \end{Bmatrix} + [\boldsymbol{J}_2] \begin{Bmatrix} \dot{\theta}_1 \\ \dot{\theta}_2 \end{Bmatrix} \tag{6.64}$$

where

$$ [\boldsymbol{J}_1] = \begin{bmatrix} -a_1s\theta_1 - a_2s(\theta_1+\theta_2) & -a_2s(\theta_1+\theta_2) \\ a_1c\theta_1 + a_2c(\theta_1+\theta_2) & a_2c(\theta_1+\theta_2) \end{bmatrix}$$

$$ [\boldsymbol{J}_2] = \begin{bmatrix} -a_1\dot{\theta}_1c\theta_1 - a_2(\dot{\theta}_1+\dot{\theta}_2)c(\theta_1+\theta_2) & -a_2(\dot{\theta}_1+\dot{\theta}_2)c(\theta_1+\theta_2) \\ -a_1\dot{\theta}_1s\theta_1 - a_2(\dot{\theta}_1+\dot{\theta}_2)s(\theta_1+\theta_2) & -a_2(\dot{\theta}_1+\dot{\theta}_2)s(\theta_1+\theta_2) \end{bmatrix}$$

The solution of the inverse kinematic problem to get θ_1 and θ_2 from x_T and y_T can be found in Eq. (6.63) as

$$\left.\begin{aligned} \theta_1 &= \tan^{-1}\left(\frac{-K_2 \pm \sqrt{K_2^2 - K_3^2 + K_1^2}}{K_3 - K_1}\right) \\ \theta_2 &= \tan^{-1}\left(\frac{y_T - a_1 s\theta_1}{x_T - a_1 c\theta_1}\right) - \theta_1 \end{aligned}\right\} \tag{6.65}$$

where K_1, K_2, and K_3 are dependent intermediate parameters as

$$\left.\begin{aligned} K_1 &= 2a_1 x_T \\ K_2 &= 2a_1 y_T \\ K_3 &= a_2^2 - (a_1^2 + x_T^2 + y_T^2) \end{aligned}\right\} \tag{6.66}$$

The solution of the inverse kinematic problem for the velocity and acceleration relation is found from Eq. (6.63) as

$$\begin{aligned} \dot{\theta} &= \begin{Bmatrix} \dot{\theta}_1 \\ \dot{\theta}_2 \end{Bmatrix} = [\boldsymbol{J}_1]^{-1} \begin{Bmatrix} \dot{x}_T \\ \dot{y}_T \end{Bmatrix} \\ \ddot{\theta} &= \begin{Bmatrix} \ddot{\theta}_1 \\ \ddot{\theta}_2 \end{Bmatrix} = [\boldsymbol{J}_1]^{-1} \begin{Bmatrix} \ddot{x}_T \\ \ddot{y}_T \end{Bmatrix} - [\boldsymbol{J}_1]^{-1}[\boldsymbol{J}_2] \begin{Bmatrix} \dot{\theta}_1 \\ \dot{\theta}_2 \end{Bmatrix} \end{aligned} \tag{6.67}$$

where $[\bullet]^{-1}$ is the operation of an inverse of matrix $[\bullet]$.

Example 6.5 Figure 6.31 shows the structure of a PUMA 560. Find (i) the solution of the forward kinematic problem and (ii) the solution of the inverse kinematic problem.

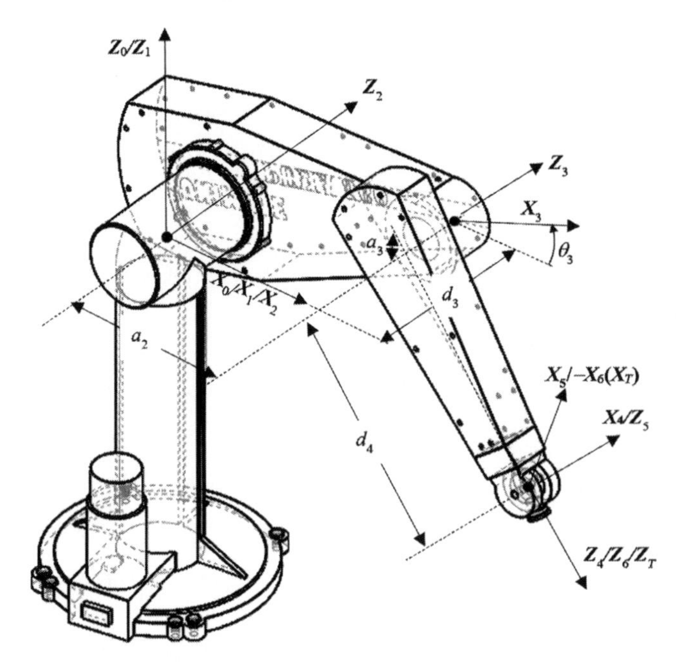

Figure 6.31 Structure of Puma 560.

The D-H parameters of the Puma 560 can be determined as follows:

Link	θ_i	d_i	α_{i-1}	a_{i-1}
1	θ_1	0	0°	0
2	θ_2	0	−90°	0
3	θ_3	d_3	0°	a_2
4	θ_4	d_4	−90°	a_3
5	θ_5	0	90°	0
6	θ_6	0	−90°	0

Using Eq. (6.48) gives the forward kinematic solution as (Singh 2014)

$$ {}^0_1\boldsymbol{A} = \begin{bmatrix} c\theta_1 & -s\theta_1 & 0 & 0 \\ s\theta_1 & c\theta_1 & 0 & 0 \\ 0 & 0 & 1 & 0 \\ 0 & 0 & 0 & 1 \end{bmatrix} \qquad {}^1_2\boldsymbol{A} = \begin{bmatrix} c\theta_2 & -s\theta_2 & 0 & 0 \\ 0 & 0 & 1 & 0 \\ -s\theta_2 & -c\theta_2 & 0 & 0 \\ 0 & 0 & 0 & 1 \end{bmatrix} $$

$$ {}^2_3\boldsymbol{A} = \begin{bmatrix} c\theta_3 & -s\theta_3 & 0 & a_2 \\ s\theta_3 & c\theta_3 & 0 & 0 \\ 0 & 0 & 1 & d_3 \\ 0 & 0 & 0 & 1 \end{bmatrix} \qquad {}^3_4\boldsymbol{A} = \begin{bmatrix} c\theta_4 & -s\theta_4 & 0 & a_3 \\ 0 & 0 & 1 & d_4 \\ -s\theta_4 & -c\theta_4 & 0 & 0 \\ 0 & 0 & 0 & 1 \end{bmatrix} $$

$$ {}^4_5\boldsymbol{A} = \begin{bmatrix} c\theta_5 & -s\theta_5 & 0 & 0 \\ 0 & 0 & -1 & 0 \\ s\theta_5 & c\theta_5 & 0 & 0 \\ 0 & 0 & 0 & 1 \end{bmatrix} \qquad {}^5_6\boldsymbol{A} = \begin{bmatrix} c\theta_6 & -s\theta_6 & 0 & 0 \\ 0 & 0 & 1 & 0 \\ -s\theta_6 & -c\theta_6 & 0 & 0 \\ 0 & 0 & 0 & 1 \end{bmatrix} $$

$$ {}^0_1\boldsymbol{T} = \begin{bmatrix} {}^0_1\boldsymbol{n} & {}^0_1\boldsymbol{o} & {}^0_1\boldsymbol{a} & {}^0_1\boldsymbol{p} \\ 0 & 0 & 0 & 1 \end{bmatrix} = \begin{bmatrix} {}^0_1n_x & {}^0_1o_x & {}^0_1a_x & {}^0_1p_x \\ {}^0_1n_y & {}^0_1o_y & {}^0_1a_y & {}^0_1p_y \\ {}^0_1n_z & {}^0_1o_z & {}^0_1a_z & {}^0_1p_z \\ 0 & 0 & 0 & 1 \end{bmatrix} = {}^0_1\boldsymbol{A} $$

$$ = \begin{bmatrix} c\theta_1 & -s\theta_1 & 0 & 0 \\ s\theta_1 & c\theta_1 & 0 & 0 \\ 0 & 0 & 1 & 0 \\ 0 & 0 & 0 & 1 \end{bmatrix} \tag{6.68} $$

$$
{}_{2}^{0}\boldsymbol{T} = \begin{bmatrix} {}_{2}^{0}\boldsymbol{n} \; {}_{2}^{0}\boldsymbol{o} \; {}_{2}^{0}\boldsymbol{a} & {}_{2}^{0}\boldsymbol{p} \\ 0\;0\;0 & 1 \end{bmatrix} = \begin{bmatrix} {}_{2}^{0}n_x & {}_{2}^{0}o_x & {}_{2}^{0}a_x & {}_{2}^{0}p_x \\ {}_{2}^{0}n_y & {}_{2}^{0}o_y & {}_{2}^{0}a_y & {}_{2}^{0}p_y \\ {}_{2}^{0}n_z & {}_{2}^{0}o_z & {}_{2}^{0}a_z & {}_{2}^{0}p_z \\ 0 & 0 & 0 & 1 \end{bmatrix} = {}_{1}^{0}\boldsymbol{T} \cdot {}_{2}^{1}\boldsymbol{A}
$$

$$
= \begin{bmatrix} c\theta_1 c\theta_2 & -c\theta_1 s\theta_2 & -s\theta_1 & 0 \\ s\theta_1 c\theta_2 & -s\theta_1 s\theta_2 & c\theta_1 & 0 \\ -s\theta_2 & -c\theta_2 & 0 & 0 \\ 0 & 0 & 0 & 1 \end{bmatrix} \tag{6.69}
$$

$$
{}_{3}^{0}\boldsymbol{T} = \begin{bmatrix} {}_{3}^{0}\boldsymbol{n} & {}_{3}^{0}\boldsymbol{o} & {}_{3}^{0}\boldsymbol{a} & {}_{3}^{0}\boldsymbol{p} \\ 0 & 0 & 0 & 1 \end{bmatrix} = \begin{bmatrix} {}_{3}^{0}n_x & {}_{3}^{0}o_x & {}_{3}^{0}a_x & {}_{3}^{0}p_x \\ {}_{3}^{0}n_y & {}_{3}^{0}o_y & {}_{3}^{0}a_y & {}_{3}^{0}p_y \\ {}_{3}^{0}n_z & {}_{3}^{0}o_z & {}_{3}^{0}a_z & {}_{3}^{0}p_z \\ 0 & 0 & 0 & 1 \end{bmatrix} = {}_{2}^{0}\boldsymbol{T} \cdot {}_{3}^{2}\boldsymbol{A}
$$

$$
= \begin{bmatrix} c\theta_1 c\theta_{23} & -c\theta_1 s\theta_{23} & -s\theta_1 & a_2 c\theta_1 c\theta_2 - d_3 s\theta_1 \\ s\theta_1 c\theta_{23} & -s\theta_1 s\theta_{23} & c\theta_1 & a_2 s\theta_1 c\theta_2 + d_3 c\theta_1 \\ -s\theta_{23} & -c\theta_{23} & 0 & -a_2 s\theta_2 \\ 0 & 0 & 0 & 1 \end{bmatrix} \tag{6.70}
$$

$$
{}_{4}^{0}\boldsymbol{T} = \begin{bmatrix} {}_{4}^{0}\boldsymbol{n} & {}_{4}^{0}\boldsymbol{o} & {}_{4}^{0}\boldsymbol{a} & {}_{4}^{0}\boldsymbol{p} \\ 0 & 0 & 0 & 1 \end{bmatrix} = \begin{bmatrix} {}_{4}^{0}n_x & {}_{4}^{0}o_x & {}_{4}^{0}a_x & {}_{4}^{0}p_x \\ {}_{4}^{0}n_y & {}_{4}^{0}o_y & {}_{4}^{0}a_y & {}_{4}^{0}p_y \\ {}_{4}^{0}n_z & {}_{4}^{0}o_z & {}_{4}^{0}a_z & {}_{4}^{0}p_z \\ 0 & 0 & 0 & 1 \end{bmatrix} = {}_{3}^{0}\boldsymbol{T} \cdot {}_{4}^{3}\boldsymbol{A} \tag{6.71}
$$

where $\theta_{23} = \theta_2 + \theta_3$

$$ {}_{4}^{0}n_x = c\theta_1 c\theta_{23} c\theta_4 + s\theta_1 s\theta_4 $$

$$ {}_{4}^{0}n_y = s\theta_1 c\theta_{23} c\theta_4 - c\theta_1 s\theta_4 $$

$$ {}_{4}^{0}n_z = -s\theta_{23} c\theta_4 $$

$$ {}_{4}^{0}o_x = -c\theta_1 c\theta_{23} s\theta_4 + s\theta_1 c\theta_4 $$

$$ {}_{4}^{0}o_y = -s\theta_1 c\theta_{23} s\theta_4 - c\theta_1 c\theta_4 $$

$$ {}_{4}^{0}o_z = s\theta_{23} s\theta_4 $$

$$ {}_{4}^{0}a_x = -c\theta_1 s\theta_{23} $$

$$ {}_{4}^{0}a_y = -s\theta_1 s\theta_{23} $$

$$ {}_{4}^{0}a_z = -c\theta_{23} $$

$${}^{0}_{4}p_x = a_3 c\theta_1 c\theta_{23} - d_4 c\theta_1 s\theta_{23} + a_2 c\theta_1 c\theta_2 - d_3 s\theta_1$$

$${}^{0}_{4}p_y = a_3 s\theta_1 c\theta_{23} - d_4 s\theta_1 s\theta_{23} + a_2 s\theta_1 c\theta_2 + d_3 c\theta_1$$

$${}^{0}_{4}p_z = -a_3 s\theta_{23} - d_4 c\theta_{23} - a_2 s\theta_2$$

$${}^{0}_{5}\boldsymbol{T} = \begin{bmatrix} {}^{0}_{5}\boldsymbol{n} & {}^{0}_{5}\boldsymbol{o} & {}^{0}_{5}\boldsymbol{a} & {}^{0}_{5}\boldsymbol{p} \\ 0 & 0 & 0 & 1 \end{bmatrix} = \begin{bmatrix} {}^{0}_{5}n_x & {}^{0}_{5}o_x & {}^{0}_{5}a_x & {}^{0}_{5}p_x \\ {}^{0}_{5}n_y & {}^{0}_{5}o_y & {}^{0}_{5}a_y & {}^{0}_{5}p_y \\ {}^{0}_{5}n_z & {}^{0}_{5}o_z & {}^{0}_{5}a_z & {}^{0}_{5}p_z \\ 0 & 0 & 0 & 1 \end{bmatrix} = {}^{0}_{4}\boldsymbol{T} \cdot {}^{4}_{5}\boldsymbol{A} \tag{6.72}$$

where

$${}^{0}_{5}n_x = (c\theta_1 c\theta_{23} c\theta_4 + s\theta_1 s\theta_4) c\theta_5 - c\theta_1 s\theta_{23} s\theta_5$$

$${}^{0}_{5}n_y = (s\theta_1 c\theta_{23} c\theta_4 - c\theta_1 s\theta_4) c\theta_5 - s\theta_1 s\theta_{23} s\theta_5$$

$${}^{0}_{5}n_z = -s\theta_{23} c\theta_4 c\theta_5 - c\theta_{23} s\theta_5$$

$${}^{0}_{5}o_x = -(c\theta_1 c\theta_{23} c\theta_4 + s\theta_1 s\theta_4) s\theta_5 - c\theta_1 s\theta_{23} c\theta_5$$

$${}^{0}_{5}o_y = -(s\theta_1 c\theta_{23} c\theta_4 - c\theta_1 s\theta_4) s\theta_5 - s\theta_1 s\theta_{23} c\theta_5$$

$${}^{0}_{5}o_z = s\theta_{23} c\theta_4 s\theta_5 - c\theta_{23} c\theta_5$$

$${}^{0}_{5}a_x = c\theta_1 c\theta_{23} s\theta_4 - s\theta_1 c\theta_4$$

$${}^{0}_{5}a_y = s\theta_1 c\theta_{23} s\theta_4 + c\theta_1 c\theta_4$$

$${}^{0}_{5}a_z = -s\theta_{23} s\theta_4$$

$${}^{0}_{5}p_x = -d_4 c\theta_1 s\theta_{23} + a_2 c\theta_1 c\theta_2 - d_3 s\theta_1$$

$${}^{0}_{5}p_y = -d_4 s\theta_1 s\theta_{23} + a_2 s\theta_1 c\theta_2 + d_3 c\theta_1$$

$${}^{0}_{5}p_z = -a_3 s\theta_{23} - d_4 c\theta_{23} - a_2 s\theta_2$$

$${}^{0}_{6}\boldsymbol{T} = \begin{bmatrix} {}^{0}_{6}\boldsymbol{n} & {}^{0}_{6}\boldsymbol{o} & {}^{0}_{6}\boldsymbol{a} & {}^{0}_{6}\boldsymbol{p} \\ 0 & 0 & 0 & 1 \end{bmatrix} = \begin{bmatrix} {}^{0}_{6}n_x & {}^{0}_{6}o_x & {}^{0}_{6}a_x & {}^{0}_{6}p_x \\ {}^{0}_{6}n_y & {}^{0}_{6}o_y & {}^{0}_{6}a_y & {}^{0}_{6}p_y \\ {}^{0}_{6}n_z & {}^{0}_{6}o_z & {}^{0}_{6}a_z & {}^{0}_{6}p_z \\ 0 & 0 & 0 & 1 \end{bmatrix} = {}^{0}_{5}\boldsymbol{T} \cdot {}^{5}_{6}\boldsymbol{A} \tag{6.73}$$

where

$${}^{0}_{6}n_x = c\theta_1 (c\theta_{23} (c\theta_4 c\theta_5 c\theta_6 - s\theta_4 s\theta_6) - s\theta_{23} s\theta_5 c\theta_6) - s\theta_1 (s\theta_4 c\theta_5 c\theta_6 + c\theta_4 s\theta_6)$$

$${}^{0}_{6}n_y = s\theta_1 (c\theta_{23} (c\theta_4 c\theta_5 c\theta_6 - s\theta_4 s\theta_6) - s\theta_{23} s\theta_5 c\theta_6) + c\theta_1 (s\theta_4 c\theta_5 c\theta_6 + c\theta_4 s\theta_6)$$

$${}^{0}_{6}n_z = -s\theta_{23} (c\theta_4 c\theta_5 c\theta_6 - s\theta_4 s\theta_6) - c\theta_{23} s\theta_5 c\theta_6$$

$${}^0_6o_x = c\theta_1(-c\theta_{23}(c\theta_4 c\theta_5 s\theta_6 + s\theta_4 c\theta_6) + s\theta_{23} s\theta_5 s\theta_6) - s\theta_1(-s\theta_4 c\theta_5 c\theta_6 + c\theta_4 c\theta_6)$$

$${}^0_6o_y = s\theta_1(-c\theta_{23}(c\theta_4 c\theta_5 s\theta_6 + s\theta_4 c\theta_6) + s\theta_{23} s\theta_5 s\theta_6) + c\theta_1(-s\theta_4 c\theta_5 s\theta_6 + c\theta_4 c\theta_6)$$

$${}^0_6o_z = s\theta_{23}(c\theta_4 c\theta_5 s\theta_6 + s\theta_4 c\theta_6) - c\theta_{23} s\theta_5 s\theta_6$$

$${}^0_6a_x = -c\theta_1(c\theta_{23} c\theta_4 s\theta_5 + s\theta_{23} c\theta_5) - s\theta_1 s\theta_4 s\theta_5$$

$${}^0_6a_y = -s\theta_1(c\theta_{23} c\theta_4 s\theta_5 + s\theta_{23} c\theta_5) + c\theta_1 s\theta_4 s\theta_5$$

$${}^0_6a_z = s\theta_{23} c\theta_4 s\theta_5 - c\theta_{23} c\theta_5$$

$${}^0_6p_x = c\theta_1(-d_4 s\theta_{23} + a_3 c\theta_{23} + a_2 c\theta_2) - d_3 s\theta_1$$

$${}^0_6p_y = s\theta_1(-d_4 s\theta_{23} + a_3 c\theta_{23} + a_2 c\theta_2) + d_3 c\theta_1$$

$${}^0_6p_z = -d_4 c\theta_{23} - a_3 s\theta_{23} - a_2 s\theta_2$$

Using the results of Eqs. (6.68) to (6.73) in Eq. (6.55) of the Jacobian matrix gives

$$[\boldsymbol{J}_v] = \begin{bmatrix} \frac{\partial \boldsymbol{p}_{0,6}}{\partial \theta_1} & \frac{\partial \boldsymbol{p}_{0,6}}{\partial \theta_2} & \frac{\partial \boldsymbol{p}_{0,6}}{\partial \theta_3} & \frac{\partial \boldsymbol{p}_{0,6}}{\partial \theta_4} & \frac{\partial \boldsymbol{p}_{0,6}}{\partial \theta_5} & \frac{\partial \boldsymbol{p}_{0,6}}{\partial \theta_6} \end{bmatrix} \tag{6.74}$$

where

$$\frac{\partial \boldsymbol{p}_{0,6}}{\partial \theta_1} = \begin{Bmatrix} -s\theta_1(-d_4 s\theta_{23} + a_3 c\theta_{23} + a_2 c\theta_2) - d_3 c\theta_1 \\ -c\theta_1(-d_4 s\theta_{23} + a_3 c\theta_{23} + a_2 c\theta_2) - d_3 s\theta_1 \\ 0 \end{Bmatrix} \qquad \frac{\partial \boldsymbol{p}_{0,6}}{\partial \theta_2} = \begin{Bmatrix} c\theta_1(-d_4 c\theta_{23} - a_3 s\theta_{23} - a_2 s\theta_2) \\ s\theta_1(-d_4 c\theta_{23} - a_3 s\theta_{23} - a_2 s\theta_2) \\ d_4 s\theta_{23} - a_3 c\theta_{23} - a_2 c\theta_2 \end{Bmatrix}$$

$$\frac{\partial \boldsymbol{p}_{0,6}}{\partial \theta_3} = \begin{Bmatrix} c\theta_1(-d_4 c\theta_{23} - a_3 s\theta_{23}) \\ s\theta_1(-d_4 c\theta_{23} - a_3 s\theta_{23}) \\ d_4 s\theta_{23} - a_3 c\theta_{23} \end{Bmatrix} \qquad \frac{\partial \boldsymbol{p}_{0,6}}{\partial \theta_4} = \frac{\partial \boldsymbol{p}_{0,6}}{\partial \theta_5} = \frac{\partial \boldsymbol{p}_{0,6}}{\partial \theta_6} = \begin{Bmatrix} 0 \\ 0 \\ 0 \end{Bmatrix}$$

$$[\boldsymbol{J}_\omega] = \left[{}^0_1\boldsymbol{Z}_1 \ {}^0_2\boldsymbol{Z}_2 \ {}^0_3\boldsymbol{Z}_3 \ {}^0_4\boldsymbol{Z}_4 \ {}^0_5\boldsymbol{Z}_5 \ {}^0_6\boldsymbol{Z}_6 \right] = \left[{}^0_1\boldsymbol{a}_1 \ {}^0_2\boldsymbol{a}_2 \ {}^0_3\boldsymbol{a}_3 \ {}^0_4\boldsymbol{a}_4 \ {}^0_5\boldsymbol{n}_5 \ {}^0_6\boldsymbol{n}_6 \right] \tag{6.75}$$

Using the Jacobian matrix to map the velocities of joint motion into that of the end-effect tool gives

$$\begin{Bmatrix} {}^0_6\dot{p}_x \\ {}^0_6\dot{p}_y \\ {}^0_6\dot{p}_z \\ {}^0_6\omega_x \\ {}^0_6\omega_y \\ {}^0_6\omega_z \end{Bmatrix} = \begin{bmatrix} \boldsymbol{J}_v \\ \boldsymbol{J}_\omega \end{bmatrix} \begin{Bmatrix} \dot{\theta}_1 \\ \dot{\theta}_2 \\ \dot{\theta}_3 \\ \dot{\theta}_4 \\ \dot{\theta}_5 \\ \dot{\theta}_6 \end{Bmatrix} \tag{6.76}$$

6.6 Dynamic Modelling

Dynamics is the study of the cause and effect relationship of the forces and motions of machines. A dynamic model describes the relationship between applied forces/torques and the resulting motions of a machine. Dynamic modelling is required in the design, simulation, analysis, and control of a machine. Similar to kinematic modelling, an *inverse dynamic problem* is used to obtain driving forces of joints when the motion of the end-effector is given and *a forward dynamic problem* is used to find the motion of the end-effector when the driving forces at joints are given.

In a forward dynamic problem, the time evolution of $\ddot{\boldsymbol{q}}$ (t) (and then of $\boldsymbol{q}$(t) and $\dot{\boldsymbol{q}}$ (t)) is computed when the vector of generalized forces (torques and/or forces) $\boldsymbol{\tau}$(t) applied to the joints and end-effectors are given. Assuming that the initial conditions are $\boldsymbol{q}(t = t_0)$, $\dot{\boldsymbol{q}}(t = t_0)$, the forward dynamic model is described as

$$\ddot{\boldsymbol{q}} = f(\boldsymbol{q}, \dot{\boldsymbol{q}}, \boldsymbol{\tau}) \tag{6.77}$$

where $\ddot{\boldsymbol{q}}$ is the vector of motion accelerations, $\boldsymbol{\tau}$ is the vector of torques or forces at actuators and end-effector, $\boldsymbol{q}$ and $\dot{\boldsymbol{q}}$ are the vectors of the motion displacements and velocities in the current moment of time t. The displacements and velocities are then updated as

$$\dot{\boldsymbol{q}} = \int \ddot{q}\, dt, \; q = \int \dot{\boldsymbol{q}} dt \tag{6.78}$$

In an inverse dynamic problem, the generalized forces (torques and/or forces) $\boldsymbol{\tau}(t)$ are computed when the machine motion $(\boldsymbol{q}, \dot{\boldsymbol{q}}, \ddot{\boldsymbol{q}})$ are given. The inverse forward dynamic model is described as

$$\boldsymbol{\tau} = f^{-1}(\boldsymbol{q}, \dot{\boldsymbol{q}}, \ddot{\boldsymbol{q}}) = g(\boldsymbol{q}, \dot{\boldsymbol{q}}, \ddot{\boldsymbol{q}}) \tag{6.79}$$

6.6.1 Inertia and Moments of Inertia

When the relationship between force and motion is considered, the amount of materials has to be quantified. *Inertia* is used to quantify materials, which is a material property by which it continues in its existing state of rest or uniform motion in a straight line, unless that state is changed by an external force.

As shown in Figure 6.32a, when a force (F) acts on a mass, the mass will be linearly accelerated proportionally, and the proportion is the *inertia* of the mass (m). As shown in Figure 6.32b, when a force acts on a mass at a distance, it turns into a torque (T) on the mass. The mass will be angularly accelerated proportionally, where the proportion is the *moment of inertia* (I), which is determined by both the mass and the spatial position of the mass and acting force, i.e.

$$I = m \cdot r^2 \tag{6.80}$$

where I is the moment of inertia, m is the mass, and r is the distance of the mass to the axis of the acting torque.

Equation (6.80) shows that the moment of inertia relates to the distances of a mass to the axis of the acting torque. When the mass of the body ($\rho(r)$) is distributed, as shown

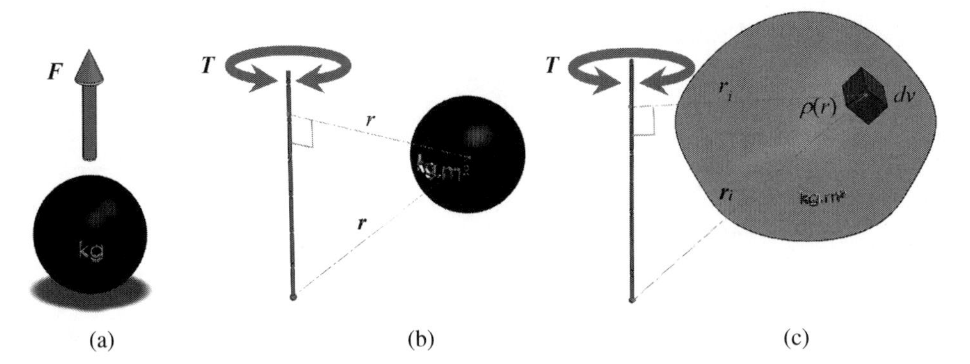

Figure 6.32 Inertia and moment of inertia of mass and body. (a) Inertia for linear motion. (b) Moment of inertia of particle for angular motion. (c) Moment of inertia of body for angular motion.

in Figure 6.32c, every mass particle in the body has its own distance to the acting torque. Equation (6.80) is modified as

$$I = \int \rho(r) \cdot r^2 dv \tag{6.81}$$

where I is the moment of inertia of the body with respect to the acting axis of torque, $\rho(r)$ is the density function, r is the distance of the mass particle to the acting axis of torque, and dv is the volume of the mass particle.

Equation (6.80) or (6.81) indicates when the moment of inertia varies with the axis of the acting torque. When a torque is applied on an arbitrary axis, it is convenient to determine the vector of angular accelerations using a tensor of the moment of inertia ($\boldsymbol{I}$). In a Cartesian coordinate system {O-XYZ}, a tensor of the moment of inertia ($\boldsymbol{I}$) is defined as

$$[\boldsymbol{I}] = \begin{bmatrix} I_{xx} & -I_{xy} & -I_{xz} \\ -I_{xy} & I_{yy} & -I_{yz} \\ -I_{xz} & -I_{yz} & I_{zz} \end{bmatrix} = \int_V \rho(x,y,z) \begin{bmatrix} y^2+z^2 & -xy & -xz \\ -xy & x^2+z^2 & -yz \\ -xz & -yz & x^2+y^2 \end{bmatrix} dx\,dy\,dz \tag{6.82}$$

where I_{xx}, I_{yy}, and I_{zz} are called the *mass products of inertia* along X, Y, and Z axes, respectively, and I_{xy}, I_{xz}, and I_{yz} are called the *cross-products of inertia* over XY, XZ, and YZ planes, respectively.

The tensor of the moment of inertia ($\boldsymbol{I}$) in Eq. (6.82) include six independent parameters (I_{xx}, I_{yy}, I_{zz}, I_{xy}, I_{xz}, and I_{yz}). The values of these parameters depend on the selection of the coordinate system {O-XYZ}. There exists a special coordinate system called a *principal coordinate system*, where all of the cross-products of inertia are zeros ($I_{xy} = I_{xz} = I_{yz} = 0$). These axes in a principal coordinate system are called the principal axes of the body.

Figure 6.33 shows the impact of the coordinate transformation on the tensor of the moment of inertia ($\boldsymbol{I}$). For the block, when the origin of the coordinate system is moved to the geometric centre of the block, all of the terms I_{xy}, I_{xz}, I_{yz} in the tensor become zero. The coordinate system established at its geometric centre is a principal coordinate system for the block.

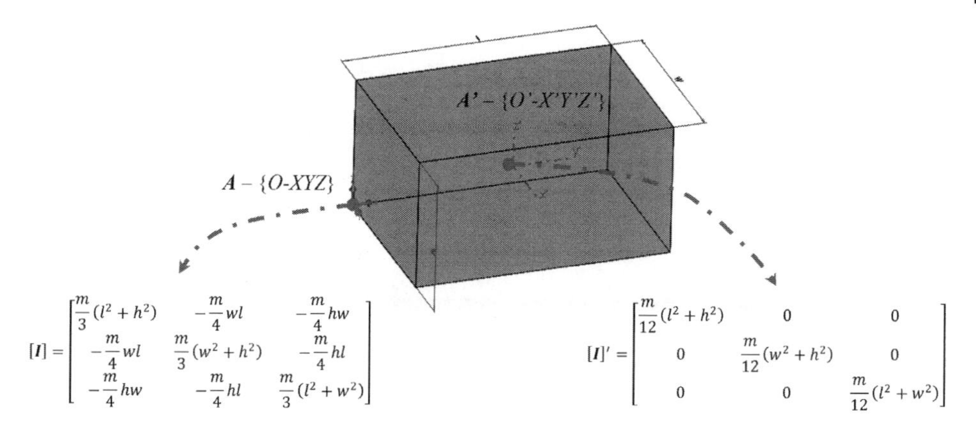

Figure 6.33 Example of principal coordinate system.

6.6.2 Newton–Euler Formulation

A machine is designed to fulfil a task by the end-effector and the end-effector is driven and moved by a number of actuators at the joints. It is necessary to know what forces or torques are required to generate the expected motion of the end-effector subjected to specified loads. A dynamic model can be expressed by a set of differential equations called *equations of motion*. Two common methods for dynamic modelling are (i) the Newton–Euler formulation and (ii) the Lagrangian formulation. In the *Newton–Euler formulation*, a machine is decomposed into rigid bodies, and Newton's second law and the Euler law are applied for each rigid body to establish the relations between driven forces and motion accelerations.

Figure 6.34 shows the Newton and Euler laws for the relations of applied force/torque and linear/angular accelerations, respectively. Note that the force (***F***) and toque (***T***) are net quantities when all of the internal and external forces/torques are combined. The Newton and Euler laws can be applied to any physical object in a system. Before these laws are applied to a body, its free-body diagram (FBD) must be drawn to identify (i) the external forces exerted on the body and (ii) the internal reactional forces at the contacts from other bodies in the system.

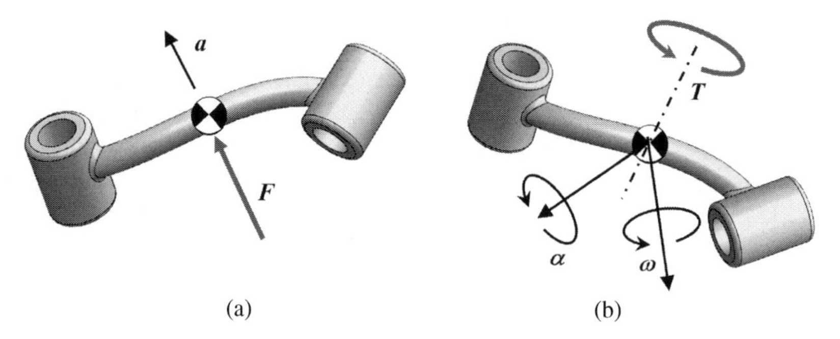

Figure 6.34 Newton and Euler laws. (a) Force causes linear acceleration ($\boldsymbol{F} = m \cdot \boldsymbol{a}$). (b) Torque causes angular acceleration ($\boldsymbol{T} = I\boldsymbol{\alpha} + \boldsymbol{\omega} \times I\boldsymbol{\omega}$).

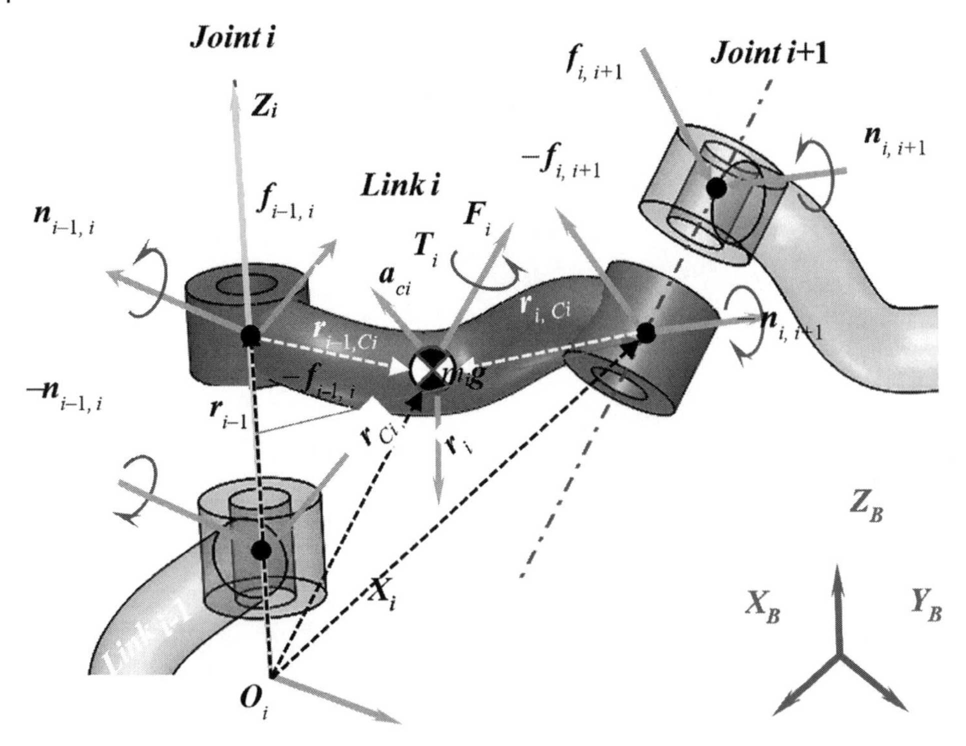

Figure 6.35 Free-Body-Diagram (FBD) of link *i* in frame *i*.

Figure 6.35a shows the FBD of link i in a manipulator, which includes an external force/torque ($\boldsymbol{F}_i$ and $\boldsymbol{T}_i$), a gravity force ($m_i \cdot \boldsymbol{g}$) and torque ($\boldsymbol{T}_i$), the reactional force ($\boldsymbol{f}_{i-1,\,i}$) and the moment ($\boldsymbol{n}_{i-1,i}$) from link $i-1$, and the reactional force ($-\boldsymbol{f}_{i,\,i+1}$) and moment ($-\boldsymbol{n}_{i,\,i+1}$) from link $i+1$. All of the above variables are given in the base coordinate systems. By applying Newtonand Euler laws in FBD, the dynamic model of link i can be found as

$$\boldsymbol{f}_{i-1,i} - m_i\boldsymbol{g} - \boldsymbol{f}_{i,i+1} + \boldsymbol{F}_i = m_i\boldsymbol{a_i} \tag{6.83}$$

$$\begin{aligned} &{}^{B}_{i}\boldsymbol{R}^{-1}(\boldsymbol{n}_{i-1,i} - \boldsymbol{n}_{i,i+1}) + [{}^{B}_{i}\boldsymbol{R}^{-1}\boldsymbol{f}_{i-1,i}] \times \boldsymbol{r}_{i-1,Ci} - [{}^{B}_{i}\boldsymbol{R}^{-1}\boldsymbol{f}_{i,i+1}] \times \boldsymbol{r}_{i,Ci} + {}^{B}_{i}\boldsymbol{R}^{-1} \cdot \boldsymbol{T}_i \\ &= \boldsymbol{I_i}\,[{}^{B}_{i}R^{-1}\dot{\boldsymbol{\omega}}_i] - [{}^{B}_{i}R^{-1}\boldsymbol{\omega}_i] \times \boldsymbol{I_i} \cdot [{}^{B}_{i}R^{-1}\boldsymbol{\omega}_i] \end{aligned} \tag{6.84}$$

Note that in Eq. (6.84), the tensor of the moments of inertia is usually defined in a local principal coordinate system built at the centre of the mass. The variables under the base coordinate system have to be transformed to the LCS by an inverse transformation matrix from frame i to the base reference coordinate system ${}^{B}_{i}R^{-1}$.

The system model is assembled from the dynamic models of all individual bodies. Variables included in the equations of motion from the Newton–Euler method include internal reaction forces and driving forces occurring to joints. Assume that as the information of all displacements, velocities, and accelerations of bodies are known, the dynamic model of a system corresponds to a set of linear equations about the variables of internal reactional forces and external driving forces.

Example 6.6 Table 6.3 gives the values of structural dynamic parameters using the example mechanism in Figure 6.36 (Bi and Kang 2014).

6.6.2.1 Inertia Force/Moment

As shown in Figure 6.36, LCSs $\{\boldsymbol{O}_e\text{-}\,\boldsymbol{x}_e\,\boldsymbol{y}_e\,\boldsymbol{z}_e\}$, $\{\boldsymbol{F}_i\text{-}\,\boldsymbol{x}_i\,\boldsymbol{y}_i\,\boldsymbol{z}_i\}$, and $\{\boldsymbol{D}_i\text{-}\,\boldsymbol{x}_{Di}\,\boldsymbol{y}_{Di}\,\boldsymbol{z}_{Di}\}$ are assumed to be part of an inertia coordinate system of the end-effector, support bar i, and slide i, respectively. According to Newton's law and Euler's equation, the inertia force/moments of the elements could be calculated as

$$\left.\begin{aligned} \boldsymbol{h}_e &= m_e\boldsymbol{R}_e^{-1}\mathrm{a}_e \\ \boldsymbol{n}_e &= \boldsymbol{I}_e(\boldsymbol{R}_e^{-1}\boldsymbol{\epsilon}_e) + (\boldsymbol{R}_e^{-1}\boldsymbol{\omega}_e)\times(\boldsymbol{I}_e(\boldsymbol{R}_e^{-1}\boldsymbol{\omega}_e) \end{aligned}\right\} \tag{6.85}$$

$$\left.\begin{aligned} \boldsymbol{h}_i &= m_l\boldsymbol{R}_i^{-1}\boldsymbol{a}_{F_i} \\ \boldsymbol{n}_i &= \boldsymbol{I}_l(\boldsymbol{R}_i^{-1}\,\epsilon_i) + (\boldsymbol{R}_i^{-1}\,\boldsymbol{\omega}_i)\times(\boldsymbol{I}_l(\boldsymbol{R}_i^{-1}\boldsymbol{\omega}_i)) \end{aligned}\right\}\quad (i = 1, 2, 3) \tag{6.86}$$

Table 6.3 Dynamic parameters.

	Platform	Support bar	Slide
Mass (kg)	1.348	0.161 65	0.213 5
Centre of mass	O_e	F_i ($l_c = 0.3l_i$)	D_i
Moment of inertia (N m) (I_{xx}, I_{yy}, I_{zz})	(0.002 706, 0.002 030, 0.004 736)	(0.000 602, 0.000 602, 0.000 032)	N/A

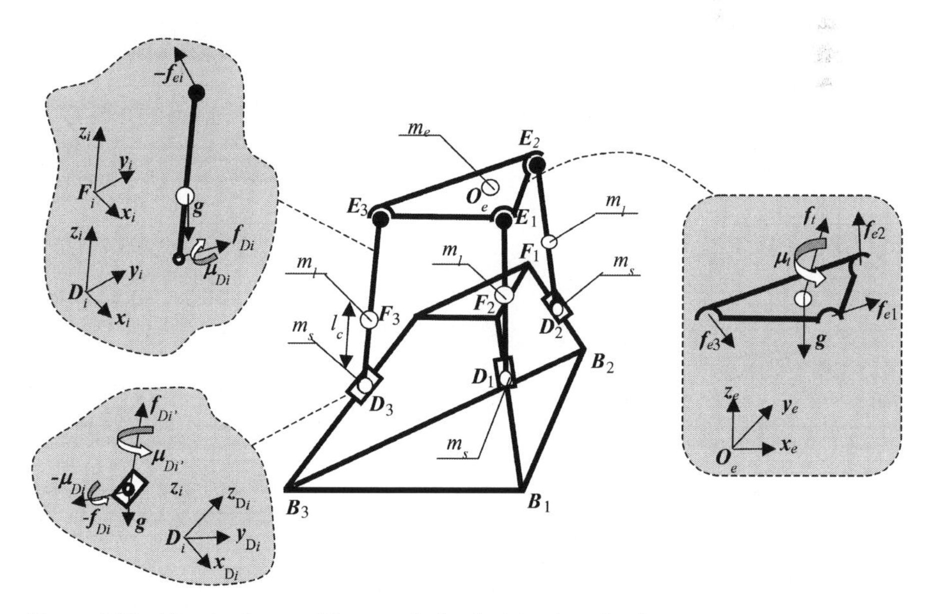

Figure 6.36 Mass centres and force analysis of a tripod mechanism.

where

$$R_i = [x_i \ y_i \ z_i] = \begin{bmatrix} -s\alpha_i & -c\phi_i c\alpha_i & s\phi_i c\alpha_i \\ c\alpha_i & -c\phi_i s\alpha_i & s\phi_i s\alpha_i \\ 0 & s\phi_i & c\phi_i \end{bmatrix}$$

$$\left.\begin{aligned} h_{D_i} &= m_s R_{D_i}^{-1} a_{D_i} \\ n_{D_i} &= 0 \end{aligned}\right\} \quad (i = 1, 2, 3) \tag{6.87}$$

where

$$R_{D_i} = [x_{D_i} \ y_{D_i} \ z_{D_i}] = \begin{bmatrix} -s\alpha_i & -s\gamma c\alpha_i & c\gamma c\alpha_i \\ c\alpha_i & -s\gamma s\alpha_i & c\gamma s\alpha_i \\ 0 & c\gamma & s\gamma \end{bmatrix}$$

6.6.2.2 Force Equilibrium Equations

As shown in Figure 6.36, based on the principles of the selection of reference coordinate systems, LCSs $\{O_e\text{-}x_e y_e z_e\}$, $\{D_i\text{-}x_i y_i z_i\}$, and $\{D_i\text{-}x_{Di} y_{Di} z_{Di}\}$ are assumed to be the reference coordinate system used to derive the equilibrium equations for the end-effector, support bar i, and slide i, respectively.

The equilibrium equations for the end-effector are derived as

$$\left.\begin{aligned} & f_t + \sum_{i=1}^{3} R_e^{-1} f_{e_i} + m_e R_e^{-1} g = h_e \\ & \sum_{i=1}^{3} r_i \times (R_e^{-1} \cdot f_{e_i}) + \mu_t = n_e \end{aligned}\right\} \tag{6.88}$$

where

f_t and μ_t are the given external force/torque with respect to the end-effector coordinate system $\{O_e\text{-}x_e y_e z_e\}$

g is the gravity acceleration with respect to the world coordinate system

f_{ei} is the unknown force from support bar i with respect to the world coordinate system

h_e and n_e are given in Eq. (6.87)

The equilibrium equations for support bar i are derived as

$$f_{D_i} - R_i^{-1} f_{e_i} + m_i (R_i^{-1} g) = h_i \quad (i = 1, 2, 3) \tag{6.89}$$

$$\mu_{D_i} - r_{D_i E_i} \times (R_i^{-1} f_{e_i}) + r_{D_i F_i} \times (R_i^{-1} g) - r_{D_i F_i} \times m_i (R_i^{-1} a_{F_i}) = n_i \quad (i = 1, 2, 3) \tag{6.90}$$

where

$$r_{D_i F_i} = l_c (0, 0, 1)^T$$

$$r_{D_i E_i} = l_i (0, 0, 1)^T$$

$$\mu_{D_i} = (0, \mu_{D_{iy}}, \mu_{D_{iz}})^T$$

f_{Di} is the unknown force from slide i with respect to the system $\{D_i\text{-}x_i y_i z_i\}$

h_i and n_i are given in Eq. (6.86)

The equilibrium equations for slide i are derived as

$$-(\boldsymbol{R}_{D_i}^{-1}\boldsymbol{f}_i)\boldsymbol{f}_{D_i} + \boldsymbol{f}_{D_i'} + m_s\boldsymbol{R}_{D_i}^{-1}\mathbf{g} = \boldsymbol{h}_{D_i} \quad (i = 1,\ 2,\ 3) \tag{6.91}$$

where

$\boldsymbol{f}_{D'i}$ is unknown force from the base with respect to the coordinate system $\{\boldsymbol{D}_i\text{-}\ \boldsymbol{x}_{Di}\ \boldsymbol{y}_{Di}\ \boldsymbol{z}_{Di}\}$
$\boldsymbol{h}_{Di}$ is given in Eq. (6.87)

The reaction force $\boldsymbol{f}_{D'i}$ includes the contribution of the driving force of joint motor i. Therefore, the driving force of joint i is the projection of $\boldsymbol{f}_{D'i}$ on the translational direction of the prismatic joint

$$\zeta_i = \boldsymbol{f}_{D_i'} \cdot \boldsymbol{z}_{D_i} \quad (i = 1, 2, 3) \tag{6.92}$$

where

ζ_i is the force of joint i

6.6.2.3 Dynamic Model and Solution

A combination of Eqs. (6.88) and (6.90) gives a group of 15 linear equations with 15 unknown parameters, i.e. 3×3 parameters for $\boldsymbol{f}_{ei}$ ($i = 1, 2, 3$) and 2×3 parameters for $\boldsymbol{\mu}_{D_i}$ ($i = 1, 2, 3$). This group of linear equations can be solved so that $\boldsymbol{f}_{ei}$ and $\boldsymbol{\mu}_{D_i}$ are obtained.

Reaction forces $\boldsymbol{f}_{Di}$ and $\boldsymbol{f}_{D'i}$ can be derived easily using Eqs. (6.88) and (6.91) sequentially once $\boldsymbol{f}_{ei}$ are known, and Eq. (6.92) is used finally to calculate the driving forces of joints.

Figure 6.37 shows an example of the required acceleration for a given path. Using the dynamic model, the required joint forces are calculated as shown in Figure 6.38.

A validating experiment is conducted and the actual joint forces are measured. The discrepancy of theoretical and measured results is illustrated in Figure 6.39. The maximum discrepancy is 6.1% and the average discrepancy is 1.83%. Note that some simplifications, such as the assumption of a centralized mass of a link body, have been made in dynamic modelling and the discrepancy is very reasonable and acceptable.

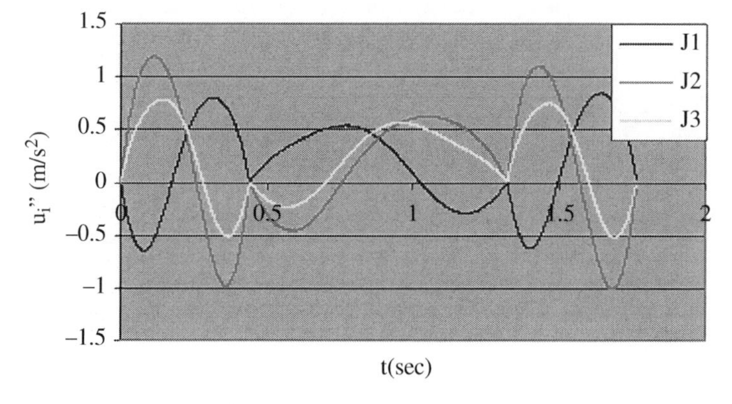

Figure 6.37 Joint accelerations.

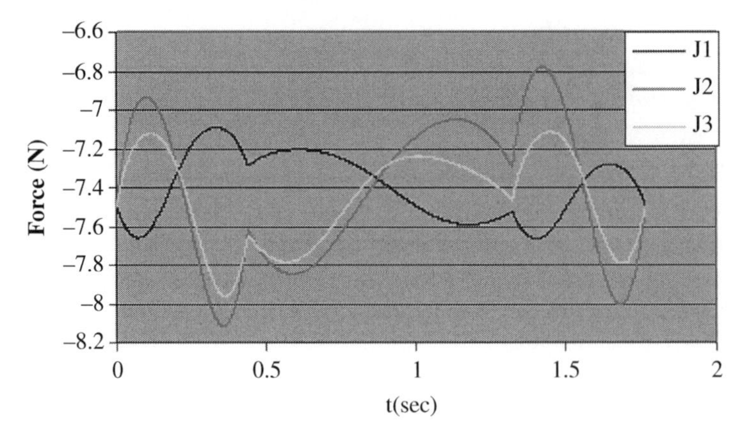

Figure 6.38 Joint forces calculated from the dynamic model.

6.6.3 Lagrangian Method

The Lagrangian formulation describes the dynamic behaviour of a system in terms of work and energy. The internal forces at joints are automatically eliminated and the closed-form dynamic models can be derived in the generalized coordinates. Assume that the generalized coordinates of the system are $q_1, q_2, \ldots, q_n$ and the Lagrangian term L is calculated from the kinetic and potential energy (K and P) of the system as

$$L(q_i, \dot{q}_i) = K(q_i, \dot{q}_i) - P(q_i) \tag{6.93}$$

where the potential energy P is a function of the generalized coordinates q_i and the kinetic energy is the function of both of q_i and $\dot{q}_i$. The equations of dynamic motion of the system can be defined as

$$\frac{d}{dt}\frac{\partial L}{\partial \dot{q}_i} - \frac{\partial L}{\partial q_i} = Q_{i,} \quad i = 1, 2, \ldots, n \tag{6.94}$$

where Q_i is the generalized force corresponding to the generalized coordinate q_i. Taking into consideration the virtual work by non-conservative forces yields the generalized forces exerted on the system.

In the following, the 2D robot in Figure 6.40 is used as an example to illustrate the application of the Lagrangian method in dynamics modelling. Figure 6.40 gives all system parameters and dynamic variables.

The angular velocities of the links are

$$\left.\begin{aligned} \omega_1 &= \dot{\theta}_1 \\ \omega_2 &= \dot{\theta}_1 + \dot{\theta}_2 \end{aligned}\right\} \tag{6.95}$$

The linear velocities of the links at their CoGs are

$$\boldsymbol{V}_{c1} = \begin{Bmatrix} -l_{c1}\sin\theta_1 \\ l_{c1}\cos\theta_1 \end{Bmatrix} \dot{\theta}_1 \tag{6.96}$$

$$\boldsymbol{V}_{c2} = [\boldsymbol{J}_{c2}]\{\dot{\boldsymbol{q}}\} = \begin{bmatrix} -a_1\sin\theta_1 - l_{c2}\sin\theta_{12} & -l_{c2}\sin\theta_{12} \\ a_1\cos\theta_1 + l_{c2}\cos\theta_{12} & l_{c2}\cos\theta_{12} \end{bmatrix} \begin{Bmatrix} \dot{\theta}_1 \\ \dot{\theta}_2 \end{Bmatrix} \tag{6.97}$$

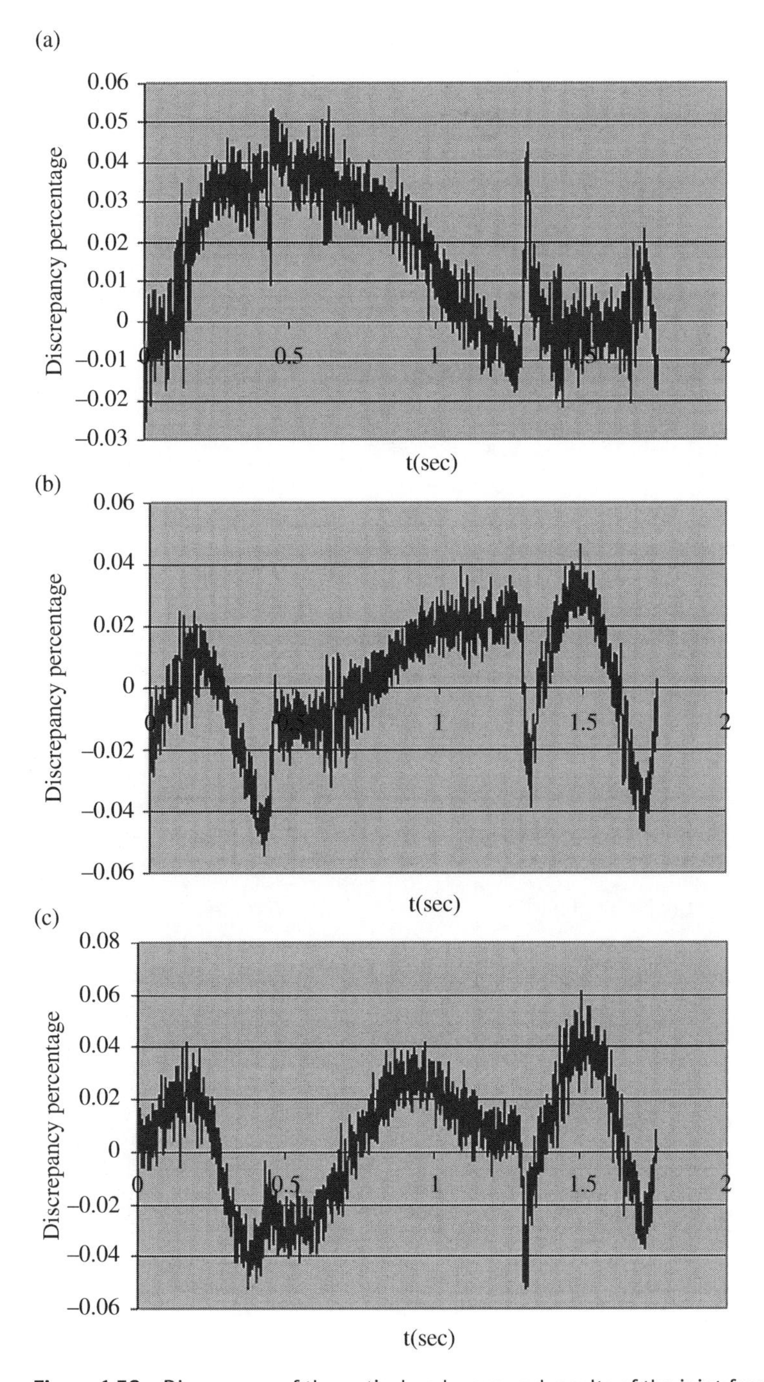

Figure 6.39 Discrepancy of theoretical and measured results of the joint forces. (a) Joint 1. (b) Joint 2. (c) Joint 3.

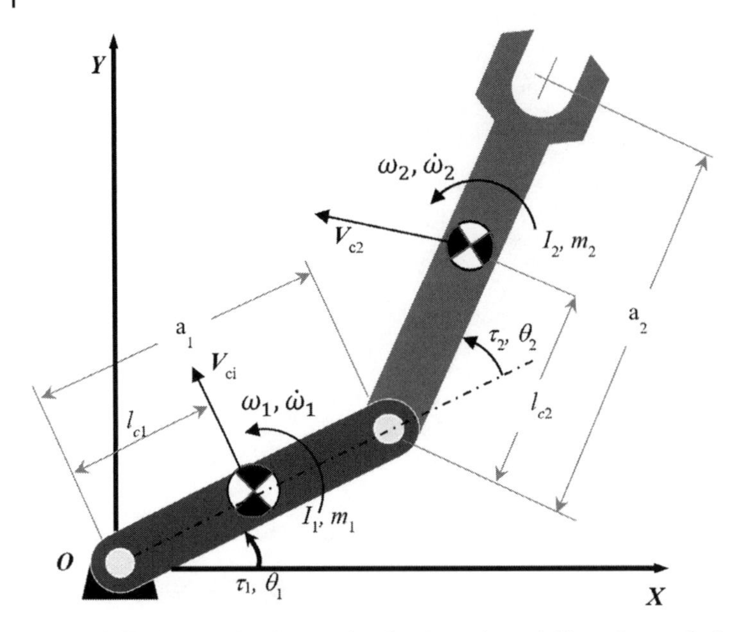

Figure 6.40 Lagrangian formulation for dynamic modelling of a two DoF robot.

where $\theta_{12} = \theta_1 + \theta_2$, $\{\dot{\boldsymbol{q}}\} = \begin{Bmatrix} \dot{\theta}_1 \\ \dot{\theta}_2 \end{Bmatrix}$ is the vector of angular velocities and $[\boldsymbol{J}_{c2}]$ is 2×2 Jacobian matrix relating to the velocity of center of gravity (CoG) of link 2. Note that both linear and angular velocities are the functions of angular displacements and velocities and are dependent.

The total kinematic energy is calculated by

$$\boldsymbol{K} = \sum_{i=1}^{2} \left(\frac{1}{2} m_i |\boldsymbol{V}_{ci}|^2 + \frac{1}{2} I_i {\omega_i}^2 \right) = \frac{1}{2} H_{11} \dot{\theta}_1^2 + H_{12} \dot{\theta}_1 \dot{\theta}_2 + \frac{1}{2} H_{22} \dot{\theta}_2^2$$
$$= \tfrac{1}{2} \left(\dot{\theta}_1 \ \dot{\theta}_2 \right) \begin{bmatrix} H_{11} & H_{12} \\ H_{12} & H_{22} \end{bmatrix} \begin{Bmatrix} \dot{\theta}_1 \\ \dot{\theta}_2 \end{Bmatrix} \tag{6.98}$$

where

$$\left. \begin{aligned} H_{11} &= H_{11}(\theta_2) = m_1 l_{c1}^2 + I_1 + m_2(a_1^2 + l_{c2}^2 + 2a_1 l_{c2} cos\theta_2) + I_2 \\ H_{22} &= m_2 l_{c2}^2 + I_2 \\ H_{12} &= H_{12}(\theta_2) = m_2(l_{c2}^2 + a_1 l_{c2} cos\theta_2) + I_2 \end{aligned} \right\} \tag{6.99}$$

The potential energy stored in the two links is given by

$$P = m_1 g l_{c1} \sin\theta_1 + m_2 g\{a_1 \sin\theta_2 + l_{c2} \sin\theta_{12}\} \tag{6.100}$$

Applying the Lagrangian formulation for the first and second joints gives

$$\left.\begin{aligned}\tau_1 &= \frac{d}{dt}\frac{\partial L}{\partial \dot{\theta}_1} - \frac{\partial L}{\partial \theta_1} = H_{11}\ddot{\theta}_1 + H_{12}\ddot{\theta}_2 + \frac{\partial H_{11}}{\partial \theta_2}\dot{\theta}_1\dot{\theta}_2 + \frac{\partial H_{12}}{\partial \theta_2}\dot{\theta}_2^2 - (m_1 g l_{c1}\cos\theta_1 + m_2 g l_{c2}\cos\theta_{12}) \\ \tau_2 &= \frac{d}{dt}\frac{\partial L}{\partial \dot{\theta}_2} - \frac{\partial L}{\partial \theta_2} = (H_{12}\ddot{\theta}_1 + H_{22}\ddot{\theta}_2) + \frac{\partial H_{12}}{\partial \theta_2}\dot{\theta}_1\dot{\theta}_2 - (m_2 g(a_1\cos\theta_2 + l_{c2}\cos\theta_{12}))\end{aligned}\right\} \tag{6.101}$$

where τ_1 and τ_2 are the driving torques at joints 1 and 2, respectively, H_{11}, H_{12}, H_{22} are expressed in Eq. (6.99), and

$$\frac{\partial H_{11}}{\partial \theta_2} = -2m_2 a_1 l_{c2}\sin\theta_2, \quad \frac{\partial H_{12}}{\partial \theta_2} = -m_2 a_1 l_{c2}\sin\theta_2$$

6.7 Kinematic and Dynamics Modelling in Virtual Design

Machine design is a complicated process. With computer implementation of the aforementioned kinematic and dynamic modelling methods, computer aided design (CAD) tools are able to perform a number of critical activities for designers to design and optimize a machine for the expected functional requirements. Figure 6.41 shows that a designer should interact and fully utilize the capabilities of a CAD tool in virtual machine design (Markkonen 1999). *Firstly*, the design process begins with preparation of a CAD model of a machine, including the geometry, dimensions, material properties, mating relations, and boundary conditions. *Secondly*, a simulation job should be defined for a computer aided tool to establish a mathematic model for the targeted simulation. *Thirdly*, designers should specify the model parameters, such as the properties of motors, the profiles of expected motions, the duration of simulation, and design variables to be investigated. *Fourthly*, the simulation is performed to find the solution to the formulated mathematic model numerically. *Fifthly*, the simulation result is analysed and verified, and an iterative process is repeated to precedent steps until the design and analysis goal is achieved. It should be noted that even CAD tools are available to take over many critical tasks, such as developing a mathematical model or solving a system model, and sufficient interaction between a designer and a computer aided design tool is essential to a successful machine design.

For a machine design at different stages, design scopes and goals are different, and different design tools are needed at different stages. SolidWorks provides a comprehensive toolset to support virtual machine design (Markkonen 1999). Figure 6.42 shows a number of main functional modules that can be fully utilized in designing and analysing a machine.

6.7.1 Motion Simulation

SolidWorks *Motion Analysis* is to create a simulation model to study the position, velocity, acceleration, and torque of a mechanism subject to external loads. It can be used to design, analyse, and optimize a machine design. Motion analysis in a virtual environment brings significant benefits to machine design: (i) the number of physical prototypes can be minimized since the simulation helps to identify potential design errors and omissions;

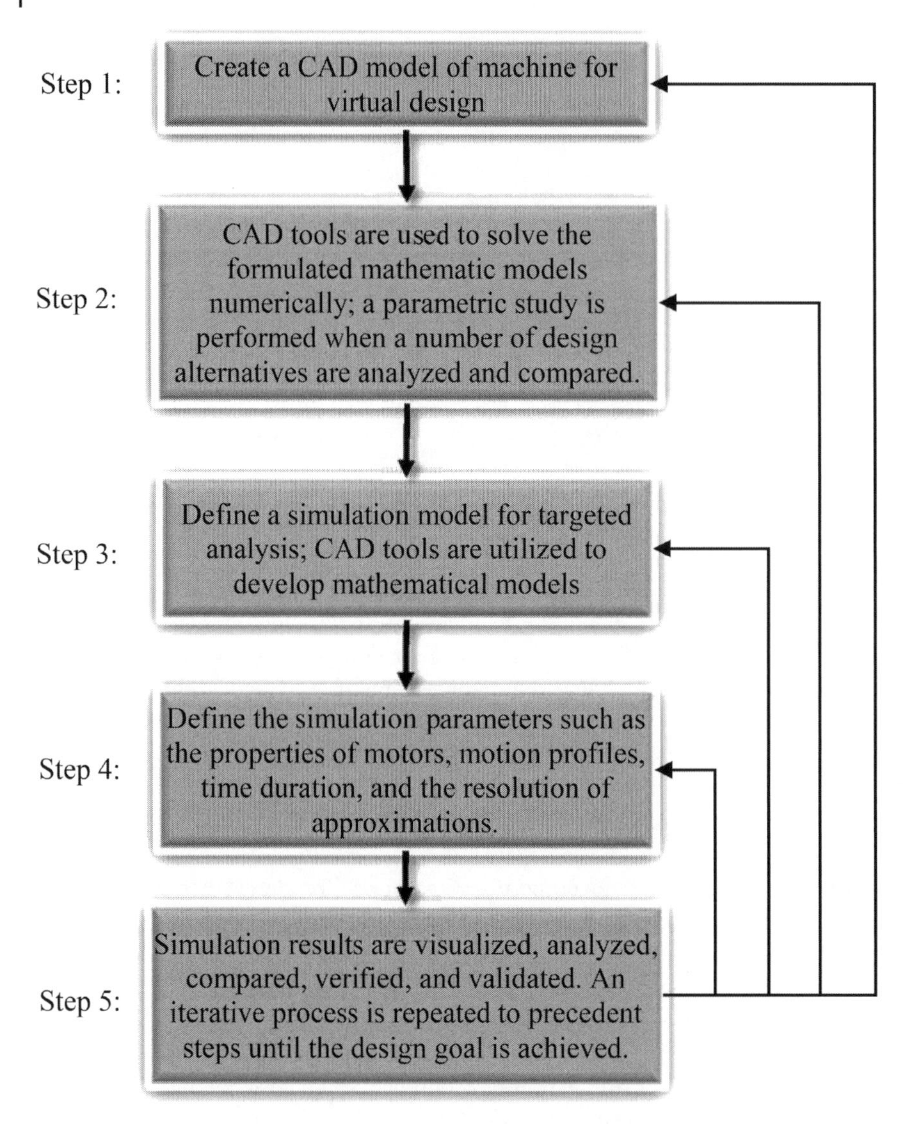

Figure 6.41 Steps of using a computer aided tool in a virtual machine design.

(ii) virtual analysis takes much less time than experiments and supports parametric studies, which allows more design options to be explored at a very early stage; (iii) the simulation-based optimization can be utilized to improve machine design; (iv) the simulation is integrated with all other engineering analyses and makes it possible to gain quantified insights ibto additional engineering analyses and look into the feasibility of design over the entire product lifecycle.

A motion analysis of a machine is capable of answering all of the questions related to kinematic and dynamic models of a machine, such as *What is the cycle time to machine operation? What force is required for a given operation? What sizes of motors should be used for the given task of the machine? What shapes and dimensions will generate expected motions? What*

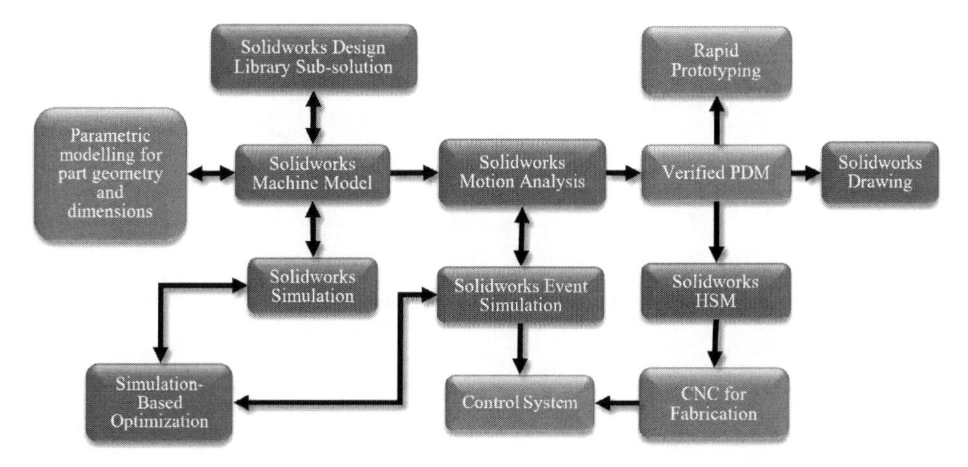

Figure 6.42 Main functional modules for a virtual machine design in SolidWorks.

are the expected velocities and accelerations subject to the given driving forces and torques? What accuracy of motion can one expect? Will parts fail? How long can the machine last subject to the given operating conditions?

6.7.2 Model Preparation

Create a design scheme where the shapes of parts are designed and the nature of their connections are presented in a simplified form and the materials are selected to determine allowable stresses accounting for all the factors that affect the part strength. The forces acting on parts in the process of machine operation are then determined. All parts or components must be assigned proper material properties.

After modelling each part, the next step is to model the assemblies of parts. If two parts have a static relation in space, these two parts should be grouped since no relative motion is allowed between them. If two parts have a relative motion, a correct joint type must be selected. Assembly relations of a machine are represented by mates in SolidWorks. Figure 6.43 shows the main mate types, which are catalogued into (a) standard mates, (b) advanced mates, and (c) mechanical mates. A bonded relation or a joint of two parts can mainly be defined as a combination of mates in (a) and (b); other mates occurring to more than two entities or a relative motion with a coupling are defined by the mates in (b) and (c).

Figure 6.44 shows an example of assembly modelling of a yumi collaborative robot. Yumi has two 7 DoF arms. Any one of 14 joints can be defined as a combination of one '*coincident*' mate of two planes and one '*concentric*' mate of two cylindrical surfaces from two joined parts.

6.7.3 Creation of a Simulation Model

SolidWorks has three functional modules for motion study: *Animation*, *Basic Motion*, and *Motion Analysis*. The animation tool studies the kinematic behaviours of models without

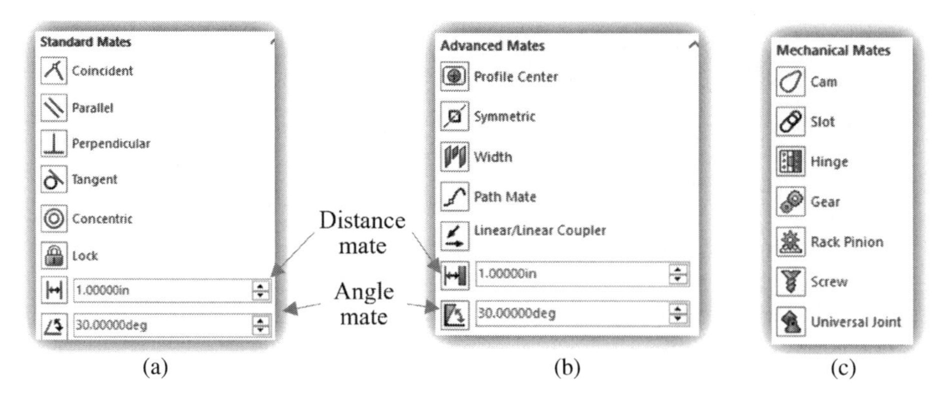

Figure 6.43 Mates in assembly modelling of SolidWorks. (a) Standard mates. (b) Advanced mates. (c) Mechanical mates.

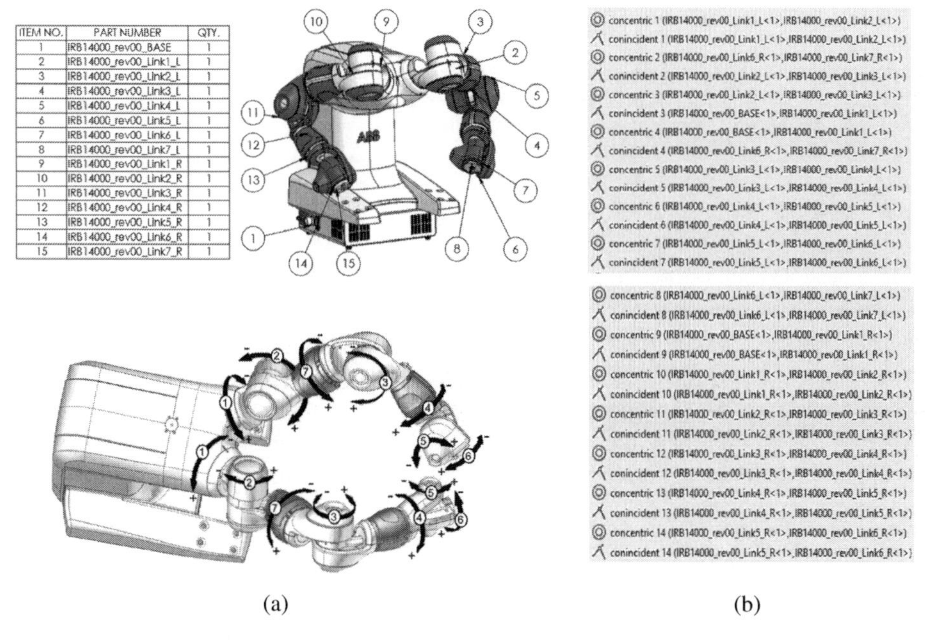

ITEM NO.	PART NUMBER	QTY.
1	IRB14000_rev00_BASE	1
2	IRB14000_rev00_Link1_L	1
3	IRB14000_rev00_Link2_L	1
4	IRB14000_rev00_Link3_L	1
5	IRB14000_rev00_Link4_L	1
6	IRB14000_rev00_Link5_L	1
7	IRB14000_rev00_Link6_L	1
8	IRB14000_rev00_Link7_L	1
9	IRB14000_rev00_Link1_R	1
10	IRB14000_rev00_Link2_R	1
11	IRB14000_rev00_Link3_R	1
12	IRB14000_rev00_Link4_R	1
13	IRB14000_rev00_Link5_R	1
14	IRB14000_rev00_Link6_R	1
15	IRB14000_rev00_Link7_R	1

Figure 6.44 Assembly modelling of an ABB yumi robot. (a) 14 joints (ABB 2018). (b) Assembling modelling.

considering dynamics. Users can use animation to visualize the models. Galliera (2010) gave a comparison of three tools in SolidWorks in Table 6.4.

As shown in Figure 6.45a, *Motion Analysis* is included in Add-Ins. A user has to activate the tool before it can be used. A new Motion Study can be created by right-clicking the Motion Study tab (Figure 6.45b). Since the Motion Analysis has been activated, the options under animation include Animation, Basic Motion, and Motion Analysis.

Table 6.4 Comparison of SolidWorks Animation, Basic Motion, and Motion Analysis.

Types	Solvers	Description
Animation	3D dimensional constraint manager (3DDCM) by D-cube	The 3DDCM solver is capable of positioning parts in an assembly or mechanism. Users can use it to build, modify, and animate the model to visualize the changes of comprehensive geometry, dimensions, and design constraints. The animation time can be used to create simple animations where the interpolations are used to specify point-to-point motion of parts in assemblies.
Basic Motion	Ageia PhysX	The Ageia PhysX is a physics solver primarily for games. The tool simulates how objects behave, move, and react for life-like motion and interaction. Adopting the Ageia PhysX in Basic Motion makes the simulation look real but not the precise one. The Basic Motion tool can be used to approximate the effects of motors, springs, collisions, and gravity on assemblies. The physics-based simulation can be generated quickly with less computation. It is best suited for presentation-worthy animation.
Motion Analysis	ADAMS solver	The ADAMS solver is a sophisticated tool used to analyse the complex behaviour of mechanical assemblies. The Motion Analysis in SolidWorks aims at analysing the forces, torques, contact forces, and power consumption accurately. Users can export any kinematic and dynamic quantities over time from analysis. Motion Analysis can be used to simulate and analyse the motion of a machine with consideration of driving forces, springs, dampers, and frictions. The kinematic solver takes into account motion constraints, material properties, mass, and component contacts. The result of dynamic loads can be exported directly for discrete event simulation of the machine.

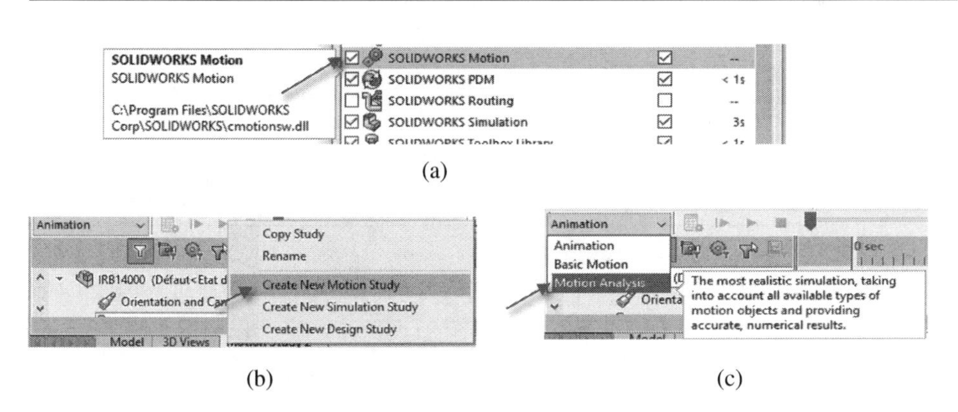

Figure 6.45 Create a Motion Study for a machine model. (a) Activate *Motion* in *Adds-Ins*. (b) Create a *New Motion Study* by right-clicking the blank area above the *status bar*. (c) Select *Motion Analysis* in the drop-down menu of *Animation*.

6.7.4 Define Motion Variables

As shown in Figure 6.46, a motion study includes the motion variables for *motor*, *spring*, *damper*, *force*, *contact*, and *gravity*. For example, motors have to be defined for all of the active joints in a machine. The Motion Analysis allows users to set up many parameters for a motor.

Figure 6.47 shows the interface for the definition of a motor. A motor motion can be *translational* or *rotational*, and the motion occurs to a moving body relative to a reference body along a specified direction. In defining a motor, the moving body, the reference body, and the motion direction must be specified. The profile of a motion can be one of *Constant Speed*, *Distance*, *Oscillation*, *Segments*, *Data Points*, *Expression*, or *Servo Motor* from the drop-down list of *Motion*. The direction of the motion must be specified.

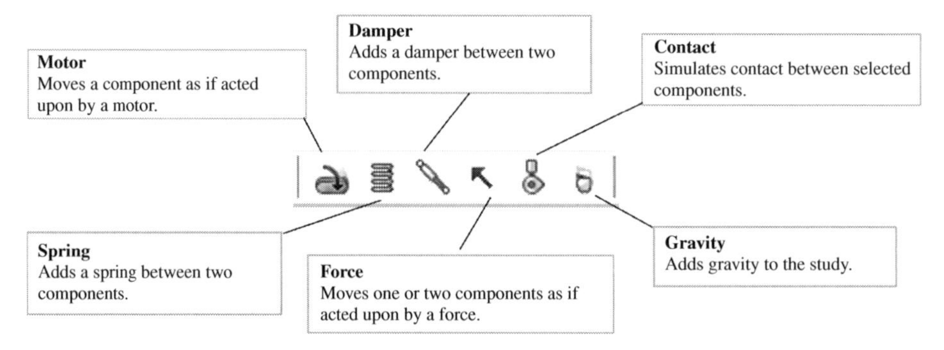

Figure 6.46 Types of motion variables in a Motion Study.

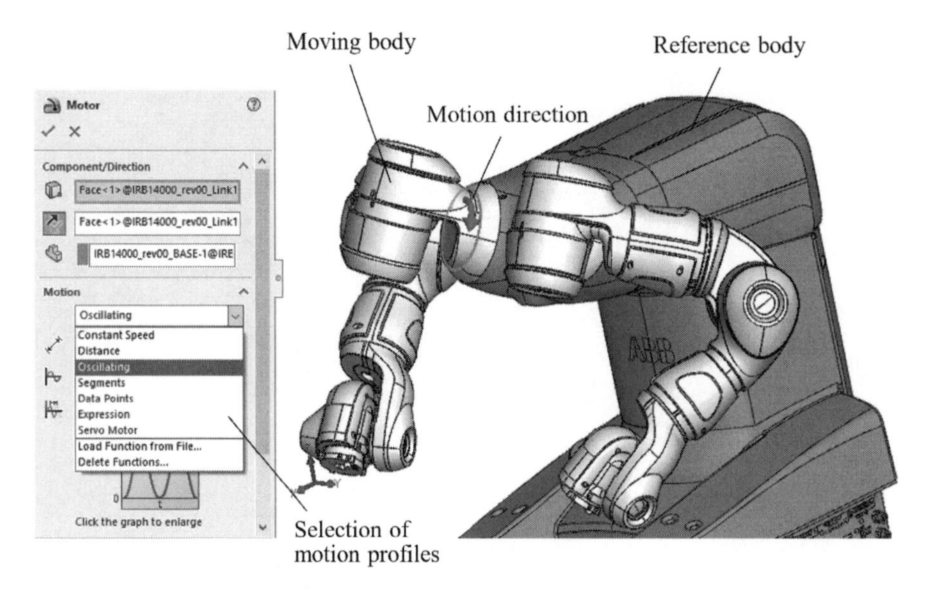

Figure 6.47 Defining a motor in a SolidWorks Motion Study.

6.7.5 Setting Simulation Parameters

Other than motion variables, other simulation parameters of the model can be customized via the *Motion Study Properties* shown in Figure 6.48a. The user can set the number of calculation steps per second, decide whether or not to run the simulation during the calculation, refine the accuracy of 3D contact or of the representation of solid geometry, specify the cycle settings, and select the analysis algorithms and tolerance of terminations, as shown in Figure 6.48b.

6.7.6 Run Simulation and Visualize Motion

The motion analysis does the calculations automatically for the simulation model. Note that the calculations will not be updated unless the user clicks the calculation icon shown in Figure 6.49. After the calculation is completed, SolidWorks keeps the results and the user can use the animation tools to review the animation of motions over time. In addition, the result from the simulation can be saved externally in .AVI or other formats.

Special attention should be paid when the user makes some changes on motion variables or simulation settings after the previous simulation is completed. The user has to run the calculations again to update the simulation results.

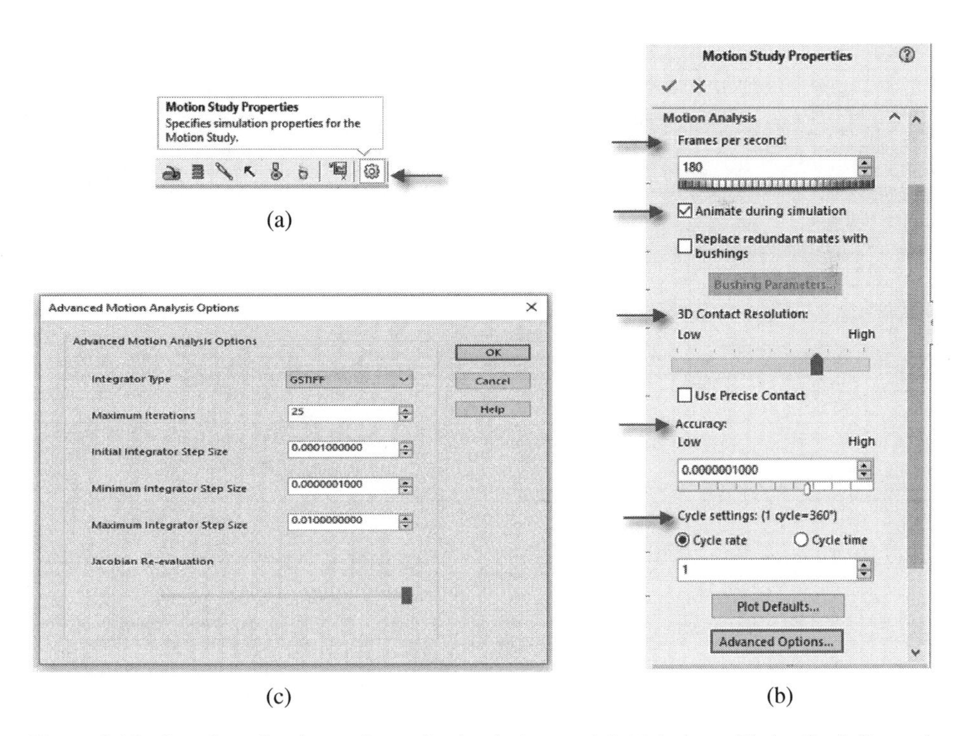

Figure 6.48 Interfaces for the settings of a simulation model. (a) Activate *Motion Study Properties*. (b) Define simulation properties. (c) Advanced setting for motion analysis.

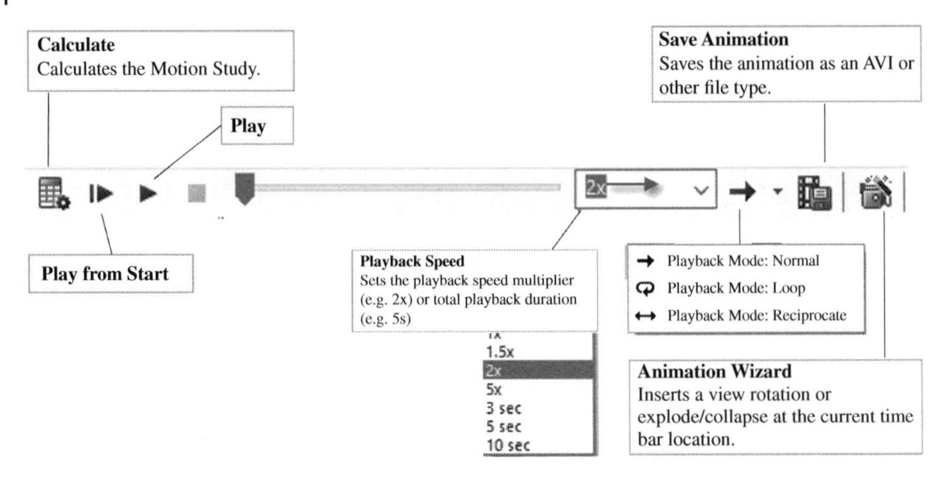

Figure 6.49 Run simulation and visualize motion.

6.7.7 Analyse Simulation Data

A Motion Analysis model involves a large number of motion variables and simulation parameters. It would be a very rare case where a user defines all of the simulation parameters appropriately at his or her first iteration. The user must possess a knowledge of kinematics and dynamics, be able to understand the simulation results, and make the engineering judgements to determine whether the simulation model is defined correctly. Figure 6.50 shows a wide selection of kinematic and dynamic variables in kinematic or dynamic models whose values in simulation with respect to time can be readily retrieved for visualization and analyses. Figure 6.51 shows an example plot for the change of torque over time for a specific motor in the mechanism.

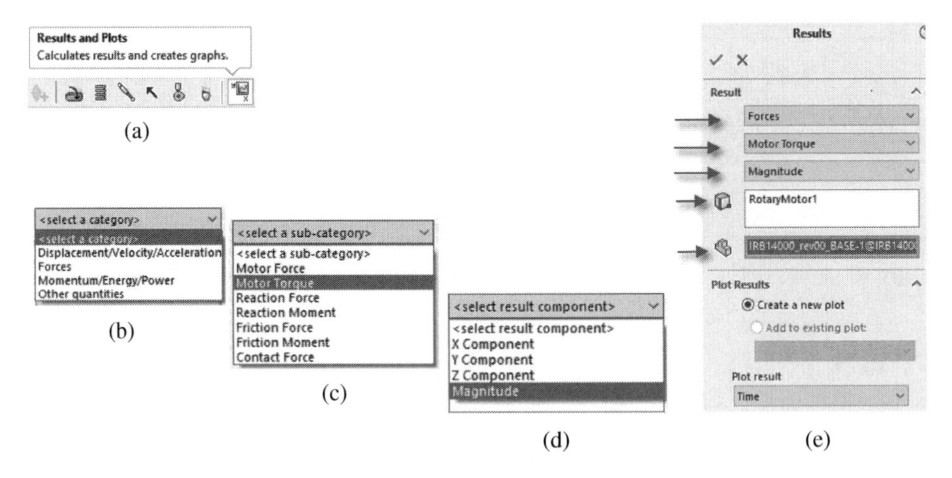

Figure 6.50 Review the results of kinematic and dynamic variables. (a). Activate *Results and Plots*. (b) Quantity types. (c) Quantity sub-types. (d) Component types. (e) Select an object and the coordinate system.

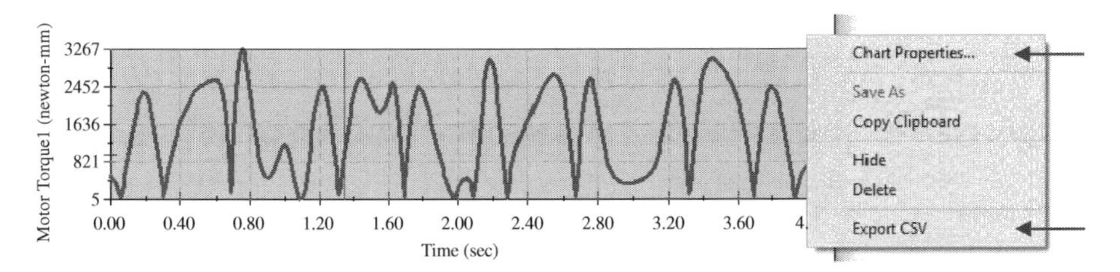

Figure 6.51 Review the results of kinematic and dynamic variables.

6.7.8 Structural Simulation Using Motion Loads

For a machine with motion, the exerted forces on its constitutive bodies vary over time, and stress distributions over the bodies are changed accordingly. For a safe design, it is helpful to determine the maximum stress on a body during the motion. The historical loads from a Motion Analysis simulation can be utilized for structural analysis of the bodies in a machine.

In a structural analysis using motion loads, the stress, factor of safety, or deformation of components can be analysed without setting up loads and boundary conditions. The loads are imported automatically from the results of a Motion Analysis study. The user is able to investigate the effects of motion loads on the stress distributions and deformations for one or more components. Figure 6.52 shows an example where the stress and deflection of a component in a robot has been analysed at the specified timeframe, where the loads and boundary conditions are automatically defined using a motion analysis model.

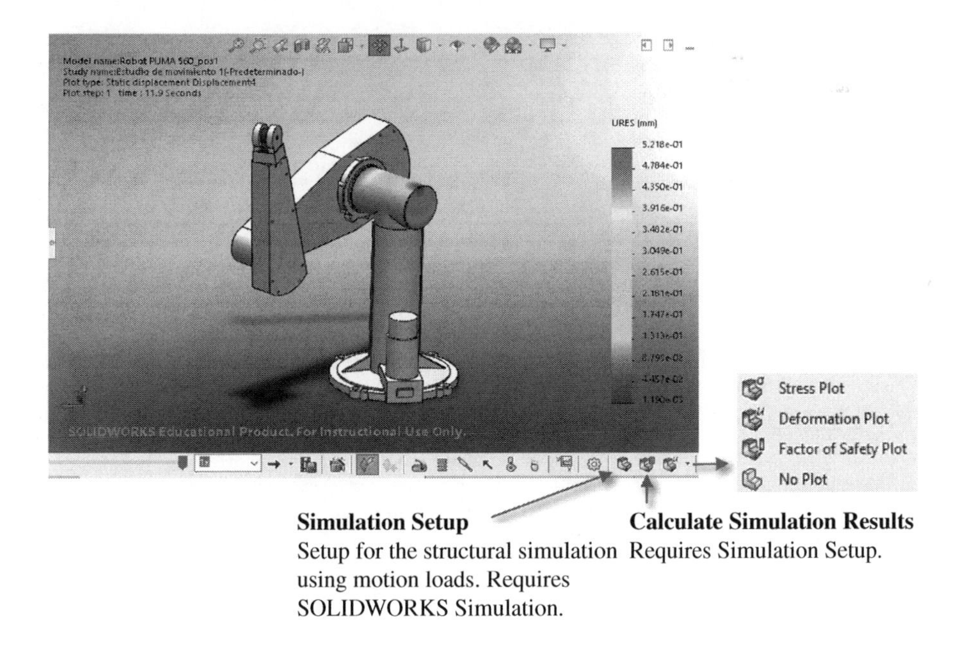

Figure 6.52 Mechanical event simulation in a motion analysis model.

The analysis can be performed at specified times and time ranges. The flow of data from a Motion Analysis Study to Structural Simulation is one-directional. The results of the stress analysis do not change the results of Motion Analysis. A more detailed stress analysis should be performed in SolidWorks Simulation as a combined rigid–flexible body analysis where the loads are exported from motion analysis one at a time (SolidWorks 2019).

6.8 Summary

Kinematics and dynamics lay the foundation to machine design since the motions of a machine are driven by force and power. Kinematics is the study of the relation of the motions of rigid bodies in a machine and dynamics is the study of the causes of motions. Machine design involves design analysis and synthesis. Design analysis is used to predict outcomes for a given configuration of a machine and design synthesis is used to analyse and compare a number of design alternatives and optimize the machine design for a given set of functional requirements (FRs). Kinematic and dynamic modelling for a machine is not a trivial task. A computer aided machine design tool automates the modelling process and will allow designers to predict machine motions when a virtual model of the machine becomes available. A computer aided machine tool also makes it possible to utilize the results of a Motion Analysis study for structural analysis and optimization.

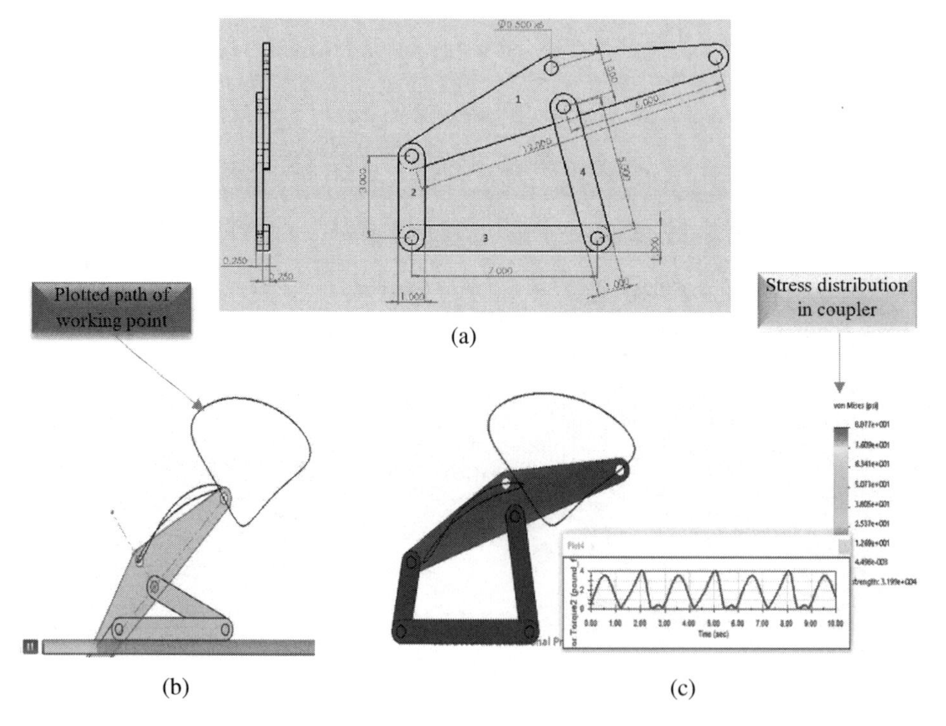

Figure 6.53 Example of a *Motion Analysis* design project. (a) Dimensions of the four-bar mechanism. (b) Motion simulation. (c) Discrete event simulation.

6.9 Design Project

6.1 Apply SolidWorks Motion Analysis and Discrete Event Simulation for computer aided design and analysis of mechanisms.

1. Select a device/machine/machine element with motion.
2. Create the parts and assembly models for the product.
3. Use the Motion Analysis to determine and plot (i) the trace path of the working point (i.e. tool-tip) and (ii) the driving force/torque of the motor.
4. Conduct Discrete Event Simulation to evaluate and visualize the deformation, stress, and safety factor of the critical link/body with an external load in the product.
5. Document your work in your project report (Figure 6.53).

References

ABB (2018). Robotics Product specification IRB 14000, Workspace R18-2 version a10, Published 2018-10-10, Skribenta version 5.3.008. http://library.e.abb.com/public/c9f66b2b80bc42bba49c20970bed1b0d/3HAC052982%20PS%20IRB%2014000-en.pdf?x-sign=09i961Kw9WYypn8AUDHOiC8hUT/kTjLE+VudbTeATcD5UcH0biKarKLuRwEL+rKw.

Bi, Z.M. and Kang, B. (2014). An inverse dynamic model of over-constrained parallel kinematic machine based on Newton–Euler formulation. *ASME Journal of Dynamic System, Measurement, and Control*, paper no. 041001. DOI: https://doi.org/10.1115/1.4026533.

BrainyQuote (2019). Steve Jobs quotes. https://www.brainyquote.com/quotes/steve_jobs_416936

Corke, P. (2011). *Robotics, Vision & Control*. Springer. ISBN: 978-3-319-54412-0.

Craig, J. (2018). *Introduction to Robotics Mechanics and Control*, 4e. Pearson. ISBN: 9780133489798.

Denavit, J. and Hartenberg, R.S. (1955). A kinematic notation for lower-pair mechanisms based on matrices. *American Society of Mechanical Engineers* 23: 215–221.

Galliera, J. (2010). Solvers used for animation, basic motion, and motion analysis. https://forum.solidworks.com/community/simulation/motion_studies/blog/2010/02/08/solvers-used-for-animation-basic-motion-motion-analysis.

Gosselin, C. and Merlet, J.-P. (1994). On the direct kinematics of planar parallel manipulators: special architectures and number of solutions. *Mechanism and Machine Theory* 29 (8): 1083–1097.

Hayes, J. (2018). Type, number, and dimensional synthesis. http://faculty.mae.carleton.ca/John_Hayes/5507Notes/Ch3JH.pdf.

Markkonen, P. (1999). On multi body systems simulation in product design. Doctoral thesis. Royal Institute of Technology, KTH, Stockholm.

Reuleaux, F. (1876). *The Kinematics of Machinery: Outlines of a Theory of Machines* [Theoretische kinematik: Grundzüge einer Theorie des Maschinenwesens](trans. A.B.W. Kennedy). New York, NY, USA: Macmillan and Company.

Singh, S. (2014). Kinematics. http://robotics.itee.uq.edu.au/~metr4202/2014/lectures/L3.Kinematics.pdf.

SolidWorks (2019). Introduction to motion studies. http://help.solidworks.com/2019/English/SolidWorks/motionstudies/c_introduction_to_motion_studies.htm.

Part II

Computer Aided Manufacturing (CAM)

7

Group Technology and Cellular Manufacturing

7.1 Introduction

Modern products are becoming more and more complex and the variety and number of the resources required to manufacture products in a plant are increasing continuously. Planning and scheduling of manufacturing activities are more and more complicated and challenging. The physical plant must be organized to run manufacturing businesses efficiently to avoid unnecessary messes such as the machine shop shown in Figure 7.1.

Platforming technologies help to increase the utilization rate of manufacturing resources, reduce product costs, and respond to market changes promptly. The philosophy of using platform technologies is that *similar things should be done in similar ways*. The similarities of products should be explored at any phases of product lifecycles. Taking an example of a manufacturing phase, products made at the same plant might show similarities of many aspects such as:

1. Materials, features, functions, sizes, markets, and costs of products,
2. Manufacturing processes, planning, scheduling, and controlling,
3. Machine tools such as fixtures, moulds, dies, inspecting tools,
4. Material handling processes, storages, supply chain management, transportation, and
5. Administrative functions.

Depending on the categories of considered design factors, the types and levels of the similarities are different. A manufacturing system should be organized based on the most dominated types of product similarities. In this chapter, the methods to identify the similarities of products are discussed and these methods are incorporated to optimize the layout design of a manufacturing system. Note that the similarities of products may occur to one, two, or more product families. We focus on the similarities of the correspondences of products and machining tools of manufacturing processes.

7.2 Manufacturing System and Components

A manufacturing system is an arrangement and operations of *machines tools*, *materials handling systems*, *positioning devices*, *fixtures*, *inspecting devices*, *people*, and *computer systems*. The system is functioned to design, manufacture, and assemble value-added

Computer Aided Design and Manufacturing, First Edition. Zhuming Bi and Xiaoqin Wang.
© 2020 John Wiley & Sons Ltd. This Work is a co-publication between John Wiley & Sons Ltd and ASME Press.
Companion website: www.wiley.com/go/bi/computer-aided-design

Figure 7.1 Example of unorganized machine shop.

physical, informational, or service products. A manufacturing system is measured by a number of evaluation criteria such as *market share of products*, *cost*, *profit*, *quality*, *responsiveness*, and *lead time* of products (Sysdesign 2019; Groover 2010).

Figure 7.2 gives a general description of a manufacturing system. The system consists of a hardware system and a software system. *The hardware system* consists of all tangible manufacturing resources such as production machines, cutting tools, material-handling devices, fixtures, humans, and other auxiliary tools. *The software system* consists of the virtual models for all tangible manufacturing resources and software tools for planning, scheduling, and controlling a system at different levels and domains. The manufacturing system is driven by *customers' requirements* (CRs) and the aim is to transform starting materials into value-added products to meet CRs and generate revenues.

Manufacturing processes are essential steps to transform starting materials to finished products. The cost reduction in manufacturing contributes greatly to the reduction of product cost. Black (1991) indicated that an average 40% of the total product cost was invested at the manufacturing stage. The manufacturing cost can be broken down as shown in Figure 7.3, which includes *direct labour* (12%), *indirect labour* (12%), *energy and depreciation of plant and machine tools* (26%), *and raw materials and parts* (50%), while the costs for direct labour, indirect labour, and energy and depreciation of plant and machine tools depend greatly on the organization and layout design of a manufacturing system.

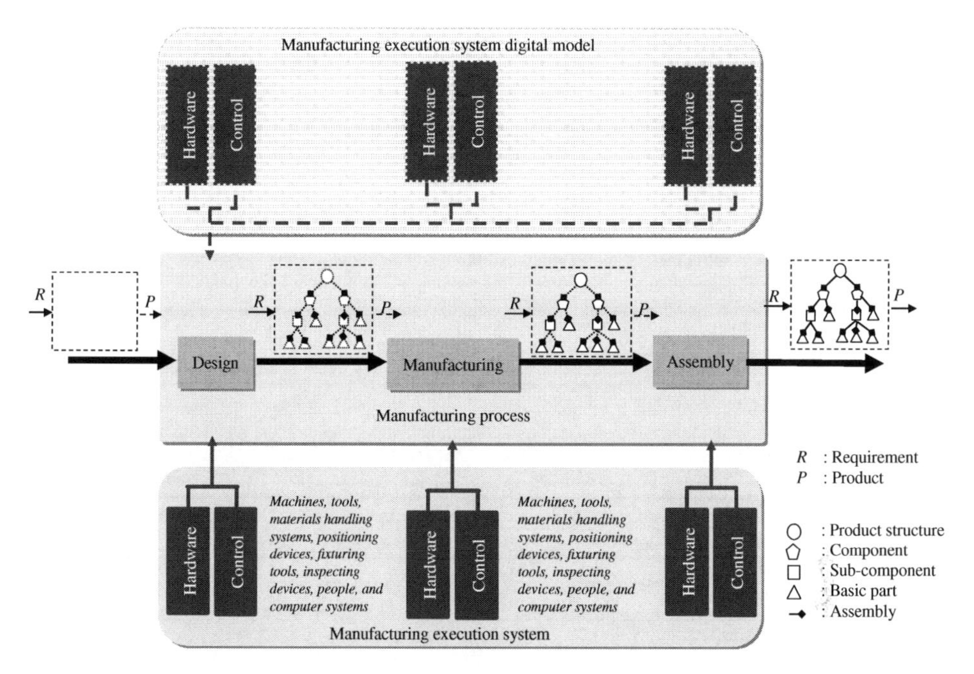

Figure 7.2 Manufacturing system model (Bi et al. 2008).

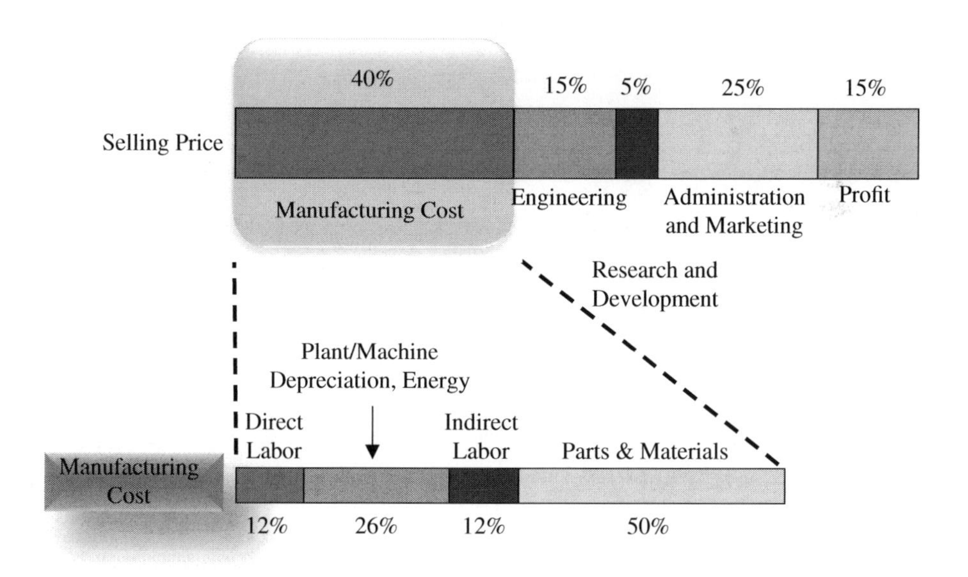

Figure 7.3 Breakdown of costs for a manufactured product (Black 1991).

Example 7.1 A product can be made in three processes (A, B, C) with different cost structures as follows:

Process	Fixed cost	Variable cost
A	\$120 000	\$3.00
B	\$90 000	\$4.00
C	\$80 000	\$4.50

What is the most economical process for a volume of 8000 units per year? Will the selection be different if the company makes a volume of 40 000 units per year?

Solution

The total cost for 8000 units consists of a fixed cost for (i) energy and depreciation of plant, (ii) machine tools, and (iii) a variable cost for raw materials and parts:

$$TC = FC + v(Q) \tag{7.1}$$

where TC, FC, and $v(Q)$ denote the total cost, fixed cost, and variable cost for Q units, respectively.

Equation (7.1) can used to estimate the annual cost for three processes as follows:

Process	TC ($Q = 8\,000$)	TC ($Q = 40\,000$)
A	\$144 000	\$240 000
B	\$122 000	↑\$250 000
C	\$116 000	\$260 000

Therefore, the most economic process for 8000 and 40 000 units are Process A and Process C, respectively.

Some early discussions on the relation of products and the organization of a manufacturing system are limited to two main factors, i.e. *volumes* and *variants* of products. As shown in Figure 7.4, an ideal paradigm of a manufacturing system varies with a combination of volume and variants of products.

In fact, design of a manufacturing system is usually a very complex optimization problem. In design optimization, each design variable corresponds to the changing directions towards favourable and adversarial performance based on a specified evaluation criterion. However, when multiple design criteria, variables, and constraints are involved, some approaches such as the *Pareto set* should be applied to trade off the solutions for the optimization of overall system performances. Sequentially, Figure 7.4 shows that different manufacturing paradigms exist in today's business environments and that some common paradigms are *mass production*, *cellular manufacturing (CM)*, *flexible manufacturing systems (FMSs)*, and *process-focused manufacture*.

However, Figure 7.4 does not give the technical guidance in designing plant layouts or organizing machine tools and other manufacturing resources. These tasks must be involved

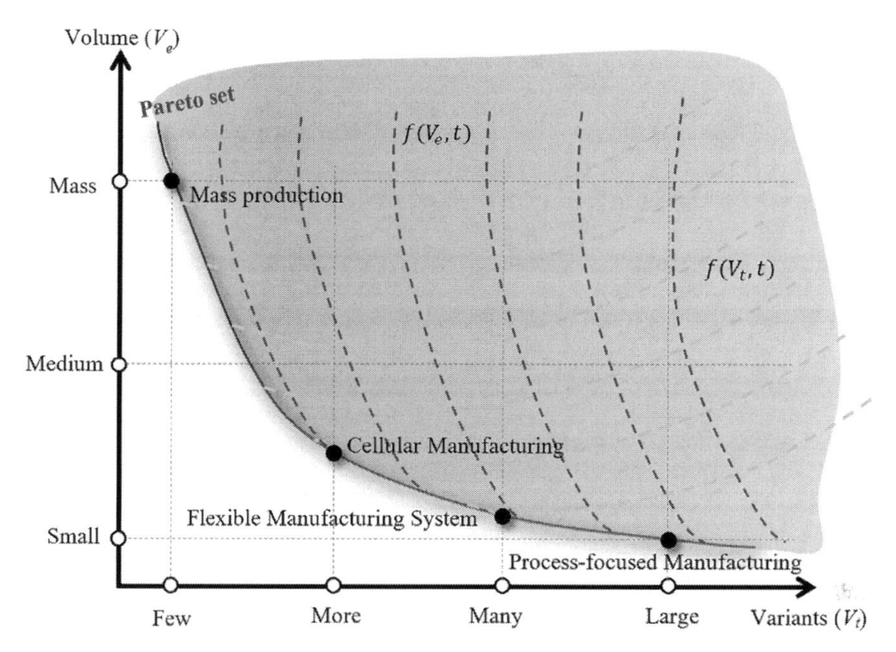

Figure 7.4 Ideal system paradigm versus volumes and variations of products. (*See color plate section for color representation of this figure*).

in designing and implementing a system paradigm. This chapter aims to organize manufacturing resources based on the similarities of products and manufacturing processes.

To organize manufacturing resources, the main components of a manufacturing system are discussed first. Figure 7.5 shows the main types of resources in the *material flow* and the *information flow*. Tangible components in the material flow are machine tools, fixtures, material handling systems, assembly equipment, storages, and inspecting systems. Tangible components in the information flow are computers, communication infrastructure, and other hardware facilitates for data acquisition, processing, planning, scheduling, and control of a system at different levels and domains.

7.2.1 Machine Tools

As shown in Figure 7.6, a traditional machine tool falls into one of four types: *machine tools for machining*, *metal forming machines*, *plastic forming machines*, and *other non-traditional machines* such as electro-discharge machining (EDM). Corresponding to product varieties and volumes in Figure 7.4, machine tools can be classified based on the production capabilities into *general purpose machine tools* for a wide range of parts with low-level productivity, *production machine tools* for a limited range of parts with medium-level productivity, *special purpose machine tools* for mass production and specific products, and *single purpose machine tools* for highly automated processes with a high level of productivity.

A machine tool for material removal operations is a machine that is used to shape a workpiece by cutting, boring, grinding, shearing, forming, or deforming. Unwanted material is removed from the workpiece to generate the desired shape. A machine tool

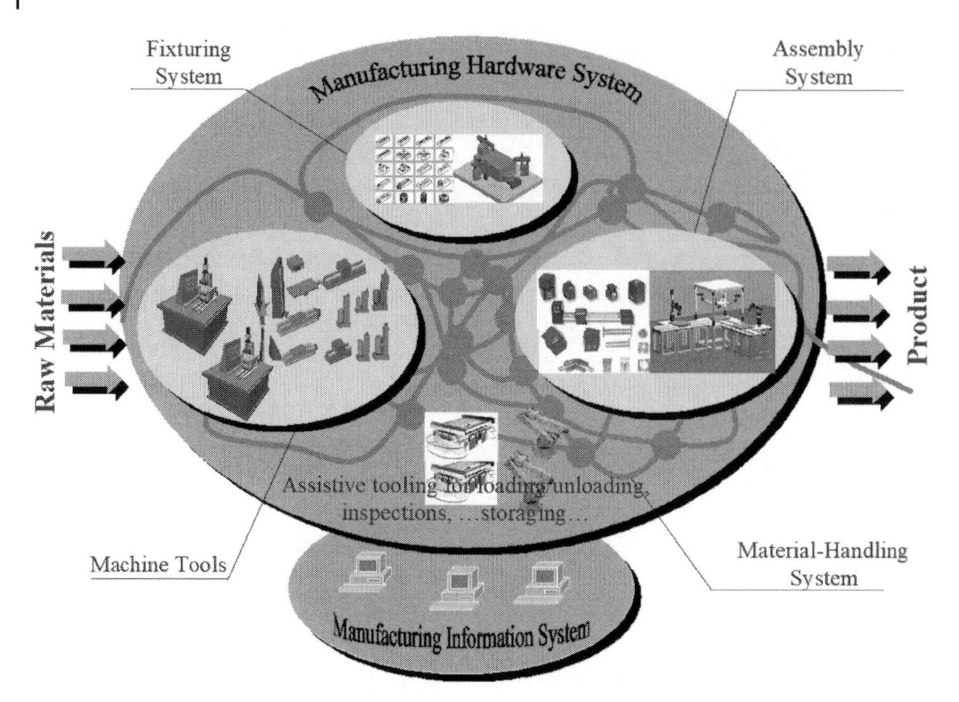

Figure 7.5 Main components in a manufacturing system (Bi et al. 2008). (*See color plate section for color representation of this figure*).

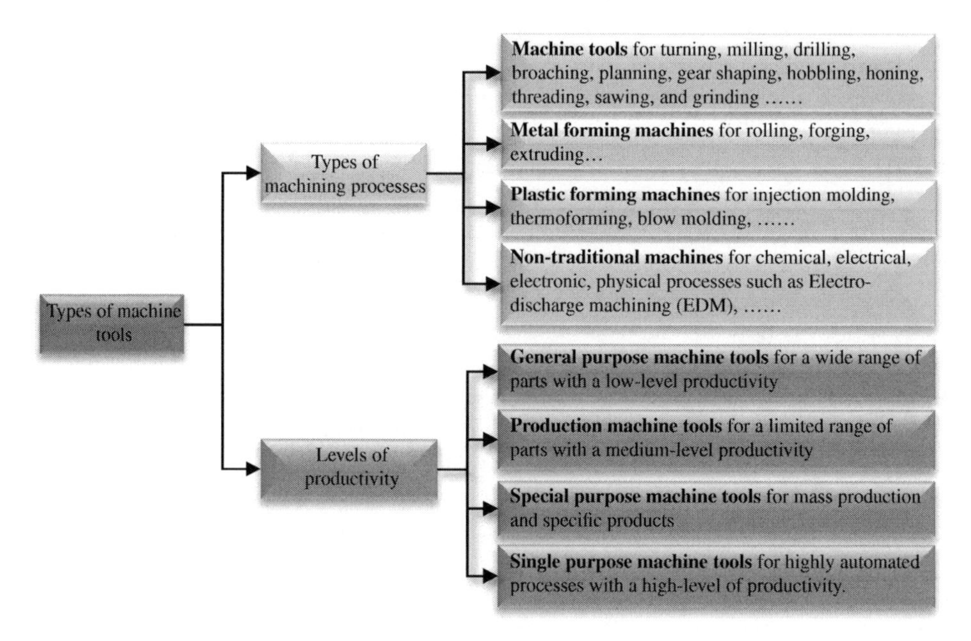

Figure 7.6 Typical types of machine tools.

operates together with auxiliary tools. Some examples of auxiliary tools in material removal operations are milling cutters, drill bits, and shapers. The machine tool is equipped to locate the workpiece and provide a guided motion of the cutting tool over the workpiece. The relative motion of the workpiece and the cutting tool is called a *toolpath*. The motion of the cutting tool along a toolpath is controlled autonomously or semi-autonomously by the machine.

7.2.2 Material Handling Tools

Material handling (MH) tools are used for the transportation and storage of materials, tooling, and other accessories in the material flow of a manufacturing system. A *material handling system* makes the connections of products in a series of manufacturing processes from raw materials to final products. MHI (2019) classifies material handling tools into the following five types.

1. *Transport equipment* for the transportation of objects from one location to another; for example, the movement of a workpiece between two workstations or between a loading dock and a storage site. Common transport equipment types are *conveyors*, *cranes*, and *industrial trucks*. Note that objects can also be transported manually without any equipment.
2. *Positioning equipment* for handling objects at a single location, so that the materials can be placed appropriately for subsequent handling, machining, transporting, or storage. Positioning equipment differs from transport equipment in the sense that the former is used to handle parts at a single workplace.
3. *Unit load formation equipment* for restricting objects, so that the objects maintain the integrity when handling a single load during transport and for storage. If the objects are self-restraining (e.g. a single part or interlocking parts), then they can be formed into a unit load with no equipment.
4. *Storage equipment* for holding or buffering objects for a certain period of time. Some storage equipment may be integrated with the equipment for the transportation of objects such as automated storage and retrieval systems (AS/RSs) and storage carousels. No storage equipment is needed for block stacked materials on the floor.
5. *Identification and control equipment* for collecting and transmitting data that is used to coordinate the materials flow in a system. Sometimes objects can be identified and controlled manually with no specialized equipment.

7.2.3 Fixtures

Fixtures are work-holding or support devices in a manufacturing system. A fixture aims to securely position the part when it is subjected to a certain operation. Using a fixture allows the smooth operation and quick transition from part to part, reduces the needs for skilled labour, simplifies the setup of the workpiece, increases the quality consistence of products, and improves the economy of production (Wikipedia 2019). Bi and Zhang (2001) gave the taxonomy of fixtures, while computer aided fixture design will be covered in the following chapter.

7.2.4 Assembling Systems and Others

Depending on the product types, a manufacturing system may include other types of value-added or non-value-added processes such as assembling, prototyping, and inspecting. Correspondingly, the hardware and software systems are needed to mechanize or automate these processes. Sequentially, the plant layout design involves planning, scheduling, and controlling of these resources.

7.3 Layouts of Manufacturing Systems

Manufacturing systems can be distinguished from each other from different perspectives such as structures and physical arrangements. Figure 7.7 shows the classification of manufacturing systems from the perspective of physical arrangements. A system layout can be a *rigid* system layout such as *job shop*, *flow shop*, *project shop*, or *continuous process*, or a *flexible* system layout such as *cellular manufacturing*, *flexible manufacturing*, *distributed manufacturing (DM)*, or *virtual manufacturing (VM)*. In this section, the features, the advantages and disadvantages of these layout designs are discussed.

7.3.1 Job Shops

In a *job shop*, product variants are manufactured in small batch sizes based on specific customer orders. To be able to perform a wide variety of manufacturing processes, a job shop usually consists of a number of general-purpose machines, where highly skilled human operators are needed to deal with the high variety of work arrangements.

As shown in Figure 7.8, the production machines are grouped according to the types of manufacturing processes the machines can perform; for example, all of the lathes are in one department for turning operations and all of the drill presses are in another department for drilling operations. In a job shop, each part requiring a sequence of manufacturing processes is routed through different departments in the proper order. The sequence of manufacturing processes is specified by a *route sheet*. Such a layout is called a *functional or*

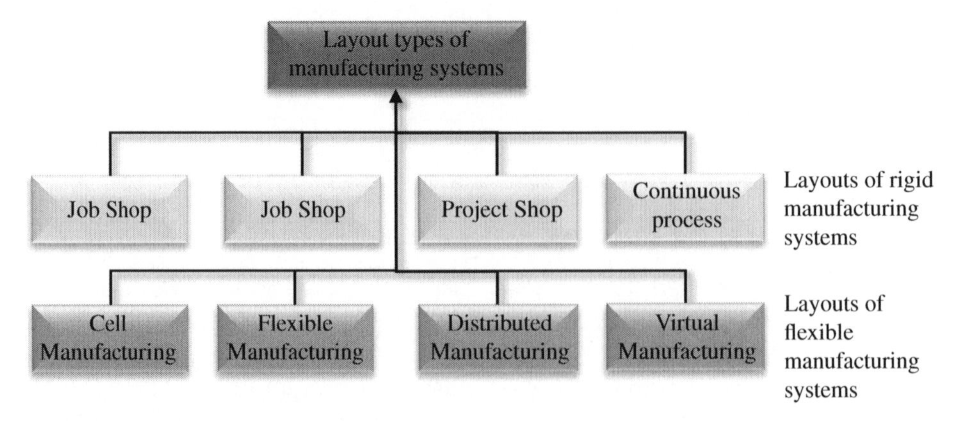

Figure 7.7 Layout types of manufacturing systems.

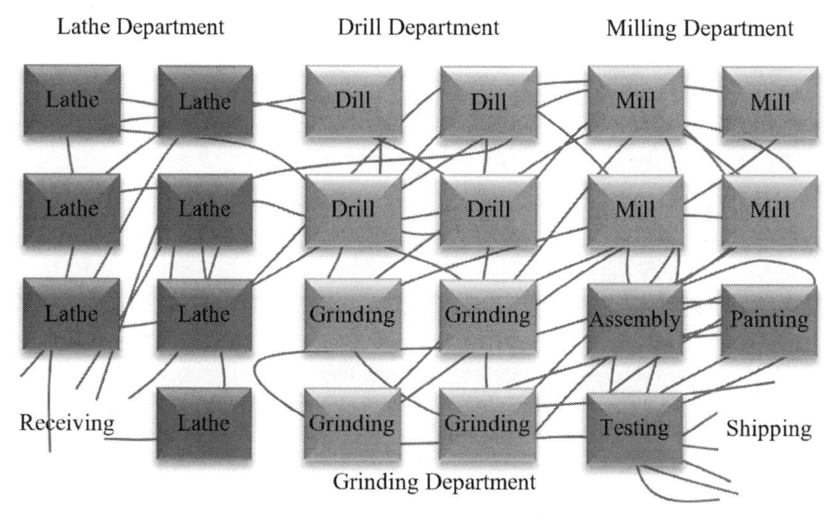

Figure 7.8 Functional or process layout.

Table 7.1 Characteristics of process layout (job shop).

Advantages	Disadvantages
• Can handle a variety of processing requirements • Not particularly vulnerable to equipment failures • Equipment used is less costly • Possible to use individual incentive plans	• In-process inventory costs can be high • Challenging routing and scheduling • Equipment utilization rates are low • Material handling is slow and inefficient • Complexities often reduce span of supervision • Special attention for each product or customer • Accounting and purchasing are more involved
Examples: Machine shops, foundries, press working shops, plastic industries.	

process layout. The examples of using a process layout include machine shops, foundries, press working shops, and plastic industries. Table 7.1 gives the characteristics of a job shop.

7.3.2 Flow Shops

A *flow shop* is product-oriented. It consists of a number of flow lines for given product families. A flow shop aims at achieving a high production rate; the layout is designed for making particular products or a family. Generally, special purpose machines are used rather than general purpose machines. In addition, the skill level of human operators is not as critical as that in a job shop. When the volume of production becomes large, it can evolve into *mass production*. A material handling system is used in the material flow to transfer parts from one workstation to another in a sequence of manufacturing. The time that a part spends in each station or location is fixed and equal as much as possible. As shown in Figure 7.9, the workstations are arranged in a line according to the sequence of manufacturing processes for given products. Table 7.2 gives the characteristics of a flow shop.

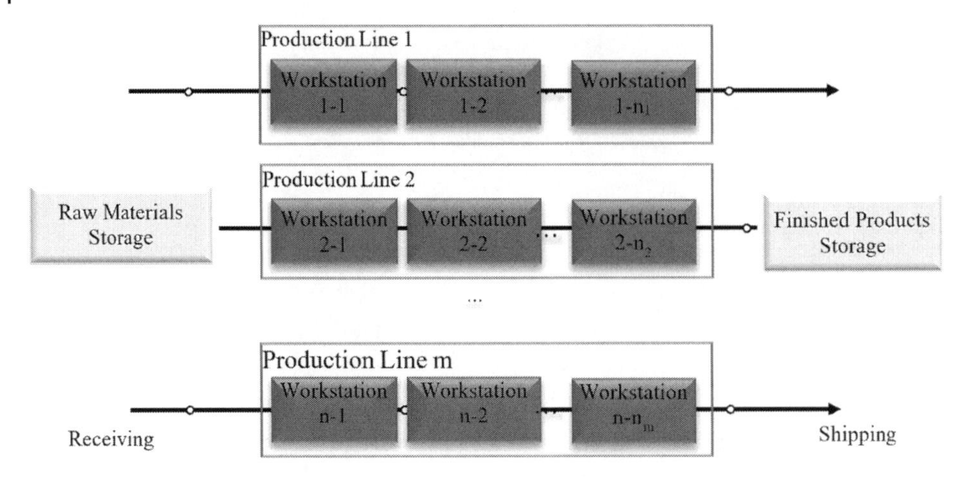

Figure 7.9 Product layout.

Table 7.2 Characteristics of product layout (flow shop).

Advantages	Disadvantages
• High rate of output • Low unit cost • Labour specialization • Low material handling cost • High utilization of labour and equipment • Established routing and scheduling • Routing accounting and purchasing	• Creates dull, repetitive jobs • Poorly skilled workers may not maintain equipment or quality of output • Fairly inflexible to changes in volume • Highly susceptible to shutdowns • Needs preventive maintenance • Individual incentive plans are impractical
Examples: Automated assembly lines and television production lines.	

7.3.3 Project Shops

In a *project shop*, a product is usually of a large scale and overweight. As shown in Figure 7.10, the product remains at a fixed site while the manufacturing and assembling processes are performed. The materials, machines, and human operators are brought to the site to perform manufacturing processes. The layout is also called a *fixed position layout*. Some examples of project shops are locomotive manufacturing, large aircraft assembly, and ship building. Table 7.3 gives the characteristics of a project shop.

7.3.4 Continuous Production

Continuous production is also called a *continuous process* or a *continuous flow process*. In continuous production, the materials in dry bulk or fluids undergo chemical reactions, mechanical deformations, or heat treatments. The materials being processed are continuously in motion. Continuous production is contrasted with batch production since the product flows in a continuous process physically. In a continuous process, the materials do

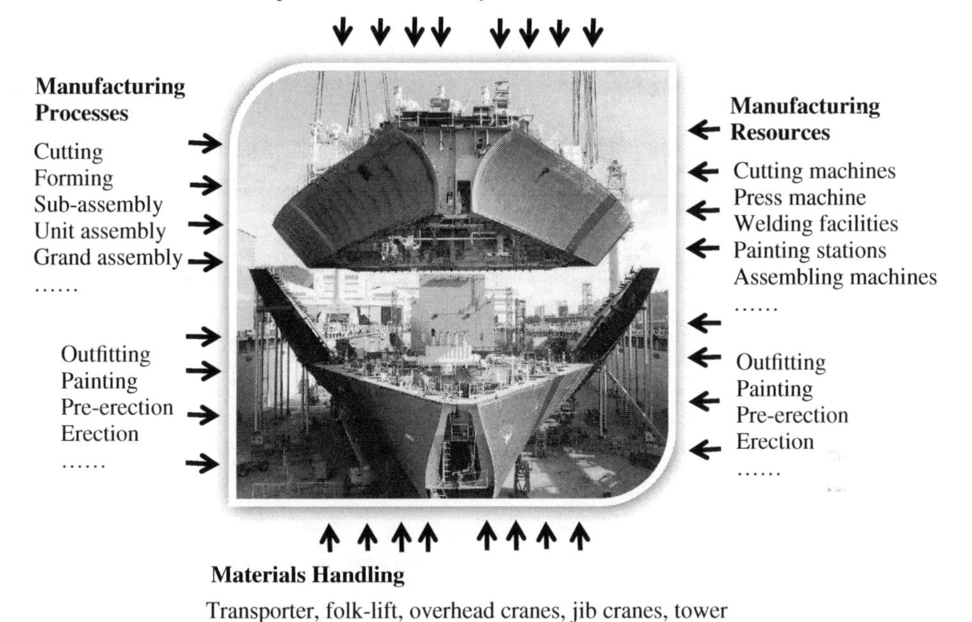

Figure 7.10 Example of project shop (AQT Solution 2016; Song and Woo 2013).

Table 7.3 Characteristics of fixed position layout (project shop).

Advantages	Disadvantages
• Minimum capital investment • Continuity of operation • Less total production cost • Offers greater flexibility • Allows the change in production design • Permits a plant to elevate the skill of its operators	• Machines, tools, and workers take more time to reach the fixed position • Highly skilled workers are required • Complicated jigs and fixtures (work holding device) may be required
Examples: Automated assembly lines and television production lines.	

flow since they are liquids, gases, or power. Figure 7.11 shows an example of a continuous process layout. It usually has the leanest and simplest production system since the control system is simplified. The production system involves the least work-in-progress (WIP). The examples of continuous products are oil refineries, chemical process plants, food processing industries, and manufacturing of metal sheets and rolls. Table 7.4 gives the characteristics of continuous production.

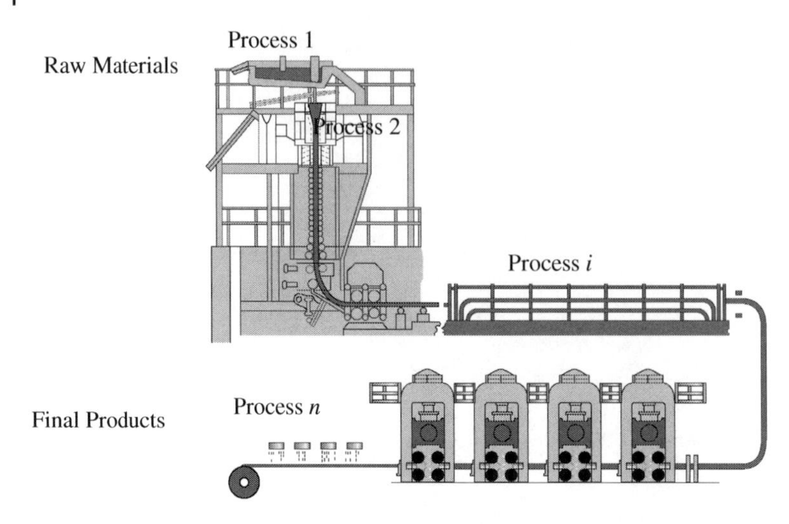

Figure 7.11 Example of continuous products (HIBA 2019).

Table 7.4 Characteristics of continuous production (Knowledgiate Team 2017).

Advantages	Disadvantages
• The quality of output is kept uniform because each stage develops skill through repetition of work • Any delay at any stage is automatically detected. As a result, there is automatic control of time and the direct labour content is reduced • Work-in-progress is minimum on account of sequence balancing • Handling of materials is reduced due to the set pattern of production line • Control over materials, costs, and output is simplified. The repetitive nature of processes makes production control easier • Overhead cost per unit is reduced due to spreading of large fixed costs of specialized equipment over a large volume of output • There is quick return on capital employed	• Is very rigid and if there is a fault in one operation, the entire process is disturbed • It becomes necessary to avoid piling up of work or any blockage on the line • Unless the fault is cleared immediately, it will force the preceding as well as the subsequent stages to be stopped
Examples: Oil refineries, chemical process plants, and food processing industries.	

7.3.5 Cellular Manufacturing

Cellular manufacturing (CM) is a hybrid system to synergy, having the advantages of (i) *job shops* for the capability in producing a wide variety of products and (ii) *flow lines* for the efficiency of product flows and high productivity. A cellular manufacturing system (CMS)

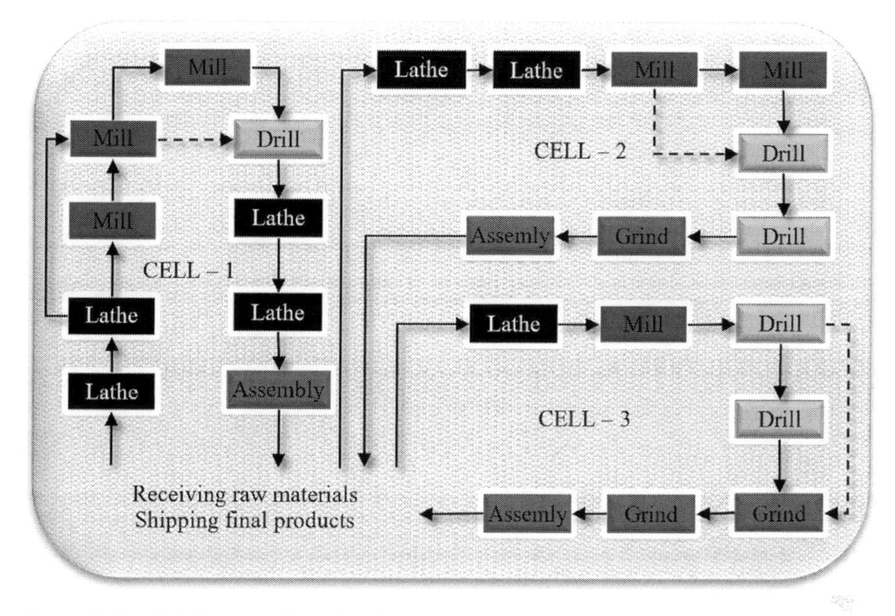

Figure 7.12 Cellular manufacturing layout.

is composed of '*linked cells*'. Figure 7.12 shows the main structure of a CMS. In the cells, the workstations are arranged as a flow shop: the machines can be modified, retooled, and regrouped for different product lines within the same 'family' of parts. This system has some degree of automatic control for loading and unloading of raw materials and work pieces, changing of tools, and transferring of work pieces and tools between workstations. The cells can be fully automated or assisted by human operators. In the latter, multifunctional operators can be used to move parts from machine to machine and run several machines simultaneously to improve efficiency of production. Table 7.5 gives the characteristics of cellular manufacturing (CM).

7.3.6 Flexible Manufacturing System (FMS)

An FMS integrates all of the major manufacturing elements into a highly automated system. An FMS is flexible in the sense that it is capable of dealing with a variety of part configurations and producing them in any order. Figure 7.13 shows an example of an FMS; it consists of a number of programmable resources such as computer numerical control (CNC) machines, material handling systems, automated storage and retrieval systems, and control systems.

An FMS usually involves a major capital investment. Therefore, the *return on investment* (RoI) is feasible only when the manufacturing resources in the system can be utilized efficiently. In addition, planning, scheduling, and controlling are critical since an FMS aims at no setup time to shift from one manufacturing operation to another. An FMS should be capable of performing various operations in different orders and on different machines.

Table 7.5 Characteristics of cellular manufacturing (Weber 2004).

Advantages	Disadvantages
• Cells are designed to make products for a family; it reduces setup time since no change is needed for machines and tools within cells to process similar parts • With reduced setup times, the amount of work in progress can be reduced • Each part is processed in a single cell which reduces part travelling distance and time. No effort is wasted to store, protect, and control materials • Single machine can be used to manufacture one or more products in each cell for high machine utilization • Lead time can be used due to the reduction of setup time, work in process, and the increase of machine utilization • Manufacturing processes become simple and incentive and simplified processes boost worker morale	• It reduces the manufacturing flexibility • It is challenging to balance cells • High mixes of low-volume production can make cells impractical • Job rotation is common in CM but it can cause the problems due to changes • Hard for operators to adapt cells and there will be some resistance from operators since no rest period when parts are made • Easily underestimate training needs
Examples: U-shaped, inverted U-shaped, or straight-line cells for making discrete parts.	

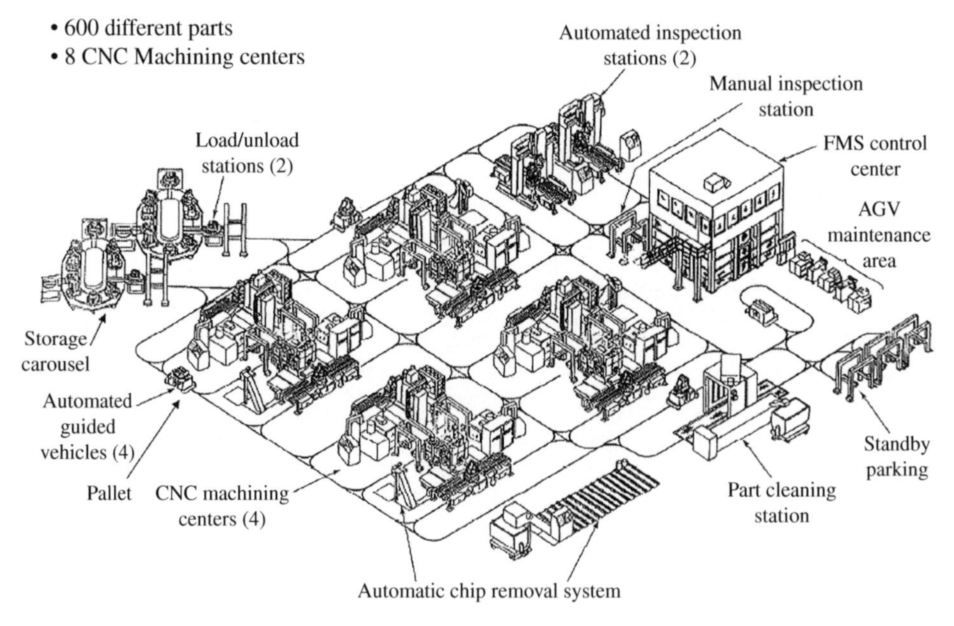

Figure 7.13 Example of flexible manufacturing system (O'Sullivan 2019; BrainKart 2019).

Table 7.6 Categories of FMS layout (O'Sullivan 2019; BrainKart 2019).

Layout	Description
Progressive	The machines and handling system are arranged in a straight line as shown in Figure 7.14a. The parts progress from one workstation to the next in a well-defined sequence with work always moving in one direction and with no back-flow.
Loop	Workstations are organized in a loop that is served by a looped parts handling system as shown in Figure 7.14b. The parts usually flow in one direction around the loop with the capability to stop and be transferred to any station. Each station has secondary handling equipment so that a part can be brought to and transferred from the workstation to the material handling loop. Load/unload stations are usually located at one end of the loop.
Ladder	A ladder layout in Figure 7.14c consists of a loop with rungs upon which workstations are located. The rungs increase the number of possible ways of getting from one machine to the next, and obviates the need for a secondary material handling system. It reduces average travel distance and minimizes congestion in materials handling to reduce transport time between stations.
Robot centre	The layout is robot centralized, as shown in Figure 7.14d; one or more robots work with the conveyer to deal with all material handling activities.
Open field	An open field layout consists of multiple loops and ladders; it may include sidings as well. This type of layout can be used to process a large family of parts. It may have a limited number of machine types and parts are usually routed to different workstations,depending on which one becomes available first see Figure 7.14e.

Based on the layout designs of material flows, FMSs can be classified into five categories. These types of FMSs are described and illustrated in Table 7.6 and Figure 7.14, respectively. As a summary, Table 7.7 gives the characteristics of FMSs (CPV 2019).

7.3.7 Distributed Manufacturing and Virtual Manufacturing

Distributed manufacturing (DM) is also referred to as *distributed production*. In DM, enterprises use a network of geographically dispersed manufacturing facilities to make products. Manufacturing resources are coordinated by an integrated information system (Wikipedia 2019). Figure 7.15 shows the scenario of distributed manufacturing. A group of networked enterprises work cooperatively to make products; each enterprise contributes certain value-added processes to products through an integrated supply chain. The coordination and interoperation of manufacturing processes are performed over the Internet. DM provides the high flexibility needed since manufacturing resources can be configured dynamically to catch emerging business opportunities.

To understand how DM can be successfully applied in practice, Rauch and Dallasega (2017) argued that it was critical to determine a framework for the network models. Table 7.8 gives five network models, including (i) *owner-based micro production networks*, (ii) *contract manufacturing networks*, (iii) *mobile factory networks*, (iv) *production franchise networks*, and (v) *collaborative cloud manufacturing*.

Virtual manufacturing (VM) is an integrated manufacturing environment with a mixture of real and virtual objects, business activities, and manufacturing processes. VM supports

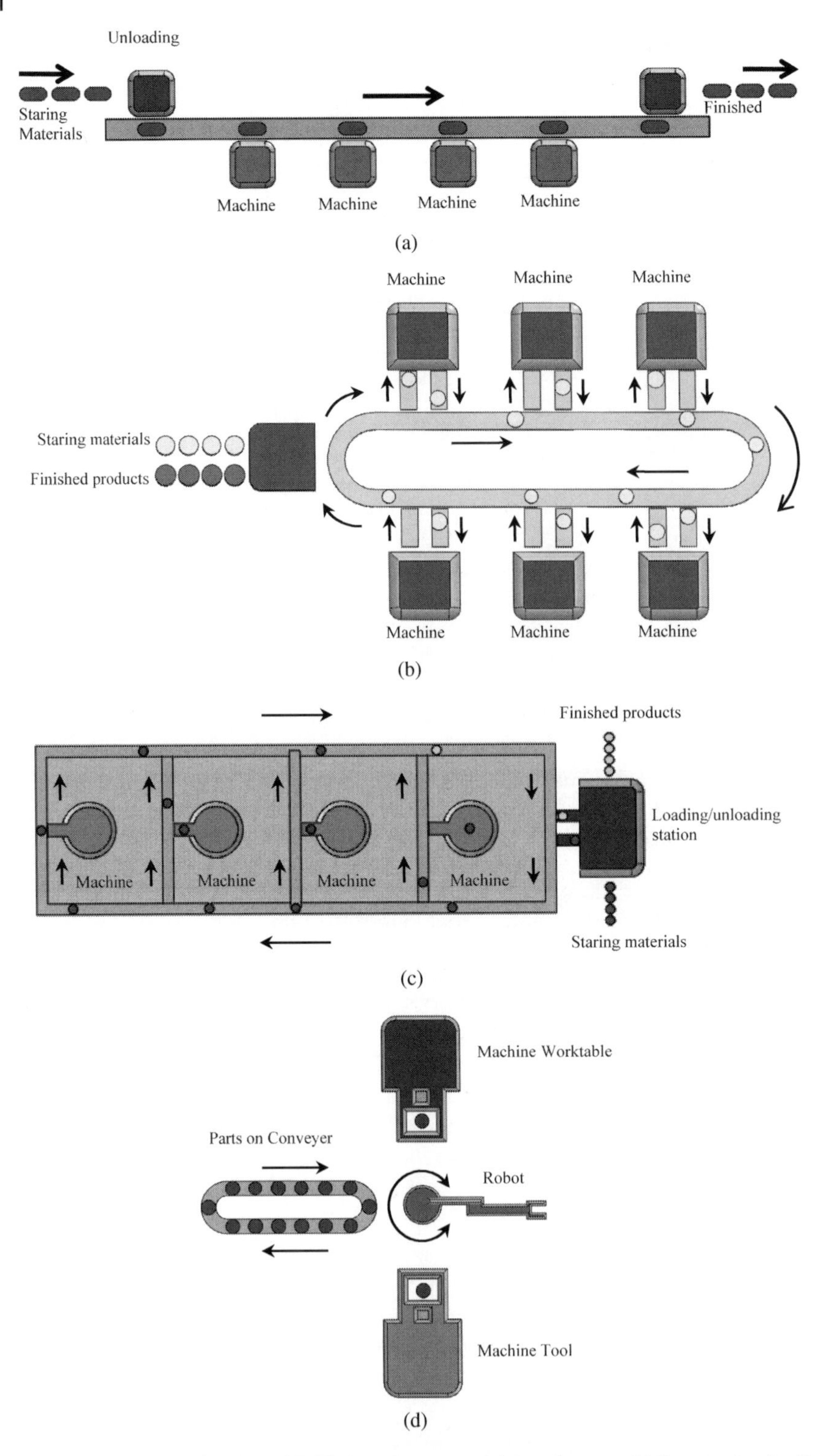

Figure 7.14 Classification of FMSs based on material flow layouts. (a) Progressive (in-line) layout. (b) Loop layout. (c) Ladder layout. (d) Robot-centralized layout. (e) Open-field layout.

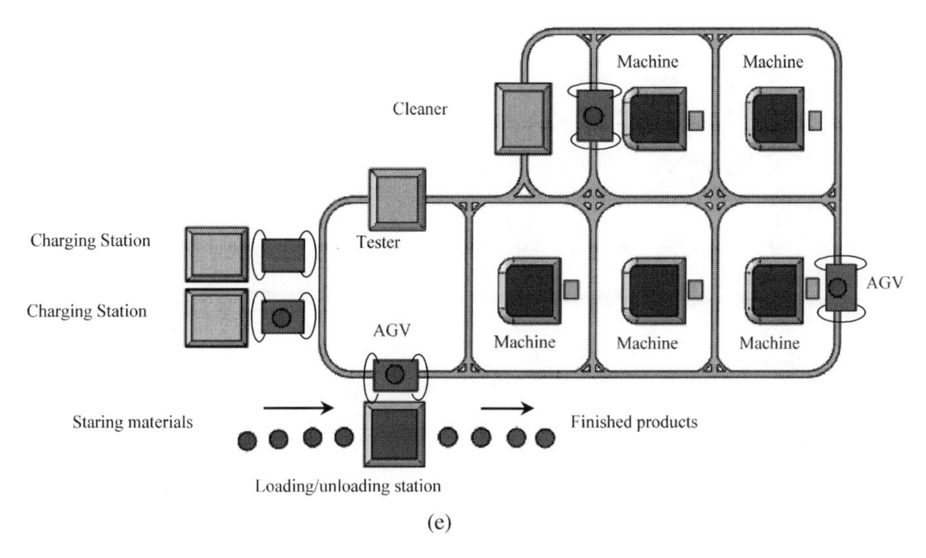

Figure 7.14 (*Continued*)

Table 7.7 Characteristics of flexible manufacturing systems (CPV 2019).

Advantages	Disadvantages
• Parts can be produced randomly in batch sizes, as small as one, and at lower cost • Lead times of parts are reduced and thus reduce work-in-progress parts • Labour and inventories are reduced • Production is more reliable due to the self-correcting functions of system • Product quality is uniform • Increased machine utilization • Fewer machines required • Reduced factory floor space • High flexibility gives a great responsiveness to change	• It requires extensive planning for detailed design and scheduling • The implementation in production processes can be complicated • It requires highly skilled employees to operate the machinery • Due to the complexity, a different set of skilled workers is needed for maintenance and repairs • Purchasing and adapting machinery is expensive • It mainly applies to large companies with enough revenue for initial investments
Examples: FMSs with progressive, loop, ladder, robotic centre, and open-field layouts.	

the construction and utilization of simulation models by a collection of computer aided tools for optimized decision-making supports. VM aims to increase the value, accuracy, and validity of decision-making supports. VM is applicable to various domains and levels of an enterprise organization factory (Rahdarmanan 2002).

DM or VM rapidly enables configuration of a multidisciplinary network of small, process-specific enterprises to meet new business opportunities in designing and making specific products. Using DM and VM technologies, a group of entrepreneurs and enterprises can integrate their expertise and resources, capitalize on market opportunities, and implement business plans to meet emerging customer needs with minimized costs.

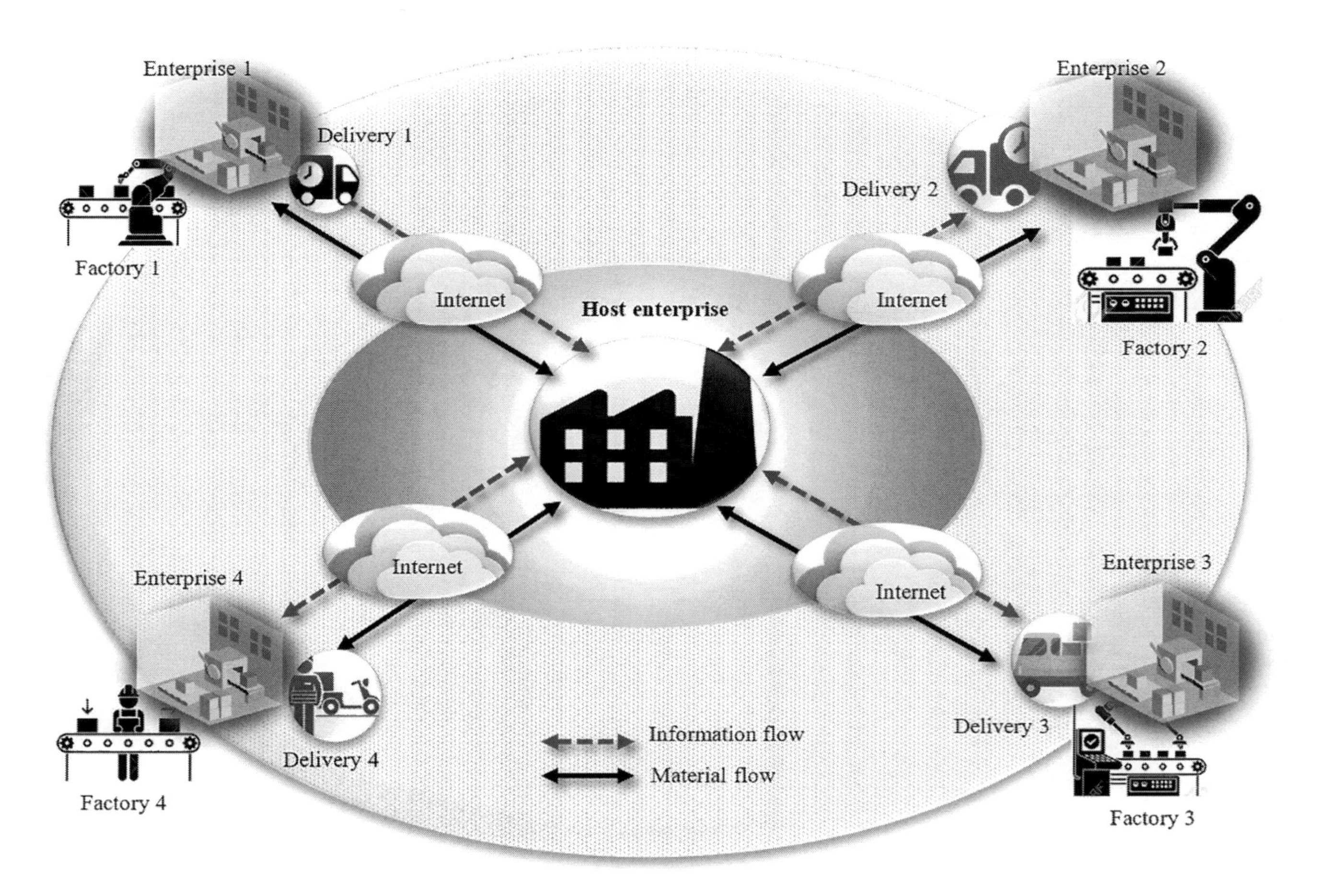

Figure 7.15 Scenario of distributed manufacturing (Mahdabi et al. 2007).

Table 7.8 The types of networked models for distributed manufacturing (Rauch and Dallasega 2017). (*See color plate section for color representation of this figure*).

Network models	Description
Reach of individual company Network of small businesses (a) Owner-based micro production networks	An owner-based micro production network consists of a number of small or medium enterprise to product goods to meet customers in local or regional markets. Such a network model provides the opportunities for local manufacturers to develop/upgrade their own businesses without large investment.
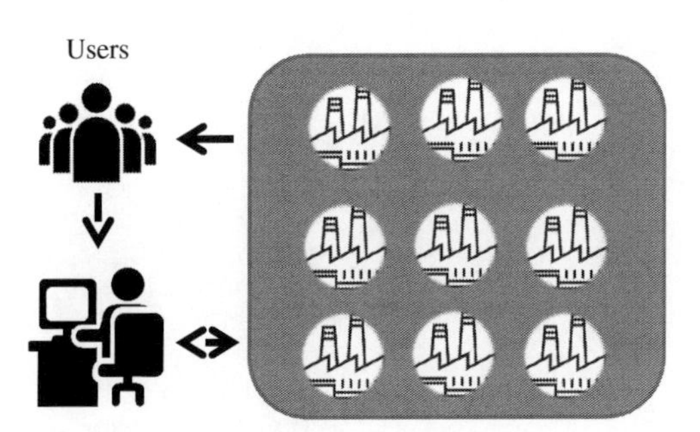 (b) Contract manufacturing networks	In a contract manufacturing network, enterprises perform manufacturing processes as the services accessible by users and production intermediaries. The network provides the high flexibility for geographically distributed enterprises to contract out manufacturing resources and production capacities. Intermediaries bring global users and locally distributed manufacturer together over the network.
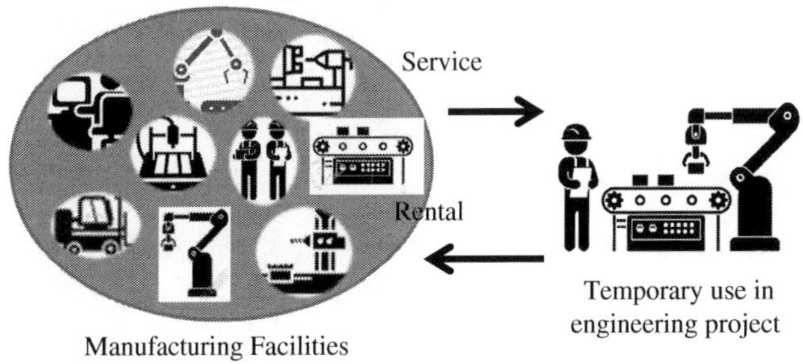 (c) Mobile factory networks	A mobile factory network supports the rental-based approach for commercialization. It consists of a pool of standard functional modules and it allows configuration of mobile factories in order to meet different business needs.

(*continued*)

Table 7.8 (Continued)

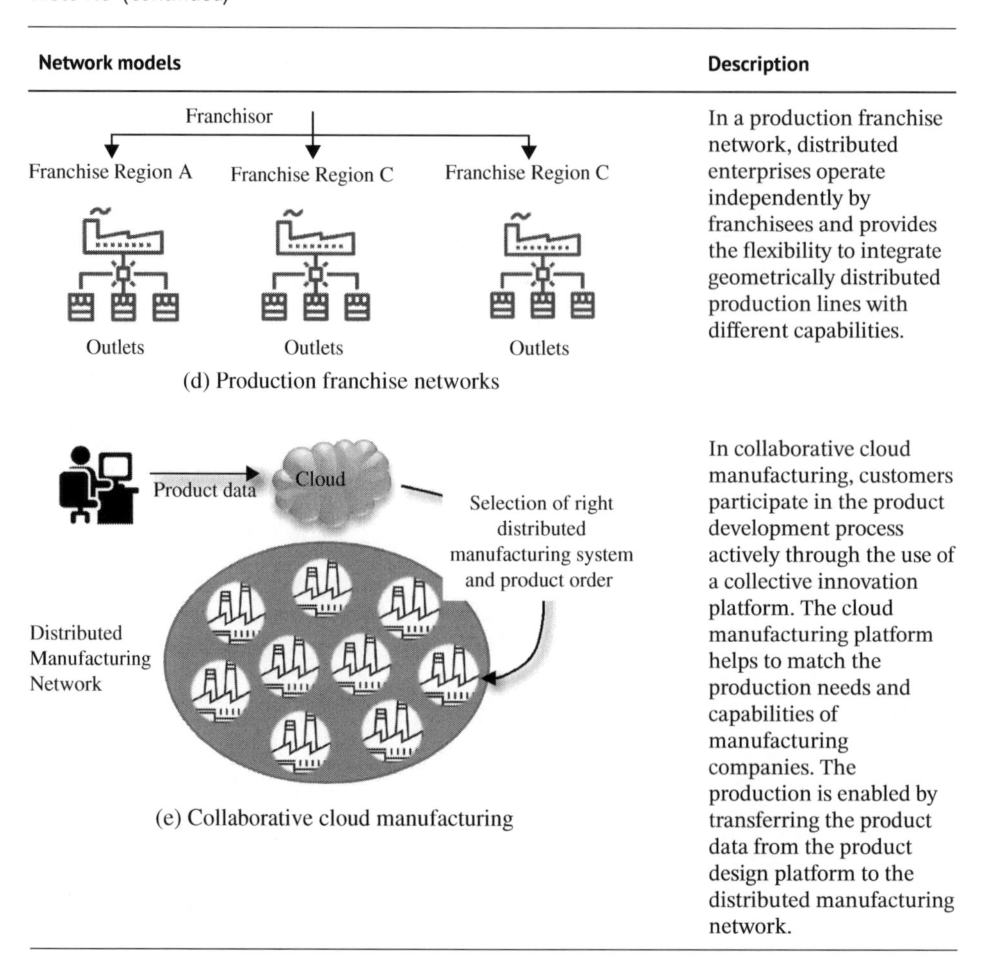

Network models	Description
(d) Production franchise networks	In a production franchise network, distributed enterprises operate independently by franchisees and provides the flexibility to integrate geometrically distributed production lines with different capabilities.
(e) Collaborative cloud manufacturing	In collaborative cloud manufacturing, customers participate in the product development process actively through the use of a collective innovation platform. The cloud manufacturing platform helps to match the production needs and capabilities of manufacturing companies. The production is enabled by transferring the product data from the product design platform to the distributed manufacturing network.

However, DM or VM operates over networks and the Internet, and demands reliable and secure communication as well as high computation. In addition, enterprises have to deal with a high risk to collaborate with dynamic partners in highly turbulent business environments.

7.3.8 Hardware Reconfiguration Versus System Layout

Eight system layouts have been discussed in this section. Among these layouts, job shops, flow shops, project shops, and continuous processes are rigid systems, which implies that hardware systems are not reconfigured for products to be manufactured. On the other hand, system layouts FMS, DM, and VM are reconfigurable but the reconfiguration is achieved by means of control software or dynamic alliances, and the hardware reconfiguration of a machine cell or a distributed enterprise is beyond the scope of system layout design.

From this perspective, CM is a unique layout type since both hardware and software components in the system have to be reconfigured to make product variants. Therefore, in the rest of the chapter, Group Technologies (GTs) will be discussed in detail for system layout designs.

7.4 Group Technology (GT)

Each system layout in Section 7.3 corresponds to certain scenarios when the overall performance of the manufacturing system can be optimized. In designing or operating a manufacturing system, it is critical to map the production scenarios and system layouts appropriately. The general guide in identifying the optimized system layout is to maximize the utilization of manufacturing resources based on the similarities of products, processes, the correspondences of products and processes, as well as other activities involved in product lifecycles. In this section, the similarity of products is firstly discussed. *Group Technology* (GT) is a manufacturing philosophy where similar products are identified and grouped to take advantage of the similarities in manufacturing. The grouped products are called 'product family'. As shown in Table 7.9. Similarities occur to design and products to be manufactured.

GT aims to identify similar features occurring to products and group them as product families to take advantage of the similarities in product design, manufacturing, assembling, and other business activities. Typically, a successful GT project leads to a number of manufacturing cells that specialize in production of certain product families. GT can be implemented by manual or automated techniques. GT is ideal to product variants with medium quantity ranges that are made in batches. In contrast to traditional batch productions with downtimes for changeovers and high inventory carrying costs, GT overcomes these advantages by identifying and grouping similar products to maximize the utilizations of manufacturing resources. If GT leads to a permanent solution of a manufacturing cell for a product family, such a solution becomes an FMS.

A product designed for manufacturing is usually made by a series of succeeding manufacturing processes. If there is a large spectrum of products to be produced, it becomes beneficial for different products with similar features to share manufacturing processes.

Table 7.9 Design and manufacturing attributes for similarities.

Design attributes	Manufacturing attributes
• Major dimensions • Length/diameter ratio • Basic external shape • Basic internal shape • Material type • Part function • Tolerances • Surface finish	• Major process • Operation sequence • Batch size • Annual production • Machine tools • Cutting tools • Material type

This is achieved by grouping products together as the families either according to their geometric similarities or to similar fabrication methods. Three practical ways in identifying product families are visual inspection, product classifications and coding, and production flow analysis (PFA). These methods are discussed, respectively, in the following sections.

7.4.1 Visual Inspection

In a visual inspection method, designers use the best judgements on product variants to identify similarities and group them into product families based on the materials, geometries, shapes, and features of physical parts or virtual models of products. Here, a *part family* is a collection of products that possess similarities in materials, geometries, shapes, features, and sizes, or types of manufacturing processes.

There are always differences among products in a family, but the similarities should be close enough to identify a set of common processing machines. Therefore, different products with identical shapes and sizes are unnecessary in a one-part family. Figure 7.16 shows an example that, even with the same geometry, it might not be appropriate to group them into a product family if the materials are different, since the shaping processes would heavily rely on the strengths of materials to be processed and the selected processing machines would be significantly different. Based on similar reasons, it is inappropriate to treat (i) 1 000 000 units per year with 1015 CR steel for a tolerance of $\pm$0.010 inch and (ii) 100 units per year with nickel plate for a tolerance of $\pm$0.001 inch in one product family where there are significant differences of materials, batches, and manufacturing tolerance.

To use a visual inspection method, designers must gain a good understanding of machined features on products and what and how these features are made on processing machines. Figure 7.17 shows an example where unorganized products may have an appropriate level of similarities to be grouped into product families.

It is worth emphasizing the geometries and shapes that are important measures in grouping products, but the similarities can be exploded beyond these factors. Figure 7.18 shows examples where the products in one family may or may not have similarities on geometries and shapes.

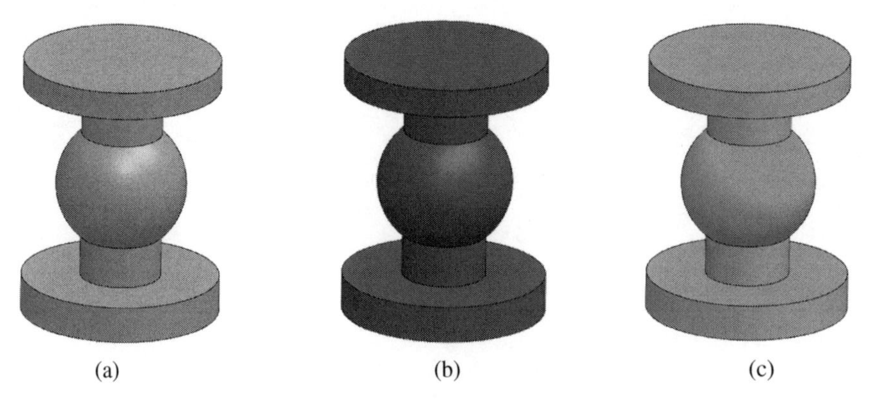

Figure 7.16 Example of inappropriate part family by visual inspection. (a) Plastic. (b) Iron. (c) Wood. (*See color plate section for color representation of this figure*).

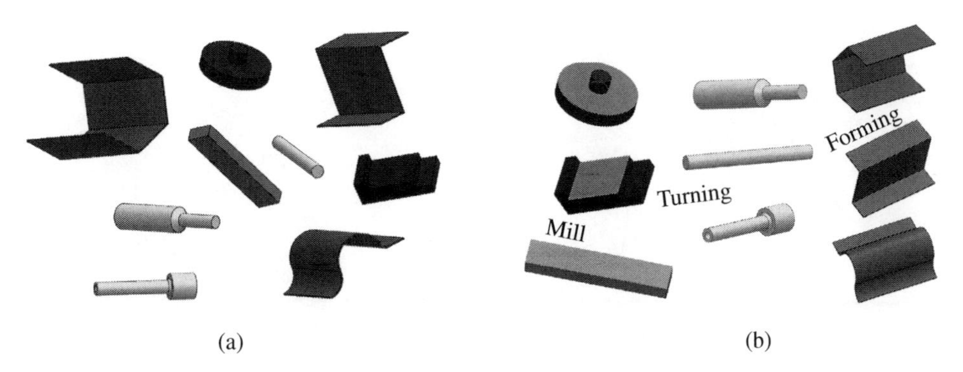

Figure 7.17 Example of organizing products into families. (a) Unorganized products. (b) Organized product families.

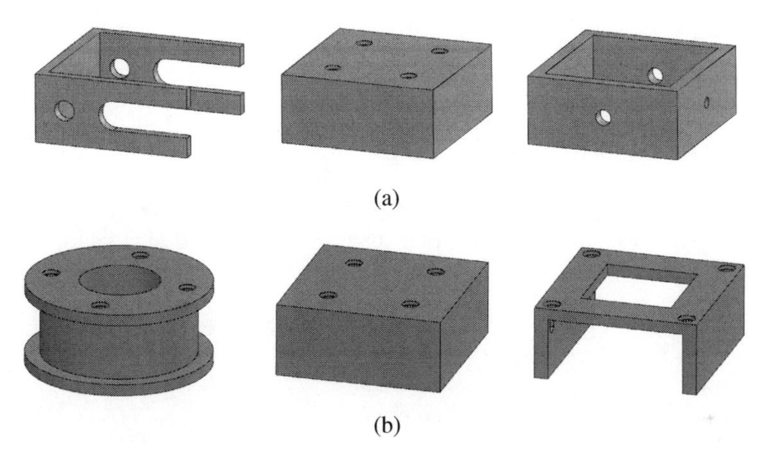

Figure 7.18 Examples of product families with and without the geometric similarities. (a) A group of prismatic parts with similarities of milling operations. (b) A group of dissimilar parts with similarities of milling operations.

7.4.2 Product Classification and Coding

In a product classification and coding method, product variants are coded based on a certain coding scheme and the product codes are analysed to maximize the coded similarities in product groups. Coded products can be analysed manually or automatically.

To classify product variants and machines, product geometries and features are described by codes and coded products are analysed to identify the similarities. Using a given coding scheme, products can be coded manually or by computers. One challenge in coding products for classification is to decide what and how product features are coded. There is no generic rule that is applicable to a wide spectrum of products. For product geometries, a common practice is to classify products into *rotational products* and *non-rotational products*. The industry survey can be applied to rank the main parameters in products based on their importance to manufacturing processes. The most important ones would be the candidates to be included in a classification system. A similar study was made with sheet metal parts. The major shape, the material, and the material specification had the highest

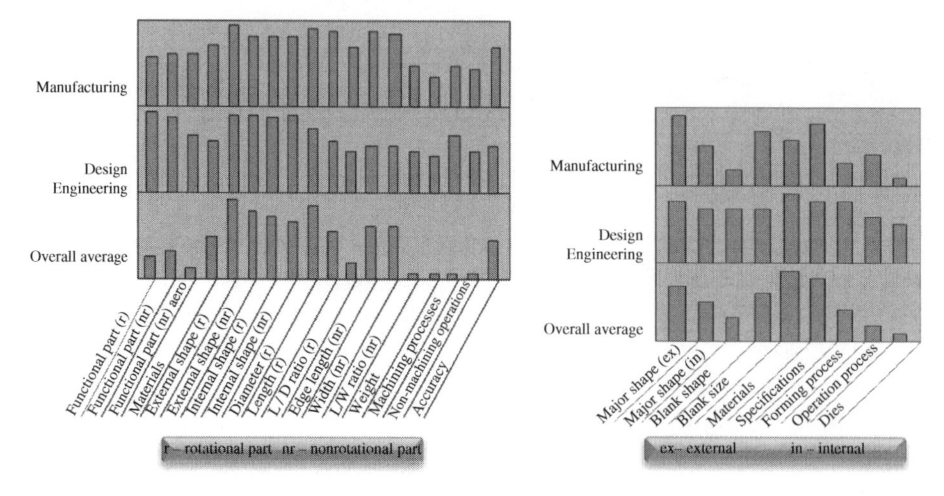

Figure 7.19 Comparative rankings for major parameters of sheet metal products (Zeng 2009).

rankings. Figure 7.19 gives an example of comparative rankings for products from metal sheets for determination of main parameters in a coding system.

Many companies customize existing coding systems to their specific needs. If GT is developed in a specified company, designers may take into consideration other design factors in developing a coding system:

1. The types of processes and limits a company can carry out in the factory;
2. Processing costs relevant to tool changes and setups;
3. The number of setups for machining operations;
4. The balance of utilization rates of available machine tools.

However, it is not uncommon for a new product to be wrongly handled in an existing coding system. In such cases, an additional economic study is needed either to eliminate products from machining cells by GT or to expand existing coding systems for more features.

The basic requirements for a coding system are (i) the code must be flexible enough to describe, classify, group current products as well as potential future products, and handle future as well as current parts. (ii) The scope of product types to be included must be specified, for example, whether the coding system is applicable for rotational, prismatic, or sheet metal products. (iii) The code must be able to distinguish products with different values for critical attributes in manufacturing processes such as material properties, tolerances, and the types of machine tools. As a general guide, all information necessary for grouping products should be included in a code system if it is possible; this information covers geometric features such as outside shape, end shape, internal shape, holes, and overall dimensions.

A coding system consists of four main components, i.e. (i) components, (ii) details, (iii) structures, and (iv) digital representations. Code detail is crucial and should be concise but identify every product type uniquely and fully represent design and manufacturing features of products. Cumbersome codes cause a waste of resources in data collection and computation. As shown in Figure 7.20, existing code structures can be classified into a hierarchical

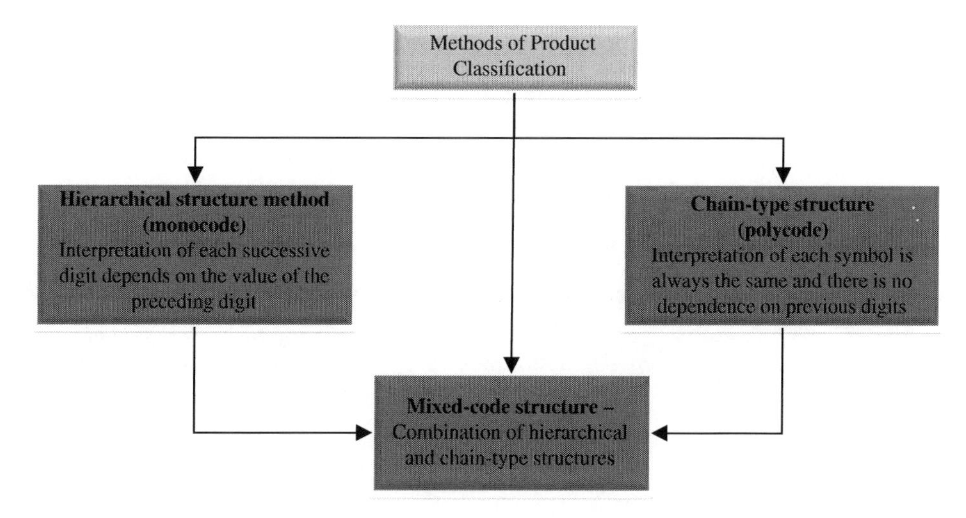

Figure 7.20 Three types of coding structures (Askin and Standridge 1993).

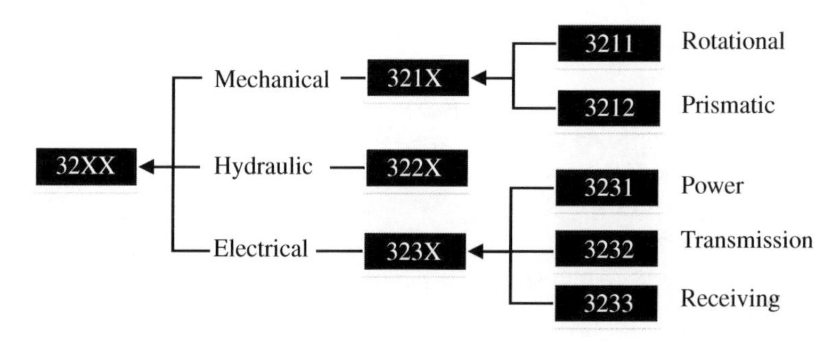

Figure 7.21 Hierarchical structure of a monocode.

type or *monocode*, chain type or *polycode*, and *hybrid* type for a combination of monocode and polycode.

7.4.2.1 Monocodes

A monocode uses a tree-like hierarchical structure. Figure 7.21 shows an example of a monocode with four digits. The third digit represents implementation of a driving mechanism, which can be mechanical, hydraulic, or electrical. The fourth digit represents the variety of driving mechanism after the implementation is determined. Different implementations correspond to different variants at this level.

The information of each digit at an upper level is amplified in the next level, in such a way that a monocode uses a limited number of digits to represent a large amount of product information. A monocode is efficient in storing and retrieving the information relevant to a product such as part geometry, material, and size. However, it shows its limitations in capturing the information of manufacturing sequences. The applicability of monocodes in manufacturing is confined.

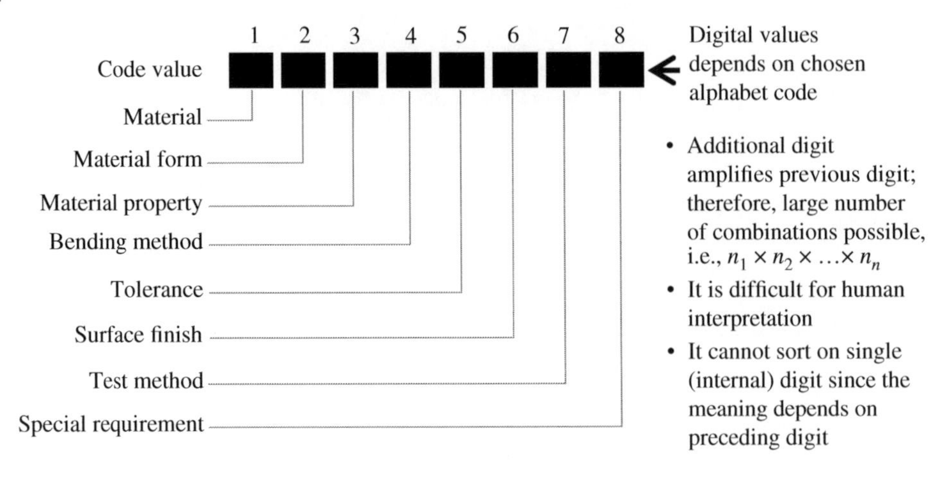

Figure 7.22 Details of a monocode.

Figure 7.22 shows an example of the details of a monocode for sheet metal products. It consists of eight digits, which stand for the variants of product features for material, material form, properties, bending method, tolerance, surface finish, testing method, and specific requirements. In the hierarchical structure, the next digit amplifies the previous digit. Therefore, a monocode is capable of representing a large number of combinations, $n_1 \times n_2 \times \cdots \times n_n$, where n_i is the number of possible choices of the digital at position i ($i = 1, 2, \ldots, n$) and n is the total number of digits. The contents of one digit depend on the value of the digit in the precedent position.

Example 7.2 Given a monocode system shown in Figure 7.23a, determine the code for the product shown in Figure 7.23b.

Solution

The coding system is a four-level hierarchical structure: the first level of an overall geometric shape (*cylindrical* or *block*), the second level and the third level for the ratio of the dimensions in different axes, and the fourth level for the tolerance of inside features. The shapes, dimensioning, and tolerances of the product in Figure 7.23 are examined, so that the digital values for product features in the monocode system can be determined as

1. The first digit '0' is for a cylinder.
2. The second digit '0' is for $L/D = 1/1.25 = 0.8 < 1$.
3. The third digit '1' is for $I/D = 1/1.25 = 0.8 > 0.5$.
4. The fourth digit '0' is for tolerance $= 0.00005 < 0.0001$.

The product on the left has a code '0010'.

7.4.2.2 Polycodes

In a chain-type code (polycode), the values of a digit at a given coding place have their consistent meanings. For example, the value of '3' at the third place of a code has the same meaning for all products. A chain-type code is relatively easier to learn but is less efficient.

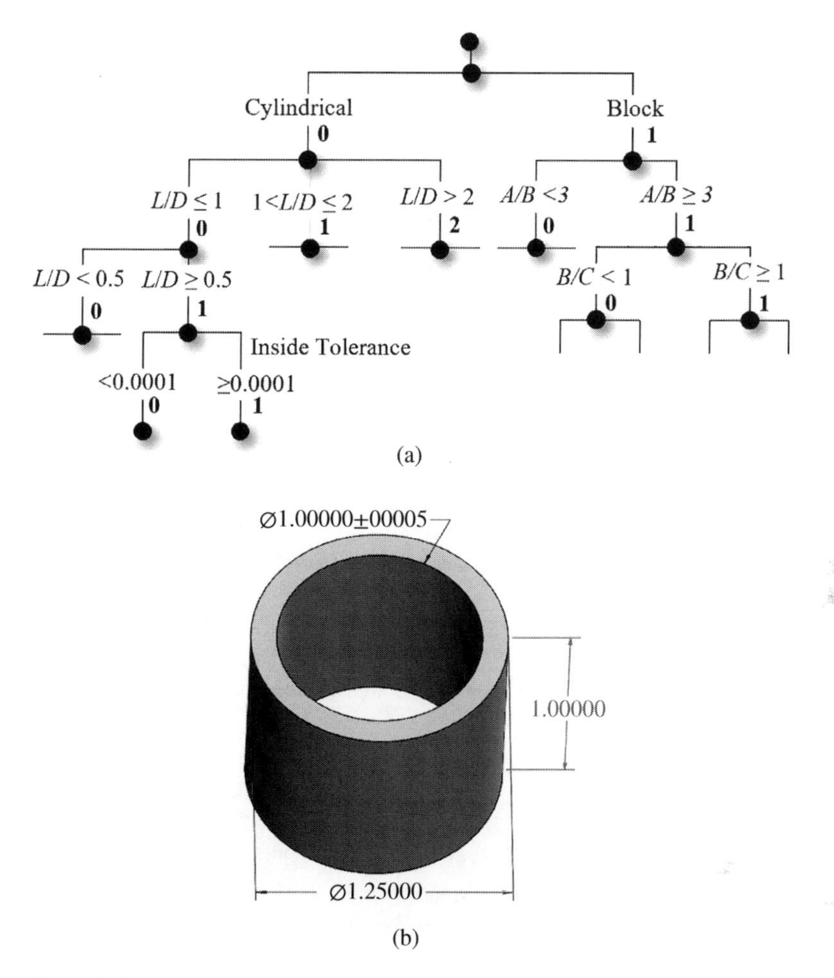

Figure 7.23 Example of using a coding system to code products. (a) Example of the monocode. (b) Drawing of an example product.

In addition, not all of the digits in a polycode are meaningful to certain products; some digits may become meaningless to these products.

As shown in Table 7.10 for a polycode example, the symbols in a polycode are independent of each other; each digit at the specific place represents a unique property of the product. A polycode is fairly simple to learn. However, if a product involves a large number of manufacturing processes, the corresponding polycode may become too lengthy and excessive to include all combinational features.

Figure 7.24 shows an example of converting a three-level polycode scheme (two-dimensional) into the digital representation (one-dimensional). Since a digit at the certain place has a clear meaning in a coding system, the digital representation of products is easily understood when the coding system (such as shown in Table 7.10) is available.

Example 7.3 Given a polycode system shown in Table 7.11 for Nissan passenger cars, determine the code for a product shown in Figure 7.25 (Kirby 2019).

Table 7.10 Example of polycode.

Digit	Class of feature	Possible values of a digit			
		1	2	3	4
1	External shape	Cylindrical without deviations	Cylindrical with deviations	Boxlike	...
2	International shape	None	Centre hole	Brind centre hole	...
3	Number of holes	0	1~2	3~5	...
4	Type of holes	Axial	Cross	Axial cross	...
5	Gear teeth	Worm	Internal spur	External spur	...

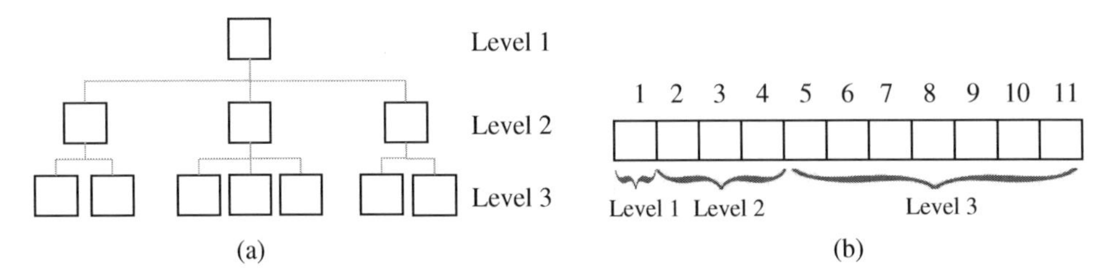

Figure 7.24 Example of converting a polycode scheme into digital representation. (a) Polycode scheme. (b) Polycode.

Table 7.11 Polycode structure for Nissan passenger cars.

Size (Digit 1)	Package (Digit 2)	Colour (Digit 3)	Interior (Digit 4)	Radio (Digit 5)	Tyre size (Digit 6)
1 for 'Maxima'	1 for 'GXE'	1 for 'White'	1 for 'Black'	1 for 'AM/FM'	1 for '15'
2 for 'Altima'	2 for 'XE'	2 for 'Black'	2 for 'Grey'	2 for 'CD'	2 for '17'
3 for 'Sentra'	3 for 'SE'	3 for 'Gold'	3 for 'Brown'	3 for 'CD changer'	
	4 for 'GLE'	4 for 'Blue'	4 for 'Leather'	4 for 'Premium'	
		5 for 'Red'			
		4 for 'Dark Grey'			

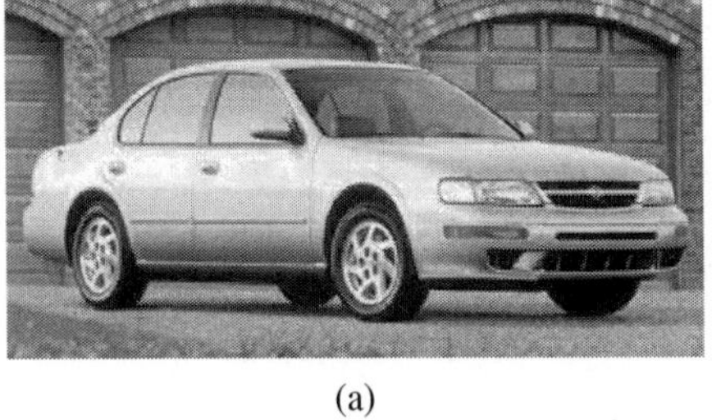

(a)

Features/Attributes		
Digits		Description
1	Type of car	1 for "Maxima"
2	Car package	1 for "GXE"
3	Car color	3 for "Gold"
4	Interior	4 for "Leather"
5	Radio	1 for "AM/FM"
6	Tire size	1 for "15 inch"

(b)

Figure 7.25 Example of coding products by a polycode. (a) Car model. (b) Code for a car model (a).

Solution

The polycode in Table 7.11 consists of six digits for the product features on type, package, colour, interior finish, radio, and tyre size. Assigning values for these product features according to the car photo in Figure 7.25a gives the code for this car model as '1 1 3 4 1 1'.

Different coding systems may be proposed for the same set of product variations. It is worth noting the capacity difference of information storages between the monocode and the polycode.

Example 7.4 Assume that a code consists of six digits. Each digital place has possible values from 0 to 9. Determine the number of product variants the monocode and the polycode can represent, respectively.

Solution

The number of product variants in a monocode is: $10^1 + 10^2 + 10^3 + 10^4 + 10^5 + 10^6 = 1,111,110$, while the number of product variants in a polycode is: $10 + 10 + 10 + 10 + 10 + 10 = 60$.

7.4.2.3 Hybrid Codes

A hybrid code is a mixture of monocode and polycode. A hybrid code is proposed to retain the advantages of the two basic coding systems. As shown in Figure 7.26, a hybrid code consists of several sub-groups that contain either hierarchical or chain structures. Hybrid codes are generally versatile since they can be tailored to represent both general product attributes (monocode) as well as process-specific or company-specific features (polycode).

Most practical coding systems use hybrid codes. Some popular code systems include Opitz from the Technology University of Aachen in Germany, the Brisch System from Brisch-Birn Inc., the CODE from Manufacturing Data System, Inc., CUTPLAN from

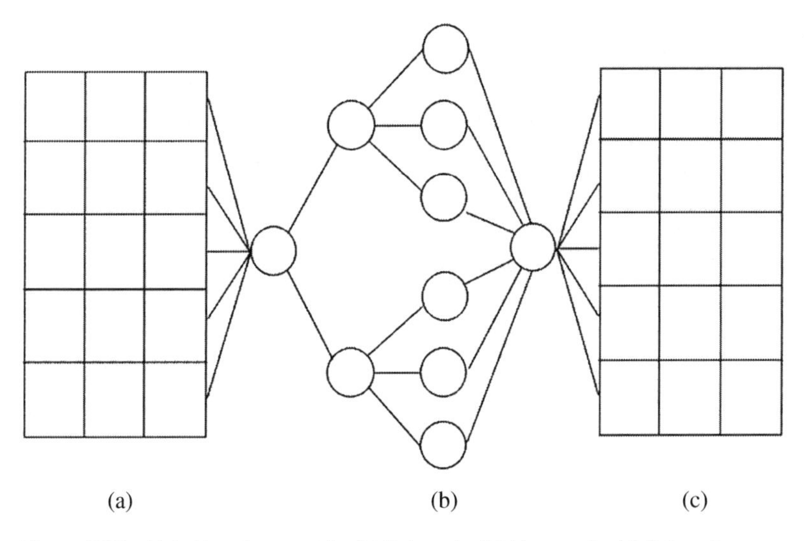

Figure 7.26 Hybrid code example. (a) Polycode. (b) Monocode. (c) Polycode.

Metcut Associates, DCLASS from the Brigham Young University, MultiClass from the Organization for Industrial Research (OIR), and the Part Analog System from Lovelace, Lawrence & Co. (Khan 2013). Among these coding systems, Opitz is still the most popular one and is used to illustrate the structure of the hybrid code and the application.

7.4.2.4 Opitz Coding System

Opitz is a hybrid coding system developed by Opitz at the Technical University of Aachen (Haworth 1968). It is proven as a basic framework for understanding the classification and coding process. Opitz is widely used to group machined, non-machined (both formed and cast), and purchased products. It covers both design and manufacturing information of products. The Opitz coding system consists of three groups of digits shown in Table 7.12.

The roles of nine digits in the Opitz coding system are explained in Table 7.13. The structure of the Opitz coding system is shown in Figure 7.27, where the form code (digital 1–5) is a monocode and the supplementary code (digital 6–9) is a polycode. Moreover, Table 7.14 gives the criteria for coding products based on the design and manufacturing features of the products.

Table 7.12 Constitutions of the Opitz code.

Digits in Opitz classification												
1	2	3	4	5	6	7	8	9	A	B	C	D
Form code					Supplementary code				Secondary code			

Table 7.13 Roles of digits in the Opitz coding system.

Digit	Description
1	Differ rotational shapes from non-rotational shapes. Products with a rotational shape are classified by the length-to-diameter ratio and the products with a non-rotational shape are classified by length, width, and thickness.
2	Differ external shape features.
3	Differ products with the machining needs for internal features such as holes or threads or other rotational features, and differ non-rotational products with rotational shape features.
4	Differ products with machined surfaces such as flats and slots.
5	Differ products with auxiliary holes, gear teeth, and other features.
6	Differ products based on overall dimensions.
7	Differ products based on working materials such as steel, aluminium, and iron.
8	Differ products based on original shapes of starting raw materials.
9	Differ products based on accuracy requirements.

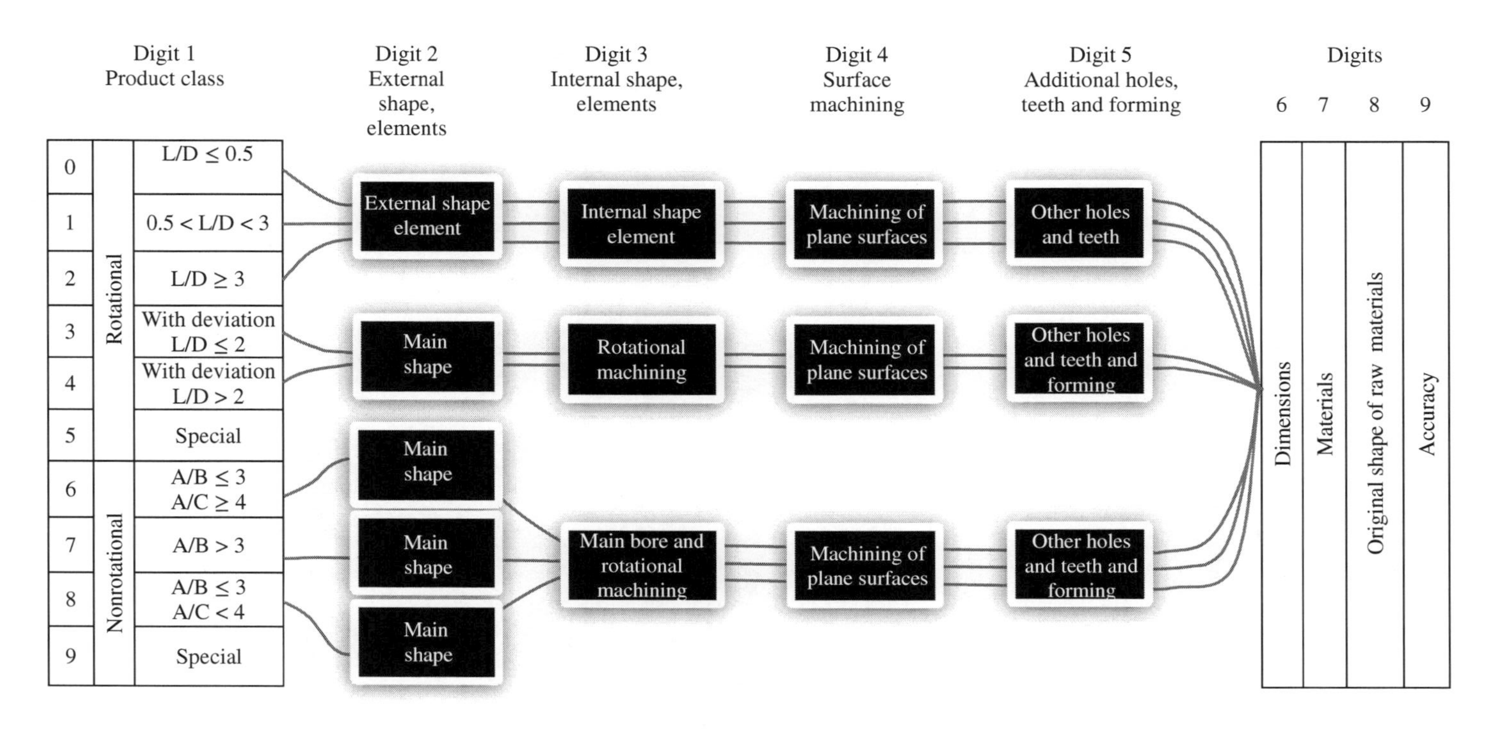

Figure 7.27 Structure of the Opitze coding system.

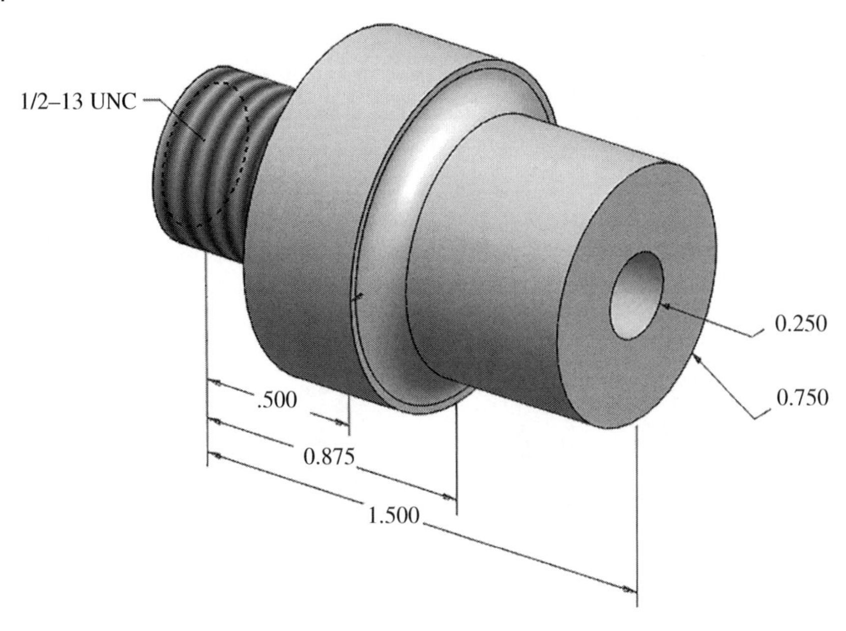

Figure 7.28 Product drawing for Example 7.5.

Example 7.5 Given the product shown in Figure 7.28, determine five digits of the 'form code' using the Opitz coding system in Table 7.14.

Solution

The dimensions in Figure 7.30 are used to obtain the values of design variables in Table 7.7 as follows:

1. The length (L) of the product is 1.75 and the overall diameter (D) is 1.25, L/D = 1.4 → (code 1).
2. The external shape of the rotational product is stepped on both with one thread → (code 5).
3. The internal shape of the rotational product has a through hole → (code 1).
4. The product has no surface machining → (code 0).
5. The product has no auxiliary hole or gear teeth → (code 0).

Therefore, the Opitz code of the product is '15100'.

Example 7.6 Given the product shown in Figure 7.29, determine nine digits of the 'form code' and 'supplementary code' using the Opitz coding system.

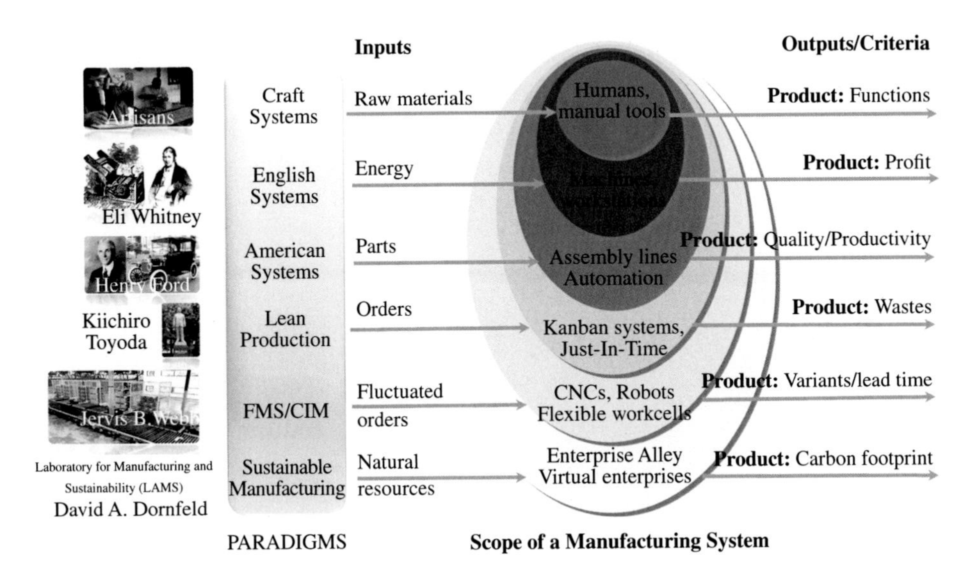

Figure 1.3 The growth of scale and complexity of manufacturing systems (Bi et al. 2014).

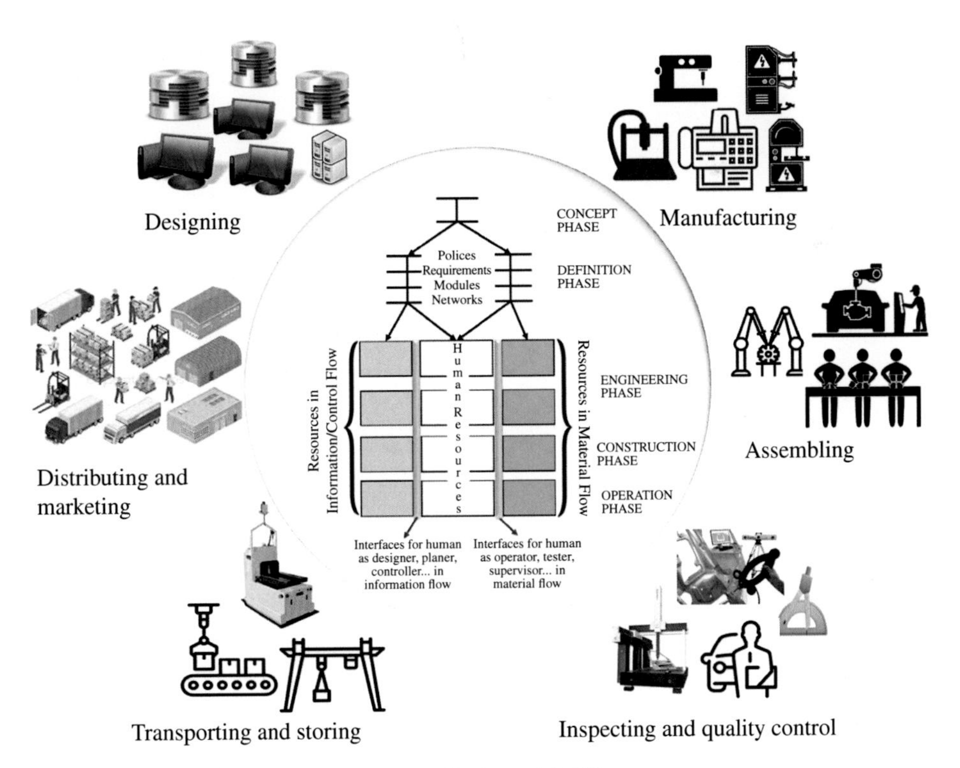

Figure 1.4 Human's role in manufacturing (Ortiz et al 1999).

Computer Aided Design and Manufacturing, First Edition. Zhuming Bi and Xiaoqin Wang.
© 2020 John Wiley & Sons Ltd. This Work is a co-publication between John Wiley & Sons Ltd and ASME Press.
Companion website: www.wiley.com/go/bi/computer-aided-design

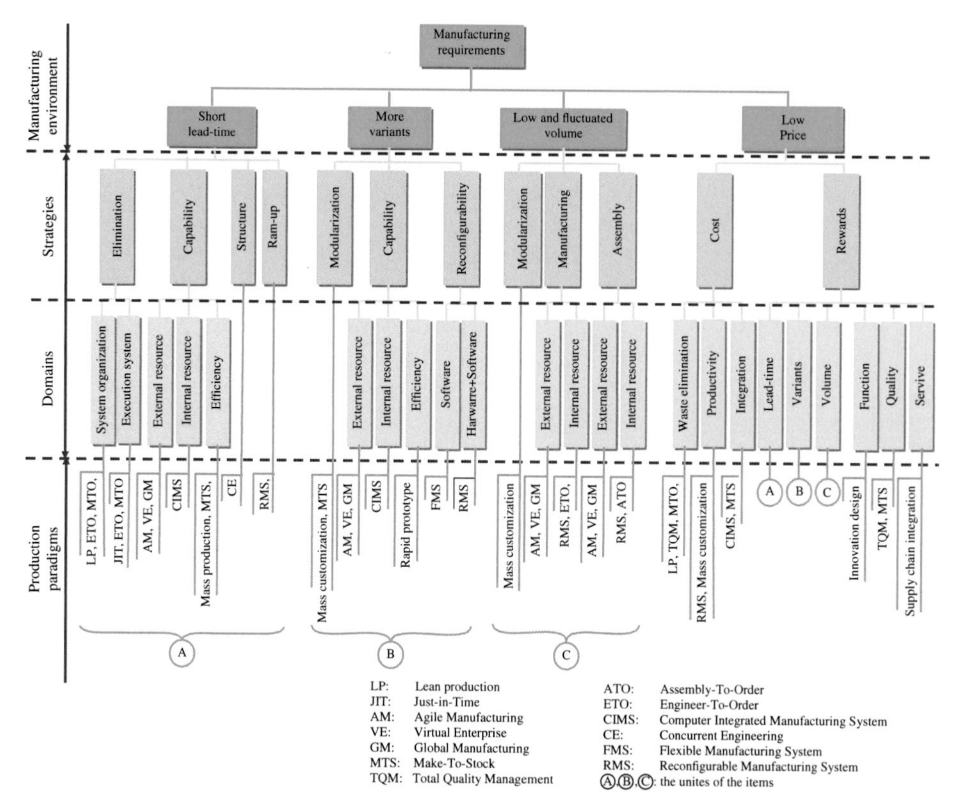

Figure 1.5 The strategies, domains, and production paradigms of advanced manufacturing technologies (Bi et al 2008).

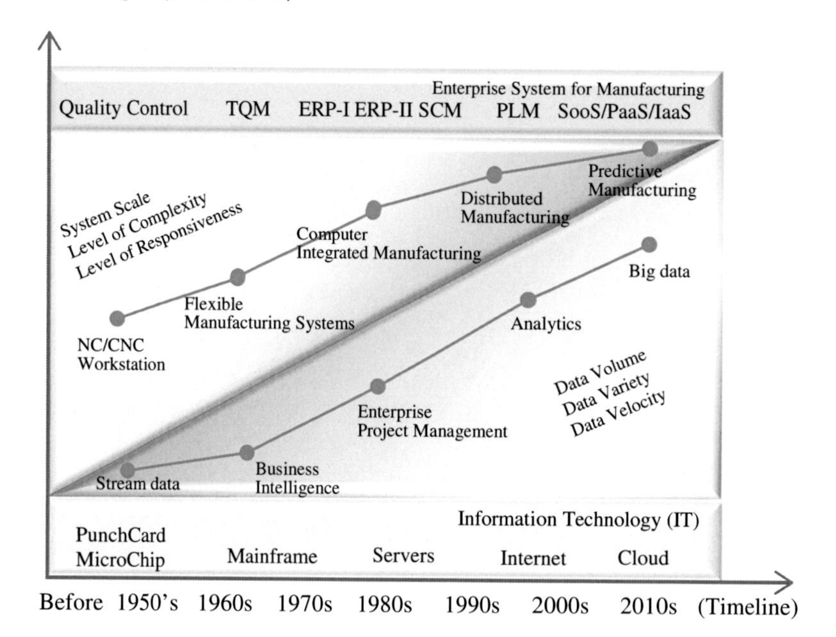

Figure 1.6 Evolution of computer aided technologies in manufacturing (Bi and Cochran 2014).

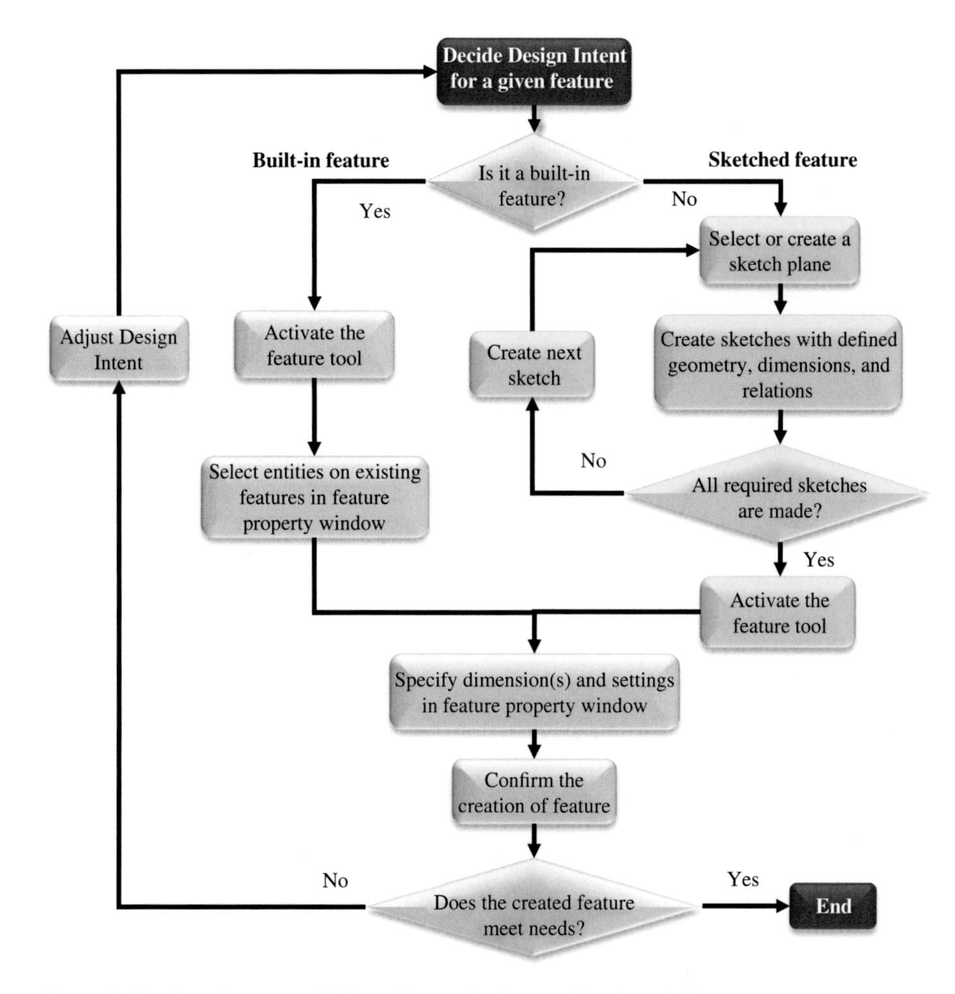

Figure 2.40 Creating or modifying a feature in feature-based modelling.

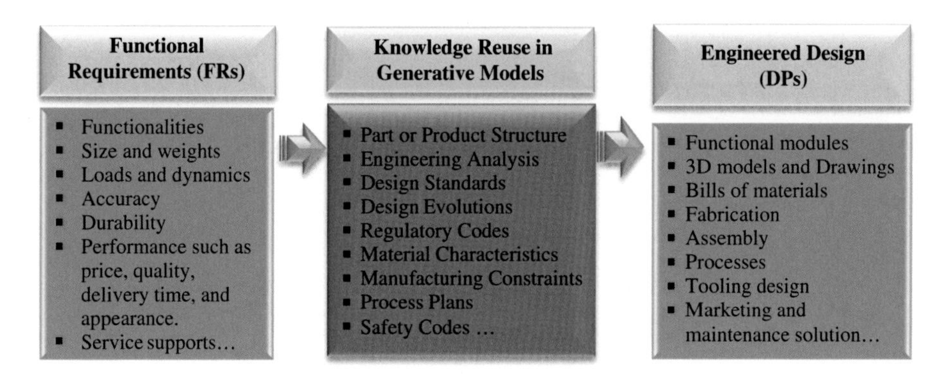

Figure 3.1 Engineering design process.

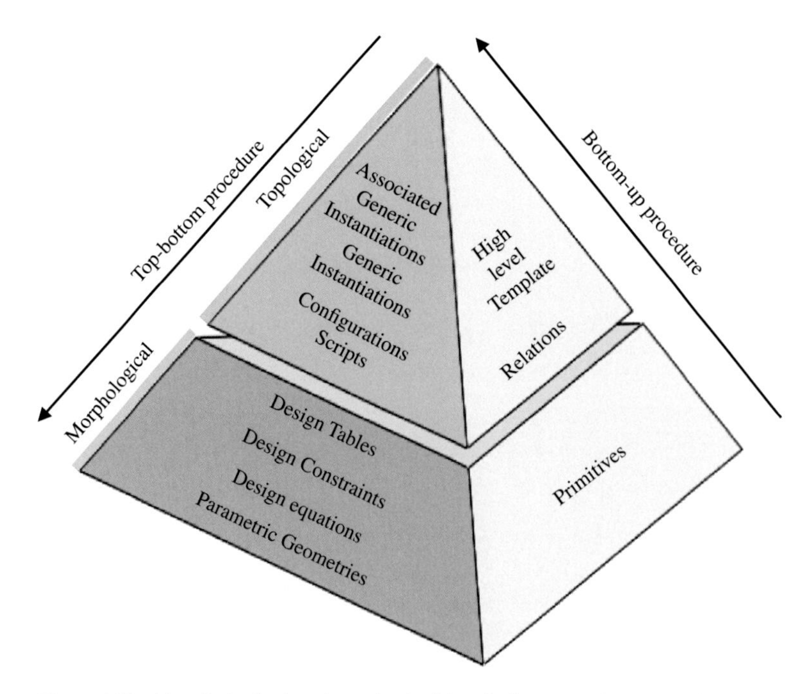

Figure 3.2 Morphological and topological level of geometric automation.

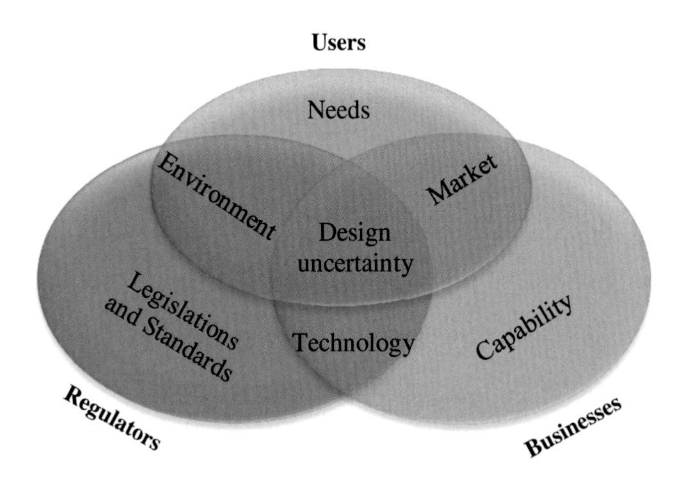

Figure 4.4 Challenges in CE practice (Nadadur et al. 2012).

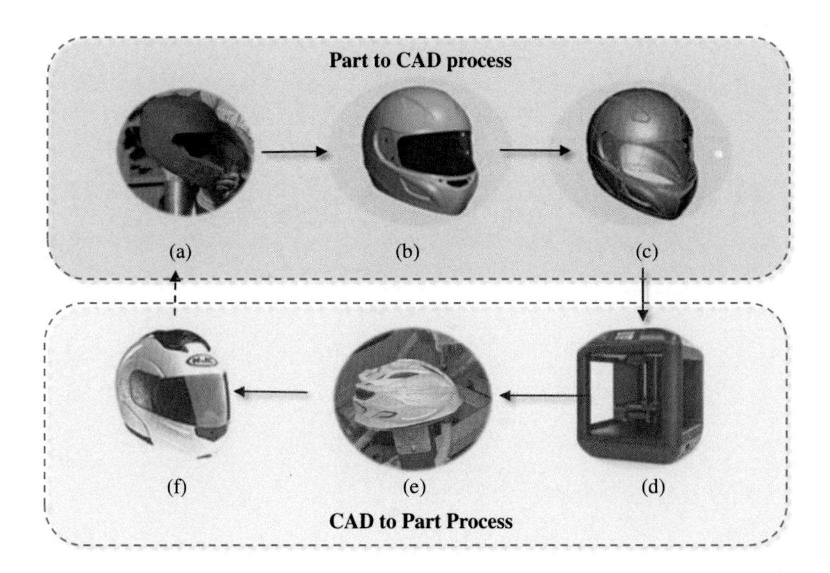

Figure 5.2 Part-to-CAD and CAD-to-part processes in RE. (a) Clay model. (b) Scanned points cloud. (c) Parametric model. (d) Rapid prototyping. (e) Tooling and manufacturing. (f) Physical product.

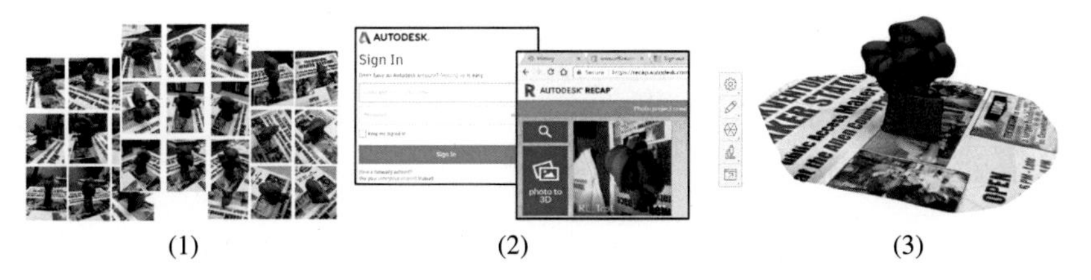

Figure 5.29 The steps of using Recap Pro for an RE project. (1) Capture sufficient photos of objects: take photos from different locations, directions, and altitudes of objects. (2) Create an RE project. Create or access a user account at https://accounts.autodesk.com/users, launch ReCap Pro, create a new project, select the mesh quality, the export file types, upload the photos, and confirm the submission. (3) Wait for the completion of surface reconstruction and download the polygonal model in the *.rcm* format. The uploaded images will be processed via the online service by cloud computing. After processing, one can review the model online or download the model for use in other software platforms.

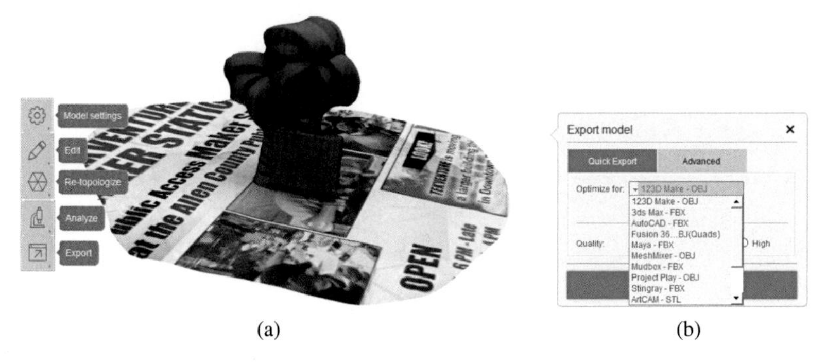

Figure 5.30 Processing and exporting a polygonal model in Recap Pro. (a) Editing a polygonal model in Recap Pro. (b) Exporting a polygonal file.

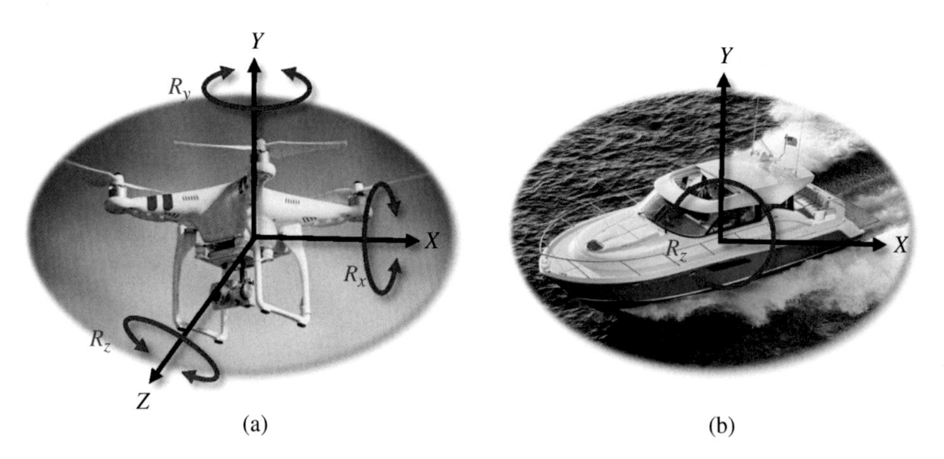

Figure 6.6 Free object and its degrees of freedom (DoF) of motion. (a) Six DoFs for an object in 3D space. (b) Three DoFs for an object in 2D space.

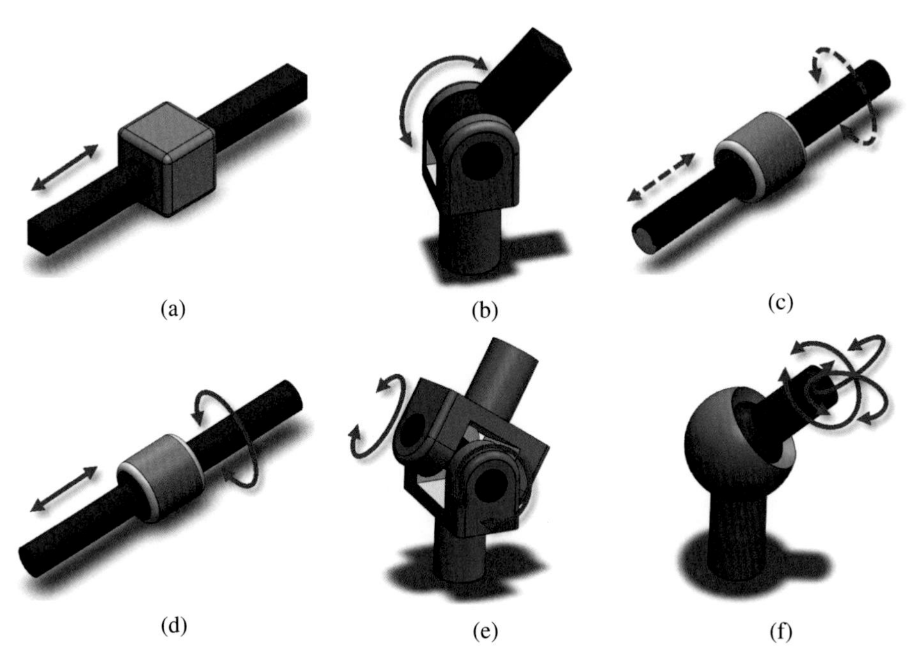

Figure 6.7 Classification of joints and the degrees of freedom of motion. (a) Prismatic joint. (b) Revolute joint. (c) Screw joint. (d) Cylindrical joint. (e) Universal joint. (f) Spherical joint.

Table 7.8 (Continued)

Network models	Description
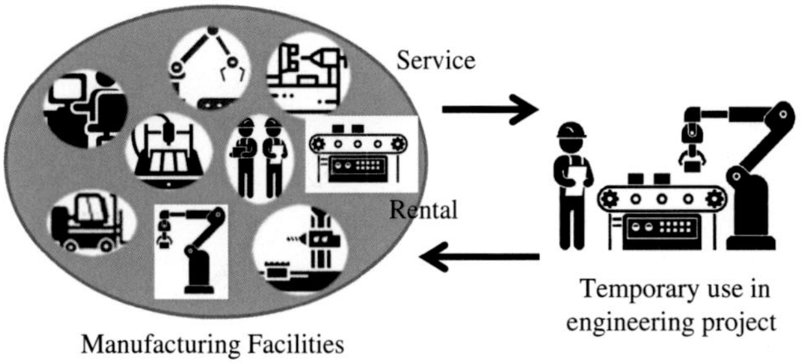 Service Rental Temporary use in engineering project Manufacturing Facilities (c) Mobile factory networks	A mobile factory network supports the rental-based approach for commercialization. It consists of a pool of standard functional modules and it allows configuration of mobile factories in order to meet different business needs.
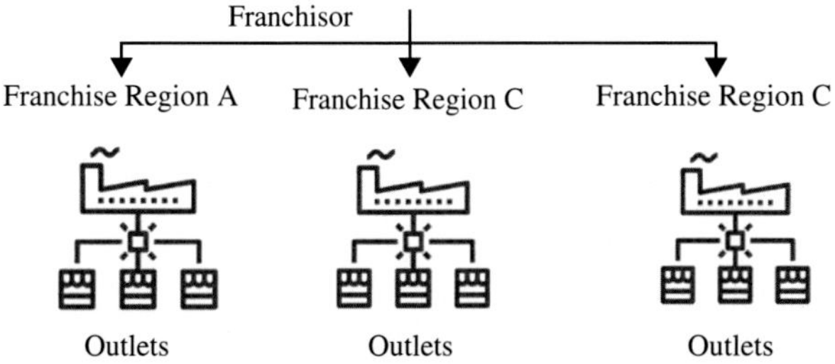 Franchisor Franchise Region A Franchise Region C Franchise Region C Outlets Outlets Outlets (d) Production franchise networks	In a production franchise network, distributed enterprises operate independently by franchisees and provides the flexibility to integrate geometrically distributed production lines with different capabilities.
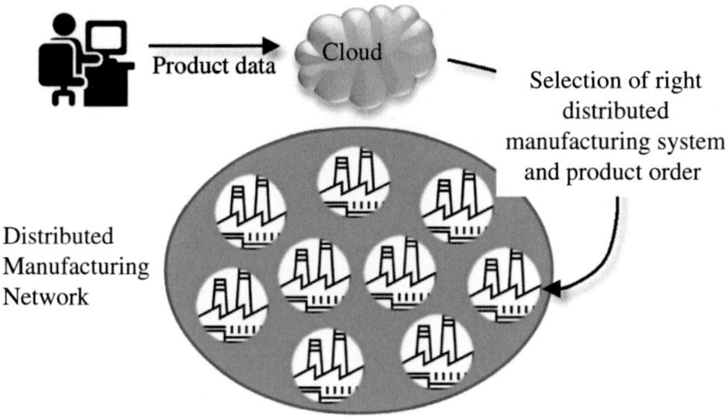 Product data Cloud Selection of right distributed manufacturing system and product order Distributed Manufacturing Network (e) Collaborative cloud manufacturing	In collaborative cloud manufacturing, customers participate in the product development process actively through the use of a collective innovation platform. The cloud manufacturing platform helps to match the production needs and capabilities of manufacturing companies. The production is enabled by transferring the product data from the product design platform to the distributed manufacturing network.

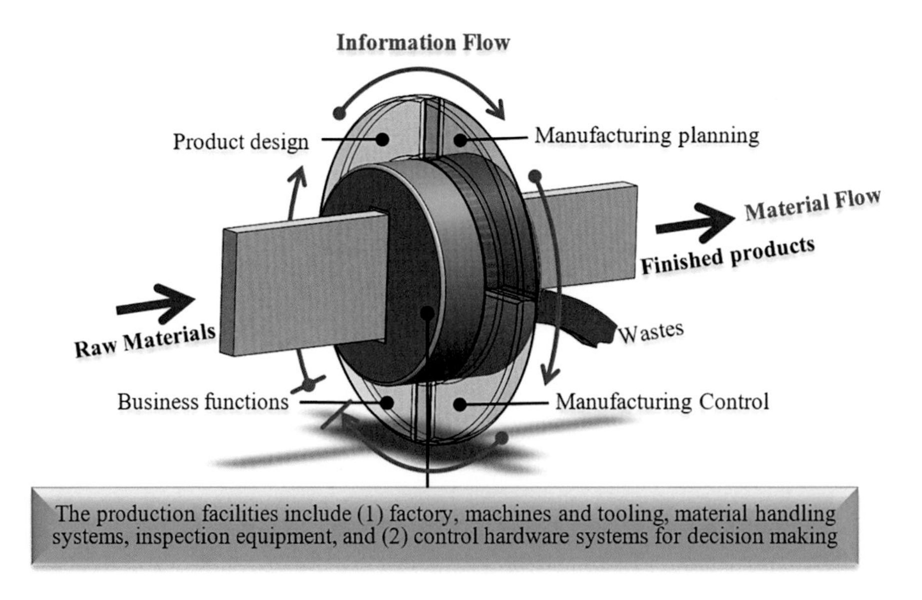

Figure 9.1 Production facilities in a manufacturing system (Groover 2007).

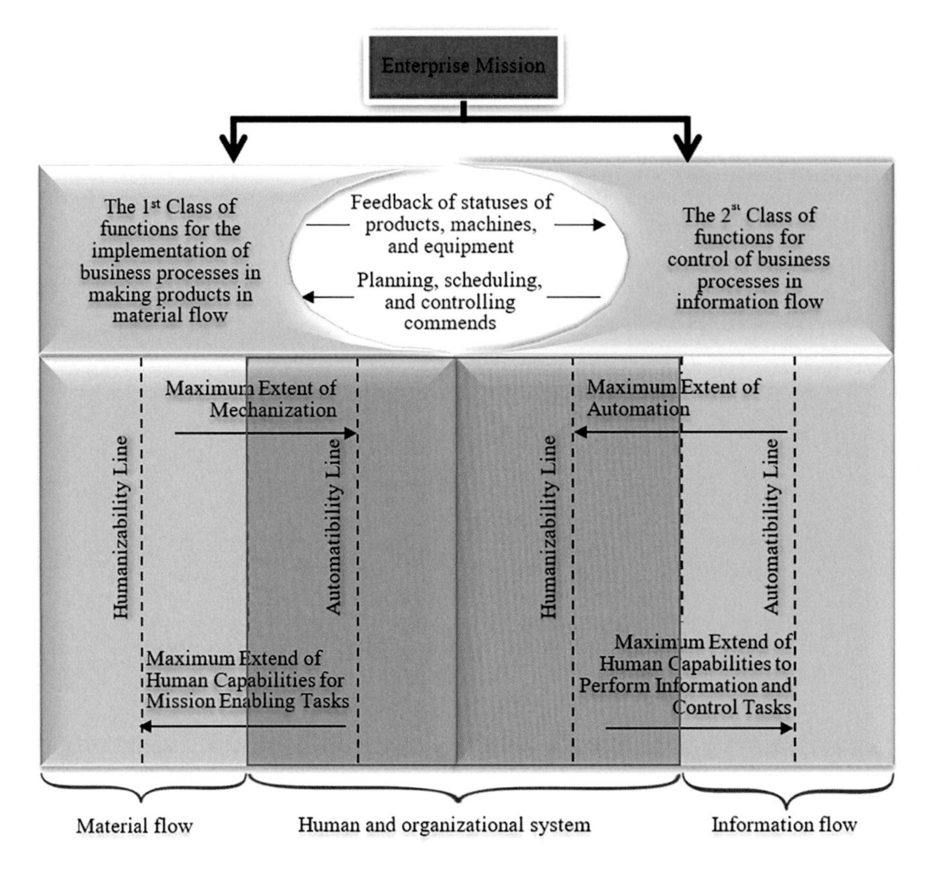

Figure 9.2 Humans and machines in a manufacturing system (Williams and Li 1998).

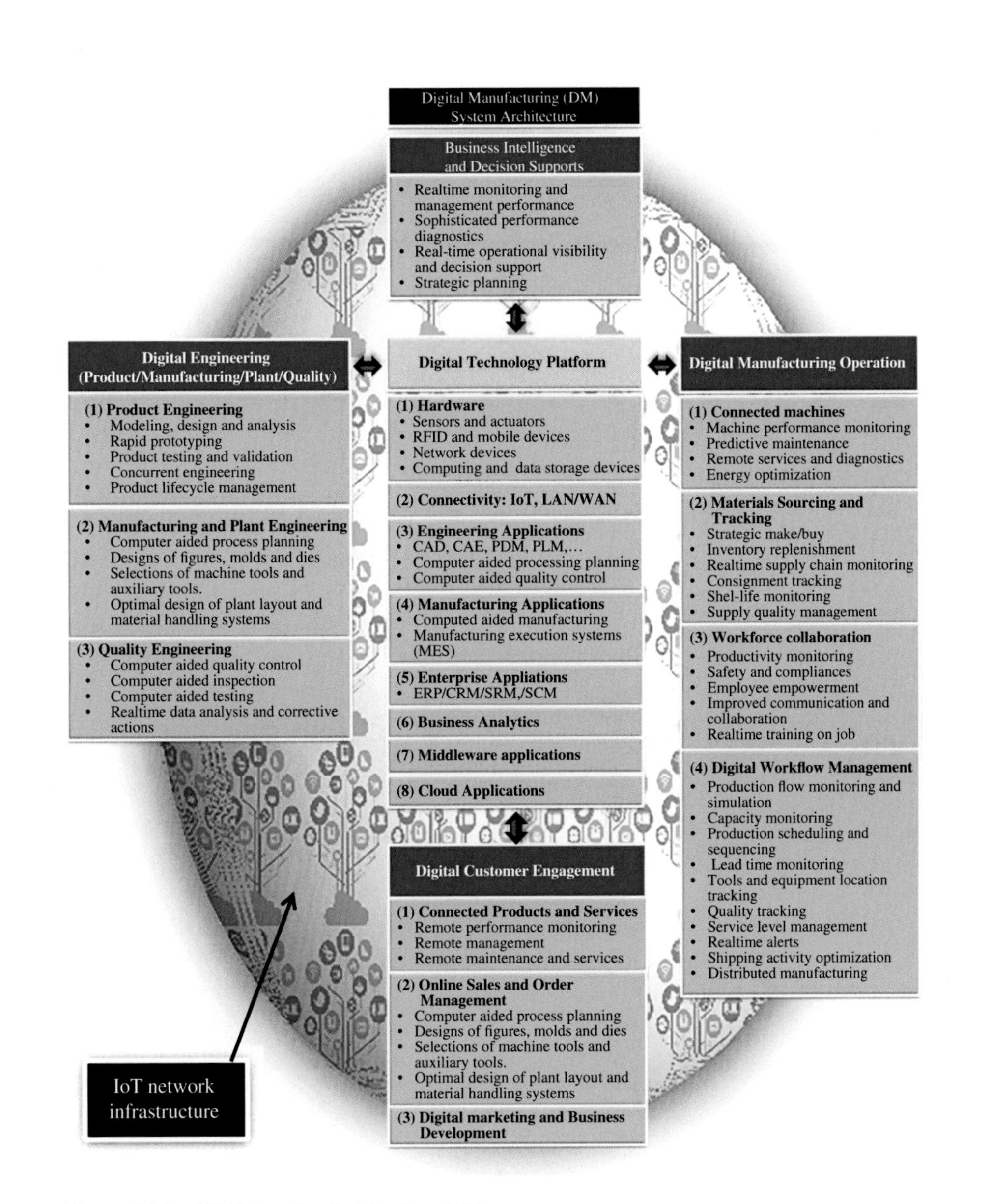

Figure 12.4 DM Enterprise Architecture (EA).

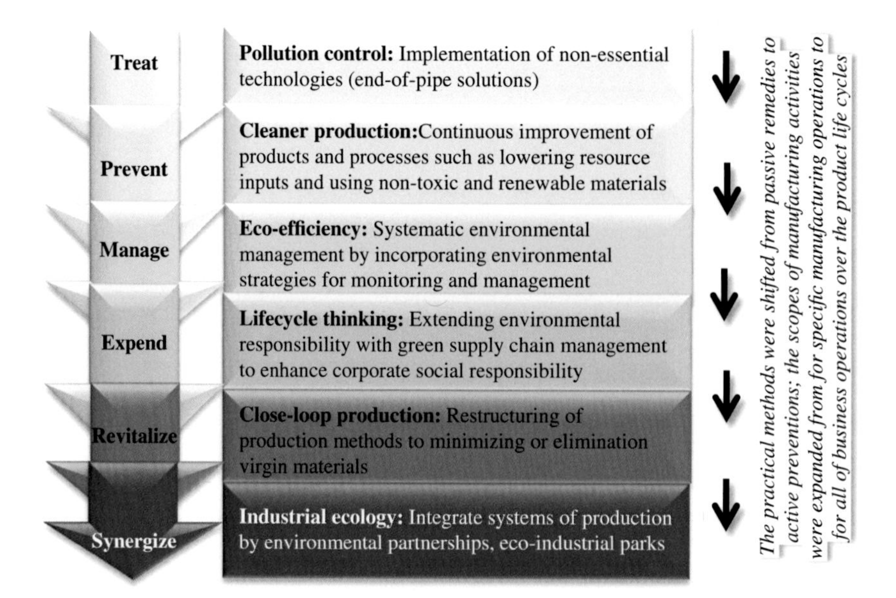

Figure 14.3 Evolution of sustainable manufacturing and concepts.

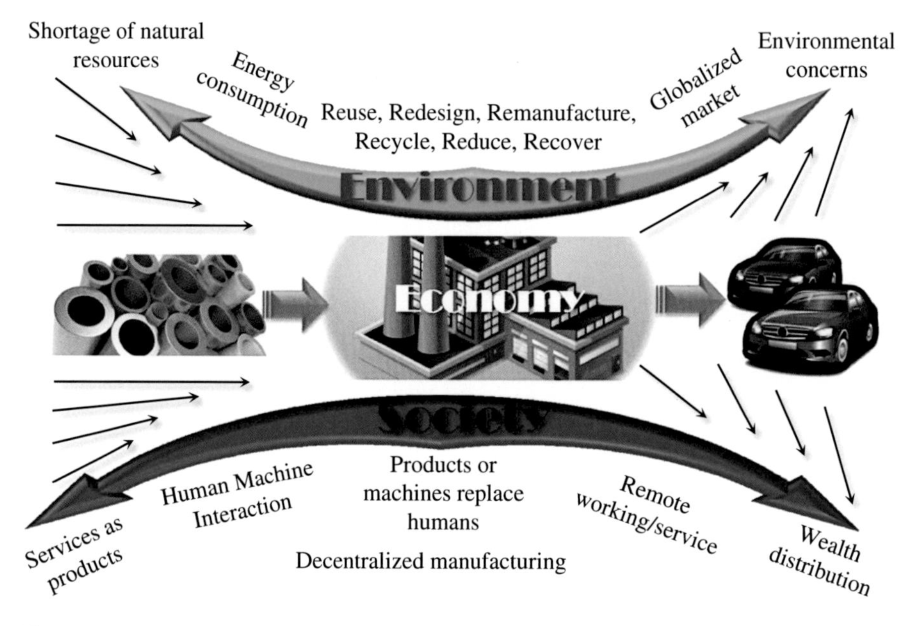

Figure 14.4 Changes of manufacturing business environment.

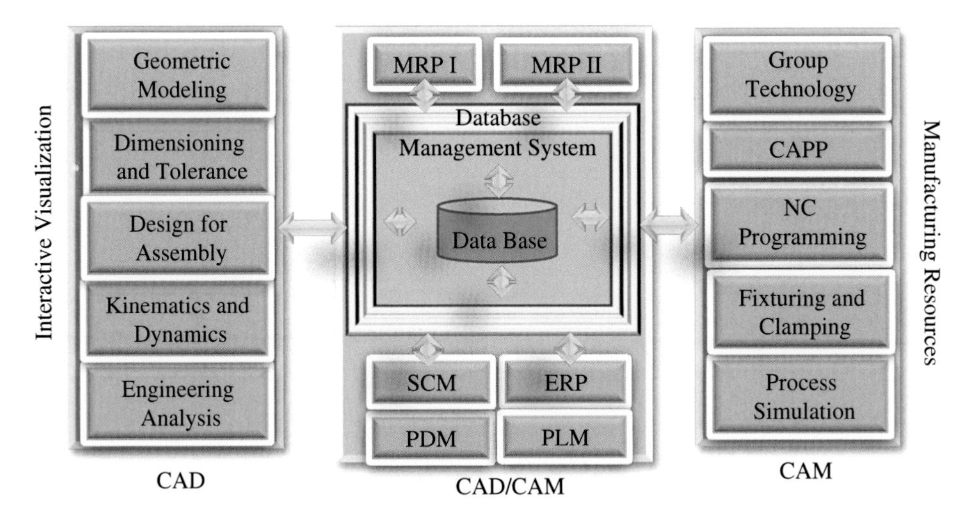

Figure 1.19 Typical computer aided tools in CAD, CAM, and CAD/CAM.

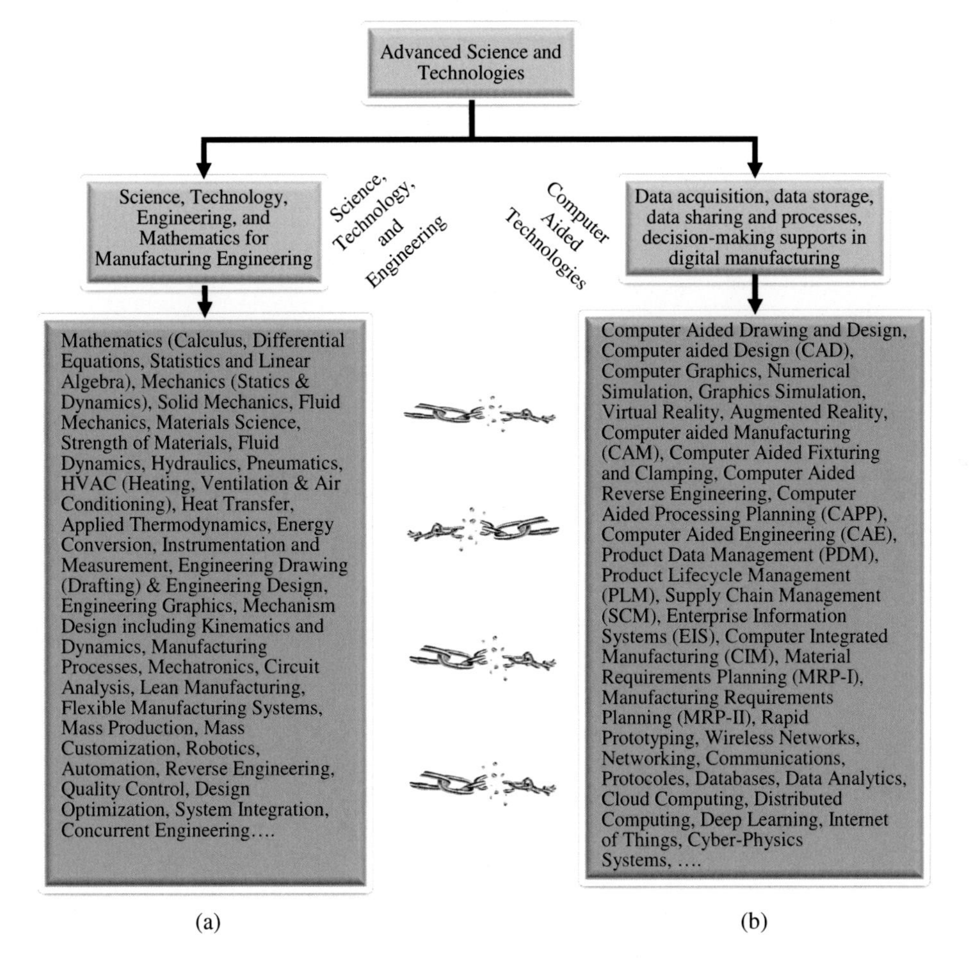

Figure 1.21 Mismatch of sub-disciplines and computer-aided tools in manufacturing engineering. (a) Subdisciplines in manufacturing engineering and (b) computer aided tools in digital manufacturing.

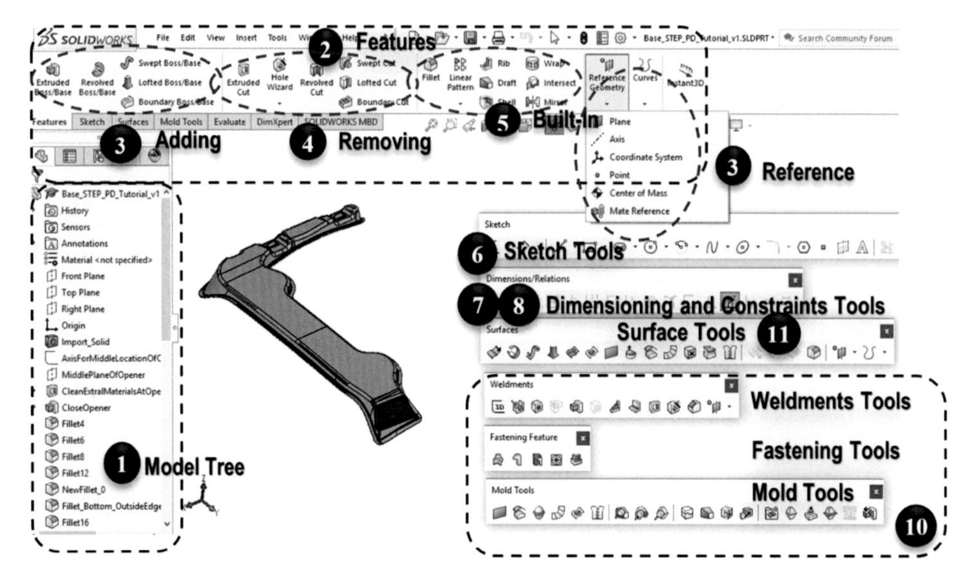

Figure 2.39 Feature-based modelling tools in SolidWorks.

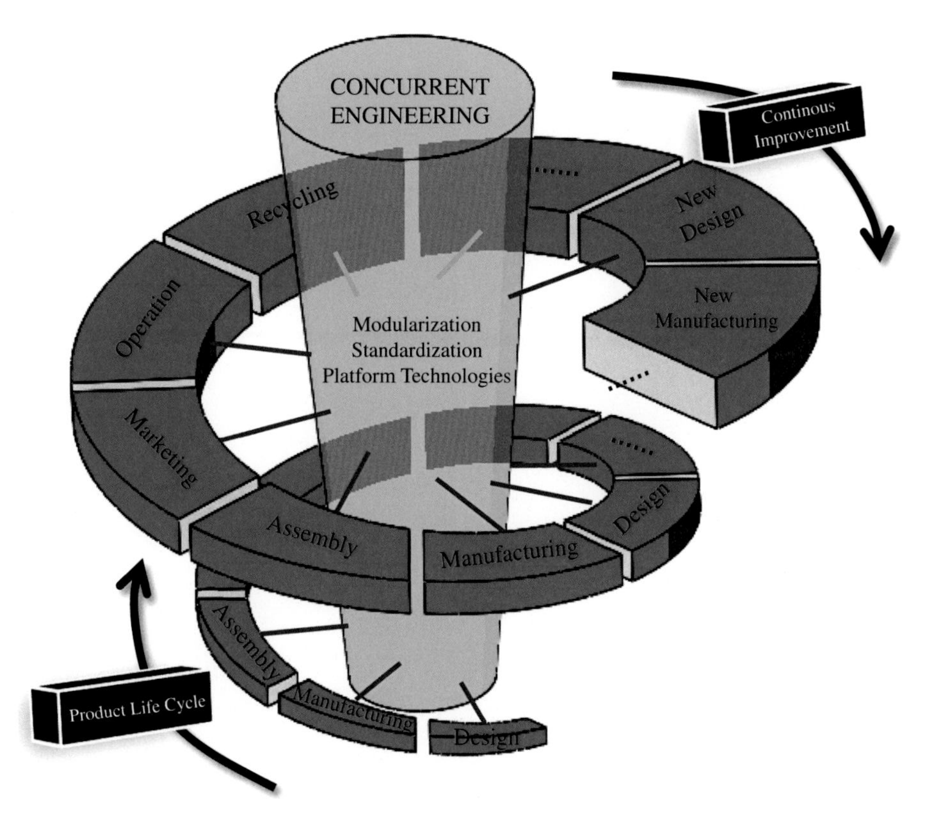

Figure 4.5 CE and Continuous Improvement (CI) in product design cycle.

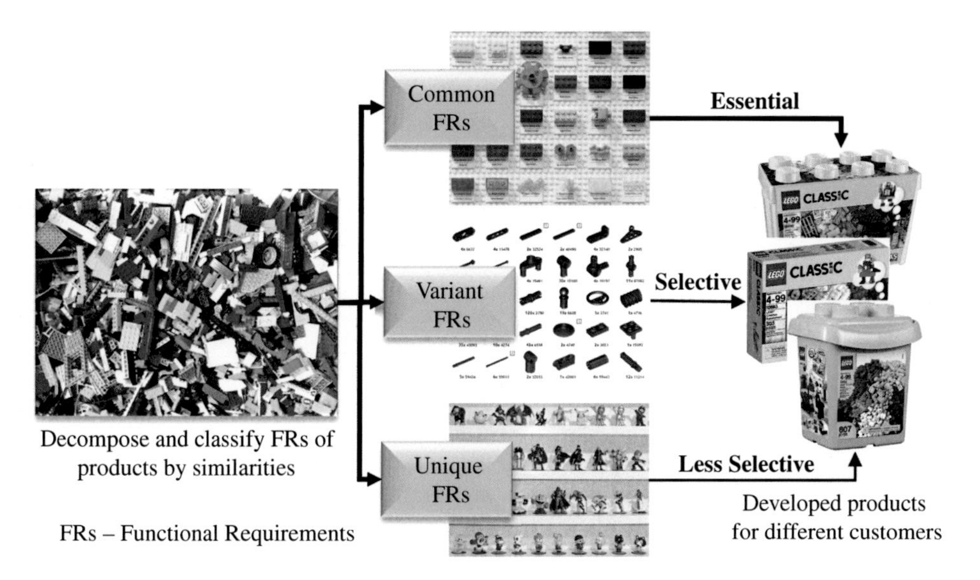

Figure 4.19 Modularization of a product family.

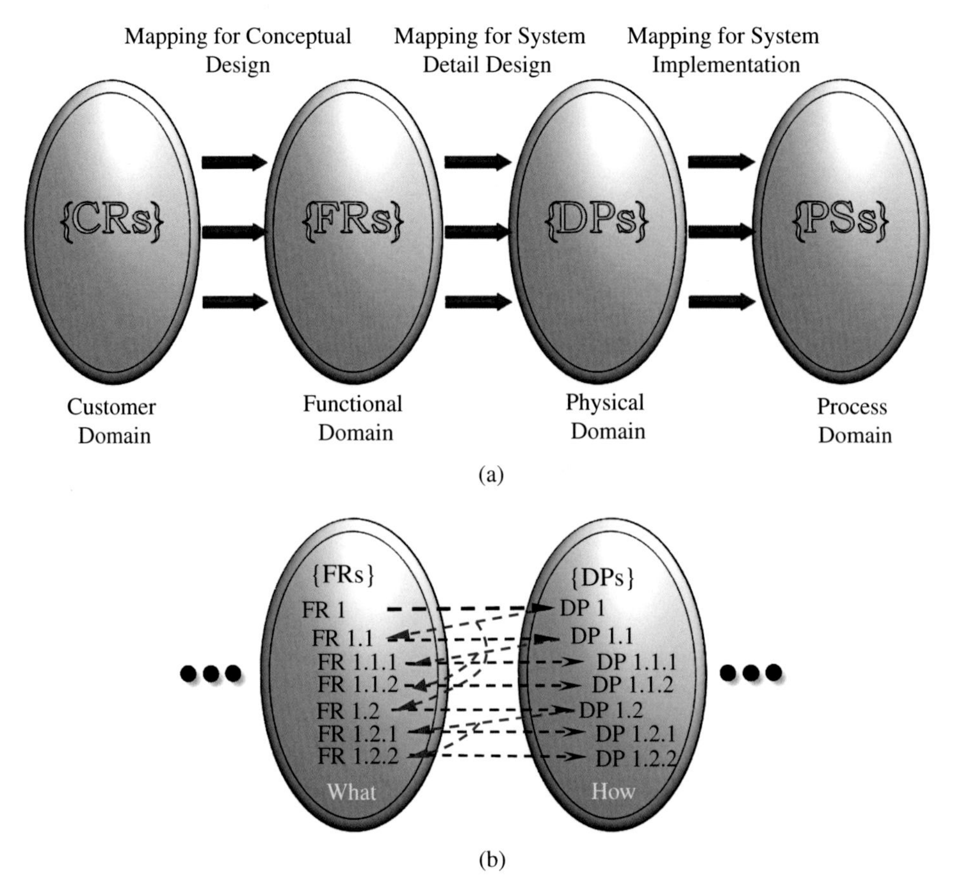

Figure 4.20 Zigzagging decomposition in axiomatic design theory (ADT). (a) Mapping in four domains of axiomatic design. (b) Zigzagging decomposition.

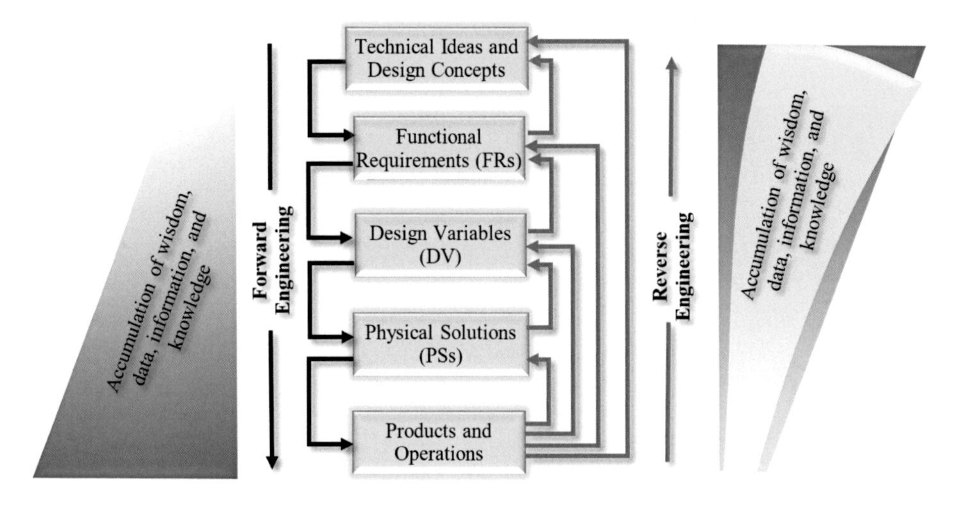

Figure 5.1 Forward engineering (FE) and reverse engineering (RE).

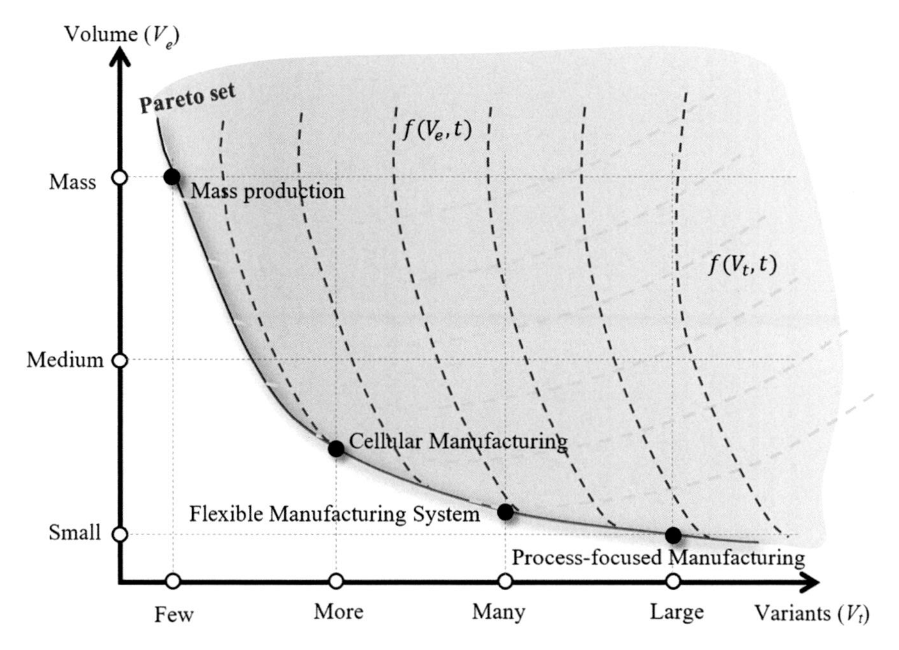

Figure 7.4 Ideal system paradigm versus volumes and variations of products.

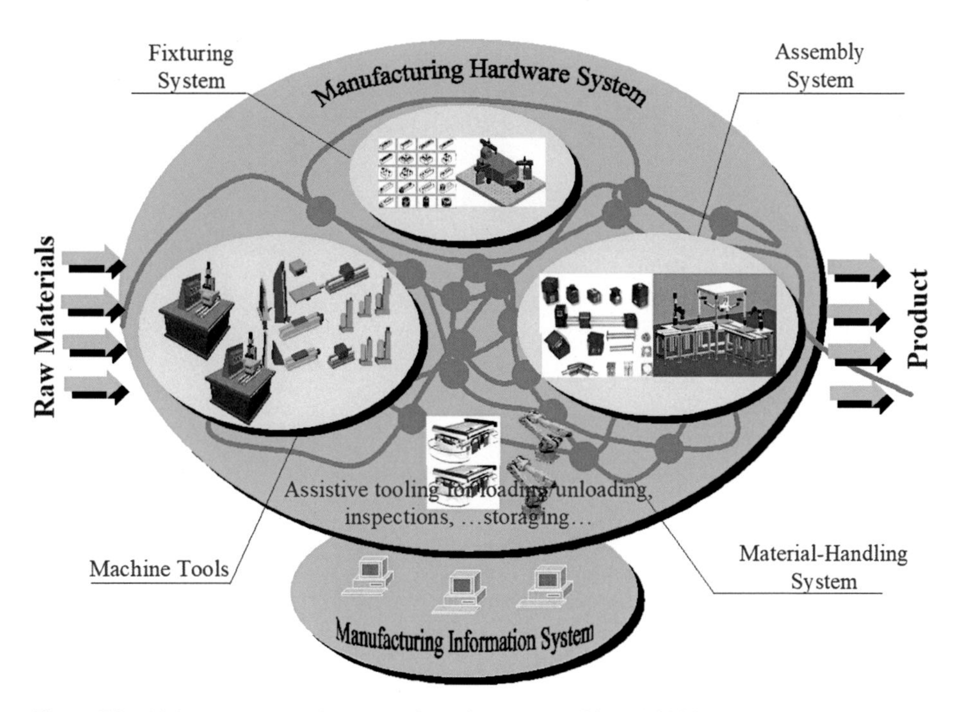

Figure 7.5 Main components in a manufacturing system (Bi et al 2008).

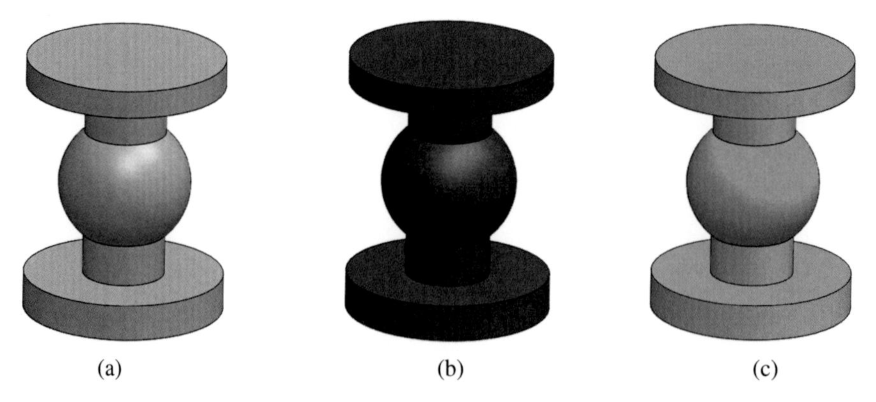

Figure 7.16 Example of inappropriate part family by visual inspection. (a) Plastic. (b) Iron. (c) Wood.

Table 7.8 The types of networked models for distributed manufacturing (Rauch and Dallasega 2017).

<table>
<tr><th>Network models</th><th>Description</th></tr>
<tr><td>Reach of individual company
Network of small businesses
(a) Owner-based micro production networks</td><td>An owner-based micro production network consists of a number of small or medium enterprise to product goods to meet customers in local or regional markets. Such a network model provides the opportunities for local manufacturers to develop/upgrade their own businesses without large investment.</td></tr>
<tr><td>
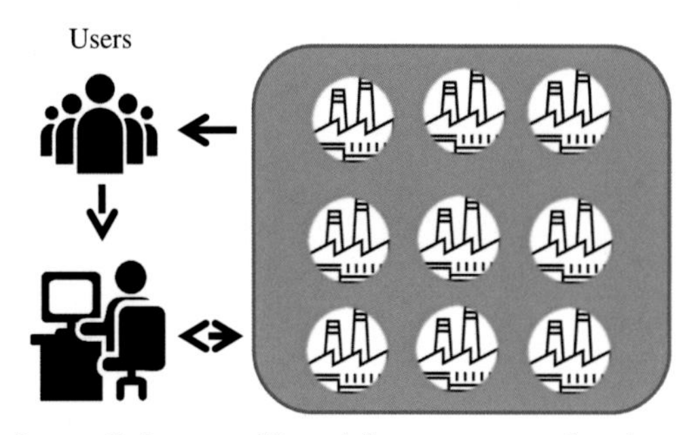

(b) Contract manufacturing networks</td><td>In a contract manufacturing network, enterprises perform manufacturing processes as the services accessible by users and production intermediaries. The network provides the high flexibility for geographically distributed enterprises to contract out manufacturing resources and production capacities. Intermediaries bring global users and locally distributed manufacturer together over the network.</td></tr>
</table>

(*continued*)

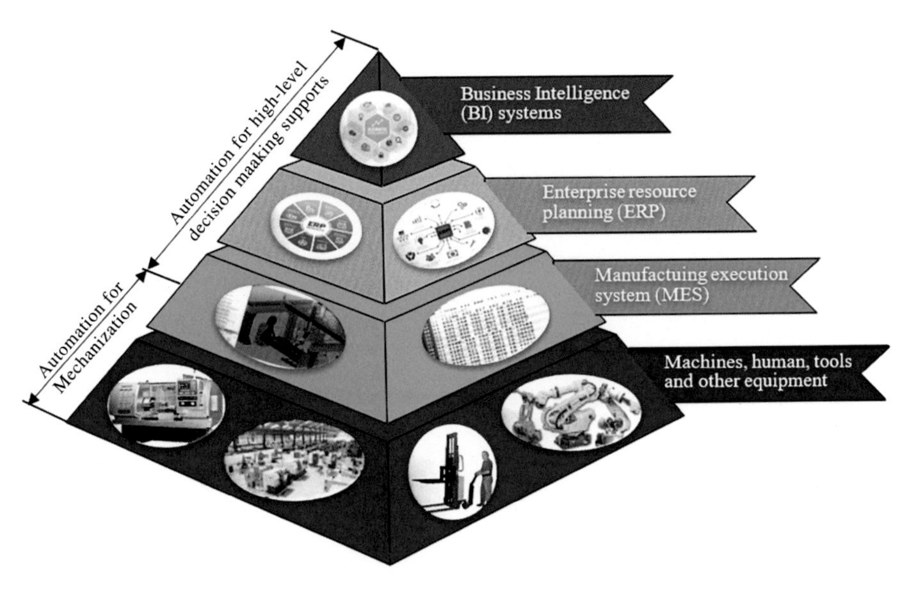

Figure 9.3 Automated decision-making supports in manufacturing.

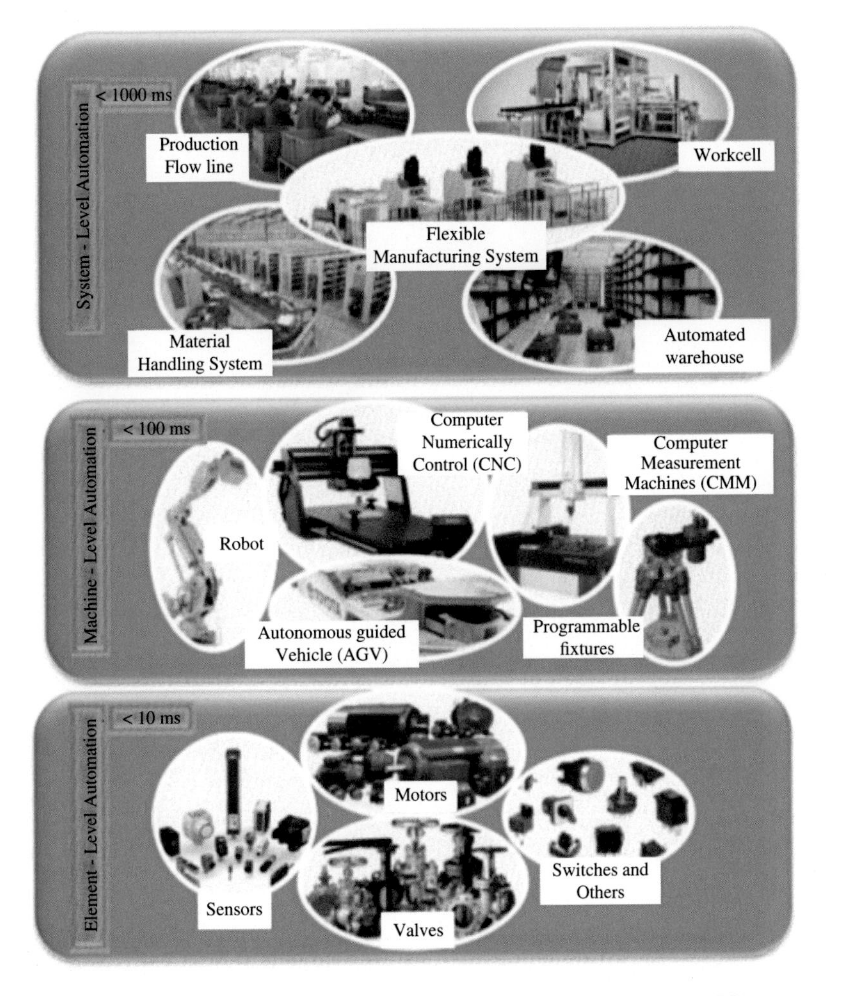

Figure 9.4 Level of automation in a manufacturing execution system (MES).

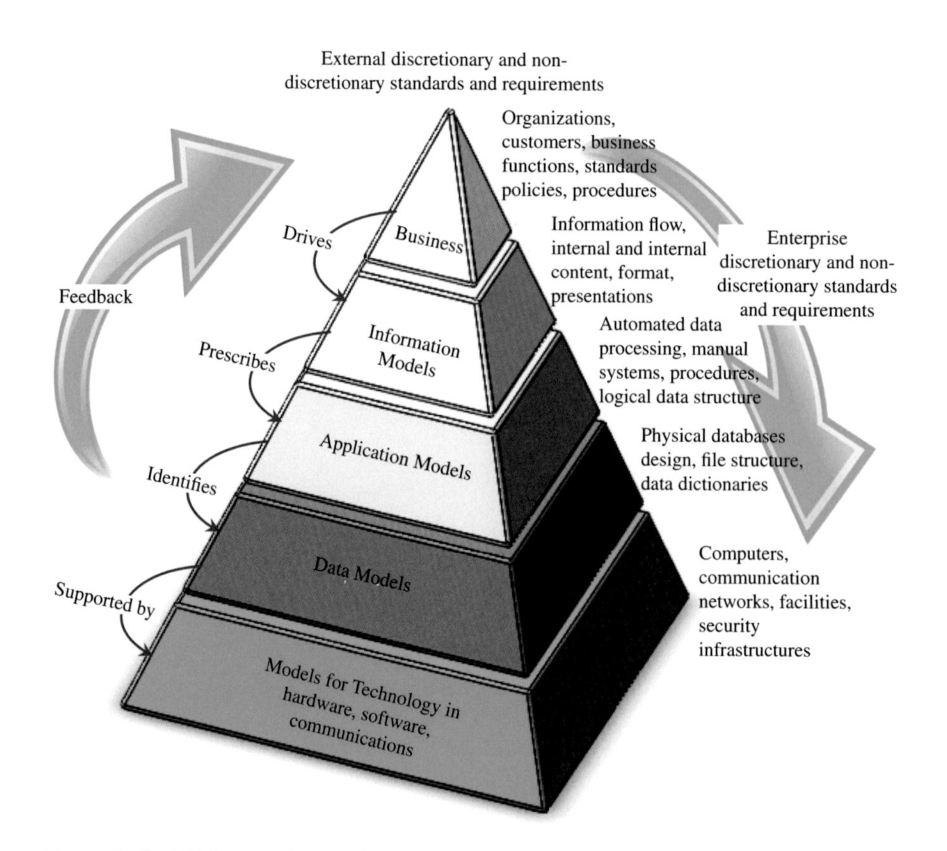

Figure 12.3 NIST enterprise architecture model (Wikipedia 2019a).

Table 7.14 Coding product features in Opitz coding system.

First digit			Second digit				Third digit			Fourth digit		Fifth digit	
Product class			External shape and elements				Internal shape and elements			Surface machining		Auxiliary holes and gear teeth	
0		$L/D \leq 0.5$	0	Smooth, no shape elements			0	No hole no breakthrough		0	No surface machining	0	No auxiliary hole
1	Rotational	$0.5 < L/D < 3$	1	Stepped one end		No shape elements	1	Smooth or stepped on one end	No shape elements	1	Surface plane/curved	1	Axial, not on pitch circle diameter
2		$L/D \geq 3$	2		or smooth	Thread	2		Thread	2	External plane surface, circular graduation	2	Axial on pitch circle diameter
3		With deviation $L/D \leq 2$	3			Groove	3		Groove	3	External groove and/or slot	3	Radial, not on pitch circle diameter
4		With deviation $L/D > 2$	4	Stepped both ends		No shape elements	4	Smooth or Stepped on both ends	No shape elements	4	External spline (polygon)	4	Radial, on pitch circle diameter
5		Special	5			Thread	5		Thread	5	External plane surface/ slot spline	5	Axial and/radial and/ other direction
6		$A/B \leq 3$ $A/C \geq 4$	6			Groove	6		Groove	6	Internal plane surface or slot	6	Spur gear teeth
7	Non-rotational	$A/C \geq 4$	7	Functional cone			7	Functional cone		7	Internal spline (polygon)	7	Bevel gear teeth
8		$A/B \leq 3$ $A/C < 4$	8	Operating speed			8	Operating speed		8	Internal or slot/external polygon	8	Other gear teeth
9		$A/B \leq 3$	9	All others			9	All others		9	All others	9	All others

(*continued*)

Table 7.14 (Continued)

Sixth digit		Seventh digit		Eighth digit		Ninth digit	
Diameter *D* or length of edge A (mm)		Material		Initial shape		Accuracy in coding digital	
0	≤20	0	Grey cast iron	0	Round bar	0	No accuracy specified
1	>20 & ≤ 50	1	Nodular graphitic cast iron and malleable cast iron	1	Bright drawn round bar	1	2
2	>50 & ≤ 100	2	Steel < 42 kg mm^{-2}	2	Triangular, square, hexagonal, or other bar	2	3
3	>100 & ≤ 160	3	Steel ≥ 42 kg mm^{-2}	3	Tubing	3	4
4	>160 & ≤ 250	4	Steel 2 + 3 heat-treated	4	Angled U.-T. and similar sections	4	5
5	>250 & ≤ 400	5	Alloy steel	5	Sheet	5	2 + 3
6	>400 & ≤ 600	6	Alloy steel heat-treated	6	Plates and slabs	6	2 + 4
7	>600 & ≤ 1000	7	Non-ferrous metal	7	Cast or forged component	7	2 + 5
8	>1000 & ≤ 2000	8	Light alloy	8	Welded group	8	3 + 4
9	>2000	9	Other materials	9	Pre-machined component	9	(2 + 3) + 4 + 5

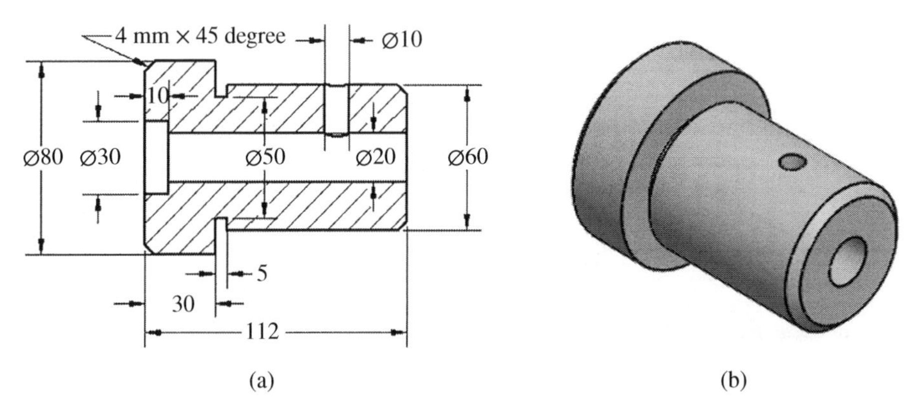

Figure 7.29 Product drawings for Example 7.6. (a) Dimension. (b) 3D part view.

Solution

The dimensions in Figure 7.29a are used to obtain the values of design variables in Table 7.7 as follows:

1. The length (L) is 112 mm and the diameter (D) is 80 mm, so L/D is 1.4 (the first digit is set as '1').
2. The external shape is rotational and is stepped to one end (the second digit is set as '1').
3. The product has an internal shape element, which is smooth or stepped to one end (the third digit is set as '1').
4. The product has no surface machining (the fourth digit is set as '0').
5. The product has auxiliary holes in the radial direction (the fifth digit is set as '5').
6. The diameter $D = 80$ mm is within the range of (50 mm, 100 mm) (the sixth digit is set as '2').
7. The product uses mild steel (the seventh digit is set as '2').
8. The initial shape of the product is a round bar (the eighth digit is set as '0').
9. No accuracy requirement has been specified (the ninth digit is set as '0').

Therefore, the product code is '111052200'.

The Opitz coding system has a total of 13 digits. Other than five digits (1, 2, 3, 4, and 5) as form codes and four digits (6, 7, 8, and 9) as supplementary codes, the last four digits (A, B, C, and D) in Table 7.12 are the secondary codes for more manufacturing attributes. Each digit has the options of 10 different values. Therefore, the Opitz coding system has the capacity to represent a considerably large number of product variants.

7.4.3 Production Flow Analysis

In a PFA method, the correspondences of products and processing facilities are modelled as relational matrices. The resulting matrices are analysed and converted to a set of product-machine groups. Each group corresponds to the set of products in one family as well as the machines to perform all manufacturing processes for this product family.

PFA uses production route sheets for grouping rather than product design data, and products with identical or similar route sheets are classified into product families. Route sheets help to deal with the similarities for (i) products with different geometrics but requiring the same or similar manufacturing processes and (ii) products with similar geometrics but requiring different manufacturing processes.

PFA consists of the following steps in the implementation:

1. Collect the data for the sequence of manufacturing processes and the routes of machines for all products.
2. Short the products and machines based on the routing similarities of manufacturing processes and arrange the identified groups of products and machines into 'packs'.
3. Create a PFA chart where all packs are displayed and each pack corresponds to a product-machine incidence matrix.
4. Conduct a cluster analysis to convert the packs into groups with similar routings.
5. Transfer each group of machines into a machine cell.

A number of algorithms have been developed to classify products and machines into groups. Two popular algorithms are the *single-linkage clustering (SLC) algorithm (SLCA)* and the *Rank-Order Clustering (ROC) Algorithm (ROCA)*. In the following, the ROCA by King (1980) is introduced to illustrate the procedure of using a product-machine matrix to group products and machines.

Assume that the correspondence of products and machines is defined as a relational matrix $[M]_{n \times m}$ where n and m are the numbers of products and machines, respectively. ROC is based on sorting rows and columns of the machine-part incidence matrix as follows:

1. Step 1. Assign a binary weight and calculate a decimal weight for each row and column using the formulas:
   i) Decimal weight for row $i = \sum_{p=1}^{m} b_{ip} 2^{m-p}$.
   ii) Decimal weight for column $j = \sum_{p=1}^{n} b_{pj} 2^{n-p}$.
2. Step 2. Rank the rows in order of decreasing decimal weight values.
3. Step 3. Repeat steps 1 and 2 for each column.
4. Step 4. Continue the preceding steps until there is no change in the position of each element in each row and column.

Example 7.7 Consider an eight-part-and-six-machine problem shown in the following table and form the part family and machine group.

		Products							
	M_{ij}	1	3	4	7	2	5	6	8
Machines	A	1	1			1			
	E				1				1
	C		1	1			1	1	
	F				1				1
	D			1	1		1	1	
	B	1	1			1			

Solution

Step 1. We assign column 8 place value 1, column 7 place value 2, column 6 place value 4, and so on. Row A receives a value of $128 + 64 + 8 = 200$ for its 1s in the first, second, and fifth columns. Evaluating all rows produces the values shown.

	Part									
	M_{ij}	1	3	4	7	2	5	6	8	$\sum_{j=1}^{j=8} 2^{n-j} M_{i,j}$
Machines	A	1	1			1				200
	E				1				1	17
	C		1	1			1	1		102
	F				1				1	17
	D			1	1		1	1		54
	B	1	1			1				200
	$2^{(n-j)}$	2^7	2^6	2^5	2^4	2^3	2^2	2^1	2^0	

Step 2. Rank the row in order of decreasing decimal weight values. The rows are reordered to A, B, C, D, E, and F.

	Part									
	M_{ij}	1	3	4	7	2	5	6	8	$\sum_{j=1}^{j=8} 2^{n-j} M_{i,j}$
Machines	A	1	1			1				200
	B	1	1			1				200
	C		1	1			1	1		102
	D			1	1		1	1		54
	E				1				1	17
	F				1				1	17
	$2^{(n-j)}$	2^7	2^6	2^5	2^4	2^3	2^2	2^1	2^0	

Step 3. Repeat steps 1 for each column by assigning the value to each row with bottom row value 1, 2nd bottom row 2, and so on. This produces the following result:

	Products									
	M_{ij}	1	3	4	7	2	5	6	8	$2^{(m-j)}$
Machines	A	1	1			1				2^5
	B	1	1			1				2^4
	C		1	1			1	1		2^3
	D			1	1		1	1		2^2
	E				1				1	2^1
	F				1				1	2^0
	$\sum_{i=1}^{i=6} 2^{m-i} M_{i,j}$	48	56	12	7	48	12	12	3	

In addition, the column is reordered in decreasing values from left to right as 3, 1, 2, 4, 5, 6, 7, and 8.

Step 4. Repeat steps 1, 2, and 3, and finally the result is given by:

	Products									
	M_{ij}	3	1	2	4	5	6	7	8	$2^{(m-j)}$
Machines	A	1	1	1						2^5
	B	1	1	1						2^4
	C	1			1	1	1			2^3
	D				1	1	1	1		2^2
	E							1	1	2^1
	F							1	1	2^0
	$\sum_{i=1}^{i=6} 2^{m-i} M_{i,j}$	56	48	48	12	12	12	7	3	

Next, repeat Step 1 for further ordering. However, the row ordering for the repeat step is unchanged and we stop.

The outcome of ROC depends on the given incidence matrix. ROC may run into some issues in grouping products and machines as follows:

1. It is not uncommon for the iterative steps in ROC clustering to lead to an oscillation; it can be solved by introducing duplicated machines.
2. When the finished clusters include outliers and/or voids, an outlier can be addressed by a machine replication and no action is required for a void since the corresponding machine can just skip the product in a cell.
3. A clustering algorithm is generally needed to convert existing routes into machine groups for planning and scheduling of productions. It requires an engineering study prior to developing a series of routers on a core sample of products that represent most of the possible production scenarios in machine shops.

7.5 Cellular Manufacturing

GT is applied to divide manufacturing facilities into groups or cells of machines. Once the machine cells are formed, *cellular manufacturing* is used to design, plan, schedule, and control manufacturing processes in machine cells. Each of these cells is dedicated to a specified family or set of products. Therefore, once products are grouped into the families by GT, the next challenge is to arrange the machines in the factory by CM. If the machines are poorly located, manufacturing costs may increase significantly.

Cellular manufacturing design can be referred to as a *cell formation problem* (CFP). The goal to solve a CFP is to group products into families based on the processing requirements, with heterogeneous machines grouped into machine cells. Subsequently, designate product families to machine cells. A CFP can be solved in three ways: (i) product families are first determined and the machines are clustered into cells based on the requirements of manufacturing processes of product families. (ii) The machine cells are recognized and product families are allocated to machine cells sequentially. (iii) The product families and machine cells are developed concurrently. CFP is a non-deterministic polynomial (NP) design issue.

The *composite product approach* was proposed to design cellular manufacturing. A composite product is an actual or imagined product with merged primitives of all products

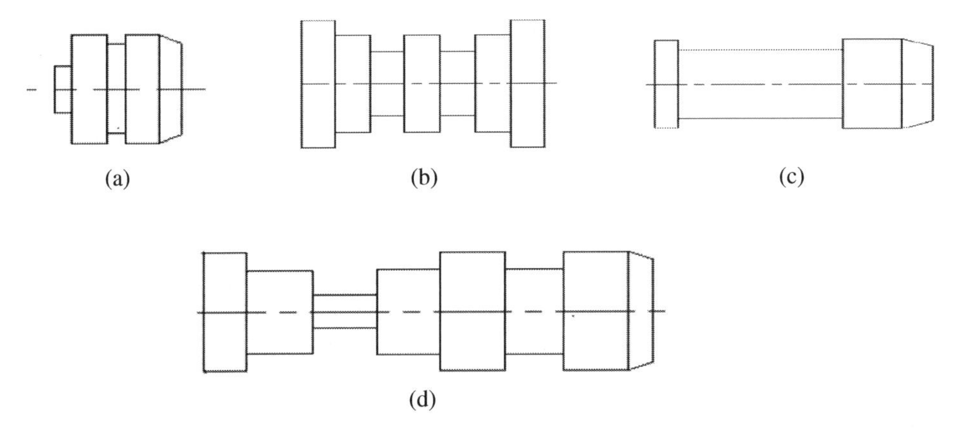

Figure 7.30 Merging product features for a composite product. (a) Instance 1 (Chamber, slot, left shoulder). (b) Instance 2 (Slot, left shoulder, right shoulder). (c) Instance 3 (Face, chamber, left shoulder). (d) Composite product (Chamber, slot, left shoulder, right shoulder, right face).

in a product family. Cellular manufacturing is designed to ensure that the composite product can be completely processed in the machine cell. If a new product is assigned to the machines in a cell, the dissimilarities of the product from the hypothetical composite product must be minimized; ideally, the composite product possesses all of the manufacturing features of a new product. Therefore, manufacturing resources can be planned based on the processing needs of the composite product.

Figure 7.30 shows an example of merging the manufacturing features of individual products as features of a hypothetical composite product. The features from three product instances (Figure 7.30a, b, and c) are face, chamber, left shoulder, right shoulder, and slot. The composite product with all of these features is shown in Figure 7.32d.

The composite product helps to select machine tools and auxiliary tools in a CM machine cell. As shown in Figure 7.31, the processing variables of products in one family have the same allowable ranges; each product in one family requires machines and tools available in a CM machine cell. In addition, raw materials for products should be reasonably consistent; the auxiliary tools such as fixtures should be flexible to support all manufacturing processes required by the composite product. It is ideal to standardize machine setups to minimize changeover costs from one product to another.

GT and CM can be used together to design the layout of a manufacturing system, especially when the product flow analysis is used in GT. By integrating GT and CM, the machines are clustered to form machine cells, each cell consisting of all machines/tools needed to make any product in the family. CM design follows a logical sequence of steps; each step can be formulated into a multiobjective optimization sub-problem where the designers have to make tradeoffs between sets of conflicting requirements subjected to specified technical limitations. To optimize CM, designers must possess a deep and profound knowledge of products, manufacturing processes, and the interactions in machine workcells as well as the elements of a workcell, their functions, and their interactions. Strategosinc (2019) shows the flowchart of some main tasks involved in the design of CM in Table 7.15, which consists of the identification of product families, the design of manufacturing processes, and the layout design.

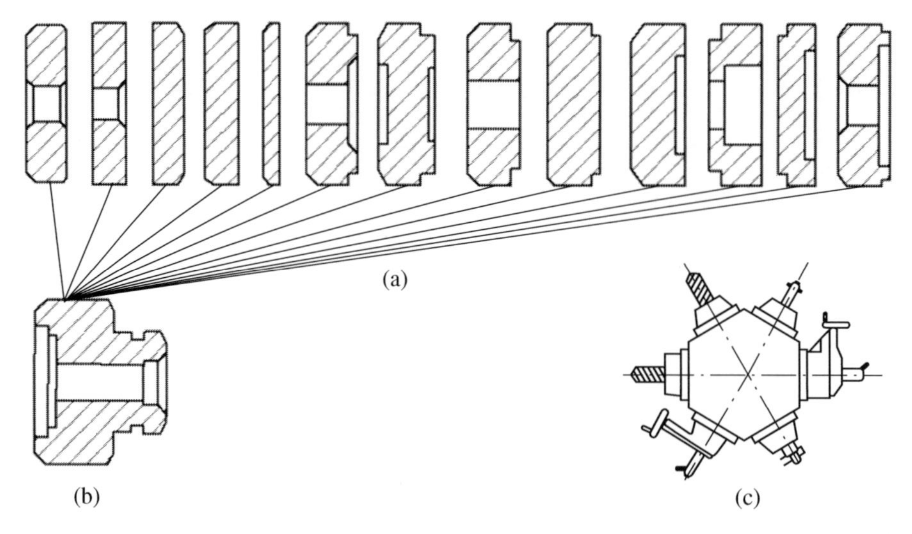

Figure 7.31 Composite product for the determination of machines and tools. (a) Product variants in a family. (b) Composite product with all manufacturing features of the product family. (c) Toolset for machining of all products/features for a family.

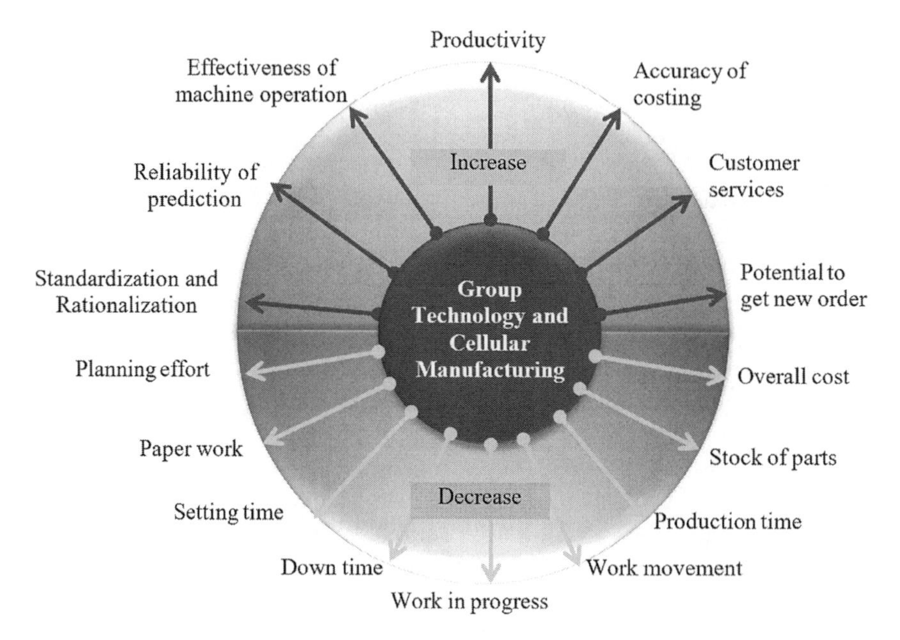

Figure 7.32 Potential benefits of implementing GT and CM.

Table 7.15 Main tasks in design of CM (Strategosinc 2019).

<table>
<tr><th>Step</th><th>Flowchart and main tasks</th></tr>
<tr><td>Step 1. Identification of product families. This step aims to find product families, where each family corresponds to a group of machines that can be processed without changeovers or other setups. The more products are analysed, the more challenges the identification process will be. The GT methods such as the process mapping method can be used in the identification process.
The following basic questions should be answered in the solution: (i) How are all product variants classified into groups? (ii) What will be the utilization rates of machines? (iii) Is there a need to have reserve capacity?</td><td>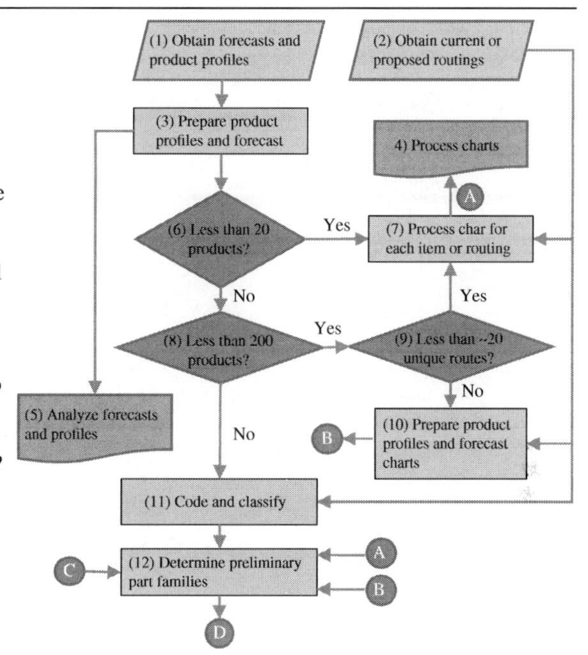
</td></tr>
<tr><td>Step 2. Determination of manufacturing resources (machines, tools, and labours) for identified families.
For each product, the required manufacturing processes are analysed to identify the needs of machines, tools, and labour. The capacities of machine cells are calculated based on the requirements of product variants at these aspects.
The following basic questions should be answered for each product: (i) What process steps are needed to make the product? (ii) What is the best sequence of steps? (iii) What machine tools are needed? (iv) Is any labour assigned to the operation or programming? (v) What lot size is appropriate and for each machine tool, any or how many replications are appropriate?</td><td>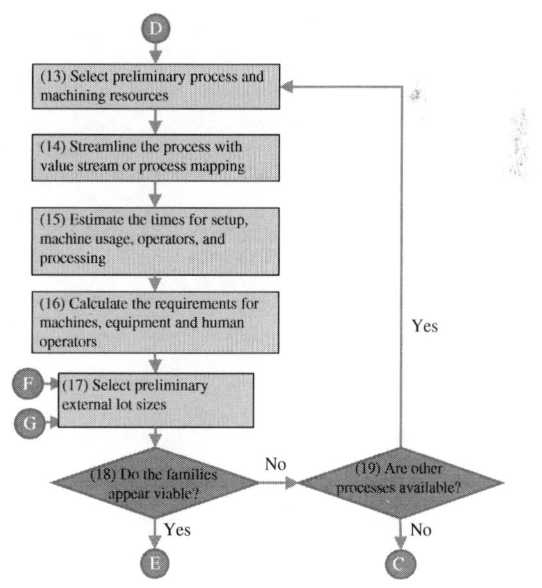
</td></tr>
</table>

(*continued*)

Table 7.15 (Continued)

Step	Flowchart and main tasks
Step 3. Detailed design of machine workcells. An infrastructure of CM for a collection of machine workcells is defined with consideration of indirect manufacturing resources such as containers, scheduling, and balance methods. Note that the infrastructure is intangible but lacking in awareness of infrastructure may cause the design fail of CM workcells. The following basic questions should be answered in defining an infrastructure: (i) What are the means to deal with material handling from cell to cell? (ii) What methods are used to balance the workloads of machines? (iii) What methods are used to plan and schedule production? (iv) Would it be work-in-process workpieces? How to deal with inventories if needed? (v) What methods are used to ensure quality? (vi) What methods are used to motivate operators in the system?	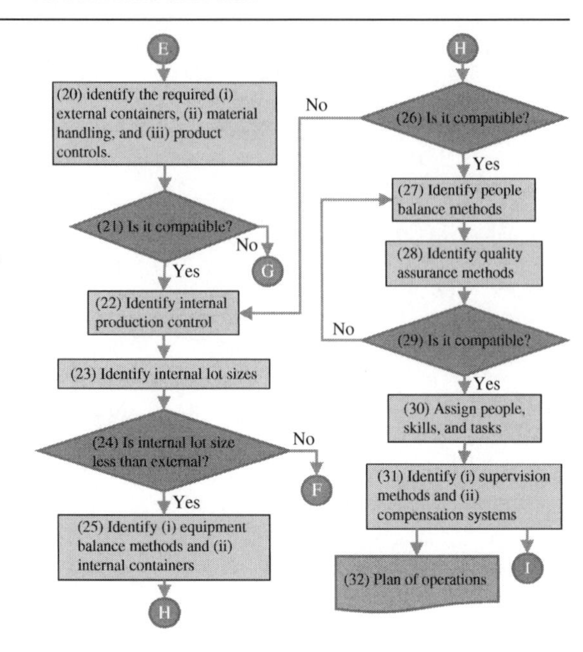
Step 4. Design of workcell layout. Layout design is straightforward when all tasks in the previous three steps are accomplished thoroughly. CM layout includes: (i) *inter-cell layouts* for the arrangement of machines to minimize the inter-cell movement of parts and (ii) *intra-cell layouts* for the arrangement of cells to minimize the transportation overall cost of CM. The design of inter-cell layouts can begin with process charts for product families. Note that basic layout types of CM have been discussed in Section 7.3.5. Three basic questions in layout design are: (i) How can the best arrangement of machines and tools be achieved for optimized overall system performance? (ii) What mechanisms are used to deal with external constraints? (iii) How can network layouts be integrated?	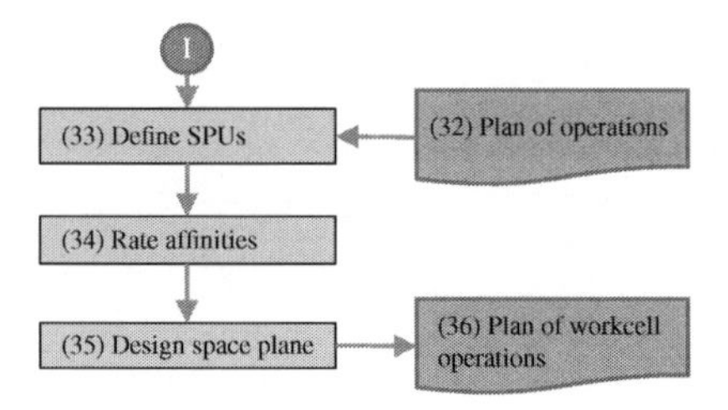

Table 7.16 Statistic survey of benefits of CMs (Hyer and Wemmerlov 2001).

	Wemmerlov and Johnson (1997) – 46 Firms			Wemmerlov and Hyer (1989) – 32 Firms		
Performance measure	**Average (%)**	**Minimum (%)**	**Maximum (%)**	**Average (%)**	**Minimum (%)**	**Maximum (%)**
Reduction of move distances/move times	61.3	15.0	99.0	39.3	10.0	83.0
Reduction in throughout time	61.2	12.5	99.5	45.6	5.0	90.0
Reduction of lead time	50.1	0.0	93.2	—	—	—
Reduction of work in progress (WIP)	48.2	10.0	99.7	41.4	8.0	80.0
Reduction of setup times	44.2	0.0	96.6	32.0	2.0	95.0
Reduction of finished goods inventory	39.3	0.0	100.00	29.2	10.0	75.0
Improvement in part and product quality	28.4	0.0	62.5	29.6	5.0	90.0
Reduction of unit costs	16.0	0.0	60.0	—	—	—

Table 7.15 provides a generic procedure and tasks to be accomplished for GT and CM. The practical implementation takes into consideration the types, varieties, and volumes of products as well as available manufacturing resources in enterprises. The effectiveness of GT and CM has been proved in many case studies. For example, Hyer and Wemmerlov (2001) conducted a survey showing the benefits for companies to use CM in their businesses and Table 7.16 shows the statistics for performance improvement of systems on reductions for transportation, throughput, lead time, WIP, setup times, product unit cost, and improvement in product quality.

7.6 Summary

A manufacturing system consists of many different manufacturing resources such as machines, tools, fixtures, and human operators. The system layout should be designed to maximize the utilization of manufacturing resources and reduce waste caused by no-value-added activities such as transportation and idle states of machines or tools. A system layout must be designed appropriately to tailor to manufacturing needs of products in enterprises. Conventional system layouts such as job shops, flow shops, project shops, and continuous production are efficient only in certain circumstances, such as a flow shop layout for mass production. In this chapter, the main factors and procedures of layout designs are discussed, different layout types and applicable business conditions are introduced, and the main factors and the importance of using GT and CM in today's manufacturing environment is outlined.

Small and medium sized enterprises (SMEs) usually deal with a wide scope of product varieties with limited volume. A high flexibility to deal with the varieties, changes, and uncertainty becomes crucial to the manufacturing systems at SMEs. GT and CM turn into effective methods to support layout designs for SMEs. Designers should be familiar with the approaches and tools for the implementation of GT and CM. Note that the system layout is for all manufacturing resources in the material flow. It would affect the operations of a manufacturing system at any level and domain, from the lowest level for machine elements to the highest level for strategic partnerships with virtual enterprise over the Internet. The influence of GT and CM to SMEs is radical. Figure 7.32 gives a summary of potential advantages for SMEs by implementing GT and CM in system design, operation, and continuous improvement.

7.7 Design Problems

7.1 Use Matlab or other languages to program the ranking of clustering (ROC) algorithm and use the program to find the machine groups for the following two machine-part incidence matrices. Alternatively, if you have limited programming skills, use ROC by hand to find the solution of GT.

(a) Case 1

Machine	**Part number**					
	1	**2**	**3**	**4**	**5**	**6**
M1		1		1		1
M2		1		1		1
M3	1		1		1	
M4	1		1		1	

(b) Case 2

Machine ID	**Part number**					
	1	**2**	**3**	**4**	**5**	**6**
A			1		1	
B		1	1			
C	1			1		
D		1	1		1	
E	1			1		1

7.2 Consider a six-part-and-nine-machine problem shown in the following table. Use ROC to form the part family and machine group.

	Product						
	M_{ij}	1	2	3	4	5	6
Machine	A	1					1
	B		1			1	
	C		1	1		1	
	D	1			1		
	E	1			1		1
	F			1		1	
	G	1		1	1		1
	H		1			1	
	I	1			1		

7.3 Consider a 20-part-and-10-machine problem shown in the following table. Use ROC to form the part family and machine group.

	Product																				
	M_{ij}	1	2	3	4	5	6	7	8	9	10	11	12	13	14	15	16	17	18	19	20
Machine	A					1		1				1							1		
	B		1			1						1					1		1		
	C		1			1		1				1									
	D			1							1				1						
	E										1				1						1
	F								1				1				1			1	
	G				1								1				1			1	
	H	1			1				1								1			1	
	I						1			1		1		1		1					
	J						1			1						1		1			

7.4 Create a composite product for a product family with the instances shown in Figure 7.33.

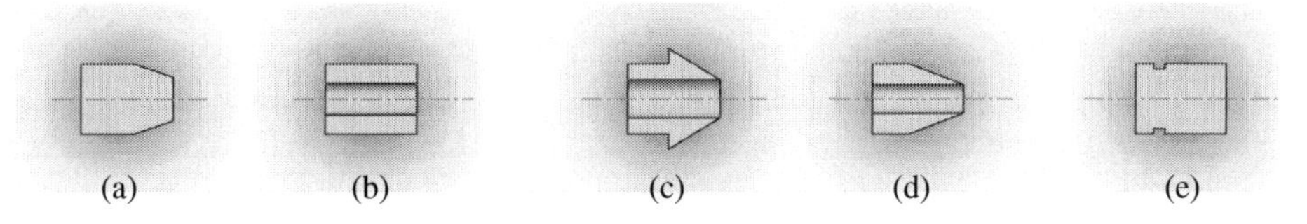

Figure 7.33 Product variants for Design Problem 7.4. (a) Chamber. (b) Through hole. (c) Through hole, step, and chamber. (d) Through hole and chamber. (e) Slot.

References

AQT Solution (2016). Spotlight on Hyundai: the biggest shipbuilding company in the world. http://www.aqtsolutions.com/the-biggest-shipbuilding-company-in-the-world.

Askin, R.G. and Standridge, C.R. (1993). *Modeling & Analysis of Manufacturing Systems.* Wiley. ISBN: 0-417-51418-7.

Bi, Z.M. and Zhang, W.J. (2001). Flexible fixture design and automation: review, issues, and future directions. *International Journal of Production Research* 39 (13): 2867–2894.

Bi, Z.M., Lang, S.Y.T., and Shen, W.M. (2008). Reconfigurable manufacturing systems: the state of the art. *International Journal of Production Research* 46 (4): 967–992.

Black, J.T. (1991). *The Design of Factory with a Future.* McGraw-Hill. ISBN: 10: 0070055505.

BrainKart (2019). Flexible manufacturing systems (FMS) applications and benefits. https://www.brainkart.com/article/Flexible-Manufacturing-Systems-FMS--Applications-and-Benefits_6426.

CPV Manufacturing (2019). Advantages and disadvantages of flexible manufacturing systems. https://www.cpvmfg.com/news/advantages-disadvantages-flexible-manufacturing-system.

Groover, M.P. (2010). *Fundamentals of Modern Manufacturing – Materials, Processes, and Systems*, the Fourth Version. Wiley. ISBN: 978-0470-46700.

Haworth, E.A. (1968). Group technology using Opitz system. *Production Engineering* 47 (1): 25–35.

HIBA (2019). Operations management. https://www.slideshare.net/Joanmaines/process-and-layout-strategies.

Hyer, N. and Wemmerlov, U. (2001). *Reorganizing the Factory, Competing Through Cellular Manufacturing.* Portland, OR: Productivity Press ISBN-10: 1563272288.

Khan, N. (2013). Computer integrated manufacturing. https://www.slideshare.net/NoumanKhan2/9-oct-2013-lec-13-1415161718.

King, J.R. (1980). Machine–component grouping in production flow analysis: an approach using a rank order clustering algorithm. *International Journal of Production Research* 18 (2): 213–232.

Kirby, K. (2019). Focused factories and group technology. http://web.utk.edu/~kkirby/IE527/Ch9.pdf.

Knowledgiate Team (2017). Advantages and disadvantages of continuous production systems. https://www.knowledgiate.com/wp-content/cache/wp-rocket/http://www.knowledgiate.com/advantages-disadvantages-continuous-production-system/index.html_gzip.

Mahdabi, I., Shirazi, B., Solimanpur, M., and Ghobadi, S. (2007). Designing an e-based real time quality control information system for distributed manufacturing shops. *Proceedings of the Ninth International Conference on Enterprise Information Systems*, vol. EIS, Funchal, Madeira, Portugal (12–16 June 2007). DOI: https://doi.org/10.5220/0002356701590163.

MHI (2019). Material handling taxonomy. http://www.mhi.org/cicmhe/resources/taxonomy.

O'Sullivan, D. (2019). Flexible manufacturing system. http://www.nuigalway.ie/staff-sites/david_osullivan/documents/unit_15_flexible_manufacturing_systems.pdf.

Rahdarmanan, R. (2002). Virtual manufacturing: an emerging technology. https://peer.asee.org/virtual-manufacturing-an-emerging-technology.pdf.

Rauch, E. and Dallasega, P. (2017). Distributed manufacturing network models for smart and agile mini-factories. *International Journal of Agile Systems and Management* 10 (3/4): 185–205.

Song, Y.J. and Woo, J.H. (2013). New shipyard layout design for the preliminary phase & case study for the green field project. *International Journal of Naval Architecture and Ocean Engineering* 5: 132–146. http://dx.doi.org/10.2478/IJNAOE-2013-012.

Strategosinc (2019). Design of workcelland micro layouts – cellular manufacturing. http://www.strategosinc.com/celldesign.htm.

Sysdesign (2019). Manufacturing system design decomposition. http://sysdesign.org/msdd/webdefinitions.htm.

Weber, A. (2004). The pros and cons of assembly cells. https://www.assemblymag.com/articles/83136-the-pros-and-cons-of-cells.

Wemmerlov, U. and Hyer, N.L. (1989). Cellular Manufacturing in the U.S. industry: a survey of users. *International Journal of Production Research* 27 (9): 1511–1530.

Wemmerlov, U. and Johnson, D.J. (1997). Cellular manufacturing at 46 user plants: implementation experiences and performance improvements. *International Journal of Production Research* 35 (1): 29–49.

Wikipedia (2019). Fixure (tool). https://en.wikipedia.org/wiki/Fixture_(tool).

Zeng, B.C. (2009). Group technology (GT) in manufacturing. http://www.me.nchu.edu.tw/lab/CIM/www/courses/Computer%20Integrated%20Manufacturing.htm.

8

Computer Aided Fixture Design

8.1 Introduction

A manufacturing system transforms raw materials to final products through a series of manufacturing processes. Manufacturing companies are classified into process manufacturers and discrete manufacturers. A *process manufacturer* builds something that is not supposed to be taken apart; for example, liquid and powder products. A *discrete manufacturer* makes discrete goods such as screws, nuts, and handles that can be taken apart and used on something else when needed. Design variables and system parameters in process manufacturing and discrete manufacturing are quite different. Table 8.1 gives the comparison of the characteristics of these two types of manufacturing processes (Sophen 2019).

Continuous manufacturing and discrete manufacturing use some common engineering principles for technical innovation and new product development. For example, mass production is preferable to improve productivity and reduce cost in a sellers' market when the demands are larger than the supplies of products. The enterprises for either continuous manufacturing or discrete manufacturing use the stage-gate process to guide new product development, which consists of several design stages. There is a gate between to two stages; the move from one stage to the next must ensure all of the targeted goals are met.

From a technical or product lifecycle perspective, there are significant differences in fabrication processes and tooling. For example, a food manufacturer usually highlights technical complexities on ingredient management, recipes, marketing, labelling, and packages; sophisticated software tools are sufficient to manage the activities in these disciplines. On the other hand, an auto manufacturer highlights technical complexities of manufacturing techniques, assembling processes, planning, scheduling, and controlling of manufacturing processes. Both hardware and software tools are configurable in order to deal with changes and uncertainties occurring to business operations (Sophen 2019; Batka 2011).

In discrete manufacturing, each manufacturing process involves a number of design variables, which have to be determined in process planning before the process can be performed. The objective of *process planning* in a manufacturing enterprise is to generate the optimized sequence of necessary manufacturing processes in order to produce the expected products in a technologically feasible and economically effective way. Process planning is based on the information defined in a product model such as *geometrical features*, *material properties*, *tolerances* (accuracy), and *surface roughness*. The information of product design relevant to manufacturing processes are referred to as *manufacturing*

Computer Aided Design and Manufacturing, First Edition. Zhuming Bi and Xiaoqin Wang.
© 2020 John Wiley & Sons Ltd. This Work is a co-publication between John Wiley & Sons Ltd and ASME Press.
Companion website: www.wiley.com/go/bi/computer-aided-design

Table 8.1 Comparison of continuous manufacturing and discrete manufacturing.

	Continuous manufacturing	**Discrete manufacturing**
Definition	A manufacturing process of assembling pieces into a standardized solution that cannot be disassembled back into their original components	A manufacturing process of making distinct items that can be disassembled back into the original components
Examples	Sugar; Paint; Chemical materials; Medical pills	Machine elements; Tools; Computers; Car
Characteristics	• Variable ingredients for different products • Attribute driven • Units of materials are material specific • Use multiple recipes and formulas (requirement) • Need lot, grades, potency, shelf-life • Mixed, blends, transforms (operation) • Brand refresh and reframe are critical to get maintain consumer attention • Brand loyalty is critical • Portfolio management is primarily based on balancing line extensions with new to market • Makes 'stuff'	• Standard parts, components • Part number driven • Units of materials (UOMs) are each and piece • Use complex multilevel bills of materials (BOMs) • Needs the identities for classification • Builds, assembles, fabricates • Manufacturing process is assembly driven with higher R&D spending • Customers are looking for regular cadence of new product releases • The 'next new thing' rules • Portfolios planning is heavily dependent platform and technology planning and capital investments • Makes 'thing'

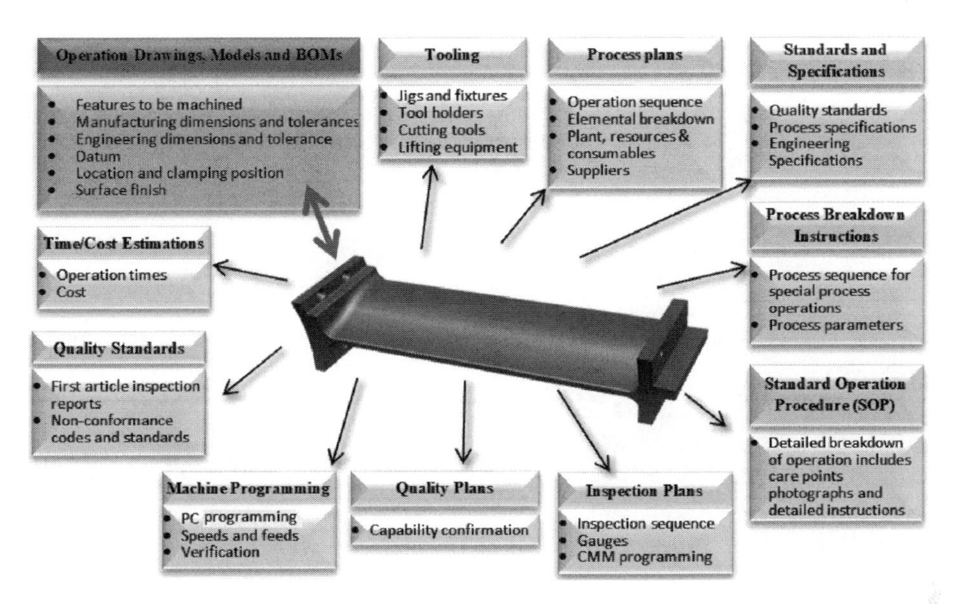

Figure 8.1 Typical tasks in process planning of discrete manufacturing.

features. Manufacturing features present the connection of computer aided design (CAD) and computer aided manufacturing (CAM) (Majstorovic et al. 2013). Process planning is essential to use physical products based on virtual product models. In process planning, the inputs are the design information of products such as dimensions, material properties, and tolerances. The outputs are the sequence of manufacturing processes and the selections or designs of machines and auxiliary tools for all of the required manufacturing processes from raw materials to finished products (Wakhare 2016).

Figure 8.1 shows the typical tasks and outcomes in computer aided processing planning (CAPP) (Buerer 2017; Wysk 2008). The top left on the operation drawing shows the models and bills of materials (BOMs), which give the inputs of CAPP, which will also be the part of the outputs of CAPP. Other typical outputs include *tooling*, *process plans*, *standards and specifications*, *process breakdown instructions*, *standard operation procedure* (SoP), *inspection plans*, *quality plans*, *machine programming*, *quality standards*, and *time and cost estimations*. While it is unnecessary to cover the applications of computer aided techniques for all of the aforementioned aspects, computer aided machining programming will be discussed in Chapter 9, while computer aided fixture design (CAFD) for tooling will be discussed in details in this chapter.

8.2 Fixtures in Processes of Discrete Manufacturing

In discrete manufacturing, product geometries are usually shaped through a series of manufacturing processes subjected to intensive mechanical or thermal loads. Fixtures are essential tools to position and secure workpieces precisely when they perform manufacturing operations (Bi and Zhang 2001; Wang et al. 2010; Ivanov et al. 2018). In a sharping process, mechanical or other-form external forces are large enough to cause the separation of

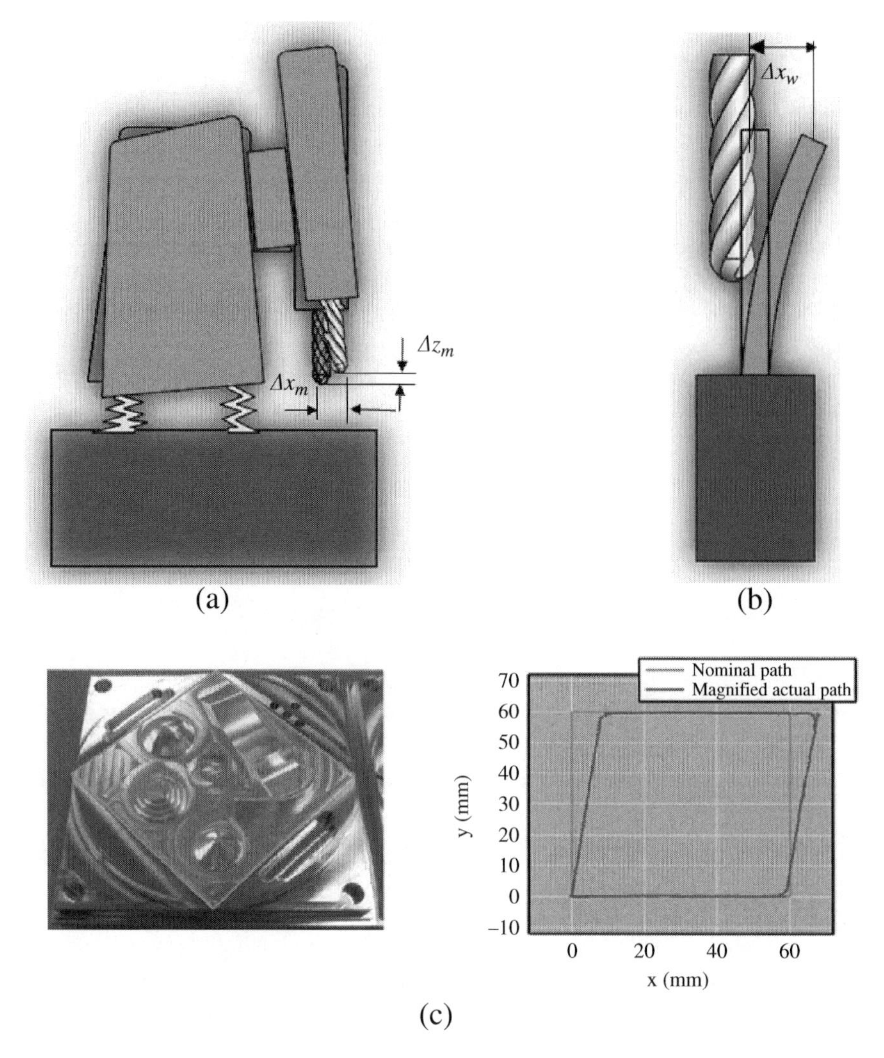

Figure 8.2 Discrepancy of actual and planned paths caused by the cutting force. (a) Machine deformation. (b) Workpiece deformation. (c) Total error occurring to the workpiece.

unwanted materials from workpieces and considerable deformations over the workpiece. Figure 8.2a and b shows examples of exaggerated deformation over the machine tool and workpiece caused by the cutting forces. This could cause a discrepancy between actual and planned cutting paths, and eventually dimensional and orientation errors of workpieces.

The discrepancy of actual and planned cutting paths relates to many factors, such as the material properties of machine components and workpieces, processing parameters, and, more importantly, the fixtures to position and secure workpieces. In discrete manufacturing, fixtures have a direct impact on product quality, productivity, and manufacturing cost. The costs associated with the design and fabrication of fixtures have been estimated to be

10–20% of the total manufacturing cost. Moreover, approximately 40% of the rejected products were due to dimensional errors attributed to poor fixturing solutions (Bi and Zhang 2001). A great deal of attention has been paid to the study of CAFD and many achievements have been reported on the development of flexible fixturing systems and software tools for CAFD (Wang et al. 2010).

8.3 Fixtures and Jigs

A *jig* is a special type of fixture that provides the guidance of machining operations other than securing the workpiece. Since both a fixture and a jig are functioned to secure workpieces, a jig is easily confused with a fixture. The primary purpose of a jig is to control the location and motion of the tool during the operations to ensure the required repeatability, accuracy, and interchangeability. Therefore, a device that does both holding the work and guiding a tool is called a jig.

Figure 8.3 shows a jig used to hold the workpiece and guide the drilling operation simultaneously. The jig has drill guides for horizontal and vertical holes. The position and orientation of a drill tool is guided by these components. Jigs and fixtures are commonly used to hold the workpieces in manufacturing processes, but they have some significant differences, which are summarized in Table 8.2 (MechanicalBooster 2019).

The main difference between a jig and a fixture is in the guidance of machining operations. In a jig design, the guide element is usually integrated with a clamping mechanism to simplify the fixture design. Figure 8.4 gives two main types of jig solutions, i.e. leaf jigs and box jigs (MechanicalBooster 2019).

In this chapter, we emphasize the design and implementation of work holding. Therefore, it is not critical to distinguish fixtures and jigs from this perspective.

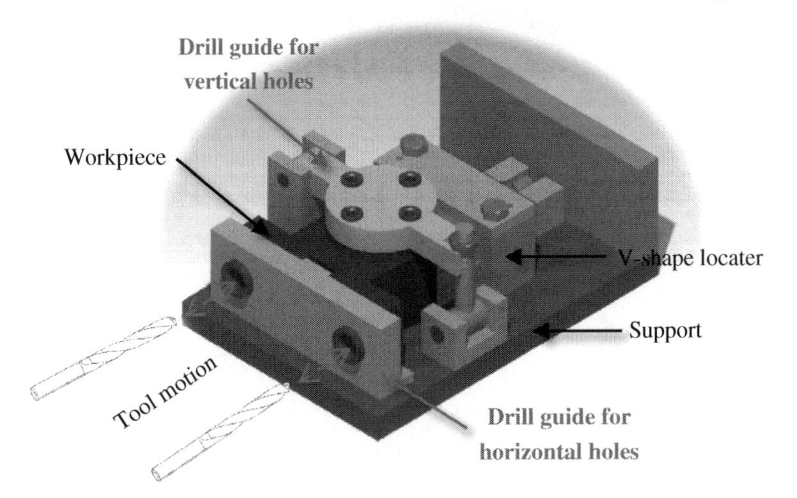

Figure 8.3 Drill jig example.

Table 8.2 Difference of jigs and fixtures.

	Jigs	Fixtures
1.	It is a work-holding device that holds, supports, and locates the workpiece and *guides the cutting tool* in an operation.	It is a work-holding device that holds, supports, and locates the workpiece for a specific operation, but does *not guide the cutting tool.*
2.	Jigs are *light* in weight.	Fixtures are *rigid and bulky.*
3.	Gauge blocks are *not necessary.*	Gauge blocks may be needed for effective handling.
4.	The jigs are special tools particularly used in *unidimensional machining* such as drilling, reaming, tapping, and boring operation.	Fixtures are specific tools used particularly in *multidimensional machining* such as milling machine and shapers and slotting machine.
5.	Usually it is *not fixed* to the machine table.	It is *fixed* to the machine table.
6.	Its *cost is more.*	Its *cost is less* as compared with the jig.
7.	Their designing is *complex.*	Their designing is less complex.

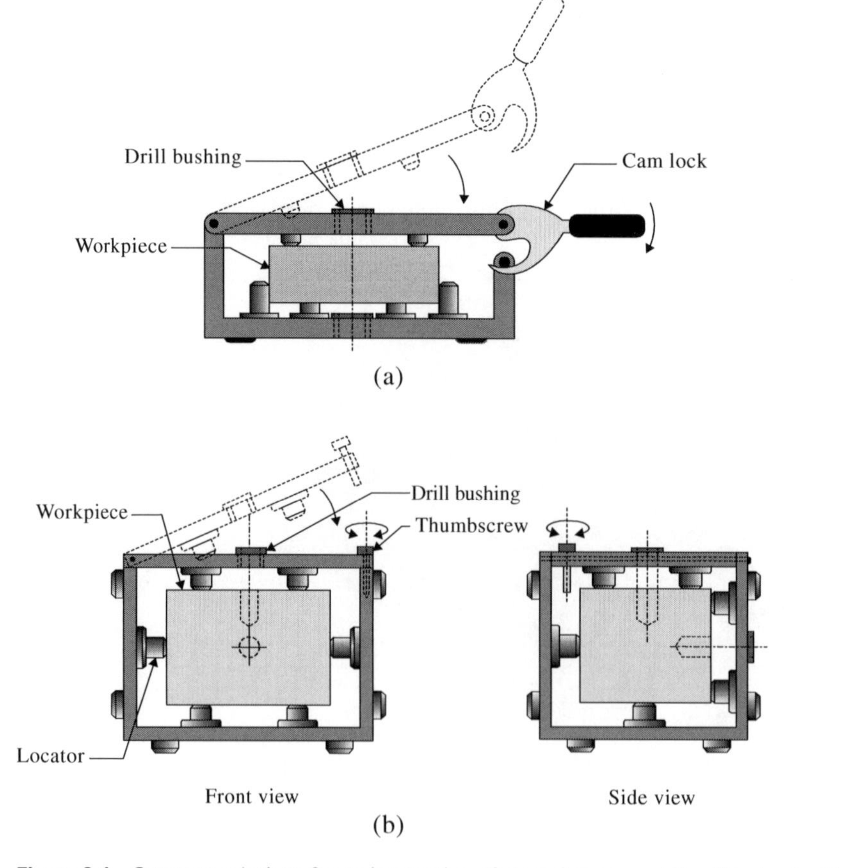

Figure 8.4 Common solutions for an integration of clamping and guiding (MechanicalBooster 2019). (a) Leaf jig. (b) Box jig.

8.4 Functional Requirements (FRs) of Fixtures

A *fixture* is used to maintain the stability of a workpiece during a machining process. An operational fixture has to satisfy the following three basic requirements to perform its functions fully as a work-holding device:

1. *Positioning* to accurately position and orient a workpiece relative to the cutting tool,
2. *Supporting* to increase the stiffness of compliant regions of a workpiece, and
3. *Clamping* to secure the workpiece in its desired location rigidly.

The fixture is used to fix all possible motions with respect to the cutting tool in a given coordinate system. As shown in Figure 8.5, if no constraint applies to an object in a three-dimensional space, the object may have a total of 12 possible directions of motion along 6 degrees of freedom. To secure a workpiece in a machining process, the fixture must be designed to eliminate a rigid motion along any of 12 possible motion directions subjected to cutting forces. To enable fixturing, a fixture system usually consists of three main types of components, including *base supports*, *locators*, and *clampers*, as shown in Figure 8.6.

Figure 8.5 Degree of freedoms (DoF) of a solid body.

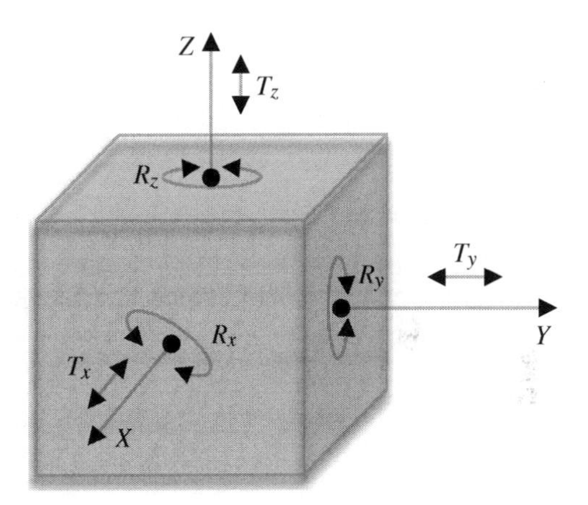

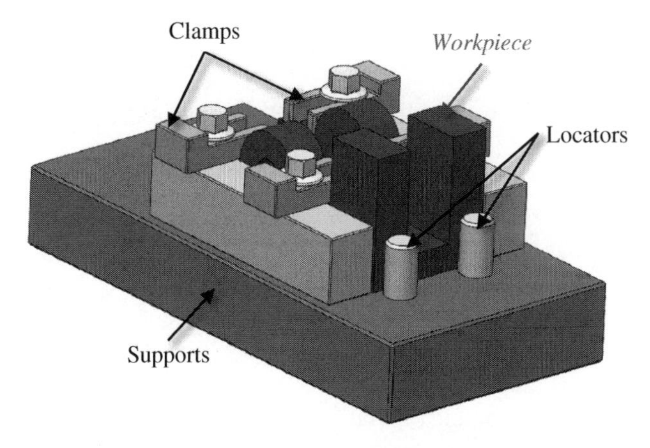

Figure 8.6 Three types of main components in a fixture system.

Table 8.3 Other quantifiable evaluation criteria in fixture design.

Stiffness:	**is the capability of a fixture element to remain in its original shape within a specified range of deflection subjected to given external loads.**
Accuracy:	is the closeness of actual and ideal positions of the locators in the given coordinate frame.
Repeatability:	is the closeness of the locations from the repeated operations of locators in fixturing workpieces.
Flexibility:	is the capability of reconfiguring fixtures for different products and different manufacturing processes.
Reconfiguration time:	is the total amount of time required for reconfiguration.
Design and deployment time:	is the total amount of time spent on design and installation of the fixture solution.
Capital cost:	is the investment cost required for implementation of a fixture design.

To meet the above functional requirements (FRs), the following three design constraints should be taken into consideration in fixture design (Basha and Salunke 2013).

1. *Geometric constraints* that guarantee all elements of a fixture are placed on datum surfaces. There is no interference between any fixture components and a cutting tool during a machining process. In addition, other desirable characteristics for a fixture are quick loading and unloading, minimized number of components, accessibility, supports of multiple cutting operations, and low cost.
2. *Deflection limit* occurs to workpieces since the deformation over a workpiece adversely affects the accuracy of machining processes. Due to elastic or plastic properties of workpiece materials, clamping force, cutting forces, and the cutting tool motion relative to the machine, it is unavoidable to eliminate the deformation of a workpiece completely. However, the deformation has to be controlled within the acceptable range to achieve tolerance specifications.
3. *Deterministic location* of a workpiece is constrained by the locators of the fixture. The dimensional accuracy of finished products relates not only to the deflection in machining processes but also to the fact that the locating errors are due to locators or locating surfaces. The workpiece should be accurately positioned to achieve better dimensional accuracy of products in the machine coordinate frame.

In designing, evaluating, and comparing fixture options, other quantified design criteria are included in Table 8.3 (Erdem 2017).

8.5 Fundamentals of Fixture Design

Figure 8.6 has shown that a fixture consists of only three main types of elements that have relatively independent functionalities. Therefore, the axiomatic design theory (ADT) can be used to decompose a fixture design problem into the sub-problems of designing the fixture elements for positioning, supporting, and clamping (Bi and Zhang 2001). Sequentially, the

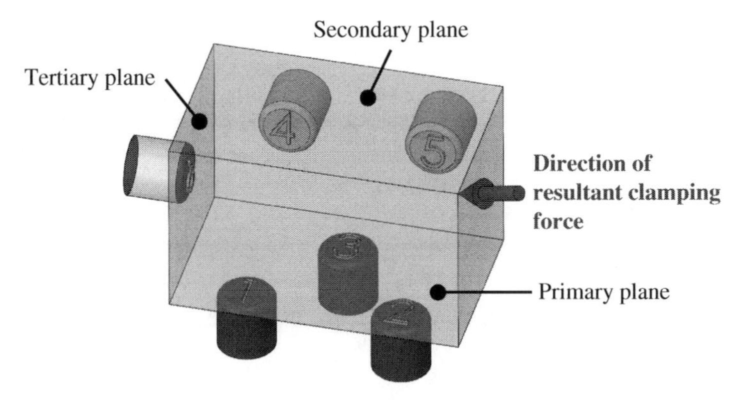

Figure 8.7 The 3-2-1 principle in fixture design.

fundamental design principles for fixturing elements can be converted into the axioms used in the design process (Singh 2019).

8.5.1 3-2-1 Principle

As shown in Figure 8.5, a free object in a three-dimensional space has 6 degrees of freedom (DoF) with 12 directions of possible motions. To immobilize the object, the fixture must provide the constraints for all of 6 DoF. Conventionally, it is achieved by *the* 3-2-1 *principle*. In the 3-2-1 principle, three mutually perpendicular planes are used to confine an object. These planes are called the *primary plane*, *secondary plane*, and *tertiary plane*, respectively. The primary plane confines 3 DoF and the secondary and tertiary planes confine 2 and 1 DoF of the object, respectively. As shown in Figure 8.7, three planes are represented by 3, 2, and 1 locators on the corresponding plane, respectively. In addition, the primary and secondary planes should be selected to mainly counteract external machining forces and clamping forces.

The 3-2-1 principle is widely used to design dedicated fixtures for prismatic parts. The '3' in the principle means that three passive fixture elements are used as the locators on the primary plane, which is used as the primary datum surface. As shown in Figure 8.8, the locators on the primary plane eliminate 3 DoF and the motion along five directions for the object. As shown in Figure 8.9, the locators on the secondary and tertiary planes eliminate 2 DoF and 1 DoF and the motion along 3 and 1 directions for the object, respectively. Finally, a clamping force towards the negative directions of the primary, secondary, and tertiary planes confines the motion along the last three directions, as highlighted in Figure 8.9b.

8.5.2 Axioms for Geometric Control

The 3-2-1 principle can be interpreted as a set of axioms for the geometric control of workpiece as:

1. Only six locators are necessary to completely locate a rigid prismatic workpiece. More locators are redundant and may give rise to uncertainty.
2. One locator removes one DoF.

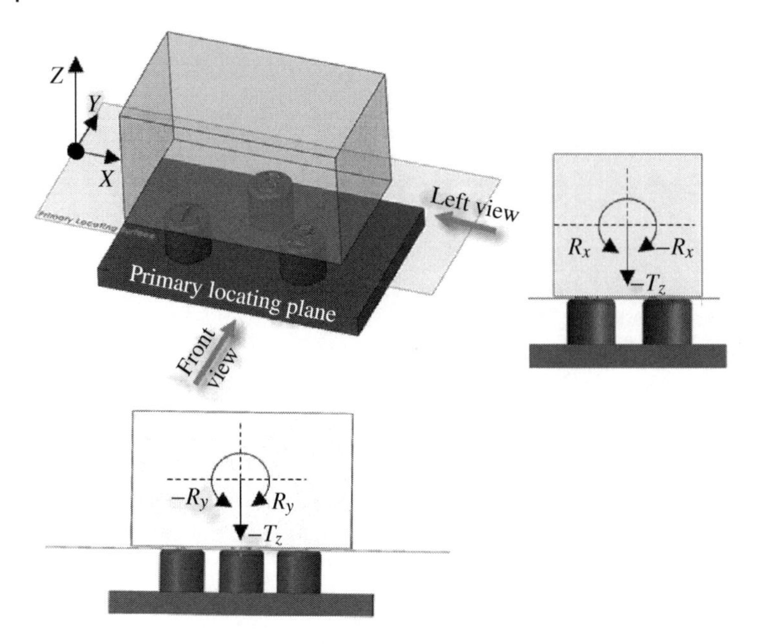

Figure 8.8 Three locators on the primary plane eliminate 3 DoF and the motion along five directions.

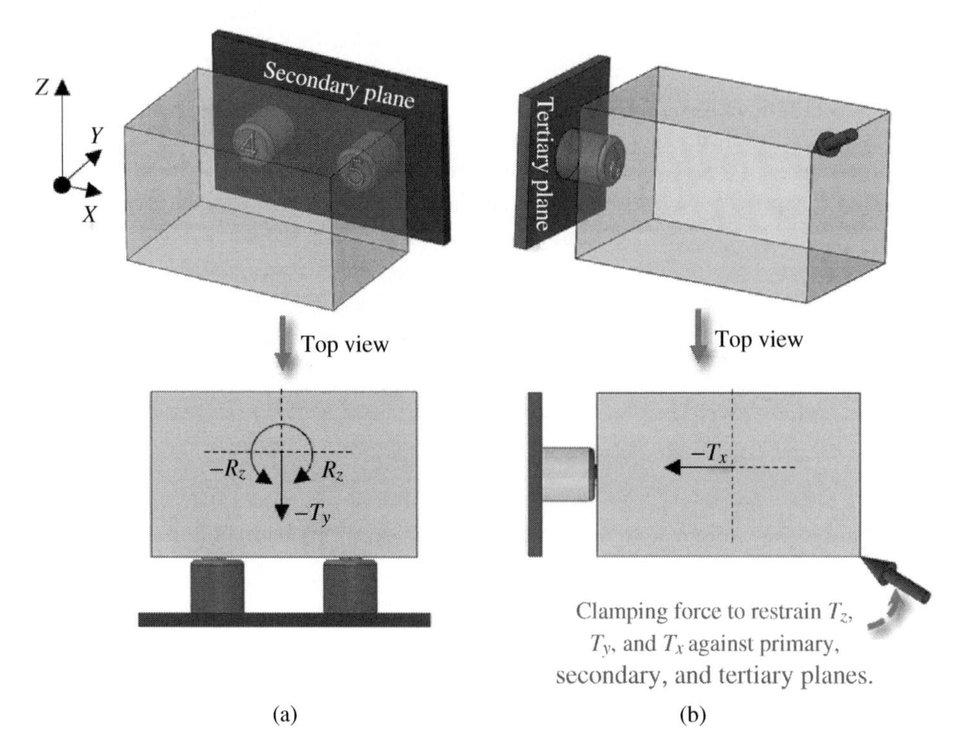

Figure 8.9 The secondary and tertiary planes remove 2 and 1 DoF on the object, respectively. (a) Two locators on the secondary plane. (b) One locator on the tertiary plane and clamping force.

3. Two directions of each DoF have to be restrained to immobilize an object completely.
4. Three locators define a primary plane, two locators for a secondary plane, and the last one on a tertiary plane.
5. Without the consideration of deformation by self-gravity, six locators should be placed as widely as possible to stabilize the object.
6. For the best possible accuracy, locators should contact the workpiece on its most accurate surfaces.

Note that although point contact would yield the best positioning accuracy, most locators have a planar contact surface to minimize the damage to the workpiece due to potentially high-pressure contact points.

8.5.3 Axioms for Dimensional Control

To achieve the best possible accuracy of the dimensions via fixture design, designers have to follow the axioms below.

1. To prevent tolerance stacks, a locator should be placed on one of the reference surfaces from where the workpiece is dimensioned.
2. When two planar surfaces are used to define a geometrical tolerance such as *parallelism* or *perpendicularity*, the primary plane must be selected as the reference planar surface, which is defined by three locators, as shown in Figure 8.10.
3. When an axiom for geometric control conflicts with that for dimensional control, the precedence is given to the axiom for dimensional control.
4. To locate the centreline of a cylindrical surface, the locators must straddle the centreline.
5. If possible, locators should be placed on machined surfaces to achieve better dimensional accuracy.

8.5.4 Axioms for Mechanical Control

Taking into consideration the deformations of workpiece and fixture elements on the dimensional accuracy of machining processes, designers need to follow the axioms below.

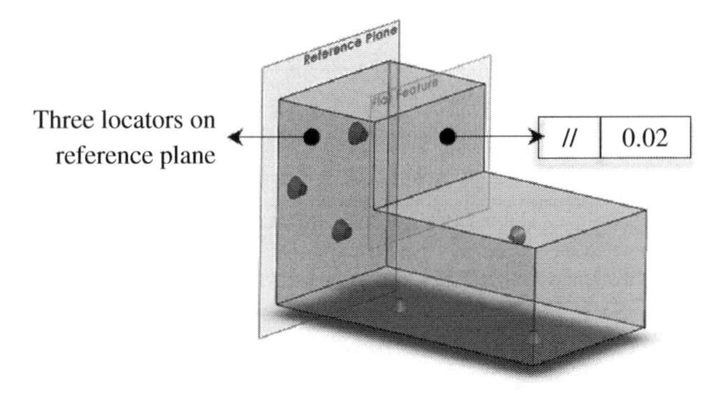

Figure 8.10 Example of using the primary plane for geometric tolerance.

1. Place a locator directly opposite the cutting force to minimize the deformation on the locator.
2. Place a locator directly opposite the clamping force to minimize the deformation on the locator.
3. If an external force cannot react directly from any locator, place fixed supports (no locators) opposite to the force to minimize the deflection on the locators.
4. Avoid an initial contact of a fixed support with the workpiece.
5. Ensure that a clamping or holding force is applied towards the locators.
6. The moment of a clamping force about any possible rotational axes must be sufficient to overcome that caused by the cutting force.
7. The cutting force should be in the direction that will force the workpiece to remain in contact with the locators.

8.5.5 Fixturing Cylindrical Workpiece

Many machined parts have symmetric geometric features, which lead to ambiguity in selecting locators. One typical example is a cylindrical object for turning operations. The 3-2-1 principle for prismatic objects may be revised to fix a cylindrical object.

Since a cylindrical object is axially symmetric, only the primary and secondary planes are needed to fix the object. Figure 8.11 shows a common fixture setup where an end plenary surface perpendicular to the main motion axis is selected as the primary plane to constrain one translational and two rotational DoF along the motion directions of T_z, R_x, and R_y. A V-block makes the contact with the lateral cylindrical surface to constrain two translational DoF along the motion along two directions of T_x and T_y. The rotation along the Z-axis in not critical due to its axially symmetric nature.

8.5.6 Kinematic and Dynamic Analysis

Conventional approaches to fixture design and planning rely heavily on the thumb rules, axioms, and prior experiences. A valid fixture design often needs the trial and error

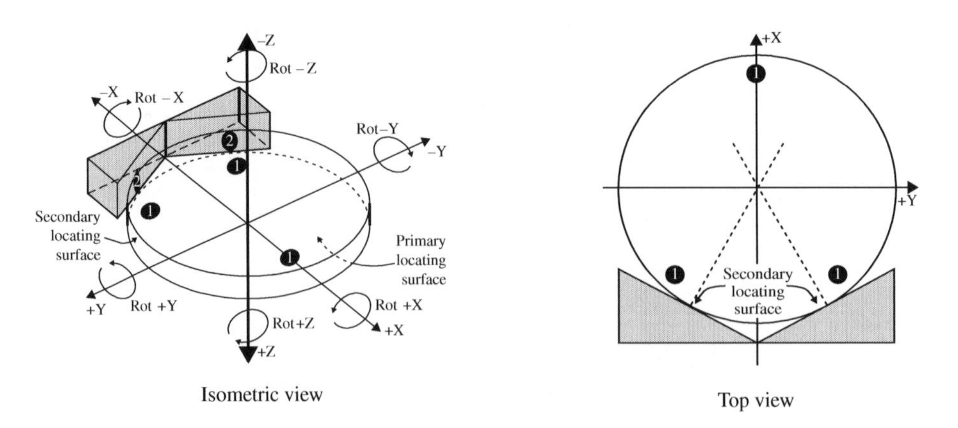

Figure 8.11 Common fixture setup for a cylindrical object.

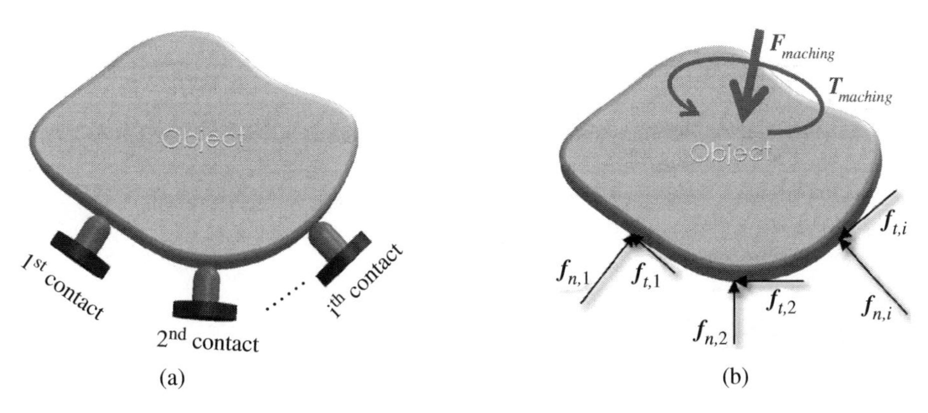

Figure 8.12 The object restricted by a number of contacts and the free-body diagram (FBD). (a) Workpiece by contacts. (b) Force analysis of an object subjected to external load.

iterations. It is desirable to use more scientific methods in fixture analysis and design. *Kinematic and dynamic modelling* for mechanical structures can be an alternative.

In kinematic and dynamic analysis, a fixture is modelled as a set of *rigid contacts* that are fixed in a reference frame. The workpiece is modelled as *a rigid body* whose motion is completely restricted by the contacts of the fixture elements. Figure 8.12a shows the case where an object is restricted by a number of contacts from the fixture elements. Figure 8.12b shows the free-body diagram (FBD) of the object when it receives the machining operation. All of the forces, including external cutting forces, the reactional forces at the contacts, and the clamping force, must meet the force-closure condition for immobilization of the object.

Kinematic and dynamic analysis aims to determine possible motions and deformation of the object that is constrained by fixture elements. A possible motion refers to an instantaneous rigid motion and a possible deformation refers to a displacement that an elastic object undergoes when sufficient constraints are applied to restrain the object by fixture elements. The instantaneous motion of an object is affected by (i) the shape and number of contacts, (ii) the relative locations and orientations of the contacts, and (iii) the friction forces at the contacts.

In regards to contacts, fixture elements involve three common contact conditions: (i) *point contacts* (e.g. point-on-plane, plane-on-point, or line-on-line), (ii) *line contacts* (e.g. line-on-plane or plane-on-line), and (iii) *planar contacts* (e.g. plane-on-plane). A contact only removes possible translational motion(s) at a point, line, or plane. However, the reactional force at the contact will induce the friction force that may restrain the motion along the tangent direction of the contact.

A fixture design can be evaluated by the criteria for *form closure* and *force closure*. To meet the condition of form closure, the contacts with the object must resist any external forces/moments occurring to the machining process, so that the object is form closed or, equivalently, the fixture provides the form closure to the object. In other words, the set of contacts provides the form closure of an object if these contacts eliminate all DoF of the object purely based on the geometrical placement of the contacts. To meet the condition of force closure, the object maintains the expected contacts to fixture elements subjected to external forces and moments. In practice, most of the fixtures for machining processes are force closure since they rely on the frictional forces to immobilize an object completely.

In summary, a kinematic and dynamic modelling method is used to evaluate whether a fixture design meets the necessary and sufficient conditions: (i) no movement subjected to the constraints from fixture elements, (ii) ease of loading/unloading of workpieces for good accessibility and detachability, and (iii) appropriate positioning and minimized deformations for better dimensional accuracy.

8.6 Types and Elements of Fixture Systems

Figure 8.6 has shown that the fixture elements are classified as *locators*, *clampers*, and *supports*. A locator is used to position an object, a clamper is used to apply a restrained force on an object, and a support is mainly a base plate for the moment plate of a machine tool on which all locators and clamping mechanisms are attached.

Under fixturing conditions where the conditions for both form closure and force closure are satisfied, the numbers and sizes of fixture elements should be minimized so that more surfaces of the object can be exposed to receive machining operations. In such a way, the number of fixture setups to perform a series of manufacturing processes can be reduced. However, as the number of locators and clamps decrease, the required force per unit area increases. This may cause damage to fixture elements or the object. Therefore, depending on the complexity of the object, the numbers and locations of locators and clamps may be increased.

Figure 8.13 shows the classification of basic fixture elements. A *tool body* is the base frame on which other fixture elements are mounted. A tool body contributes to the main mass

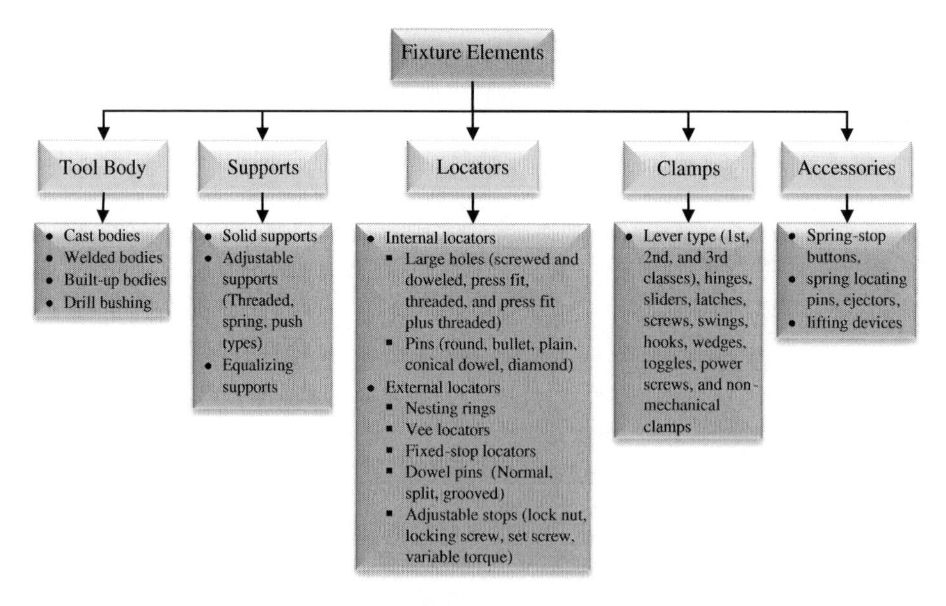

Figure 8.13 Typical elements used in a fixturing system (Keyvani 2008).

of a fixture system and the materials for the body should have high strengths. Different techniques such as welding, casting, assembling, and drill bushing can be applied to build tool bodies. *Supports* make direct contacts to the workpiece and support workpieces to endure external forces. Supports can be fixed or adjustable. *Locators* are used for positioning the workpiece in the course of machining operations. Locators and supports work together to restrain products in the right position and orientation. *Clampers* are used to provide additional restrained forces to ensure that the workpiece is kept in the fixture safely in the manufacturing processes. There are other types of fixture elements such as spring-stop buttons, spring locating pins, ejectors and lifting devices for loading/uploading, and the flexibility for reconfiguration. Each type of fixture element in Figure 8.13 consists of a set of sub-groups who have similar functionalities for supporting, positioning, and clamping.

8.6.1 Supports

A support usually has a flat surface that can be used as a primary or secondary plane. Supports are used in rough machining where the external force on an object is large, and the corresponding machining process is to remove unwanted materials at a high material removal rate. Figure 8.14a and b shows the supports to a horizontal surface or inclined surface, respectively Figure 8.14c supports an object on a vertical plane. A set of parallels in Figure 8.14d supports an object with stepped heights. Vee blocks in Figure 8.14e supports an object with a round surface by line contacts. The pins in Figure 8.14f provide point supports to an object.

8.6.2 Types of Fixture Systems

Figure 8.15 gives a classification of the types of fixturing systems. Based on the component types and lifecycle times of fixturing systems, fixtures can be classified as (i) *general-purpose fixtures* such as mechanical vises and lathe chucks, shown in Figure 8.16, (ii) *flexible fixtures* whose components are modularized or adjustable for different tasks, such as modular

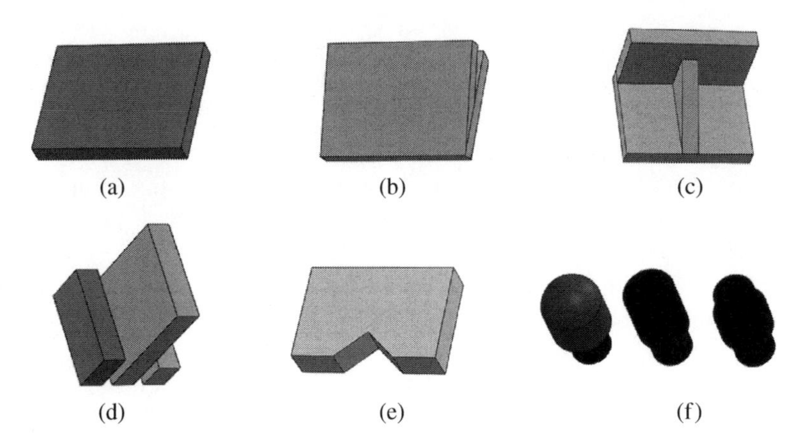

Figure 8.14 Common component supports. (a) Sub plate. (b) Sine plate. (c) Right angle plate. (d) Parallels. (e) Vee blocks. (f) Pins with spherical cap, round cap, and shoulder.

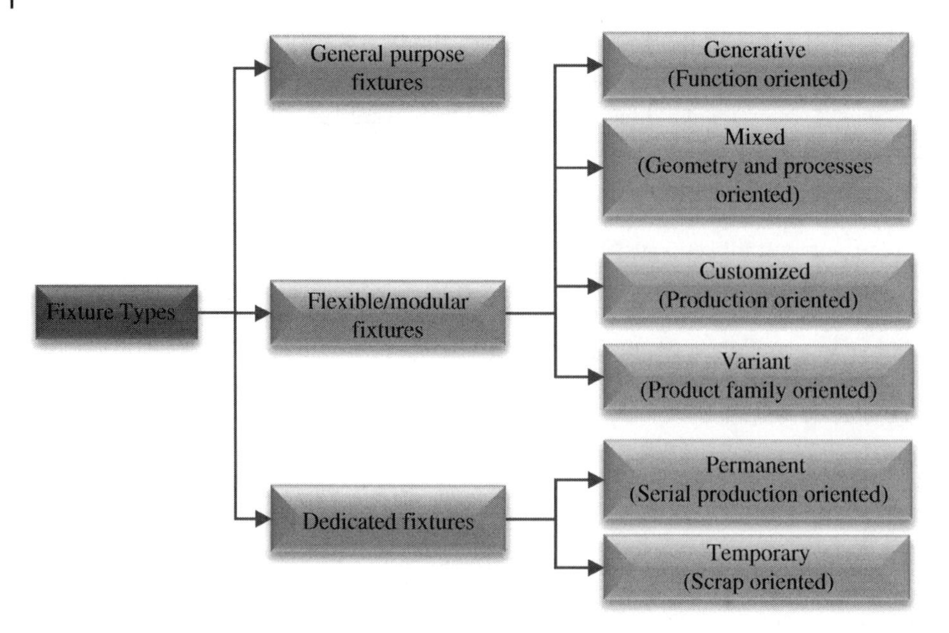

Figure 8.15 Fixture types based on generality of uses.

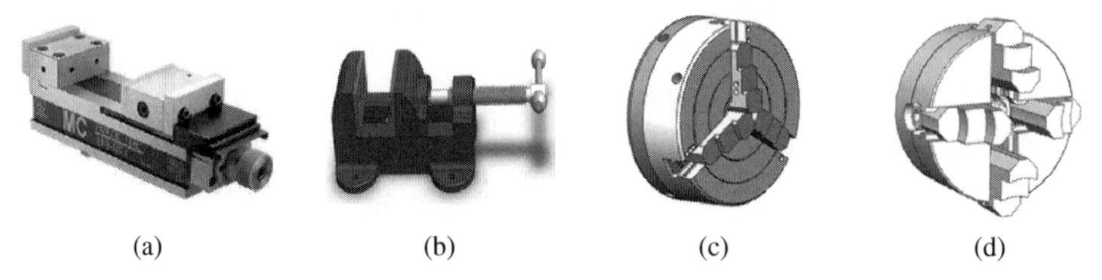

Figure 8.16 Examples of general-purpose fixtures. (a) Vise. (b) Versatile vise. (c) Three-jaw chuck. (d) Four-jaw chuck.

fixtures, pin array, and phase change, which will be discussed in Section 8.6.5, and (iii) *dedicated fixtures* that work only for given product types and a limited number of manufacturing operations (Figure 8.17).

General-purpose fixtures are often used together with general-purpose machines in job shops; fixture elements are versatile for different objects. General-purpose fixtures are applicable only to objects with simple geometries and where the productivity is low. Under light machining loads, work-holding mechanisms can be simplified. Vises in Figure 8.16a and b have limited accuracy due to finite rigidity and the clearances among moving members. Chucks in Figure 8.16c and d are applicable to cylindrical objects. They have multiple jaws that move radially to locate and clamp objects simultaneously.

Dedicated fixtures are usually used in mass production, where the costs on fixtures are shared by a high quantity of products and dedicated fixtures lack the flexibility to deal with different product types. Flexible fixtures are popular in cellular manufacturing (CM) for product families with low to medium quantities. Flexible fixture systems can be

Figure 8.17 Examples of dedicated fixtures. (a) Tombstone fixture. (b) Milling fixture. (c) Custom fixture.

reconfigured by selecting different elements and assembling them in different ways for a wide scope of product variants. Depending on the strategies to modularize systems, flexible fixture systems can be *functional oriented*, *geometric and process oriented*, *production oriented*, or *product family oriented*.

8.6.3 Locators

A locator makes direct contact with the object to be fixed. The smaller the contact area, the less dimensional uncertainty there is for the object. However, a locator also provides support to the object, and it deforms under the contact force. The smaller the contact area, the larger is the deformation. In determining the size and shape of a locator, a designer has to balance the dimensional errors from two sources: the positioning error and the error caused by the deformation at the contact.

Figure 8.18 shows the geometries of six common locators. The locator in Figure 8.18a has a round cap at the contact and a small contact area for better positioning accuracy when the operational force on the object is small. The locator in Figure 8.18b has a flat

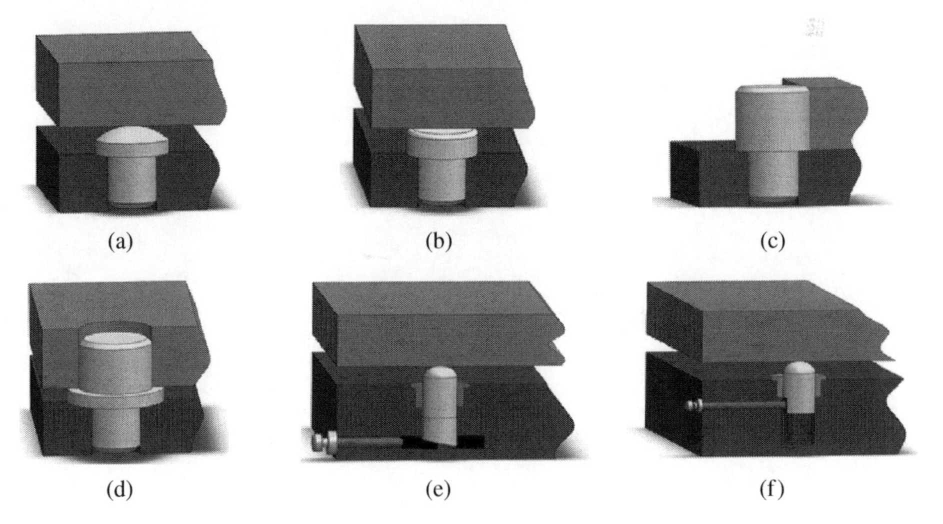

Figure 8.18 Fixture types based on generality of uses. (a) Round locator. (b) Flat locator. (c) Radical locator. (d) Concentric locator. (e) Adjustable locator – pushing. (f) Adjustable locator – spring.

surface at the contact in order to minimize the deformation when the operational force on the object is large. The locators in Figure 8.18c and d indicate the positioning on lateral flat and round surfaces, respectively; line contacts are involved. Figure 8.18e and f both integrate an adjustable mechanism in a locator; the height at the contact can be adjusted by a wedge or spring to make a contact at different heights.

8.6.4 Clamps

A clamp makes the contact to an object at the opposite side of locators or supports. A clamp is a mechanism that applies a clamping force to immobilize the object in machining processes. The criteria of designing a clamp are to (i) minimize the footprint for a large accessible space, (ii) optimize the contact location with a minimized clamping force, and (iii) avoid damaging the object. Figure 8.19 shows some common mechanisms to generate mechanical forces for a clamping purpose (Hazen 1990). As shown in Figure 8.19e and f, fixture elements can be actuated for both clamping and locating functions. Special jaws can be inserted that conform to irregular shapes of the objects. Figure 8.19e is a commonly used solution to prismatic objects on computer numerical control (CNC) machines. The vise consists of two halves, one that is fixed and the other towards the fixed half. When the vise jaws have a shoulder and one additional stop, all degrees of freedom are eliminated.

8.6.5 Flexible Fixtures

Either general-purpose fixtures or dedicated fixtures have significant limitations as the solutions to a wide scope of products variants with small or medium quantities. For small and medium size enterprises (SMEs), fixture systems should be reconfigurable to serve different products with minimal setup times. In other words, the same set of hardware devices can

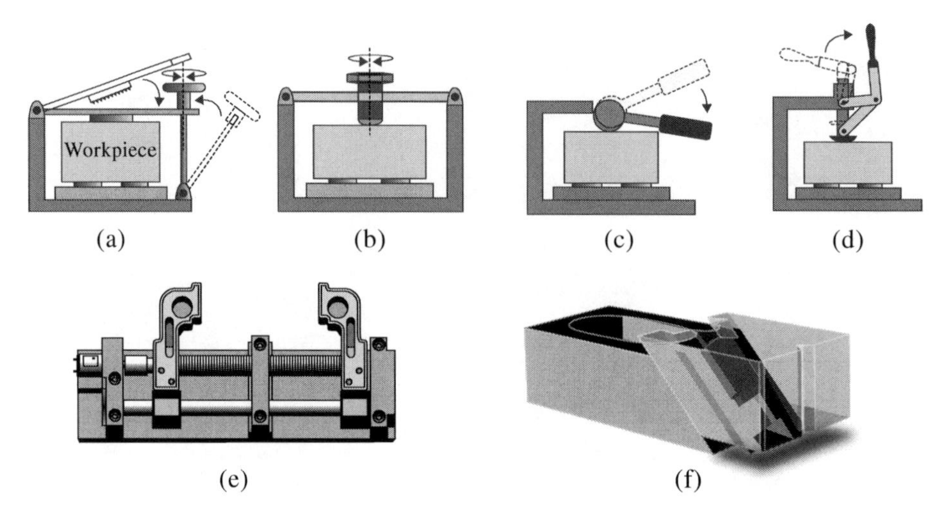

Figure 8.19 Examples of clamping mechanisms. (a) Strap clamp. (b) Screw clamp. (c) Eccentric clamp. (d) Toggle clamp. (e) Actuated parallel clamp. (f) Toe clamp.

be used for different tasks. System flexibility can be achieved in two basic ways: (i) using adjustable components whose controllable variables can be changed to adapt changes and (ii) modularizing system architecture so that the system consists of standardized modules, where different modules can be selected and assembled in different ways to generate various system configurations for different tasks. In addition, the system flexibility can be pursued in both hardware and control aspects (Bi and Zhang 2001).

As shown in Figure 8.20, the evolution of work-holding techniques is mainly driven by the needs for high-level flexibility and automation. Early innovations on fixtures occurred to new elements, strategies, and techniques for locating and holding. The innovations, such as *adjustable fixtures*, *modular fixtures*, *phase-change work-holding*, *conformable fixtures*, and *fixtureless operations*, increased the system flexibility in the sense that the same hardware system could be used for product families in a variety of machining processes (Bakker et al. 2019). Fixtures are flexible due to the improved compliance to complex geometries of objects. Recent innovations have been extended to the control systems where fixture elements are instrumented. The reconfiguration, setup, and operation of a fixture system can be automated. The fixture system is *smart*, either to monitor a machining process (open-looped) or to respond to task changes (closed-looped) by reconfiguration (Gameros et al. 2017).

8.6.5.1 Adjustable Fixtures

An adjustable fixture is still an integral fixture; however, a number of components in the fixture are adjustable to deal with different tasks. Figure 8.21 shows an example of adjustable fixtures developed by Walczyk et al. (1999). It served as a reconfigurable discrete die for large sheet metal parts and consisted of an array of pins, each pin being equipped with a hydraulic actuator to adjust the vertical position. Once all of the pins were set, the entire matrix was clamped as a work-holding device. Chan and Lin (1996) developed a flexible gripper, shown in Figure 8.22, which had a group of standardized elements to locate, support, and clamp planar objects. Each element has four fingers and eight DoF to make the contacts to any arbitrary convex surface. The fingers were controlled by one motor with two mechanical clutching systems. In Figure 8.23, Ivanov et al. (2018) introduced some ideas to make a clutch adjustable to deal with geometric variants.

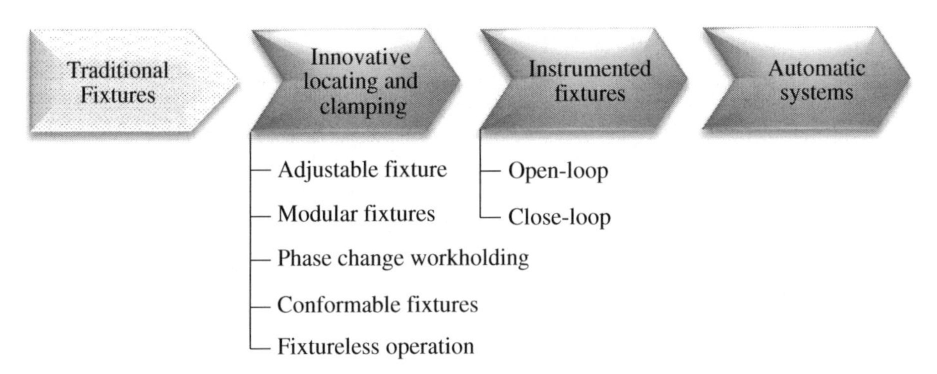

Figure 8.20 Evolution of work-holding techniques (Gameros et al. 2017).

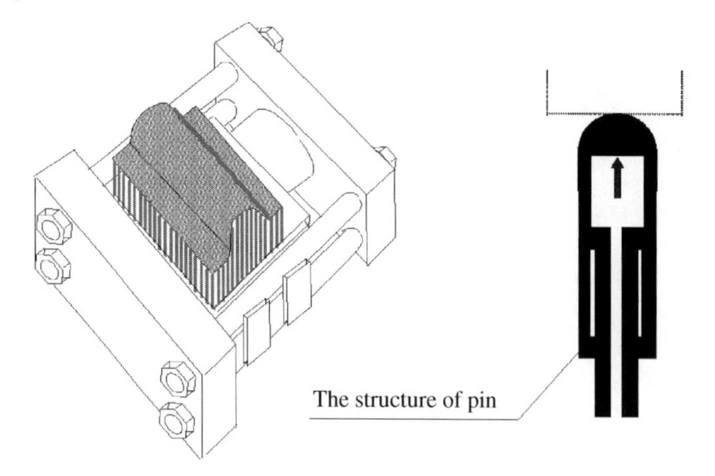

Figure 8.21 Adjustable work-holding by Walczyk et al. (1999).

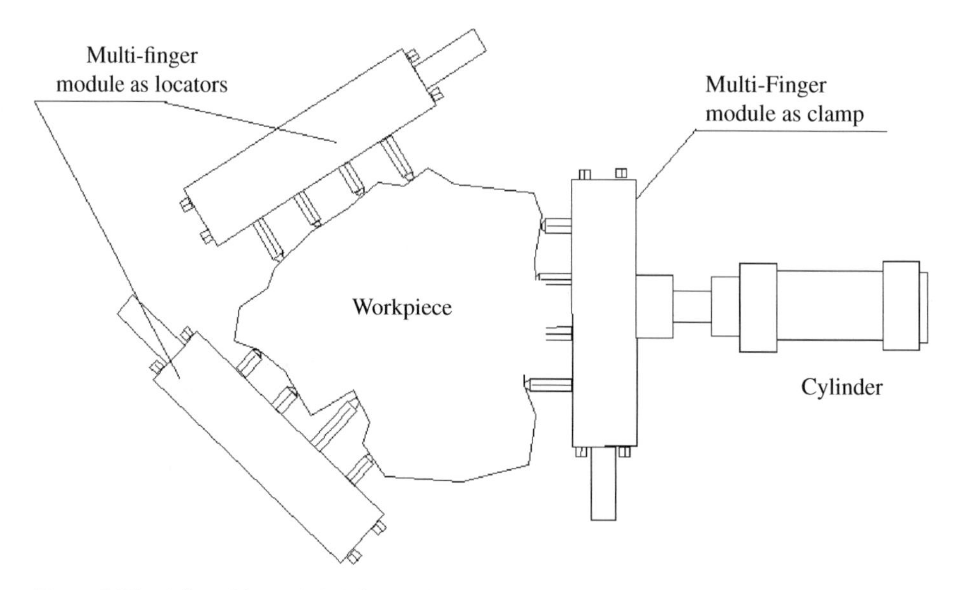

Figure 8.22 Adjustable work-holding by Chan and Lin (1996).

8.6.5.2 Modular Fixtures

The idea of modularity is correlated to the concept of interchangeability conceived during World War II. Modular fixtures began to be used in numerical control (NC) machines from the late 1960s, and their applications expanded with the advent of multiple-axis CNC machines and Flexible Manufacturing Systems (FMSs). Figure 8.24 shows two examples of modular fixtures. It is seen that a modular fixture system consists of many standard modules. These modules fall into four groups: base plate, locators, clamps, and connections. Flexibility is achieved by choosing different modules and assembling them in different ways.

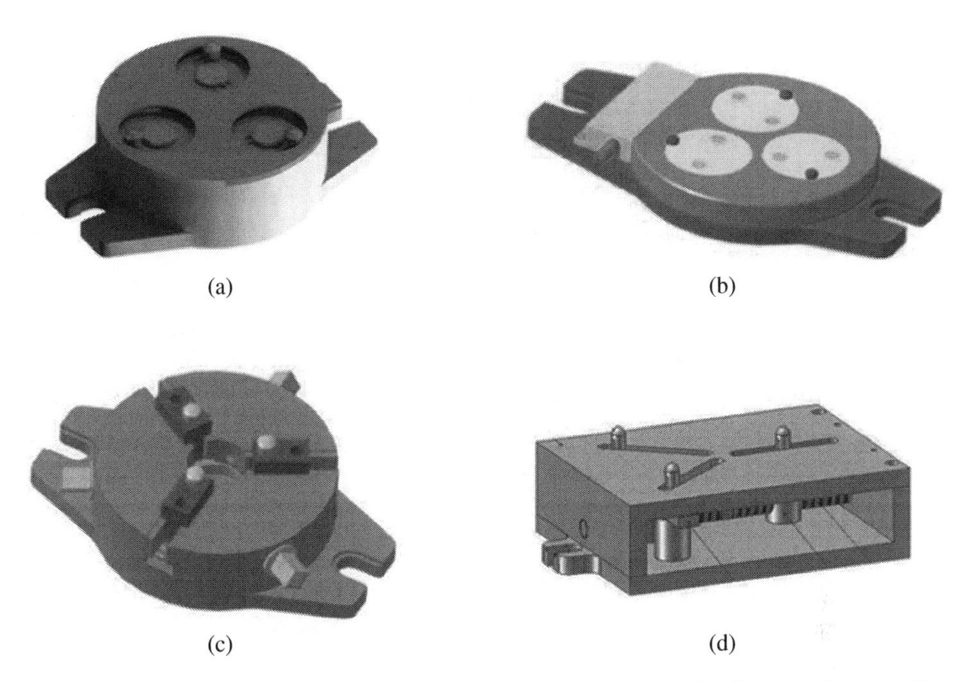

Figure 8.23 Examples of adjustable fixtures. (a) Circular adjustment I. (b) Circular adjustment II. (c) Line adjustment I. (d) Line adjustment II.

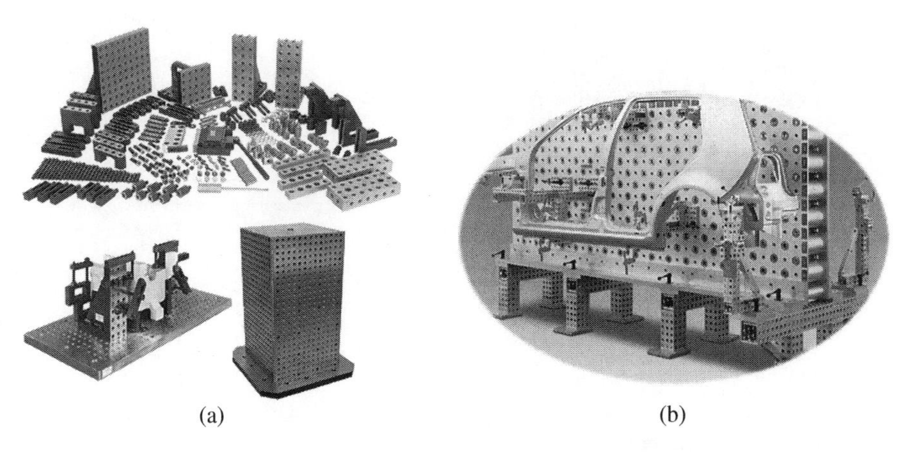

Figure 8.24 Two examples of modular fixtures. (a) Modular fixture example 1: http://www.stevenseng.com/images/stevens_home_page2.jpg. (b) Modular fixture example 2: https://media.horst-witte.com/bilder/produkte/system/sandwichplatten/vorrich2.jpg.

System architecture is critical when developing a modular system. To fix different objects, the positions and orientations of contacts by locators, supports, and clamps must be adjustable. The system must be modularized to enable the required adjustments. Figure 8.25 shows three modularized architectures that locate and clamp different objects, which are adjusted by T-slots, dowel-pin connections, and the inserts in grid holes, respectively.

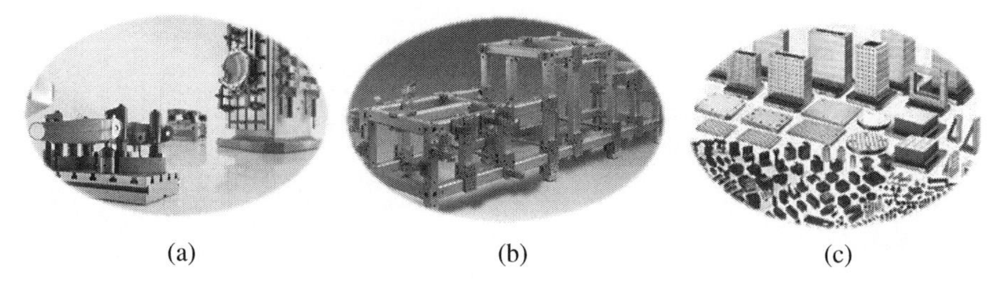

Figure 8.25 Types of locating techniques in modular fixtures (Bi and Zhang 2001; Li 2009). (a) T-Slot or Tenon-Slot system: (i) Erwin Halder Modular Jig and Fixture System, USA, (ii) Warlton Unitool, UK, (iii) CATIC (China National Aeronautical Technology Import and Export Corporation) System, China, and (iv) Gridmaster System, UK (Tenon Slot). (b) Dowel pin system: (i) Bluco Technik, Germany, (ii) SAFE (Self Adapting Fixture Element) System, USA, and (iii) Write Alufix System, Germany. (c) Grid hole system: (i) Venlic Block Jig System, IMAO Corporation, Japan, (ii) Yuasa Modular Flex System, USA, and (iii) Kipp Modular Flexible Fixturing System, Germany.

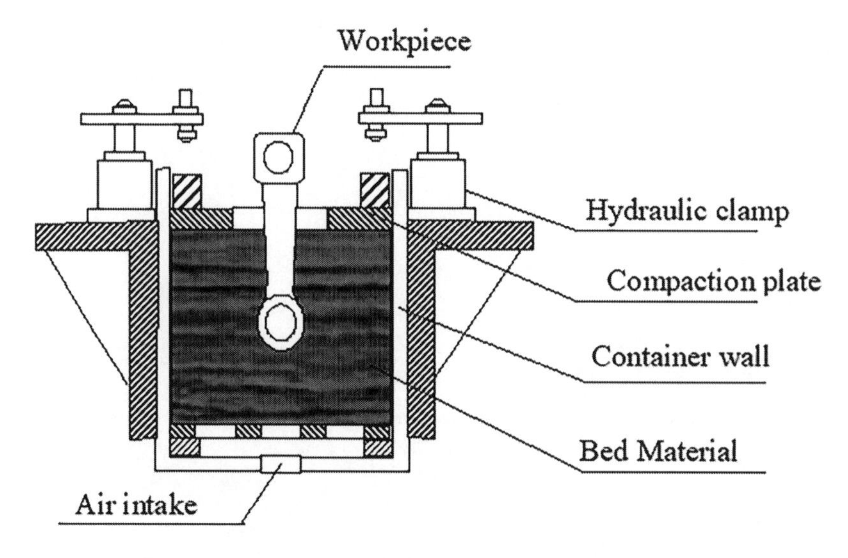

Figure 8.26 Example of a flexible fixture using phase-change materials (Hazen and Wright 1990).

8.6.5.3 Phase-Change Work-Holding

Phase-change material properties can be utilized to develop a flexible fixture. Material phase changes can be temperature induced, electrically induced, or a combination of the two. The fixtures based on temperature-induced phase changes were traditionally used to hold an object for special-purpose precision machining. Figure 8.26 shows an example of such a flexible fixture that is used to hold a turban blade for milling operation (Hazen and Wright 1990).

Theoretically, any material with a solid and liquid phase could be used to hold objects. In practice, the selection of phase-change materials is very restrictive since it is not allowed to cause damage to the workpiece, such as contamination, chemical reaction, and metallurgical modifications. In addition, the phase change needs an additional energy source that should be tolerable to the workpiece. Other criteria in selecting materials include phase-change times, ease of cleaning the encapsulation materials from the workpiece, and the material strengths at working temperature.

8.6.5.4 Conformable Fixtures

A conformable fixture adjusts the contacts to workpieces to adapt changing geometries mechanically. The adjustments in a conformable fixture are active. In other words, the fixture can actively change to adapt the geometry of the object rather than forcing the workpiece to fit the fixture by clamping. This reduces the deformation over the fixture and workpiece. From this perspective, the adjustable fixture in Figure 8.20 is also a conformable fixture, since it consists of an array of independently controllable pins that conform to the surface of the products. Figure 8.27 gives another example of a conformable fixture that was developed at MIT. It consisted of a 4 × 4 array of fingers that were normally locked by the squeezing actions of heavy-duty Belleville springs. The shape memory alloy wires were used to activate the motions (Hazen and Wright 1990).

8.6.5.5 Fixtureless Operations

A fixtureless operation usually involves a robot that is used for assembling without fixtures. Therefore, it is also called a robotic fixtureless assembly (RFA). RFA has been used to perform assembling processes for different products including automotive, aircrafts, and cameras. The operations without fixtures can reduce the cost and setup time of tooling greatly. General Motors of Canada Ltd. is developing a prototype of RFA, which is used for picking and accurately locating a large number of complex-shaped 3-D sheet metal (and sheet plastic) parts in their assembly processes. For example, Bandyopadhyay et al. (1993) introduced a 'fixture-free' machining centre to make components from this starting

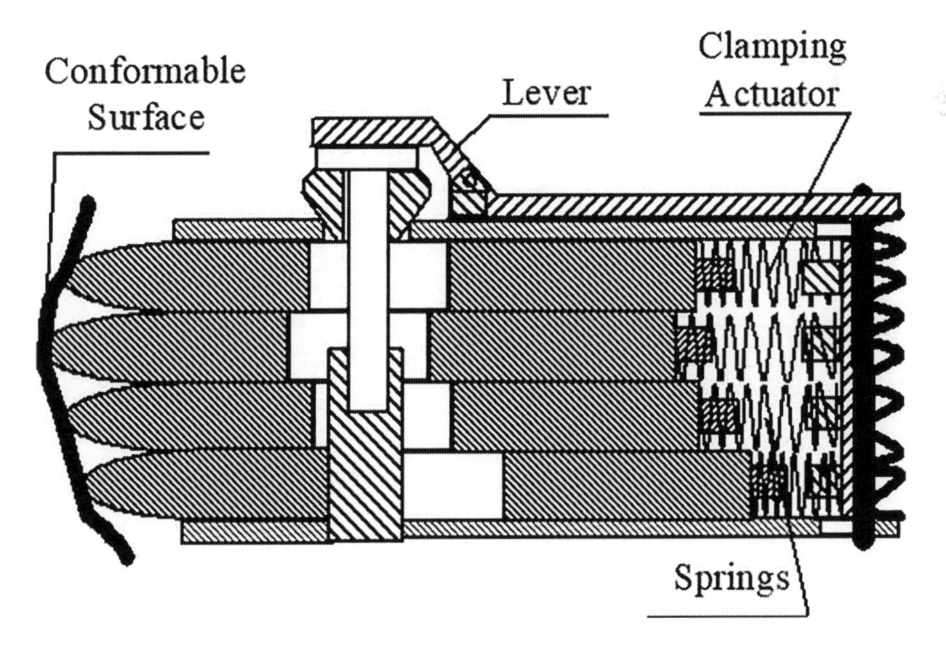

Figure 8.27 Example of a conformable fixture.

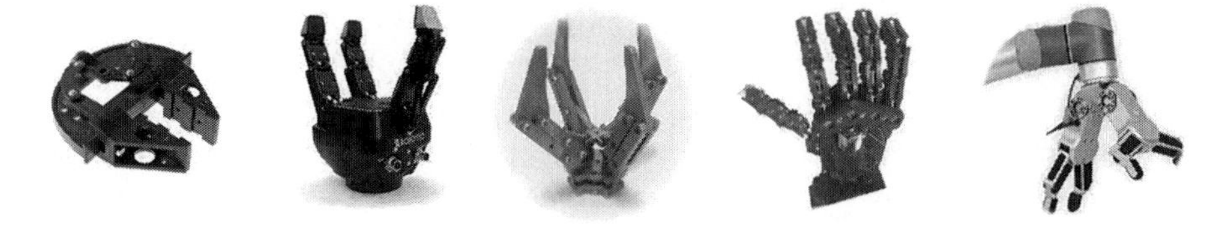

Figure 8.28 Grippers for fixtureless operations.

material in a block shape. Figure 8.28 shows a few examples of robotic grippers that can be used to hold and position an object for light assembly operations.

8.7 Procedure of Fixture Design

Figure 8.29 shows the relationship of a fixture design with other planning tasks in the process planning of products. The tasks involved in a fixture design are highlighted in the middle. Other than fixture design, machine tools, cutting tools, process parameters, cutting paths, and machining programs must be determined for each manufacturing process

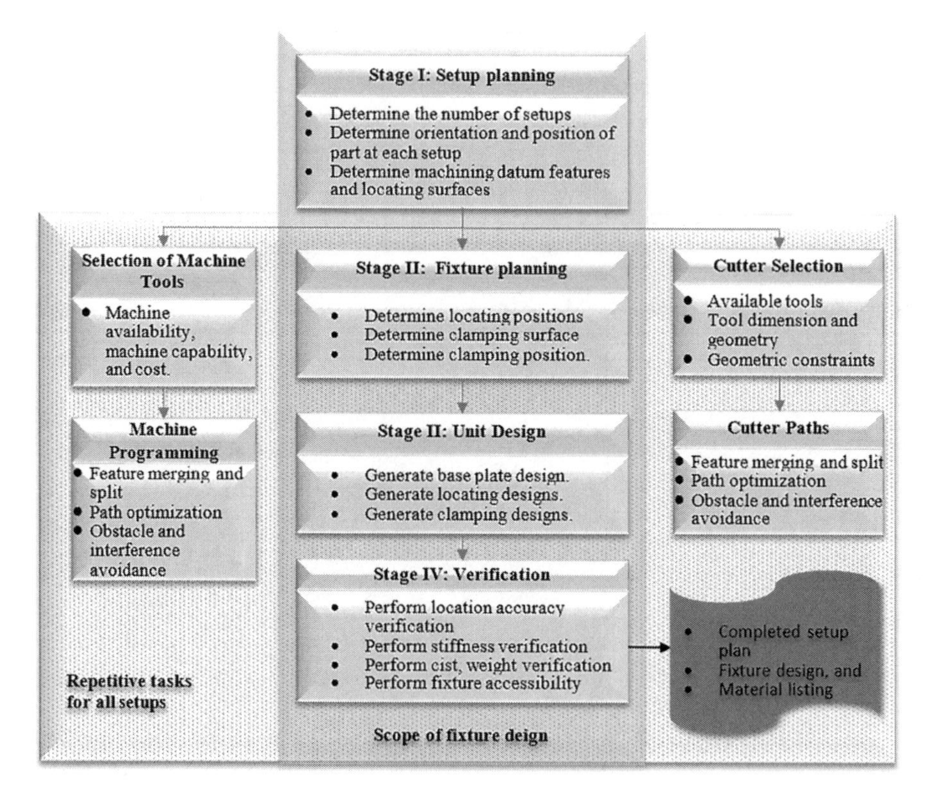

Figure 8.29 Relations of a fixture design with other planning tasks in the process planning (Buerer 2017; Wysk 2008).

Table 8.4 The tasks at four stages of fixture design.

Stage	Objective	Tasks
I	Setup planning	1. Analyse all manufacturing processes and determine the number of setups to perform all manufacturing processes for the workpiece. A setup is defined for one or more manufacturing processes that are performed by a single machining tool without the positional or orientation change of workpiece. 2. Correspond each manufacturing process to a setup with specified position and orientation of the workpiece.
II	Fixture planning	1. For each setup, determine the number and positions of locators for a work-holding purpose. 2. Identify the surfaces where the locators and clamps make contact and determine corresponding positions and orientations on the workpiece.
III	Fixture unit design	1. Design or select functional modules for the supports, locators, and clamps.
IV	Verification	1. Model and simulate to verify all design constraints of manufacturing processes are satisfied. 2. Evaluate the fixture design to ensure other constraints such as cost, weight, and setup time are satisfied. 3. Repeat stages II–IV until the solutions of the fixtures for all of the setups are obtained.

of a workpiece. The procedure of fixture design consists of a number of key activities that are logically followed one after another. However, an iterative process is expected since the design constraints at following steps are not verified when the solution at the current step is developed.

The fixture design for each setup is to (i) determine the types and positions where the supports, locators, and clamps are applied and (ii) design or select corresponding fixture elements to fulfil the specified functions. As shown in Figure 8.29, the procedure of the fixture design involves four stages, and the corresponding tasks at these stages are described in Table 8.4 (Buerer 2017; Wysk 2008).

All relevant information about products must be available when the fixtures for the manufacturing processes of products are considered. Figure 8.30 shows that the important data for fixture design relate to *products*, *machine tools*, *manufacturing processes*, and *verification* and *quality control* (Wang et al. 2010). The more detailed information can be obtained from those sources, the better the chance that the designed fixture avoids a conflict with design constraints.

After a fixture is designed, it should be verified thoroughly since the fixture makes direct contacts to a workpiece and is relevant to many manufacturing resources, including machine tools, cutters, and other auxiliary tools such as lubrication or coolant supplies. Verification is an integrated step in the fixture design process due to a number of reasons as follows.

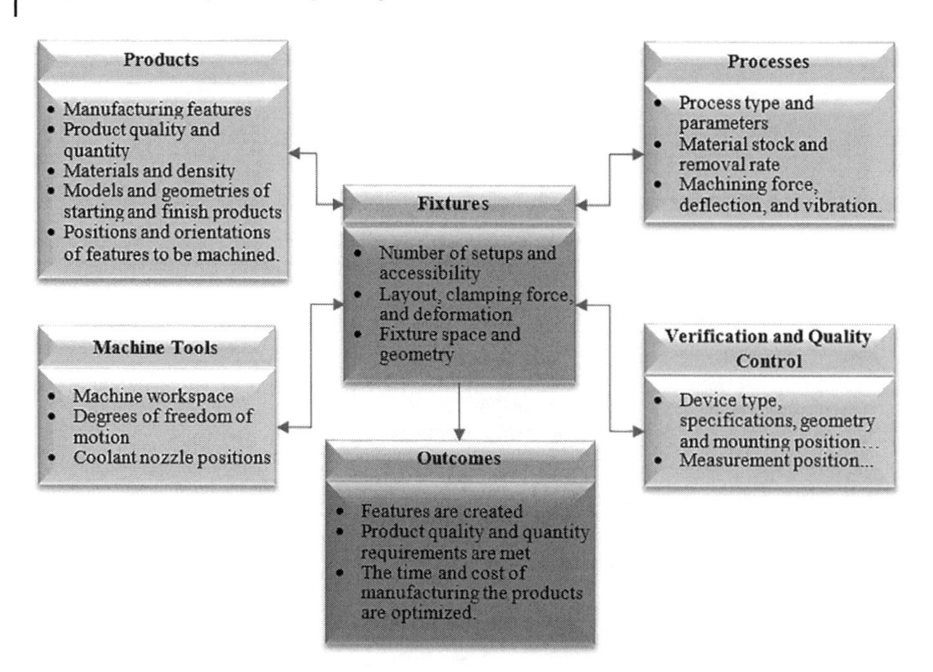

Figure 8.30 Key design factors and criteria of a fixture design.

1. As shown in Figure 8.30, many design factors are involved in the process of fixture design and many design variables have their uncertainties. However, it is difficult to establish accurate analysis models to include these design factors.
2. Only the portion of design constraints are taken into consideration at each step. It is possible that the design constraints in one step may be violated in another step; for example, the fixture design must be verified to avoid any interference in the assembly process or actual application.
3. As shown in Figure 8.29, a fixture design is closely affected by the activities for the fulfilments of other tasks, such as programming of machine tools and path planning of cutting tools. Therefore, a designed fixture must be feasible and practical to implement given manufacturing processes.
4. In application, verification is required to justify whether the fixture system is in a good working condition. Also needed is the use of a fixture system to justify whether the system is in a good condition.

Wang et al. (2010) introduced a framework for the verification of a fixture design in Figure 8.31. The suggested verifications include *analysis of geometric constraints*, *tolerance analysis*, *stability analysis*, *stiffness analysis*, and *accessibility analysis*. It is clear that a fixture design in process planning of products is not a trivial task. A knowledge-based engineering (KBE) approach should be applied to accelerate the design process and eliminate design defects.

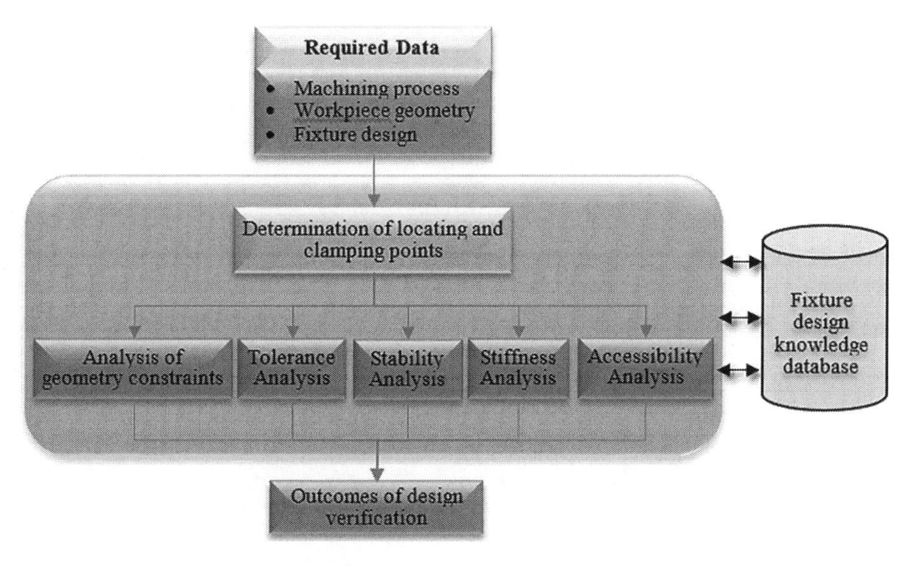

Figure 8.31 Overview of fixture verification systems (Wang et al. 2010).

8.8 Computer Aided Fixture Design

Fixture design tends to be tedious and time-consuming. Traditionally, the design and manufacture of a fixture might take several days or even longer to complete, and the performance of the fixture system relies heavily on designers' experiences and skills. To accelerate new product development, it is highly expected to utilize computer aided techniques in all aspects of fixture design. CAFD technology was developed to meet this need in past decades. Keyvani (2008) gave a summary of the historical development on using computer aided techniques for fixture design. As shown in Figure 8.32, early computer applications focused on virtual modelling and standardization of fixture elements, while recent applications were expanded to all design aspects, especially on heuristic or expert fixture design systems, kinematic and dynamic modelling, and quantitative evaluation and verification.

Gmeiner (2014) argued that CAFD should be an essential component of the integrated information system from product concept design, CAD, CAPP, CAM, CNC, and MRP to the deliveries of finished products. As shown in Figure 8.33, with data collection and exchange in the integrated system, CAFD is allowed to access all of the relevant information for fixture design. The information about manufacturing processes and product features are directly input to CAFD from CAD and other CAPP modules, while the outputs of CAFD are utilized for programming machine tools and planning cutting paths in the following CNC module. In this section, computer aided techniques for some specific tasks in fixture design are discussed.

8.8.1 Fixture Design Library

Fixtures are needed for nearly all manufacturing processes in discrete manufacturing. A fixture system is nothing more than supports, locators, and clamps. Even though the

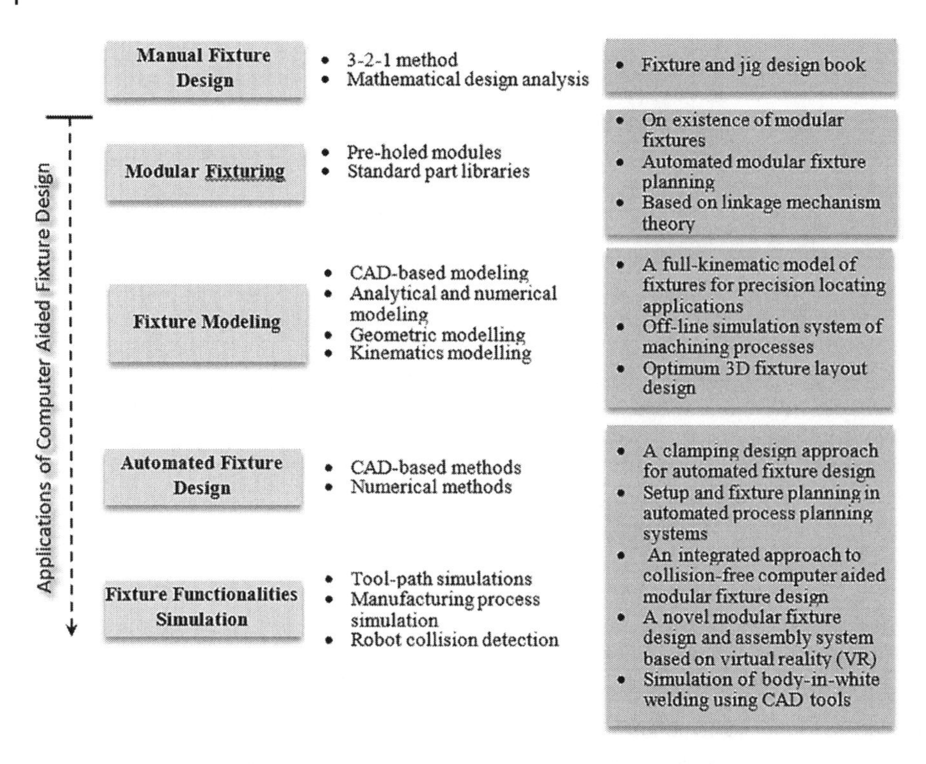

Figure 8.32 History of the applications of a computer aided fixture design.

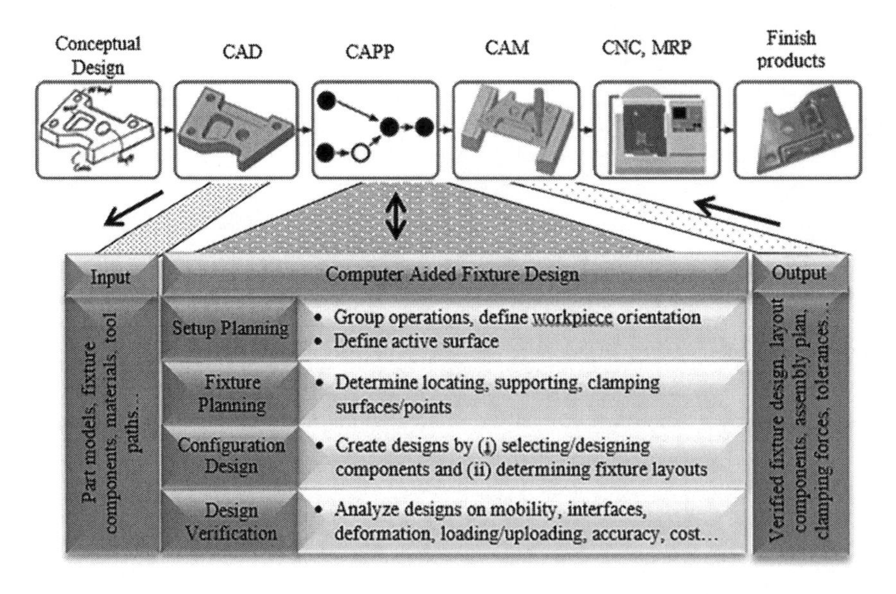

Figure 8.33 Framework of a computer aided fixture design.

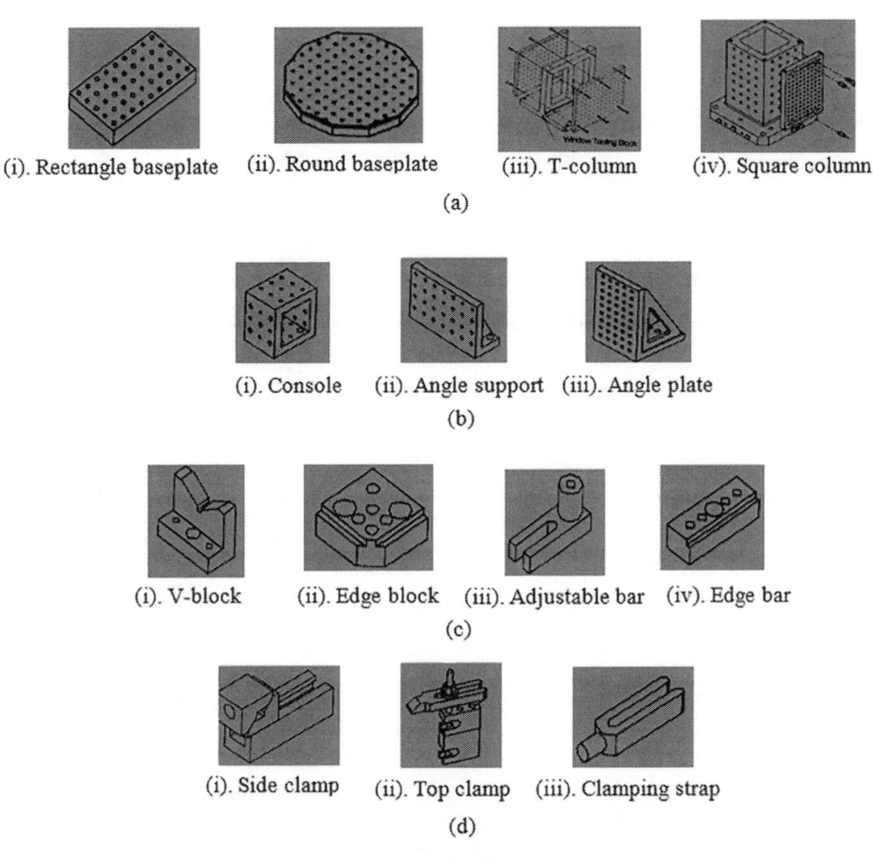

Figure 8.34 Design library example of functional modules for CAFD (Li 2009). (a) Fixture bases. (b) Supports. (c) Locators. (d) Clamps.

work-holding solution differs from one setup to another, many fixture elements, such as the locators in Figure 8.18, are reusable. Commonly used fixture elements should be standardized. The parametric models of these functional modules should be modelled and stored in the design libraries for product families. Reusing functional modules helps to identify fixture elements for new manufacturing process or new product in existing product families efficiently. Figure 8.34 shows an example of the design library of commonly used supports, locators, and clamps for CAFD by Li (2009).

8.8.2 Interference Detection

Any interference is not allowed in assembling a fixture. Virtual analysis should be conducted to ensure that a fixture is free of interference. With the virtual models of the fixture system and machine tool and workpiece, mainstream commercial CAD systems are capable of running interference checks to detect potential interferences before the fixture is prototyped. Figure 8.35 shows an example of an assembly of fixture design where the potential interference for a given sequence of assembling operations can be virtually simulated and detected (Li 2009).

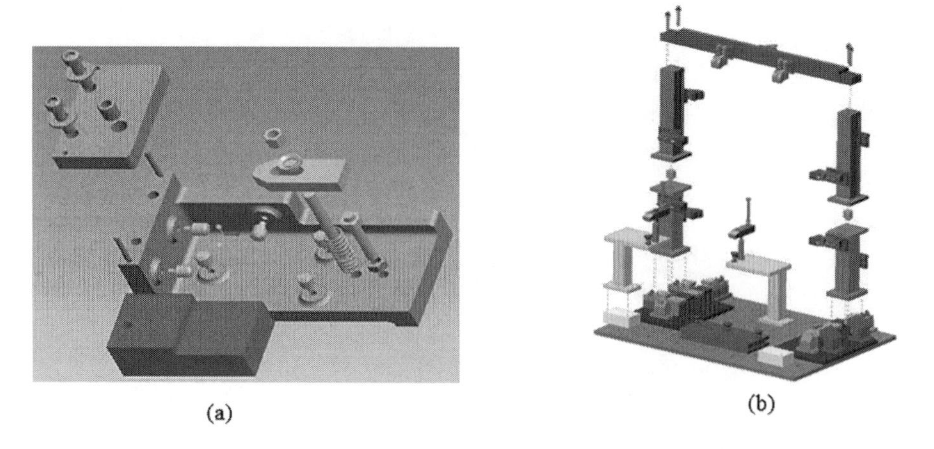

Figure 8.35 Example of interference detection (Li 2009). (a) Exploded view. (b) Visualization of the assembling part.

8.8.3 Accessibility Analysis

Accessibility analysis concerns the potential collisions of the fixture elements with surrounding objects in (i) loading and uploading a workpiece from a fixture and (ii) moving a cutting tool along the path. In accessibility analysis, the motions of machine elements, cutting tool, and workpiece with respect to fixture elements are taken into consideration. Motion simulation tools in CAD systems can be used to evaluate the accessibility of a workpiece or cutting tool for a given fixture design (Ghappande 2008). Figure 8.36 shows five examples of virtual fixturing systems in the application environment that can be utilized to run an accessibility analysis.

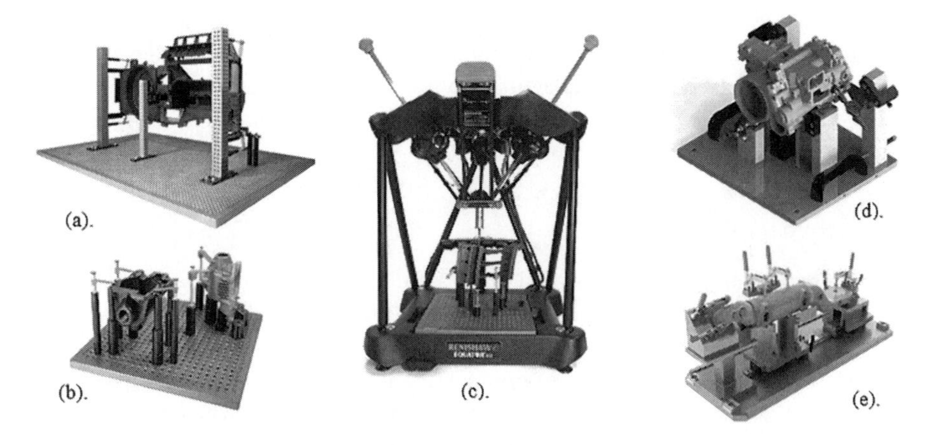

Figure 8.36 Examples of an accessibility study (Ghappande 2008).

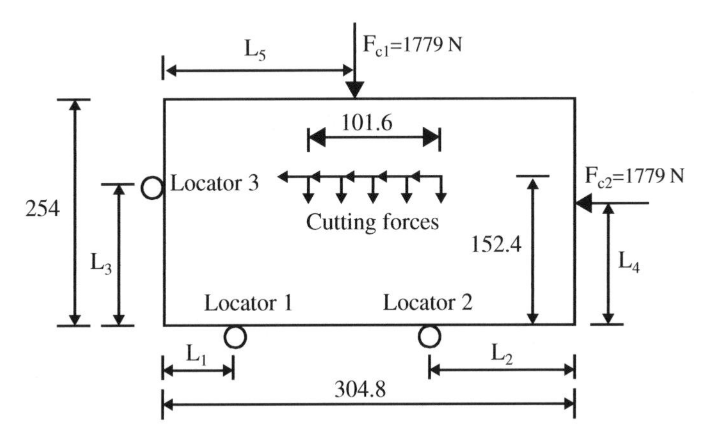

Figure 8.37 Example of deformation and accuracy analysis (Kaya 2006).

8.8.4 Analysis of Deformation and Accuracy

A material removal process in discrete manufacturing usually involves a large mechanical force; any object would be deformed when subjected to these loads. To verify the dimensional accuracy, the deformations occurring to the fixture elements and workpiece subjected to the cutting forces need to be analysed. A computer aided engineering (CAE) tool can be used to analyse the deformation and accuracy of a fixture system subjected to mechanical loads. Kaya (2006) provided a comprehensive discussion on using CAE for the optimization of a fixture design.

Figure 8.37 shows an example of formulating a deformation analysis problem into a numerical simulation in CAE (Kaya 2006). The workpiece was simplified as a two-dimensional plate (204.8 mm in width and 254 mm in height). The machining forces were assumed to be constant along the cutting path as $F_x = F_y = 889.6$ N. Two clamps with the force magnitude of 1779 N are applied on the top and right surfaces, respectively. The design space for three locators and two clamps are specified as 5 mm $< L_1 <$ 148 mm, 5 mm $< L_2 <$ 148 mm, 5 mm $< L_3 <$ 249 mm, 5 mm $< L_4 <$ 249 mm, and 5 mm $< L_5 <$ 300 mm. To determine L_1 to L_5 for minimization of the discrepancy of the cutting tool along the path, a parametric study was defined in the CAE environment.

The result with the minimized positional discrepancy of cutting time from the numerical simulation led to the optimized fixture design as $L_1 = 50.4$ mm, $L_2 = 101.6$ mm, $L_3 = 101.6$ mm, $L_4 = 50.8$ mm, and $L_5 = 152.4$ mm.

8.9 Summary

A manufacturing business can be classified as continuous manufacturing and discrete manufacturing. The wo types of manufacturing have significant differences in designing fabrication processes and tooling. Fixture design is essential to all manufacturing processes in discrete manufacturing. A fixture provides positioning references to product features and immobilizes a workpiece when a manufacturing process is performed. The 3-2-1 principle, the axioms for geometric control, dimensional control, and mechanical control, are used to guide fixture design. Fixture design is an essential component of computer aided process planning.

It has been found that CAFD can be used to accelerate new product development and reduce design cycle and cost. Engineers should gain the knowledge and skill to use CAFD for (i) developing design libraries for standard functional modules, (ii) interpreting design guides and rules for knowledge-based engineering for fixture design, (iii) simulating fixture assemblies for interference detection, (iv) evaluating the accessibilities of cutting tools and workpieces, and (v) performing engineering analysis for quality improvement and fixture design optimization.

8.10 Design Projects

8.1
1. Select a machined part you can reach (or download the blade model from the Website) or create one part model with machined features such as a part shown in Figure 8.38.
2. Design a fixture system for one or a few manufacturing processes of the part.
3. Design and create an assembly model of the fixture system.
4. Estimate the machining forces.

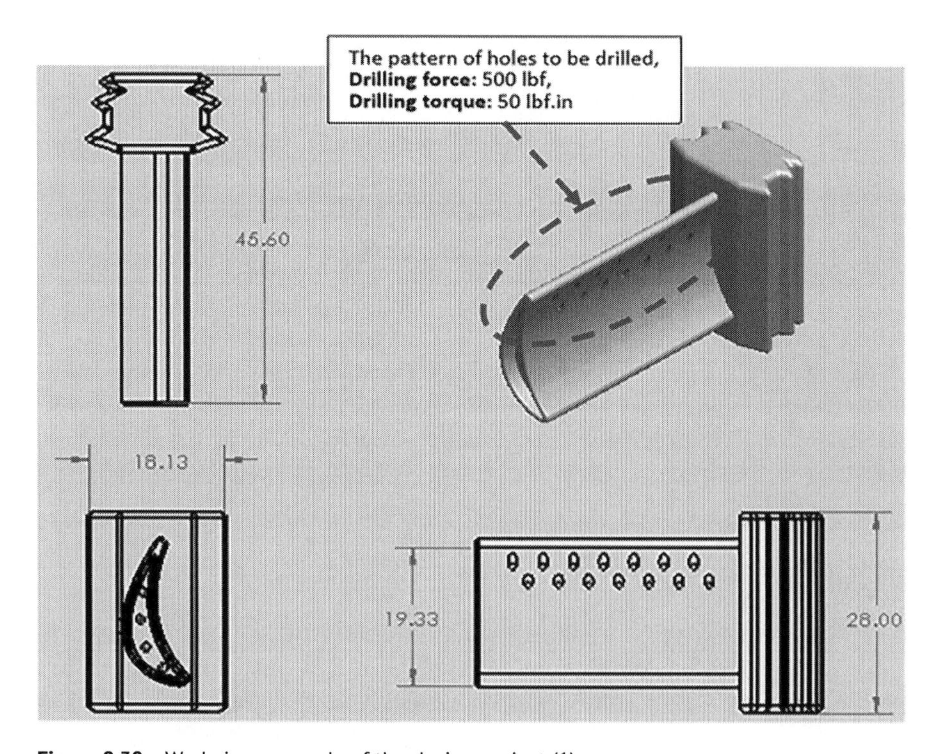

Figure 8.38 Workpiece example of the design project (1).

5. Analyse the deflection of fixture elements and workpiece using a computer aided engineering (CAE) tool.
6. Define a parametric study to optimize the fixture design for a minimized positional discrepancy of the cutting tool.

8.2 Fixture design optimization problem. As shown in Figure 8.39, the workpiece is simplified as a two-dimensional plate (300 mm in width and 90 mm in height). The machining forces are assumed to be constant along the cutting path as $F_x = 100$ N and $F_y = 286$ N. Two clamps with force magnitudes of 350 N and 200 N are applied on the top and right surfaces, respectively. The design space for three locators and two clamps are specified as 5 mm< L_1 <148 mm, 5 mm< L_2 <148 mm, 5 mm < L_3 < 85 mm, 5 mm < L_4 < 65 mm, and 5 mm < L_5 < 125 mm. Define and run a parametric study to determine L_1 to L_5 for minimization of the discrepancy of the cutting tool along the path (Kaya 2006).
(Reference solution: L_1 = 62.1 mm, L_2 = 9.11 mm, L_3 = 21.3 mm, L_4 = 8.7 mm, and L_5 = 62.2 mm.)

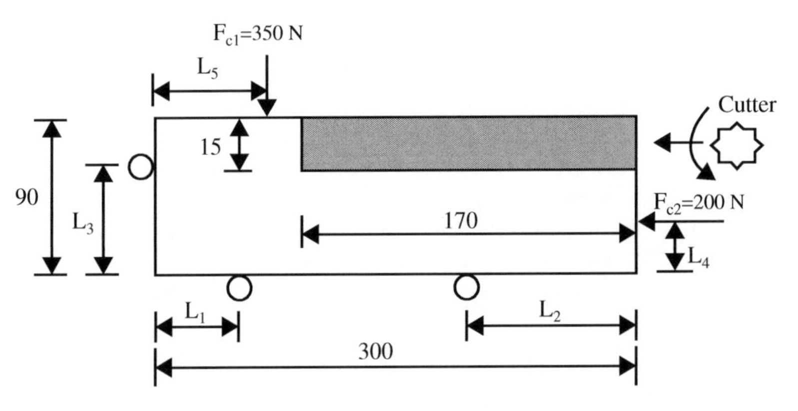

Figure 8.39 2D fixture parameters and tool path direction for the design project (2).

References

Bakker, O.J., Papastathis, T.N., Ratchev, S.M., and Popov, A.A. (2019). Recent research on flexible fixtures for manufacturing processes. http://eprints.nottingham.ac.uk/29136/1/RPME_6_2_107-121.pdf.

Bandyopadhyay, B.P., Hoshi, T., Latief, M.A., and Hanada, T. (1993). Development of a fixture-free machining center for machining block-like components. *Journal of Materials Processing Technology* 39 (3–4): 405–413.

Basha, V.R. and Salunke, J.J. (2013). An advanced exploration of fixture design. *International Journal of Engineering Research and Applications* 6 (5), part 3): 30–33.

Batka, S. (2011). Managing process manufacturing and the challenges of a hybrid process and discrete environment. https://apicsfoxriver.starchapter.com/images/downloads/Meeting_Downloads/apics_fox_river_march_2011_pdm_shared.pdf.

Bi, Z.M. and Zhang, W.J. (2001). Flexible fixture design and automation: review, issues and future directions. *International Journal of Production Research* 39 (13): 2867–2894.

Buerer, C. (2017). Integrated part manufacturing (from engineering to production). https://www.plm-europe.org/admin/presentations/2017/2021_PLMEurope_24.10.17-17-00_CORSIN-BUERER_SPLM_integrated_part_manufacturing__from_engineering_to_production.pdf.

Chan, K.C. and Lin, C.S. (1996). Development of a computer numerical control (CNC) modular fixture- machine design of a standard multifinger module. *International Journal of Advanced Manufacturing Technology* 11 (1): 18–26.

Erdem, I. (2017). *Flexible Fixtures – A Treatise on Fixture Design and Efficiency*. Gothenburg, Sweden: Chalmers University of Technology http://publications.lib.chalmers.se/records/fulltext/248405/248405.pdf.

Gameros, A., Lowth, S., Axinte, D. et al. (2017). State-of-the-art in fixture systems for the manufacture and assembly of rigid components: a review. *International Journal of Machine Tools and Manufacture* 123: 1–21.

Ghappande, P. (2008). Study of fixturing accessibilities in computer aided fixture design. PhD thesis. Worcester Polytechnic Institute.

Gmeiner, T.C. (2014). Automatic fixture design based on formal knowledge representation, design synthesis, and verification. PhD thesis. ETH Zürich, Switzerland.

Hazen, F.B. (1990). Automated setup assembly mechanisms for the intelligent machining workstation. CMU-RI-TR-90-20. https://www.ri.cmu.edu/pub_files/pub3/hazen_f_brack_1990_1/hazen_f_brack_1990_1.pdf.

Hazen, F.B. and Wright, P.K. (1990). Workholding automation: innovations in analysis, design and planning. *Manufacturing Review* 3 (4): 224–236.

Ivanov, V., Dehtiarov, I., and Zajac, J. (2018). Flexible fixtures for parts machining in automobile industry. *MMS Conference 2017*, Starý Smokovec, Slovakia (22–24 November 2018). MMS. DOI: 10.4108/eai.22-11-2017.2274155. http://eudl.eu/pdf/10.4108/eai.22-11-2017.2274155.

Kaya, N. (2006). Machining fixture locating and clamping position optimization using genetic algorithms. *Computers in Industry* 57: 112–120.

Keyvani, A. (2008). Modular fixture design for BIW lines using process simulate. http://www.diva-portal.org/smash/get/diva2:229240/fulltext01.

Li, Q. (2009). Virtual reality for fixture design and assembly. PhD thesis. University of Nottingham. http://eprints.nottingham.ac.uk/10650/1/BillyThesisFinal.pdf.

Majstorovic, V., Sibalija, T., and Ercevic, B. (2013). CAPP model for prismatic parts in digital manufacturing. *6th Programming Languages for Manufacturing*, Dresden, Germany (October 2013), pp. 190–204. https://hal.inria.fr/hal-01485815/document.

MechanicalBooster (2019). Difference between jigs and fixtures. http://www.mechanicalbooster.com/2016/11/difference-between-jigs-and-fixtures.html.

Singh, R. (2019). Fixturing/workholding. https://www.me.iitb.ac.in/~ramesh/courses/ME338/fixturing.pdf.

Sophen (2019). Discrete versus process manufacturing innovation. https://www.sopheon.com/discrete-versus-process-manufacturing-innovation.

Wakhare, W.M. (2016). Rule based setup and fixture planning for prismatic parts on 3-axis and 4-axis milling machines. MS thesis. Ohio University, USA. https://etd.ohiolink.edu/!etd.send_file?accession=ohiou1470758431&disposition=inline.

Walczyk, D.F., Lakshimikanthan, J., and Kirk, D.R. (1999). A comparison of rapid fabrication methods for sheet metal forming dies. *Transactions of the ASME, Journal of Mechanical Design* 121: 214–223.

Wang, H., Rong, Y., Li, H., and Shaun, P. (2010). Computer aided fixture design: recent research and trends. *Computer-Aided Design* 42: 1085–1094.

Wysk, R. A. (2008). Process engineering. http://www.engr.psu.edu/cim/ie550/ie550capp.ppt.

9

Computer Aided Manufacturing (CAM)

9.1 Introduction

In a manufacturing system, the raw materials are transformed into finished products through a series of manufacturing processes, which are performed at production facilities. Figure 9.1 shows the transformation of a manufacturing system (Groover 2007). A manufacturing system is represented by a number of major activities in an information flow and a material flow. An *information flow* describes all of the decision-making activities involved in a product lifecycle. A *material flow* describes the transportation from raw materials, pre-fabricates, parts, components, and integrated objects to finished products as a flow of entities. *Production facilities* are the hardware components that make *physical touches* to physical objects in the material flow. Main types of production facilities in a manufacturing system are factories, machines and tools, material handling systems, and inspection systems. In addition, product facilities also include control hardware systems to run main product facilities.

When manufacturing businesses are automated, the roles of human beings in operating machines or decision-making supports are replaced by various functional modules in an *enterprise information system* (EIS), which is the collection of computer tools for data acquisition, processing, and utilization to support decision-making activities in a manufacturing system (Bi et al. 2014). An EIS deals with some critical activities involved in an *information flow* of a manufacturing system from high-level business functions, product design, and manufacturing process planning to low-level machine control:

1. The unit for *business functions* supports the decision-making activities to respond to customer needs and deal with the changes and uncertainties in the dynamic business environment at the enterprise level. It functions at the beginning and the closing end of the information flow. Typical decision-making activities of business functions are the identification and forecasting of market needs, order entries, costing accountings, and customer billings.
2. Decision-making activities in *product design* relate to the product orders from customers or internal departments. Products can be one of the following types: (i) products with given specifications, (ii) manufacturers' proprietary products, and (iii) internal orders for proprietary products based on forecasting. When the products to be made are proprietary, the company designs products before they can be manufactured.

Computer Aided Design and Manufacturing, First Edition. Zhuming Bi and Xiaoqin Wang.
© 2020 John Wiley & Sons Ltd. This Work is a co-publication between John Wiley & Sons Ltd and ASME Press.
Companion website: www.wiley.com/go/bi/computer-aided-design

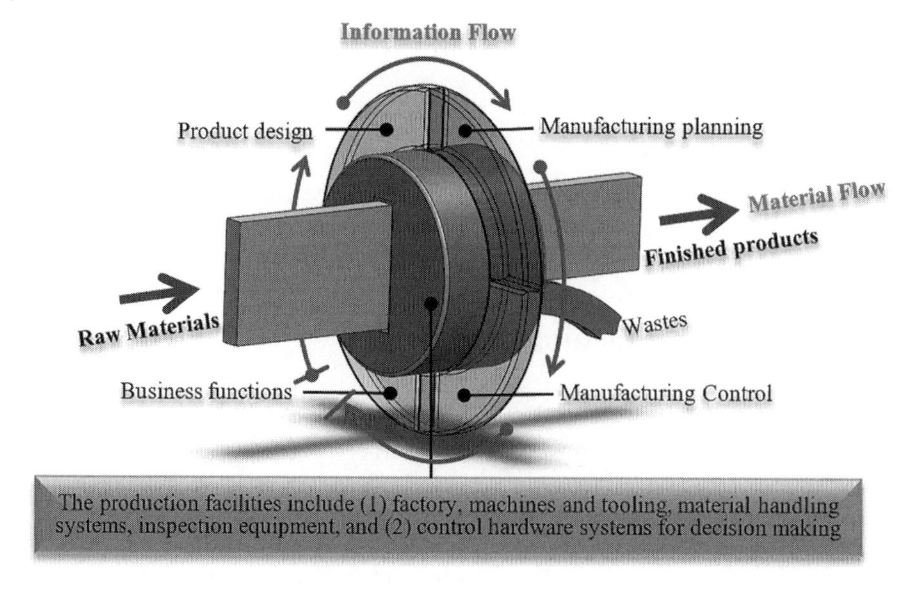

Figure 9.1 Production facilities in a manufacturing system (Groover 2007). (*See color plate section for color representation of this figure*).

3. *Manufacturing planning* is used to convert the data of designed products into the information to plan and schedule manufacturing resources and processes. The processed information includes process planning, master scheduling, requirement planning, and capacity planning. The processes are planned to determine the sequences of manufacturing, assembling, and testing processes that are needed to make products.
4. *Machine and process control* relates to managing and controlling actual operations to execute manufacturing plans in factories. The information flows from manufacturing planning to control. The information also flows back and forth between manufacturing control and factory operations (Groover 2007).

9.1.1 Human and Machines in Manufacturing

Using the concept of *the separation of functions*, the functions of a manufacturing system are classified into two classes: (i) the functions in the first class are for operations of business processes in the material flow, which are designed to make products that fulfil the enterprise's mission; (ii) the functions in the second class are to control the process executions in the information flow and include all planning, scheduling, control, and data management relating to the controls of the manufacturing processes (Williams and Li 1998). The functions in both types are bidirectionally interconnected: the collected data about the states of production facilities in the material flow are processed and utilized by the functional modules in the information flow and the operational commands from informational functional modules are transmitted from the information flow back to the material flow for execution on production facilities.

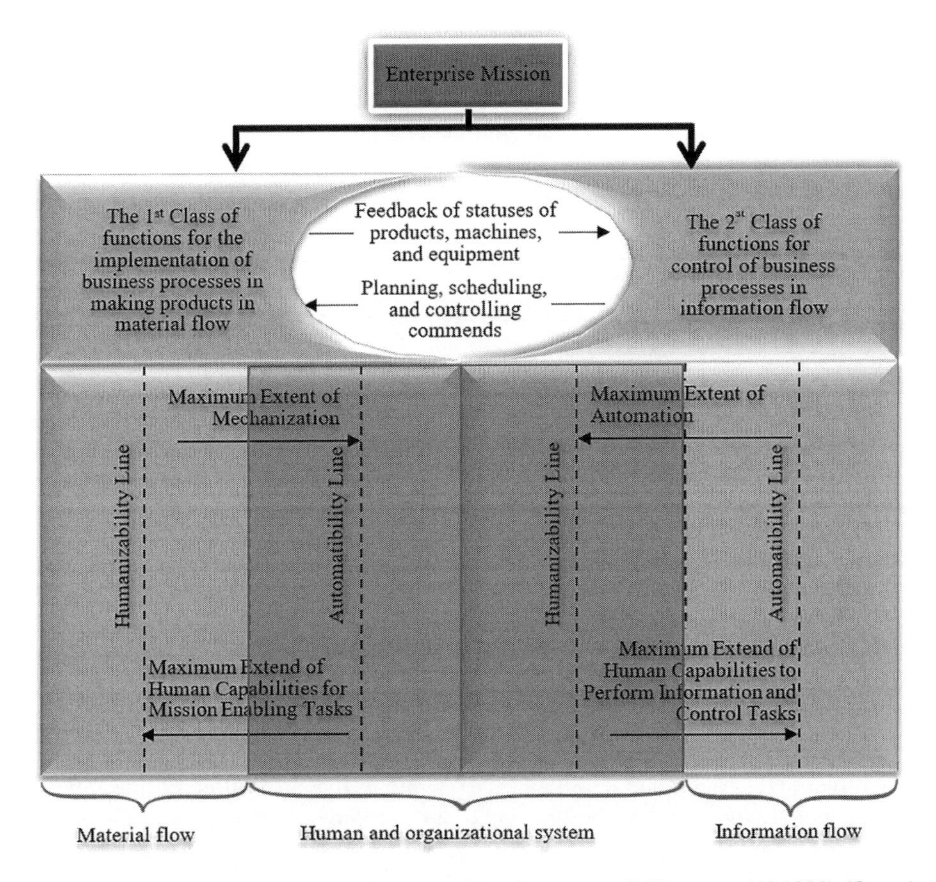

Figure 9.2 Humans and machines in a manufacturing system (Williams and Li 1998). (*See color plate section for color representation of this figure*).

In the bottom block of Figure 9.2, the central part shows the human roles in running manufacturing businesses in an enterprise. For the *material flow* on the left, human operations are productive for some intricate tasks but underperform compared with machines for simple and repetitive tasks; humans are dispensable for some tasks where the human flexibility and agility are critical. For the *information flow* on the right, humans are extremely efficient in making decisions where the manufacturing environment is very complex and involves many changes and uncertainties. Humans are not efficient where a mathematic model or a computer algorithm can reach optimum decisions. In the middle range, either humans or computers can deal with the scopes and the complexities of decision-making problems. Therefore, there are limits for the automation level in both the material flow and the information flow. Accordingly, two boundaries of automation are marked by the *'maximized extent of mechanization'* line and the *'maximized extent of automation'* line in Figure 9.2.

The mechanization of machine tools began in the eighteenth century when the Watt steam engine was invented; steam engines were used in Jacquard looms, lathes, and screw machines to relieve human labours during the first Industrial Revolution. However, those mechanisms were not fully automated since they were operated by humans (Gerovitch 2019). Recent advancements are mainly on the automation of decision-making supports at various domains and levels of businesses of manufacturing enterprises. As a general trend of technological development, more and more machines and computers are being adopted to mechanize or automate manufacturing processes. In fact, the advancement of a manufacturing system can be reflected by the *level of automation* where human resources are replaced by machines. The level of automation of manufacturing turnover has been increased from 1% to 70% (Holmes Noble 2017).

Sullivan (2009) and Groover (2007) summarized the advantage of maximizing mechanization and automation for the following aspects:

1. To increase the productivity of human resources. Automating a manufacturing process will reduce the cycle times of manufacturing processes and increase the productivity of human resources. This leads to an increased throughput of production per unit time.
2. To reduce labour needs. The ever-increasing labour cost has been and will continue to be the trend of industrialized societies in the world. Consequently, traditionally high investments in automation have become economically justifiable to replace increasingly expensive labour.
3. To mitigate the effects of labour shortages. There is a general shortage of labour in many advanced nations. This has stimulated the development of automation solutions in modern manufacturing.
4. To reduce or eliminate routine manual and clerical tasks. Automation has significant social values to relieve human resources from operations that are routine, boring, fatiguing, and possibly irksome. Automation will improve human working conditions in the manufacturing environment.
5. To improve worker safety. Automation transfers workers from participation in the actual manufacturing processes to supervisory roles to authorize and monitor processes. This has provided an impetus for automation.
6. To sustain consistent quality of the product. In contrast to manual operations, automation performs the tasks with consistent quality; this will reduce the defect rates.
7. To reduce lead time in manufacturing. The lead time is the elapsed time from the moment when a customer places the product order to the moment when the product is delivered to the customer. Automation can reduce the manufacturing time due to higher productivity. The company can gain its competitive advantage to catch more orders in the future. In addition, it helps to reduce work-in-process parts.
8. To perform complex processes that cannot be accomplished manually. Many manufacturing processes are mechanized since they are beyond human capabilities, for example generating complex geometric features on products, producing dimensional features with high precision, or making products with extremely small or large sizes. Examples of indispensable automation are rapid prototyping and fabrication of integration circuits.
9. To reduce the cost by mass production or mass customization. The initial investment for manufacturing is facilitated by the quantity of products that are made by the same set of machines and tools. The higher productivity an automated system is, the more products are made and the less cost there is for each product unit.

In summary, automation has exhibited its advantages in improving product quality, increasing sales, sustaining better relations of companies with employees, and increasing business profits in unexpected and intangible ways.

9.1.2 Automation in Manufacturing

Automation becomes essential to modern manufacturing systems (Mishev 2006). Automation has been treated as the effective means for manufacturing companies to reduce product costs, achieve consistent quality, improve productivity, and sustain a high competitive advantage in the highly turbulent business environment. *Manufacturing automation* refers either to the mechanization of manufacturing processes and operations in (i) the material flow or (ii) the automation of decision-making supports in the information flow.

Depending on the degree of substitution of humans by machines, a manufacturing automation can be classified into hard automation, programmable automation, and full automation. Accordingly, the task allocations of humans and machines vary with the level of automation.

9.1.2.1 Hard Automation

In *hard automation*, the sequence of processing operations is fixed by controllable variables in a program. The operations are executed sequentially since the order is predefined. The complication of hard automation mainly depends on the types of operations the machine can cope with; the more operations a machine can perform, the more complex and expensive the hard simulation will be. In general, hard automation is characterized by a high initial cost and a high production rate. Therefore, hard automation is appropriate for continuous production flow and discrete mass production systems where the products have a very high demand and volume. The examples of hard automation are distillation processes, paint shops, conveyers, and transfer lines. Hard automation is applicable to the following circumstances (Kharagpur 2019):

1. Large volumes but limited variants of products in terms of sizes, shapes, materials, and functionalities
2. The demands for products are stable and predictable for a period of two to five years.
3. The high throughput of production is expected.
4. How to reduce production cost is the main strategy to sustain competitiveness in the market.

9.1.2.2 Programmable Automation

Programmable automation is used for a changeable sequence of manufacturing operations where the machines can be reconfigured by control systems. The automation solutions to machines are capable of generating control programs for different manufacturing processes to make product variants. A sequence of operations refers to a collection of computer instructions that are issued to machines for execution. However, it is a non-trivial task to program machines or sequences of operations for different products. Therefore, investment on programmable equipment is economically attractive only when the manufacturing processes are not changed very frequently. Programmable automation is particularly useful to the batch production processes where the volume of the product is medium to high.

Note that at an early development of computer aided techniques, programmable automation covered a broad scope other than for programming machine tools. For example, NTIS (1984) divided programmable automation into three categories as *computer aided design* (CAD) for products and machine tools, *computer aided manufacturing* (CAM) for robots, computerized machine tools, flexible manufacturing systems, and *computer aided techniques for management*, such as management information systems and computer aided planning. An assembly of these technologies are referred to as *computer integrated manufacturing* (CIM). Examples of programmable automations are numerical control (NC) machines, industrial robots, steel rolling mills, and paper mills. In this chapter, we emphasize the programmable automation in CAM.

9.1.2.3 Full Automation

Full automation aims to make a wide range of products from one product to another with a minimized changeover time. Ideally, both hardware and software systems can be reconfigured automatically by computer programs. There is no loss of production time in reprogramming and reconfiguring an automated system for next products. A full automated system is capable of making different combinations and schedules to perform manufacturing processes for products. This is fundamentally different from the scenarios by programmable automation that product variants are made in separate batches in semi-automation paradigms. A fully automated system is equipped with some fully automated product facilities, such as automated guided vehicles (AGVs), robots, and computer numerical controls (CNCs). Full automation is applicable to the following scenarios:

1. A company has its business strategies to meet highly diversified needs from customers. A product mix requires a combination of different parts, components, and products that have to be manufactured from the same production system.
2. The product lifecycles are relatively short and the demanded product volumes are highly unpredictable and change quickly with respect to time; the product facilities need frequent upgrades to adopt new functions and technologies and to accommodate the diversified needs of new customers.
3. A company makes batch-size-one products or products with very volumes, the lead-times of products are very short, and product demands are very dynamic and unpredictable.

9.1.3 Automated Decision-Making Supports

Figure 9.2 has shown that manufacturing automations can be of two classes: mechanizations of manufacturing processes and automation of decision-making supports. The *first class of automations* is implemented for respective manufacturing processes and the automation solution is associated with hardware components for the specified manufacturing processes. The automation is for the controls of a hardware system. On the other hand, the *second class of automations* covers a very broad scope of decision-making activities, and the implementation depends greatly on the complexity of the manufacturing system. Note that system complexity can be measured by the *number and types of system components*,

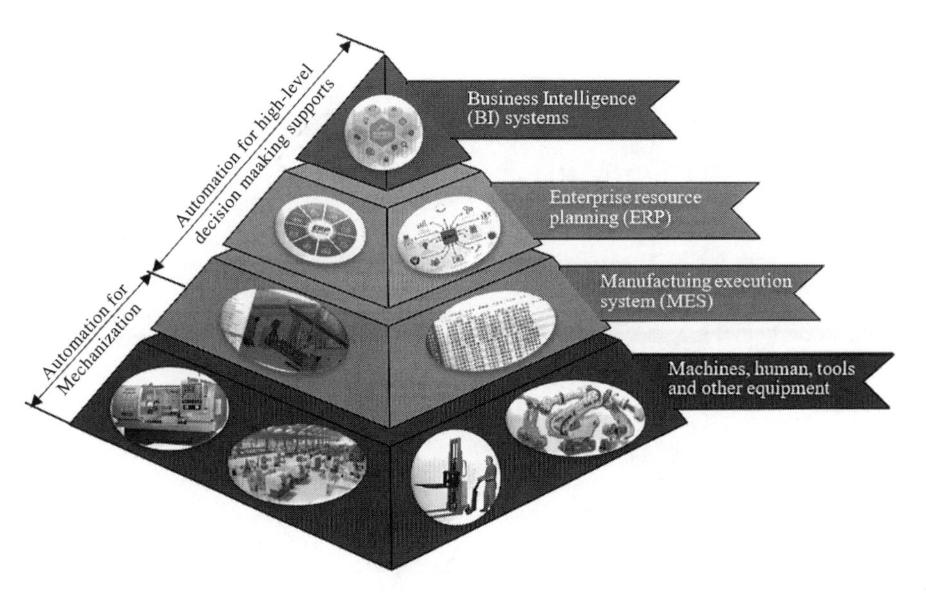

Figure 9.3 Automated decision-making supports in manufacturing. (*See color plate section for color representation of this figure*).

their interactions, the *dynamic characteristics of components*, and the *interactions with respect to time* (Bi et al. 2014).

As shown in Figure 9.3, since businesses in a modern enterprise are becoming more and more complex, sophisticated software tools are needed to support decision-making activities at different domains and levels. For example, the software tools for the controls of hardware systems are usually included in a manufacturing execution system (MES). The software tools for planning manufacturing resources, supply chain managements, marketing, and sales are included in the layer of enterprise resource planning (ERP) and the software tools for enterprise alliances, virtual manufacturing, and manufacturing as services are developed at the layer of business intelligence (BI).

9.1.4 Automation in Manufacturing Execution Systems (MESs)

For the decision-making supports in Figure 9.3, a *manufacturing execution system* (MES) is used to connect, monitor, and control complex manufacturing systems and data flows collected from the material flow. An MES ensures effective executions of manufacturing processes for high production outputs, faster deliveries, and competitive costs.

A manufacturing system consists of different types of hardware resources; at the device level, every hardware device corresponds to its own automated solution. Therefore, an MES is a comprehensive solution to automate all hardware systems in a manufacturing plant. Computer programs in MES support all of the required decision-making activities; in addition, these programs support the cooperation and coordination of different information processing units.

Depending on the scale and business scope of a manufacturing system, system components in an MES are classified in terms of the level of automation into *element-level*,

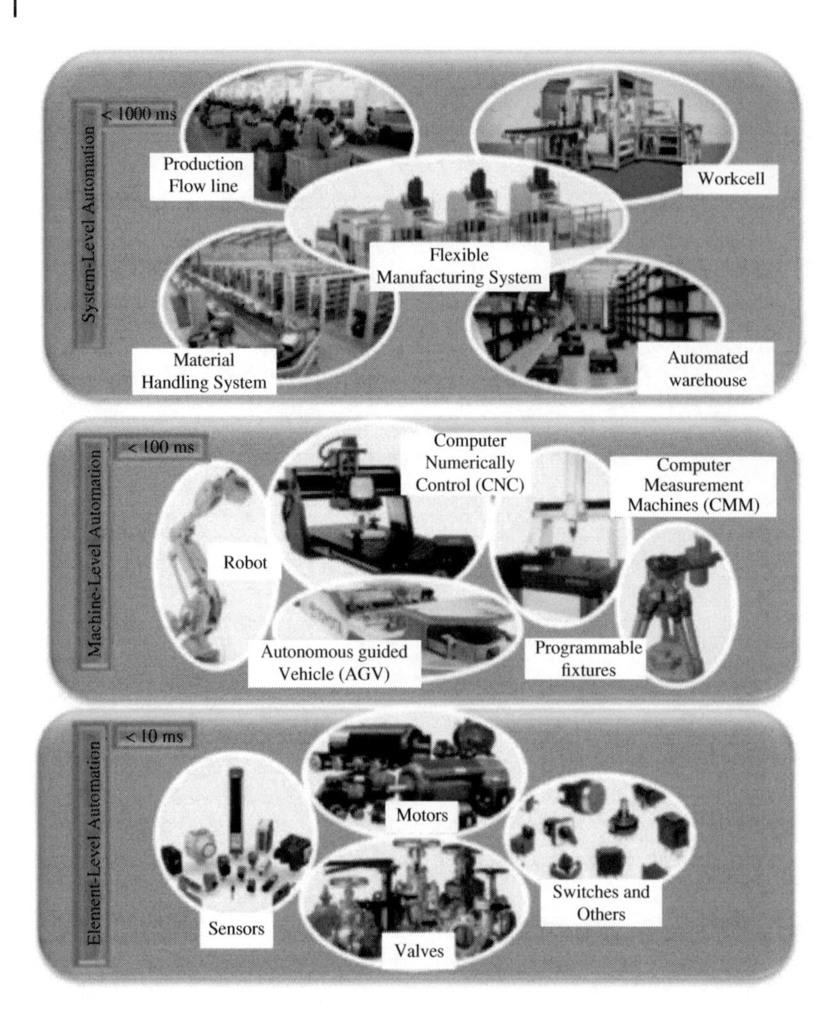

Figure 9.4 Level of automation in a manufacturing execution system (MES). (*See color plate section for color representation of this figure*).

machine-level, and *system-level*. As shown in Figure 9.4 (Calvo et al. 2013), typical enabling technologies for the implementation of MESs are *computer aided design and manufacturing* (CADM), *computer aided process planning* (CAPP), *computer numerical control* (CNCs), *flexible machining systems* (FMSs), *automated storage and retrieval systems* (AS/RS), *automated material handling systems* (AMHSs) for robots and automated cranes and conveyors, and *computerized scheduling and production control*.

One difference between the automations at different levels is the frequency of updating control commands: the real-time performance over 100 Hz is usually required for the controls at the element level, while the frequencies at the machine level and system level are normally 10 and 1 Hz, respectively. The rest of the chapter will concentrate on programming of material removal processes using CNCs in discrete manufacturing.

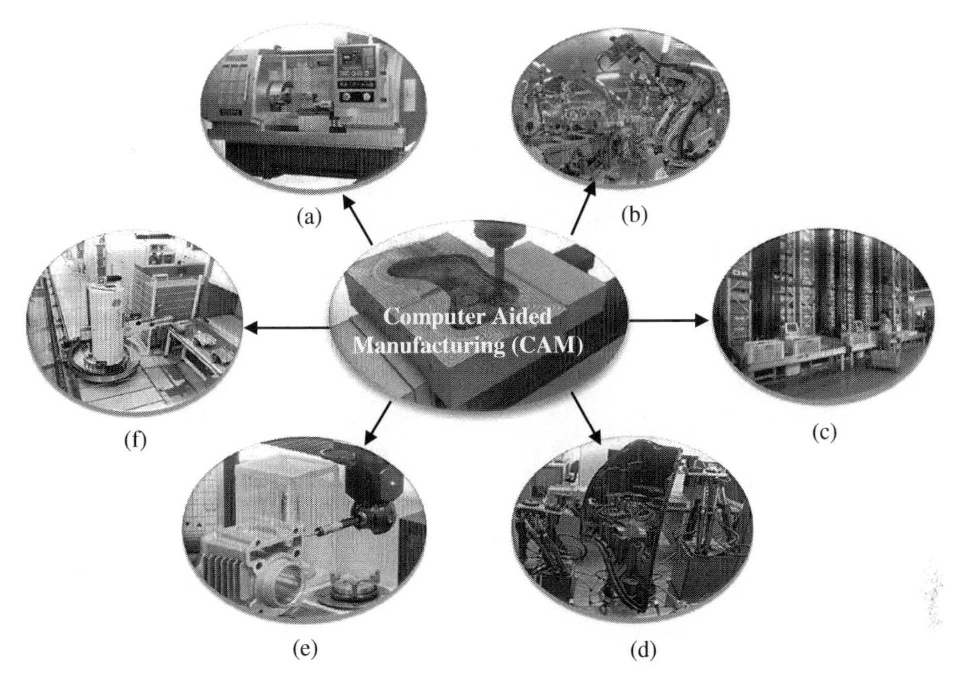

Figure 9.5 Typical production facilities supported by CAM. (a) Numerically controlled centres. (b) Industrial robots. (c) Automated storage and retrieval systems (AS/RSs). (d) Flexible fixture systems (FFSs). (e) Coordinate measurement machines (CMMs). (f) Automated material handling systems (AMHSs).

9.2 Computer Aided Manufacturing (CAM)

Computer aided manufacturing (CAM) is a widely used term in the literature although researchers give various meanings for CAM. It is the result of computer aided processes after CAD and computer aided engineering (CAE) were applied widely.

After a product is modelled in CAD and the product design is analysed and verified in CAE, the product is ready to proceed to the production stage. CAM is applied to create control programs and run machine tools in manufacturing processes. Therefore, CAM is often used together with CAD and CAE for the design and manufacture of products. Here, CAM is defined as the technology for the use of software to control machine tools and other auxiliary tools in production. CAM falls into the category of programmable automations, which are used primarily in manufacturing systems to make products. Figure 9.5 shows some common production facilities where CAM tools are needed for control implementation. These production facilities are now discussed briefly.

9.2.1 Numerically Controlled (NC) Machine Tools

A *numerically controlled* (NC) machine is usually a machining system consisting of a machine tool, a number of motors to move the cutting tool along the path, and a controller that is responsible to generate and execute NC commands. An NC is used to cut away the unwanted materials from a workpiece in order to produce the desired product features and dimensions.

Machine tools are critical product facilities to the metalworking industry. Conventional machine tools are operated manually; in other words, all of the control parameters such as the feed rate, feed depth, cutting speed, and the coolant flow are determined and input by human operators. The programs of early NC machines were written as punched holes on paper or mylar plastic tape. A control command was represented by a set of holes, and was issued and transmitted to the motors by relays and other switches to move the machine. An NC was programmable and semi-automated in the sense that the machine tool could be re-programmed to make different parts. In addition, the machine tool was operated automatically to move the cutting tool and adjust the coolant system without human intervention. The controllers of CNC machines allow the operators to edit the programs at the machine interfaces. There is no need to send the tapes back and forth when some changes are occurring to the manufacturing processes. Therefore, CNC machines are substantially more reliable in comparison with the ordinary NC machines and no mylar tap is needed to store and input the control commands.

9.2.2 Industrial Robots

Robots are programmable devices to move end-effector tools around in order to perform given tasks. Industrial robots are robotic systems used in manufacturing industries. Industrial robots are widely applied in different operations such as loading/unloading, welding, debugging, painting, packaging, labelling, assembling, inspecting, and testing. Robots can substitute for humans to do unpleasant jobs as these job positions often have a high turnover of workers. Moreover, manual operations may cause inconsistent quality of products. Industrial robots perform very well to many repetitive processes such as spray painting and spot welding. These tasks are relatively easy to be automated because the paths for operations are predictable.

Industrial robots can be programmed in three different ways: (i) the *manual teaching method* where the end-effector positions are taught and recorded manually via a teach pendant, (ii) the *lead by the nose technique* where a user moves a robot along a robot path in the de-energized state by hand and the software logs the working positions into the memory, and (iii) *offline programming* where workpieces, robots, auxiliary tools and instruments are modelled, and the robotic motions are simulated, verified, and finally converted into control programs.

9.2.3 Automated Storage and Retrieval Systems (ASRS)

An automated storage and retrieval system (ASRS) consists of a variety of automated devices for placing and retrieving loads in a storage system. ASRSs are typically applied where (i) a high volume of goods are moved into and out of storages frequently; (ii) the space utilization is critical due to the space constraints; (iii) no value-added processing is needed but the temporary storage and transportation are essential to run the manufacturing systems; and (iv) no defect is expected in storage and transportation. Enterprises can benefit greatly from ASRSs to reduce labour in transporting products, reduce parts in the inventory, and track work-in-process parts accurately (Wikipedia 2019a). An ASRS can be viewed as a *discrete*

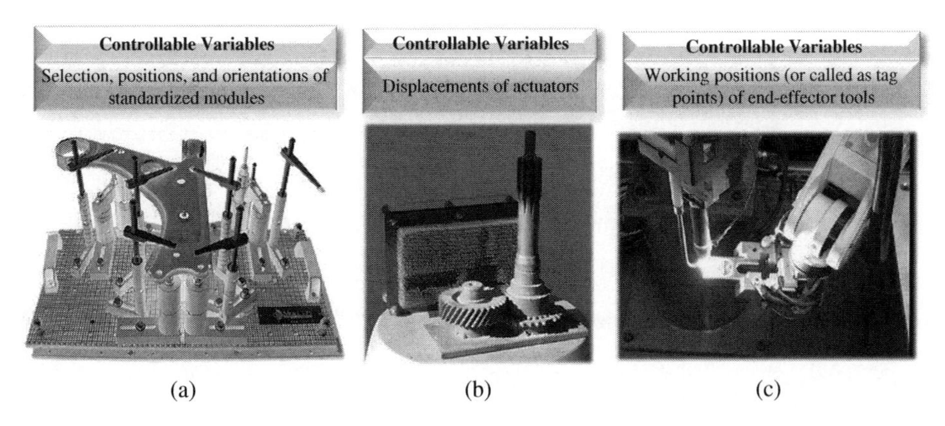

Figure 9.6 Controllable variables in three FFSs. (a) Modular fixture. (b) Adaptive fixture. (c) Fixtureless welding system.

event dynamic system (DEDS) and the controlling methods for DEDSs, such as programming logical controls (PLCs), Petri Nets, Matlib/Simulinks, rule-based controls and fuzzy controls, can be used to implement the controls of ASRS (Soyalan et al. 2012; Pohjalainen 2015).

9.2.4 Flexible Fixture Systems (FFSs)

Fixtures are required to support, locate, and clamp workpieces accurately subjected to the mechanical forces in manufacturing processes. Traditionally, fixtures are dedicated and tailored to the geometry and manufacturing processes of certain products; such fixture are very expensive. A flexible fixture system (FFS) has the flexibility to accommodate the variants, geometries, and manufacturing processes of products, and FFS is capable of self-reconfiguration to deal with different tasks by controls (Bi and Zhang 2001). Chapter 8 gives a detailed introduction to fixture design. Figure 9.6 shows the main controllable variables in three types of FFSs, i.e. *modular fixture systems*, *adaptive fixture systems*, and *fixtureless systems*.

Control of an FFS is different from that of other machine tools in the sense that the fixture is usually immobilized in the operation. Therefore, control of an FFS is often viewed as part of the fixture setup where the controllable parameters are set up in an off-line or on-line way. Depending on the types of adjustable mechanisms, controllable variables vary greatly in different FFSs. As shown in Figure 9.6, controllable variables for a modular fixture system are the selections, positions, and orientations of standardized functional modules. Controllable variables for an adaptive fixture are the displacements of actuators. Controllable variables for a robotic fixtureless system are the working positions (or called tag points) of end-effector tools.

9.2.5 Coordinate Measurement Machines (CMMs)

A coordinate measuring machine (CMM) is used to measure the geometry of physical objects by probing and sensing a number of discrete points on the surface of an object. The coordinates of the positions of interest can be measured by mechanical devices, optical devices, or laser or white light. The positions to be probed can be controlled manually or by computer programs. Typically, a CMM defines a probing position in terms of the displacements with respect to a global coordinate system. Some CMMs have more than three degrees of freedom and are capable of sensing angular coordinates of some features on the object.

9.2.6 Automated Material Handling Systems (AMHSs)

An AMHS has similar material handling functions as ASRS for storage, except that an AMHS can be applied in a production line and should be flexible enough to deal with various material handling tasks for flow controls, product routings, and the distribution of products (Azizi et al. 2018; Heragu and Ekren 2019). Some common facilities in an AMHS are conveyers, robots, gantries, cranes, hoists, and automated guided vehicles (AGVs). Each type of material handling device can be viewed as an automated system that consists of different motors and an actuator. From this perspective, the robotic control theory and methods can be applied to program and control automated systems.

9.3 Numerical Control (NC) Machine Tools

An NC or a CNC are controlled by numerical programs. Numerical programs are used to control different types of machine tools such as lathes, mills, routers, grinders, and many others. Control programs are changed and customized to product features to be manufactured. A control program consists of a series of instructions to control motion and operating parameters of a machine tool, such as feed rates, coordinates, locations, and speeds. In this chapter, NCs and CNCs are used to illustrate computer aided machining programs. Therefore, we will expand our discussions to the following four aspects: (i) the differences between NCs and CNCs, (ii) the programming methods, (iii) the advantages and limitations of computer aided programming tools, and (iv) the application of the commercial module of high-speed machining (HSM) for computer aided machining programming.

Figure 9.7 shows the difference between the programming methods for NCs and CNCs. The NC machine in Figure 9.7a consists of a program input unit, a machine control unit (MCU), and the processing equipment. A program is a sequence of planned operations and is written as numerical codes by human operators. Early NC programs were punched into punch cards, which were also called tape readers. The codes were sent to the MUC line by line and control commands were converted into the instructions to move the processing equipment. NC programs were eventually executed by the processing equipment.

Similar to an NC machine, a CNC machine is controlled by programs with numerical codes, which are usually written as G-codes and M-codes. However, a CNC machine usually has better machining capabilities as the programs can be written by a variety of input devices and are then fed to the MCU directly. Figure 9.7b shows the block diagram of a CNC machine. A computer can be used as an input device to create and store NC programs for different manufacturing processes. The programs can be easily edited when they are needed and the computer allows integration with a CAD software tool, where a CAM tool can generate NC programs directly based on the manufacturing features of a CAD model.

The driving system and the feedback devices are individual function modules in the machine tool. The driving system receives the control commands from the MCU and executes the commands to move the end-effector tool. The feedback devices report the states of

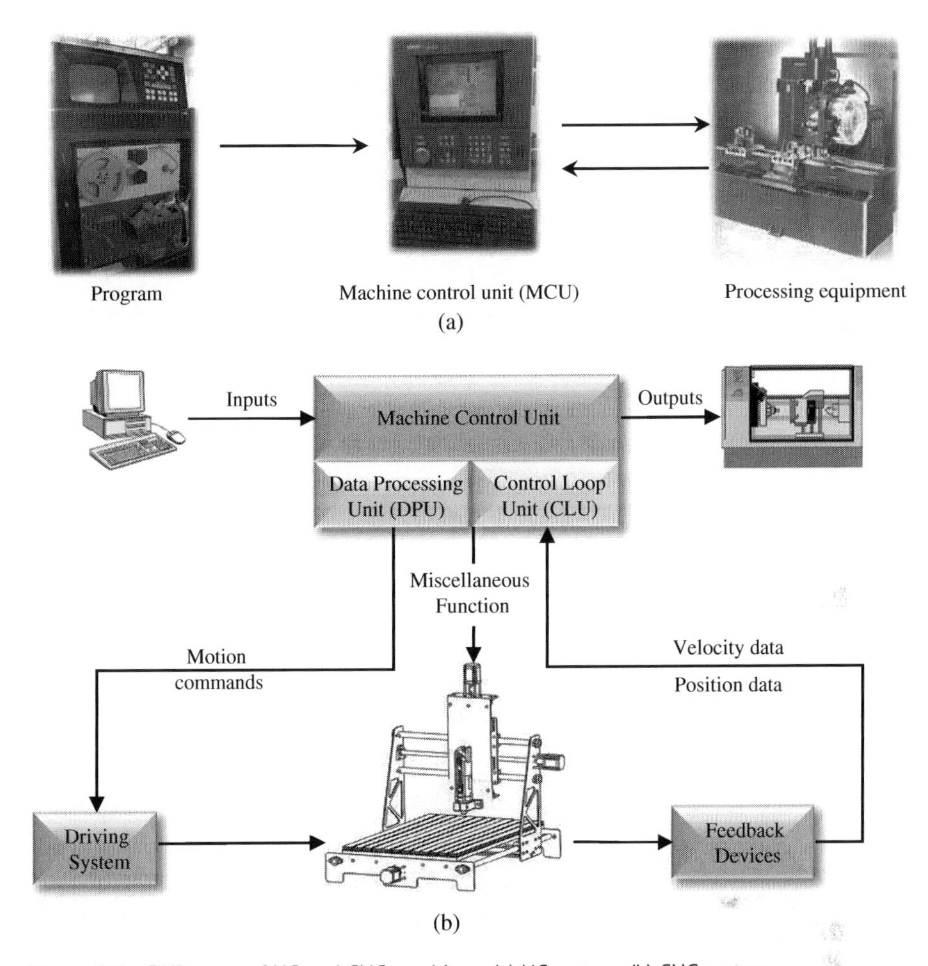

Figure 9.7 Difference of NC and CNC machines. (a) NC system. (b) CNC system.

the machine tool, such as displacements and velocities of joints. The MCU takes the feedback in the closed-loop control to adjust motion commands at sequential steps. Table 9.1 gives a brief comparison of NC and CNC machines to show their differences on definition, program storage, numerical codes, flexibility, maintenance, program execution, machining accuracy, the need for operator skills, running time, and integration (Engineeringinsider 2019).

The capabilities of NC machines can be expanded as *direct numerical controls* (DNCs) and *machining centres*. A DNC in Figure 9.8 is a collection of NC machine tools, where the control programs are stored and shared by the set of NC machines. The programs are provided to the networked machines by on-demand distribution of data to machines. With the network, one computer can be used to control a number of NC machine tools, which eliminates the need for substantial hardware for individual controllers of each NC machine tool.

A machining centre is capable of performing multiple machining operations and processes in a single setup. As shown in Figure 9.9, a machining centre usually has multiple axes and is equipped with an automatic mechanism to change tools. All of the motion

Table 9.1 The comparison of NC and CNC machines (Engineeringinsider 2019).

Features	NC machines	CNC machines
Definition	Programs are stored in punch cards; an MCU processes the basic commends for the movement of the machine tool	Programs are written through the computer and stored in the memory for machine tool operation; an MCU processes basic and advanced commends for the operation of the machine tool
Storage	Punch cards are used	Directly stored in computer memory
Numerical codes	Numeric, symbols, letters	G-codes and M-codes
Flexibility and computational ability	Low	High
Maintenance and cost	Less maintenance and less cost	More maintenance and high cost
Program execution	Requires a long time to execute a program for the set of given features	Requires less time to execute a program for the set of given features
Accuracy	Less	High
Need for operator skills	Operators are highly skilled	Operators are marginally skilled
Running time	Cannot run continuously	Can run for 24 h
Integration	No capability of integration	Support CAD and CAM integration

axes are driven by servomotors that are equipped with positioning feedbacks for precise closed-loop controls of machine motions.

9.3.1 Basics of Numerical Control (NC)

Figure 9.10 illustrates the working principle of NC. MCU consists of a data processing unit (DPU) and the control loop unit (CLU). The DPU decodes the information contained in the part-program, processes the information, and provides the instructions to the CLU. The CLU operates the drives attached to the machine by lead screws. The CLU provides the feedback about the actual position and velocity of the motion axes.

The drive units are actuated by voltage pulses. The number of pulses transmitted to each axis determines the required incremental motion and the frequency of these pulses represents axial velocities. Each incremental motion is called a *basic length unit* (BLU); accordingly, one pulse is equivalent to 1 BLU. One BLU represents the resolution of the NC machine tool, i.e. one BLU corresponds to the positional resolution of the axis of motion. For example, if one BLU is set as 0.0001 inch, this implies that the axis will move 0.0001 inch in response to one electrical pulse received by the motor. A BLU is also referred to as a *bit* (binary digital), i.e. one pulse → one BLU → one bit.

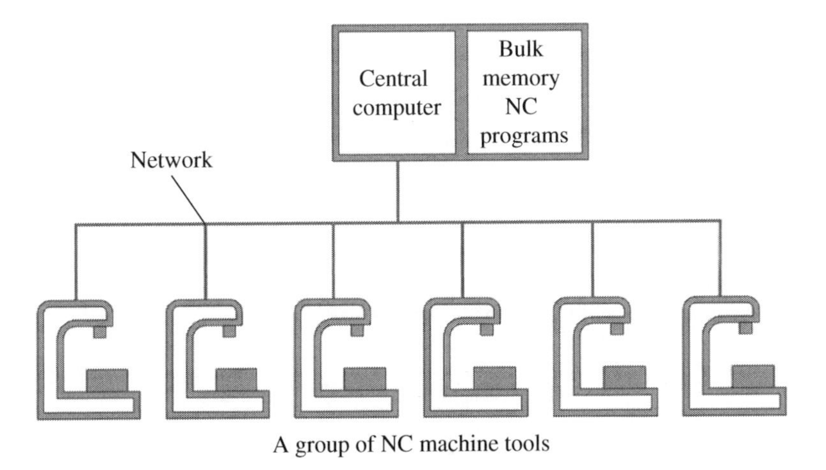

Figure 9.8 A central computer for the controls of DNCs.

Figure 9.9 Example of machining centre (Kasugai et al. 2003).

Example 9.1 A machine has 1 BLU = 0.001 inch. To move the distance d of 5 inches of the table on the x-axis at a speed rate V_L of 6 inches per minute, determine (i) the pulse rate and (ii) the total number of pulses for the required distance.

Solution

(i) The pulse rate R_p can be determined by the required moving speed V_L and BLU as

$$R_p = \frac{V_L}{BLU} = \frac{6}{0.001} = 6000\,(\text{pulse})/\text{minute}$$

(ii) The total number of pulses N_p for the required distance can be found as

$$N_p = \frac{d}{BLU} = \frac{5}{0.001} = 5000\,(\text{pulse})$$

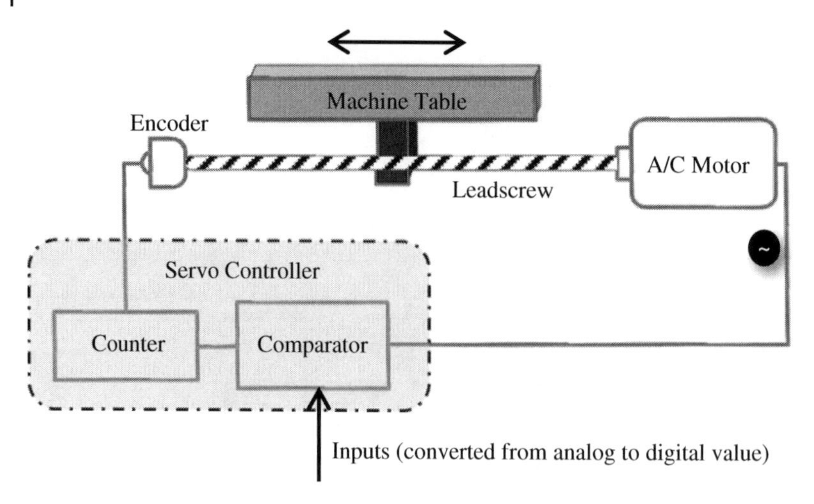

Figure 9.10 Numerical control working principle.

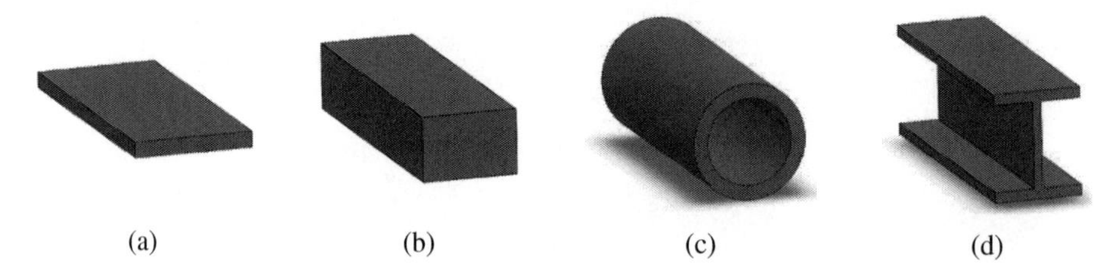

Figure 9.11 Examples of stock shapes. (a) Sheet. (b) Block. (c) Tube. (d) Beam.

9.4 Machining Processes

Before the machining programming of a machine tool is discussed, it is helpful to understand the operations and controllable parameters of machining processes. Note that we limit our discussion on programming of the machine tools for material removal processes.

Machining operations refer to a variety of material removal processes in which a cutting tool removes unwanted materials from a workpiece to produce the desired features and dimensions. The starting materials are usually cut from a larger piece of stock in a simple shape such as flat sheets, solid bars, hollow tubes, or construction beams shown in Figure 9.11. Machining operations can also be used to perform secondary operations on some products from other shaping processes such as casting or forging. Primary operations generate rough and inaccurate dimensions for these products.

As shown in Figure 9.12, machined parts are usually *prismatic* (Figure 9.12b and c) or *cylindrical* (Figure 9.12a) in their overall shapes. Some common features to be machined are *holes*, *slots*, *pockets*, *flat surfaces*, and *complex surface contours*. Machining is considered as the most versatile manufacturing process and can be applied to all commonly used engineering materials such as metals, plastics, composites, and wood.

Machining processes are usually expansive due to several reasons: (i) machining is a not net-shape operation and unwanted materials have to be removed to shape the right part of

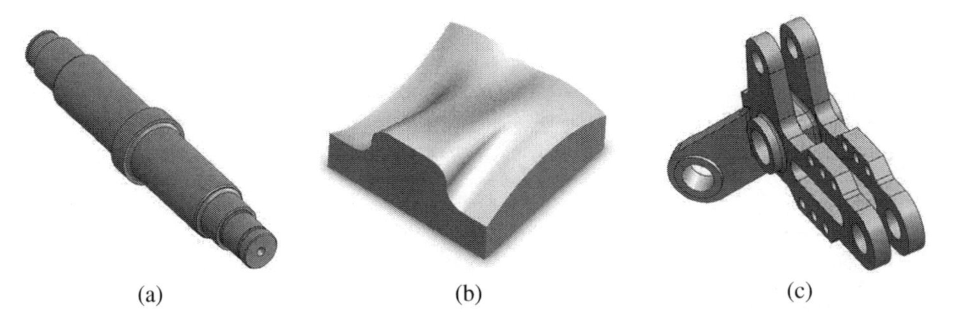

Figure 9.12 Examples of machined parts. (a) Part with turned features. (b) Part with milled features. (c) Part with mixed features.

the geometry; (ii) the initial investment to a high-performance machine tool is very high; (iii) and even though the cost to set up the tooling is low, it takes a long machining time and may be cost prohibitive for large quantities. Therefore, machining is often used to fabricate prototypes or custom tooling with a limited quantity. The parts are machined as moulds or tools for other manufacturing processes such as forging and injection moulding. Machining operations allow a tight tolerance and high surface finish; therefore, they can be used as a secondary process to add new features (such as side holes on a moulded part) or improve the accuracy of existing features on parts.

Machining operations can be performed in different ways as long as unwanted materials can be removed from starting parts. Depending on the methods for material removal, machining processes are classified into *conventional machining* and *non-conventional machining*. In conventional machining, unwanted materials are cut away by cutting tools subjected to large mechanical forces. In non-conventional machining, the materials are removed by chemical or thermal means.

Depending on the types of cutting tools and the cutting paths over workpieces, conventional machining processes are classified into single-point cutting, multi-point cutting, and abrasive machining. *Single-point cutting* involves one cutting edge, *multi-point cutting* involves multiple cutting edges, and *abrasive machining* removes material particles by wear. Figure 9.13 shows examples of machining operations under these categories. Turning operations are single-point cutting, turning and drilling operations are multi-point cutting, and grinding operations are abrasive cutting.

Other than the types of features a machine tool can make, another important measure for the capability of a machine tool is the *material removal rate* (R_{MR}). It is defined as the volume of removed materials in unit time. Therefore, the material removal rate is evaluated based on four process parameters, as shown in Figure 9.14:

1. Cutting speed (V) is the relative speed where the cutting tooltip passes through the workpiece. It depends on the motions of the workpiece and the cutting tool at the contact. The material removal rate is greatly affected by the cutting speed. The higher the cutting speed, the larger R_{MR} is and the better the productivity is. Tool manufacturers often recommend the cutting speed based on the combined material selections of the cutting tool and workpiece.

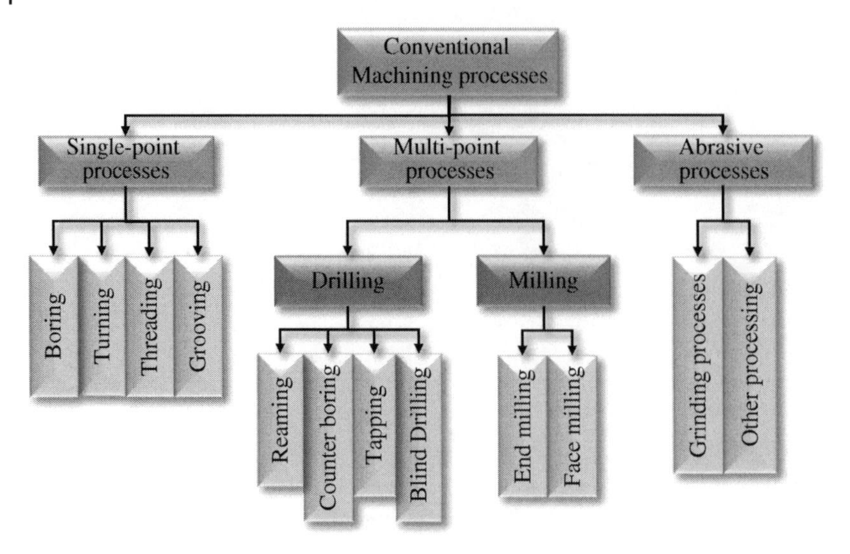

Figure 9.13 Types of conventional machining processes.

2. Spindle speed (N) is the angular velocity of the motion axis expressed in the number of revolutions per minute (RPM). It can be determined as

$$N = \frac{V}{\pi \cdot D} \tag{9.1}$$

   where V is the cutting speed, D is (i) the diameter of the workpiece at contact in a turning and boring process or (ii) the diameter of a cutting tool in drilling or milling operations.
3. The depth of cut (d) measures how far the tool penetrates into the workpiece to remove material in the current pass.
4. The feed rate (f) is the velocity at which the cutter is fed as it advances against the workpiece. It is expressed in units of distance per revolution for turning and boring, where the common units are inches per revolution and millimetres per revolution.

Accordingly, the material removal rate R_{MR} can be computed as

$$R_{MR} = v \cdot f \cdot d \tag{9.2}$$

9.5 Fundamentals of Machining Programming

Figure 9.15 shows the basic structure of a machine tool. The main moving components to be controlled are worktables, spindles, axes, cutting tools, and other auxiliary equipment attached to motion axes in machining operations.

9.5.1 Procedure of Machining Programming

CAM generates control programs for a machine tool to create geometric features on parts. CAM requires the information about machined features on a part. The information is usually defined in the CAD model of a part. Therefore, CAM is driven by the CAD model of

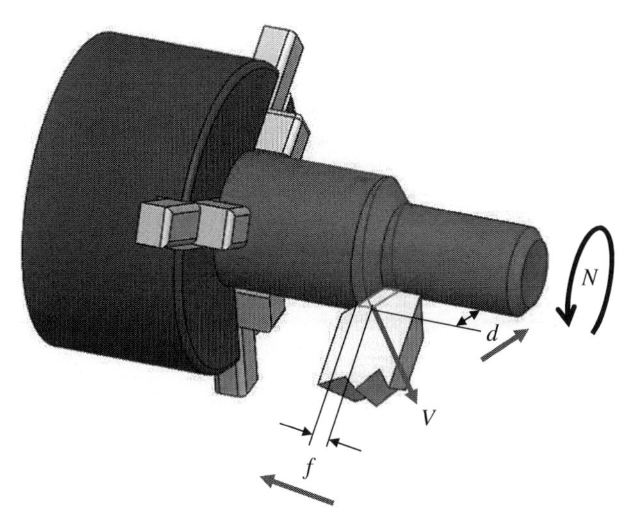

Figure 9.14 Process parameters in a turning operation.

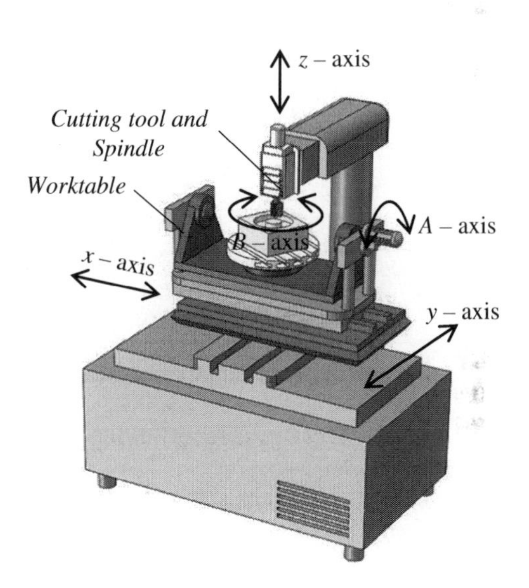

Figure 9.15 Examples of the moving components in a CNC machine tool.

the part to be made. Figure 9.16 shows the workflow of machining programming, where the procedure of machining programming consists of eight sequential steps. The main activities involved in these steps are elaborated in Table 9.2.

9.5.2 World Axis Standards

Computer programming follows a set of rules called *syntax* to combine symbols and words as statements. A program consists of valid statements that are legal constitutions or fragments. For machining programming, the motion nomenclatures and coordinate systems follow the EIA267-C standards by the Electronic Industries Association (EIA) (Global Engineering 2019). EIA267-C aims to eliminate inconsistencies and

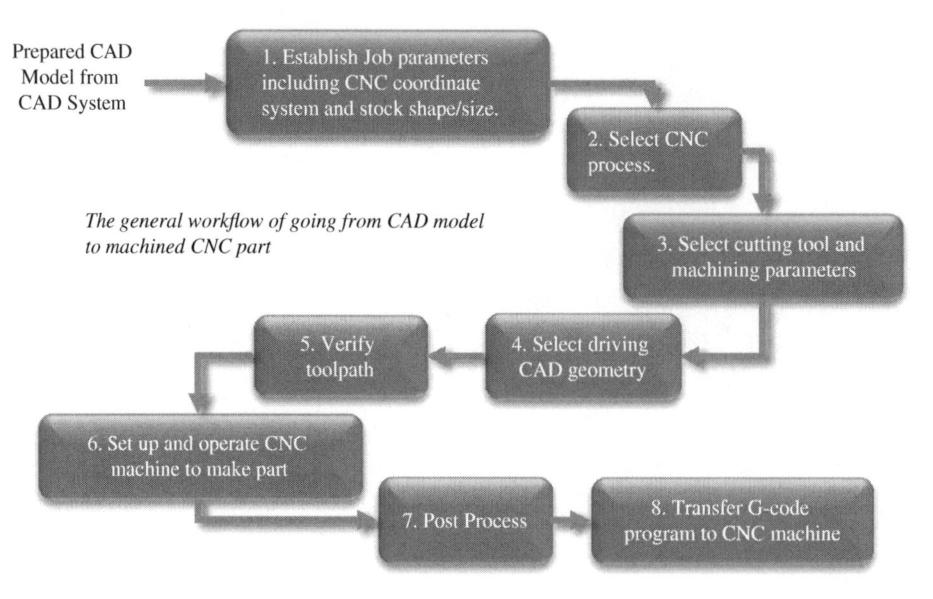

Figure 9.16 The workflow of CNC programming.

Table 9.2 Main activities involved in CNC programming.

Step	Activities
1	The job parameters, such as coordinate systems and the sizes and shapes of stock, are defined
2	The types of CNC operations, such as turning, drilling or milling, are selected
3	The cutting tools are selected and the machining parameters are determined
4	The part feature is analysed to specify a number of working points for each path and the tool path is planned
5	The feasibility of toolpath(s) is verified to ensure all working points are accessible and without interference
6	The toolpaths are converted to control programs that are downloadable to set up the machine tool
7	The program is post-processed to take into consideration the global coordinate system, the reference coordinate systems for the cutting tool and the workpiece. It is desirable to verify CNC codes by simulation
8	The verified program is downloaded as the G-codes to control the machine tool to perform the machining operation

misunderstandings among CNC manufacturers, programmers, and users. It simplifies machining programming and facilitates the interchangeability of the programs on different machine tools.

EIA267-C specifies the coordinates and movements of machine tools, so that a programmer can model the machining operations even before the motion of the cutting tool relative to the workpiece is defined. EIA defines 14 standard axes for the motions and positions of the workpiece, cutting tools, and other objects involved in the machining processes (SAE

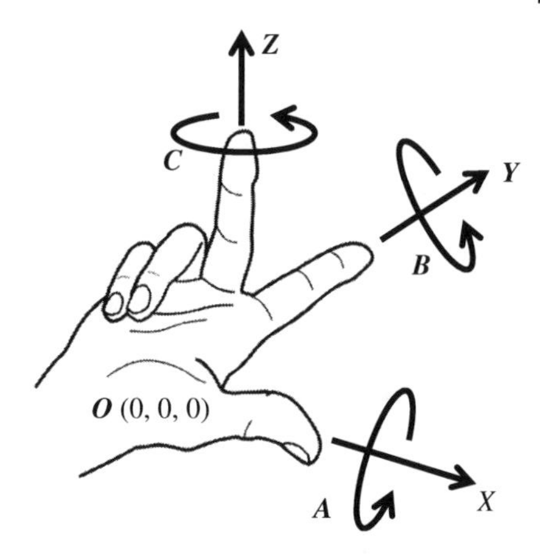

Figure 9.17 Right-hand rule for the definition of coordinate systems and motions.

2016). Nine axes are sufficient to define most of the machine tools in industry, while some multiplexed machine tools have more auxiliary motion axes. However, machine tools will continue to evolve, since the central processors are able to deal with more and more computations per nanosecond. This allows a machine tool to perform more and more functions simultaneously. For the simplicity of our discussion, we confine the discussions on a machine tool with no more than nine axes: three primary linear axes (***X***, ***Y***, and ***Z***), three primary rotary axes (***A***, ***B***, and ***C***), and three secondary linear axes (***U***, ***V***, and ***W***).

The motion of the cutting tool is defined with respect to one or more reference coordinate systems. All of the coordinate systems follow the right-hand rule shown in Figure 9.17:

1. The direction of each finger represents the direction of the translational motion.
2. The axis of the main spindle is always defined as the *Z*-axis (the direction of the middle finger) and the positive direction is into the spindle.
3. The longest travel slide is designated as the *X*-axis (the direction of the thumb) and is always perpendicular to the *Z*-axis.
4. If one rotates the hand for 90° by looking into the middle finger, the forefinger represents the *Y*-axis. The base of the hand is set as the origin *O* (0, 0, 0) of the coordinate system {*O* - *XYZ*}.
5. The right-hand rule for the positive direction of an axis is used to determine the clockwise rotary motion about the *X*, *Y*, *Z* axes.
6. To determine the positive (or clockwise) direction about an axis, close the hand with the thumb pointing to the positive direction of the axis. Note that the thumb may represent any one of the *X*, *Y*, *Z* axes, and the curl of the fingers represents the positive (or clockwise) rotation about the axis. The rotary motions about the *X*, *Y*, *Z* axes are known as *A*, *B*, and *C*, respectively.

Using the aforementioned rules, the motion nomenclatures for conventional machine tools are illustrated in Table 9.3.

9.5.3 Default Coordinate Planes

A motion path of the cutting tool mostly consists of line and arc segments on a plane. It is helpful to use a plane in the Cartesian coordinate system to define a motion path. The

Table 9.3 Motion nomenclatures of machine tools.

Type	Description	Illustration
Lathe	On most CNC lathes, the Z-axis is parallel to the spindle, and the longer motion axis is defined as the X-axis.	
Mill	In the shown mill, the spindle travels along the Z-axis. The other axis with a longer travel stroke is defined as the X-axis and the other axis with a shorter travel stroke is defined as the Y-axis.	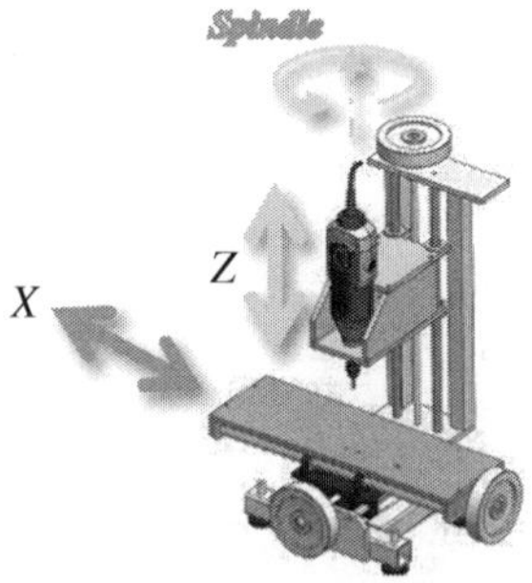
Knee mill	On a common vertical knee CNC mill the spindle is stationary and the direction of the spindle axis is defined as the Z-axis. The other two axes are defined as X- and Y-axes, respectively, based on their travel distance.	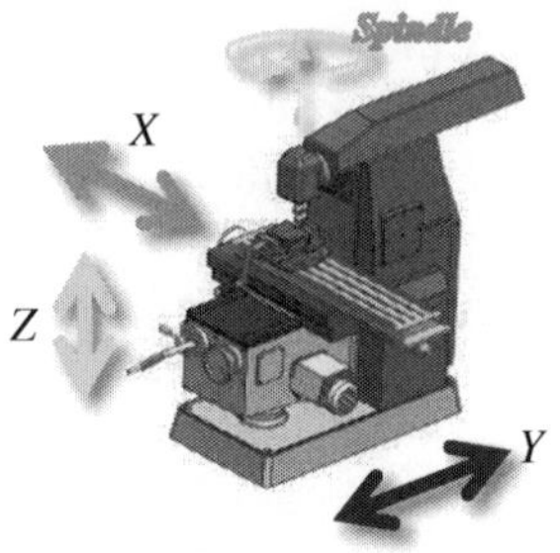
Contour mill	On this five-axis horizontal contour milling machine, note the orientation of the X- and Y-axes in relation to the Z-axis. The rotary axes for both the X- and Y-axes are designated by the A and B rotary tables.	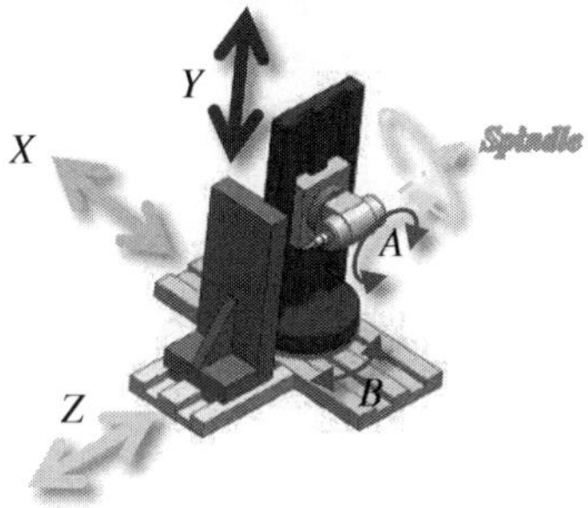

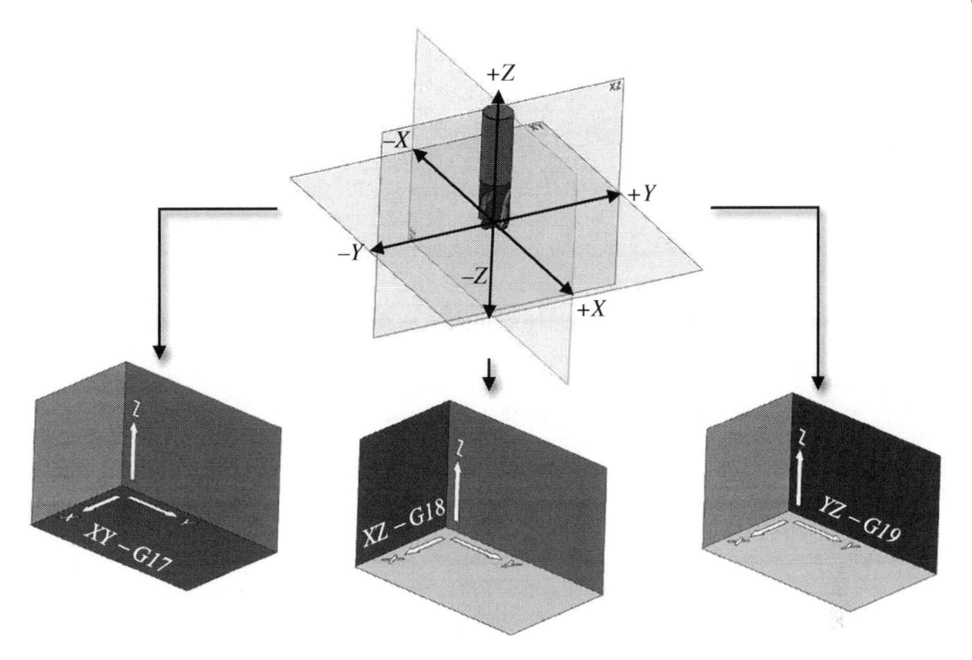

Figure 9.18 Default coordinate planes for a mill machine.

default Cartesian planes *XY*, *XZ*, and *YZ* are defined based on the given coordinate axes *X*, *Y*, and *Z*. Figure 9.18 shows the correspondence of these default planes and their notations for a mill machine.

As shown in Table 9.3, a turning machine only has a primary axis (*Z*-axis) and a secondary axis (*X*-axis). The turning machine does not have a motion in the third axis (*Y*-axis) except the depth of cut by the cutting tool. All of the CNC lathes use a default two-axis coordinate system, shown in Figure 9.19. This allows for the transfer of CNC programs or measurements among different lathes. Note that the cutting tool on a CNC lathe differs from a conventional lathe in the sense that the cutting tool is usually placed on the top or back side of the machine tool instead of being placed at the front side horizontally on a conventional lathe.

In the default *XZ* plane, a central reference point *O* is used to measure the coordinates along the *X* and *Z* axes, and the coordinates of *O* are set as *O* (0, 0). For consistence, it is convenient to set the right-hand central endpoint of the workpiece as the origin of the ***XZ*** plane. However, keep in mind that at times the centre left-hand end-point of the workpiece might be used occasionally.

A CNC lathe makes axially symmetric parts, while corresponding NC programs can use diametrical programming or radical programming. *Diameter (or diametrical) programming* relates the *x*-coordinate to the diameter of the workpiece at a specified location. For example, if a feature on the workpiece has a 5 inch outside diameter and the NC program uses absolute coordinates to move the cutting time to the outside diameter, the *x*-coordinate of the motion commended is X5.0. *Radius (or radial) programming* relates the *x*-coordinate to the radius of the workpiece at a specified location. For the move

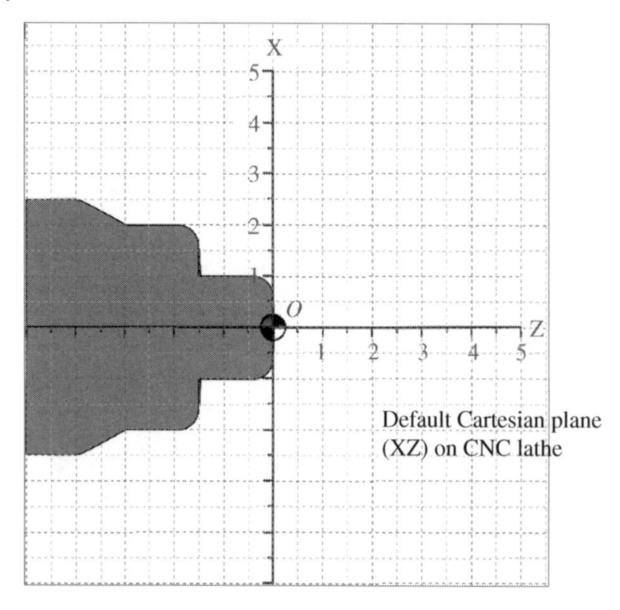

Figure 9.19 Default two-axis coordinate system on a CNC lathe.

commended for the feature with a 5 inch outside diameter, the *x*-coordinate of the motion comment becomes X2.5.

9.5.4 Part Reference Zero (PRZ)

To determine the relative position of an object in different coordinate systems, the following points are usually taken as the references (Figure 9.20): (i) Machine Reference Zero (*M*), (ii) Part Reference Zero (PRZ) (*W*), and (iii) Tool Reference (*R*). All coordinates of the working positions of a cutting tool are defined based on *M* and *W*:

1. All CNC machine tools require a reference point (*M*) from which to determine the coordinates of working points.
2. It is generally easier to use a point on the workpiece itself for a reference, since the coordinates apply to the part anyway – thus the PRZ designation (*W*).
3. The PRZ (*W*) is defined as the lower left-hand corner and the top of the stock of each part.

9.5.5 Absolute and Incremental Coordinates

The working points can be represented by absolute coordinates and incremental coordinates with respect to a reference coordination system. *Absolute coordinates* of a point are determined when the origin of the coordinate system is used as the reference point for measurement. As shown in Figure 9.21, the coordinates of the points *A*, *B*, and *C* are measured from the reference point *O*, i.e. the origin of the coordinate system {*O-XY*}.

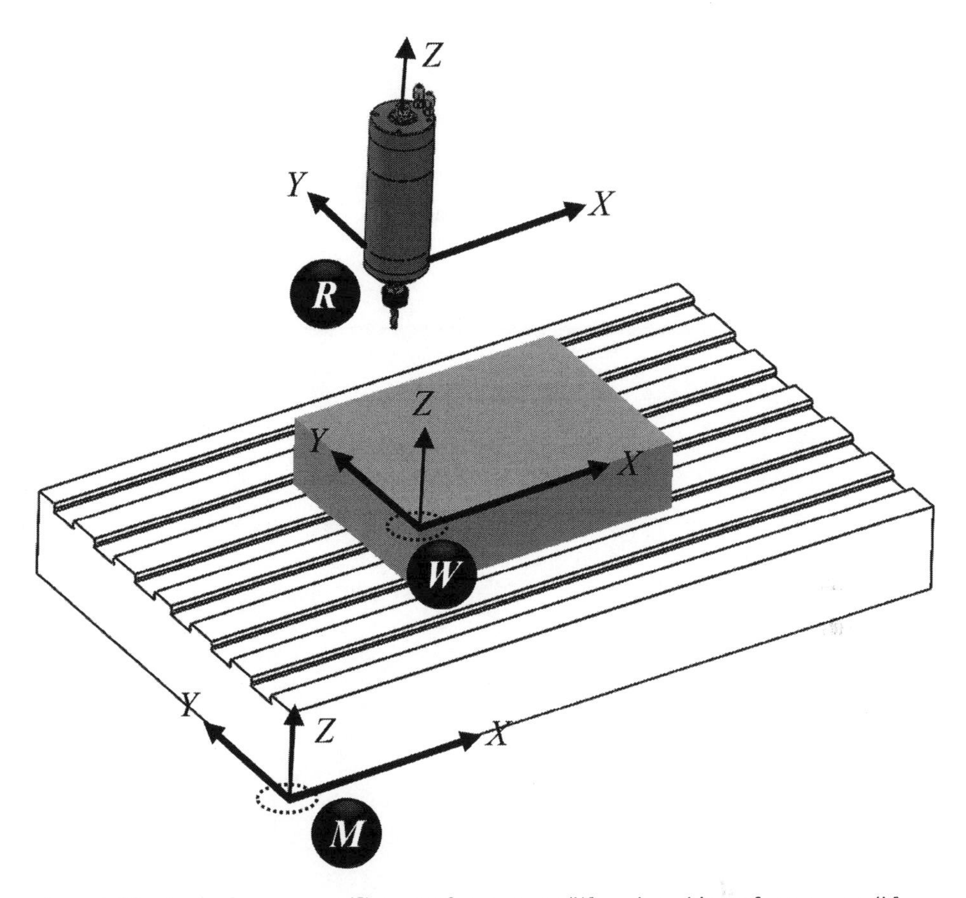

Figure 9.20 Tool reference zero (*R*), part reference zero (*W*), and machine reference zero (*M*).

Example 9.2 Determine the absolute coordinates of the tag points on the path shown in Figure 9.22a.

Solution

The coordinates are determined by taking the origin *O* as the reference for the measurement. The coordinates for these tag points are shown in Figure 9.22b.

Absolute coordinates apply to any coordinate planes including the default *XZ* plane for a CNC lathe. In programming the operations for a lathe, absolute coordinates always start at the origin (X_0, Z_0). As shown in Figure 9.23a, to determine the coordinates for the next point, track the path along the *Z*-axis until the next point is reached on the *z*-axis right below the point; read the *Z* coordinate and then go up to reach the point and the vertical distance will be the *z*-coordinate. The absolute coordinates for the points P_1 to P_6 are determined in Figure 9.23b.

Incremental coordinates use the present position as the reference point for the next movement. It implies that any point in the Cartesian coordinate system can be plotted accurately by measuring relative distances of points by pairs in a sequence starting at the

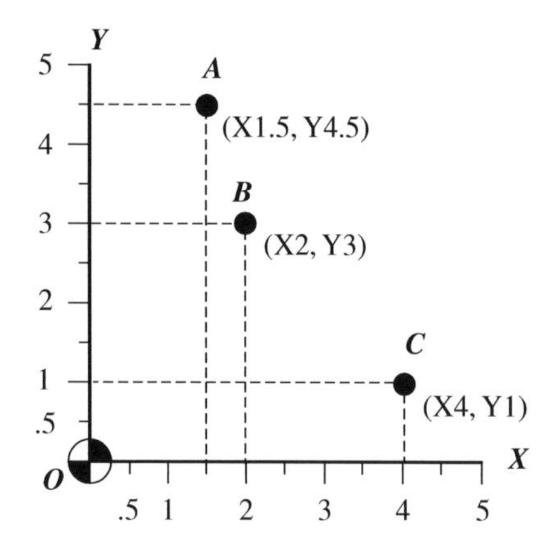

Figure 9.21 Absolute coordinates measured from the origin of the coordinate system.

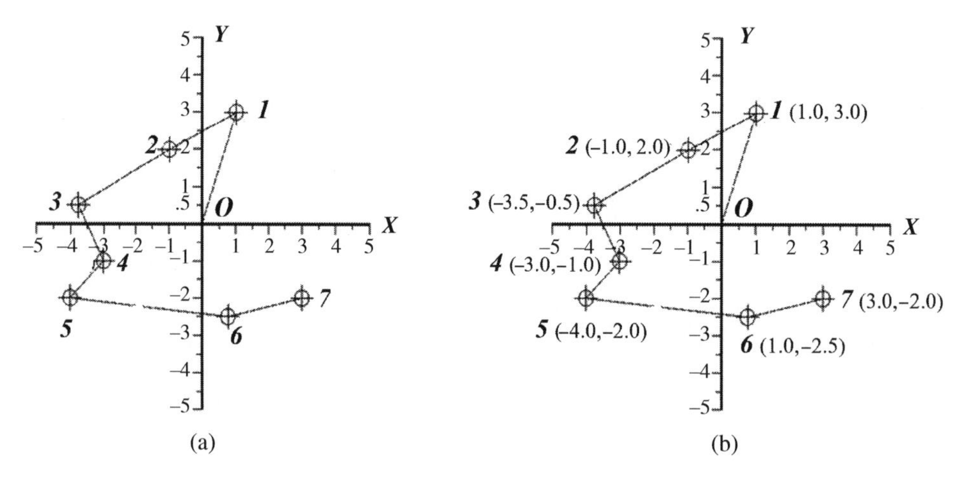

Figure 9.22 Example of absolute coordinates. (a) Tag points on path. (b) Absolute coordinates.

origin. Figure 9.24 shows the scenario of defining incremental coordinates for tag points on the path.

Example 9.3 Determine the incremental coordinates of the working points in Figure 9.25a.

Solution

The incremental coordinates are determined sequentially from the current point to the next point. The coordinates for the working points for *O*, *A*, *B*, *C*, *D*, *E*, and *F* are shown in Figure 9.25b.

9.5.6 Types of Motion Paths

In the machining process, the path of the cutting tool is planning to generate part features. The type of motion path depends on the feature type to be generated. Figure 9.26 shows four main types of motion paths. Figure 9.26a is a *point-to-point motion* where the cutting tool

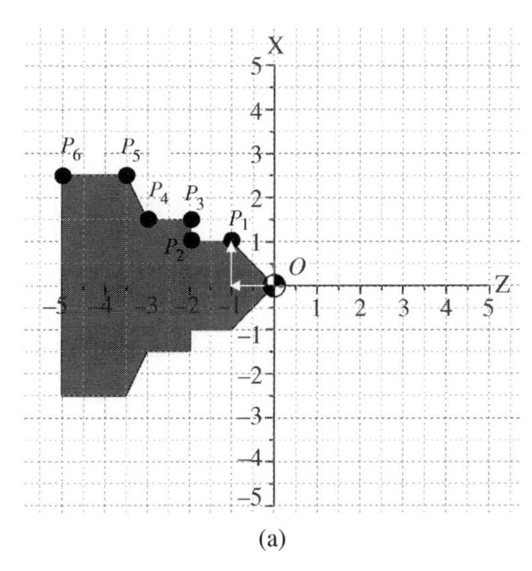

(a)

Position	Z	X_R	X_D
O	0.0	0.0	0.0
P_1	−1.0	1.0	2.0
P_2	−2.0	1.0	2.0
P_3	−2.0	1.5	3.0
P_4	−3.0	1.5	3.0
P_5	−3.5	2.5	5.0
P_6	−5.0	2.5	5.0

X_R and X_D are for radical and diametrical programming, respectively

(b)

Figure 9.23 Example of absolute coordinates for lathe. (a) Working points on a turned part. (b) Absolute coordinates on the *XZ* plane.

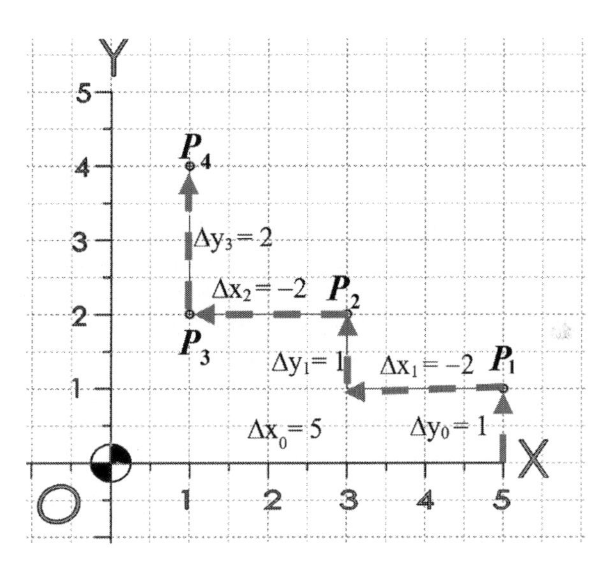

Figure 9.24 Incremental coordinates measuring relative displacements of points in sequence.

operates at a set of discrete points but is not significant for the connection of two working points since no operation occurs to the cutting tool. A point-to-point motion is applicable to drilling or boring operations. Figure 9.26b shows a *point-to-point straight motion*. The path consists of a set of line segments, where each segment is formed by connecting two working points in a straight line. The cutting tool operates continuously from one point to another and the connection of the two points is straight. This applies to a frame milling to remove unwanted materials on lateral sides. Figure 9.26c shows a *two-dimensional continuous path*. It is formed as the smooth connection between a set of tag points on a plane that is switchable, depending on the orientation of the feature. It is applicable to 2D contouring mills. Figure 9.26d shows a *three-dimensional continuous path* that is used to generate a curvy

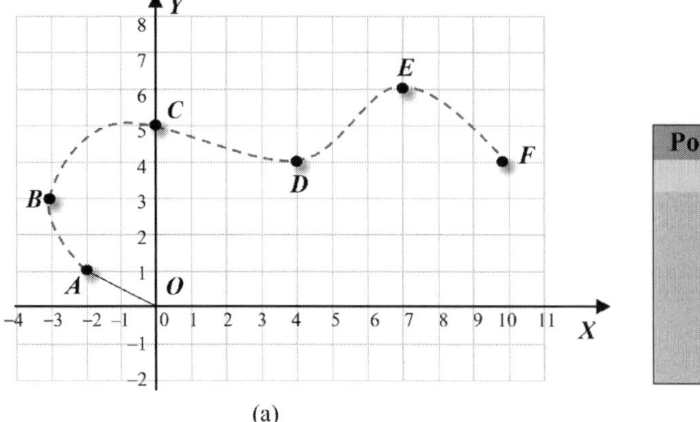

Position	ΔX	ΔY
O	0.0	0.0
A	−2.0	1.0
B	−1.0	2.0
C	3.0	2.0
D	4.0	−1.0
E	3.0	2.0
F	3.0	−2.0

(b)

Figure 9.25 Example of incremental coordinates. (a) Tag points on path. (b) Incremental coordinates.

surface in space. The path is formed by following the nodes on the grid by zigging-zagging. It is applicable to 3D contouring mills.

9.5.7 Programming Methods

NC machine tools are controlled by computer programs. To generate the designed features on a workpiece, CNC machining operations must be programmed first so that MCU can control the motion of the cutting tool to generate the required features. Three common ways to program a machine tool are *manual (NC) programming*, *conversational (shop-floor) programming*, and *CAM system programming*. Each of these three methods has its niche in the manufacturing industry (Lynch 1994).

Manual programming works well if the part is simple since no advanced programming tool is needed to create programs. Since a manual program is written in the same language the CNC machine is equipped with, the program can be as concise as possible to achieve efficiency of the execution. However, manual programming could be very tedious and error-prone if a large number of operations are involved. Manual programming is preferable when (i) a program will be for the same parts with a high volume and (ii) the priority of a machining program is the efficiency of machining operations. In addition, CNC users are expected to understand the techniques of manual programming, so that they are able to check, verify, and make corrections for NC programs when needed.

Conversational programming is supported by CNC machines where graphic and menu-driven functions are used to create programs for parts. A user can check the inputs of the program, visualize the cutting tool path graphically, and simulate what would happen when the program is used to run the machining process. Conversational programming is widely used in small or medium sized enterprises (SMEs). In SMEs, CNC machine operators usually take full responsibility for the setups of machine tools, fixtures, and tooling, preparing, verifying, and optimizing programs, and finally running CNC machines. In contrast to manual programming, conversational programming helps them to reduce the programming time dramatically. Note that the capability of conversational

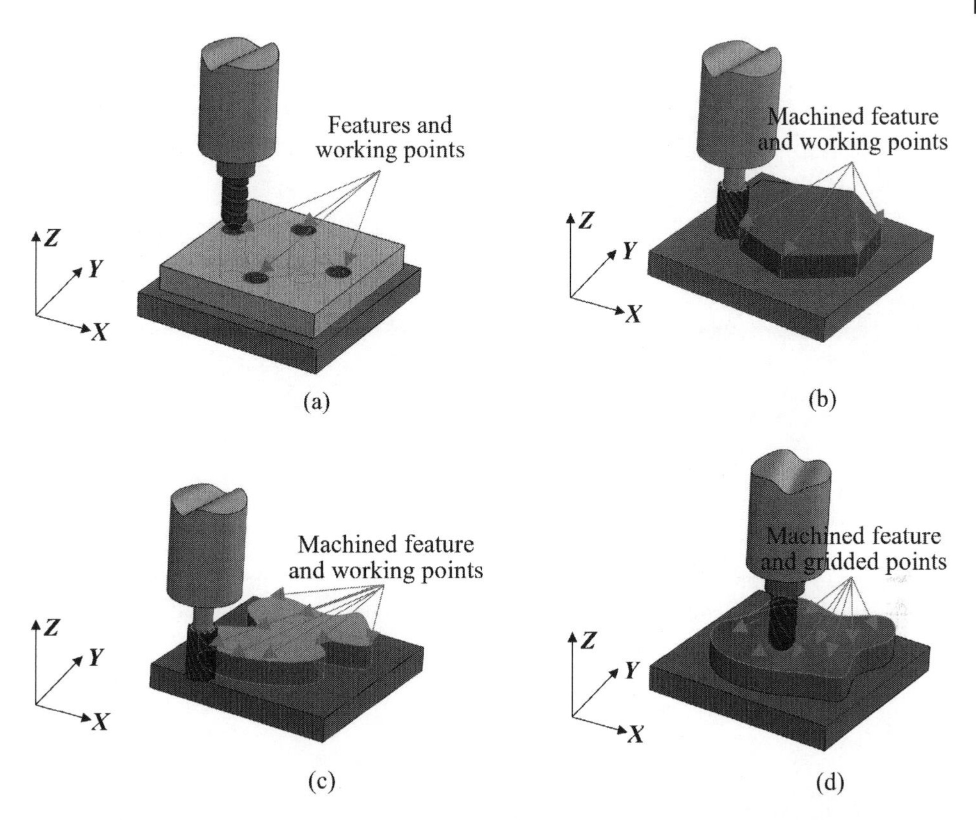

Figure 9.26 Common types of motion paths. (a) Point to point motion (drilling and boring). (b) Point to point straight motion (frame milling). (c) Two-axis contouring with a switchable plane (2D contouring milling). (d) Three-axis contouring with a continuous path (3D contour milling).

programming is provided by CNC machines and varies greatly from one machine vendor to another. From this perspective, a CNC machine with the capability of supporting conversational programming can be treated as a single-purpose CAM system since it provides a convenient way to write programs for parts from the same type of machines. Conversational programming used to be the only way to program some legacy machine tools, but modern CNC machines support both conversational programming and offline programming.

A CAM system prepares CNC programs on a much higher level than manual programming and conversational programming. It has gradually gained popularity. A CAM system facilitates CNC programming mainly on three aspects: (i) do mathematical calculation of tool paths automatically, (ii) create generic programs that can be compatible with different machine types, and (iii) provide the library with reusable functional modules and routines for common machining operations. In programming by a CAM system, users create NC programs based on CAD models of parts in an off-line mode. The CAM system is capable of generating the programs in G-code, which is similar to manually created NC programs. The verified programs in a CAM system can be downloaded and directly transferred to machine tools. CAM can be non-graphic-based or graphic-based. Early CAM systems were

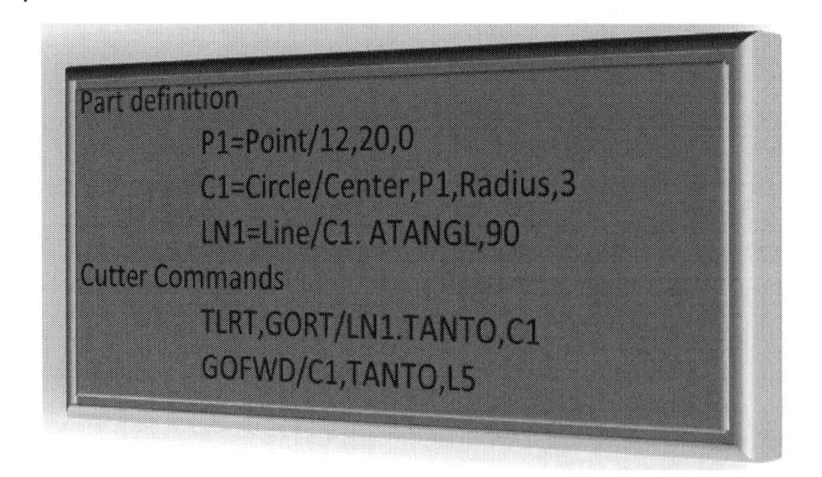

Figure 9.27 NC example code from APT.

non-graphic-based and created the programs in BASIC, C Language, or other languages; those systems were not user-friendly. Recent CAM systems are graphic-based and CNC programs can be created interactively. Users can get feedback visually at every programming step. Off-line programming by a CAM system can minimize programming and debugging time for the preparation of CNC programs.

9.5.8 Automatically Programmed Tools (APT)

An Automatically Programmed Tool (APT) is a high-level computer programming language, which is most commonly used to generate the instructions for NC machine tools. APT was developed by Douglas T. Ross at MIT in 1956. APT is the language and system that has made NC practical. APT is capable of generating complex machine codes such as one for a five-axis machine tool. It is still used today and accounts for about 5~10% of all programming in the defence and aerospace industries.

An NC program from APT consists of four types of statements or declarations, i.e. (i) *geometric statements* for part geometry, (ii) *motion statements* for the paths of cutting tools, (iii) *post-processor statements* for the specifications of feeds, speeds, and feature actuations, and (iv) *auxiliary statements* for identifications of parts, tools, or tolerances. As shown in Figure 9.27, a programmer defines a series of *lines*, *arcs*, and *points* that define the overall part geometry locations. These features are then used to generate a *cutter location* (CL) *file*.

Example 9.4 Write an NC program for the milling operation on the part shown in Figure 9.28.

Solution
Table 9.4 gives the result of NC programming. The program consists of a series of lines. The first three lines are written to label the program, specify machining type, and select the tool. The second part of the program defines all geometric entities. The third part of the program sets up the operating conditions for cutting speed, federate, and coolant supply. The fourth

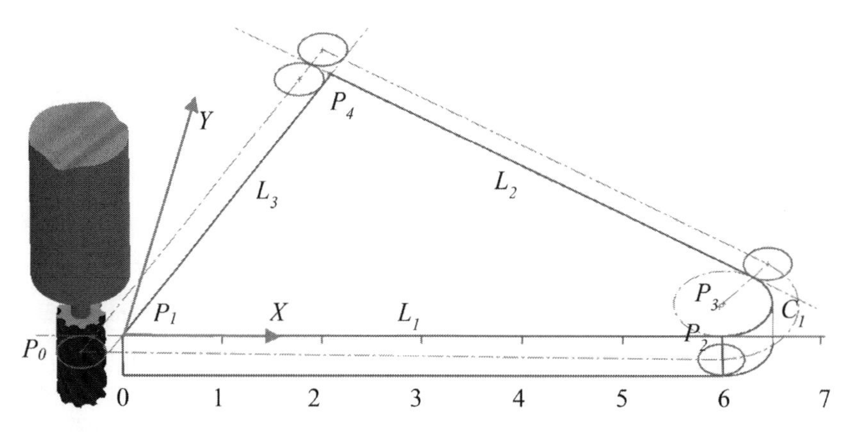

Figure 9.28 Example of a milling path.

Table 9.4 NC program for the milling operation of part in Figure 9.28.

NC code	Explanation
PARTNO/EXAMPLE9-3	; label the program as 'EXAMLE9-4'
MACHIN/MILL, 1	; select the target machine and controller type
CUTTER/ 0.5000	; specifies the cutter diameter
P0 = POINT/−1.0, −1.0, 0.0	
P1 = POINT/ 0.0, 0.0, 0.0	; geometry statement to specify the pertinent surface of part
P2 = POINT/ 6.0, 0.0, 0.0	
P3 = POINT/ 6.0, 1.0, 0.0	
P4 = POINT/ 2.0, 4.0, 0.0	
L1 = LINE/P1, P2	
C1 = CIRCLE/CENTER, P3, RADIUS, 1.0	
L2 = LINE/P4, LEFT, TANTO, C1	
L3 = LINE/P1, P4	
PL1 = PLANE/P1, P2, P3	
SPINDLE/573	; set the spindle speed to 573 RPM
FEDRAT/5.39	; set the feed rate to 5.39 inch per minute
COOLNT/ON	; turn the coolant on
FROM/P0	; gives the starting position for the tool
GO/PAST, L3, TO, PL1, TO, L1	; initialize contouring motion, drive, part, and check surfaces
GOUP/L3, PAST, L2	
GORGT/L2, TANTO, C1	; motion statements to contour the part in the clockwise direction
GOFWD/C1, ON, P2	
GOFWD/L1, PAST, L3	
RAPID	; move rapidly once cutting is down
GOTO/P0	; return the tool to the home position
COOLNT/OFF	; turn off the coolant
FINI	; terminate the program

part of the program is for all motion statements. The last part of the program resets the tool position and terminates the machining operation.

9.6 Computer Aided Manufacturing

APT offers the advantages for manual programming. However, it involves defining part geometries and tool positions that tend to be very error-prone. To address this concern, graphics-based CAM was introduced and became popular in the 1980s. This made it possible to describe part geometry by points, lines, and arcs directly rather than requiring a translation to a text-oriented notation. CAM is driven by the graphic representation of a part to generate NC programs semi-automatically. It can usually be run on workstations or personal computers. There is a need when using CAM to understand the fundamentals of CNC programming and machining processes, such as the selection of speeds, feed rates, and feed depths based on the materials, tolerances, and tool specifications. Therefore, CAM needs the users to have some built-in expertise about manufacturing processes. As a powerful computer aided technique, CAM provides the following functionalities:

1. Make standardized tools and materials available from the built-in design libraries.
2. Support the simulation of motion paths of cutting tools.
3. Be capable of editing and optimizing tool paths.
4. Estimate cutting times and machining costs.

CAM allows minimum information from users in programming machines for parts. The default settings in CNC programming follow the EIA standards or common practices, especially the following:

1. It is assumed that the tool moves relative to the workpiece no matter what actual motion occurs to the machine tool. Taking as an example of the operations on a shaper, the cutting tool is kept still and the workpiece is moved towards the cutting tool.
2. The positions of the objects are described in a Cartesian coordinate system.
3. If the reference position (0.0, 0.0, 0.0) is flexible, it can be specified by the programmer; such a position is called a floating zero.

9.6.1 Main Tasks of CNC Programming

To implement the procedure of CNC programming in Figure 9.2, a CAM system needs to accomplish a series of tasks, which are explained in Table 9.5.

9.6.2 Motion of Cutting Tools

In CNC programming, the motion of a cutting tool can be specified using one of the following five methods:

1. A rapid travel to a reference location without engaging a machining operation.
2. A straight linear motion actuated by a single axis or multiple axes.
3. A circular motion that is defined in a planer surface.

Table 9.5 Steps and tasks for CNC programming in CAM.

Tasks	Description
1	Prepare the CAD model of a part. Even though imported geometric model or reversely generated model is acceptable, it is ideal to generate a parametric model so that all of the machined features can be selected by the CAM system automatically.
2	Identify the features to be manufactured. Common features are contours, pockets, hole patterns, free-form surfaces, planar surfaces. Select the type of machining operation and tooling for each feature.
3	Define the cutting and processing parameters, mainly the information of the machine tool such as type, revolutions per minute (RPM), federate, and cut depth, and path planning methods such as zigg-zagging, spiralling, inside-out, and the parameters for rough/finish cutting.
4	Run the simulation of cutting operation, visualize the motion of the cutting tool along the cutting path, and modify or delete the cutting sequences if it is needed.
5	Identify the machine and cutting tool to be used and post-process the current location (CL) for machine-specific NC programs.
6	Filter out the CL information and format it into an NC code based on machine-specific parameters such as work envelope, limits of federate, tool changer, RPM, and G&M function capabilities.
7	Output the NC code. An NC code is a series of commands that controls the motion of the cutting tool. A machine tool can download and run the verified NC codes for actual machining operations.

4. A combination of planar and linear motions that is commonly called as 2½-D motion. A planar motion is implemented by two motion axes, which move simultaneously in one plane. This plane is perpendicular to the motion direction of the third axis at the feeding direction.
5. Any complex motion that can be treated as a combination of the aforementioned motions; for example, the motion along an arc in an arbitrary plane can be approximated by a set of straight lines and the arc path can be followed by three simultaneous motion axes.

Due to advanced technologies in computing and controls, most recent CNCs are able to support true 3D motion planning along any path and orientation in the workspace of the motion tool.

9.6.3 Algorithms in NC Programming

To program interference-free tool-paths for part features, CAM is equipped with various algorithms to calculate tag points, generate paths, and detect interferences. To convert part features into an executable program, the algorithms in a programming system must be able to:

1. Perform the trigonometry calculation to determine tag points, path segments, and their connections.

2. Computer offset contours based on specified part outlines (or tool path) and cutter dimensions.
3. Post-process a generic program to accommodate the differences of reference zeros, features, and motion capabilities.
4. Convert all motion instructions into the motion commands of individual motion axes.
5. Visualize and monitor the machine operations and, especially, the motion path of a cutting tool.
6. Verify the feasibility of CNC programs. Sometimes verification can be performed by actual tests on a workpiece with substituted materials such as light metals, plastics, foams, wood, laminates, and other low-cost materials for NC proofing.

9.6.4 Program Structure

A CNC program consists of a series of blocks. Each line in a program corresponds to a block. A block consists of one or several instructions or words that are separated by spaces or tab characters. Therefore, it is not possible to use a space within a word. A word is a character for a single function; for example; '*X*' is the displacement along the *X*-axis and '*F*' is the feed-rate.

A block in a CNC program is similar to a sentence in the English language. The sentences in English are separated by periods while the blocks in a CNC program are separated by the end-of-block (EOB) character. A block includes all the information contained between successive EOB characters. The EOB character can be easily generated by pressing the *Enter* or *Return Key* on the data-entry terminal. A collection of English sentences makes up an essay and the collection of blocks makes up a program. The EOB character signals the controller to execute the commands contained in the block. The same EOB character that ends one block also signals the start of the next block.

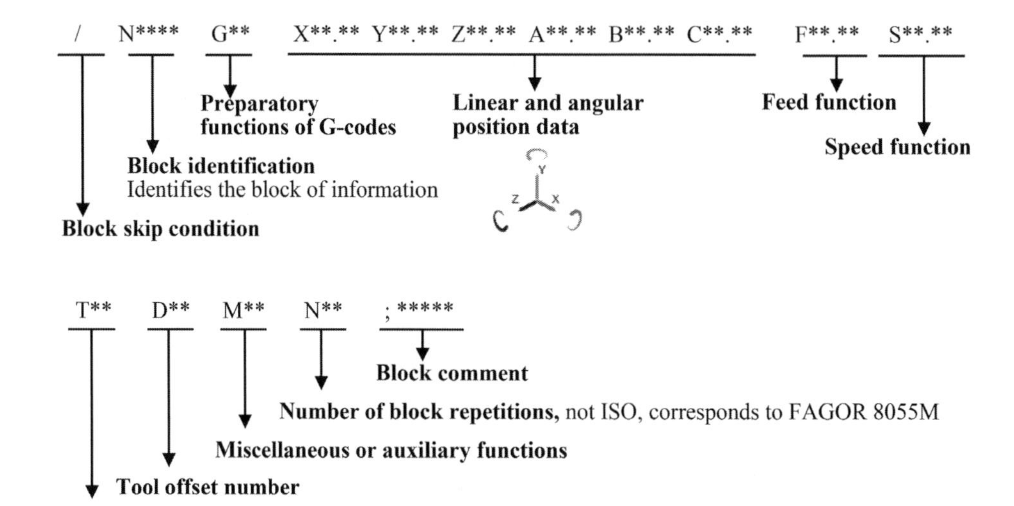

Figure 9.29 The block structure of an NC statement.

Figure 9.29 shows the block structure of a statement in an NC program with some commonly used words: '/' specifies whether or not the block is skipped; '*N*' is the block identification, where a block might not have the block identification if it is not referenced by other blocks such as a jump statement; '*G*' is for the preparatory functions provided for next operations; '*X*', '*Y*', '*Z*' are for the primary displacements in the *X*, *Y*, *Z* directions; '*U*', '*V*', '*W*' are for the secondary displacements in the *X*, *Y*, *Z* directions; '*A*', '*B*', '*C*' are for the angular displacements in the *X*, *Y*, *Z* directions; '*I*', '*J*', '*K*' are for the distances to the arc centre or thread leads parallel to the *X*, *Y*, *Z* directions, respectively; '*F*' is for federate; '*S*' is for the cutting speed; '*T*' is for tool number; '*D*' is for tool offset; and '*M*' is for other miscellaneous functions.

9.6.5 Programming Language G-Code

The language for CNC programming is known as the G-code, which has been standardized by several organizations. The main version used in the United States was version D of RS274, which was developed by the Electronic Industries Alliance (EIA) in 1979 (Kramer et al. 2000). Globally, DIN 66025 is used in Germany, PN-73M-55256 and PN-93/M-55251 are used in Poland, and the standard ISO 6983 is used by other counties (Wikipedia 2019a, 2019b). However, different standards of G-code use the same block structure and similar words in defining the motion of cutting tools and specify process parameters. All G-code versions have five types of words, which are listed in Table 9.6.

The tool motions are defined by G words. Table 9.7 provides the set of commonly used G-words.

Each G word has its specific format in NC programming. In the following, the G words for some common tool motions are introduced as examples:

1. Rapid traverse motion (G00) is to take the shortest route to move the tool at a rapid feed rate to the new position with specified coordinates:

Table 9.6 Five types of words in G-code.

Words	Function	Block example
N *G*	is to specify block identity is used as the preparatory commands for motion control	N010 X70.0 Y85.5 F175 S500 (EOB) where *N*-word = a sequence number (010) *X*-word = *x* coordinate position (70.0 mm) *Y*-word = *y* coordinate position (85.5 mm) *F*word = feed rate of 175 mm min^{-1} *S*-word = spindle speed of 500 RPM (EOB) = end of block
S, *F*, *T*, *D*, …	are to specify cutting speed, federate, tool, offset, …	
X, *Y*, *Z*, *U*, *V*, *W*, *A*, *B*, *C*, …	are to specify motion displacements	
M	is to set up miscellaneous parameters for machine control	

Table 9.7 Commonly used G words.

G-word	Description
G00	Rapid point-to-point movement
G01	Linear motion between two points
G02	Clockwise circular motion
G03	Counterclockwise circular motion
G04	Tool dwell
G10	Tool offset
G17	Selection of *XY* plane
G18	Selection of *XZ* plane
G19	Selection of *YZ* plane
G20	Input data in inches
G21	Input data in millimetres
G28	Go to reference point
G90	Absolute coordinates
G91	Incremental coordinates
G94	Feed per minute in milling and drilling
G95	Feed per revolution in milling and drilling
G98	Feed per minute in turning
G99	Feed per revolution in turning

Format:	G00 X*** Y*** Z***
where:	X, Y, Z are the words for the displacements along three primary motion axes, respectively, and *** is the value of a displacement.

2. Straight line motion (G01) is to move the tool from its current position to a new position with specified coordinates at the given feed rate. Figure 9.30 shows that G01 generates a straight line path from its current position to the next position:

Format:	G01 X*** Y*** Z*** F***
where:	X, Y, Z are the words for the displacements along three primary motion axes, respectively, F is the word for federate, and *** is the value of a displacement or feed rate.

3. Close-wise arc motion (G02) is to move the tool along with a circular arc in the clockwise direction until it reaches new position coordinates at the given feed rate. Figure 9.31 shows a close-wise arc path in XY, XZ, and YZ planes, respectively:

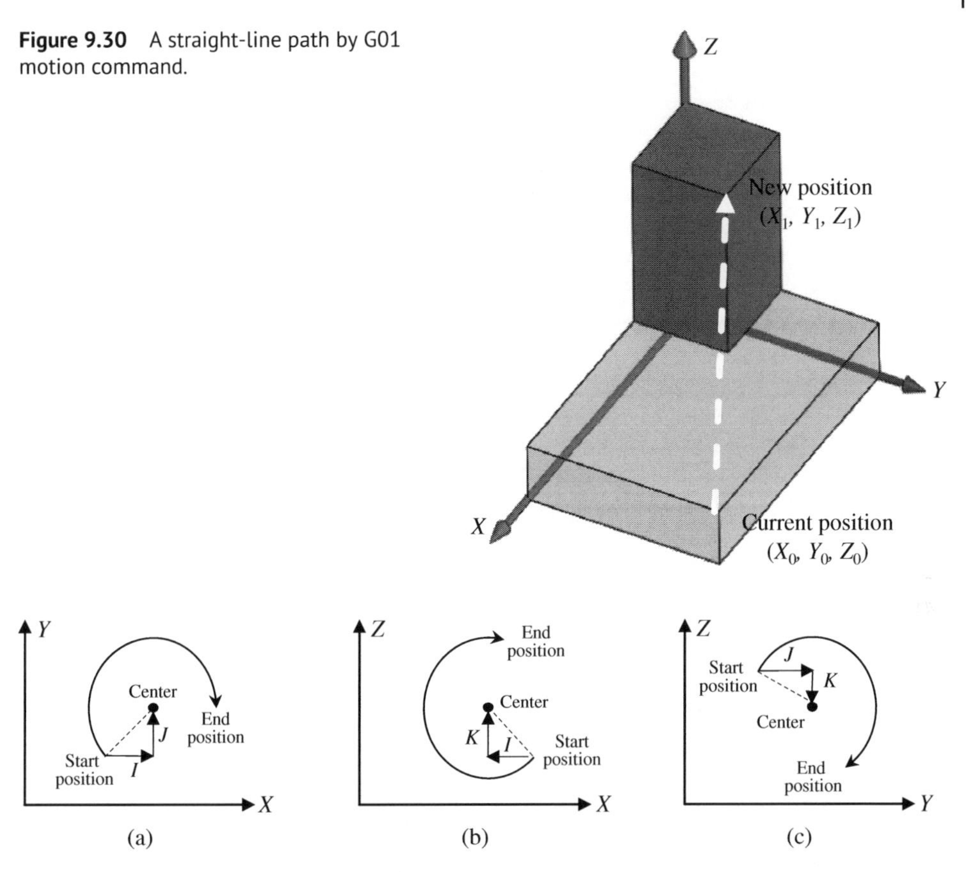

Figure 9.30 A straight-line path by G01 motion command.

Figure 9.31 A clock-wise arc path by G02 in *XY*, *XZ*, and *YZ* planes. (a) *XY* plane. (b) *XZ* plane. (c) *YZ* plane.

Format:	G17 G02 X*** Y*** I*** J*** F*** ;Arc on XY plane
	G18 G02 X*** Z*** I*** K*** F*** ;Arc on XZ plane
	G19 G02 Y*** Z*** J*** K*** F*** ;Arc on YZ plane
where:	G17, G18, or G19 represents the plane where the arc locates,
	X, Y, Z are the words for the coordinates of a designated point,
	I, J, K are the words of the displacements from the starting point to the arc centre along X, Y, Z axes, respectively,
	F is the word for feed rate, and
	*** is the value of a coordinate, displacement, or feed rate.

Example 9.5 Write a program for the motion control of the tool to drill a hole at (1.0 inch, 1.0 inch) with the depth of 0.75 inch for the part shown in Figure 9.32.

Solution

For manual programming, a designer begins to plan the steps of machining operations and then interprets these steps into statements in an NC program. Figure 9.33 shows the main

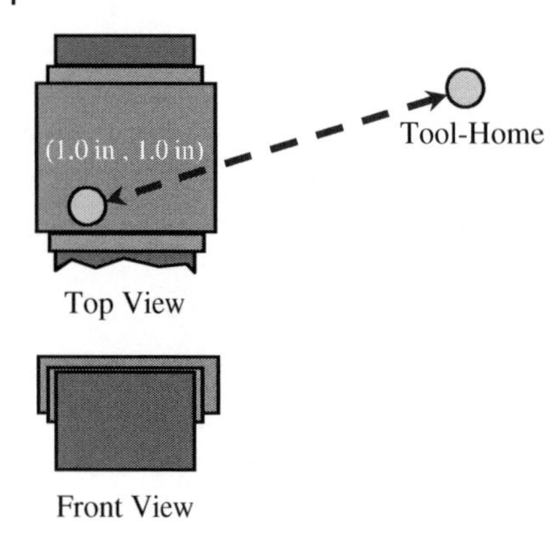

Figure 9.32 Example of NC programming for a drilling operation.

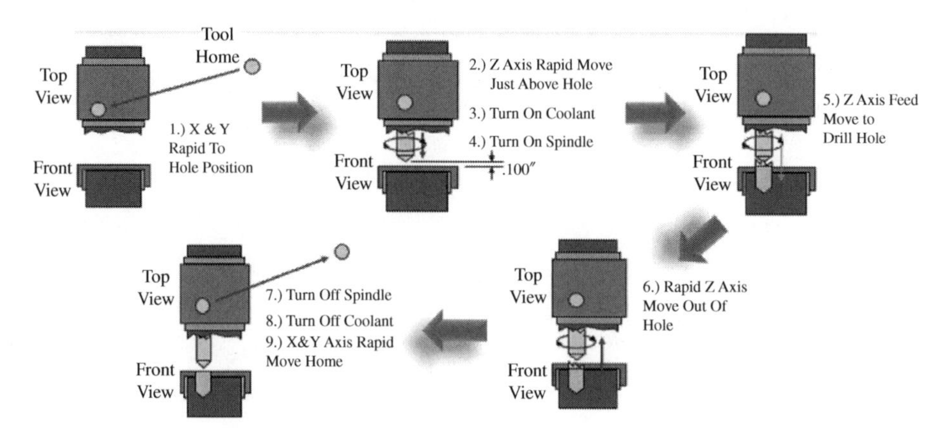

Figure 9.33 The steps of drilling operations in Example 9.5.

steps the machine tool has to take to perform the drilling operation and Figure 9.34 shows the concise version of an NC program with detailed explanations for motion instructions.

Example 9.6 Write a program with the motion commands for the contouring operation on the part shown in Figure 9.35a.

Solution

For the operation for contouring the profile of the workpiece, the steps for the motion of the cutting tool are straightforward. They begin at the home position S and follow the order of $S \to A \to B \to C \to D \to E \to F \to G \to H \to I \to A \to S$. There are two arc motions; the first one $B \to C \to D$ is in the counter-clockwise direction and the second one $G \to H$ is in the clockwise direction. Accordingly, the NC program for this series of motions are given in Figure 9.35b.

'*M*' stands for 'machine' and *M* words in G-code are mostly for miscellaneous functions to activate or deactivate machine functions. Table 9.8 shows the functions of some commonly used M words.

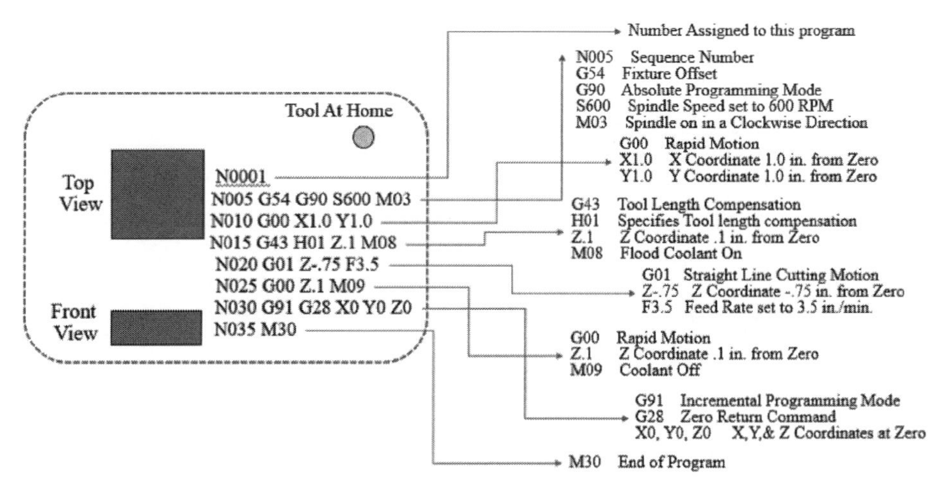

Figure 9.34 The NC program for the drilling operation in Example 9.5.

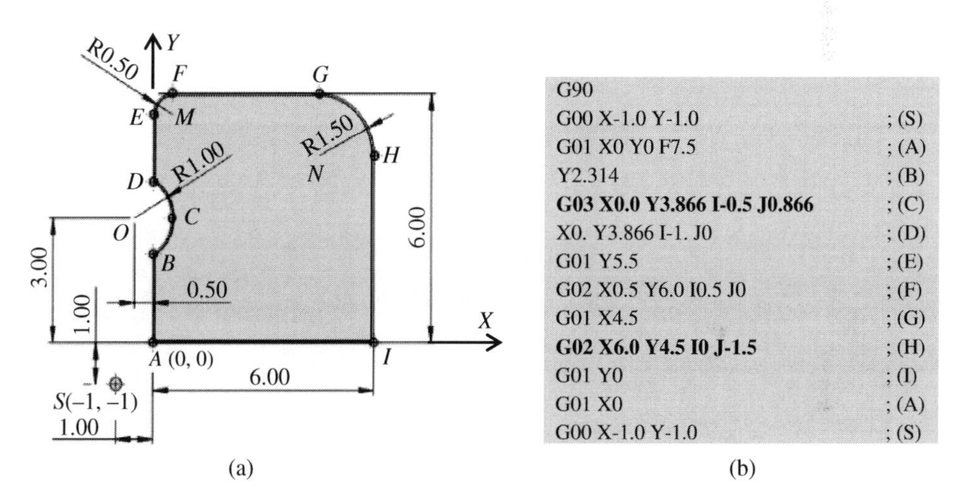

Figure 9.35 Example of CN programming with arc motions. (a) Contouring operation. (b) NC programming.

9.7 Example of CAM Tool – HSMWorks

HSMWorks is a CAM add-in tool for SolidWorks. It is designed to generate tool paths for parts with manufacturing features automatically. HSMWorks aims to produce the smoothest paths with a reduced machining time, an improved surface finish, and less tool wears. HSMWorks is equipped with all of the conventional machining strategies, including parallel, contour, pocket, scallop, radial, spiral, and pencil, and expands these strategies with the smooth links of path segments (ECAD Inc 2019; Bebjaminsen 2015). Figure 9.36 gives the collections of main features of HSMWorks (Autodesk 2017). It supports the NC programming for milling and turning operations, is very powerful in supporting the

Table 9.8 Commonly used M words.

M-word	Description
M02	End of program and machine stop
M03	Clockwise spindle start (CSS)
M04	Counterclockwise spindle start (CCSS)
M05	Spindle stop
M06	Tool change
M07	Turn cutting fluid on, flood mode
M08	Turn cutting fluid on, mist mode
M09	Turn off cutting fluid
M10	Automatic clamping of fixture
M11	Automatic unclamping of fixture
M13	CSS and turn on cutting fluid
M14	CCSS and turn on cutting fluid
M17	Turn off spindle and cutting fluid
M19	Turn off spindle at oriented position

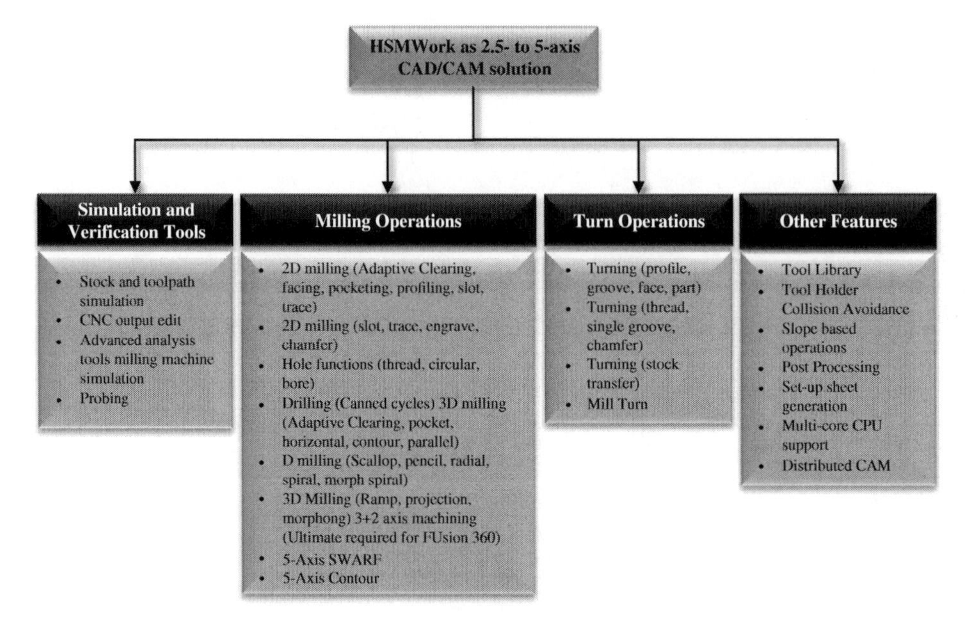

Figure 9.36 The features of HSMWorks as a 2.5- to 5-axis CAD/CAM solution.

simulation and verification of CNC programs, and has seamless integration with CAD and the CNC machine.

The use of HSMWorks is straightforward. Figure 9.37 shows the graphic-based user interfaces. The tool can be accessed by turning on the add-in of ‘CAM’. The part CAD model with manufacturing features can be directly imported. The operation for each feature can be programmed individually. A tool path can be simulated and visualized

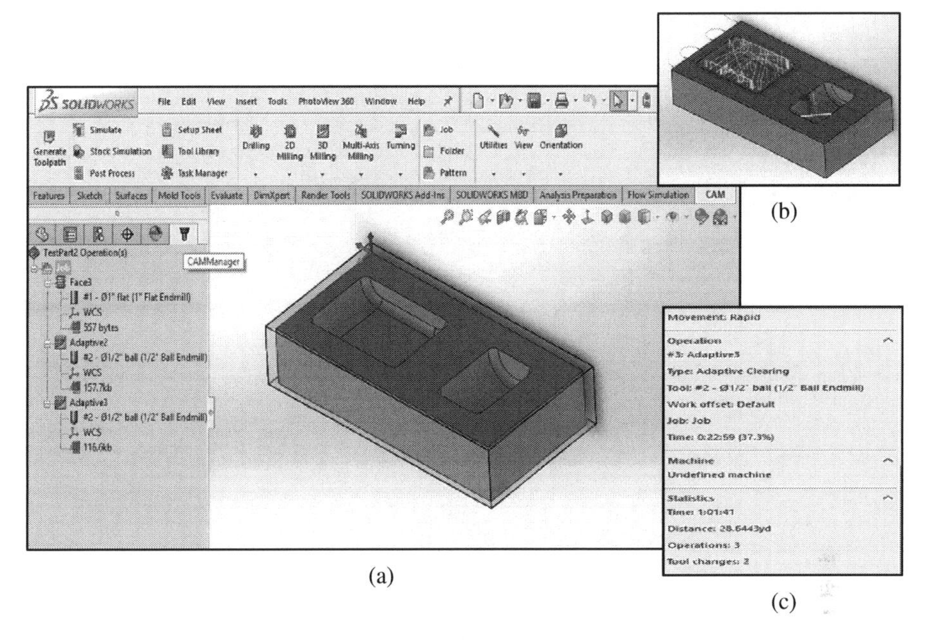

Figure 9.37 Programming interface of HSMWorks. (a) User-friendly programming interface. (b) Tool path simulation. (c) Post-processing and statistics.

at any time (Figure 9.37b). The software is equipped with design libraries for users to choose process types, machines, and tools (Figure 9.37a). If the information about the machine is known, the NC program can be post-processed and downloaded to the machine for operation. The statistics of machining operations in an NC program are generated automatically for evaluation (Figure 9.37c).

9.8 Summary

CAM is an automation process to convert a product CAD model into a code format that is readable by machines in manufacturing processes. CAM was evolved from an NC technique in the early 1950s that used the coded instructions on punched paper to control machining processes. CAM makes it possible for engineers to model and analyse manufacturing processes, detect design defects in the early stages, and evaluate the performance of production systems in a virtual environment. CAM systems can be utilized to maximize the productivity of production facilities such as CNC machines, robotics, and high-performance machine centres.

CAM is an important constitution in CIM to make the connection between the virtual world and the physical world. The CNC programs from CAM can be post-processed and downloaded to machine tools for the control of actual machining operations where products are made. Therefore, it is critical to simulate and verify CNC programs before they can be deployed in a physical manufacturing environment. Manual programming for machining operations is very tedious and error-prone, but it is advantageous for engineers to master

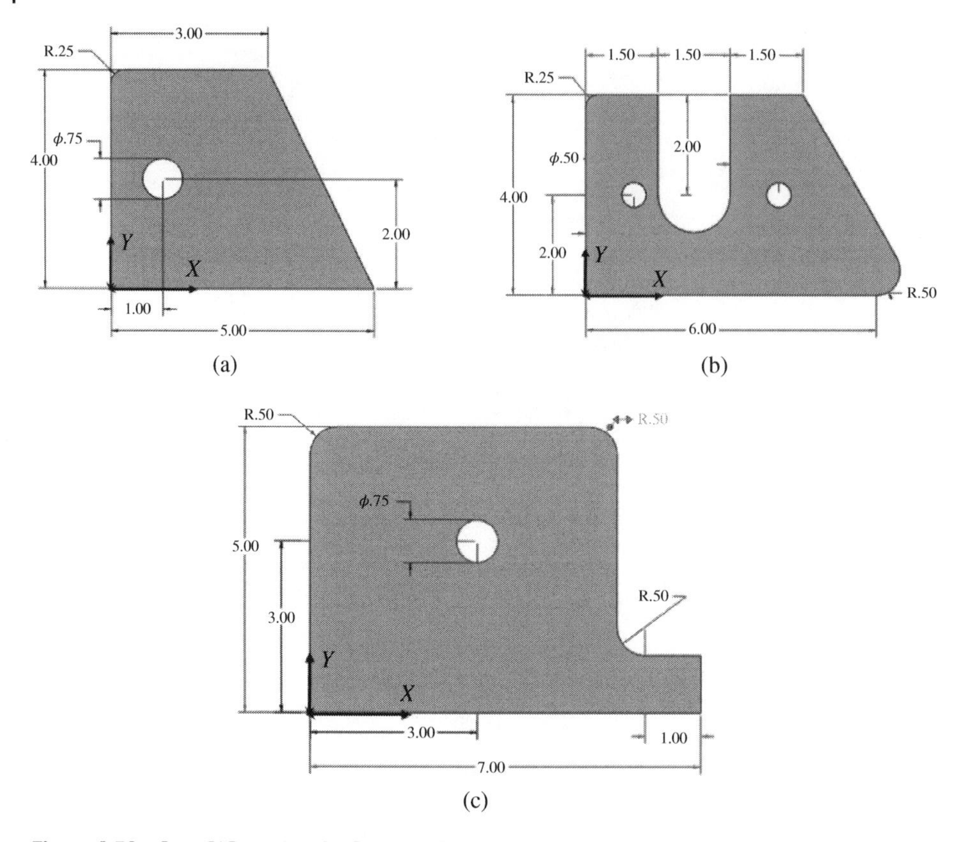

Figure 9.38 Part CAD models for Problem 9.1.

the skills of using CAM systems so that CNC programs for complex products can be created, simulated, and verified in a reduced time in a seamlessly integrated virtual environment.

9.9 Design Problems

9.1 Assume that the part material is low carbon steel and the cutters are HSS. The part thickness is 0.500 inch (tolerances are 0.020 for all dimensions, with units in inches). Generate the NC part programs for the parts shown in Figure 9.38. For each part, write two programs: (i) for cutting holes and (ii) for performing contour milling.

9.2 Figure 9.39 shows the NU logo and the toolpath required to engrave it is a rectangular plate. A milling operation is used to cut the plate and a feed rate is chosen as 10 in min^{-1}. The plate thickness is 0.500 inch and the cut depth is 0.0625 inch. Write an NC program to mill the logo.

Figure 9.39 Part CAD model for Problem 9.2.

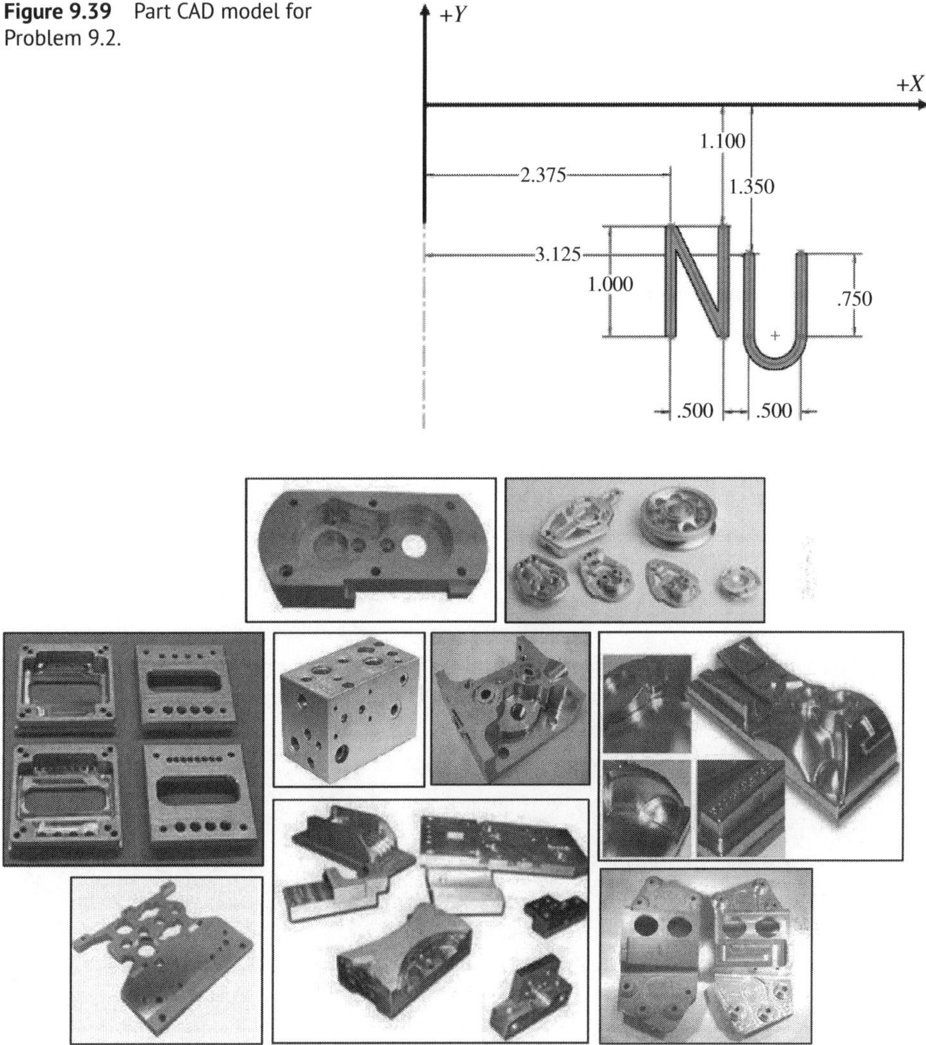

Figure 9.40 Exemplifying machined parts for design project.

9.10 Design Project

1 Find a machined part such as the one shown in Figure 9.40, build its CAD model, figure out the block material, and program the machining program for the part.
  1. Create a CAD model by reverse engineering.
  2. Fix a design for all of the machined features.

3. Create an assembly model for the fixture setup(s).
4. Create a CNC program(s) for all features.
5. Conduct a simulation of the CNC program for the basic statistics of machining (number of setups, machining tools, the number of tools, cost, etc.)
6. Post-process to generate an NC code to a generic three-axis Hass machine tool.
7. Conduct an engineering analysis to predict deflection (tolerance).
8. (Optional for bonus) make the part if CNC is accessible to you.

References

Autodesk (2017). Autodesk HSM 2.5 to 5-axis CAD/CAM solution. https://www.hungarocad.hu/wp-content/uploads/2017/05/hsm-2018-feature-comparison-matrix-en.pdf.

Azizi, A., Yazdi, P.G., Humairi, A.A. et al. (2018). Design and fabrication of intelligent material handling system in modern manufacturing with industry 4.0 approaches. *International Robotics and Automation Journal* https://medcraveonline.com/IRATJ/IRATJ-04-00121.pdf.

Bebjaminsen, T. K. (2015). Rapid prototyping CNC, http://www.robotikk.com/student/projects/Thomas%20Benjaminsen_RP_CNC/Benjaminsen-master.pdf

Bi, Z.M. and Zhang, W.J. (2001). Flexible fixture design and automation: review, issues and future directions. *International Journal of Production Research* 46 (4): 2867–2894.

Bi, Z.M., Xu, L.D., and Wang, C. (2014). Internet of Things for enterprise systems of modern manufacturing. *IEEE Transactions on Industrial Informatics* 10 (2): 1537–1546.

Calvo, I., Perze, F., Etxeberria-Agiriano, I., and de Alberniz, O.G. (2013). Designing high performance factory automation applications on top of DDS. *International Journal of Advanced Robotic Systems* https://doi.org/10.5772/56341.

ECAD Inc (2019). HSMWorks for SolidWorks. http://ecadinc.com/Products/Autodesk/Industry/Product?ProductID=60.

Engineeringinsider (2019). Difference between NC and CNC machine. https://engineeringinsider.org/difference-nc-cnc-machine/2.

Gerovitch, S. (2019). Automation, pp. 122–126. http://web.mit.edu/slava/homepage/articles/Gerovitch-Automation.pdf.

Global Engineering (2019). Standards and engineering publications. https://global.ihs.com/images/ENGL/EIA_Master_Num_Index.pdf.

Groover, M.P. (2007). *Automation, Production Systems, and Computer Integrated Manufacturing*, 3e. Prentice Hall, ISBN-10: 0132393212.

Heragu, S.S., and Ekren, B. (2019). Materials handling system design. https://pdfs.semanticscholar.org/c8c3/f4b7b4588be91fd52f3ecd9aba1422210f1b.pdf.

Holmes Noble (2017). Automation – the collaboration between human and machine. https://www.holmesnoble.com/wp-content/uploads/2017/10/Holmes-Noble-Automation-Discussion-Paper.pdf.

Kasugai, H., Fujiwara, A., Sasai, H., and Kajiyama, M. (2003). The starting point for 'creation' – 'machining centers' otherwise known as 'mother machines' yesterday, today, and tomorrow, Mitsubishi Heavy Industries, Ltd. *Technical Review* 40 (1): 1–6. www.mhi.co.jp/technology/review/pdf/e401/e401078.pdf.

Kharagpur (2019). Introduction to industrial automation and control. https://nptel.ac.in/courses/108105063/pdf/L-01(SM)(IA&C)%20((EE)NPTEL).pdf.
Kramer, T.R., Proctor, F.M., and Messina, E. (2000). The NIST RS274NGC interpreter – version 3. https://ws680.nist.gov/publication/get_pdf.cfm?pub_id=823374.
Lynch, M. (1994). Ket CNC concept #6 methods for CNC programming. https://www.mmsonline.com/articles/key-cnc-concept-6methods-for-cnc-programming.
Mishev, G. (2006). Analysis of the automation and the human worker, connection between the levels of automation and different automation concepts. Department of Industrial Engineering and Management, Jönköping School of Engineering, Sweden. https://www.diva-portal.org/smash/get/diva2:228706/FULLTEXT01.pdf.
NTIS (1984). Computerized manufacturing automation: employment, education, and the workplace. Washington, DC: US Congress, Office of Technology Assessment, OTACIT-235 (April 1984). https://www.princeton.edu/~ota/disk3/1984/8408/8408.PDF.
Pohjalainen, A. (2015). Control policies of an automated storage and retrieval system. MS thesis. Aalto University. https://aaltodoc.aalto.fi/bitstream/handle/123456789/18646/master_Pohjalainen_Antti_2015.pdf?sequence=1&isAllowed=y.
SAE (2016). Axis and motion nomenclature for numerically controlled machines (stabilized May 2016). http://standards.sae.org/eia267c.
Soyalan, M., Fenerioglu, A., and Kozkurt, C. (2012). An approach of control system for automated storage and retrieval system (AS/RS). *International Journal of Mechanical and Mechatronics Engineering* 6 (9): 1950–1954.
Sullivan, D.O. (2009). Industrial automation, course notes. https://pdfs.semanticscholar.org/9198/9bc786ff72868772d19d703acb450dc57bc7.pdf.
Wikipedia (2019a). Automated storage and retrieval system. https://en.wikipedia.org/wiki/Automated_storage_and_retrieval_system.
Wikipedia (2019b). G-code. https://en.wikipedia.org/wiki/G-code.
Williams, T.J. and Li, H. (1998, 1998). PERA and GERAM - enterprise reference architectures in enterprise integration. In: *Information Infrastructure Systems for Manufacturing II, IFIP* (eds. J. Mills and F. Kimura). Kluwer Academic Publishers. http://citeseerx.ist.psu.edu/viewdoc/download?doi=10.1.1.194.6112&rep=rep1&type=pdf.

10

Simulation of Manufacturing Processes

10.1 Introduction

In a manufacturing system, the transformation from starting materials to finished products corresponds to a series of manufacturing processes. Different manufacturing processes serve different purposes; for example, to *shape*, *transport*, *assembly*, or *change* geometries, properties, or the constitutions of products. All value-added processes are necessary to make parts, assemble parts into components, and finally integrate parts and components into finished products or systems. Manufacturing processes are usually complex since a large number of design factors and process parameters are involved; therefore, designs of manufacturing processes are traditionally treated as the art of science, since manufacturing processes involve so many uncertainties and randomness, and the designs of manufacturing processes rely greatly on experiences and trial and error methods. With the rapid development of information technologies, computer aided technologies make it possible (i) to explore design options of manufacturing processes in the virtual environment, (ii) to correlate design variables to manufacturing processes in order to predict outcomes, and (iii) to optimize processing parameters and tooling in manufacturing processes.

In this chapter, conventional manufacturing processes are overviewed, a number of forming processes are discussed in detail, and the applications of computer aided tools in manufacturing processes are exemplified by modelling and simulations of injection modelling.

10.2 Manufacturing Processes

Manufacturing processes are the steps through which raw materials are transformed into a finished product; in a manufacturing process, one of a group of workpieces is continually machined by machine tools and operators until a final product is made. *A manufacturing process* or *operation* is the subunit of the entire transformation process from starting materials to the finished product. A *manufacturing process* or operation should be accomplished at one setup. It does not involve changes of manufacturing tools or operating conditions such as changes of speed, depth, and feed rate in a turning operation (Zhang 1994).

In discrete manufacturing, a process or operation applies to parts, and the starting working material for a part is transformed in order to change the geometry, properties, and

Computer Aided Design and Manufacturing, First Edition. Zhuming Bi and Xiaoqin Wang.
© 2020 John Wiley & Sons Ltd. This Work is a co-publication between John Wiley & Sons Ltd and ASME Press.
Companion website: www.wiley.com/go/bi/computer-aided-design

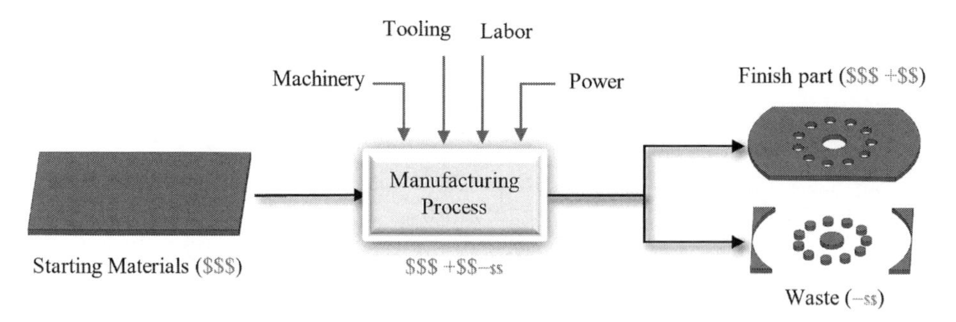

Figure 10.1 Description of a manufacturing process.

appearance of the materials. Some processes such as painting operations are applicable to components or assemblies. Figure 10.1 describes the input and output of a manufacturing process, which begins with the workpiece in its starting state, then accepts the operation at a *machinery* or workstation until it turns into the finished part as an output. Other than main machinery, a manufacturing process is assisted by other resources such as *tooling*, *labour*, and *power supply*. The manufacturing process also produces wastes other than the finished part. A manufacturing process can also be understood from an economic perspective. A company runs manufacturing businesses to achieve an economic goal, and a manufacturing process has the purpose to add value on products. The uses of any tangible objects and manufacturing resources involve costs. Therefore, the transformation of starting materials to a finished part in the material flow can be modelled as an economic transformation in the value flow.

As shown in Figure 10.2, manufacturing processes can be classified in terms of the change a process made on the workpiece of materials (Groover 2012). Firstly, manufacturing processes are classified into *processing operations*, *assembly operations*, and other non-value added processes, as shown in Figure 10.2 (Groover 2012). A processing operation applies to part of an assembly operation applied to a group of parts or components.

Processing operations can be further classified into *shaping processes*, *surface treatment processes*, and *property enhancement*. Raw materials used to make products, tools, and machines are originally from ores in nature. Suitable ores are reduced, refined, and cast as ingots. Ingots are then processed by rolling mills to produce the materials in market forms such as bloom, billets, slabs, and rods. These forms of material supply are used to make functional products with different shapes and sizes. Processing operations at different stages can be primary or secondary shaping processes, surface treatment processes, or property enhancement.

A primary shaping process such as a casting and deforming process is to make products from amorphous materials. The common processes used for primary shaping are *casting*, *powder metallurgy*, *injection modelling*, *metal sheet forming*, and *forging*. *A secondary shaping process* such as grinding makes a finished product with the desired shape, dimensions, and tolerance. Commonly used secondary processes include *turning*, *drilling*, *milling*, *threading*, *boring*, *shaping*, *planning*, *sawing*, *broaching*, *hobbing*, and *grinding*. The shaping processes also include some unconventional machining processes such as *electrochemical*

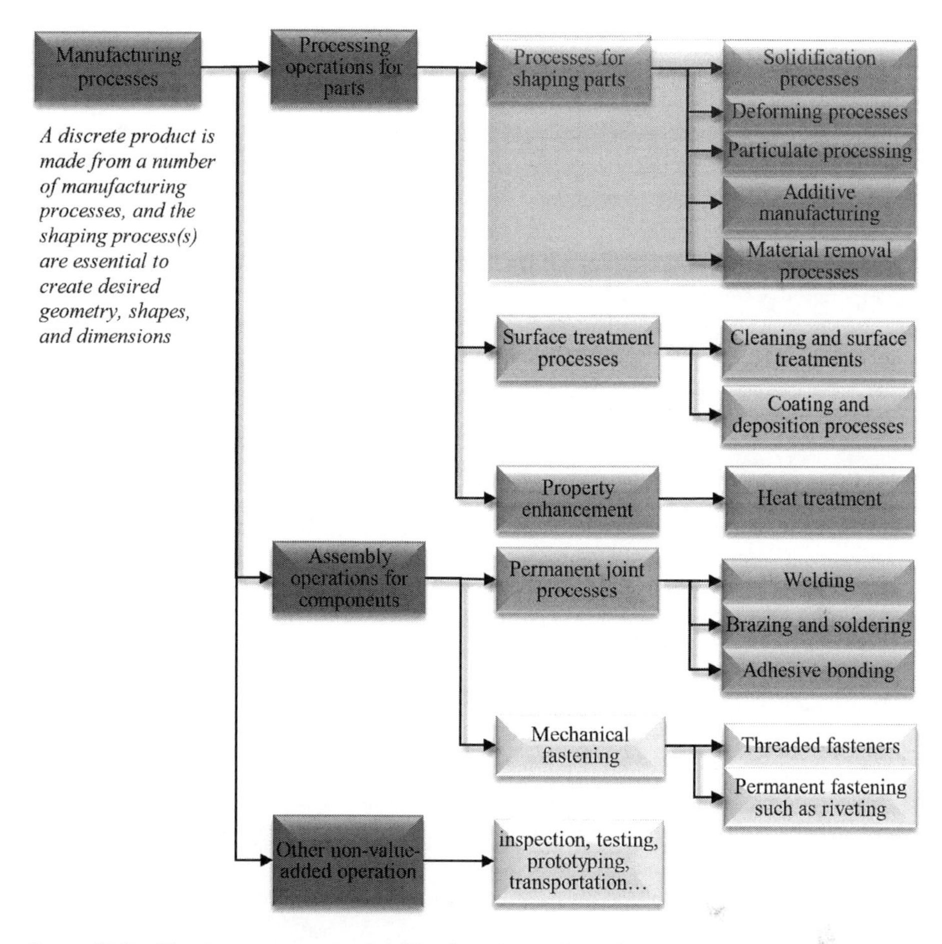

Figure 10.2 Shaping processes in classification of manufacturing processes.

machining (ECM), *laser beam machining* (LBM), *abrasive jet machining* (AJM), and *ultrasonic machining* (USM).

Surface treatment processes impart desired surface finishes on workpieces. Surface finishes are enhanced by adding or removing a very negligible amount of materials on the surface and such processes do not change the dimensions of workpieces functionally. Surface treatment is different from a material removal process since it is primarily to improve surface finish or provide a decorative or protective coating on surfaces. Commonly used surface treatments include honing, lapping, super finishing, belt grinding, polishing, tumbling, organic finishes, sanding, deburring, electroplating, buffing, metal spraying, painting, inorganic coating, anodizing, galvanizing, plastic coating, metallic coating, and sand blasting (Singh 2006).

Property enhancements are used to change material properties of workpieces suitable for some specific applications. Material properties can be changed by hardening, softening, or grain refinement in heat treatments, since temperature has its impact on physical properties with an internal crystalized structure. Metal-forming processes also have an effect on the physical properties of objects. Therefore, some commonly used processes of property enhancements are annealing, normalizing, hardening, tempering, shot peeing, and grain refining.

In an assembly operation, two or more parts or components are joined to create a new entity. The new entity is called *assembly*, *subassembly*, or *component* based on its state in the product. If the entity is an intermediate state of the product, it is called a subassembly or component. In addition, the joining processes are broadly classified into two groups – one that produces *non-permanent joints* and the other that constructs *permanent joints*. The first group includes some of the common mechanical joining processes such as screws and bolts, snap fits, and shrink fits. The permanent joining processes can be classified into four groups as *mechanical*, *solid state*, *liquid state*, and *liquid–solid state* joints. Moreover, permanent mechanical joining processes include welding, brazing, soldering, riveting, press fitting, sintering, adhesive bonding shrink filling, stitching, and stapling. In principle, these joints are heterogeneous in nature since no atomic bonding takes place along the original joint interface.

Other manufacturing operations do not change workpieces and thus do not add value to the products. Examples of these manufacturing processes are inspection, testing, prototyping, transportation, automated material handling, and even packaging.

10.3 Shaping Processes

As shown in Figure 10.2, a *shaping process* refers to a process of generating a desired geometric shape or features of a workpiece subjected to thermal, mechanical, or other types of loads. A part geometry can be shaped in many different ways. From the materials perspective, the desirable material properties for a shaping process are *low strength* and *high formability*. Since the temperature affects both the strength and formability significantly, the working temperature of workpieces is the most important factor in a shaping process: the higher the temperature, the higher the level of energy the material is and the lower the strength and the higher the formability of the material, the easier it is for the shaping process to be implemented. Accordingly, shaping processes can be classified based on the temperature levels in contrast to melting temperatures of the materials that the shaping processes are performing.

Assume that a material has the melting temperature of T_m; it becomes feasible to shape an object by *recrystallization* or *solidification* of the material when the working temperature is over $0.5T_m$. The working temperature can be used to distinguish shaping processes from one another. Figure 10.3 shows a classification of the shaping processes based on working temperatures. *Cold working*, *warm working*, *hot working*, *sintering*, and *casting* are performed at ranges of temperatures below $0.3T_m$, $(0.3T_m, 0.5T_m)$, $(0.5T_m, 0.75T_m)$, $(0.75T_m, T_m)$, and over T_m, respectively.

Different shaping processes begin with different starting materials, as shown in Figure 10.3. The starting materials of a *casting process* are in the molten state; the material does not hold the shape until it has been solidified in a cavity. The starting material of a *particulate process* is powder; the material in the powder state is pressed in a mould and heated until sintering occurs to form the crystalline state for the desired geometry.

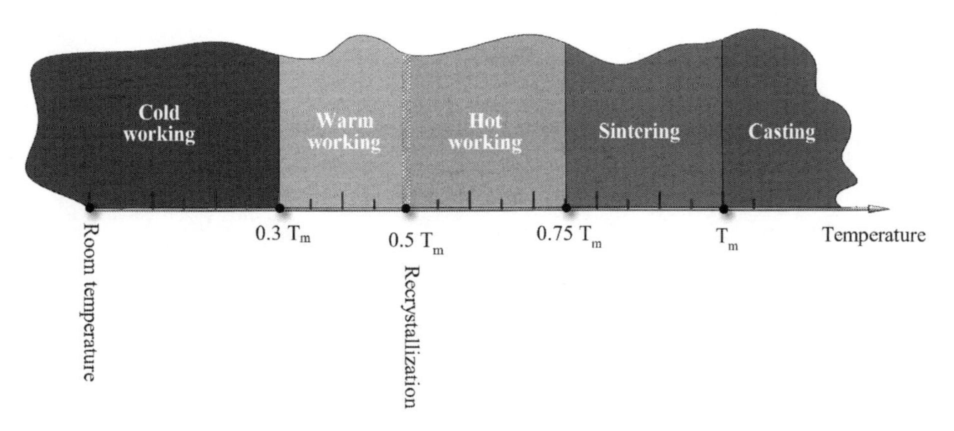

Figure 10.3 Classification of shaping processes based on working temperatures.

The starting materials of an *injection moulding* are small particles; the particles are heated and pressed into moulds and then solidified to get the finished products. The starting materials of a *deforming process* have bulk or sheet forms from primary shaping processes; the materials are deformed by mechanical with/without thermal loads to shape products. The starting materials in *additive manufacturing* are in a powder, strip, or fluid state, and are then heated, solidified, or sintered layer by layer until complete part geometry is produced.

10.4 Manufacturing Processes – Designing and Planning

A manufacturing process design aims to determine a manufacturing method and the details of the implementation by which products can be manufactured economically and competitively from a raw materials stage to finished stages of products in desired forms. Manufacturing process planning is the transformation of the specifications and features of products into detailed operation instructions including machine selections, design of machining tools, and the determinations of process parameters. Due to the diversity of manufacturing processes in Figure 10.2, the contents of a manufacturing process planning vary greatly from one type of manufacturing process to another. In this chapter, we focus on the design and analysis of some shaping processes. The basics of casting processes, injection moulding processes, and thermoforming processes are overviewed, the design challenges are discussed, and computer aided techniques are introduced to support manufacturing processing planning.

Design of manufacturing processes are heavily embedded with complexity and manufacturing processes are affected by other design activities for the products. As shown in Figure 10.4, the design issues relevant to manufacturing processes should be taken into account at the stages of product design, tooling design, and manufacturing to reduce the possibility of changes made at later stages of the product lifecycle. Similar to the roles of other computer aided techniques we use in the product lifecycle, computer aided designing and planning of manufacturing processes provide the opportunities to identify design defects at an early stage. Bringing the simulation of manufacturing processes early into

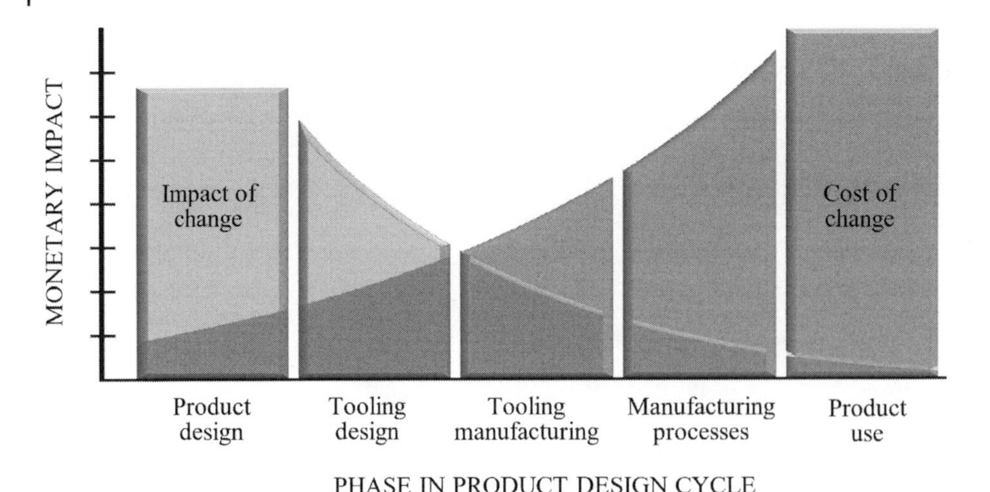

Figure 10.4 Impact of computer aided techniques on cost reduction of manufacturing processes.

the design cycle reduces costs, helps find mistakes early, and allows more exploration of the design space. This produces far better products (Reifschneider 2000; Engineering 2017; SolidWorks 2015).

10.5 Procedure of Manufacturing Processes Planning

Designing and planning a manufacturing process can be formulated as a system design problem, as described in Figure 10.5. The inputs of manufacturing process planning are product models, the properties of starting materials, available machinery, and tools at factories. The outputs of process planning are fixture designs, mould and tooling designs, and determined process variables.

The procedure for planning a manufacturing process consists of the following steps.

1. The inputs, outputs, constraints, variables, and design criteria of the manufacturing process planning are clarified. The product model and manufacturing features are analysed to identify processing types, raw materials, dimensions and tolerances, surface conditions, strengths, hardness, and other characteristics. It is important to verify whether the product features are feasible from the manufacturing point of view. If the verification leads to a doubtable conclusion, the process planner needs to ask for an improved

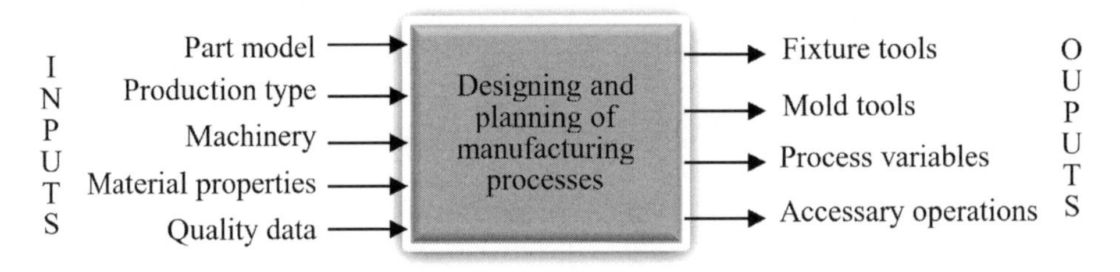

Figure 10.5 Inputs and outputs of designing and planning of a manufacturing process.

design of products until all of the potential manufacturing concerns are addressed. It is not rare that some criteria taken at the manufacturing stage conflict with those at the product design stage. To shorten the iterative process, enterprises are encouraged to take the design of manufacturing processes as an integrated component of product design for concurrent engineering (CE) practice, so that design constraints from a manufacturing perspective can be considered and satisfied at the stage of product design.

2. The form of raw materials is selected for the manufacturing process. On the one hand, the form of raw materials determines applicable types of manufacturing processes on workpieces that have been elaborated in Figure 10.6. On the other hand, the selection of raw material greatly affects the performance of a manufacturing process. Taking an example of material removal processes, the geometric shapes of selected stocks determine the amount of materials to be removed as well as the moving paths of cutting tools. Some general guides for stock selections are (i) match the quality of stock with the final product, (ii) minimize machining needs, (iii) maximize the utilization rate of materials, and (iv) reduce the manufacturing cost and machining time.
3. The machinery, fixtures, tooling, and other auxiliary resources are selected or designed. Different resources are required for different processes. Taking examples of tooling that make direct contact with the workpieces, moulds are needed for casting or injection moulding processes, the punch and dies are needed for forging, extrusions or other deforming processes, and the fixtures are needed for material removal processes. Resource planning also covers clamping, inspecting, and other auxiliary tools.
4. The process parameters for machinery operations are determined. Depending on the types of machinery, process parameters vary from one manufacturing process to another. For example, the pressure and ramp speed are critical parameters for an extrusion, and the speed, feed rate, and cut depth are important in a turning operation.

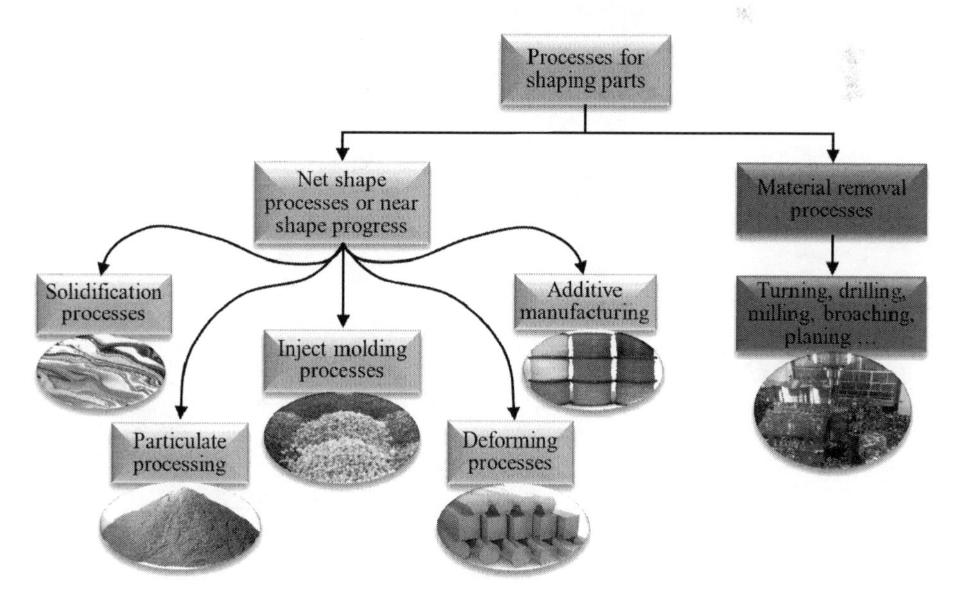

Figure 10.6 Types of starting materials in different shaping processes.

In the aforementioned steps, not all of the relations between input and output variables can be mathematically modelled. Even though a model can be developed, it would be too complicated for designers to use the model in the design of manufacturing processes manually. In addition, the uncertainties of processing parameters increase the level of difficulty in predicting the outcomes of manufacturing processes. Therefore, the simulation-based optimization of manufacturing processes becomes the only practical approach to use to support concurrent engineering.

In the following, casting processes and injection modelling are used as examples to show the challenges involved in manufacturing processes and illustrate how computer simulation can be applied to support the design optimization of manufacturing processes.

10.6 Casting Processes

Casting is a manufacturing process where the liquidized raw material is poured into a mould and then solidified in the hollow cavity of the mould to form the desired shape. A part from the casting process is also known as a *casting*. Casting processes are often used to make products with complex shapes that would otherwise be difficult or uneconomical to make by other shaping or forming methods (Wikipedia 2019).

10.6.1 Casting Materials and Products

Casting processes are applicable to most of engineering materials. Applicable materials include *metals* or other time-setting materials that cure after mixing two or more components together, such as *epoxy*, *concrete*, *plaster*, and *clay*. Figure 10.7 provides some statistical data about the raw materials and the applications of casting products (ESP International 2019). The dominant raw materials are ferrous metals such as grey iron (40%), ductile iron

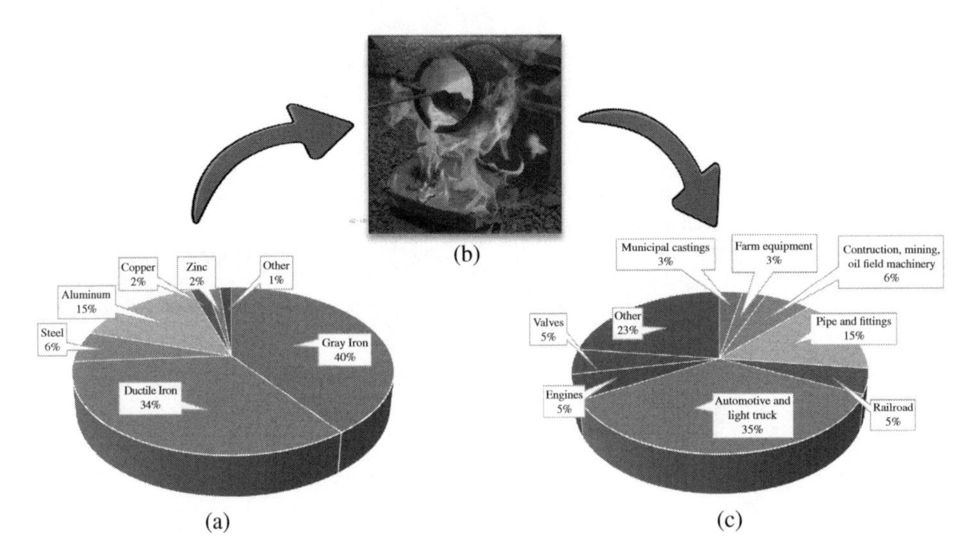

Figure 10.7 Raw materials and products of casting processes.

(34%), and steel (6%). Casting processes are often applied to other metals such as aluminium (15%), copper (2%), and zinc (2%). Casting products are widely applied in many sectors, especially in the automotive industry (35%), pipe and fitting (15%) and construction and mining oil field machinery (6%).

Figure 10.8 gives some product examples that are made from the casting processes. These examples include train wheels, engine bodies, motors, machine bases, statues, prosthetic teeth, jewellery, and casting bells. The noticeable features of these casting products are: (i) The dimensional tolerances are relatively low or the final dimensions of products are determined by secondary machining operations. (ii) The sizes of products can be massive or small, which may not be suitable for conventional machining processes. (iii) The geometry and shapes are often complex with many detailed features on the surfaces. (iv) The batch sizes of products vary from small to large volumes.

Similar to many other shaping processes, such as forging, extrusion, and machining in Figure 10.6, a casting process aims to create the shape of a part. Before the casting process is selected as the shaping method, designers should fully understand some benefits of using casting processes (Thomasnet 2019):

1. A casting process is able to create very complex geometries of parts since liquid metal facilitates the construction of intricate features no matter how complex the geometries.
2. A casting process provides a fast production cycle. Once the casting tools are made, there is minimum maintenance and downtime. Therefore, casting can be an option for mass production.
3. A casting process is applicable to hard materials; in some cases, casting may be the only viable manufacturing process for hard metals since those materials are not malleable enough to be shaped in their solid states.
4. A casting process may eliminate the assembling process of multiple parts since casting can create items in a single complete component if the materials of parts are the same.
5. A casting process can create extremely small to extremely large parts.
6. A casting process can be used to create different surface textures, i.e. smooth, semi-smooth, or rough surface textures.

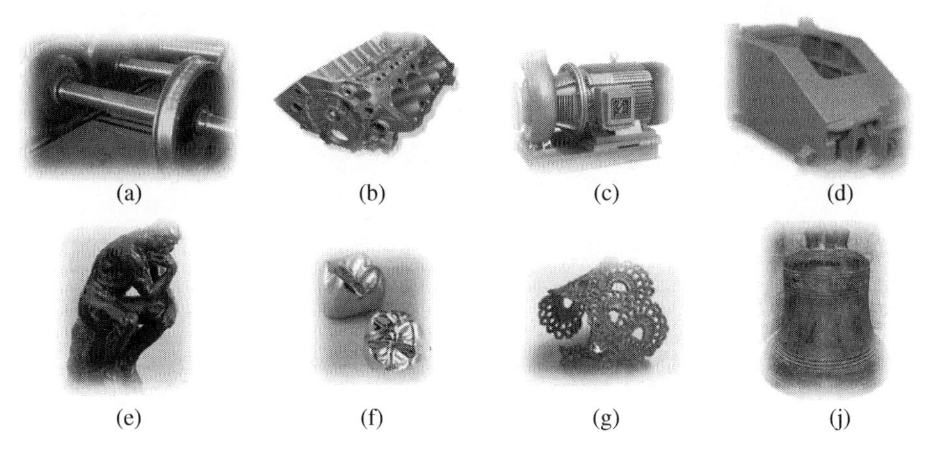

Figure 10.8 Examples of casting products.

Table 10.1 Casting processes – advantages and disadvantages.

Advantages	Disadvantages
1. Net shape or near net shape processes 2. Simple or complex geometries 3. Both external/internal shapes 4. Large or small parts 5. Applicable to small or large volumes	1. Limited mechanical properties (impurity) 2. Poor dimensional accuracy and surface finish for some processes; e.g. sand casting (shrinking) 3. Environmental issues 4. Safety hazards to workers due to hot molten metals

In addition, Table 10.1 summarizes the advantages and disadvantages of casting processes in comparison with other shaping methods.

10.6.2 Fundamental of Casting Processes

Figure 10.9 shows the setup of sand casting. A casting process involves three essential steps: the raw materials are heated, the molten metal is transported to fill a mould, and the materials in the mould are then solidified to give a casting with a desired shape. In practice, there are many types of casting processes that are different from one another in terms of (i) the selection of mould materials such sand, plaster, ceramics, and metals, (ii) the selection of pattern materials such as wood, wax, and polystyrene, and (iii) the ways used to fill the cavities, such as gravity, pressure-assisted, and centrifugal force-driven.

Taking into consideration of three aspects in Figure 10.9, available casting processes can be classified accordingly. As shown in Figure 10.10, the moulds used in the casting process can be *expandable* or *permanent*.

Expandable moulds are not reusable; each mould can be used for one product. Depending on the types of mould and pattern materials, expendable mould castings include *sand casting*, *shell mould casting*, *expanded polystyrene casting*, *investment casting*, *plaster mould*

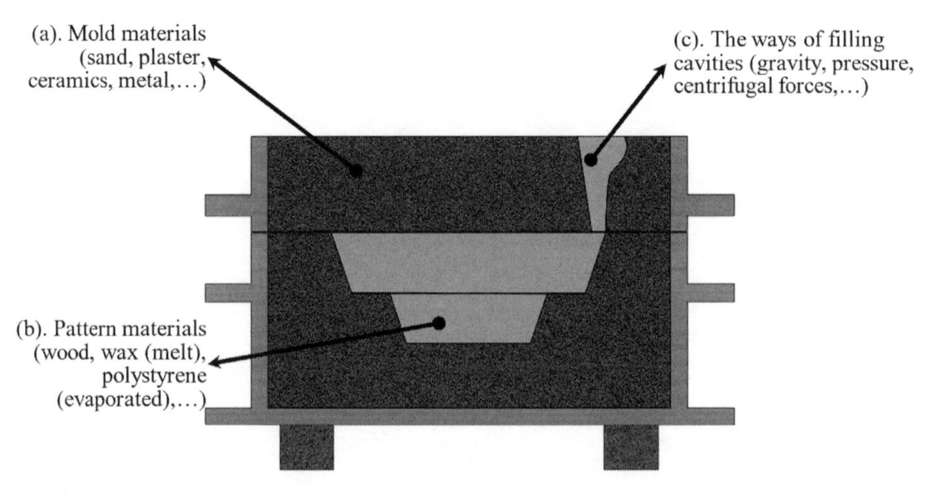

Figure 10.9 Typical setup of casting processes.

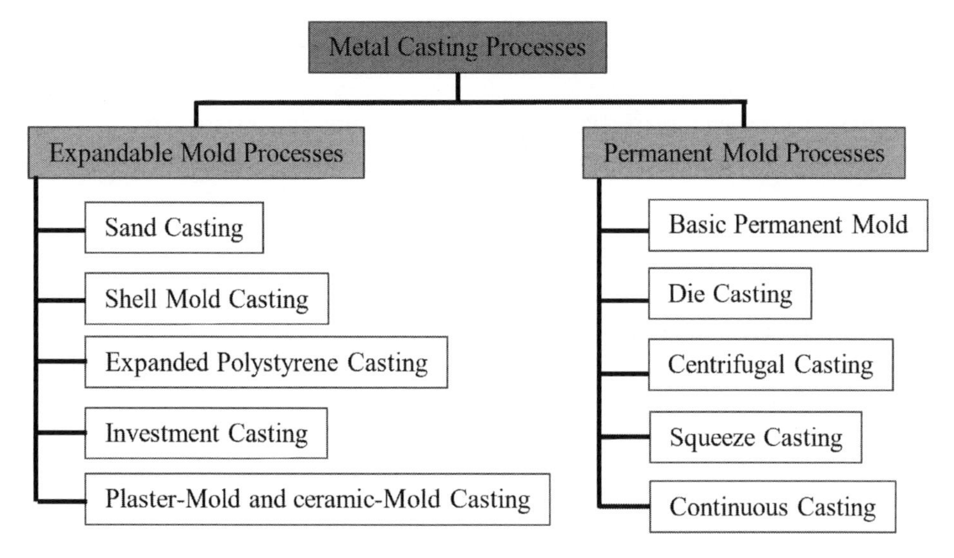

Figure 10.10 Classification of casting processes.

casting, and *ceramic mould casting*. After the molten metal in the mould cavity solidifies, the expandable mould is broken to take out the solidified cast. Expendable mould casting processes are suitable for very complex shaped parts and materials with a high melting point temperature. However, the rate of production is often limited by the time to make the mould rather than the casting itself. Permanent moulds are usually made of metallic materials, so the moulds are permanent and can be used repeatedly. Metal moulds are also called dies; they provide a superior surface finish and closer tolerance than typical sand moulds. The permanent mould casting processes broadly include pressure *basic permanent moulds*, *die-casting*, *centrifugal casting*, *squeeze casting*, and *continuous casting*.

To design and analyse a casting process, one has to understand a few of the principles that govern the casting process, as discussed below.

10.6.2.1 Energy Consumption

The raw materials for a casting process are heated over to the molten temperature. The energy cost for the casting process is mainly determined by the heating processes as

$$H = \rho V[C_s(T_m - T_0) + H_f + C_l(T_p - T_m)] \tag{10.1}$$

where

- H is the total heat required to raise the temperature of the metal to a pouring temperature (Btu)
- ρ is the density of raw materials, g cm^{-3} (lbm in^{-3})
- V is the volume of metal being heated, cm^3 (in^3)
- $C_{s,}$ C_l are the weight specific heat for the solid metal, J g^{-1} °C (Btu lbm^{-1} °F)
- T_0, T_m, T_p are the room temperature, melting temperature, and pouring temperature, respectively
- H_f is the heat of fusion J g^{-1} (Btu lbm^{-1})

Equation (10.1) shows that the required energy to heat raw materials to the pouring temperature consists of three parts: (i) the heat for raising the temperature of raw materials from room temperature to the melting point, (ii) the heat of fusion to state the change of materials from solid to fluid occurring at the melting temperature, and (iii) the heat for raising the temperature of the molten metal from the melting point to the pouring temperature.

Example 10.1 Pure copper is heated to cast a large rectangular plate in an open mould. The plate's length = 20 in, width = 10 in, and thickness = 2 in. Compute the amount of heat energy to heat the metal from ambient temperature (75 °F) to a pouring temperature of 2100 °F. Assume that the amount of metal heated will be 10% more than what is needed to fill the mould cavity. The properties of copper are as follows: the density $\rho = 0.324$ lbm in^{-3}, $C_s = 0.092$ Btu lbm^{-1} °F, $C_l = 0.090$ Btu lbm^{-1} °F, $T_m = 1981$ °F, $T_0 = 75$ °F, $H_f = 80$ (Btu lbm^{-1} °F).

Solution

Equation (10.1) can be used to calculate the required heat energy as

$$H = \rho V[C_s(T_m - T_0) + H_f + C_l(T_p - T_m)] = 37\,930 \text{ Btu}$$

10.6.2.2 Governing Equations in Pouring Operation

In the pouring operation, three types of governing equations are applicable.

Firstly, the fluid flow should be laminate to avoid turbulence. Therefore, the Reynolds number (Re) can be used to determine if the flow is laminate. Re is calculated as

$$Re = \frac{\rho V d}{\mu} \tag{10.2}$$

where μ is the viscosity, V is the velocity of flow, d is the diameter of an equivalent circular area of cross-section, and ρ is the density of the fluid. The fluid is laminate when the Reynolds number is $Re < 2000$.

Secondly, the energy is conservative in the fluid flow. Energies are presented in the form of *flow head*, *pressure*, *kinetic*, and *friction*, and the total energy is conserved. The energy conservation is described by the *Bernoulli theorem* as

$$h_1 + \frac{p_1}{\rho} + \frac{v_1^2}{2g} + F_1 = h_2 + \frac{p_2}{\rho} + \frac{v_2^2}{2g} + F_2 \tag{10.3}$$

where

h_1 and h_2	are the heads in cm (in) at the cross-sections of 1 and 2
ρ	is the density of raw materials, g cm^{-3} (lbm in^{-3})
p_1 and h_2	are the pressures of fluid in N cm^{-2} (lb in^{-2}) at the cross-sections of 1 and 2
v_1 and v_2	are the velocities of fluid in cm s^{-1} (in s^{-1}) at the cross-sections of 1 and 2
F_1 and F_2	are the head losses due to friction in cm (in) at the cross-sections of 1 and 2
g	is the gravitational acceleration constant (981 cm s^{-2} or 381 in s^{-2})

Thirdly, the fluid flow follows the continuity law Thus, the volume rate of flow remains constant as

$$Q_1 = v_1 A_1 = v_2 A_2 = Q_2 \tag{10.4}$$

where Q_1 and Q_2 are the volumetric flow rates (cm^3 s^{-1}, in^3 s^{-1}), v_1 and v_2 are the velocities (cm s^{-1}, in./s^{-1}), and A_1 and A_2 are the cross-section areas (cm^2, in^2) of locations 1 and 2.

Example 10.2 Figure 10.11 shows a partial structure of sprue and runner design. Assume the density and viscosity of the melted metal are ρ and μ, respectively. The round gate of the horizontal runner is attached directly under the bottom sprue. Find the answers to the following three questions.

1. What is the most allowable gate diameter to maximize the productivity when h_1 is given?
2. What is the size of sprue at location 2?
3. What is the estimate time to fill a mould with a volume of V_m?

Solution

1. The higher velocity of the fluid flow, the higher the productivity can be, but the flow velocity cannot be too high to cause a turbulent flow. The condition for a laminate flow can be verified at the gate as follows.
   Assume that the velocity of fluid flow at location 1 as well as the impact of pressure and friction can be ignored, i.e. $v_1 = 0$, $p_1 = p_2$, and $F_1 = F_2$ Applying Eq. (10.3) for the locations at 1 and 3 gives the velocity at location 3 as

   $$v_3 = \sqrt{2gh_1} \tag{10.5}$$

   Applying Eq. (10.3) for the condition for a laminate flow (Re < 2000) gives the diameter of gate d_3 at location 3 as

   $$d_3 \leq \frac{2000\mu}{\rho V} = \frac{2000\mu}{\rho\sqrt{2gh_1}} \tag{10.6}$$

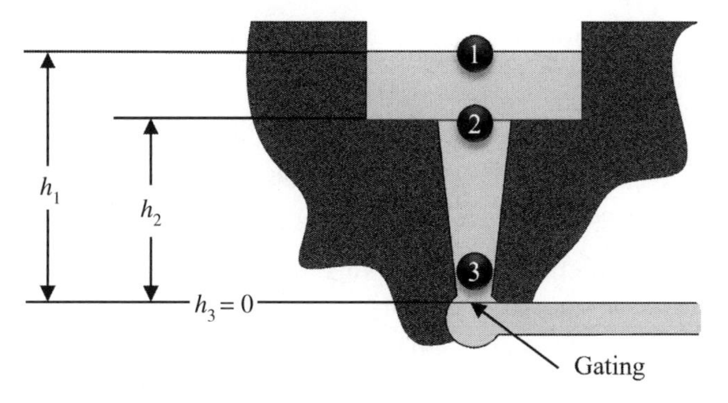

Figure 10.11 Partial structure of sprue and runner designs for Example 10.2.

The fluid flow at location 3 (Q_3) is determined as

$$Q_3 = v_3 \cdot A_3 = v_3 \cdot \frac{\pi d_3^2}{4} = \frac{\pi \mu^2 \sqrt{2gh_1} \times 10^6}{\rho^2} \tag{10.7}$$

2. Similarly, the velocity at location 2 (v_2) can be determined by the Bernoulli Eq. (10.3) as

$$v_2 = \sqrt{2g(h_1 - h_2)} \tag{10.8}$$

Further, using the continuity law Eq. (10.4) gives the diameter of sprue at location 2 (d_2) as

$$d_2 = \sqrt{\frac{v_3}{v_2}} d_3 = \sqrt{\frac{h_1}{h_1 - h_2}} \frac{2000\mu}{\rho\sqrt{2gh_1}} = \frac{2000\mu}{\rho\sqrt{2g(h_1 - h_2)}} \tag{10.9}$$

3. The filling time (t) is estimated based on the fluid flow at the gate as

$$t = \frac{V_m}{Q_3} = \frac{\rho^2 V_m \times 10^{-6}}{\pi \mu^2 \sqrt{2gh_1}} \tag{10.10}$$

10.6.2.3 Solidification Time

The solidification time (T_{TS}) is the total time required for casting to solidify after pouring. T_{TS} depends on the size and shape of casting governed by the *Chvorinov Rule*,

$$T_{TS} = C_m \left(\frac{V}{A}\right)^n \tag{10.11}$$

where

- T_{LS} is the total solidification time
- V is the volume of casting
- A is the surface area of casting
- n is the exponent with a typical value of 2
- C_m is the mould constant obtained from experiments. It depends on mould materials, thermal properties of casting, and the pouring temperature

The Chvorinov Rule gives the following design guides:

1. A casting with a higher volume-to-surface area ratio cools and solidifies more slowly than one with a lower ratio.
2. To feed the molten metal into the main cavity, T_{TS} for the riser must be greater than T_{TS} for the main casting.
3. The riser should be designed to have a larger volume-to-area ratio so that the main casting solidifies first; this minimizes the effects of shrinkage.

Therefore, one important application of the Chvorinov Rule is to design risers. A riser design must consider two critical requirements: (i) after the casting process, the materials in a riser is the waste that is separated from the casting. It is desirable to minimize the volume of metal in the riser. (ii) The solidification time of metal in a riser must be longer than that of the rest of the casting. According to the Chvorinov Rule, the shape of the riser should be designed to maximize the *V/A* ratio, which minimizes the volume of a riser for waste reduction.

Example 10.3 A cylindrical riser with the diameter-to-length ratio of 1.0 ($D = H$) is used in a sand-casting mould. The casting geometry is given in Figure 10.12 in which the units are inches. If the mould constant in the Chvorinov Rule is 19.5 min in^{-2}, determine the dimensions of the riser so that the riser will take at least 0.25 minutes longer than the casting itself to freeze.

Solution

The solidification time of the casting is calculated as

$$V/A = 46.7084/259.4917 = 0.1800 \text{ in}$$

Using Eq. (10.11) to find the solidification time of the casting gives

$$T_{TS,casting} = 19.5(0.1800)^2 = 0.631\,797 \text{ min}$$

To determine the riser size, the minimum solidification time of the materials in the riser is calculated and then the Chvorinov Rule is used to find the riser size.

The minimum solidification time for the riser is $T_{TS,\ riser} = 0.631797 + 0.25 = 0.8818$ min

For a cylindrical riser with $D = H$, the volume $V = \pi D^2 H/4 = \pi D^3/4 = 0.25\pi D^3$, and the surface area $A = \pi DH + 2\pi D^2/4 = 1.5\pi D^2$; therefore, the volume-to-surface area ratio is

$$V/A = 0.25\pi D^3/1.5\pi D^2 = D/6$$

Using the Chvorinov rule Eq. (10.11) gives

$$T_{TS,riser} = C_m(V/A)^2, \text{i.e.} 0.8818 = 19.5(D/6)^2 = 0.5417\, D^2$$

Therefore, we find that the riser size is $D = H = 1.2759$ in.

10.6.2.4 Shrink Factors

The solid phase of the materials has a higher density than that in the liquid phase. Therefore, a solidification causes a reduction in volume per unit weight of metal. Design of a casting process has to take into consideration the shrinkage in solidification. The patterns should be oversized to compensate for shrinkage and thermal contraction. The amount

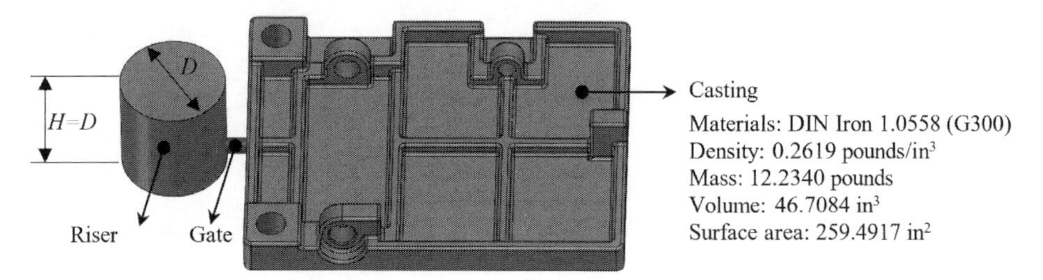

Figure 10.12 Riser design example.

by which the mould is made larger relative to the final casting size is called the *pattern shrinkage allowance*. Note that the thermal contraction might not be the only factor for determining pattern size; in particular, for the case when the secondary shaping process for surface finishing operations is required.

Table 10.2 provides the recommended values of the pattern shrinkage allowances for common casting metals (Black and Kohser 2011). To make the patterns for sand casting, the recommended shrinkage allowance is 10 ~ 20 mm m^{-1} for the dimensional compensation.

For the casting with complex geometries, uneven shrinkages may cause defects. Geometric features on casting should be refined to facilitate casting processes. Table 10.3 shows some examples where original product features are modified to facilitate a casting process.

Table 10.2 Recommended pattern shrinkage allowance.

Materials	Cast iron	Steel	Aluminium	Magnesium	Brass
Allowance (%)	0.8 ~ 1.0	1.5 ~ 2.0	1.0 ~ 1.3	1.0 ~ 1.3	1.5

Table 10.3 Modification of product features for casting processes.

Original feature	Modified feature	Original feature	Modified feature
(a) Possible area of shrinkage cavity	Hot spot removed	(g)	
(b)		(h)	
(c)		(i)	
(d)		(j)	
(e)		(k)	
(f)		(l)	

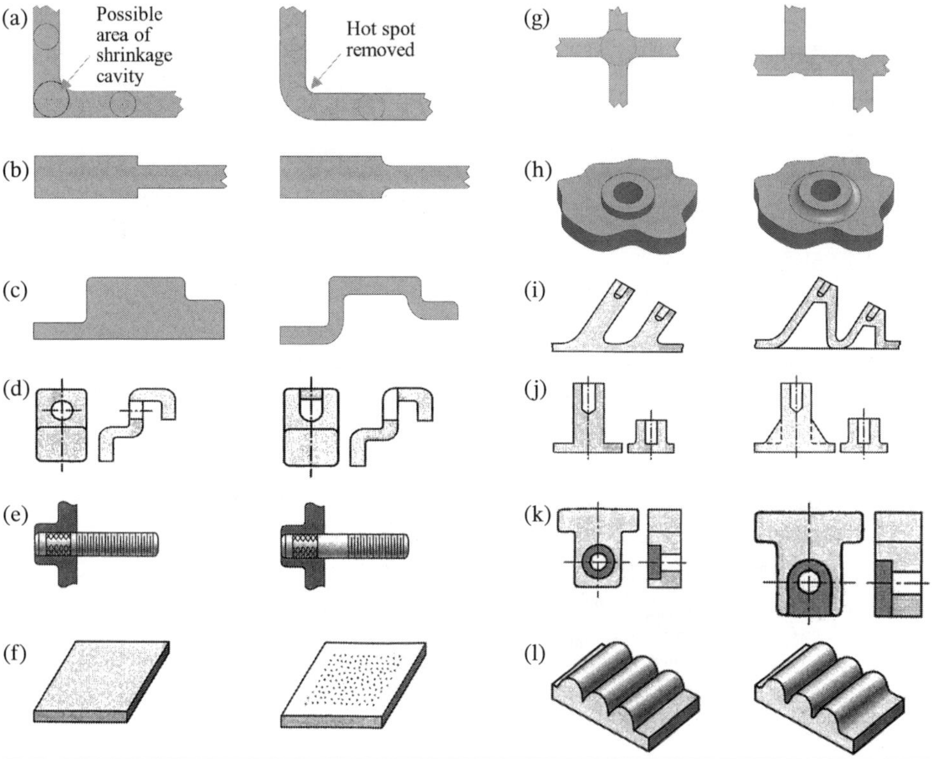

10.6.3 Design for Manufacturing (DFM) for Casting Processes

Design for manufacturing (DFM) is the engineering practice to take into consideration manufacturing requirements at the phase of product design to facilitate manufacturing processes. The materials in a casting process are involved in the temperate change in a wide range. Casting products have to be designed adequately to avoid various defects in the casting processes. Some common defects in sand casting are illustrated in Figure 10.13.

These casting defects are caused by inappropriate product designs, mould designs, or casting process parameters. To support DFM, draft features have to be added to any lateral surface for the casting purpose at a phase of product design.

As shown in Figure 10.14, a draft D is an inclined angle between a lateral face and a reference plane perpendicular to the parting plane. Draft features must be placed (i) to enable the easy removal of a pattern from the mould and (ii) to eject a part from a permanent mould. As shown in Figure 10.14, a draft angle can be within (0.5°, 2.0°). Alternatively, it can be defined in terms of the length change as 5–15 mm m^{-1}.

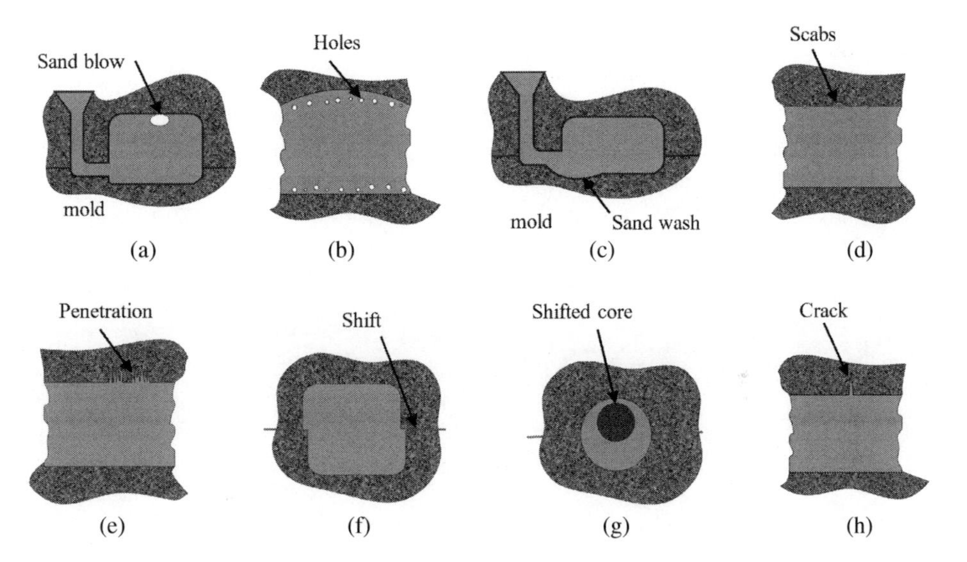

Figure 10.13 Common defect types in sand casting.

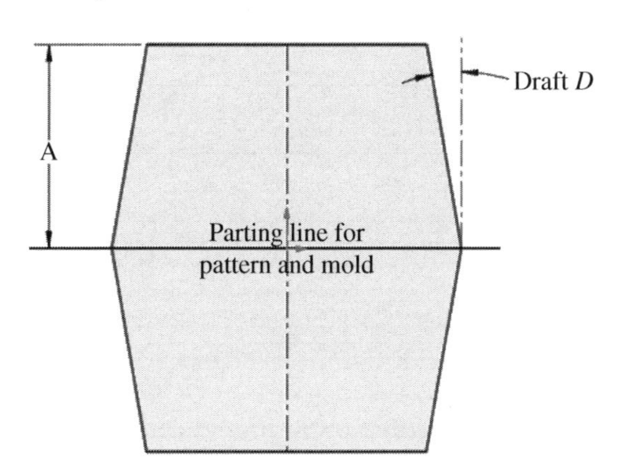

Figure 10.14 Definition of a draft for casting.

10.6.4 Steps in Casting Processes

Designing and planning a casting process involves many steps and activities: *Firstly*, the casting geometry is determined by the mould; therefore, a pattern is made to define cavity. *Secondly*, the starting material is prepared by heating the materials to pouring temperature. *Thirdly*, liquidized materials are transported and poured into the cavities at a proper rate. *Fourthly*, solidification takes place in the mould to generate the desired crystalline structure. *Fifthly*, the casting is removed out of the mould. Finally, the casting is post-processed with the necessary secondary machining operations.

The methods, the tooling, and process parameters at every step in Figure 10.15 have to be determined in the design of a casting process. The materials is changed its state from solid to liquid, and then from liquid to solid at different places. A large number of process parameters and a great deal of uncertainties are involved in a casting process. It brings the challenges to model and analyse casting processes (Figure 10.16).

10.6.5 Components in a Casting System

Other than the mould cavity for the part geometry, a casting system consists of other components to facilitate a casting process. All of these components must be designed to implement the casting process. Figure 10.16 shows that the components are directly related to the mould cavities and other elements are related to sprues, runners, gates, vents, and cores. Table 10.4 gives a brief explanation for these components and basic concepts relevant to a casting system.

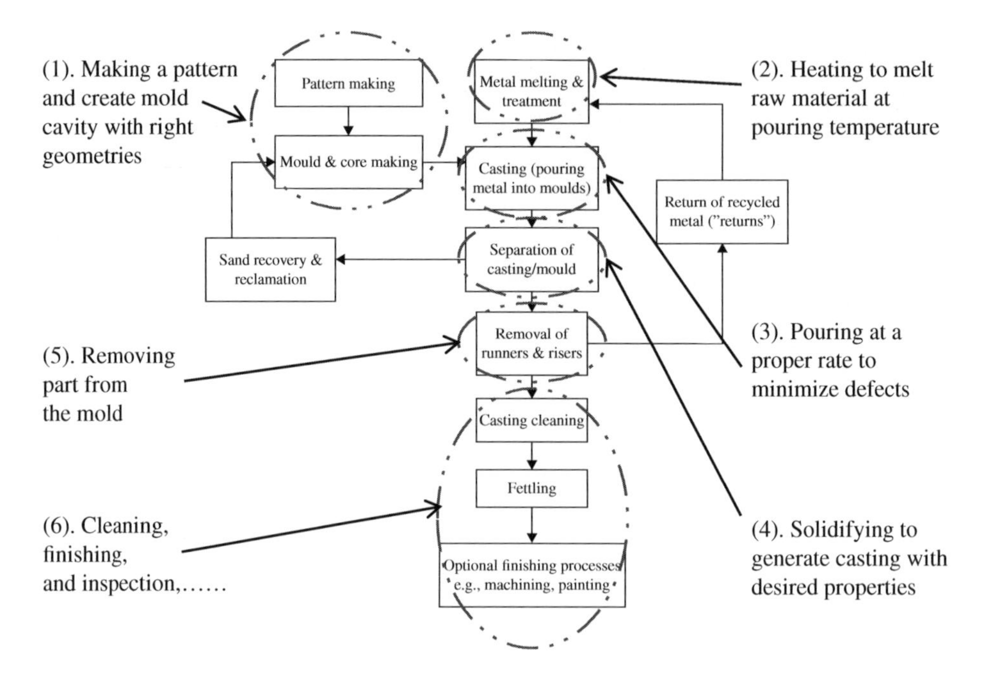

Figure 10.15 Steps and activities in a casting process.

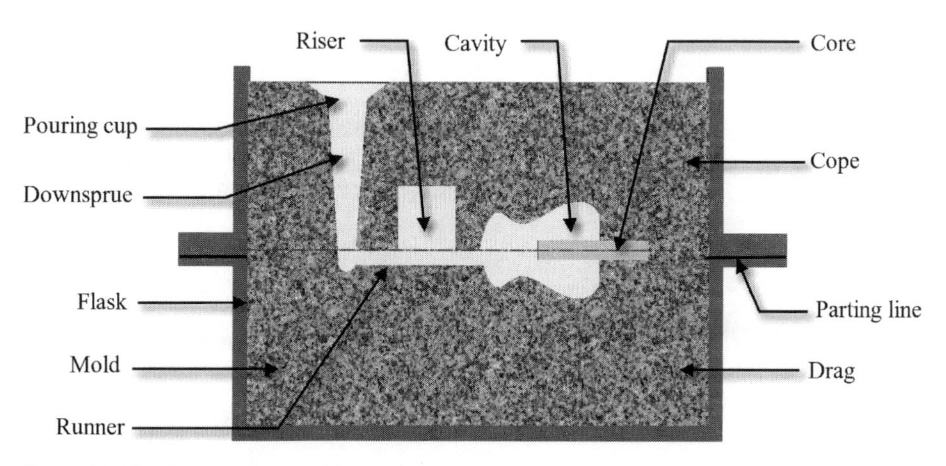

Figure 10.16 Components used in casting processes.

Table 10.4 Concepts and main components in a casting system.

Terminology	Explanation	Terminology	Explanation
Pattern:	A duplicate of final casting to form mould cavity.	Core print:	The region added to the pattern, core, or mould to locate and support the core.
Draft:	The taper on the casting or pattern that allows it to be withdrawn from the mould.	Flask:	The frame that holds the moulding material.
Mould cavity:	The combined open area of the moulding material and core, where the metal is poured to produce the casting.	Pouring cup:	The part of the gating system that receives the molten material from the pouring vessel.
Cope:	The top half of the pattern, flask, mould, or core.	Sprue:	The pouring cup attaches to the sprue, which is the vertical part of the gating system. The other end of the sprue attaches to the runners.
Drag:	The bottom half of the pattern, flask, mould, or core.	Gating system;	The network of connected channels that deliver the molten material to the mould cavities.
Parting line:	The interface between the cope and drag halves of the mould, flask, or pattern.	Runner:	The horizontal portion of the gating system that connects the sprues to the gates.
Core:	An insert in the mould to produce internal features.	Riser:	An extra void in the mould that fills with molten material to compensate for shrinkage during solidification.
Core box:	The mould or die used to produce cores.	Vents:	Additional channels that provide an escape for gases generated during the pour.

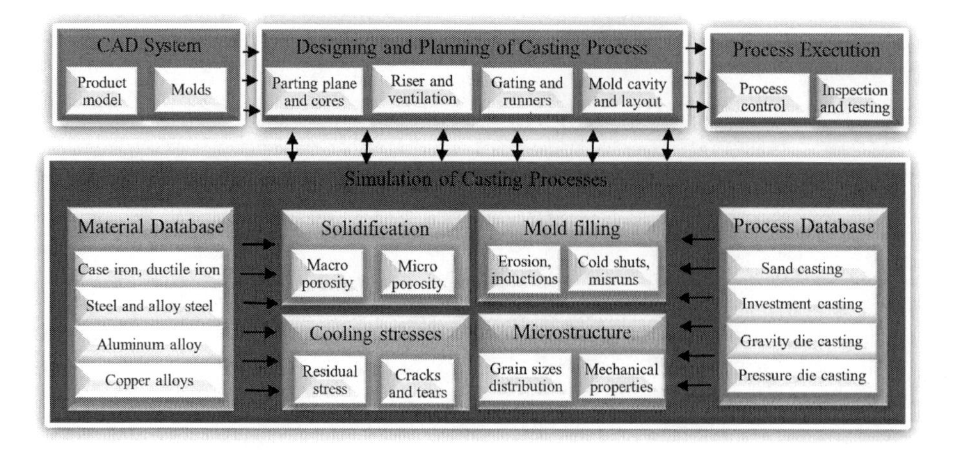

Figure 10.17 Functional models of AutoCAST (Ravi 2017).

10.6.6 Simulation of Casting Processes

Due to the complexity of casting processes, it is impractical for engineers to develop mathematical models for designs of casting processes manually. Traditionally, casting processes are designed and planned intuitively by *trial and error* methods. Computer aided tools provide scientific approaches for the design and planning of casting processes.

Casting process simulation takes into consideration *mould filling*, *solidification*, and *cooling* to predict the quality of casting quantitatively in terms of *mechanical properties*, *thermal stresses*, and *distortion*. A simulation makes it possible to describe the quality of castings virtually before an actual process starts. The verification and validation by virtual models lead to an optimal layout design of the casting system to reduce the cost of energy consumption, materials, and tooling. Casting process simulation was initially developed at universities in the early 1970s, mainly in Europe. It has been viewed as one of the most important innovations in casting technology over the last 50 years. Since the late 1980s, commercial programs (such as AutoCAST and MAGMA) have been available; software tools such as AutoCAST (Choudhari et al. 2014) and MAGMA (Hahn and Sturm 2015) are used to gain new insights into what is happening inside the mould or die during the casting process. Major modules of a typical casting simulation software are shown in Figure 10.17. A software tool for the simulation of casting processes is usually equipped with the materials and process databases and supports the analysis on solidification, mould filling, microstructure, and cooling stresses (Ravi 2010, 2017).

10.7 Injection Moulding Processes

Injection moulding is another commonly used approach to produce discrete components with a net shape. In injection moulding, polymer is heated to a highly plastic state and forced to flow under high pressure into a mould cavity where it solidifies and the *moulding* is then removed from the cavity. Injection moulding works for complex shapes and varying sizes, but it is economical only for mass production due to the high cost of the mould.

To improve productivity, a mould may contain multiple cavities, so multiple mouldings are produced each cycle.

Due to relative low strengths and low melting points in contrast to metals and ceramics, the cost of the machinery and tooling of injection modelling processes is very affordable. Therefore, injection moulding processes are very attractive to small or medium sized enterprises (SMEs) to prototype new products. Injection moulding processes have been widely adopted to produce low-cost consumer products such as the example products shown in Figure 10.18.

10.7.1 Injection Moulding Machine

In an injection moulding process, the melt is formed and delivered by heating elements subjected to pressure, and the product is shaped in a clamped mould. Therefore, an injection moulding machine consists of an injection unit and a clamping unit, as shown in Figure 10.19. *An injection unit* consists of a *barrel* to feed materials into a mould and a *screw* to (i) rotate, mix, and heat polymer and (ii) act as a ram (i.e. plunger) to inject molten plastic into the mould. The melts can be delivered by different mechanisms such as

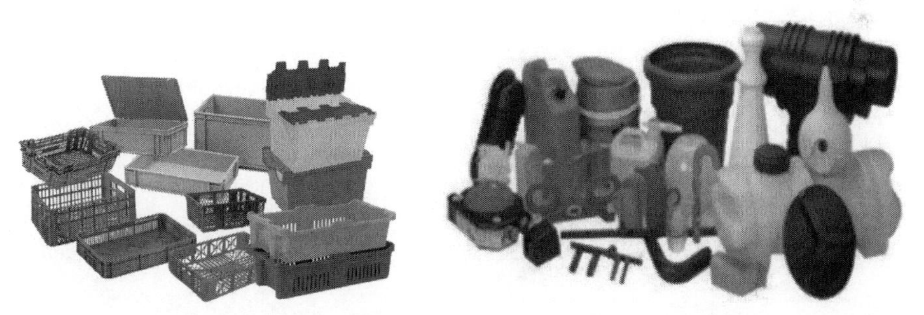

(a) Containers and storages for solid objects (b) Containers of storages for powder and liquid

Figure 10.18 Product examples from injection moulding processes.

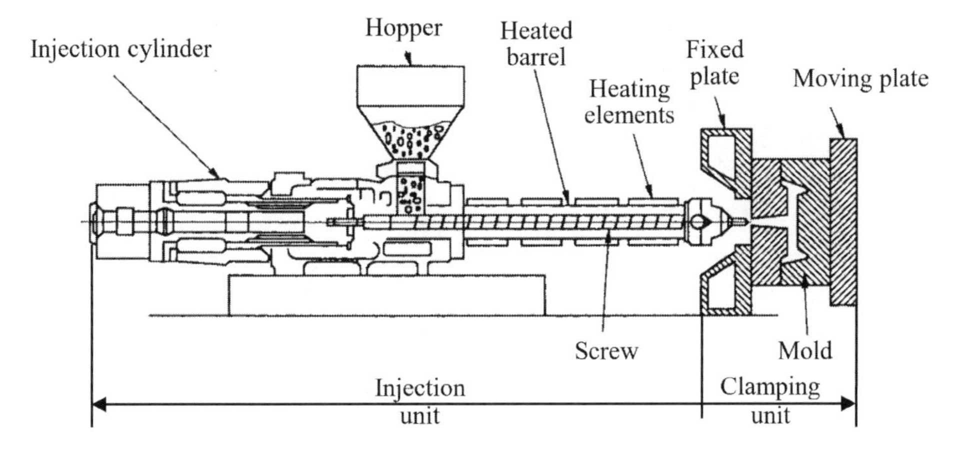

Figure 10.19 Machinery for injection moulding.

plungers, a screw pre-plasticating type, and an in-line screw type. The driving mechanism in Figure 10.19 is an in-line screw type.

The capacity of an injection moulding machine is measured by the amount of melts the machine can deliver in a full cycle of the reciprocating motion. The machine is selected based on the comparison of product weight and machine capacity (Mitsubishi Engineering Plastics Corp. 2019). Figure 10.20 shows the relation between product weight (one-shot weight) and the machine capacity. Product weight or machine capacity are in grams (g). It is necessary to select an injection moulding machine with a satisfactory capacity for the product weight. The correspondence of the machine capacity and product weight must be in area 2.

10.7.2 Steps in the Injection Moulding Process

Figure 10.21 shows the four main steps in a cycle of injection moulding. *Firstly*, the mould is closed. *Secondly*, the heated melt is injected into the mould. As the melt flows into the mould, the displaced air escapes through vents in the injection pins and along the parting plane. Runners, gates, and vents are designed to ensure all volumes of the cavities are filled adequately. *Thirdly*, once the mould is filled, the exact amount of solidification time is given to cool and harden the part. The solidification time depends on the type of resin used and the thickness of the part. To ensure uniform cooling, internal cooling or heating lines are placed, so that the water is cycled through the mould to maintain a constant temperature in the mould. While the part cools, the barrel screw retracts and draws new plastic resin into the barrel from the material hopper. The heater bands maintain the needed barrel temperature for the type of resin being used. *Fourthly*, the mould opens and the ejector rod moves the ejector pins forward to separate the workpiece from the model. After the four steps are completed in an injection moulding cycle, some post-processing activities may be needed;

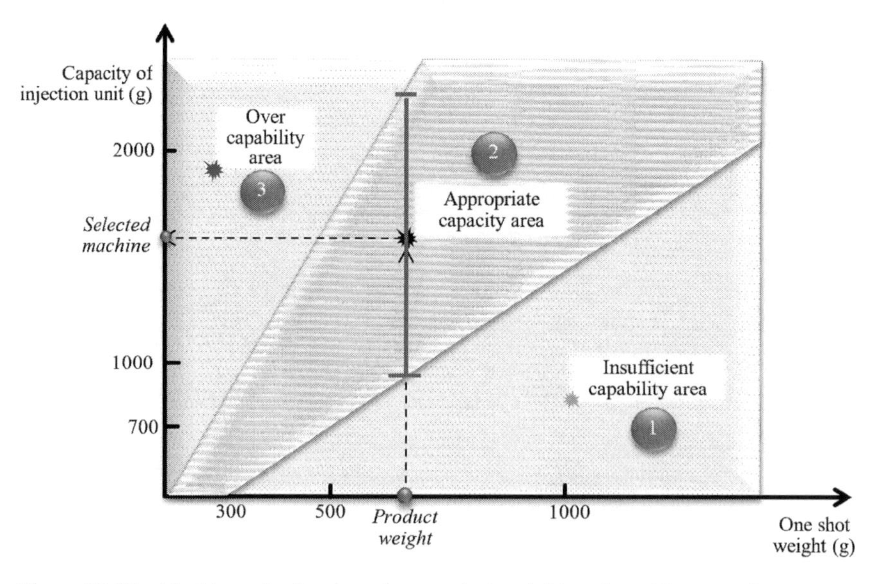

Figure 10.20 Machine selection based on product weight and machine capacity.

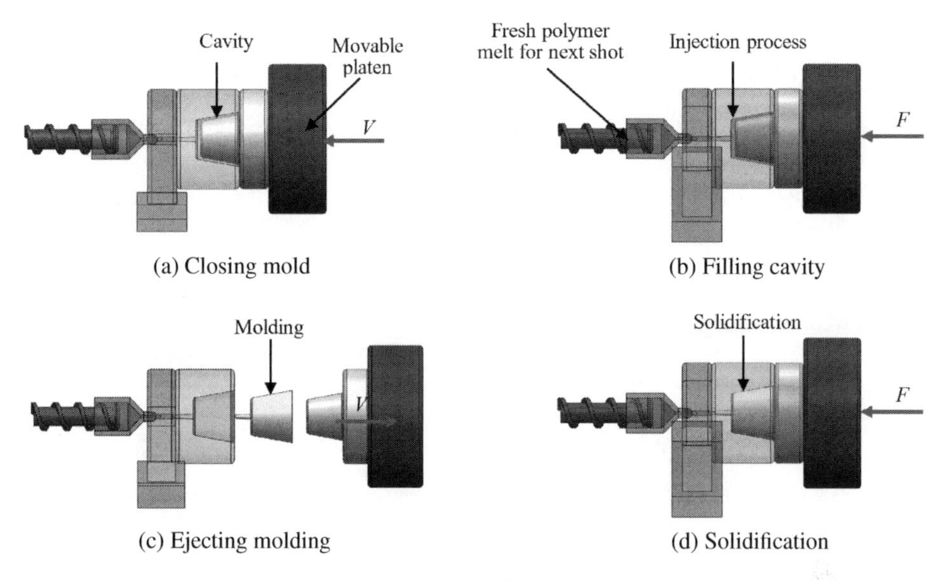

Figure 10.21 The steps in a cycle of injection moulding.

for example, separating the part from the leftovers of runners. Note that a runner is a pathway where the melt flows to fill the mould cavity. In many cases, runners are ground and recycled to reduce costs and environmental impact.

10.7.3 Temperature and Pressure for Moldability

The operating temperature and pressure are the two most criterial parameters. Figure 10.22 shows that these two parameters determine the *moldability* of the plastics in an injection moulding process (Tandon 2001). The moldability can be ensured only when the working

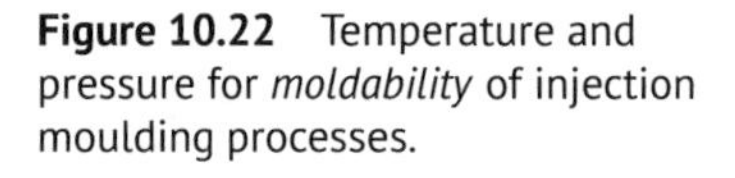

Figure 10.22 Temperature and pressure for *moldability* of injection moulding processes.

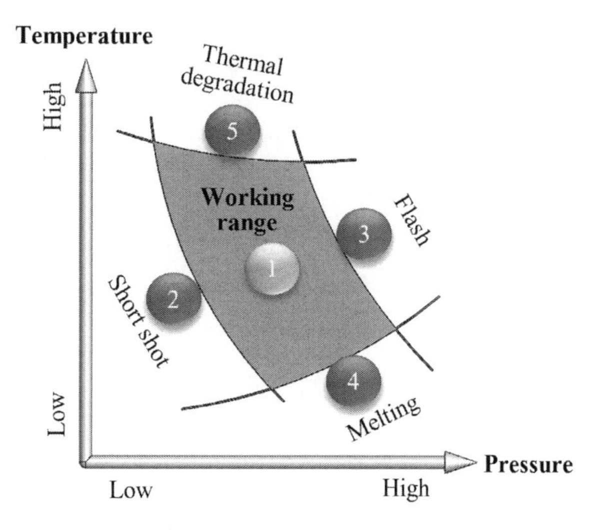

temperature and pressure are set correspondingly (area 1). On the one hand, a lower or higher pressure for a given range of the working temperature may cause the defects of *short shot* (area 2) and *flash* (area 3), respectively. On the other hand, a lower or higher temperature for a given range of the working pressure may cause the defects of *melting* (area 4) and *thermal degradation* (area 5), respectively.

A high mould temperature increases the cycle time but provides more time for crystallizing the melt. Mould temperature affects the cooling rate. The faster the melt cools, the less time is needed to form the crystallization of the melt and the less the moulded part shrinks (Fischer 2003). In addition, amorphous plastics relax internal molecular stresses when it is cooled slowly and the increased order and relaxed stresses result in greater material density and shrinkage. Therefore, rapid cooling reduces shrinkage, while too short a cooling time is prone to shrinkage and warpage after the part is moulded.

A high injection pressure must be employed in order to fill the mould rapidly and before the melt is frozen in the feed system. In addition, high clamping pressures must be used so that injection moulding machines can generate, and use, large forces. A clamping pressure must be applied in order to oppose the injection pressure. The clamping pressure is determined by the area of the part projected to the plane in the direction of the clamping force.

10.7.4 Procedure of the Injection Moulding System

Injected moulded parts are shaped by the moulds. Therefore, the mould cavity must be tailored to the part geometry. In addition, the injection moulding process involves supplying the melt from the injection unit to the clamping unit. Therefore, many components considered in a casting process are also needed in the injection moulding process. These components include cores, gates, runners, and cooling systems (Patil et al. 2005). Figure 10.23 shows the procedure of designing an injection moulding system.

1. The part model is first analysed to define the requirements of the injection modelling process, such as a one-shot weight, the need for cores, the number of cavities, the injection locations, and the projected area in the clamping direction.
2. Then the cavities of the mould are modelled, the moulding type is selected (e.g. hot runner, cold runner or conventional sprue), and the materials are selected for a mould base, inserts or mould plates, and cores.
3. Then placement of the mould cavities is considered (e.g. star, symmetrical, or in-line arrangement), and the feasibility of opening and closing a mould as well as the part ejection is verified.
4. The other components related to melt delivery, cooling, and the ejection are then designed, the type of gating system is determined (e.g. conventional, pinpoint, sub-marine, flash, tab, disc, diaphragm), the ejection element is designed (e.g. pins, stripper rings, stripper plates, air, slides, and side core), and the type of venting parting lines are specified.
5. The design of the injection moulding system is verified against the system functional requirements to ensure no design constraints are violated before the injection moulding system is fabricated.

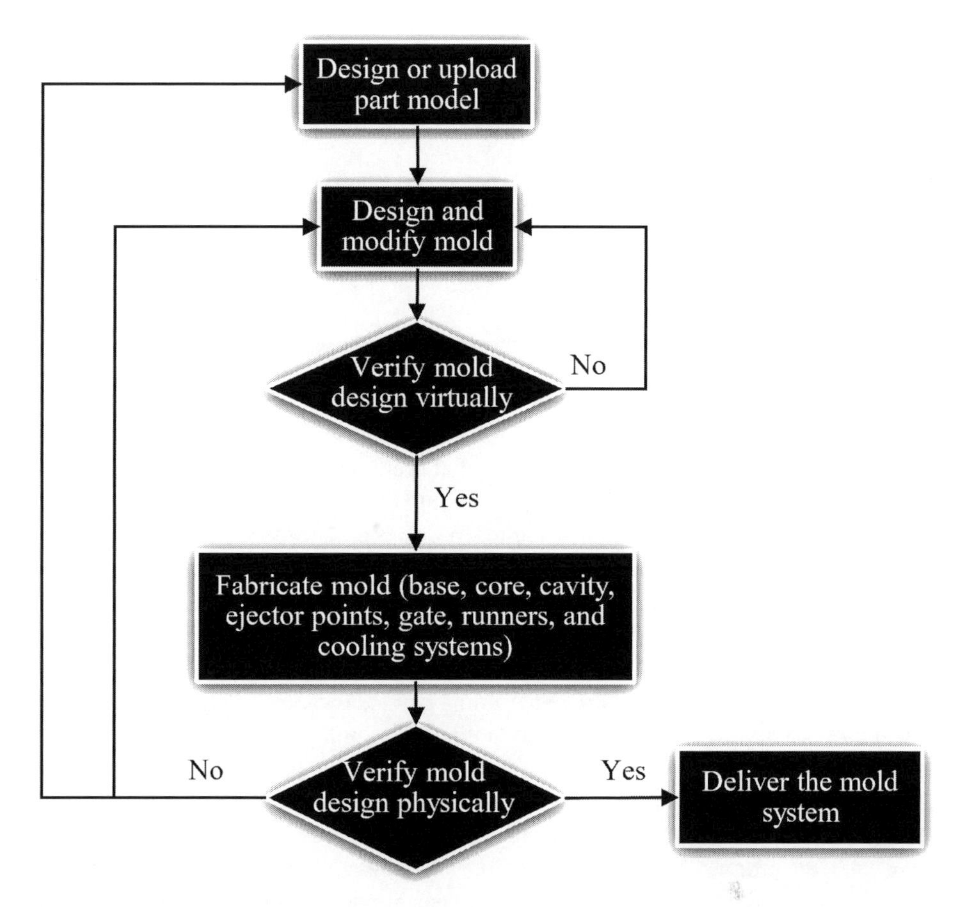

Figure 10.23 The flowchart of designing an injection moulding system.

10.7.5 Other Design Considerations

Section 10.7.3 showed that the working temperature and pressure determines mainly the moldability of parts in an injection moulding process. The following design factors should also be taken into consideration to ensure the success of the injection moulding process.

1. Shrinkage compensation
   The polymers are used as raw materials in injection moulding. Polymers have high thermal expansion coefficients. Therefore, the part dimensions are greatly varied since an injection moulding process involves a large changing range of temperatures and a significant shrinkage occurs during solidification and cooling in the moulding process. Such a dimensional shrinkage has to be compensated. Table 10.5 shows the typical compensation for shrinkage of commonly used plastics for their injection moulding processes (Groover 2012).
   A dimension in a mould cavity must be larger than the desired part dimension and can be estimated by

$$D_c = D_p(1 + S + S^2) \tag{10.12}$$

Table 10.5 Typical shrinkage compensation for commonly used plastics.

Plastics	Typical shrinkage, mm/mm (in./in.)
Nylon-6	0.020
Polyethylene	0.025
Polystyrene	0.004
Polyvinyl chloride (PVC)	0.005

where D_c is the dimension on the cavity, D_p is the dimension on part, and S is the specified value of shrinkage defined in Table 10.5. Note that the third term on the right side compensates for the second-order shrinkage that occurs in the first-order shrinkage.
To reduce the shrinkage, a designer might look into the following approaches:
(a) Increasing the injection pressure to force more material into the mould cavity.
(b) Increasing the moulding temperature to lower the polymer melt viscosity and allow more material to be packed into the mould.
(c) Prolong the compaction time to force more melts flow into the cavity.

2. Economic factors

   The mould and tooling for any type of moulding processes can be costly since each part type requires a unique set of moulds and tooling. It is applicable to injection moulding processes. A variety of injection moulding processes are available. The selection of injection moulding processes must be justified by the quantity of product, since the costs on moulds and tooling are shared by the number of parts to be made. The higher the quantity of the product, the lower is the unit cost of the product.

   Generally, an injection moulding process applies appropriately for the product with a volume over 10 000 pieces, a compression moulding process should be considered if the product has a volume less than 1000 pieces, and the transfer moulding process is a trade-off solution of injection moulding and compression moulding for the product with a volume range from 1000 to 10 000 pieces.

3. DFM

   DFM is applicable for injection moulding processes. Although a part with a high level of complexity implies a high cost of moulds and tooling, it is economically beneficial to combine multiple parts as one part. An injection modelling process makes it possible for a combination of multiple functional features when it is needed.

   Taking into consideration injection moulding processes at the product design stage is also referred to as *design for moldability*. It supports part designs from the perspective of process optimization and specific performance criteria are flow lengths, weld line locations, injection pressures, clamping requirements, scrap rate, easiness of part assembly, and the need for secondary operations such as de-gating, painting, and drilling. In addition, design for moldability aims to minimize the mould defects such as moulded-in stress, flashes, sink marks, and surface blemishes. When the design constraints of an injection moulding process are considered at the product design stage, the following product features must be paid special attention.

(a) *Wall thickness*. A thick wall causes waste, it is more likely to cause a warping defect due to uneven shrinkage, and it takes a long time to harden the materials. The wall features should be designed to be thin and ideally with even thickness.
(b) *Reinforcing ribs*. When the strengths have to be enhanced at certain sections, considering adding ribs increases the stiffness without thick walls. Reinforcing ribs should be made thinner than the walls they reinforce to minimize the defects of sink marks on the outside wall.
(c) *Corner radii and fillets*. Sharp external or internal corners are undesirable in a moulded part, since a corner affects the smooth flow of melt. The sharper a corner is, the more severe is the impact on the melt flow. Therefore, sharper corners should be eliminated to avoid surface defects and stress concentrations.
(d) *Holes*. Holes are quite feasible in plastic moulding, but the impact of the holes on the mould complexity and part ejections have to be taken into consideration.
(e) *Drafts*. Draft features are essential to side surfaces. A part should be designed with the drafts on the lateral surfaces to facilitate part ejections. The recommended drafts for thermosetting and thermoplastic materials are $\sim 1/2$–$1°$ and $\sim 1/8$–$1/2°$, respectively.
(f) *Tolerances*. Even though the shrinkage is predictable under closely controlled conditions, generous tolerances are desirable since (i) variations in process parameters can affect shrinkage and (ii) diversity of part geometries is encountered.

10.8 Mould Filling Analysis

The discussion in Section 10.7 shows that an injection moulding process is very complex, since the raw materials experience a large range of temperature changes in the injection moulding unit and clamping unit. It involves a large number of design variables and system parameters. Traditionally, the design of an injection moulding process relies heavily on designers' experiences and the trial and error methods. Computer simulation tools provide the scientific methods to design and analyse injection moulding processes.

Mould filling analysis is useful in designing and optimizing injection modelling processes. Computer aided mould filling analysis can be applied and used to simulate the injection modelling process and find the answers to most of the design problems discussed in the previous section. Figure 10.24 shows the inputs and outputs of a computer aided mould filling analysis system. The system inputs include part and mould design, material flow characteristics, heat transfer properties, melt temperature, mould temperature, and the designs of runners, gating systems, and cooling systems. The system outputs include all simulation data relevant to the injection moulding process and the quality of mould parts such as flow and fill parameters, locations of weldlines, pressure to fill, pressure patterns, clamping force, temperature patterns, shear patterns, filling, temperature distribution, shear thinning, freezing and reheating, and temporary stoppages of flow (Eastman 2019).

Using a computer aided mould filling analysis system, critical mould design variables and process parameters can be identified and computer models for parametric studies can be defined, so that the analysis system is able to support simulation-based design optimization for injection moulding processes.

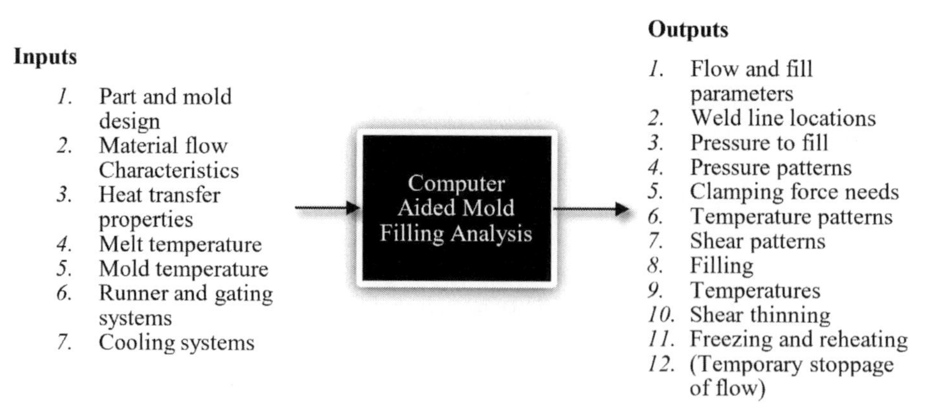

Figure 10.24 Inputs and outputs of computer aided moulding filling analysis.

10.8.1 Mould Defects

An injection moulding process relates to numerous design factors. There is no comprehensive mathematical model to guide the design of injection processes. Design of an injection moulding process is usually an iterative process used to optimize process parameters and moulding systems gradually. The design improvement can be justified by the reduction of moulding defects. Moulding defects might be caused by relevant and complicated reasons, such as design flaws in products, moulds, and tools, improper selection of moulding materials, inappropriate moulding conditions, and malfunctions of the moulding machine. Here, a few of commonly observed mould defects are discussed.

1. Weldlines

   A *weldline* is a thin line created when different flows of molten plastics in a mould cavity meet and remain undissolved. It is the boundary between the flows caused by incomplete dissolution of molten plastics. Figure 10.25 shows the scenario of how a weldline is developed by two flows. Weldline defects can be alleviated by (i) increasing the injection speed and raising the mould temperature, (ii) lowering the molten plastics temperature and increasing the injection pressure, and (iii) adjusting the gate position and the flow of molten plastics to prevent the development of weldlines.
2. Flashes

   Flashes are developed at the parting line of the mould or the location where an ejector pin is installed. Figure 10.26 shows the case where the flashes are developed at the parting plane. A flash is a phenomenon where the molten polymer smears out or sticks to the seam or gap of two objects. Some common causes of flashes are (i) the poor quality of the mould, (ii) the low viscosity of the molten polymer, (iii) a too high injection pressure, or (iv) a too low clamping force. Therefore, to reduce flashes, the following methods can be considered: (i) avoiding excessive differences in thickness in a part design, (ii) reducing the injection speed, (iii) applying well-balanced pressure to the mould for consistence of the clamping force, (iv) increasing the clamping force, and (v) enhancing the surface quality of the parting plane, ejector pins, and holes.

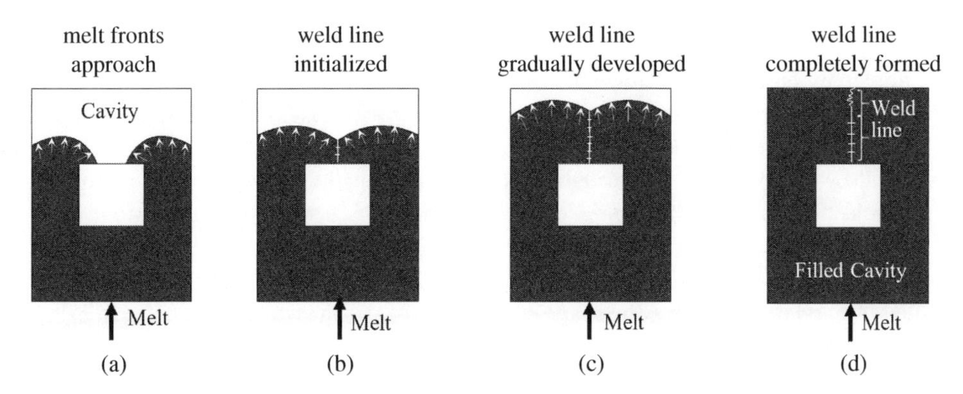

Figure 10.25 The development of a weldline.

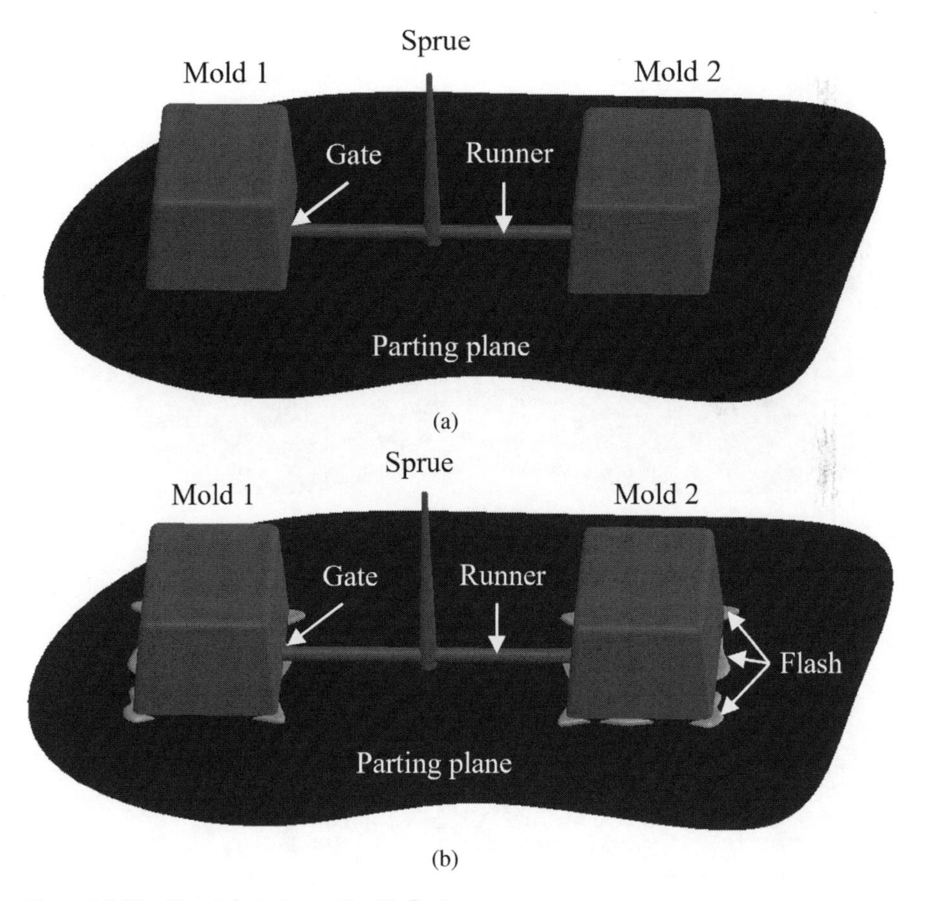

Figure 10.26 The defected mould with flashes.

3. Short shots

   A *short shot* is the phenomenon where molten plastics do not fill the mould cavity completely. Figure 10.27 shows the difference between parts with and without short shorts. Short shots usually occur at the areas with the longest flow distance to the gate. The portion of parts with short shots becomes an incomplete shape. Short shots are caused by (i) a low one-shot weight, (ii) an insufficient injection pressure, and (iii) a low injection speed leading to the case where the melt becomes frozen before it flows to the end of the mould. Short shots can be addressed by (i) applying a high injection pressure, (ii) installing air vents or a degassing device to increase pressure difference in the flow, and (iii) changing the shape of the mould or the gate position for a better flow of the melt.

4. Warpages

   A *warpage* appears when the part is removed from the mould and pressure is released. Figure 10.28 shows a part with warpage defects. Warpages are caused by (i) uneven shrinkage due to the mould temperature difference (surface temperature difference at the cavity and core) and the thickness difference in the part and (ii) a low injection

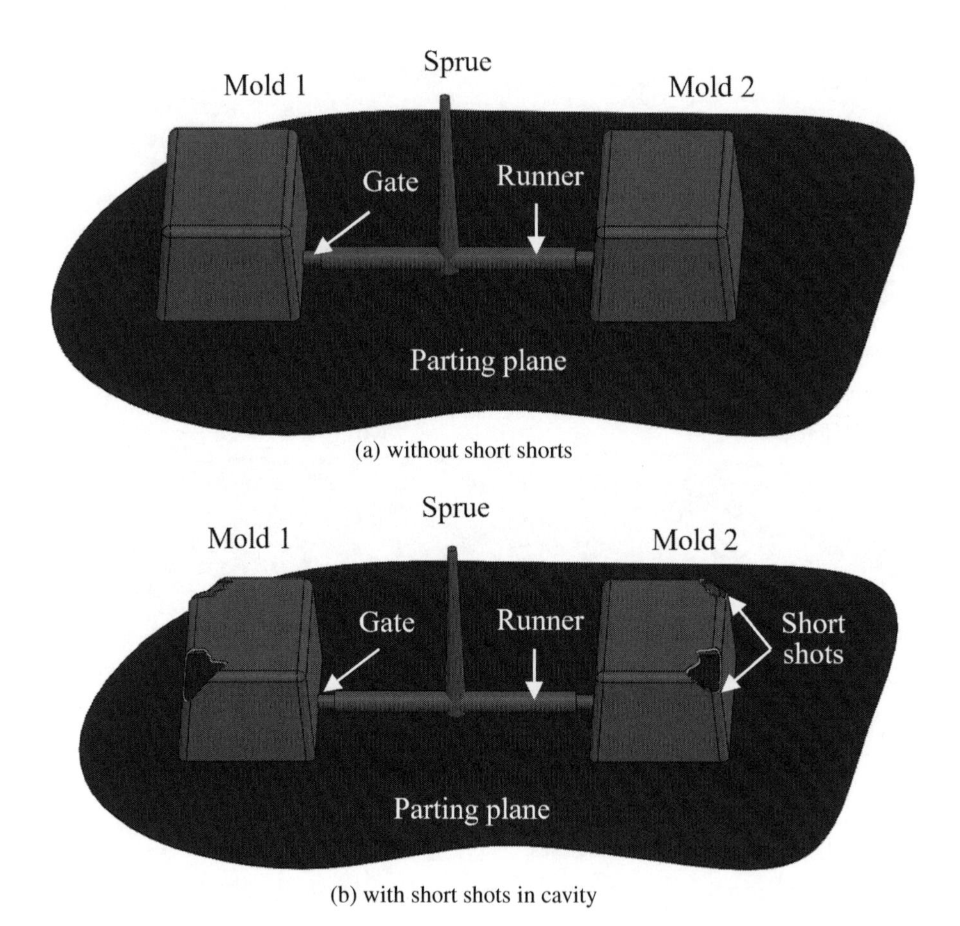

Figure 10.27 The defected mould with short shots.

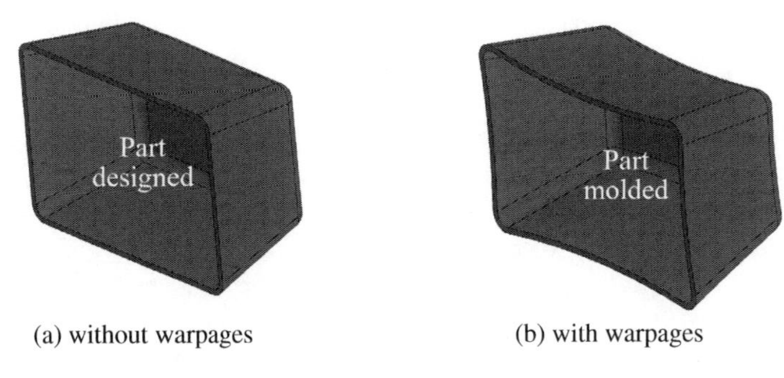

Figure 10.28 The defected mould with warpages.

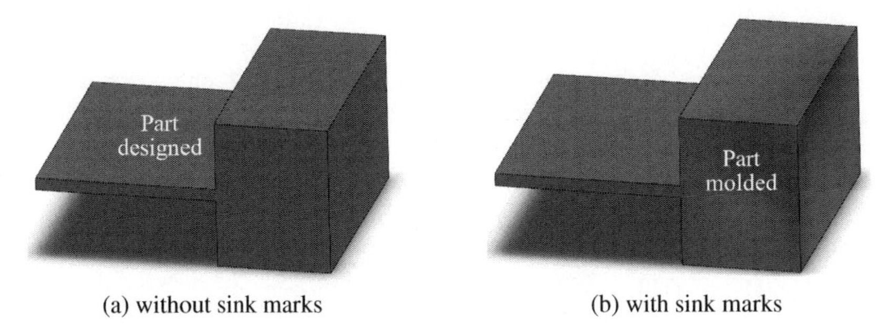

Figure 10.29 The defected mould with sink marks.

pressure and insufficient packing. To reduce warpages, the possible solutions include (i) taking a longer cooling time and lowering the ejection speed, (ii) adjusting the position of ejection pins, (iii) enlarging draft angles, (iv) balancing cooling lines, and (v) increasing packing pressure.

5. Sink marks

   Sink marks develop when material in the region of thick features such as ribs or bosses shrinks more than the material in the adjacent wall. Figure 10.29 shows the difference between parts with and without sink marks. Sink marks are caused by insufficient melts in the given area before the channel for more melt is closed due to solidification. The practical methods to reduce sink marks are to (i) increase packing and holding pressure to get more melts in the cavity, (ii) reduce the injection speed to keep the gates open longer and let more melts enter, and (iii) reduce the mould temperature to strengthen the plastic surface instead of adding more plastics.

10.9 Mould Flow Analysis Tool – SolidWorks Plastics

SolidWorks Plastics is an add-in feature to the standard SolidWorks package. It is a computer mould flow analysis program used to simulate an injection moulding process and analyse *flow patterns*, *weld lines*, and *cooling rates*. The SolidWorks Plastics allows users to quickly identify and correct manufacturing defects in parts and moulds and supports the simulation-based optimization to shorten cycle times, reduce product defects, and eliminate costly mould rework. It is fully integrated with SolidWorks modelling, assembling, and simulation modules to help part designers, mould designers, and mould makers optimize designs for manufacturability without leaving their familiar 3D design experience. With the SolidWorks Plastics, a designer does not have to be an expert to identify and address potential defects by making changes to the part or mould design, plastics material, or processing parameters, saving resources, time, and money.

SolidWorks Plastics works directly on a 3D model; no translation is required. The wizards for mould analysis and meshing are very user-friendly; the interface is intuitive and step-by-step. Guided analysis, intelligent defaults, and automated processes ensure correct setups even though users are not experts in simulation. Figure 10.30 shows the interface to activate SolidWorks Plastics. It is an add-ins, which can be simply checked to be activated. Figure 10.30b shows that once the SolidWorks Plastics is activated, the wizards for mould filling analysis and automatic meshing become available.

Creating a mould filling analysis model is straightforward. Figure 10.31 shows six steps of mould filling analysis in the SolidWorks Plastics. The analysis follows the setup of the unit system in the welcome window to generate mesh, select material, specify process parameters, add injection locations, run the simulation, and finally to review the results. As shown in Figure 10.32, the SolidWorks Plastics offers automatic meshing for either surface or solid meshes. As shown in Figure 10.32b, the SolidWorks Plastics offers automatically to add injection locations and predict a flow pattern at step 4 for injection locations. After the simulation is completed, the users can review many types of simulation data related to flow, pack, and warp. Figure 10.33 shows the list of distributed field variables that are generated from the simulation. Figure 10.34 shows some commonly used data from mould filling analysis, which include volumetric shrinkage, pressure, sink marks, cooling time, and changing the mould temperature.

Besides the wizard tools for beginners, the SolidWorks Plastics allows experienced users to access more variables and setups relevant to mould filling analysis and mould layout analysis. The interface to these tools is shown in Figure 10.35. The tools in mould filling analysis

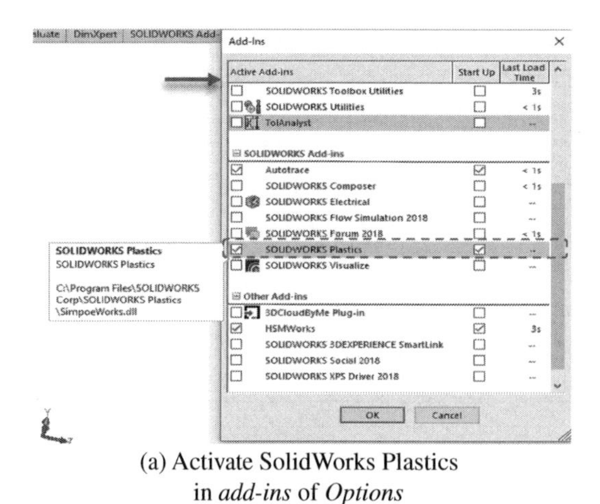

(a) Activate SolidWorks Plastics in *add-ins* of *Options*

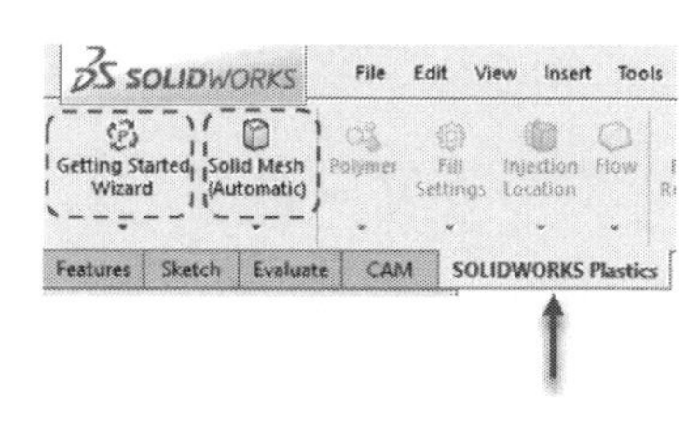

(b) CommendManager SolidWorks Plastics become available

Figure 10.30 Activating a SolidWorks Plastics tool.

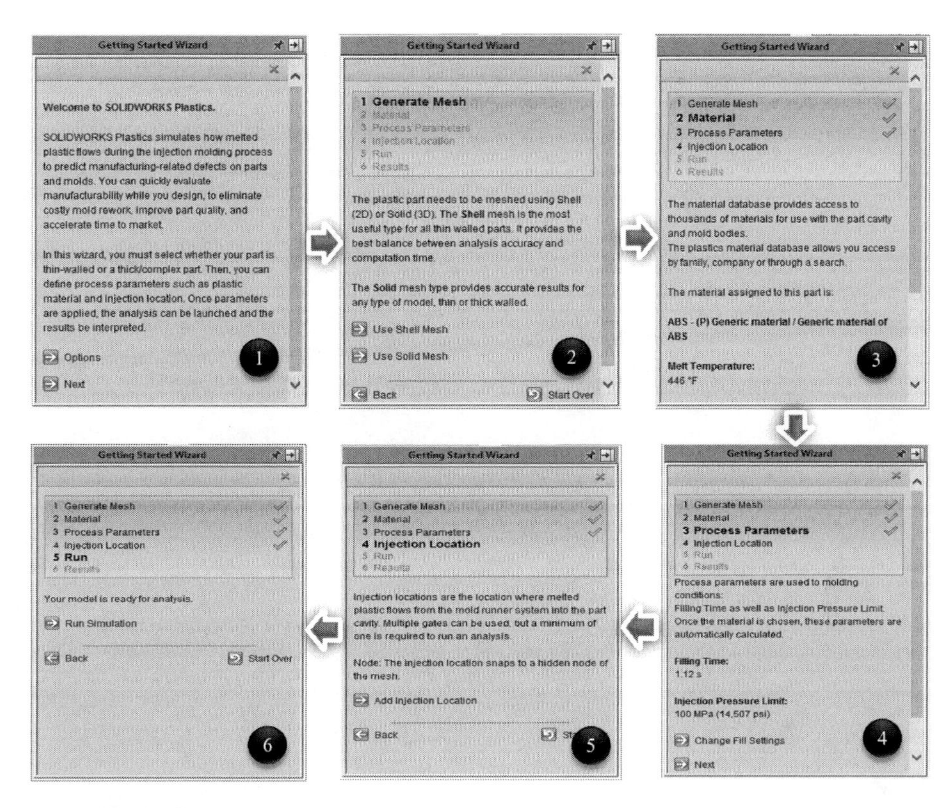

Figure 10.31 Steps in mould analysis wizard.

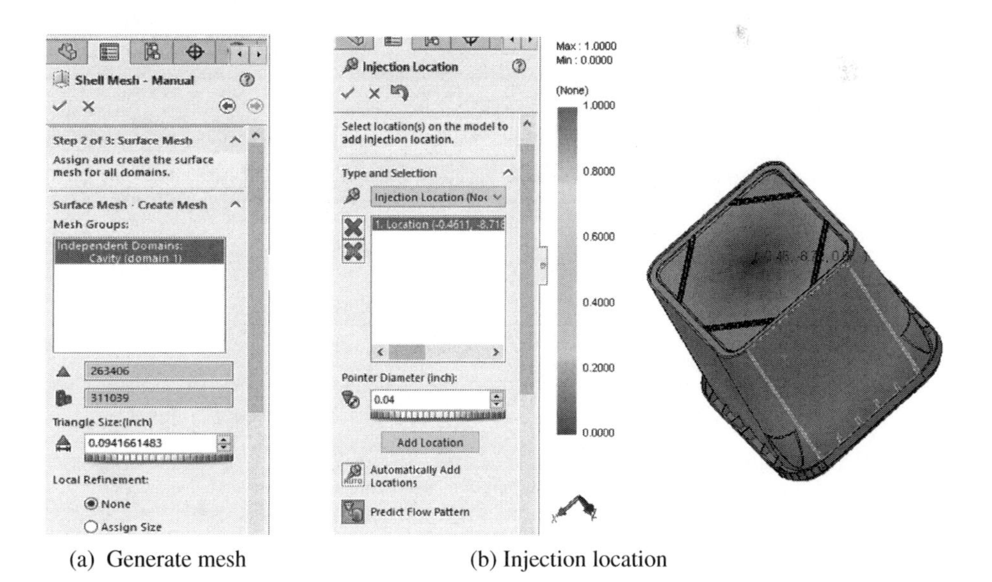

Figure 10.32 Interfaces in the steps of *generate mesh* and *injection location*.

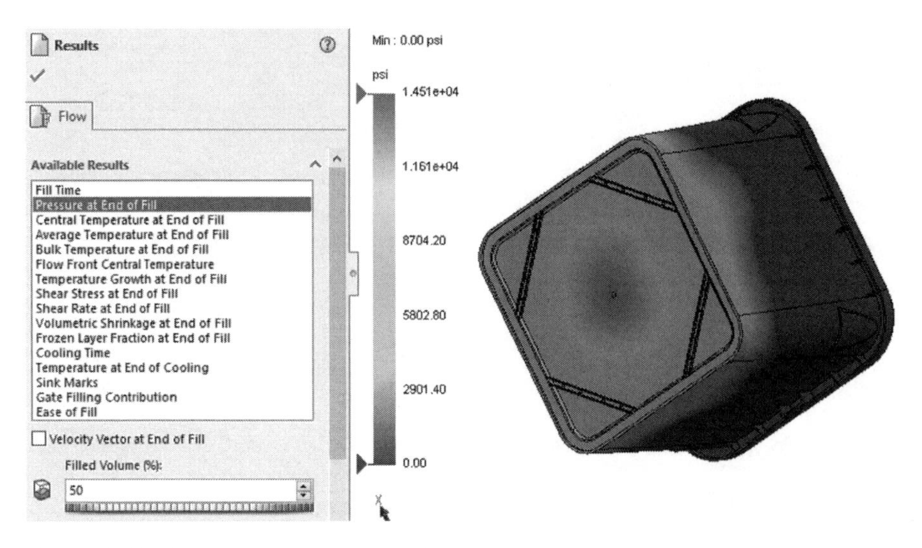

Figure 10.33 Review mould filling analysis results.

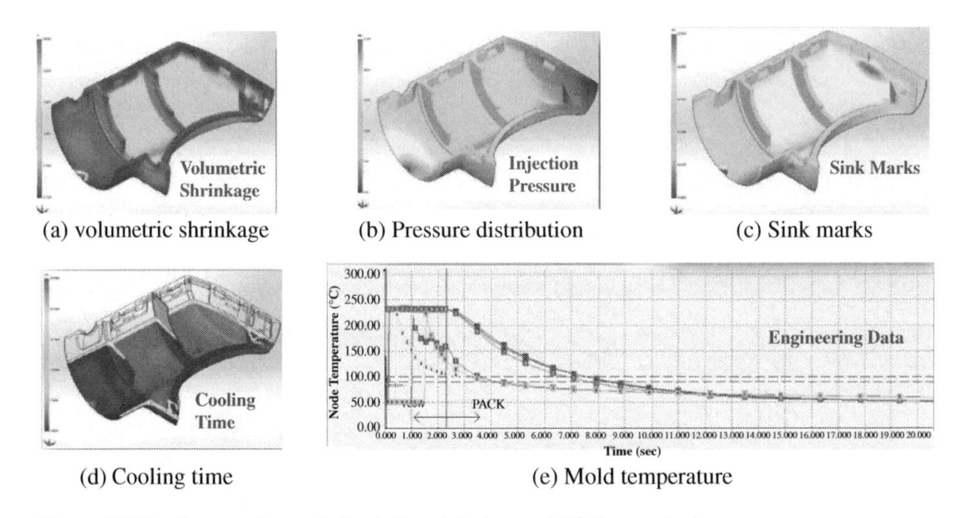

Figure 10.34 Commonly used simulation data in mould filling analysis.

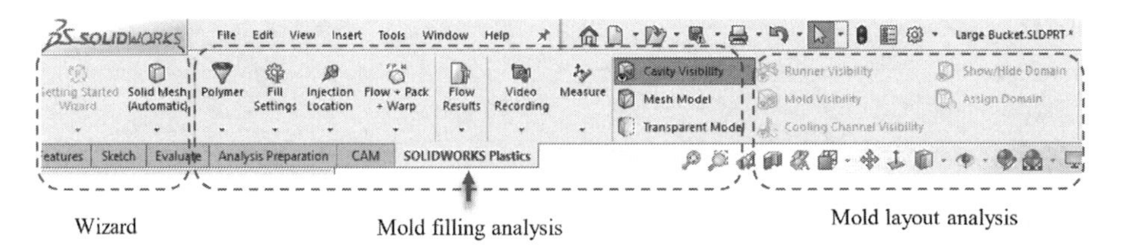

Figure 10.35 Available tools in SolidWorks Plastics.

include *polymer* to change materials, *filling settings* to setup injection moulding parameters, such as filling time, melt temperature, mould temperature, injection pressure limit, and clamp force limit, *injection location* to add or change injection locations, *flow + pack + warp* to specific the simulation type of interest, and *flow results* to review simulation results. Mould layout analysis is used to simulate the impacts of runners, mould assembly, and cooling systems.

10.10 Summary

Numerous types of manufacturing processes are available. Among these, the processes for creating the geometries of parts are shaping processes. Shaping processes such as casting and injection moulding are very complex and involve a large number of processing parameters and uncertainties.

Virtual design of manufacturing processes can assist in identifying design defects of products and custom tools and predicting outcomes of manufacturing processes. In this chapter, the designs of casting processes and injection moulding processes are discussed in detail; the challenges in designing complex shaping processes are explored; it is unrealistic to develop analytic models for shaping process design; and traditional designs of manufacturing processes heavily rely on designers' experiences and trial-and-error methods. Computer numerical simulation are introduced to conduct mould-filling analysis to predict manufacturing defects such as sink marks, warps, and residual stresses.

10.11 Design Project

10.1 1. Create a plastic part you can reach (or download and select one model from the Website) such as the one shown in Figure 10.36.

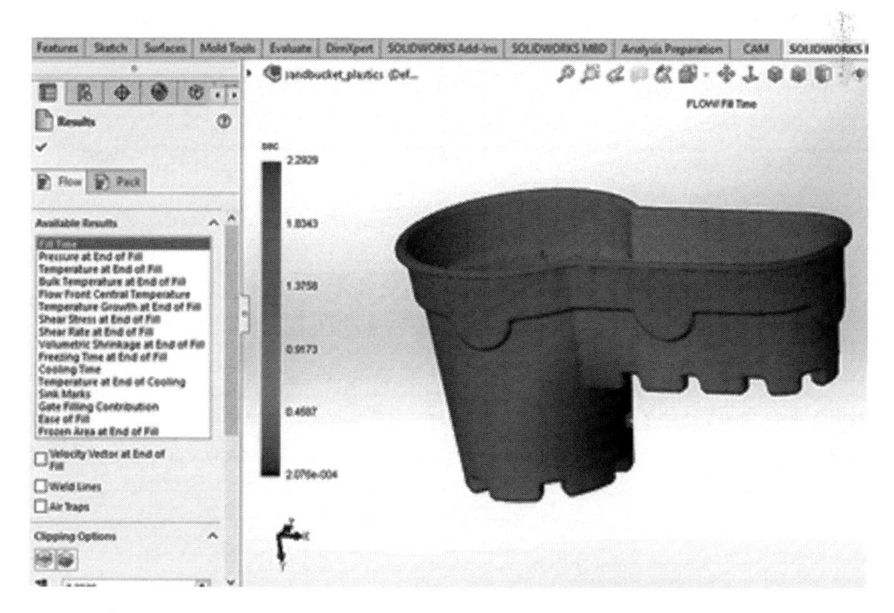

Figure 10.36 Example of product and mould filling analysis for Design Project 10.1.

2. Use an SW Plastics package and select injection locations and filling settings.
3. Run an injection simulation to predict the injection moulding time.
4. Predict defects including air traps, welding lines, and wraps.
5. 5 Document your process and result.

References

Black, J.T. and Kohser, R.A. (2011). *Degarmo's Materials and Processes in Manufacturing*, 11e, ISBN-13: 978-0470924679. Wiley.

Choudhari, C.M., Narkhede, B.E., and Mahajian, S.K. (2014). Casting design and simulation of cover plate using AutoCAST-X software for defect minimization with experimental validation. *Procedia Materials Science* 6 (2014): 786–797.

Eastman (2019). Medical devices processing guide. https://www.eastman.com/Literature_Center/S/SPMBS3689.pdf.

Engineering (2017). Engineering design platforms and simulation in-CAD benefit product development teams. https://www.engineering.com/DesignSoftware/DesignSoftwareArticles/ArticleID/14534/Engineering-Design-Platforms-and-Simulation-in-CAD-Benefit-Product-Development-Teams.aspx.

ESP International (2019). Casting design guide. https://www.espint.com/CASTING%20CATALOG2-opt.pdf.

Fischer, M.J. (2003). Cause of molded part variation: processing. In: *Handbook of Molded Part Shrinkage and Warpage*, 51–77, ISBN 978-1-884207-72-3. William Andrew Inc.

Groover, M.P. (2012). *Introduction of Manufacturing Processes*. Wiley ISBN 978-0-470-63228-4.

Hahn, I., and Sturm, J.C. (2015). Simulation evolves to autonomous optimization. https://www.magmasoft.com/export/shared/.galleries/pdfs_publications/2015_Simulation-evolves-to-autonomous-optimization.pdf.

Mitsubishi Engineering Plastics Corp (2019). Molding. https://www.m-ep.co.jp/en/pdf/product/iupi_nova/molding.pdf.

Patil, S.S., Choudhari, U.M., and Reddy, A.C. (2005). Computer integrated injection mould design, analysis and development. In: *National Conference on Computer Applications in Mechanical Engineering (CAME – 2005)*, 202–205. https://jntuhceh.ac.in/faculty_portal/uploads/staff_downloads/736_nc053-2005.pdf (last accessed on 1 October 2019).

Ravi, B. (2010). Casting simulation – best practice. http://citeseerx.ist.psu.edu/viewdoc/download?doi=10.1.1.459.3783&rep=rep1&type=pdf.

Ravi, R. (2017). Computer-aided casting design – past, present and future. http://citeseerx.ist.psu.edu/viewdoc/download?doi=10.1.1.473.1273&rep=rep1&type=pdf.

Reifschneider, L. (2000). Teaching design for manufacturability with desktop computer-aided analysis. *Journal of Industrial Technology* 16 (3): 1–5.

Singh, R. (2006). *Introduction to Basic Manufacturing Processes and Workshop Technology*. New Age International Limited Publishers, ISBN (10): 81-224-2316-7.

SolidWorks (2015). SolidWorks Plastics optimize the design of plastic parts and injection molds. www.solidsolutions.co.uk/Uploaded/Documents/Products/Datasheets/2015/solidworks%20Plastics%20Datasheet.pdf.

Tandon, R. (2001). Metal injection molding. In: *Encyclopedia of Materials: Science and Technology* (eds. K.H.J. Buschow, M.C. Fleming, E.J. Kramer, et al.). UK: Elsevier. ISBN: 978-0-08-043152-9 https://www.sciencedirect.com/topics/engineering/injection-molding-process.

Thomasnet (2019). Processes involved with casting. https://www.thomasnet.com/articles/custom-manufacturing-fabricating/casting-processes.

Wikipedia (2019). Casting. https://en.wikipedia.org/wiki/Casting.

Zhang, H.-C. (1994). Manufacturing process planning in book. In: *Handbook of Design, Manufacturing and Automation*, Chapter 29 (eds. R.C. Dorf and A. Kuziak), 587–616. Wiley.

11

Computer Aided Design of Tools, Dies, and Moulds (TDMs)

11.1 Introduction

In a manufacturing process, a machine tool makes physical contacts with workpieces. Part features are determined by (i) the specified tool geometries and (ii) the relative motions of tooling with respect to a workpiece in manufacturing processes. Different manufacturing processes or products need different types of tools; for example, a material removal process needs cutting tools, an injection moulding or casting process needs moulds, a metal forming process needs dies, and an inspecting process needs probes. In design tooling for a manufacturing process, the required functions, constraints, and objectives are driven by the features, geometry, dimensions, and design intents of products. Therefore, the knowledge-based engineering (KBE) approach is an effective means to transform product information into the requirements, design constraints, and objectives of moulds and dies.

A conventional manufacturing process such as an injection moulding or turning operation creates part features via *machinery* and *tooling*. Machinery provides the power and all the functions to enable manufacturing processes and tooling makes direct contact and changes workpieces. Table 11.1 shows the types of machinery and the methods in which product features are generated in material removal and forming processes.

In general, tooling refers to an auxiliary tool that makes direct contact to physical objects in certain manufacturing processes. Therefore, machining tools show a wide scope of variety for different manufacturing processes and different part geometries. Available tools can be classified based on their functions in different manufacturing activities; for examples, fixturing tools for immobilizing objects, cutting tooling for material removal processes, deforming tools for forming processes, and measuring and testing tools for quality controls. Design of a manufacturing process covers the design, verification, and validation of required auxiliary tools. Figures 11.1 and 11.2 show some common tools used in conventional material removal processes and forming processes, respectively.

For machining tools in material removal processes, designs of fixture tools in Figure 11.1a have been extensively discussed in Chapter 8. The geometries of cutting tools in Figure 11.1b to d are not closely related to the part features, since these features depend greatly on the motions of machineries relative to workpieces. The tools in Figure 11.1e and f are used for general purposes to pick and place objects. These types of tools should be standardized to maximize their utilization rates and reduce tooling costs.

Computer Aided Design and Manufacturing, First Edition. Zhuming Bi and Xiaoqin Wang.
© 2020 John Wiley & Sons Ltd. This Work is a co-publication between John Wiley & Sons Ltd and ASME Press.
Companion website: www.wiley.com/go/bi/computer-aided-design

Table 11.1 Machinery and tooling in conventional manufacturing processes.

Types of machinery	Working principles	Examples
Material removal processes (machine tools and tooling)	• Power-driven machines used to operate cutting tools to remove unwanted materials • Part features are defined by relative motions/paths of cutting tools with respect to workpieces and tool geometry	Lathes, mills, grinders, drills, and laser- or water-cutting tools
Forming processes (production equipment, moulds, dies, and tools)	• Production equipment used to support near net-shape manufacturing processes such as casting, compression, extrusion, drawing, injection moulding • Features of the part are defined by the cavities or geometry of moulds, dies, and tools	Casting facilitates, compression machines, presses, forge hammers, and plastic injection moulding machines

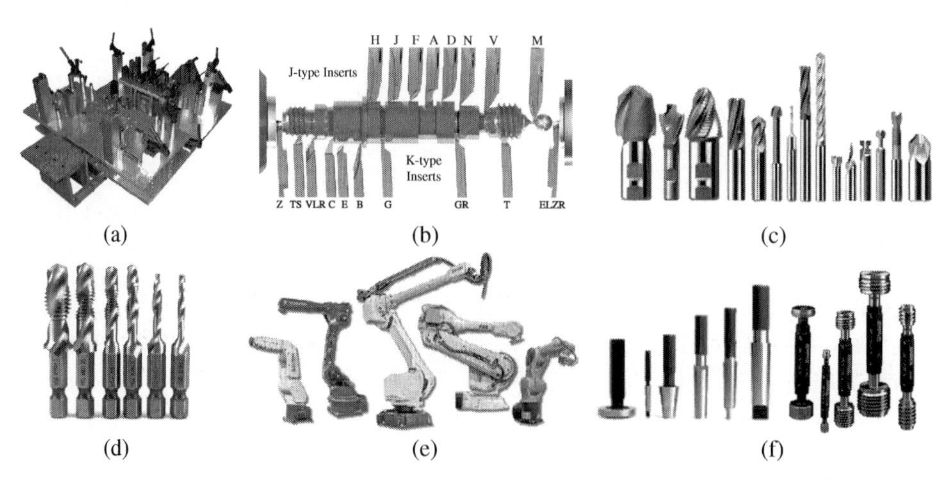

Figure 11.1 Examples of machining tools for conventional material removal processes: (a) fixture, (b) cutting tools, (c) milling tools, (d) drilling tools, (e) picking and placing tools, and (f) inspection tools.

For the machining tools in forming or deforming processes and the geometries of wire drawing or rolling tools in Figure 11.2a and b are relatively simple, since their part geometries are gradually formed, and controllable tooling parameters are two-dimensional. On the other hand, the tool geometries for manufacturing processes in Figure 11.2c to f are customized to those parts to be manufactured. Therefore, the tool geometries must be directly derived from part geometries. In the rest of this chapter, computer aided design (CAD) for these machining tools especially are investigated.

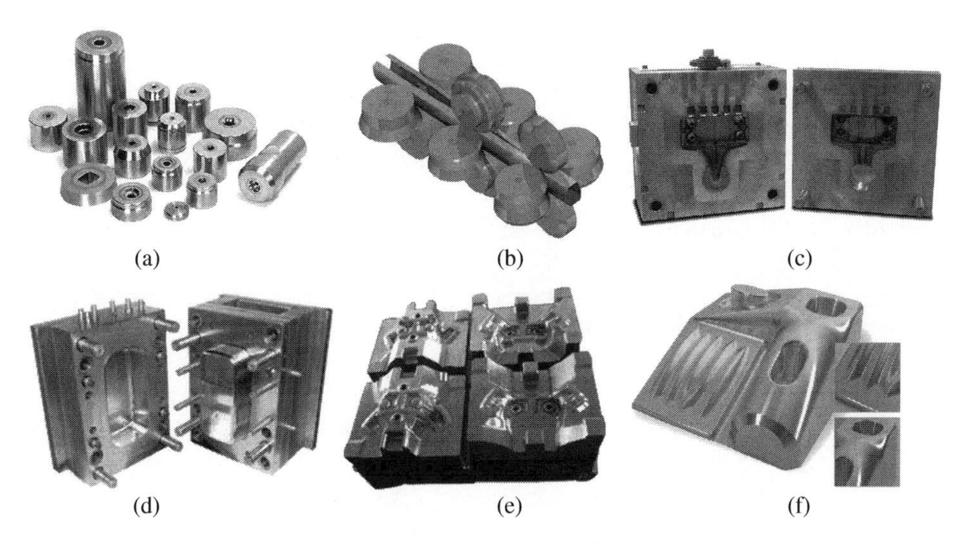

Figure 11.2 Examples of moulds and dies for forming and deforming processes: (a) wire drawing, (b) rolling process, (c) permanent mould casting, (d) injection modelling, (e) forging process, and (f) thermoforming process.

11.2 Overview of Tools, Dies, and Industrial Moulds (TDMs)

Tools, dies, and industrial moulds (TDMs) are used by industries such as metal stamping, die casting, and plastics moulding industries to give the final shape or form to the items being produced. In industry usage, tools include dies, punch tools for dies, industrial moulds, jigs, and fixtures. Moulds and dies are similar to some extent in exterior appearance as both are usually produced as reverse-representations of the objects or shapes to be manufactured. However, in operation, moulds generally come together and pull apart on a horizontal plane, whereas dies come together and pull apart on a vertical plane as the strike force of the press is aided by gravity. Industrial moulds are used to produce a wide variety of plastic, metal, rubber, glass, and mineral products. These include plastic and metal parts for motor vehicles, aircraft, appliances, electronics and electrical products, housewares, consumer products, furniture, military items, and medical products. Moulds for plastics include a variety of types, including injection, compression, blow, reinforced, transfer, forming, plunger, and rotational moulds, with the most widely used being injection moulds (United States International Trade Commission 2002).

The TDM industry refers to the manufacturing business sector that designs, fabricates, and tests the tooling, moulds, and dies. These machining tools are essential for manufacturers to shape raw materials into desirable features, geometry, and dimensions (Lanzuela 2018). *Tooling* is a broad term, which covers TDM. A *cutting tool* is used in a material removal process to cut unwanted materials away (Canis 2012). Dies and moulds are closely relevant concepts and sometimes are used alternatively. Generally, a *die* is a form of tooling, which is used to shape the materials in a deforming process, such as forging and stamping operations. On the other hand, a *mould* is used to shape the materials in a forming process, such as casting or injection modelling operations. A die can also be a metal mould, which is used to make products from plastics, ceramics, and composite materials.

Due to the low volumes and highly diversified nature of products, the majority of toolmakers are small and medium sized enterprises (SMEs). Nine out of ten toolmakers are privately owned and often family operated and employ fewer than 50 workers (Canis 2012). In addition, such companies hire highly skilled employees with many years of experience to tackle the challenges involved in the designs and fabrication of tools. However, tooling is needed for nearly all manufactured products, while nearly half of tooling consumption is for automotive or defence-related manufacturers.

11.3 Roles of TDM Industry in Manufacturing

The TDM industry provides essential services for designing, manufacturing, and testing machining tools to all manufacturing enterprises. Figure 11.3 shows the structure of the TDM industry (Lanzuela 2018): (i) as the manufacturers to produce TDM as finished products, they have specified suppliers to obtain raw materials for tooling products, the machinery to make tools, and computer software tools required to design, manufacture, and test products; (ii) as the capital products to make other products, tools, moulds, and dies have high requirements of strength, rigidity, durability, and toughness, property enhancements are essential after the products are shaped; (iii) as direct contacts to shape workpieces, TDM are needed for any manufacturing processes, and these products are highly fragmented due to the diversity of tooling needs in different processes; and, finally, (iv) the TDM end-users are the downstream enterprises who manufacture customer products.

The TDM industry plays a pivotal role in the manufacturing eco-environment. The growth of manufacturing industry is directly reflected by that of the tools, moulds, and dies industry. Table 11.2 shows the changes of import and export trades in TDM from 1997 to 2010 in the Unites States (US) (Canis 2012). The deficit of import and export trades

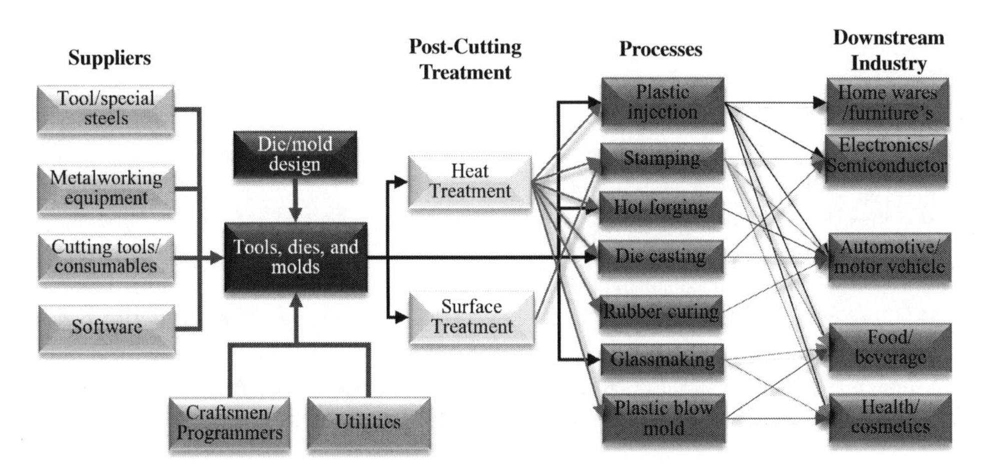

Figure 11.3 Role of TDM industry in manufacturing (Lanzuela 2018).

Table 11.2 Import and export trades of TDM from 1997 to 2010 (Canis 2012).

Product	Year	Exports	Imports	Deficit
Moulds	1997	$648 million	$1291 million	$643 million
	2010	$717 million	$4745 million	$4028 million
Tools, dies, and jigs	1997	$442 million	$696 million	$254 million
	2010	$488 million	$851 million	$363 million
Total	1997	$1090 million	$1987 million	$897 million
	2010	$1205 million	$5596 million	$4391 million

Table 11.3 The shrinkages of American manufacturing and TDM industry (Canis 2012).

Category	1998	2005	2008	2018	Percent change 1998–2010 (%)
Employment					
Manufacturing	17 616 672	14 190 394	13 382 697	11 487 497	−35
Tooling	162 032	119 308	107 187	89 661	−45
Annual payroll					
Manufacturing	$679 billion	$699 billion	$728 billion	$661 billion	−3
Tooling	$6.9 billion	$5.7 billion	$5.4 billion	$4.7 billion	−32
Number of establishments					
Manufacturing	412 453	365 351	359 844	342 844	−17
Tooling	9057	7192	6393	5789	−36

has increased from $897 million to $4391 million, an increase of nearly seven times. This implies that US manufacturers rely heavily on foreign vendors to fabricate custom dies and moulds to run manufacturing businesses.

Table 11.3 shows the correspondence between shrinkages occurring to US manufacturing and the tools, moulds, and dies industry (Canis 2012). The shrinkages of manufacturing have been evidenced by multiple aspects, including employment, annual payroll, and the number of establishments of new companies. The statistical data shows that the shrinkage of the tools, moulds, and dies industry has led to the shrinkage of the whole manufacturing industry.

For consumer products that are easy to ship (e.g. small appliances, electronics, and telecommunication devices), making products in low-cost foreign locations and shipping products to the US markets became cost-effective for manufacturers. Products such as air conditioners, radios, vacuum cleaners, power hand tools, televisions, and telephones were increasingly produced abroad. This has adversely affected US toolmakers, since the TDM sourcing has shifted to foreign locations along with the manufacturing, and toolmakers no longer supply the tooling for those products (United States International Trade Commission 2002).

Due to highly diversified products with low volumes, the manufacturing companies in TDM are facing fiercer competition than the companies in other businesses. The shifting in the TDM industry has been especially pronounced over the past decade. The National Tooling and Machining Association (NTMA) estimated that 30% of US toolmakers closed in the period between 2000 and 2003 (Canis 2012).

To boost a nation's manufacturing, it is critical to advance the TDM industry. US manufacturing industry is facing the challenge of attracting more talents for the tooling industry. NTMA surveys show that 95% of toolmakers have openings for jobs, even with relatively high US unemployment rates. The adoption of new technologies is the most vital strategy to advance manufacturing. Technologies have raised product quality and made toolmakers as much as 20% more efficient. Especially, CAD is used to design parts and blueprints for machines. Computer aided manufacturing (CAM) takes the CAD designs and converts them into sequenced instructions for cutting tools. Computer aided techniques are making significant differences to the manufacturing businesses in tools manufacturers.

11.4 General Requirements of TDM

Tooling design is a specialized area of manufacturing engineering that comprises analysis, planning, design, construction, and application of tools, methods, and procedures necessary to increase manufacturing productivity. TDM designs are highly diversified since each tooling product has to be tailored to the required function, part geometry, and manufacturing process. Figure 11.4 shows a set of diversified functional tools for different processes, including work-holding tools (jigs and fixtures), cutting tools, sheet metal dies, forging dies, extrusion dies, welding and inspection fixtures, and injecting moulds.

While tool products show significant differences in terms of functionalities tailored to part geometries and manufacturing processes, the following general requirements are applicable to manufactured products including TDM products.

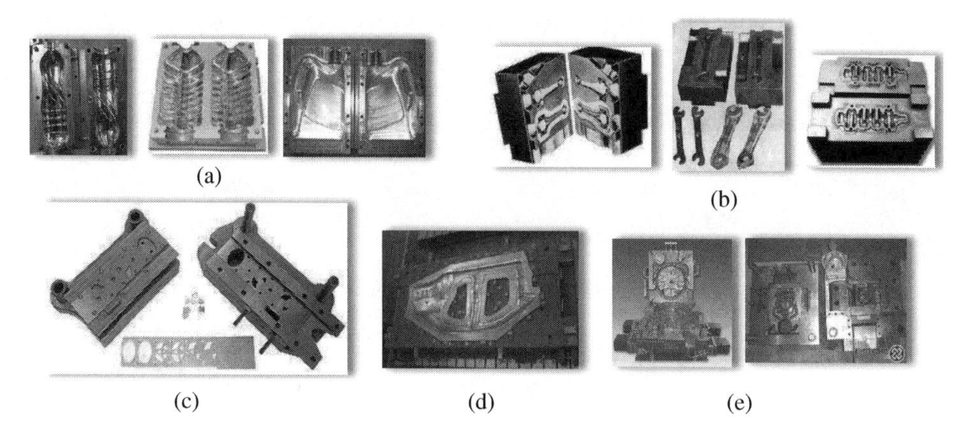

Figure 11.4 Examples of diversified functional tools for different processes. (a) moulds for plastic blow moulding, (b) dies for metal forging, (c) dies for metal stamping, (d) dies for metal stamping, and (e) mould for die casting.

11.4.1 Cost Factors

From an economic perspective, a series of manufacturing processes to transfer raw materials into finished products is viewed as a value-adding process. Therefore, the cost is an essential performance measure for the manufacturers or users of tools, moulds, and dies. Moreover, since a tool product usually has a very low volume, the cost becomes the primary design factor for product optimization. TDM end-users are seeking cost reductions wherever they can. It is especially true for the manufacturers of household appliances, power hand tools, housewares, and electronics. The retail chains for those products are growing and requesting additional pricing demands. Tooling cost was identified as the leading factor of competition cited by toolmakers (United States International Trade Commission 2002).

The total cost of tooling a product relates to all tasks that are involved in the product lifecycle. Figure 11.5 shows the cost constitutions of a tool product. Therefore, adopting new technologies to fulfil the tasks involved in designing, materials selection, manufacturing, and assembling of TDM can contribute to the cost reduction. CAD/CAM and computer numerical control (CNC) technologies in the tooling industry have shown the potential of cost reduction by improving productivity and competitiveness, increasing manufacturing capacity, and ameliorating the need for highly skilled labour.

11.4.2 Lead-Time Factors

Manufacturing of TDM products is driven by end-user orders in the pull production. From the time when the order of a product is placed to the time when the product is delivered, there is a time delay called a *lead-time* of the product. The shortened lifecycles of the products in key industries, such as automotive, appliances, electronics, and telecommunications, have forced toolmakers to shorten the lead times of TDM products that are supplied to original equipment manufacturers (OEMs).

Due to the complexity and diversity of TDM products, average lead-times are long and are measured in weeks. Figure 11.6 shows the performance evaluation of a toolmaker based on the average lead-time of its TDM products (Cimatron 2017). The average lead-times classed as laggard, average, best in class, and optimized practice toolmakers are 11.6, 9.3, 8.7, and 7.9 weeks, respectively, and the maximized difference of average lead-times is 3.7 weeks. This implies that a toolmaker with the optimized practice can gain a time window of 3.7 weeks in capturing new business opportunities than less competitive toolmakers.

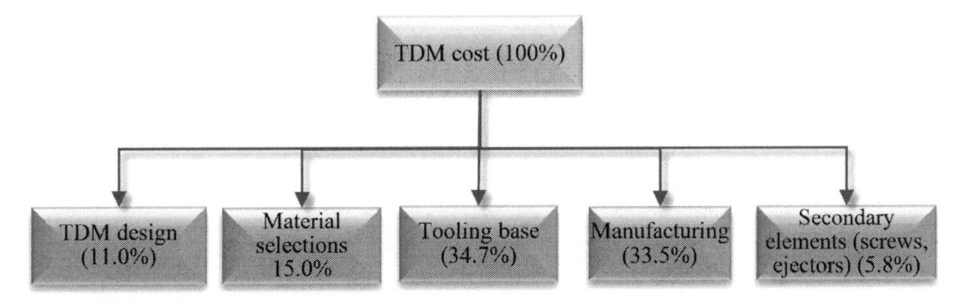

Figure 11.5 Cost constitutions of the TDM product.

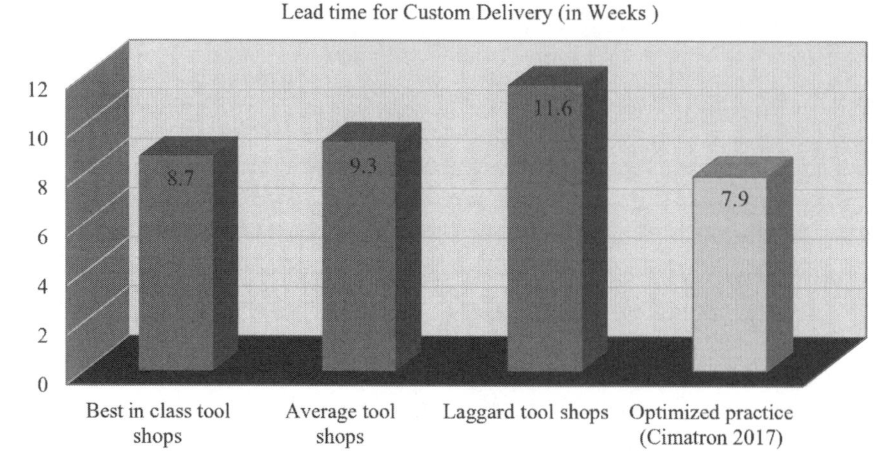

Figure 11.6 Statistic data of average lead time of dies and tools for customer delivery (Cimatron 2017).

11.4.3 Complexity

TDM products vary greatly in terms of geometries, dimensions, and served manufacturing processes, as follows. (i) The sizes of tooling vary significantly since the overall size of tooling is usually proportionate to the workpiece to be made. For example, the size of a die for stamping an automobile part is measured in feet while a mould for cell phone cases is measured in inches or millimetres. (ii) A toolmaker usually serves a group of customers and industries in different applications; accordingly, the functional requirements of TDM products are highly diversified. (iii) A toolmaker is required to encompass the complete range of tooling of a particular construction or form, from low-end, simple fabrications to highly intricate and technologically advanced products. The aforementioned varieties bring great challenges in identifying the commonalities over product variants and reusing existing knowledge for new product development.

11.4.4 Precision

Precision of a TDM product is used to describe the accuracy of tooling and the characteristics of tooling to meet specified measurement tolerances of products to be made. The accuracy of a product depends greatly on that of tooling. Alternatively, the precision requirement of tooling is defined based on the required accuracy of products to be made. The precision of tooling varies with the accuracy of moulded products. A low precision of tooling is appropriate for plastic buckets since dimensional variants do not affect the functionalities of products. An average precision of tooling is applicable to the products with average dimensional tolerances such as housings of electronic components, keyboards, or clock covers. A high precision of tooling is required for parts that are assembled with other components under tight tolerances, such as cellphone housing and closures of food containers.

The tooling accuracy is critical since any tooling error will pass over to products whose accuracy is adversely affected. The use and targeted markets determine the precision requirements and the subsequent degree of accuracy built into the tooling. A tooling for high-precision products is expected to have a repeatable dimensional tolerance of ±0.000 05 inch.

11.4.5 Quality

Quality of tooling reflects the design ideas in fulfilling the needs of end-users and the resulted manufacturing performances. Quality of tooling includes *product life*, *durability*, *performance*, and *maintainability* (United States International Trade Commission 2002). Product life and durability are measured by the lifespan of tooling in making parts to the required specifications without excess wears and premature breakdown. Good performance is reflected by a low downtime and a high productivity in making more parts in each cycle and more cycles in unit time. Maintainability is the ease with which a TDM product can be maintained to prolong the life, correct defects, prevent premature failure, meet new requirements, and improve efficiency and reliability. Designing and manufacturing TDM products with the required quality are subjected to the customers' specifications and cost limits; both creativity and craftsmanship are needed to achieve this goal.

11.4.6 Materials

The material of a TDM product must be harder than the materials of the workpiece, even at high temperatures during a manufacturing process.

Common criteria for the selection of tooling materials are *cost*, *hardness*, *hot hardness*, *strengths*, *toughness*, *ease of machining*, *dimensional stability* after heat treatment, *wear resistance*, *surface finish*, and *corrosion resistance*. Some criteria, such as ease of machining and hardness, conflict with each other. Therefore, no material can possess all of the desired properties as tooling materials. For example, if the tooling material has to be very hard, it will have lower toughness and could break easily when subjected to dynamic loads. As a trade-off selection, steel alloy and aluminium alloy are mostly selected as the materials for TDM products. Many types of tool steels are available.

11.5 Tooling for Injection Moulding

Each manufacturing process requires a certain type of tooling in the execution. A tooling design problem has its uniqueness for a specified process and product. In this section, the tooling design for injection moulding is used as an example to discuss design challenges and introduce the importance and methods of computer aided mould design.

The raw materials of injection moulding are plastic. Plastic materials are light, durable, corrosion resistant, and chemically inert. Plastic materials have a low thermal expansion of coefficient and possess good thermal and electrical insulating properties.

Figure 11.7 shows some examples of injection moulded products, such as furniture, toys, kitchen wares, parts in motorcycles, trucks, and cars, parts in appliances, cellphones, and electronic products, and the packaging parts for food, industrial, and pharmaceutical products. Since plastic is lightweight and durable, plastic products are widely used to store and

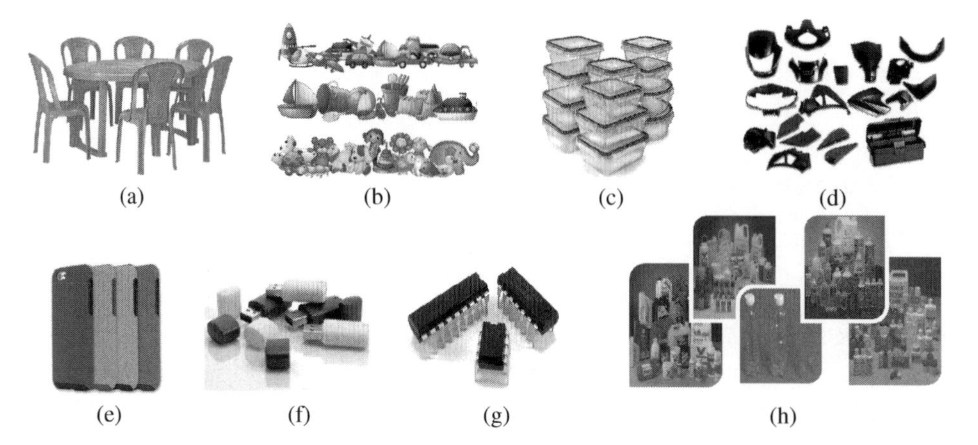

Figure 11.7 Examples of injection moulded products: (a) furniture, (b) toys, (c) kitchen wares, (d) car parts, (e) cellphone parts, (f) USB encapsulation, (g) integrated circuits (ICs), and (h) packaging for food, industrial, and pharmaceutical products.

transport goods. Transportation facilities with more plastic parts are more fuel-efficient. In contrast to metals and ceramics, plastic materials have low strengths and low melting points. Therefore, many different processes with less energy consumption can be used to shape plastic products. Plastic products are inexpensive compared to the products made of other engineering materials.

An injection moulding process is performed on an injection moulding machine, shown in Figure 10.19. The machine consists of an injection unit and a clamping unit. In this chapter, the mould assembly in a clamping unit is focused. A mould is specially designed for the part to be moulded and the complexity of the mould depends on the complexity of the part. As shown in Figure 11.8, a mould is an assembly of a number of components, such as a top plate, bottom plate, core back plate, cavity plate, cavity back plate, ejector mechanism, guide bushes, guide pillars, locating ring, and sprue bush (Custompart 2017; Nagahanumaiah and Mukherjee 2007).

11.6 Design of Injection Moulding Systems

Figure 11.8 shows that the tooling for injection moulding consists of many components for different functions. Tooling design is affected by many factors, including the number of cavities, gating systems, material viscosity, and mould venting (nttd-es.co 2019), which are discussed as follows.

11.6.1 Number of Cavities

To fully utilize the capability of an injection unit, reduce cost and improve productivity, a mould can be designed with multiple cavities, with each cavity corresponding to a part. However, multiple cavities in one mould increase the complexity of mould; this increases the cost of mould. Figure 11.9 shows the impact of the number of cavities on the product unit

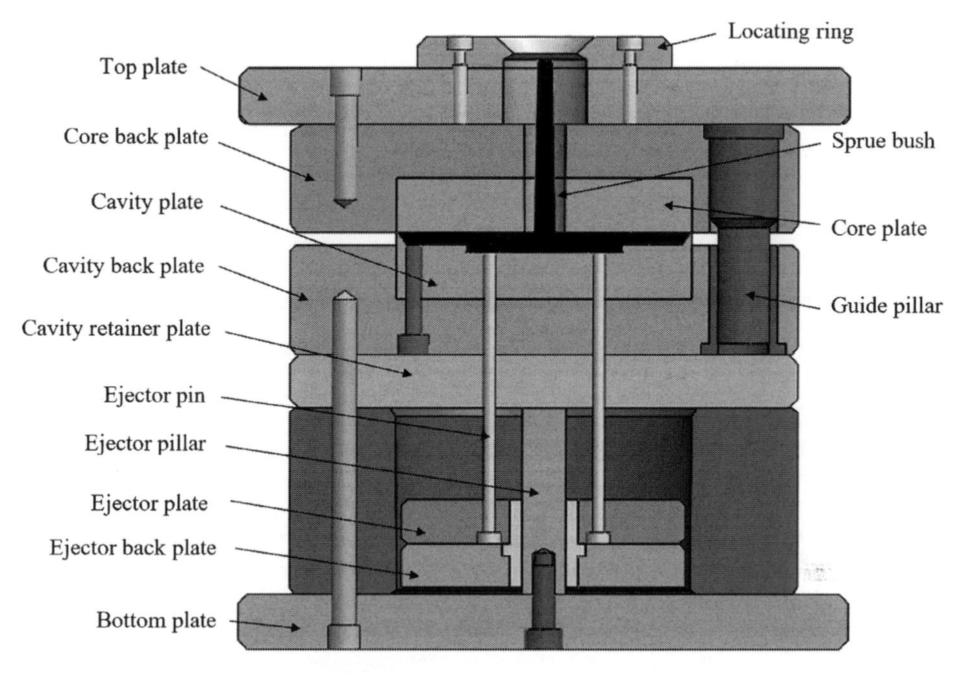

Figure 11.8 Construction of a typical injection mould.

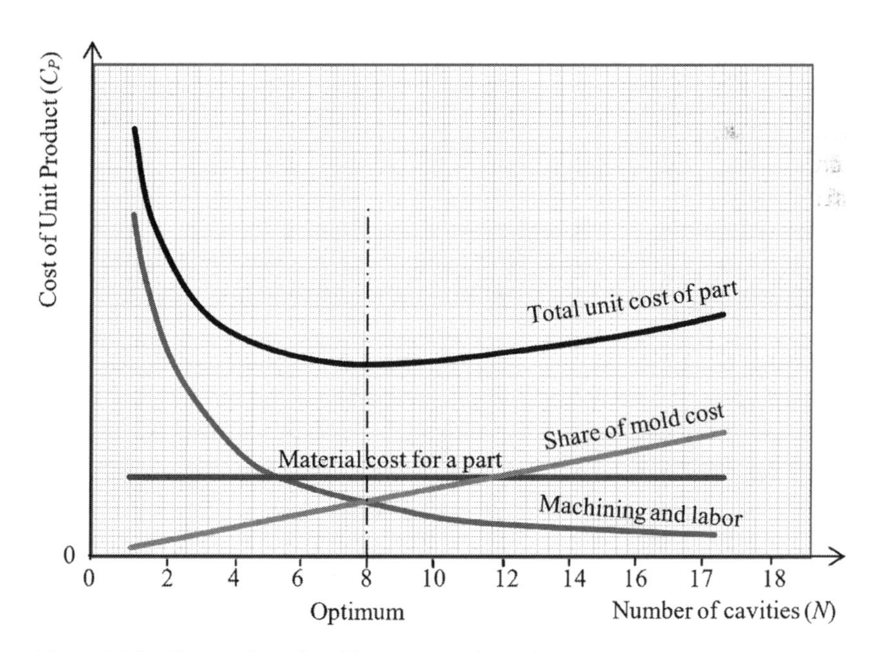

Figure 11.9 The number of cavities versus unit product cost.

cost. The total cost of the unit product (C_P) consists of (i) material cost (C_M), (ii) machining and labour cost (C_L), and (iii) the share of mould cost (C_T). The number of cavities does not affect the material cost. Increasing the number of cavities reduces the machining and labour cost due to improved productivity, but increases the share of mould cost due to the increased complexity of the mould design. Therefore, the number of cavities in a mould should not be optimized to reduce the cost of the unit product. The maximum number of cavities in a mould is also constrained by the maximum barrel capacity and the clamping force of the injection moulding machine.

A mould may include multiple cavities for different parts; in such a case, the cavity with the largest volume should be placed nearest to the sprue. Note that multiple cavities in one mould would cause quality variants of parts. If high quality and tight tolerances are required the cavities must be uniform. One must be very careful when using multiple cavities in one mould.

11.6.2 Runner Systems

An increase in the mould cost with multicavities is partially due to the need for a runner system. Similar to a runner system in a casting process, a *runner system* in an injection moulding system is the networked channels needed to carry melt pastis from injection nozzles to the cavities. A runner system can be *unheated* or *heated*. A heated runner system is called a hot runner system and includes a heated manifold and a number of heated nozzles. A *manifold* is used to distribute the melt plastic to multiple nozzles and measure the melt flow precisely at injection points.

The runner system is responsible for distributing plastic melts to these cavities as equally as possible in order to achieve a consistent quality of parts in four cavities. The running system is used to pour melted plastic into cavities. The filling process affects the performance of the injection moulding process at multiple aspects. Therefore, a running system must be designed to balance the filling of multiple cavities, minimize scraps at parting planes, reduce energy consumption, simplify the ejection operation, and provide the flexibility to control the filling and packing in a cycle of injection moulding processes. The symmetric configuration of a runner system in Figure 11.10 should be adopted to achieve these objectives.

11.6.3 Geometry of Runners

Geometries of runners affect the flow of melt plastic. The higher the surface to volume ratio runner, the more efficient it is. Therefore, the ideal cross-section for high efficiency is circular, since a circular cross-section ensures a favourable melt flow and cooling. However, using circular runners causes difficulties in fabrication since one-half of the runner must be machined in the fixed mould part and the other half in the moving mould part. Therefore, the full trapezoidal channels in one of the two mould halves can be used to simplify manufacturing processes. Moreover, a rounded off trapezoidal cross-section combines the ease of machining in one mould half with a cross-section that approaches the desired circular shape. The height of a trapezoidal runner must be at least 80% of the largest width. The correspondences of the dimensions of two trapezoidal channels with the round channel are provided in Figure 11.11 (Mold Technology 2011a).

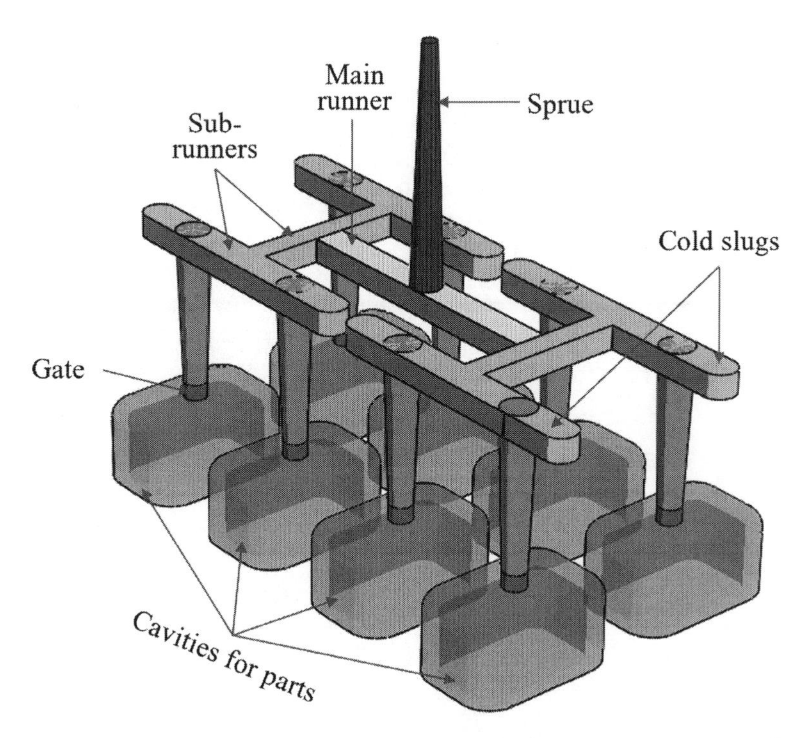

Figure 11.10 Symmetric configuration of running systems for balancing melt flows.

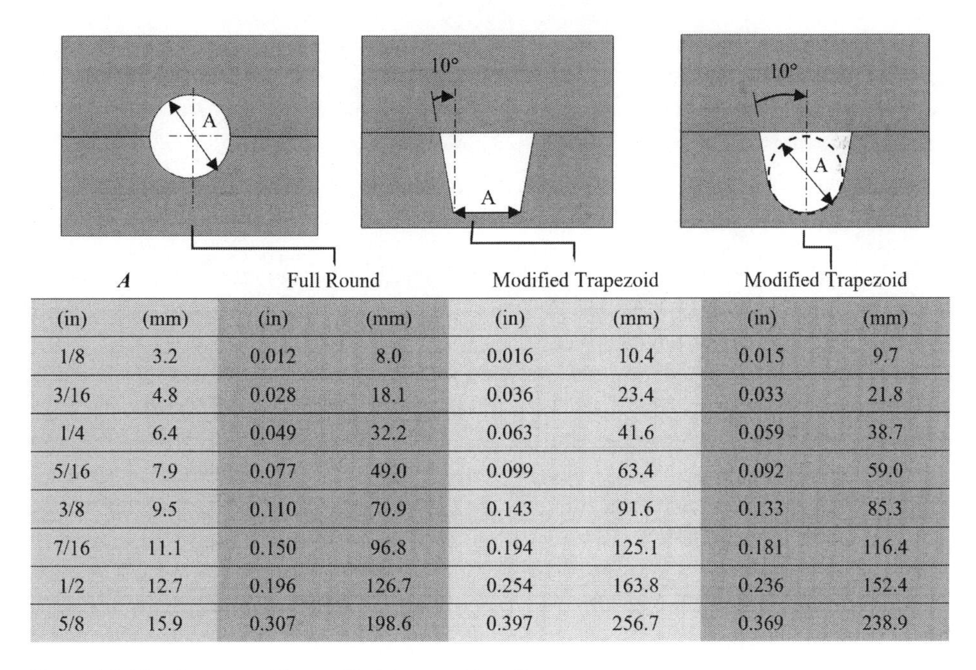

A		Full Round		Modified Trapezoid		Modified Trapezoid	
(in)	(mm)	(in)	(mm)	(in)	(mm)	(in)	(mm)
1/8	3.2	0.012	8.0	0.016	10.4	0.015	9.7
3/16	4.8	0.028	18.1	0.036	23.4	0.033	21.8
1/4	6.4	0.049	32.2	0.063	41.6	0.059	38.7
5/16	7.9	0.077	49.0	0.099	63.4	0.092	59.0
3/8	9.5	0.110	70.9	0.143	91.6	0.133	85.3
7/16	11.1	0.150	96.8	0.194	125.1	0.181	116.4
1/2	12.7	0.196	126.7	0.254	163.8	0.236	152.4
5/8	15.9	0.307	198.6	0.397	256.7	0.369	238.9

Figure 11.11 Runners with modified trapezoid profiles.

Table 11.4 Recommended runner dimensions based on maximum runner lengths.

Runner diameter		Maximum runner length			
		Low viscosity		High viscosity	
(in)	(mm)	(in)	(mm)	(in)	(mm)
1/8	3	4	100	2	50
¼	6	8	200	4	100
3/8	9	11	280	6	150
½	13	13	330	7	175

The actual diameter of a runner depends on the cavity volume, the runner length, melt flow length, machine capacity, and gate size. While a large runner assists plastic flow subjected to low pressure, it takes a longer cooling time, and demands more clamping force and produces more scrap. Therefore, minimizing runner sizes will maximize the efficiency of material use and energy consumption. However, the runner design is constrained by the moulding machine's injection pressure capability. The following equation can be used to estimate runner sizes based on the part weight and runner length (Mold Technology 2011a):

$$D = \frac{W^{0.5} L^{0.25}}{3.7} \tag{11.1}$$

where D is the runner diameter (mm), W is the part weight (g), and L is the runner length (mm).

As a rule, the diameter should be in a range of 3 to 15 mm and it must be larger than the largest wall thickness of the part. In addition, the diameter of the cold runner should be rounded up to a standard size of cutting tool. Table 11.4 shows the recommended runner dimensions (Mold Technology 2011a).

11.6.4 Layout of Runners

A runner system is usually for multiple cavities. Therefore, the layout of a runner system is also important. Figure 11.12 shows that four basic types of runner layouts are *standard* (herringbone) layouts, *H-bridge* (branching) layouts, *radical* layouts, and *star* layouts.

The layouts of a runner system should be balanced to ensure that each cavity gets the same amount of material in the cycle of the injection moulding process. An unbalanced runner system, such as the cases shown in Figure 11.13, may cause unequal filling, post-filling, and cooling of individual cavities, which will eventually lead to some failures such as incomplete filling, variants of properties from different cavities, sink marks, flashes, or poor mould releases. *A naturally balanced runner system* provides equal distance and runner size from the sprue to all the cavities, so that each cavity fills under the same conditions. To balance the runner system, the layouts in Figure 11.13 can be modified as shown by the cases in Figure 11.14, so that the plastic flows to cavities are naturally balanced.

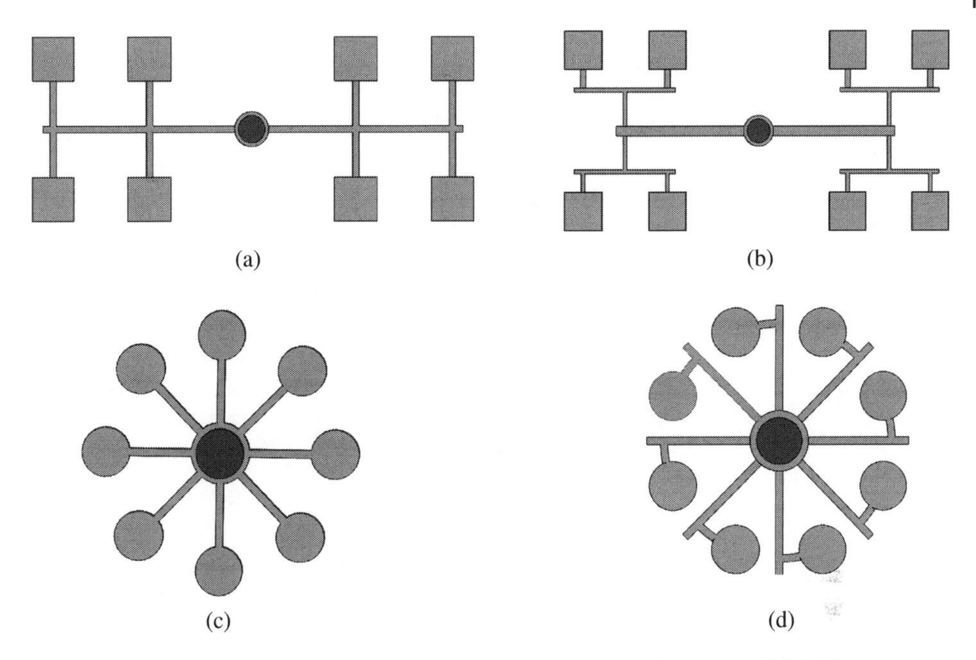

Figure 11.12 Four basic types of runner layouts: (a) standard (herringbone), (b) 'H' bridge (branching), (c) radial layout, and (d) star layout.

In practice, an unbalanced layout such as the herringbone layout in Figure 11.13b may be adopted to reduce the materials in the runner system, minimize the tooling cost, and accommodate more cavities than other naturally balanced counterparts. To tackle the issues related to an unbalance running system, a computer aided flow simulation is needed to obtain equal filling patterns by adjusting primary and secondary runner dimensions. However, it should be noted that (i) non-standard runner diameters will increase manufacturing and maintenance costs; (ii) varying runner dimensions is not applicable to extremely small parts, parts with thin section, parts with no permission to have sink marks, and a mould where the length of a primary runner is much larger than that of secondary runners.

11.6.5 Branched Runners

The law of flow continuity applies to the plastic flow in a runner. When a runner is branched, to keep the constant velocity of flow, the diameter of the branched runner should be reduced so that less material flows into it. In addition, the diametric reduction helps to reduce the materials in the runner system.

Figure 11.15 shows an example of branched runners that are divided from a main runner. Assume that there are N branched runners. The diameter of a branched runner (d_{branch}) can be determined from that of the main runner (d_{main}) as

$$d_{main} = d_{branch} N^{1/3} \tag{11.2}$$

where d_{main} and d_{branch} are the diameters of the main and branch runners, respectively.

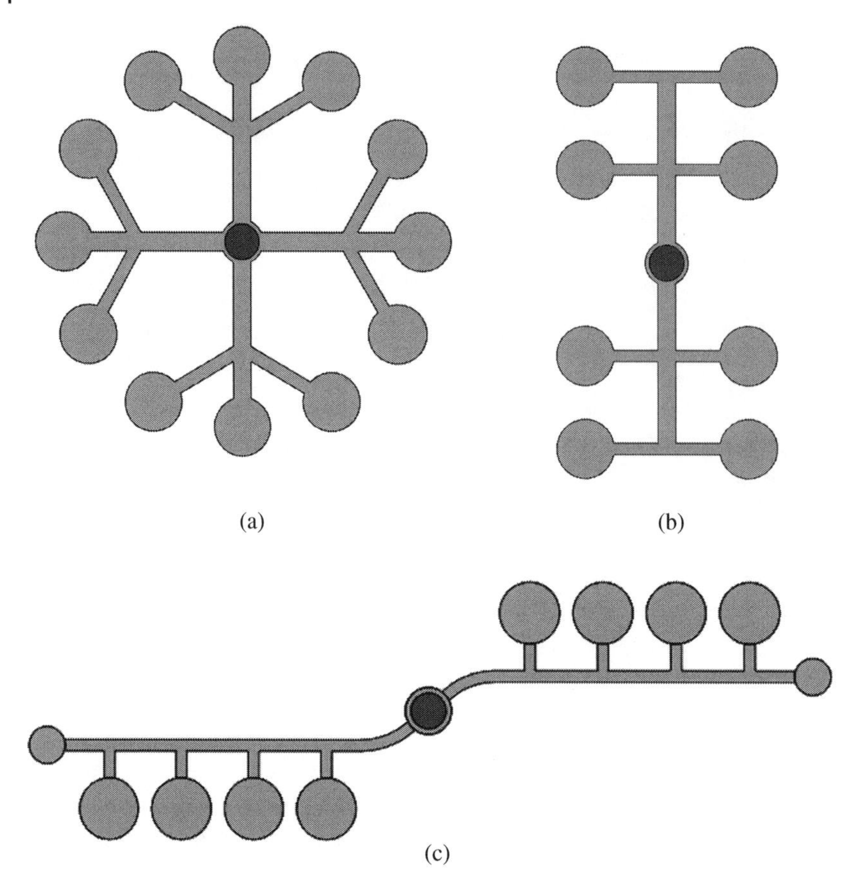

Figure 11.13 Examples of unbalanced runner systems: (a) unbalanced star layout, (b) unbalanced herringbone layout, and (c) unbalanced line layout.

Figure 11.15 also shows that at an intersection of two runners, there should be a *cold slug*. The well of a cold slug helps the material to flow past the intersection with minimal energy loss since the material with the higher viscosity moving at the forefront is trapped at the well. As shown in Figure 11.16, the length of the well (L_{well}) is usually equal to or greater than the runner diameter (d_{branch}). A clod slug is made by extending the primary runner beyond the intersection, as shown in the figure (Mold Technology 2011a).

11.6.6 Sprue Design

A typical sprue design is shown in Figure 11.17 (Mold Technology 2011a). Sprue bushing connects the nozzle of the injection unit to the runner system in a clamping unit. To ensure a clean ejection from the bushing, the bushing should have a smooth, tapered internal finish. The bushing should be polished in the pulling direction of drawing. It is recommended that a positive sprue puller is used. A cold slug well should also be included in the design; this prevents a slug of cold material from entering the feed system and finally the part.

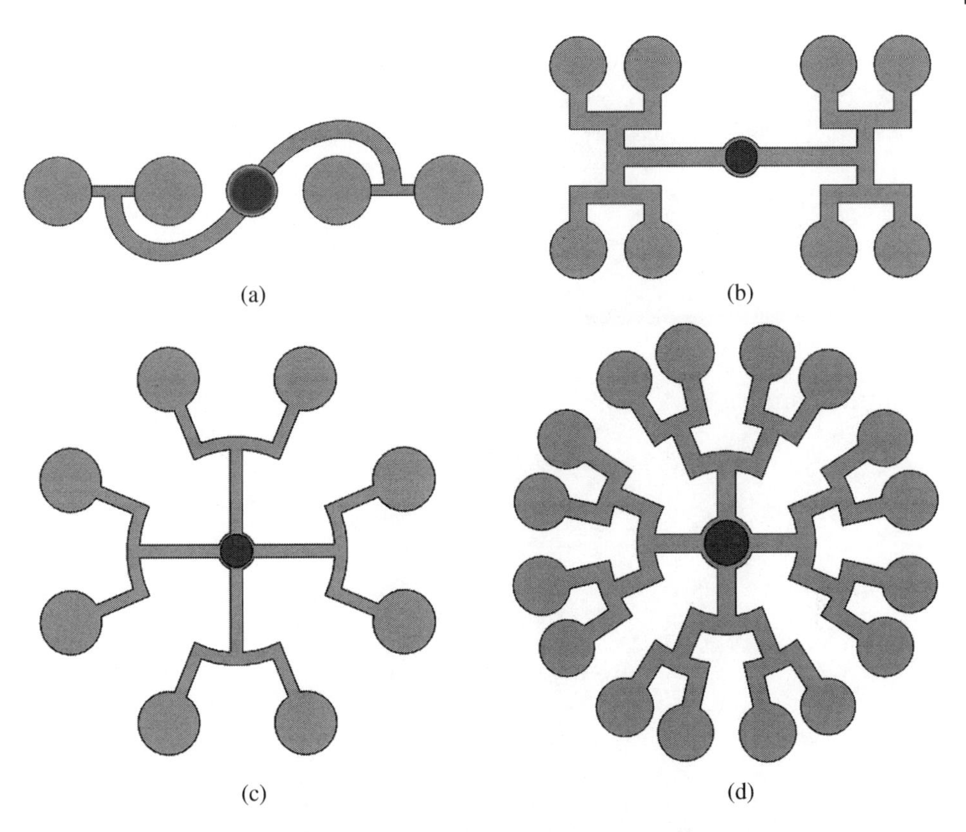

Figure 11.14 Examples of balanced runner systems: (a) standard (herringbone), (b) 'H' bridge (branching), (c) radial layout, and (d) start layout.

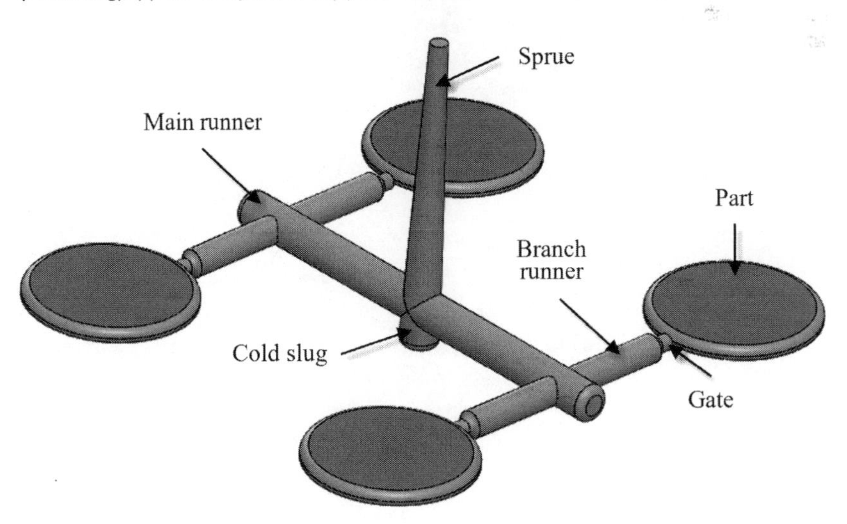

Figure 11.15 Example of branched runners and clod slug.

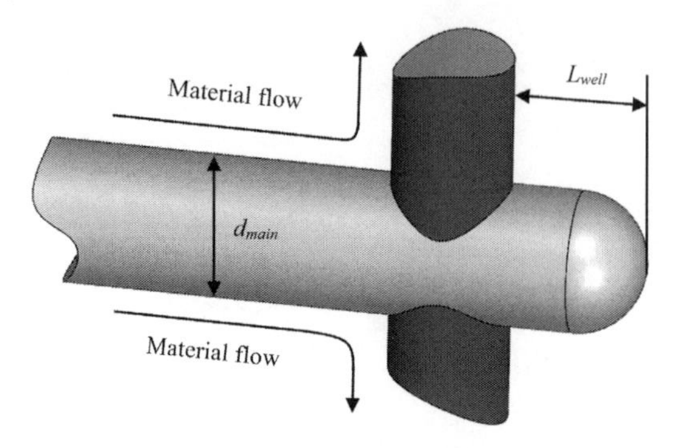

Figure 11.16 The length of clod slug ($L_{well} \geq d_{main}$).

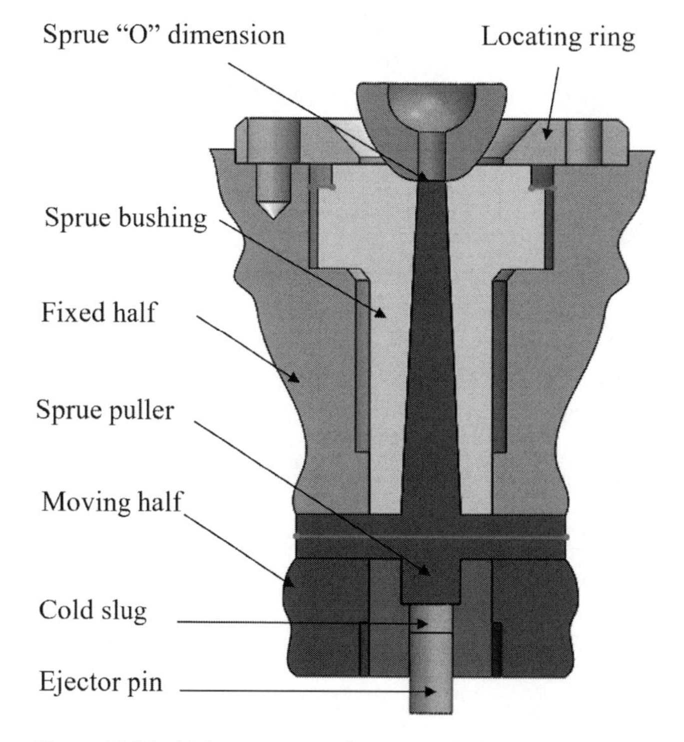

Figure 11.17 Main components in a sprue design.

The sprue size is determined mainly by the total volume of the cavities and the wall thickness of parts on the condition that no freeze occurs to the sprue before any other cross-sections in the runner system, thus sustaining sufficient transmission of the holding pressure. In addition, a sprue should be separated easily from the part without causing a defect on the part.

11.6.7 Design of Gating System

A gate in the mould is a transition zone to connect a cavity to the runner system. The gating system affects the properties and appearance of finished parts from the cavities. The gating

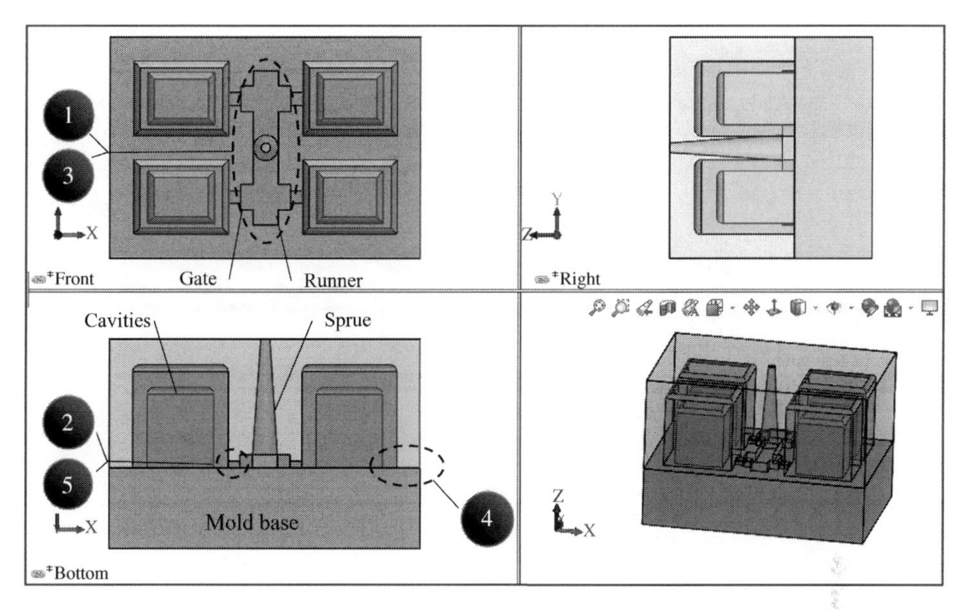

Figure 11.18 Guides to position gates in a running system.

Table 11.5 General guides to position gates.[a]

1	Set a gate position where molten plastic finished filling up each cavity simultaneously, which applies to the case with multipoint gates in Figure 11.18.
2	Place a gate to a position to the thickest area of a part. This can avoid sink marks due to mould shrinkage.
3	Position a gate at an unremarkable area of part with less impact of appearance and textures.
4	Avoid injecting from the direction where the air in the cavity or the gas generated from molten plastic is inclined to accumulate.
5	Fill up molten plastic using the wall surface to avoid the generation of jetting.

a) The numbers in the first column are marked in Figure 11.18.

system should be designed to ensure that the melt fills the cavities in the mould quickly and evenly. Figure 11.18 shows an example of a runner system with four cavities in the mould with highlighted areas related to the selection of gates. Table 11.5 gives the guides in positioning gates based on part geometries.

Other than the guides in Table 11.5, the following design factors have to be taken into consideration in determining the positions of gates (Mold Technology 2011b):

1. *Stress concentration*. Avoid the areas exposed to high external stress, since the gate area involves high residual stresses. In addition, rough surfaces left by the gate act as stress concentrators.

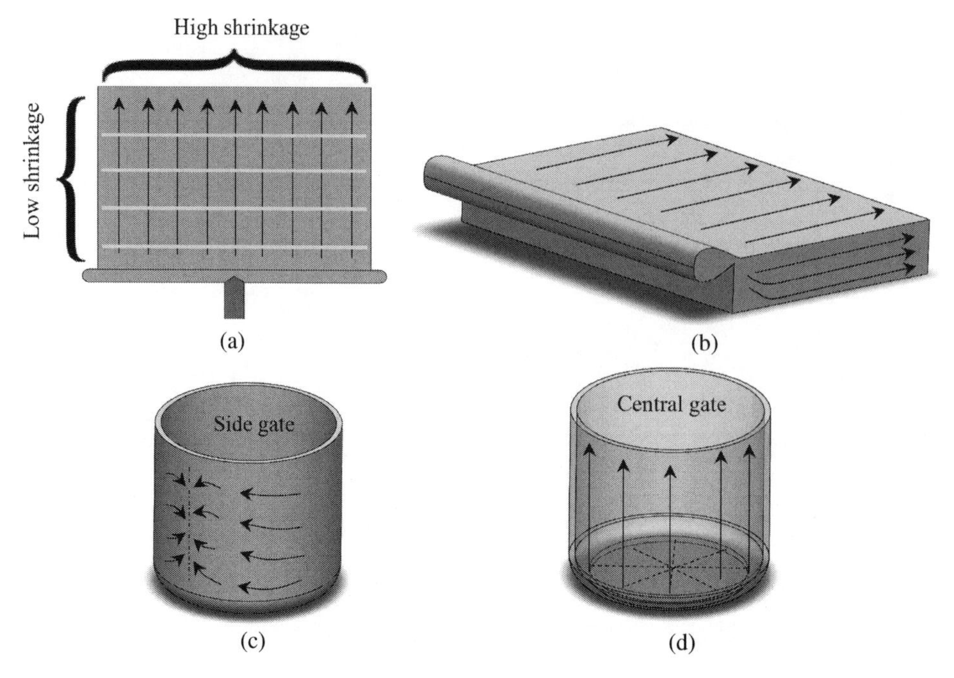

Figure 11.19 The impact of the gating system on mechanical properties: (a) influence of gating on melt orientation and shrinkage, (b) anisotropic shrinkage by film gating, (c) weld line by an unfavourable flow pattern, and (d) no weld line by a favourable flow pattern.

2. *Orientation*. Molecular orientation becomes more pronounced in thin sections as the molecules usually align themselves in the flow direction. Figure 11.19a and b show the impact of the gate position on plastic flow and the shrinkage of the materials.
3. *Weld lines*. Gates should be placed to minimize the number and length of weld lines or to direct weld lines to positions that are not objectionable to the function or appearance of the part. Figure 11.19c and d shows that the weld lines may be eliminated by placing a gate at the centre of the part.
4. *Secondary process*. The easiness of de-gating should be considered in the design of the gating system.

A gating system may include *single gate* or *multiple gates*. A single gate is preferable unless the length of the melt flow exceeds practical limits, since multiple gates always create weld lines where the flows from the separate gates meet. It is also desirable to place a gate at the centre of a part, since it generates a radial flow of the melt that gives the part symmetrical mechanical properties.

In determining the dimensions of gates, the cross-section of the gate is typically smaller than the diameter of the runner or the thickness of the part. This helps to de-gate the part from the runner system without leaving a visible scar on the part. *The gate thickness* is typically between one-half and two-thirds of the part thickness. The gate thickness controls the packing time, since the packing is terminated at the time when the material in the gate drops below the freezing temperature.

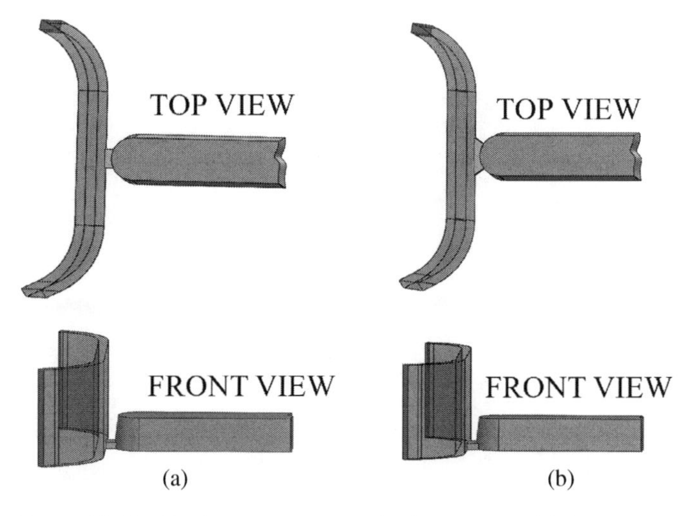

Figure 11.20 Two basic gate types: (a) edge gate and (b) fan gate.

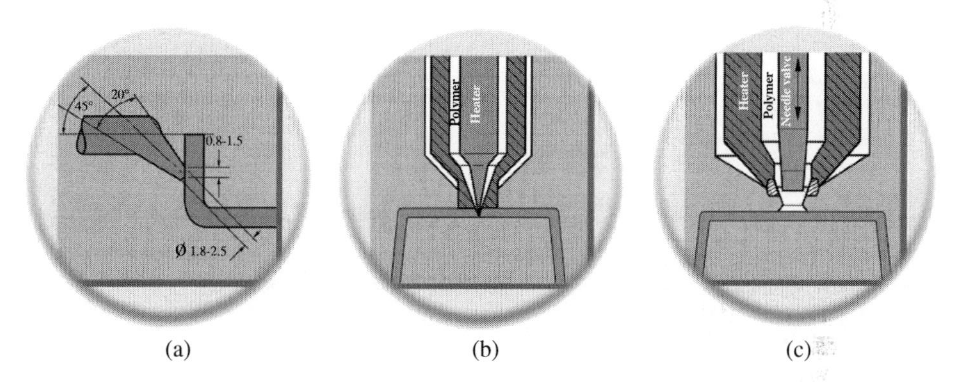

Figure 11.21 Gate types by automatic trimming (Mold Technology 2011b). (a) A submarine gate is used in a two-plate mould construction (b) Hot runner gates are also known as a sprueless gating. (c) The valve gate adds a valve rod to the hot runner gate.

Many gate types are used in injection moulding processes (Engineering Design Center 2019; Mold Technology 2011b). Two basic gate types are *edge type* and *fan type*, shown in Figure 11.20. For simplicity and ease of manufacture, edge-type gates are the most commonly used. Figure 11.21 shows three examples of gates used in manual and automatic de-gating (Mold Technology 2011b), respectively.

11.6.8 Design of Ejection System

An ejection system is used to remove moulded parts from the mould when the injection moulding process is completed. There are many types of ejection mechanisms that are different in terms of their functions and implementations. In selecting an ejection system, the designs of the mould and parts need to be taken into consideration simultaneously.

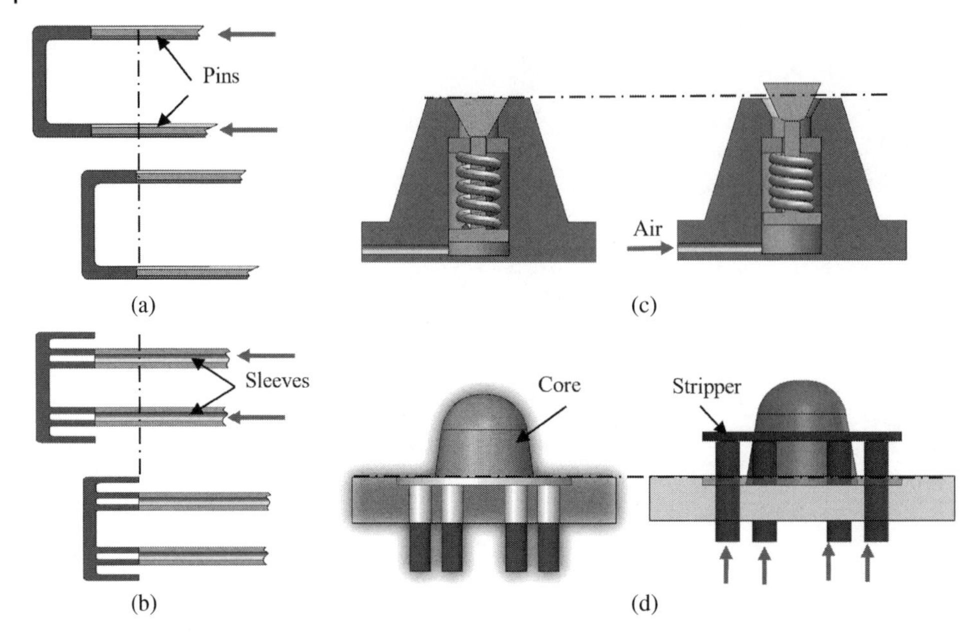

Figure 11.22 Common types of ejections: (a) pin ejection, (b) sleeve ejection, (c) air ejection, and (d) Stripper ejection.

Figure 11.22 shows some examples of ejection mechanisms. The most commonly used ejections are *pin ejections*, where an ejector pin is attached to the plate assembly to move backwards and forwards. Other ejection methods are *sleeve ejections*, *air ejections*, and *stripper plate ejections*.

11.6.9 Design of the Cooling System

In a cycle of an injecting moulding process, the cooling time takes up 70–80% of the cycle time (Injection Molding 2019). A *cooling system* in a mould serves to dissipate the heat of the moulding quickly and uniformly, where fast cooling is necessary to obtain economical production and uniform cooling is required for product quality. Therefore, a cooling system is designed to shorten the moulding time and improve the productivity.

A cooling system controls the mould temperature. Adequate control of the mould temperature is essential for consistent moulding, since the temperature and its change with respect to time greatly affect mould shrinkage, dimensional stability, deformation, internal stress, and surface quality. In designing a cooling system, the following objectives have to be taken into consideration (Rapid 2015):

1. Avoid or minimize differences in flow resistance of cooling channels, caused by diameter changes.
2. Avoid dead spots and/or air bubbles in cooling circuits.
3. Minimize heat exchange between the mould and machine.
4. Shorten cooling channels to ensure temperature differences between the inlet and outlet do not exceed 5 °C.

5. Sustain independent symmetrical cooling circuits around the mould cavities.
6. Provide effective cooling in cores using components such as baffles, bubblers, and thermal pins.

In a cooling system, the layout design of *cooling channels* is most critical to reduce the cycle time, achieve high heat transfer efficiency, and sustain the uniformness of temperature distribution in the mould. The cooling channels should be roughly drilled or milled to improve the heat transfer efficiency. Moreover, rough inner surfaces enhance turbulent flow of the coolant and a turbulent flow achieves three to five times as much heat transfer as a laminate flow.

With regard to the uniform temperature distribution, Figure 11.23 shows the cases where the layouts of a cooling system can change the temperature distribution greatly. In Figure 11.23a, the distances from cooling channels to a part surface are different, which causes a uniform temperature in the mould. This can be improved by repositioning the channels so that they are an equal distance from the part surface. In Figure 11.23c, a set of cooling channels are connected sequentially, which causes different temperature derivations between the cooling channels and the surrounding areas in the mould. To eliminate the temperature derivations caused by serial connection of channels, Figure 11.23d shows the preferable layout where a set of cooling channels are instead placed in parallel.

Cooling channels should be placed close to the mould cavity surface with equal centre distances in between. Figure 11.24 shows the definition of the distance from a cooling channel to the part surface and Table 11.6 provides the recommended distances based on the dimensions of the channels, spacing, and part.

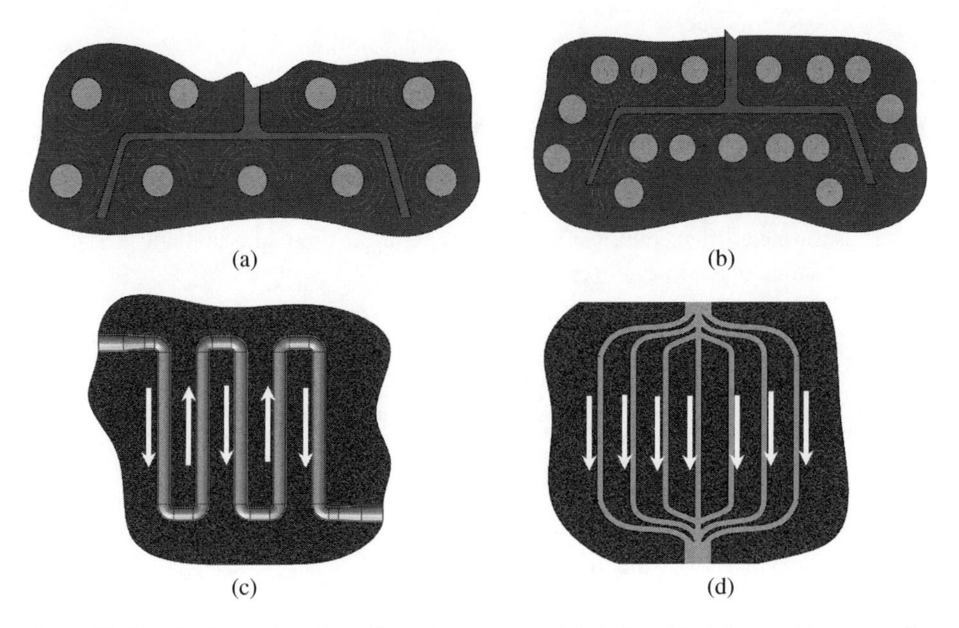

Figure 11.23 Cooling system for uniform temperature distribution. (a) Unfavourable – un-uniform temperature. (b) Favourable – uniform temperature. (c) Unfavourable – serial cooling. (d) Favourable – parallel cooling.

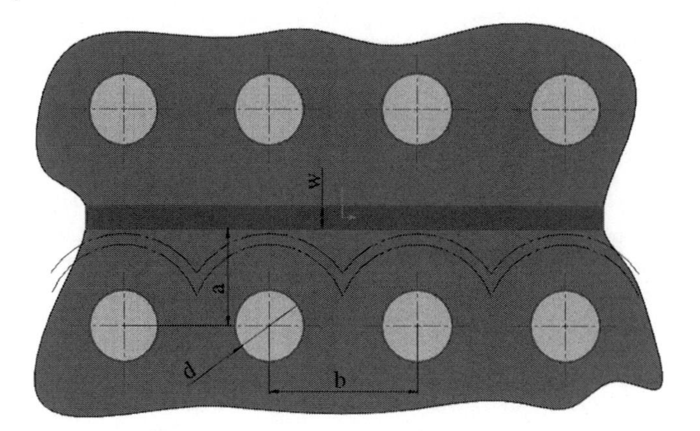

Figure 11.24 The distance of cooling channels to part.

Table 11.6 Recommended distances (Mold Technology 2011c).

Dimension	The wall thickness of the product (*w*) in mm (in)	The diameter of the cooling channels (*d*) in mm (in)	The centre distance with respect to mould cavity	The centre distances between cooling channels
Recommended distance	<2 (<0.08) 2~4 (0.08~0.16) 4~6 (0.16~0.24)	8~10 (0.31~0.40) 10~12 (0.40~0.47) 12~14 (0.47~0.55)	1.5 d~2.0 d	2.0 d~3.0 d

11.6.10 Moulding Cycle Times

The productivity of an injection moulding system is evaluated by (i) the number of cavities in a mould and (ii) the cycle time of the injection moulding process. The injection moulding cycle is a sequence of events during the injection moulding, including (i) closing the mould, (ii) injecting the melts, (iii) cooling the mould, (iv) opening the mould, and (v) ejecting parts. Therefore, the time of the injection moulding cycle (T_{cycle}) is estimated as

$$T_{cycle} = 2T_{setup} + T_{injection} + T_{cooling} + T_{ejection} \tag{11.3}$$

where

T_{setup} is the time to set up the mould (i.e. to open or close the mould)
$T_{injection} = V/R$ is the injection time determined by the volume of the cavities (V) and the filling rate (R)
$T_{cooling}$ is the cooling time
$T_{ejection}$ is the time for the ejecting part

During the moulding cycle, especially in the cooling process, it is important to maintain a holding pressure to compensate for material shrinkage, where the screw turns, feeding the next shot to the front screw.

11.7 Computer Aided Mould Design

In this section, the methods of designing moulds and dies based on part models are discussed.

11.7.1 Main Components of Mould

The parts geometry in injection moulding is defined by the geometric space of the formed cavity in the mould. Two examples in Figure 11.25 show that a mould consists of (i) the *core* (male part) and (ii) the *cavity* (female part), which are separated at (iii) the *parting line or surface* and (iv) a number of *other cores* for internal surfaces. Other cores are used to define holes, pockets, and undercut features of a part. A parting line or surface is an interface from where two main components (the core and the cavity) split and open in the respective directions.

11.7.2 Mould Tool in SolidWorks

Here the add-in *Mould* in SolidWorks is used as an example of computer aided mould tools to illustrate the procedure, steps, and tasks in designing moulds and dies based on part models for injection moulding.

Figure 11.26 shows the tools under the Mould command manager. If this is unavailable in the *Command Managers* group, it can be turned on by *Customize* in the *Options* of the top menu bar, or one can just right-click on a blank space to activate it.

The *Mould* in SolidWorks consist of three groups: (i) tools for modifying, deleting, or creating surfaces (e.g. offsets or ruled surfaces), knitting surfaces, and fixing holes by filling; (ii) tools to analyse mould features such as draft analysis, undercut analysis, and the analysis of parting lines; and (iii) tools to create mould components including cavities and the cores. In the following, these tools are used to illustrate the design procedure of a mould tool.

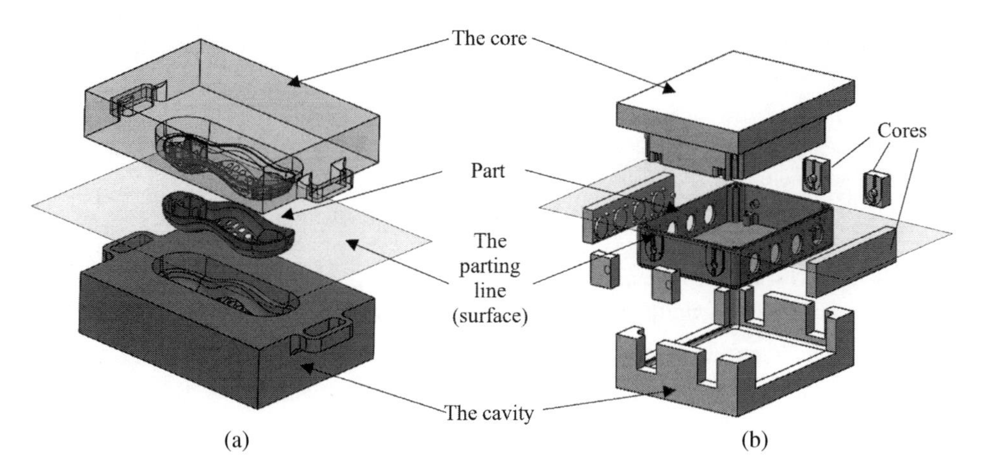

Figure 11.25 Mould components in part examples: (a) phone cover and (b) plastic container.

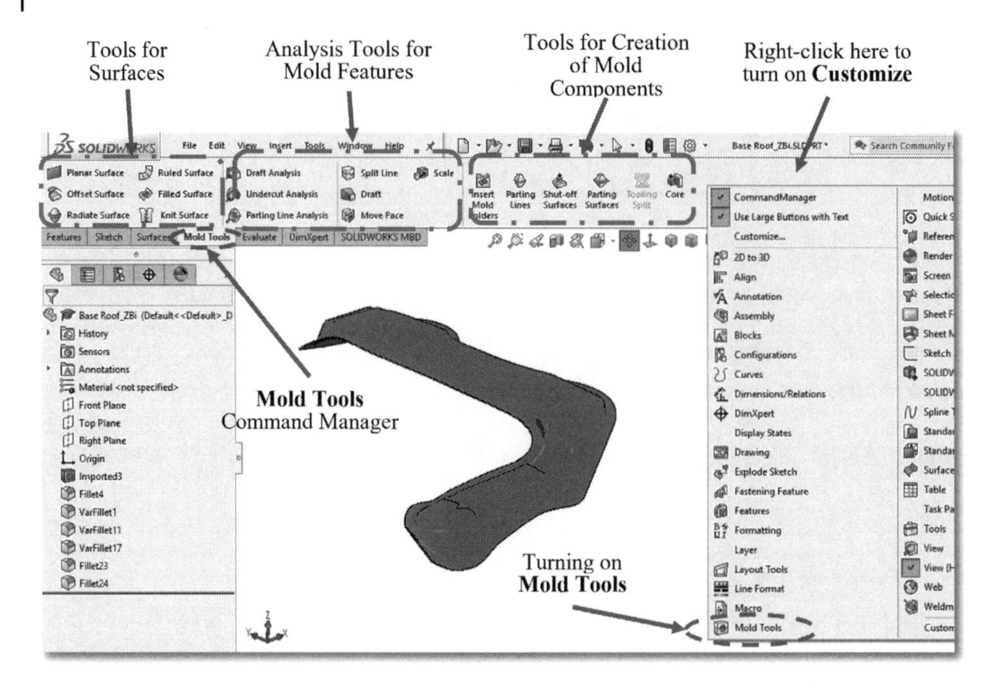

Figure 11.26 Graphical user interface of a *mould* module.

11.7.3 Design Procedure

Figure 11.27 shows the procedure of designing moulds based on part models:

1. The part model to be moulded is loaded and then scaled to compensate for shrinkage in the moulding process.
2. The draft analysis is conducted to ensure that the part can be retrieved from the mould after the moulding process.
3. The faces with inappropriate drafts or defects must be fixed to continue the design process.
4. If the part has an open space(s), the *shut-off surfaces* are created to make the core and cavity separable,
5. The *parting line* is defined where the core and cavity are separated.
6. The *parting surfaces* are generated using a parting line so that the geometries and shapes of the core and cavity are defined accordingly.
7. The defined parting surfaces are knitted with the core and cavity surfaces to create the core and cavity solids, respectively.
8. The tooling is split into core and cavity solids.
9. If the model has an undercut(s), an additional core(s) is defined to deal with the undercut.

Finally, a number of new configurations can be created to illustrate the exploded view of the assembled mould tool for the part.

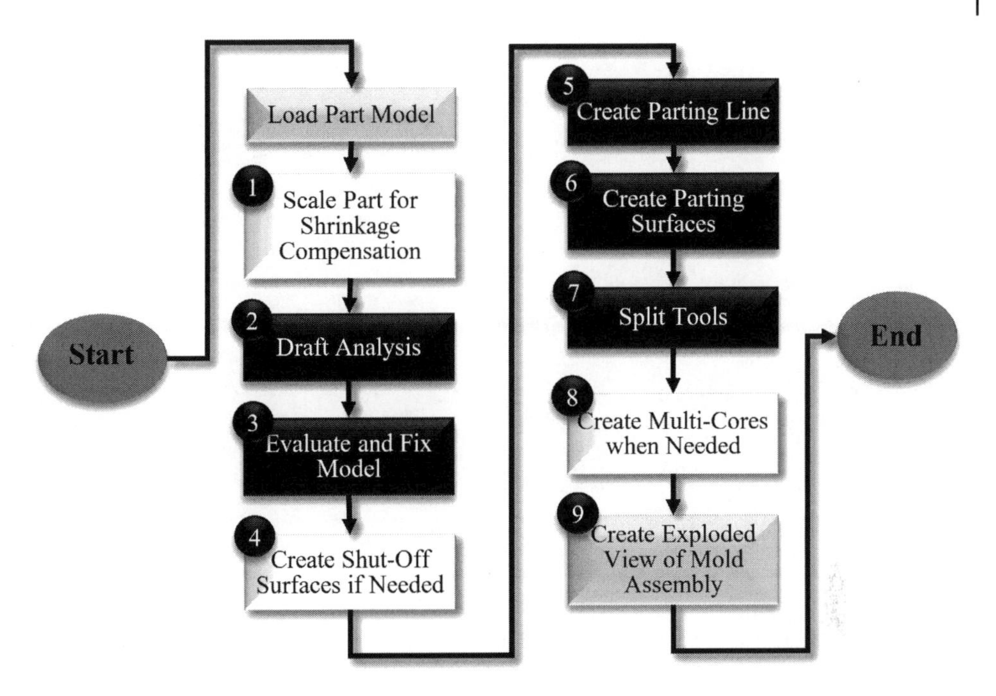

Figure 11.27 Procedure of a computer aided mould design.

11.7.4 Compensation of Shrinkage

To generate correct dimensions of part, the dimensions of the mould cavity must be adjusted to compensate for the shrinkage occurring during the cooling process. The shrinkage is represented by a *scale factor*, which is defined as the ratio of the dimensions between virtual and moulded parts. The scale factor depends on the type of materials and the shape of the part. As shown in Figure 11.28a and b, the *scale* tool can be accessed in two different ways; i.e. (i) click *Insert* > *Mould* > *Scale* and (ii) click *Scale* under the Command Manager of Mould Tools. Figure 11.28c shows that a part can be scaled by a specified factor about the *centroid*, *origin*, or *coordinate system* of the part.

11.7.5 Draft Analysis

Faces on a mould part must be drafted so that the mould tool can shape the part correctly. SW Mould Tools provide the *Draft Analysis* tools. As shown in Figure 11.29, the user can select a planar face, linear edge, or axis to define the draw direction. In addition, the user can adjust the direction of pull by selecting *Adjustment* and dragging the triad.

Along the direction of pulling, each face must have a draft angle larger than a minimum value. If not, its draft angle can be modified by the *Draft* tool. Figure 11.30 shows the inputs for adding a draft on surface(s); e.g. the value of the draft angle, the direction of pull, and the surface(s) to be drafted. It should be noted that not all surfaces can be drafted smoothly due to possible conflicts of geometries.

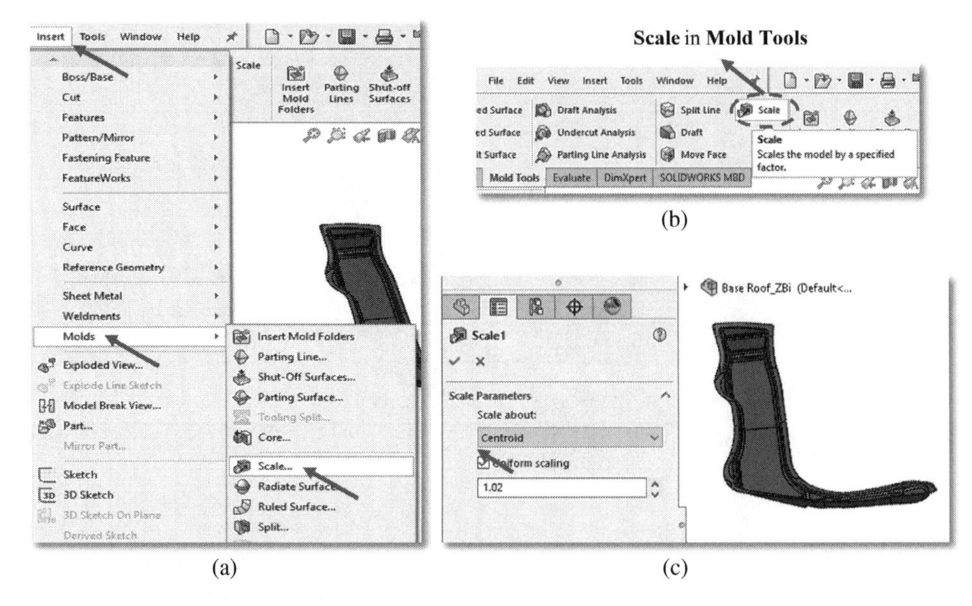

Figure 11.28 Scale a part model to compensate shrinkage: (a) access **Scale** in **Insert**, (b) access **Scale** in **Mould Tools**, and (c) **Scale** interface.

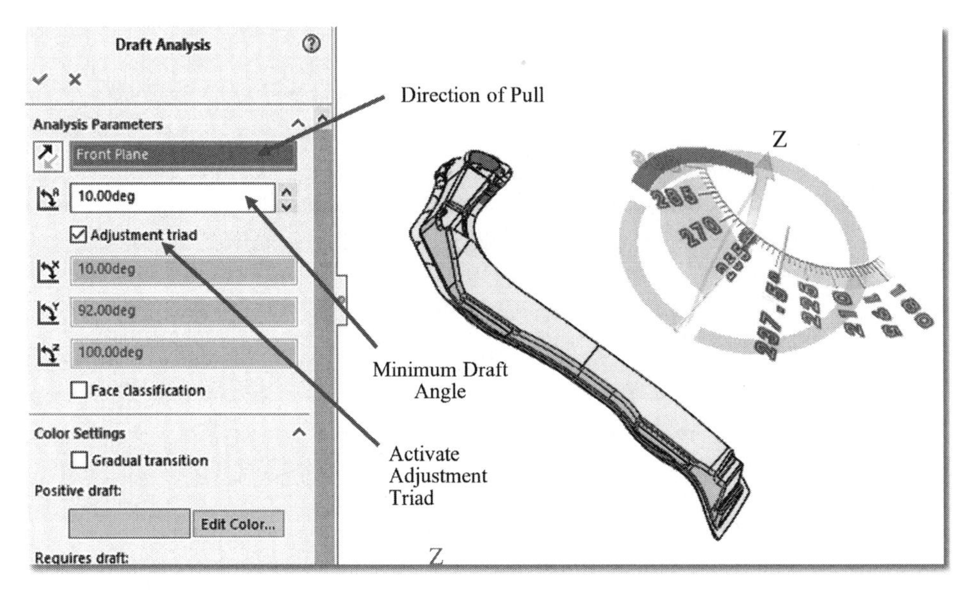

Figure 11.29 Running a *draft* analysis.

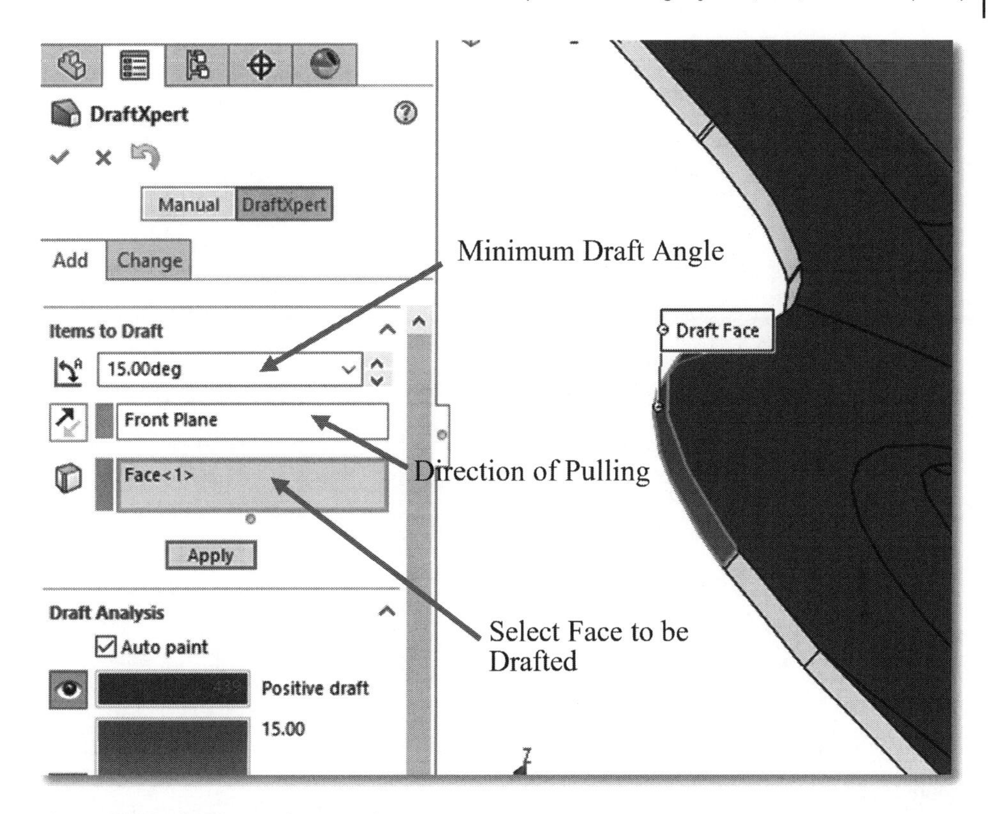

Figure 11.30 Adding drafts on surfaces.

11.7.6 Parting Line and Shut-off Planes

A mould must be able to separate into multiple parts, i.e. the core(s) and cavity. A parting line consists of boundary edges where the two parts of the mould assembly can be separated. The draft angles of the faces on the core and cavity sides are positive and negative, respectively. Figure 11.31 shows that a parting line can be defined by specifying (i) the direction of pulling, (ii) the minimum draft angle, and (iii) a collection of closed-loop edges.

In the case where a part has internal holes or openings, a shut-off plane tool can be used to make the cores and cavity separable.

11.7.7 Parting Surfaces

In the *Mould* tools, the solids of the core and the cavity are eventually knitted from a number of surfaces, including the parting surfaces where the two solids are separated. Figure 11.32 shows the three important inputs in defining a parting surface. These three inputs are: (i) the selected surface normal; (ii) the parting line; and (iii) the distance of the edge from the parting line. Note that the value for the distance should be as large as possible so that

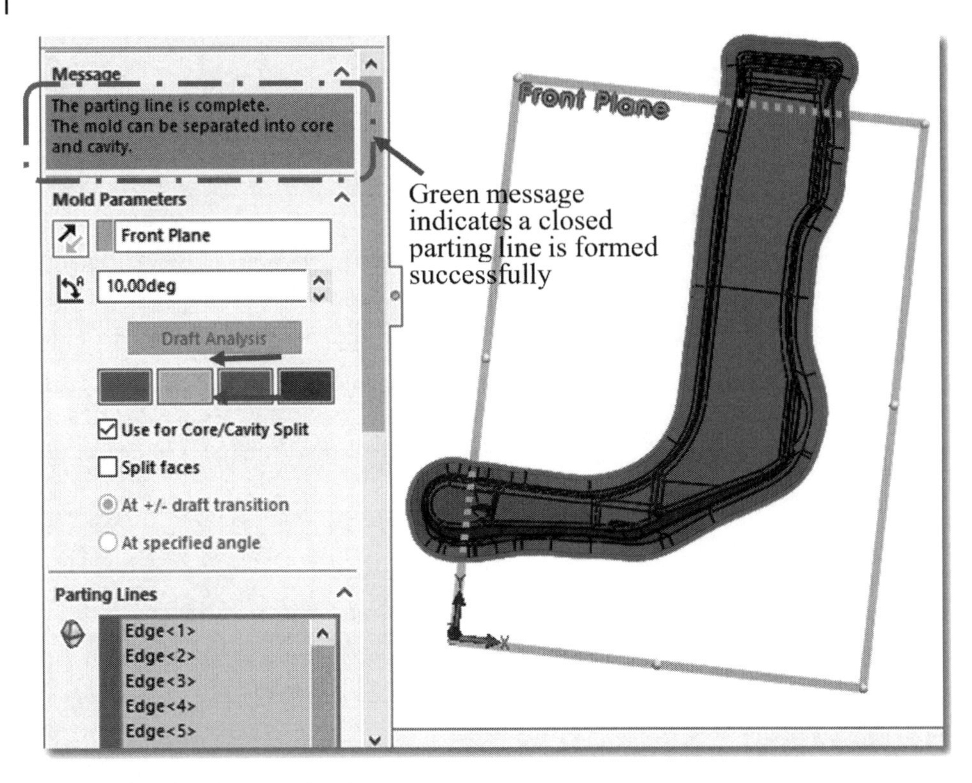

Figure 11.31 Creating a parting line.

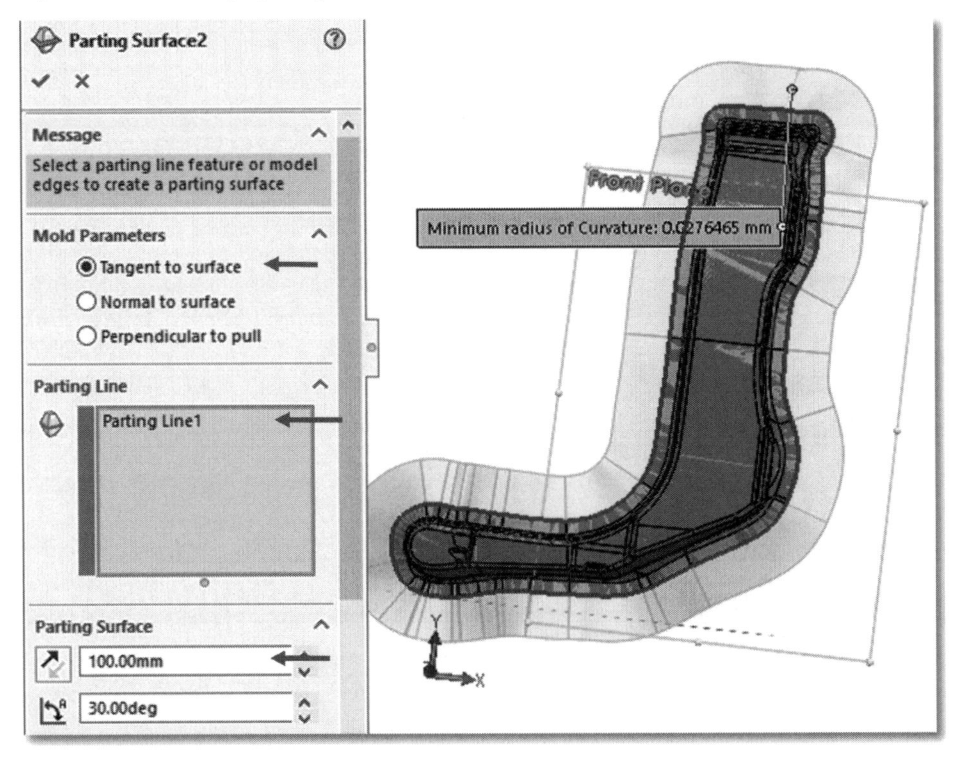

Figure 11.32 Creating a parting surface.

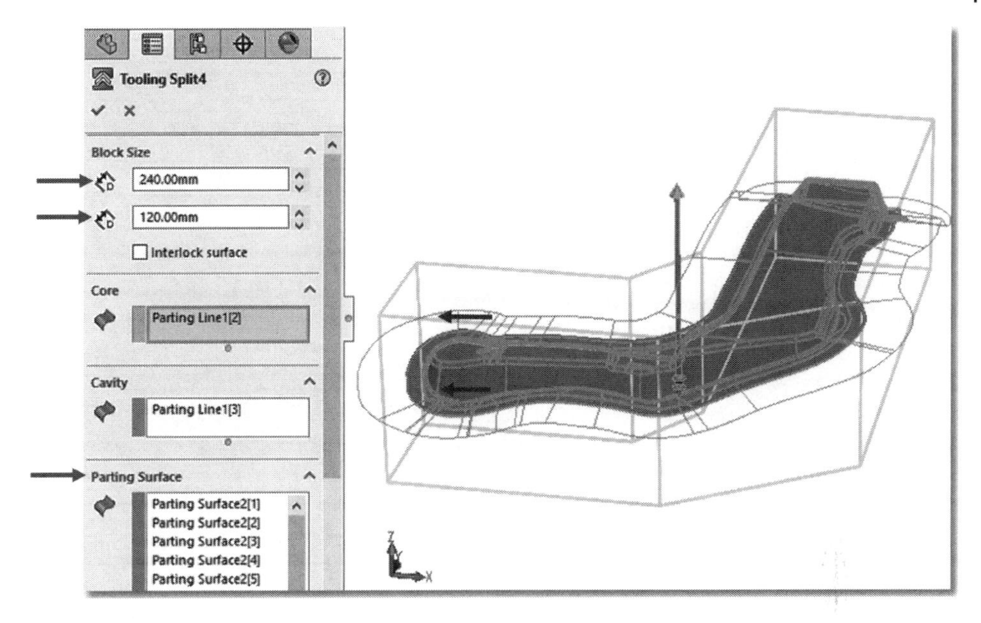

Figure 11.33 Creating *core* and *cavity* by *Tooling Split*.

the parting surface can be applied in the knitting, with an enclosed volume for the core and cavity. Otherwise, SW will give an error message about the knitting failure in the next step.

11.7.8 Splitting Mould Components

The separated solids of the core and the cavity are created by the *Tooling Split*. Before the Tooling Split tool is applied, the user has to create a sketch to define the outside boundary of the tooling. Note that the outside boundary must be completely covered by the parting surface to ensure the closeness of the knitted volume. In addition, Figure 11.33 shows other inputs when the Tooling Split tool is applied. These inputs include: (i) the depths of the core and cavity, respectively, and (ii) the specified parting surfaces. The SW tool will specify the names of the core and cavity solids in defaults, which can be renamed after they are created.

11.7.9 Assembly and Visualization of Moulds

As shown in Figure 11.34, the design process of a mould tool creates a number of reference planes, sketches, surfaces, and solids. It is helpful to hide intermediate entities after the mould tool has been created. This can be achieved by changing the hide/show settings of features and defining new configurations for the assembly of the moulding tool. In addition, additional features can be added to the core and cavity solids using the SW parametric tools.

As shown in Figure 11.35, in order to save the core or cavity solid as a separate part model, the exploded view of the mould tool needs to be collapsed and the *Save Body* tool in the Features can be applied to save the selected bodies as an individual file.

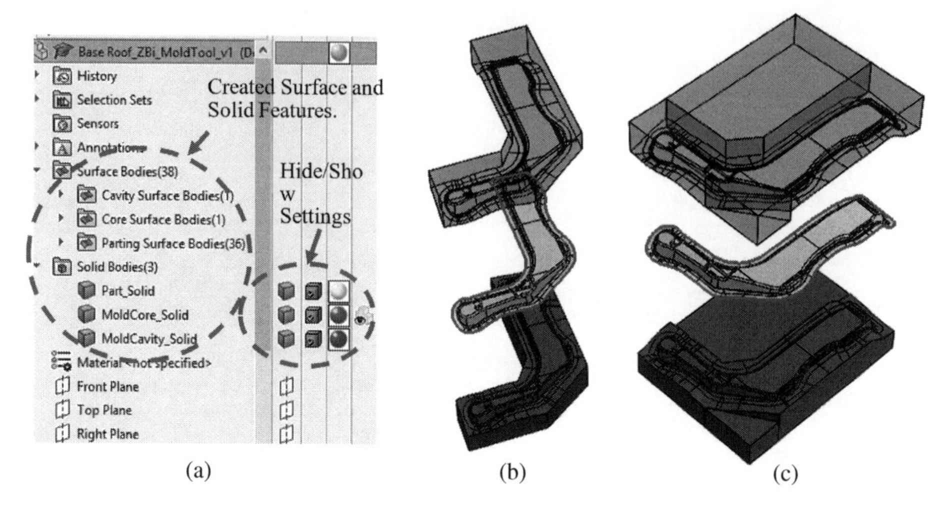

Figure 11.34 Visualization of a mould assembly. (a) Intermediate features in creating a mould tool. (b) Compact mould tool. (c) Core and cavity with additional features.

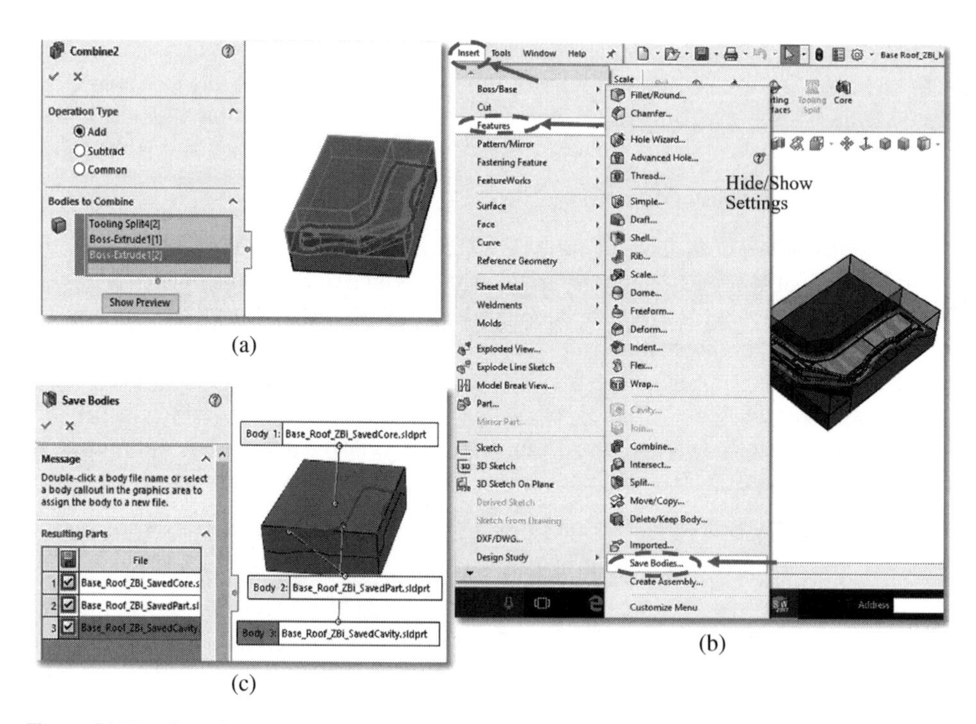

Figure 11.35 Combine and save selected bodies as an individual part model. (a) Combine multiple features as a body, (b) use save bodies, and (c) save mould components as individual parts.

11.8 Computer Aided Mould Analysis

In this section, the SolidWorks *Mould Tool* is used as an example tool to analyse the elongation occurring to the thermoformed product.

11.8.1 Thermoformable Materials and Products

Thermoforming processes are often used to make the products from composite panels. Thermoformable composite panels are thermoplastic materials, e.g. polypropylene (PP), nylon 6, polyetherimide (PEI), polyphenylenesulphide (PPS); these materials can be reinforced with some type of fibre. The materials are processed as prepreges for secondary operations, and then are thermoformed into shaped structures. Figure 11.36 shows some sheet moulding compound (SMC) products used in the automotive industry (Asadi et al. 2018).

SMC is characterized by a very high productivity, excellent part reproducibility and cost efficiency, and the possibility to build parts with complex geometries and integrated functions. Note that SMC usually has a low level of stiffness and strength because of a low fibre-volume fraction, a short fibre length, and isotropic fibre distribution (Wulfsberg et al. 2014).

11.8.2 Compression Moulding

Compression moulding is a process in which a moulding polymer, called a charge or compound, is squeezed into a pre-heated mould cavity under heat and pressure until the charge has cured. The process is a high-volume production and low-cost moulding method, suitable for parts with a complex appearance, high strength, or high impact resistance.

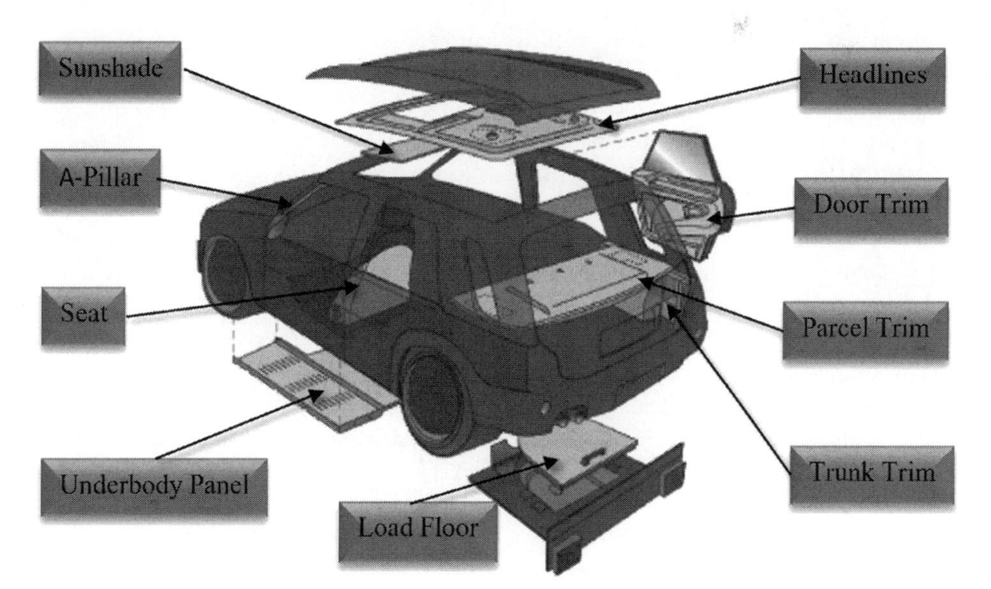

Figure 11.36 Examples of sheet moulding compound products in an automobile (Asadi et al. 2018Technologies).

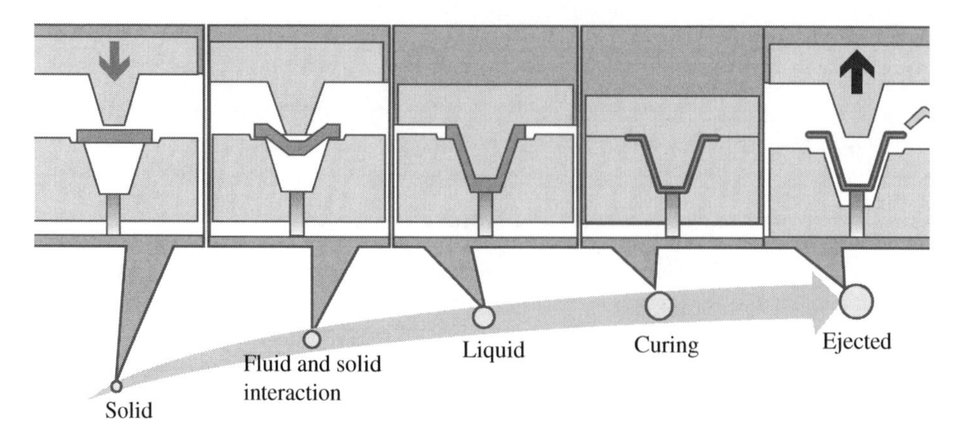

Figure 11.37 Illustration of compression moulding process.

Compression moulding, as illustrated in Figure 11.37 (Moldex3D 2019), is a common process in manufacturing complex composite components for industrial applications. SMCs, glass mat thermoplastic, and pre-impregnated composite are the most commonly used materials. With regards to possible quality issues in compression moulding, it has been found that the strain affects the inelastic properties of composite panels for all microstructures, particularly on the damage threshold. During the compression moulding, the strain distribution must be analysed to see if it exceeds the material strength (Shirinbayan et al. 2017).

11.8.3 Simulation of Compression Moulding

A number of research articles as well as commercial software tools are available for compression moulding simulation. Kim et al. (1995) investigated the effects of the process parameters on the flow and the curing of the composites in compression moulding. The flow pattern was visualized with the charges of different layers. A numerical simulation program by the thermo-viscoplastic finite-element method was developed to analyse flowing and curing and predict the curing and the fibre distributions. Kim and Im (1997) then extended their program for a 3D model using a rigid-viscoplastic approach; however, the programs worked only for simple geometry (e.g. a rectangular specimen under symmetric compression). Narasimhan (2011) developed an iFEA framework to simulate thermoforming procedures of composite headliners. It led to 80% saving of development cost in contrast to traditional methods.

Ognjen et al. (2012) discussed software tools for design and analysis of composites. Most of the mainstream computer aided engineering (CAE) tools, such as Computer-Aided Three-dimensional Interactive Application (CATIA), Unigraphics NX (NX), and Pro/Engineer (ProE), provide integrated environments for the design of composites. However, even the most comprehensive software packages do not contain all the tools necessary for design, manufacturing, and assembly of composite products. Some composite specific software tools can be used individually or as add-ons in a large software package. PAM-FORM from ESI was developed to simulate different forming processes such as

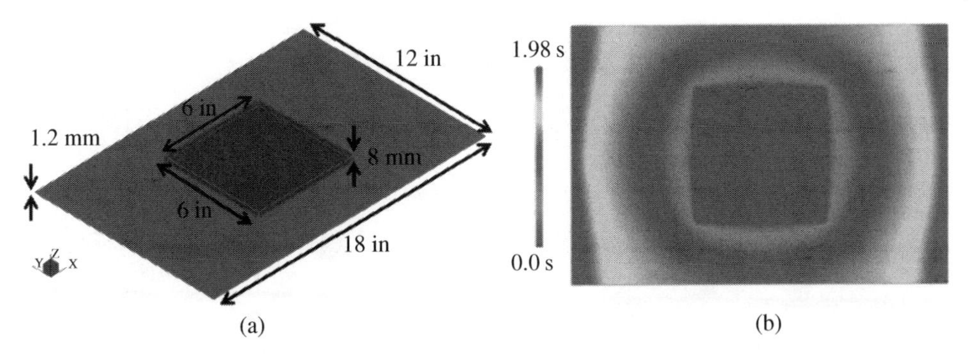

Figure 11.38 Example of compression moulding simulation in Autodesk Modflow (Li et al. 2017). (a) Geometry of the part and initial charge. (b) Fill time predictions.

stamping, diaphragm forming, thermoforming under clamping conditions and different process parameters such as tool velocity, temperature, and pressure. It can be used to predict wrinkling, bridging, thickness, optimum flat pattern, contact pressure, fibre orientation, and stress/strains (Chabin and Camanho 2012). A compression moulding process can be analysed using integrated numerical simulation that involves both Computational Structural Mechanics (CSM) and Computational Fluid Dynamics (CFD). Li et al. (2017) used the Autodesk Moldflow to simulate a compression moulding process and achieved good results for parts with chopped carbon fibre SMC. Figure 11.38 shows the initial charge and final shape of the product sample.

Hinterhölzl (2014) discussed the simulation for three types of forming processes; i.e. *draping simulation*, *automated fibre placement* (AFP) simulation, and *braiding process* simulation. As shown in Figure 11.39, the draping simulation was applied to optimize geometry and alignment of the flat preform, drapeability, and prediction of draping defects.

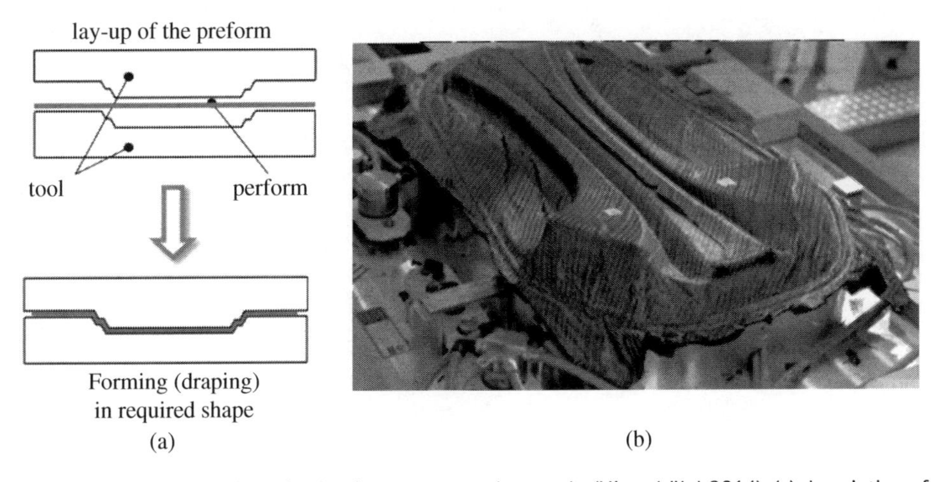

Figure 11.39 Simulation of a draping process and example (Hinterhölzl 2014): (a) description of a draping process and (b) example of a draped part.

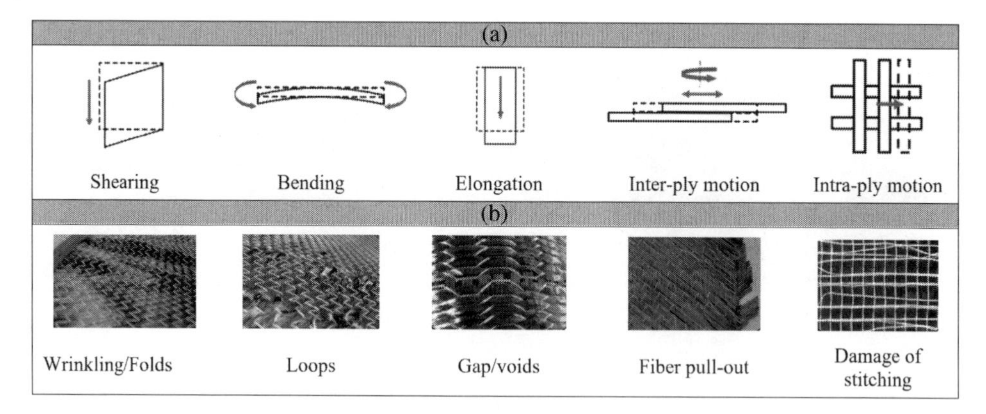

Figure 11.40 Deformations of composites in the draping process (Hinterhölzl 2014; Binetruy 2015): (a) deformation modes and (b) types of defects.

Hinterhölzl (2014) also described the deformation modes and corresponding defects of draping simulation, as shown in Figure 11.40.

Hinterhölzl (2014) classified the methods for draping simulation into (i) a coarse *kinematic approach* including a pin-join method, a purely geometric approach, computation of the fibre direction, and direct cut geometry generation, and (ii) detailed *finite element analysis* to predict deformation modes and defects in the pre-forms by modelling the pre-form on a ply-level with shell elements and simulating draping processes and tools with the consideration of forces, friction, and velocities. Two types of the approaches for draping simulation are compared in Table 11.7 (Hinterhölzl 2014). In addition, Bei (2014) has given

Table 11.7 Comparison of kinematic and finite element approaches (Hinterhölzl 2014).

	Kinematic approach	**Finite element analysis**
Advantages	• Fast and easy to use • Many software available • Good interface to structural analysis	• Accurate results even for complex shapes • Accounts for material and process influence • Prediction of draping defects possible (wrinkles, gaps, etc.)
Disadvantages	• No material behaviour • No process influence • Suspicious results for complex shapes	• Higher computational effort • Material testing required
Example tools and experiments	CATIA CPD (Kinematic Simulation) ⟺ Experiment	⟺ ESI PAM-FORM (Chabin 2013)

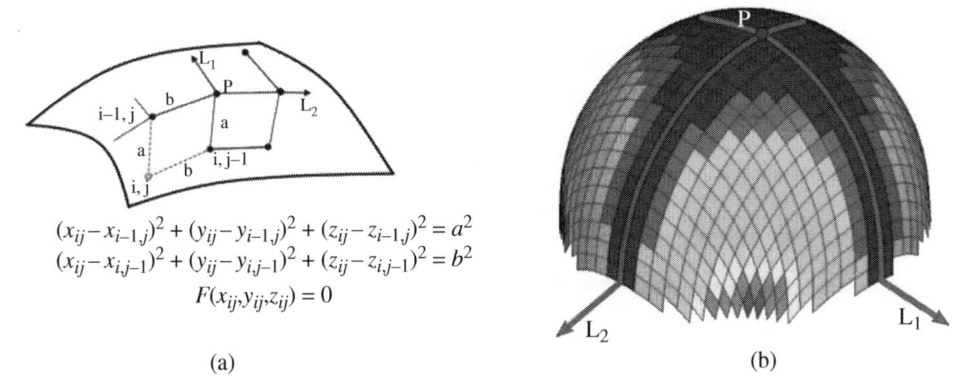

Figure 11.41 Geometric approach for evaluation of distortion (Hinterhölzl 2014): (a) determination of fibre length and (b) an arbitrary point on the surface.

an introduction to existing technologies and development trends of research on simulation of draping processes.

Figure 11.41 shows an example where the geometric approach was used to evaluate the distortion of the part (Hinterhölzl 2014).

In designing composites, the Excel sheet approach, grid approach, and solid slicing approach could be used to design ply configurations. A grid system was also utilized to optimize plies of composites based on the requirements of geometric and assembly needs. In addition, integrated composite structure analysis was available where mesh was used to analyse fibre deviations (Curtis 2012; Berardi 2016).

11.8.4 Predicating Elongation in SolidWorks

SolidWorks does not support simulation of compression moulding. However, the surface flattened tool can be used for implementation of the grid approach to predict the elongation of a part from its original and final states.

SolidWorks has various add-ins for design and engineering analysis. For example, the *Draft Analysis* in *Mould Tool* can only be used to analyse a draft without a direct relationship to elongation. Here, the *Surface Flatten* function in the *Surface Command Manager* is introduced to predict the elongation of the part. The Surface Flatten tool was originally developed mainly to create a flat template for a 3D surface. For any surface, the Surface Flatten tool can be used to determine what geometry a flat plate should start with. The tool can work for both developable and non-developable surfaces or faces. A developable surface will not have the distortion after being flattened. For a non-developable surface, the surface flatten tool can estimate the amount of stretch or contraction.

Using the Surface Flatten tool is straightforward. Figure 11.42 shows the available tools under the Surface Command Manager or we can find it from the text menu bar: i.e. *Insert > Surface > Flatten.* Once the Surface Flatten tool is activated, the user can input all of the required information to create a flattening plate, as shown in Figure 11.43. Note that the user must specify (i) one or more surfaces to be flattened and (ii) which vertex or edge of the flattening plate passes through. The other inputs are optional. Note that the input

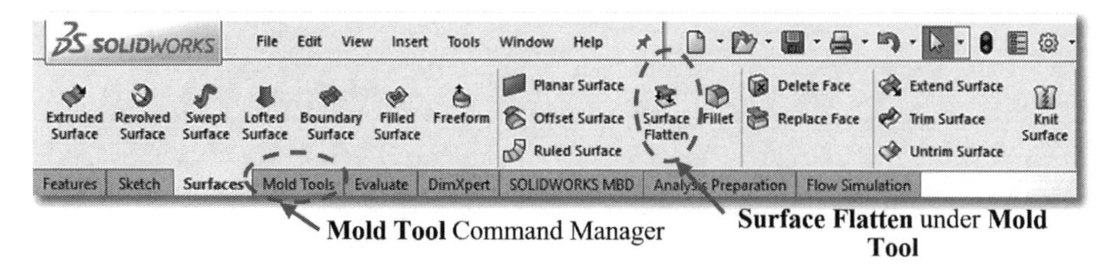

Figure 11.42 Activating the surface flatten tool in SolidWorks.

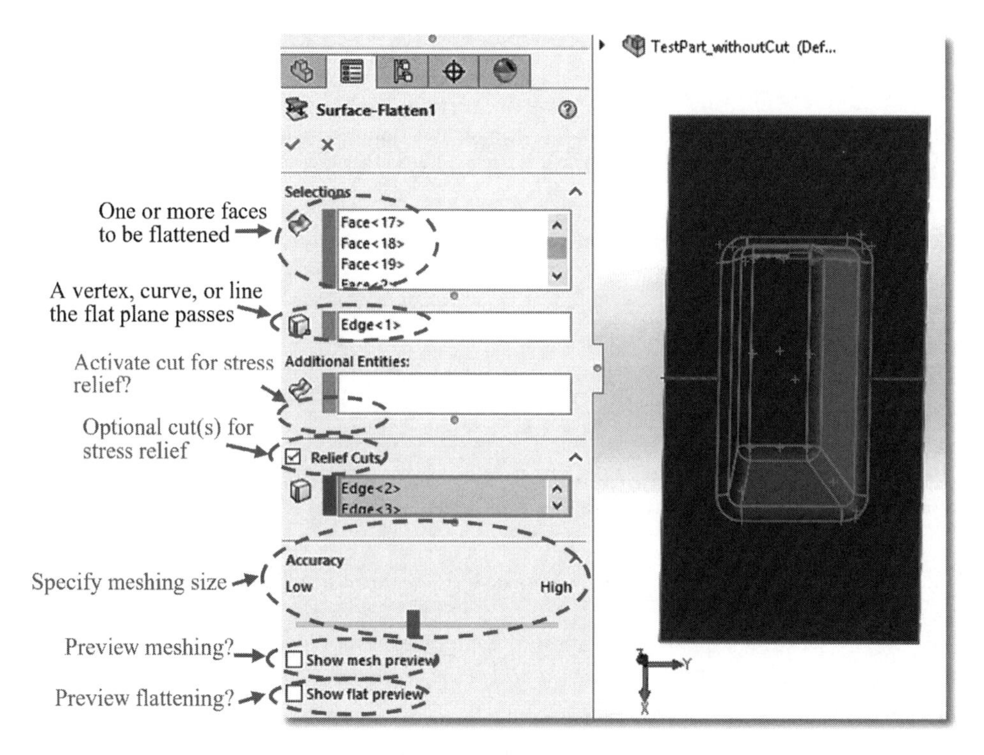

Figure 11.43 Using the surface flatten tool in SolidWorks.

for item (ii) is not critical since the result of the flattening plate is the same for whatever vertex or edge the flattening plate passes through.

Once a flatten surface is created, it will be listed as a feature in the Feature Manager tree, as shown in Figure 11.44. The simulated deformation can be viewed by right-clicking on the flatten surface and then selecting *Deformation Plot*, as shown in Figure 11.45.

In evaluating the deformation of a flattened surface, the tool has an option for introducing cut edges where stresses can be relieved. Figure 11.46 shows a comparison of deformations of a part with and without stress relief cuts. Introducing stress relief cuts reduces the amount of deformation on the majority of surface areas at the expense of a possible local fracture. These cuts can be introduced by creating split lines on the part model.

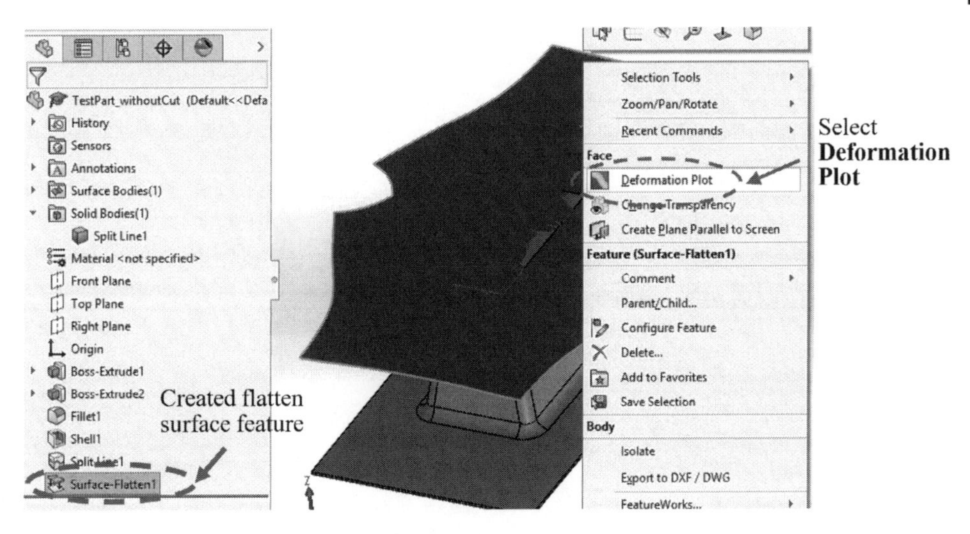

Figure 11.44 Flatten the surface as a feature of the part model.

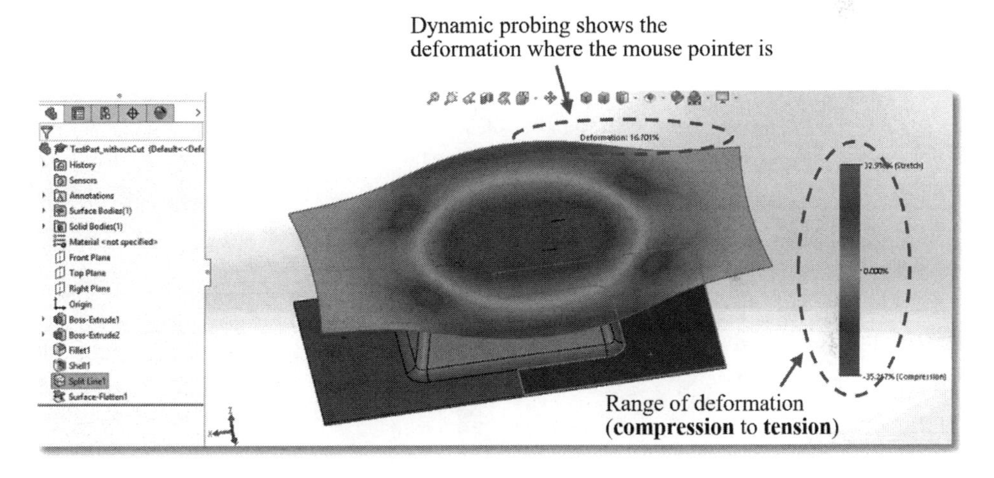

Figure 11.45 Viewing the deformation.

Similar to the grid-based approach, the Surface Flatten tool works well for some parts with smooth surfaces and relatively small curvatures. However, for a complex part with large curvatures and/or sharp corners with compound angles, the Surface Flatten tool is likely to run into the issue of *self-interference*, as shown in Figure 11.47. This is due to the fact that the Surface Flatten tool tries to maintain the connections of all the faces of a flattening plate, which includes the twist of the surface on the plane that is not part of the deformation in compression moulding. This issue can be addressed by segmenting a large surface area into several small areas and in such a way that the surface flatten feature for each surface area can be created. Figure 11.48 demonstrates the segmentation technique.

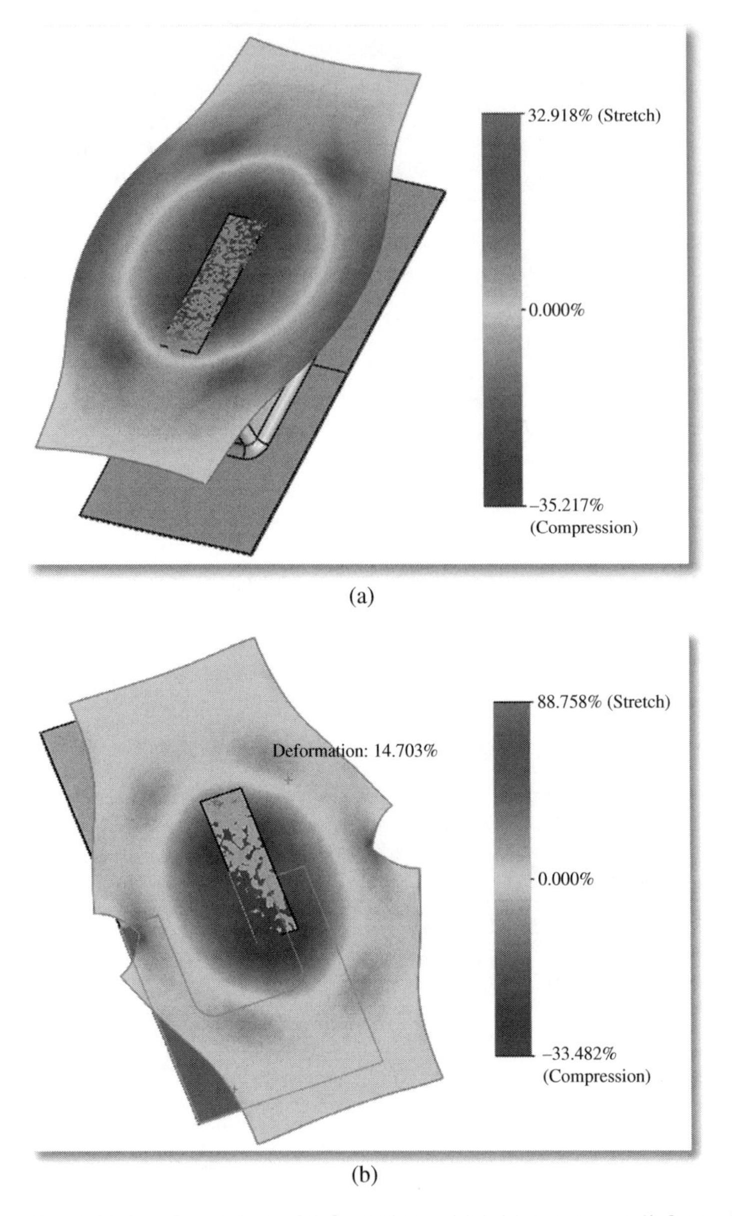

Figure 11.46 Comparison of deformations with/without stress relief cuts: (a) without stress relief cuts and (b) with stress relief cuts.

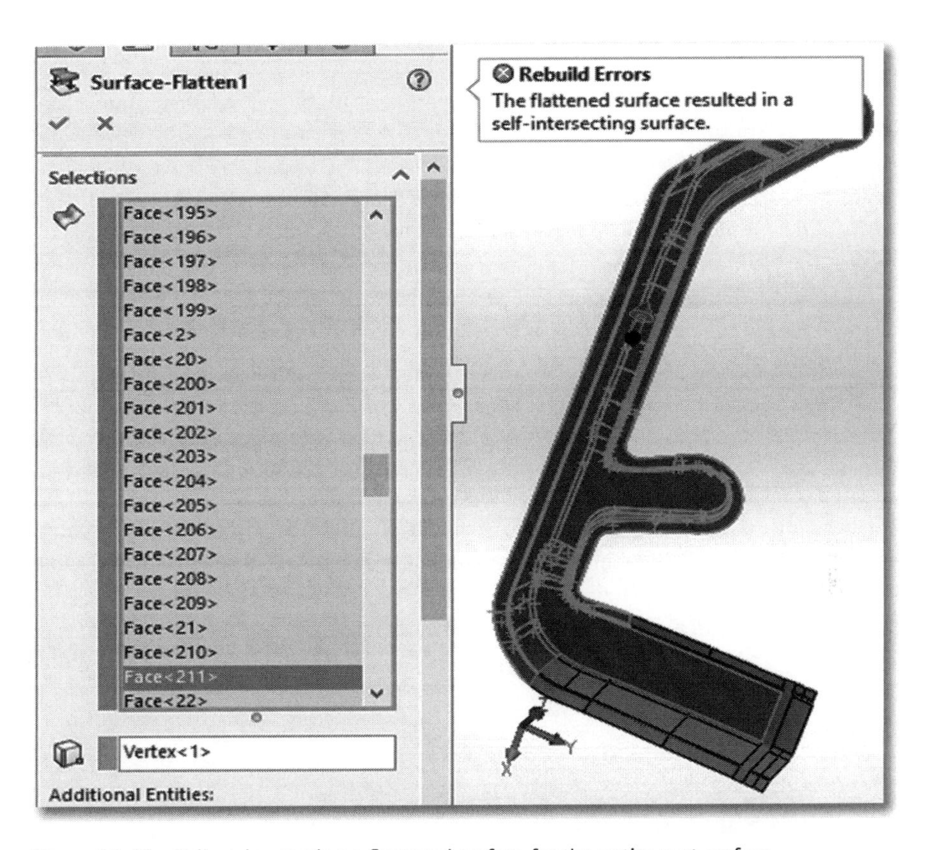

Figure 11.47 Failure in creating a flattened surface for the entire part surface.

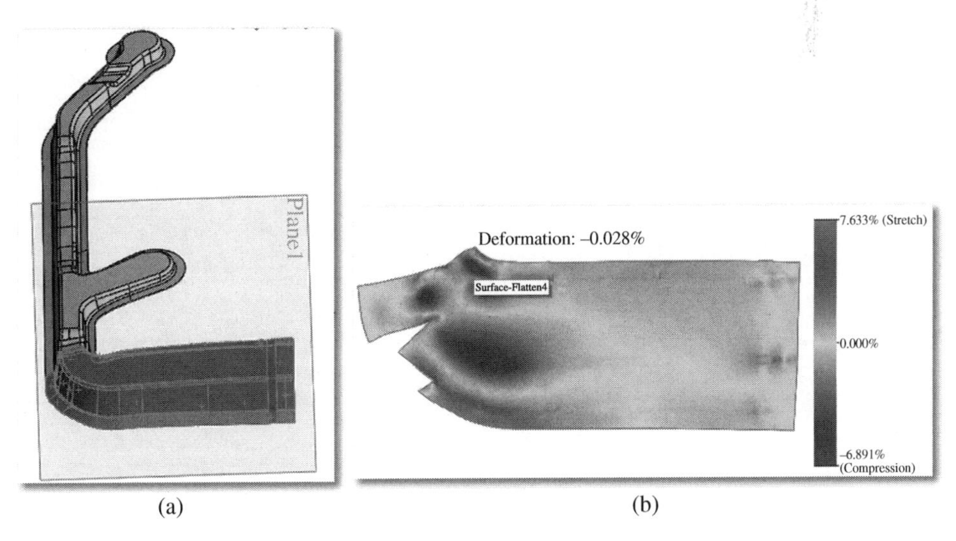

Figure 11.48 Segmentation technique applied to a complex non-smooth part: (a) segmented surface area and (b) deformation field on the segmented surface area.

11.9 Summary

TDM makes direct contacts to workpieces, which are critical to ensure the quality of products. A TDM should be designed to meet both the requirements of the product to be made and the manufacturing process to be performed. TDM design is extremely challenging since each TDM is a customized design for a specific product and manufacturing process. In addition, TDMs are usually low-volume and therefore the manufacturing businesses in the TDM industry are facing fierce competition globally.

Adopting sophisticated computer aided tools and CNCs for designing and manufacturing TDM allows manufacturing enterprises to keep competitive. In designing a TDM product, computerization can reduce the amount of time needed to develop a sufficiently skilled TDM workforce: in the past, five years of apprenticeship and five years of work experience were required to produce a skilled toolmaker. This was particularly true for die makers, since the skills related to solids are required to adjust a die during tryout before production stamping. Today, a toolmaker requires less skill; for example, within two years, a TDM trainee can acquire 70% of the knowledge of a traditionally trained toolmaker because of computerization of the production process. Less training time is needed to enable the employee to be technically proficient to produce TDMs for low and medium levels of precision, complexity, and quality. At the same time, technology has incorporated more of

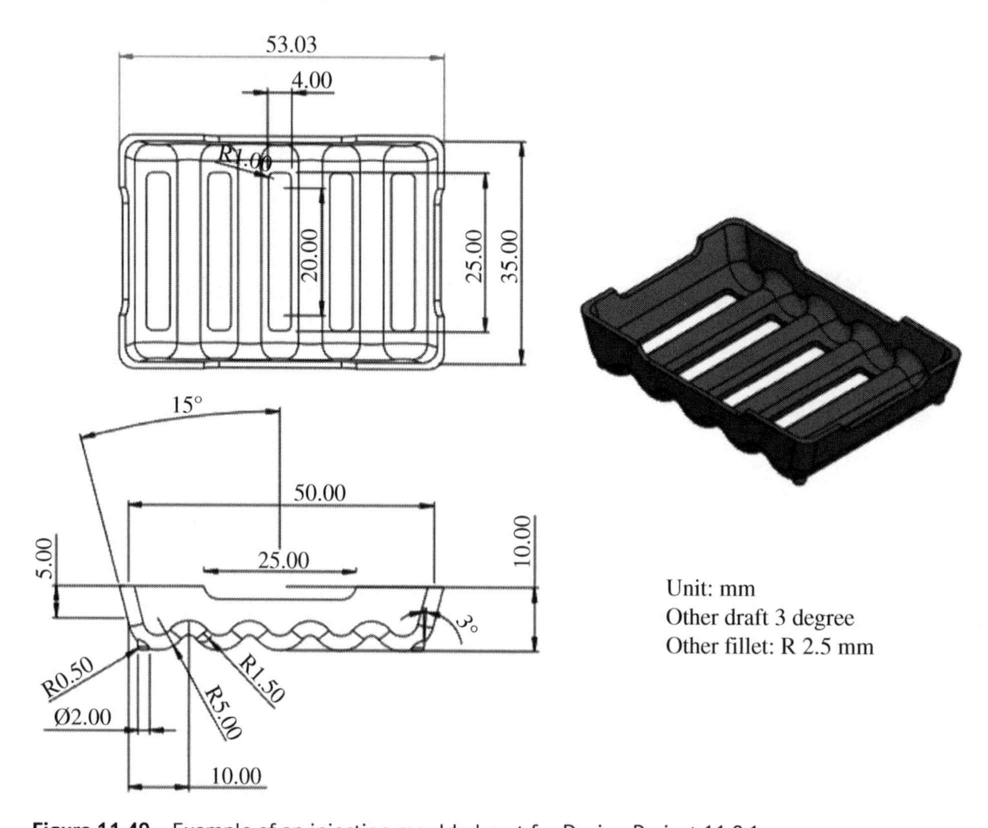

Figure 11.49 Example of an injection moulded part for Design Project 11.8.1.

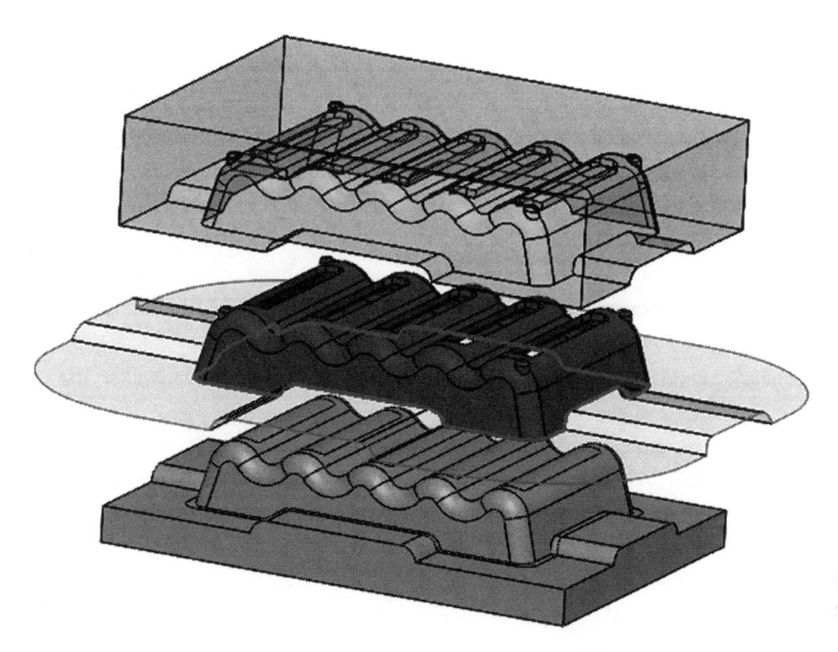

Figure 11.50 Sample mould assembly for an injection moulded part.

the knowledge base required to make TDMs, but computer aided tools cannot fully replace design creativity, talent, and experience (United States International Trade Commission 2002).

11.10 Design Projects

11.1 Create a model of the injection mould part in Figure 11.49, or find and model a plastic object by reverse engineering. Create a mould assembly (such as the sample model in Figure 11.50) for the corresponding injection moulding process.

11.2 Reverse engineer an auto part made from composite panels by thermoforming a part, such as the one shown in Figure 11.36, using the SolidWorks Surface Flatten tool to evaluate the distribution of elongation over the part in the compression moulding process.

References

Asadi, A., Baaij, F., Moon, R., and Kalaitzidou, K. (2018). Cellulose nanocrystals for lightweight sheet molding compound composites. http://speautomotive.com/wp-content/uploads/2018/03/SUST_Kalaitzidou_Georgia-Tech_Cellulose-Nanocrystals-For-Lightweight-Sheet-Molding-Compounds-Composites.pdf.

Bei, S. (2014). Draping simulation – recent achievements and future trends. www.lcc.mw.tum.de/fileadmin/w00bkg/www/PDF/Symposium/LCC_Symposium_Bel.pdf.

Berardi, D. (2016). CATIA for composites design and manufacturing preparation – design and produce: better, stronger and lighter. https://www.octima.it/wp-content/uploads/2016/01/10Nau_Dassault.pdf.

Binetruy, C. (2015). The major role of manufacturing processes on the use and properties of carbon fibre reinforced composites, Brussels – 18th November 2015 – University Foundation. http://www.seii.org/seii/documents_seii/archives/colloques/2015-11-18-Composites/4_Lecture_Binetruy.pdf.

Canis, B. (2012). The tool and die industry: contribution to U.S. manufacturing and federal policy considerations, Congressional Research Service, R 42411. http://www.ntma.org/uploads/general/Tool-and-Die-Industry.pdf.

Chabin, M. (2013). PAM-FORM 2G 2013 – thermoforming of dry textiles and prepregs. http://docplayer.net/45310029-Pam-form-2g-2013-thermoforming-of-dry-textiles-and-prepregs-july-23rd-2013-mathilde-chabin.html.

Chabin, M., and Camanho, A. (2012). Simulation for mass production of automotive composite structural components, ESI Group. www.tecnologiademateriais.com.br/mt/2012/cobertura_paineis/congresso_pr/apresentacoes/ESI.pdf.

Cimatron E. Inc. (2017). CAD/CAM solution for mold making from quoting to deliver. http://www.cimatron.com/SIP_STORAGE/files/0/1760.pdf.

Curtis, J. (2012). CATIA composite design, analysis, and manufacturing – driving innovation in the automotive industry success story: Terrafugia and the Transition roadable light sport aircraft. http://www.temp.speautomotive.com/SPEA_CD/SPEA2012/pdf/VPT/VPT12.pdf.

Custompart (2017). Die castings. http://www.custompartnet.com/wu/die-casting.

Engineering Design Center (2019). Gate types. http://www.dc.engr.scu.edu/cmdoc/dg_doc/develop/design/gate/32000002.htm.

Hinterhölzl, R. (2014). Process simulation of composites. http://www.intales.com/wp-content/uploads/presentations/2014/session1/03_Hinterhoelzl_Process_Simulation_of_Composites.pdf.

Injection Molding (2019). Injection mold cooling design. http://www.injectionmoldingplastic.com/injection-mold-cooling-design.html.

Kim, S.-Y. and Im, Y.-T. (1997). Three-dimensional finite element analysis of the compression molding of sheet molding compound. *Journal of Materials Processing Technology* 67: 207–213.

Kim, D.K., Choi, H.-Y., and Kim, N. (1995). Experimental investigation and numerical simulation of SMC in compression molding. *Journal of Materials Processing Technology* 49: 333–344.

Lanzuela, V. (2018). Competitiveness roadmap of the tool and die industry. http://boi.gov.ph/wp-content/uploads/2018/03/5th-TID-Mr.-Lanzuelas-presentation-on-Tool-and-Die.pdf.

Li, Y., Chen, Z., Xu, H. et al. (2017). Modeling and simulation of compression molding process for sheet molding compound (SMC) of chopped carbon fiber composites. *SAE International Journal of Materials and Manufacturing* 10 (2): 2017. https://doi.org/10.4271/2017-01-0228.

Mold Technology (2011a). Cold running systems. http://mold-technology4all.blogspot.com/2011/08/cold-runner-systems.html.

Mold Technology (2011b). Mold technology – gate types. http://mold-technology4all.blogspot.com/2011/08/gate-type.html.

Mold Technology (2011c). Cooling system. https://mold-technology4all.blogspot.com/search?q=cooling+system.

Moldex3D (2019). On-demand Webinar: integrated numerical simulation for SMC in compression molding process. https://www.moldex3d.com/en/on_demand_webinar/integrated-numerical-simulation-for-smc-in-compression-molding-process.

Nagahanumaiah, R.B. and Mukherjee, N. (2007). Rapid tooling manufacturability evaluation using fuzzy-AHP methodology. *International Journal of Production Research* 45 (5): 1161–1181.

Narasimhan, S. (2011). Virtual design verification and process improvement of composite sheet material products using intelligent finite element analysis (iFEA), Western Michigan University, Dissertation 442. http://scholarworks.wmich.edu/dissertations/442.

nttd-es.co (2019). Running systems. http://www.nttd-es.co.jp/products/e-learning/e-trainer/trial/en/mold/kiso/sample/step3/.runner.htm.

Ognjen, P., Simonovic, A., Stupar, S., et al. (2012). Contemporary software tools in the design process of composite structures. *5th International Scientific Conference on Defensive Technologies*, Belgrade, Serbia (18–19 September 2012), pp. 117–122.

Rapid, S. (2015). Advanced methods for making better mold tools. https://www.starrapid.com/blog/advanced-methods-for-making-better-mold-tools.

Shirinbayan, M., Fitoussi, J., Abbasnezhad, N. et al. (2017). Mechanical characterization of a low density sheet molding composition (LD-SMC): multi-scale damage analysis and strain rate effect. *Composites Part B Engineering* 131: 8–12.

United States International Trade Commission (2002). Tools, dies, and industrial molds: competitive conditions in the United States and selected foreign markets, Investigation No. 332–435. https://www.usitc.gov/publications/332/pub3556.pdf.

Wulfsberg, J., Herrmann, A., Ziegmann, G. et al. (2014). Combination of carbon fibre sheet moulding compound and prepreg compression moulding in aerospace industry. *Procedia Engineering* 81: 1601–1607.

Part III

System Integration

12

Digital Manufacturing (DM)

12.1 Introduction

The automation of decision-making supports for various manufacturing businesses in material flows have been introduced in previous chapters. An automation to specific manufacturing business used to be called an island of automation. The concept of *islands of automation* was coined in the 1980s to describe the situations where different automation solutions are isolated from each other without an integration or even communications. Since the network becomes so prevalent today, communication is no longer an issue. However, the concept of islands of automation is still used for cases where companies or organizations use automation solutions in a limited fashion without an integration.

Independence of the automation solutions can be sustained in a similar way where an object-oriented method is applied. There is no doubt that enterprises can benefit from a number of independent automation solutions by (Clifford 2006):

1. Giving the identities and defining clear boundaries of functional modules in a system. One functional module has its inputs and outputs; it is operated, maintained, and updated independently without affecting others.
2. Clarifying the functionalities and roles. Each functional module is developed for certain purposes, which are easily understood by users; for example, computer aided manufacturing for machining programming.
3. Defining inputs and outputs clearly. The movements in and out of a functional module are very controlled; users can change the outputs of the functional module only through well-defined inputs.
4. Sustaining self-sufficiency. Each functional module has the mechanisms to deal with the abnormal, uncertainties, and disturbances when performing specified functions.
5. Facilitating security enhancement. No attack can be made except through the defined protocols of communications. In a functional module, there is a clear distinction between trusted objects and outsiders. The roles of participators are clearly defined and the intruders are easily identified.

However, with ever-growing complexity of manufacturing systems, enterprises are seeking the holistic automation solution to a higher level and broader scope of manufacturing businesses than before. Islands of automation have exposed a number of the limitations. (i) Isolating the applications in a system is a haphazard approach towards information

Computer Aided Design and Manufacturing, First Edition. Zhuming Bi and Xiaoqin Wang.
© 2020 John Wiley & Sons Ltd. This Work is a co-publication between John Wiley & Sons Ltd and ASME Press.
Companion website: www.wiley.com/go/bi/computer-aided-design

technology (IT) infrastructure since the communication of functional modules are not regulated. (ii) Islands of automation is an unorganized collection of complex and unconnected setups of applications, systems, and databases; the system lacks the interdepartmental coordination, cooperation, and collaboration to support existing and new business requirements. (iii) The isolation of functional modules becomes a big hurdle to synergize segmented solutions as a holistic system-level solution, while modern enterprises need to integrate more and more diversified tools from various sources to cope with manufacturing businesses. In summary, islands of automation cause an increase in batch cycle time, material losses, delays, diminishing profits, compliance errors, and increased operational, research, and development costs (Adeptia 2017).

Digital manufacturing (DM) is the use of an integrated information system that consists of modelling and simulation, 3D visualization, analytics, and collaboration tools to design products and processes and make products. DM has been evolved from some early manufacturing initiatives such as design for manufacturability (DFM), computer-integrated manufacturing (CIM), flexible manufacturing, and lean manufacturing, but DM highlights the needs for collaborative product and process design via a digital platform (Siemens 2019a, b). In this chapter, DM is introduced as an integration of various automation solutions at different levels and scopes of manufacturing processes. The focuses are put on (i) the architecture for data collection and sharing, (ii) high-level decision-making supports, and (iii) main enabling technologies for planning, scheduling, and controlling of system-level businesses.

12.2 Historical Development

DM is an integrated approach to manufacturing that is centred on a computer system. The DM concept was inspired by the rapid development and the applications of IT. In the past few decades, IT has extensively been applied in manufacturing and this has led to numerous well-developed computer aided tools for product design, manufacturing, machine controls, and process planning. DM has developed as a virtual platform to support, network, and integrate segmented automation solutions (such as computer aided design (CAD), computer aided manufacturing (CAM), computer aided fixture design (CAFD), computer numerical controls (CNCs), computer aided processing planning (CAPP), flexible manufacturing systems (FMSs), virtual reality (VR), manufacturing execution systems (MES), material resource planning (MRP), and manufacturing resource planning (MRP-II)), and incorporate new solutions (such as computer integrated manufacturing (CIM), total quality management (TQM), just in time (JIT), concurrent engineering (CE), enterprise information systems (EIS), product lifecycle management (PLM), enterprise resource planning (ERP), customer relationship management (CRM), supply chain management (SCM), service workflows, and business intelligence (BI)), into high-level and broad scopes of manufacturing businesses.

Many researchers have discussed the application of the evolution of IT in manufacturing. The evolution of IT in manufacturing is commonly known to the research community as industry evolution. Table 12.1 shows the milestones of the industry evolution as well as the main characteristics in each stage of development (Berger 2019). The historical evolution of IT is divided into four stages from Industry 1.0 in 1784 to Industry 4.0 today.

Table 12.1 Industry evolution from Industry 1.0 to 4.0.

Year	Milestone	Description of evolution
1784	Industry 1.0	At the end of the eighteenth century, steam engines were invented In England. Steam engines were used to power a loom. This brought the dawn of mechanical products; water and steam power were used to mechanize manufacturing processes and substitute human labour.
1870	Industry 2.0	At the end of the nineteenth century, electrification enabled mass production. The manufacturing processes for products such as abattoir and automotive were broken down into a series of specialized activities performed on production lines. The division of labour began, the quality of products was improved, and product cost was reduced.
1969	Industry 3.0	The microelectronics and IT began to be applied widely in manufacturing and production automation came into reality; programmable logic controllers (PLCs) and automated solutions became prevalent. Machines competed to ever more complex tasks out of human hands with improved productivity.
Today and tomorrow	Industry 4.0	The physical system is seamlessly integrated with the information system. Adopting cyber-physical systems (CPSs) digitizes production and manufacturing systems. All of manufacturing resources such as machines, tools, workpieces, conveyers, robots, operators, and controllers are networked as *smart things* in the Internet of Things (IoT) to optimize manufacturing business over entire product lifecycles for smart products, mass customization, value-added service supports, and a secure and reliable technological infrastructure.

The fourth industrial revolution is poised to generate a new wave of wealth creation and brings industry 4.0 into reality. The main characteristics of industry 4.0 are interoperability, information transparency, technical assistance, decentralized decisions, all of which is enabled through data (Berger 2019).

From the perspective of information systems, DM is about the generation, collection, processing, sharing, mining, utilization of data from and to all manufacturing resources (Bi et al. 2014). Along the product lifecycle from raw materials, manufacturing processes, assembly processes, qualification testing and validation, marketing and sale, after-sale services, and product recycling, the amount of data is unidirectionally increased. Figure 12.1 shows the scenario where the amount of data is increased when the material flow moves forward from one manufacturing process to another. Note that the data in a manufacturing enterprise is heterogeneous and highly diversified; data can be any piece of information from or to *a smart thing* in a system, such as sensed signals or events, models, methods, algorithms, plans, commendations, information, decisions, and values.

Due to the complexity of modern manufacturing and the globalized business environment, the amount of data of a manufacturing system is usually big. To deal with big data

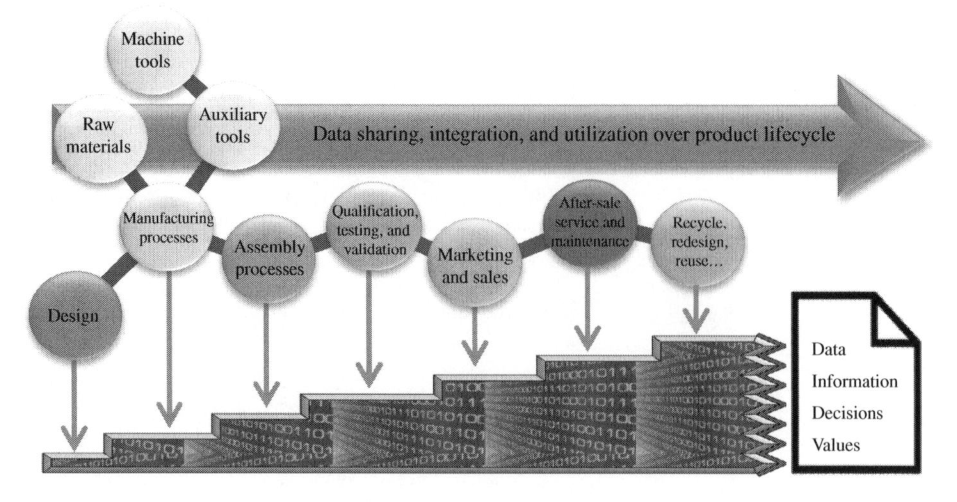

Figure 12.1 The data growth in product lifecycle (King 2018).

involved in a product lifecycle, big data analytics (BDA) is usually integrated with DM to optimize decision-making supports (Bi and Cochran 2014).

12.3 Functional Requirements (FRs) of Digital Manufacturing

In a highly competitive business environment, many companies suffer from *data rich but information poor* (DRIP) syndrome: the amount of data from products, processes, planning, and control keeps growing, while the acquired data cannot be processed to obtain the correct information for decision-making supports efficiently and promptly. DM is not just the collection of computer aided tools in respective domains and levels; DM will be equipped with advanced decision-making tools to process, utilize, mine data for the right information, and ensure that the right data is delivered from the right source to the right place at the right time. The ultimate promise of DM is the seamless integration of all enabling tools for designing, manufacturing, assembling, using, and eventually disposing of products. DM is expected to solve some practical challenges facing enterprises. Here, the main functional requirements (FRs) of DM are discussed.

12.3.1 Data Availability, Accessibility, and Information Transparency

Decision-making activities rely on reliable, sufficient, and prompt data. One of the primary goals of DM is the digitalization to ensure the right data is available where and whenever it is needed. The digitalization technologies enable the virtual product and process planning (Kritzinger et al. 2018). Enterprises expect industrial digitalization at all levels of product value chains to achieve higher competitiveness (Mäkiö-Marusik et al. 2019).

Digitalization at the system level needs to deal with the varieties, volumes, velocities of all types of data from different sources, especially the Internet of Things (IoT). DM should ensure reliable data is accessible to all functional modules; for example, when a machining process is programmed by CAM, the product model from CAD must be accessible. Acquiring data is essential to drive a functional module to perform certain manufacturing business processes. Therefore, the importance of digitization cannot be overemphasized. Digitalization makes it possible to network all of the manufacturing resources inside or outside an enterprise, and reconfigure the system when new automation or communication infrastructure is available or the mission of enterprise is changed (Buttner and Muller 2018).

12.3.2 Integration

As a complete solution to the business operations of a manufacturing enterprise, DM must seamlessly integrate all of the system components of an enterprise system, including enterprise architecture (EA), hardware and software tools for business processes at different levels and domains. System integration supports the integration of data, platforms, business processes, and applications (Fenner 2003).

1. For *data integration*, data will be integrated, maintained, and shared in the network, and the data can be accessed and distributed across database systems by using standard formats of a *component object model* (COM), *distributed component object model* (DCOM), *common object request broker architecture* (CORBA), *enterprise data integration* (EDI), *Java remote method invocation* (JavaRMI), and *extensible markup language* (XML). Full utilization of data makes a manufacturing system smart, while raw data is useless unless it is processed and transformed into the correct information. To utilize the data fully, the data lifecycle consists of the stages of data collection, transmission, storage, pre-processing, filtering, analysis, mining, visualization, and application. The data may be exploited at different stages of the lifecycle in manufacturing (Tao et al. 2019).
2. For *platform integration*, DM must integrate the processes and tools from different vendors and heterogeneous platforms. The integration supports information exchanges among heterogeneous systems securely and efficiently and data can be accessed by different applications without difficulty.
3. For *business process integration*, it is critical for an enterprise to define the processes and specify the information exchange with other processes. This allows enterprises to streamline the operations, reduce costs, and improve responsiveness to customer demands. Process integration covers process management, process modelling, and process workflow. The integrated solution to each business process includes the tasks, procedures, organizations, required input and output information, and required tools.
4. For *application integration*, data and functions of one application must be integrated with that of another application for interoperations in a real-time manner. The integration covers the enterprise's relations to the manufacturing environment: CRM systems are integrated with a company's backend applications for the interoperation of businesses across enterprise boundaries (Fenner 2003).

12.3.3 High-Level Decision-Making Supports

The computer aided systems introduced in preceding chapters are applied to the manufacturing businesses linked directly to the material flow. When a manufacturing system becomes more and more complicated, a hierarchical control architecture is needed to decompose the system control into multiple layers and domains, so that the design space for each decision-making unit is manageable. Therefore, other than the computer aided systems applying directly to the material flow, new systems for high-level decision-making supports must be incorporated in DM.

12.3.4 Decentralization

The boundaries of manufacturing enterprises are becoming vaguer and vaguer (Bi et al. 2014). The processes for making parts, components, and finished products are performed in different locations and planning, scheduling, and controlling manufacturing businesses are decentralized. Therefore, DM must provide support to make parts and components at geographically dispersed locations, bring them together via the supply chain, and assemble them into final products in a form of distributed manufacturing. Decentralized control is needed for the manufacturing processes over distributed manufacturing facilities through the network.

12.3.5 Reconfigurability, Modularity, and Composability

To make a manufacturing system sustainable, the system structure should be reconfigurable to meet unanticipated needs in a dynamic environment. Reconfigurability can be viewed as an extended system ability over the flexibility. The flexibility deals with the anticipated changes of manufacturing processes by changing plans, schedules, and reprograms, while the reconfigurability deals with unanticipated changes of the market needs by reconfiguring both hardware and software resources in a system. Reconfigurability is a comprehensive requirement and can be characterized as a set of sub-requirements, i.e. *modularity*, *compatibility*, *universality*, *mobility*, and *scalability* (Wiendahl et al. 2007; Buttner and Muller 2018).

An integral and static system is ineffective to deal with the diversities of products, manufacturing businesses, and manufacturing resources, since a manufacturing system needs the flexibility to deal with numerous changes. Different from reconfigurability measured in a system application, *modularity* is considered as the system design stage. *Modularity* measures the flexibility of choosing different system modules and assembling them in different ways, so that the same set of manufacturing resources can be reconfigured to make different products with a minimized changeover time. *Modularity* maintains the independence of information models and allows selecting and connecting information modules in multiple ways for different uses and applications. Modularity divides an information system into a collection of smaller, independent, and reusable functional modules to increase the extensibility and flexibility of the enterprise system. *Composability* is a measure for an individual functional module; composability tells how many other modules and at what level of flexibility an individual function module can be integrated with. Taking as an example a computer aided engineering (CAE) tool in DM, its composability can be measured by the number of other CAD tools that can import and export data models to CAE directly.

12.3.6 Resiliency

Manufacturing resources are networked in the digital world and so products and companies will become more vulnerable to the disruptions of supply chains or functioning of physical and cloud assets. Resiliency is the system ability to recover from an undesired state to its desired state. The attributes of resiliency of manufacturing are *persistence*, *adaptability*, *agility*, *redundancy*, *learning capability*, and *decentralization* (Kusiak 2019). The DM should be able to assess the system vulnerability to unexpected disruptions when the operations of a manufacturing system are subjected to continuous changes in the business environment.

12.3.7 Sustainability

Sustainability became a crucial factor that manufacturing enterprises must take into consideration. There is a potential of reducing manufacturing output or even shutting down company operations due to unsustainable practices. The decision factors could range from judicial and regulatory to the inability of a company to compete on a cost basis or due to shortage of materials. *Sustainable manufacturing* must ensure that products or processes are compliant with environmental standards, regulations, and guidelines. Since the manufacturing sector has the largest impact on energy consumption in comparison with other sectors (Bi and Wang 2012), DM must include effective enabling tools to design products and processes subjected to sustainability standards. Note that the sustainability is measured from three aspects: economic factors, environmental factors, and social factors. DM should include the tools to enable it to evaluate system sustainability using these factors (Gregori et al. 2017).

12.3.8 Evaluation Metrics

DM is adopted by manufacturing enterprises to gain competitiveness via digitization, cloud computing, IoT, and BDA. However, DM is not static and evolves along with time. Therefore, DM must include an evaluation system to monitor the performances of system components and take actions for continuous improvement (CI) when performances are below expectations (Castelo-Branco et al. 2019).

Although DM is an integration of many different technologies, methods, and tools, the evaluation system should focus on the functional commonality at the system level. This commonality lays the foundation to develop a set of evaluation metrics of what are applicable to any system components. Metrics can be used to define key impacts, outcomes, or issues that determine or affect the actions for CI of the manufacturing system. Macro-level metrics may be used to define key performance indicators (KPIs) that evaluate the performance of a system component and simplify the qualification of business processes. Metrics and KPIs should be developed in an understandable, meaningful, and measurable way.

12.4 System Entropy and Complexity

Converting raw materials into final products requires numerous manufacturing processes and other corresponding business operations in manufacturing. DM is an integration platform that networks all computing tools to support manufacturing businesses over product lifecycles. The complexity of DM is increasing with the number of system components, the interactions of system components, and the changes of system components and interactions with respect to time. The complexity continues to increase for all of these aspects.

System complexity can be measured by entropy. *Entropy* used to be the thermodynamic property as a measure of energy but is not available for work in a thermodynamic process. According to the second law of thermodynamics, entropy is used to express the disorder, randomness, or complexity in the system (Sönmez and Koç 2015). The complexity of a manufacturing system can be measured by the Shannon entropy (Isik 2010; Vrabič and Butala 2012) as

$$H(X) = \sum_{x \in \chi} p(x) \cdot \log_2 \left(\frac{1}{p(x)} \right) \tag{12.1}$$

where $H(X)$ is a measure of uncertainty or information and quantifies how much information is missed when one does not know random variables, $p(x)$ is the probability of the event x, $\log_2$ is a logical operation that is used to turn the probability into the binary number '0' or '1', and χ is the set of all possible events in a system.

The higher the entropy value, the more complex is the manufacturing system and the larger the amount of information or uncertainty there is in the manufacturing system. Equation (12.1) shows that entropy is determined by two main factors. *The first factor* is the number of events; an event may be an interaction of two or more system components and it can also be the change of a system component or an interaction with respect to time. The more the number of events, the higher the entropy is. *The second factor* is the probability of a certain event occurring; the higher an event is likely to happen, the less the event contributes to the entropy of system.

Figure 12.2 shows the correspondence between system-level goals and computer solutions at different levels and domains of manufacturing systems (Bi et al. 2007). The system-level goals can be defined in terms of customer needs, such as a *shortened*

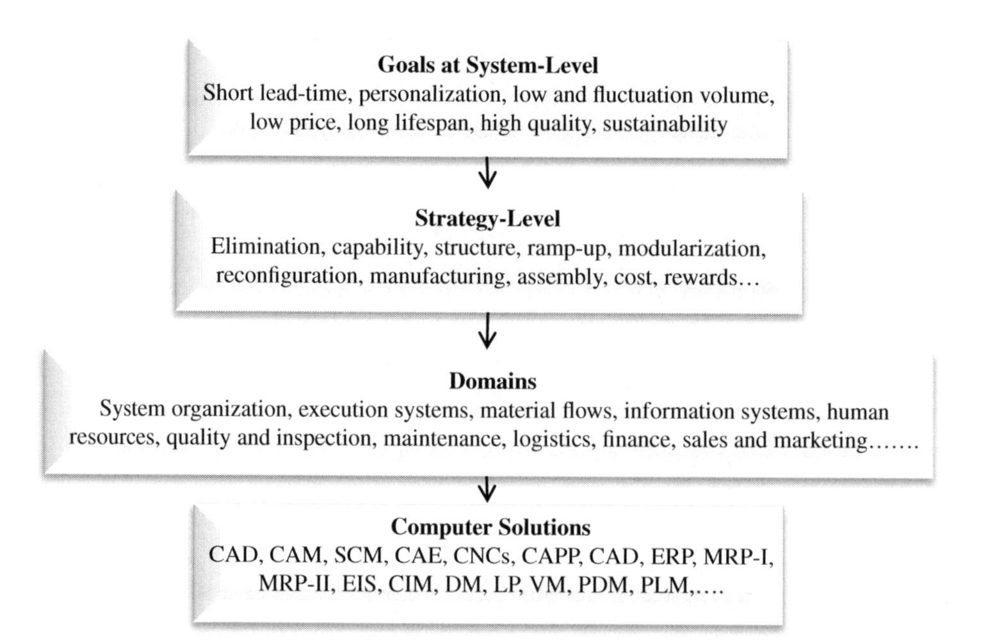

Figure 12.2 Correspondence of system-level goals and computer solution in implementations.

lead-time, *personalization*, *low and fluctuating volume*, and *cost reductions* of products. At the strategy level, a manufacturing company then derives its business strategies for a given system goal. For example, to shorten the lead-time, applicable business strategies include '*elimination*' for the reduction of non-value-added activities, '*capability*' for the high productivity of machine tools, '*structure*' for execution of tasks in parallel, and '*rampup time*' for the reduction of the setup time when a task is changed. At the structure level, the strategical needs are decomposed into the functional requirements of entities in a hierarchy with multiple domains and levels. Finally, at the implementation level, all of the technical solutions including numerous automation solutions are integrated to support manufacturing operations. Figure 12.2 shows that DM is very complex and it is going to be more and more complex with the increase in new computer aided techniques for manufacturing. A manufacturing system should have the ability to make changes in the dynamic environment.

12.5 System Architecture

A system with higher complexity tends to have more problems than that with lower complexity. Measuring manufacturing complexity provides a useful metric to compare different system solutions (Frizelle and Woodcock 1995). If the system complexity can be quantified as entropy, it can be used to (i) compare different system designs for the best solution, (ii) determine how the system has to be changed when the operation conditions are changed, and (iii) look into the impact of one system component on another in a manufacturing system (Fredendall and Gabriel 2019). A manufacturing system involves numerous manufacturing resources that heavily interact with each other. However, no matter how complex a manufacturing system is, the complexity of the corresponding enterprise systems must be *manageable* to ensure the smoothness of business operations.

12.5.1 NIST Enterprise Architecture

As shown in Figure 12.2, the components in a manufacturing system are at different levels and domains, and therefore these system components should be organized to specify the interactions between them. Note that the number and changes of the interactions affect the entropy or complexity of the system.

Enterprise architecture (EA) applies architecture principles to organize the information systems from the perspectives of business, information, process, and technology changes. More specifically, EA can be used to (Gao 2000):

1. Capture the facts of an enterprise about the mission, functions, and business foundations in an understandable manner to promote better planning and decision making.
2. Reduce system complexity by improving the communication among the business and IT organizations within the enterprise through a standardized vocabulary.
3. Provide architectural views that help communicate the complexity of large systems and facilitate management of extensive, complex environments.

4. Focus on the strategic use of emerging technologies to enable better management of the information of the enterprise and consistently insert those technologies into the enterprise.
5. Improve the consistency, accuracy, timeliness, integrity, quality, availability, access, and sharing of IT-managed information across the enterprise and highlight the opportunities for building greater quality and flexibility into applications without increasing cost.
6. Achieve the economies of scale by providing mechanisms for sharing services across the enterprise.
7. Expedite the integration of legacy, migration, and new systems.
8. Ensure legal and regulatory compliance.

EA is a comprehensive approach used to analyse, design, plan, and implement an enterprise system (ES) (FEAPO 2018). Many EAs have been proposed for CIM, where the computer aided enabling technologies in a manufacturing system are all integrated into a completely automated solution. The most popular EAs are the Open System Architecture for Computer Integrated Manufacturing (CIMOSA) by the European CIM Architecture Consortium, the GRAI Integrated Methodology (GRAI-GIM) by the GRAI Laboratory at the University of Bordeaux in France, the Purdue Enterprise Reference Architecture and the related Purdue Methodology (PEPA) by the Industry-Purdue University Consortium for CIM, and the Enterprise Architecture by the National Institute of Standards and Technology (EA-NIST) (Williams 1994). In the section, the enterprise architecture by the National Institute of Standards and Technology (NIST) is used as an example to show how an EA can be applied to simplify the organization of EISs.

To minimize the uncertainties of a manufacturing system, a hierarchical EA is often used to organize an enterprise system in a traditional manufacturing enterprise. Figure 12.3 shows a traditional EA proposed by the NIST. The EA consists of five layers from *business*, *information*, *applications*, *data*, and finally to the *technologies*. With regards to the relations of a manufacturing system and the business environment, EA is constrained by *external discretionary and non-discretionary standards*. With regards to the relations of functional modules at multiple layers, the components at the information layer are driven by the needs of components at the business layer, the components at the application layer are prescribed by those at the information layer, and the components at the application layer are used to define corresponding element at the data layers. Finally, the data components are supported by hardware, software, and communication at the technology layer. The structure is refined from the top layer to the bottom layer, and the information is fed back from the bottom layer to the top layer or in reverse.

The multilayer hierarchical structure such as EA by NIST is usually modularized. All functional components are separable, so the components at a certain layer and domain are specialized to manage, contextualize, and generate data for corresponding tasks independently. The modularization helps to maintain and upgrade an EA solution. In addition, system architecture will not be affected by local changes made to individual components.

12.5.2 DM Enterprise Architecture

Since a manufacturing enterprise is generally very complex, EA has been developed to simplify the complexity of an enterprise information system. The well-defined architecture

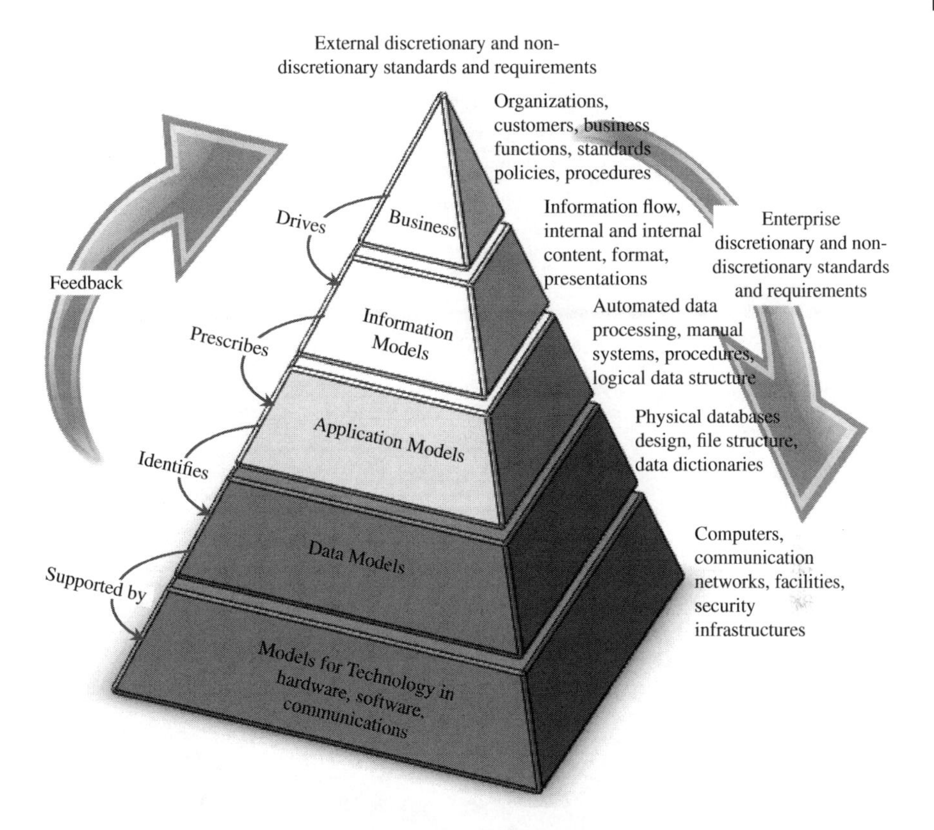

Figure 12.3 NIST enterprise architecture model (Wikipedia 2019a). (*See color plate section for color representation of this figure*).

helps an enterprise to organize a large number of functional components and relationships. Since the complexity of an enterprise is not only attributable to the types and numbers of system components, but also their relationships occurring to business processes, EA can be developed to identify and simplify the relationships (EAdirections 2015).

Enterprise architecture manages the system complexity by regulating or confining the interactions of system components by standards and laws in a layer or grid structure. For example, the interactions for the derivation and the feedback in Figure 12.3 proceed layer by layer; this might cause a number of issues such as a significant delay to communication, the resistance to necessary changes, and the complexity to evolve an enterprise information system with respect to time. Traditional EAs are able to reduce the system entropy, complexity, or uncertainties; however, such simplifications sacrifice system flexibility to certain degree. Therefore, these EAs are mostly applicable to cases where the operational conditions of manufacturing enterprises are relatively stable or changes to businesses or market needs are quite predictable. They become ineffective when an enterprise needs frequent changes on a regular base.

The higher the system entropy, the higher is the level of uncertainties in the system and the more possible it is that interactions and flexibility occur in the system components (Sönmez and Koç 2015). From this perspective, a flatten enterprise architecture that supports

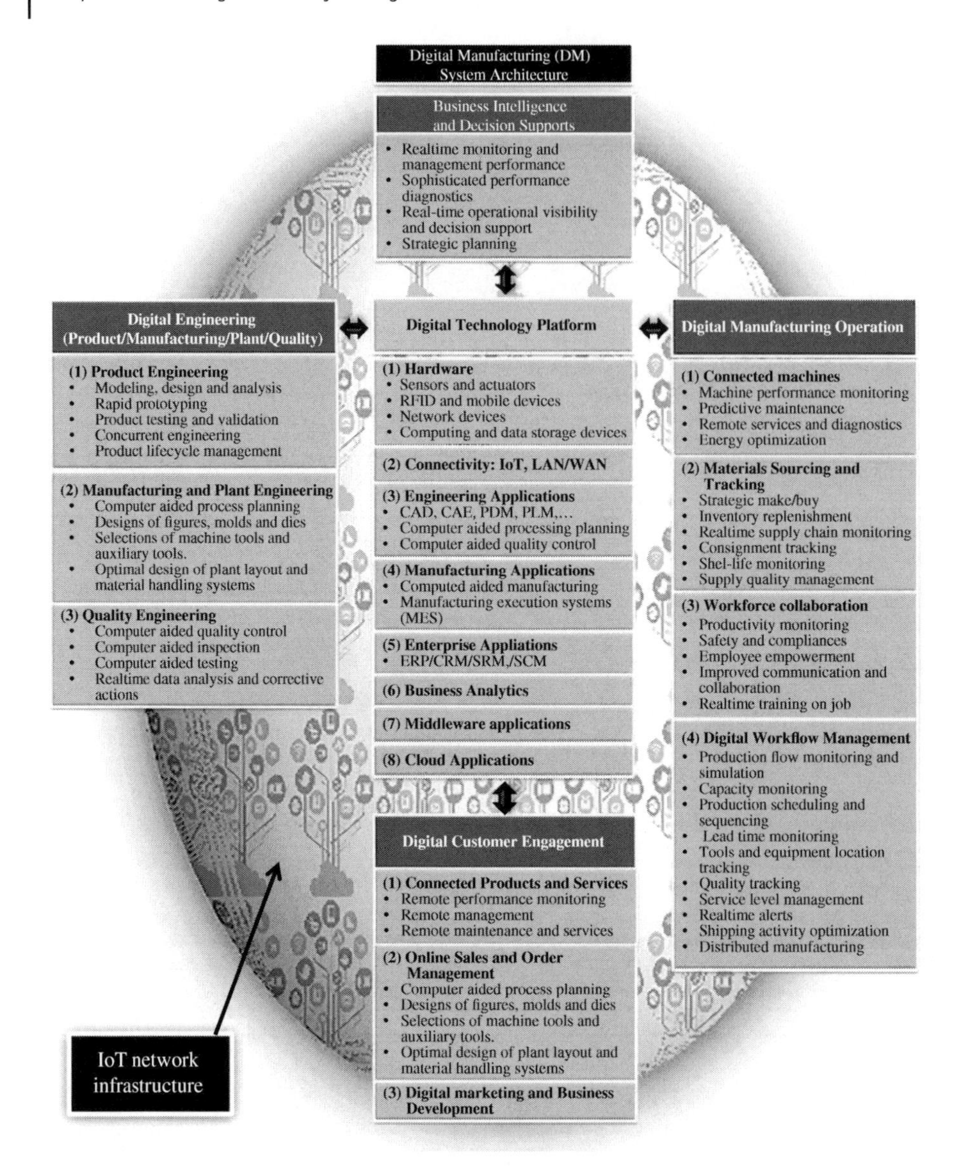

Figure 12.4 DM Enterprise Architecture (EA). (*See color plate section for color representation of this figure*).

mutual interactions of arbitrarily selected components will be ideal since it will maximize the system flexibility and adaptability to adopt anticipated or unanticipated changes and uncertainties.

Figure 12.4 shows the flatten EA for DM (Berman and Bell 2007; Pati and Bandyopadhyay 2017). All of the functional components at the same layer allow direct communication and interactions via the *infrastructure of IoT*. The EA is open to enable manufacturing flexibility, adaptability, and agility. System components are modularized and reconfigurable to meet emerging application needs. Manufacturing businesses in a system can be classified into five cohesive categories: *BI*, *digital engineering*, *manufacturing operations*, *customer engagement*, and *digital technology platform* for integration of digital technologies to support decision-making activities and automate manufacturing businesses.

1. *Digital engineering*. Engineering activities are associated with the designs or selections of manufacturing resources in the material flow, including products, processes, factories, and material handling systems. These activities are simulated and verified in virtual environments prior to the actual executions; typical enabling tools are CAD, CAE, CAPP, computer aided quality (CAQ), and discrete event dynamic system (DEDS) simulation.
2. *Manufacturing operations*. Operations refer to the actual executions of manufacturing or relevant processes, such as machining, material handling, manual operations, and transportation. These processes play critical roles in determining the cost, lead-time, and quality of products, and indirectly in system-level performance such as profits and customer satisfaction. Typical digital tools for manufacturing operations are CNCs, MRP, ERP, MES, and SCM.
3. *Digital customer engagement*. Manufacturing enterprises use various enterprise applications such as CRM to integrate with distribution channel partners and end customers. Such systems help to collect the actual sales data and predict future orders. In addition, advanced information and communication technologies (ICTs) offer the means for customers to participate in designing and developing products or services and managing the product life cycles.
4. *Business intelligence*. Manufacturing enterprises adopt BI tools for long-term survival in the highly competitive market place. This requires defining value-chain digital models before the customers even order products. To this end, digital strategies must be integrated with business strategies of enterprises. Similarly, BI processes must be connected to digital ones. Typical tools for BI are Return on Investment (ROI) and Quality Function Deployment (QFD).
5. *Digital technology platforms*. All of the activities in the above four categories are supported by the digital technology platform. Digital technologies have advanced rapidly in recent years; for example, (i) the Industrial Internet of Things (IIoT) enables machine to machine (MTM) and machine to man communications for real-time monitoring and quality controls, (ii) the Social, Mobility, Analytics, and Cloud (SMAC) lays the foundation to implement digital strategies in any organization, (iii) portable devices such as tablets, laptops, and mobile phones give access to real-time information on all events on the shop floor, and (iv) social apps provide the connectivity among people to transmit information in masses. In addition, digital enterprise architecture merges the sophisticated product lifecycle management (PLM) software and powerful automation technologies for optimization of the entire value chain.

12.5.3 Digital Technologies in Different Domains

After all, DM is an information system where any tangible resource and its behaviour in the physical world have their corresponding replicates in the virtual model. Therefore, DM can be referred to as the digital twin of the corresponding physical model. A *digital twin* is a digital replica of a living or non-living physical entity. When the physical and the virtual world

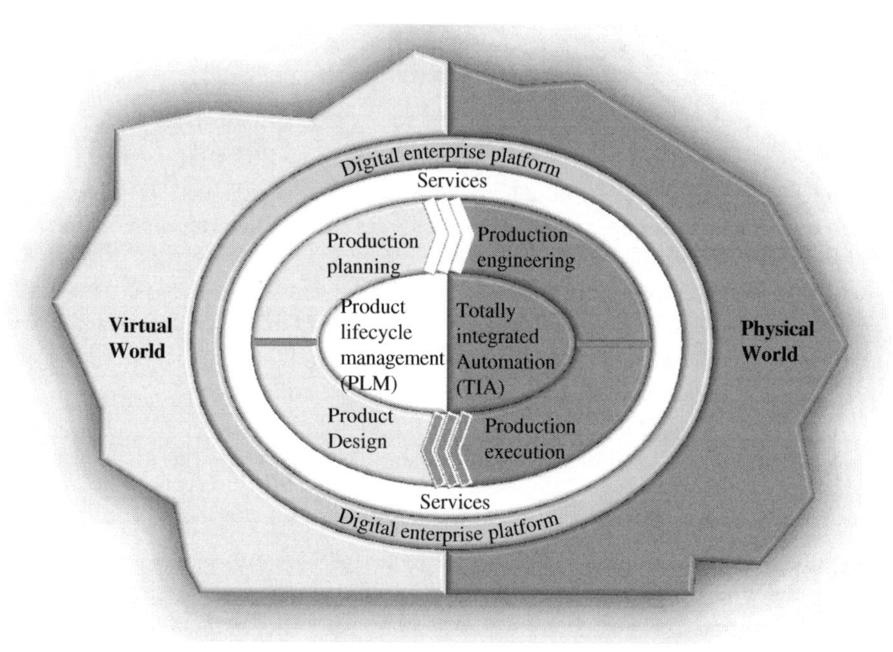

Figure 12.5 Integration of digital and physical twins.

are integrated, data is transmitted seamlessly, allowing the virtual entity to exist simultaneously with the physical entity. A digital twin is a digital replica of tangible machines, tools, processes, people, places, systems, and auxiliary devices that are used for the execution of manufacturing processes (Wikipedia 2019b). Figure 12.5 shows the mirror relation of digital and physical twins in the enterprise information platforms where both digital and physical resources are integrated seamlessly over the infrastructure of IoT (Siemens 2015).

In contrast to the traditional EAs, the flatten EA for DM in Section 12.5.3 does not classify functional components in terms of the levels and the domains where these functional components are applied. However, different virtual functional components work for different types of manufacturing resources in the physical world. To understand the variety of functional components in EA better, it is helpful to look into the correspondences of virtual functional components and applicable levels and domains of manufacturing resources in the physical world. Figure 12.6 shows the applicable levels and domains of 15 types of digital technologies in the physical word, and these 15 technologies are explained in Table 12.2.

12.5.4 Characteristics of Internet of Things (IoT) Infrastructure

As shown in Figure 12.4, EA adopts IoT as the infrastructure to support the interactions of functional components at any level and domain. IoT is the extension of the Internet connectivity into physical devices, everyday objects, and, most importantly, various digital tools for decision-making supports for DM. IoT integrates the digital world seamlessly with the physical world. In the IoT infrastructure, embedded with electronics, Internet connectivity, and other forms of hardware (such as sensors), these devices can communicate and interact with others over the Internet, and can be remotely monitored and controlled (Wikipedia

Figure 12.6 Digital enabling technologies in different domains (PWC 2017).

2019c). Figure 12.7 shows that the IoT makes all types of interactions possible in the integrated infrastructure (Vermesan and Friess 2016). The interactions occur for all combinations of manufacturing resources such as *cyber and physical systems*, *machine-to-machine*, *human-to-machine*, *human-to-human*, *machine-to-human*, *machine-to-infrastructure*, and *machine-to-environment*.

As a matter of fact, the advancements in IT have always been the key drivers for the digital transformation of manufacturing business models. Figure 12.8 shows the impact of new IT developments on the trend of DM models (Kawasmi 2018). Three main characteristics

Table 12.2 Fifteen representative digital technologies in DM (PWC 2017).

No.	Technology	Explanation
1	Digital twin of factory	Helps to design, plane, and construct the plant and infrastructure. Digital twin supports testing, simulation, and commissioning the implementation.
2	Digital twin of machine tool	A digital twin of machine tool is used to design, virtual startup, and ongoing operation. It focuses on the simulation of production operations, helps to set and optimize its key parameters, and enable concepts such as predictive maintenance or augmented reality.
3	Digital twin of product	A digital twin of a product is to link engineering and product lifecycle management with factory operation. It is engineered as part of the research and development process and helps to drive front loading in product development by making it possible to simulate and test the product at an early process stage.
4	Networked factory	It refers to the concept of connecting smart things such as resources, machines, transportation vehicles, or products through a connectivity layer for control and optimization purposes. Often leverages manufacturing execution systems integrated with an enterprise resource planning system.
5	Modular manufacturing resources	Uses modularized manufacturing resources instead of traditionally integral systems. Modular systems such as robots, autonomous guided vehicles (AGVs), and fixtures can be reconfigured to accommodate new manufacturing needs.
6	Flexible production methods	Uses digitized technologies to support the manufacturing processes for products with a high variety for high system flexibility.
7	Process visualization and automation	Such as mobile applications combined with virtual and augmented reality (AR) such as haptic devices or digital glasses to improve cooperation between people and machines via innovative user interfaces.
8	Integrated planning	From machine level, MES, ERP, to supply chain management including extended partners. Integrated planning allows an immediate reaction to changes in resources availability or demand.
9	Autonomous intra-plant logistics	Enterprises are capable of operating and performing logistics activities with minimized human intervention. These systems sense and process real-time information about digital or physical surroundings to navigate safely through indoor and outdoor environments, while simultaneously performing all necessary tasks. The technical solutions include mobile robots, autonomous guided vehicles, and aerial drones.
10	Predictive maintenance	Remote monitoring of dynamic conditions of machines with the help of sensor data and bid data analytics to predict maintenance and repair situations. This helps to increase resource availability and optimize maintenance efforts.
11	Big data-driven process/quality optimization	Helps to detect patterns in production or quality data and provide insights to optimize processes or product quality. Models ranges from pure statistical 'black box' models to expert and knowledge-based 'while box' approaches.

Table 12.2 (Continued)

No.	Technology	Explanation
12	Data-enabled resources optimization	Optimization of energy and resource consumption through intelligent data analyses and controls, e.g. energy or pressurized air management in facilities based on actual demand and supply.
13	Transfer of production parameters	Refers to a fully automated transfer of production parameters to other factories, i.e. to implement a lead plant concept where optimizations can be reproduced in other plants.
14	Fully autonomous digital factory	These can operate independently based on self-learning algorithms. People are required only for initial design and setup, and ongoing monitoring and exception handling. It applies to hazardous or remote production facilitates.
15	Track and trace	Things in the manufacturing system are tracked via sensors and integrated into a data platform connected to internal systems such as MES or ERP. It provides a full transparency for statuses of manufacturing resources and progression of productions for responsive decision-making supports.

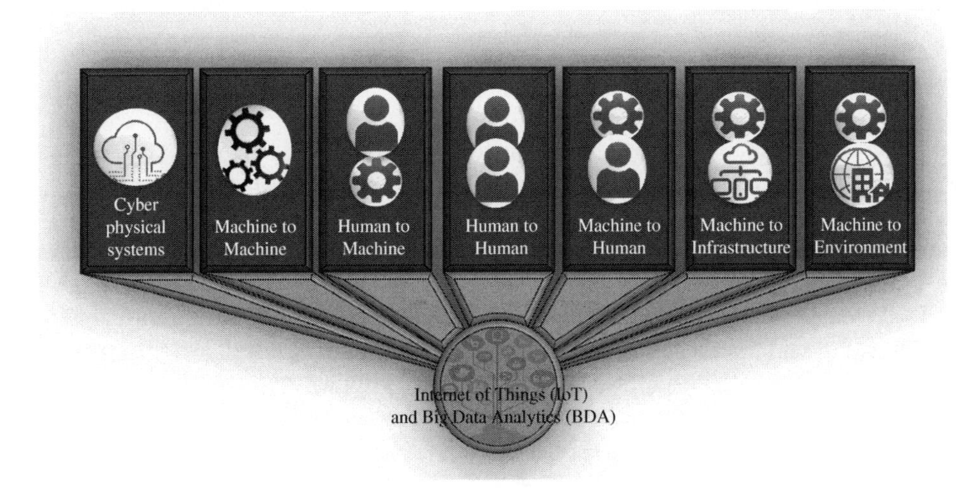

Figure 12.7 IoT and BDA make all types of interactions possible.

of IoT-based EA are (i) the *networks* that connect all of the things involved in product lifecycles, (ii) the *digitization* for data-driven business and the real-time visibility of all physical things in the digital world, and (iii) the *consumerization* for full personalization and customer experience of products. These characteristics of DM business models cannot be implemented without the new development of IT technologies, such as *in-memory computing*, *BDA*, *machine learning*, *artificial intelligence*, *cloud scalability*, *ubiquitous connectivity*, *edge computing*, *workflows*, and *blockchain technologies* (*BCT*) (Viriyasitayat et al. 2018, 2019).

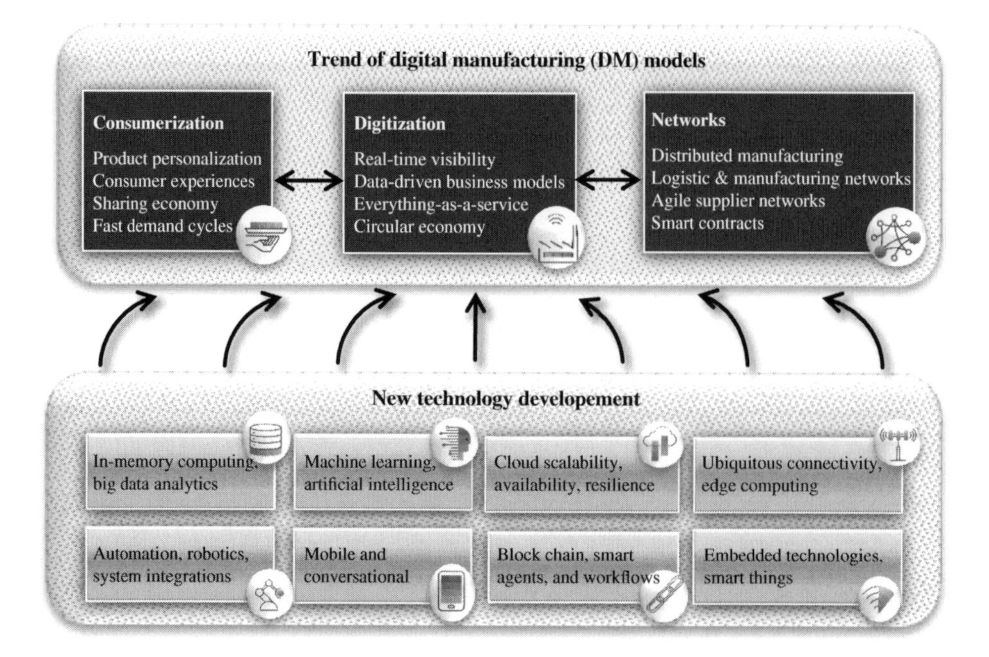

Figure 12.8 Trend of manufacturing business model versus IT development (Kawasmi 2018).

12.5.5 Lifecycle and Evolution of EA

An EA represents system components, their relations, and the evaluation of system elements with respect to time. Similar to a product, product family, or a process, an EA has its own lifecycle. Moreover, EA has evolved gradually within its lifecycle to adopt significant changes of products, processes, market needs, and information infrastructure. Note that the primary goals of an EA are to reduce the response time to market changes, trade-off analysis, redirection of strategic plans, and making tactical reactions. The EA must remain updated to reflect the reality of manufacturing enterprises, the organizations, and business environments. Maintaining an EA is as important as developing an EA. To sustain an EA, the enterprise or organization needs to oversee the EA with independent verification and assess the alignment of the EA with ever-changing business practices, funding profiles, and new technologies.

As shown in Figure 12.9, if the EA is not current, it will limit the capabilities of the manufacturing enterprise to meet its business goals and mission. On a regular time base, it will be necessary to revisit the vision of the enterprise and update it when it is needed. The current or baseline architecture must be ensured to reflect the statuses of system elements and their relations appropriately. When the predicted changes of any component in architecture are significant enough, the sequencing plans must be made to reflect the prevailing priorities of enterprise and resources that will be available. The plans should determine all of the updates to the EA and corresponding changes in respective projects. Sequentially, the EA is evolved to a target architecture that will reflect the updated business vision and adopt new technologies of the enterprise since the last release (Gao 2000).

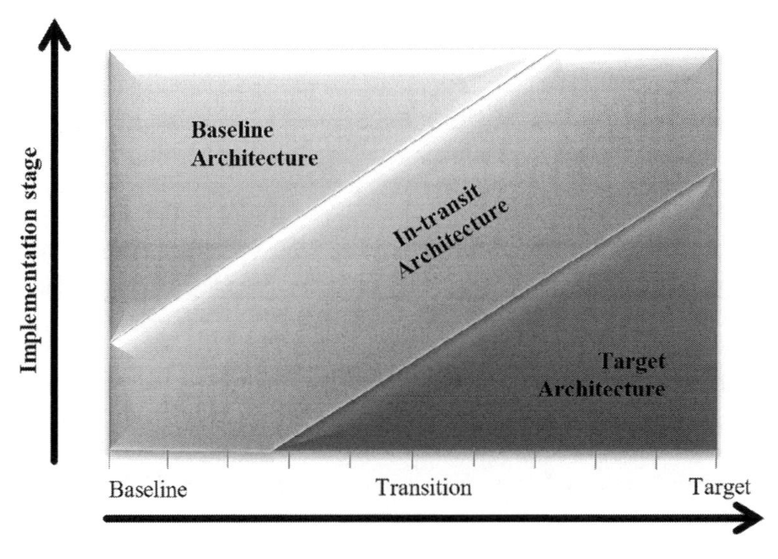

Figure 12.9 Transition of EA in its lifecycle (Gao 2000).

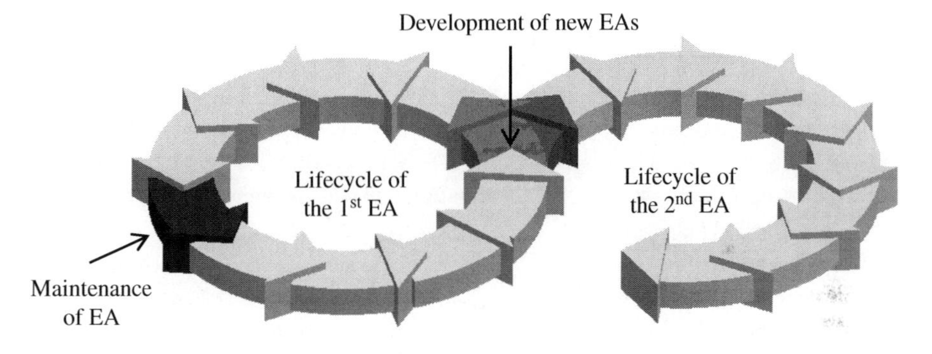

Figure 12.10 Maintenance of EA in its lifecycle.

As shown in Figure 12.10, continuous evolution occurs with EA in its lifecycle until part of the unchanged elements in a baseline architecture are not the majority of all elements in the target architecture. In such a case, the lifecycle of the EA should be terminated and the new architecture should be defined.

12.6 Hardware Solutions

The hardware and software development of IT has been a major driver for manufacturing enterprises to pursue competitiveness in the global market. In return, the applications in manufacturing are demanding more and more computing capacities to deal with the increasing complexity and dynamics of manufacturing businesses. The functional requirements discussed in Section 12.3 pose challenges for computer hardware and software systems to meet the real-time needs of decision-making supports in DM. The implementation of DM systems should be powered by the rapid advancement of information technologies. Enterprises should use sophisticated computer technologies such as high-performance computing (HPC) to deal with big data and obtain design solutions at shortened lead-times and reduced cost.

Big data poses significant challenges to both BDA and computing capability of DM as traditional workstations might not have insufficient capacities to deal with the workloads in DM and those workstations might not be scalable to larger systems. Computer platforms might reach their maximum capacities to deal with large workloads. There is a need for developing high-performance computer aided systems to deal with rapid data growth and increased workload demands.

At the device level, *sensors* and *devices* networked in the IIoT can provide enterprises with real-time data about all manufacturing resources. Data collected from smart things are integrated with other unstructured business data from a product lifecycle. Advanced computer tools such as BDA and artificial intelligence (AI) can be powered by HPCs, such as Dell EMC (Table 12.3), to analyse data and support decision-making activities. HPC-powered analytics and AI are revolutionizing engineering to help manufacturers speed the time to market with more innovative and higher quality products.

12.7 Big Data Analytics (BDA)

As an information system, DM is functioned to process and convert data such as the information of products, processes, machines and tools, design intents, and the decisions in various business operations. The system complexity depends on the amount and types of data, numbers of inputs and outputs, their correlations, as well as their dynamic characteristics over time. The DM performance can be measured by the system capabilities in dealing with the volume, variety, and velocity of data and its responsiveness in making appropriate decisions for system operations. Manufacturing enterprises today are facing unprecedented challenges of complexity and growth, since the integration of advanced technologies in DM makes it possible to develop a *digital factory*. A digital factory is an industrial facility where all products, personnel, raw materials, machines, and processes are connected, and where abundant real-time data is collected, processed, and utilized to optimize production operations and achieve higher levels of efficiency continuously (SME 2018).

The flatten EA in Section 12.5 has ruled out the simplification methods where the system elements, activities, interactions, communications, processes, and relevant data are structured. Therefore, DM is much more complex than traditional CIM in the sense that (i) a large number of system elements are included, (ii) the communications of system elements are not structured by layers, grids, or their combinations, (iii) the boundaries of manufacturing enterprises are vague, and manufacturing businesses involve more and more changes and dynamics, (iv) IoT networks everything in the Internet that continuously generates a large amount of data in the real-time mode, and (v) enterprises have higher functional requirements from DM (Bi and Cochran 2015).

Table 12.3 Technical specifications of Dell EMC Ready Solutions (Dell 2019).

Servers or processors	Head node	PowerEdge R640	
	Compute nodes	Choice of: PowerEdge R640, PowerEdge C6420, PowerEdge R840	
	Processors	Intel® Xeon® 8200, 8100, 6200, 6100, 5200, 5100, 4200, 4100, 3200, and 3100 series	Intel Xeon SKL-F only on C6420 Intel Xeon E7–4800 v3, E7–8800 v4
Operating systems	Head nodes	RHEL (2- or 4-socket)	
	Compute nodes	RHEL for HPC Compute Node (2- or 4-socket)	
Software		OpenHPC	
Networking			
OPA (Open-source programming)	OPA HFI	Intel Omni-Path Host Fabric Interface Adapter 100 Series 1 Port PCIe x16	
	OPA switches	Dell EMC Networking H1000 Edge series: H1048 and H1024 Dell EMC Networking H9100 series	
	OPA IFS drive	10.9	
IB	IB host channel adapters	Rack: Mellanox ConnectX-5 EDR single port or Mellanox ConnectX-3 FDR	Blade: Mellanox ConnectX-3 SFF: FDR dual port or FDR10 mezzanine cards
	IB switches: FDR and RDR	Rack: Mellanox SwitchX 6xxx series Mellanox MSB 78xx series	Blade: Mellanox M4001F (supported on Mellanox SB 77xx and 78xx series M640 blades)
	Drivers	Mellanox OFED 4.5	
Ethernet	NICs	1, 10, 40GbE (full and low profile)	
	Dell EMC Networking	Z and S series, 1, 10, 40GbE	
Storage	NFS	Dell EMC Ready Solutions for HPC NFS Storage	
	Lustre	Dell EMC Ready Solutions for HPC Lustre Storage	
	SAS RAID controller	PERC 10	
System management		Dell EMC Deployment Toolkit (DTK) Dell EMC OpenManage (OM)	

12.7.1 Big Data in DM

As shown in Figure 12.11, a DM system can be analogous to a manufacturing system. The acquired data can be viewed as ‘raw materials’ and the system outputs are knowledge, wisdom, or decisions.

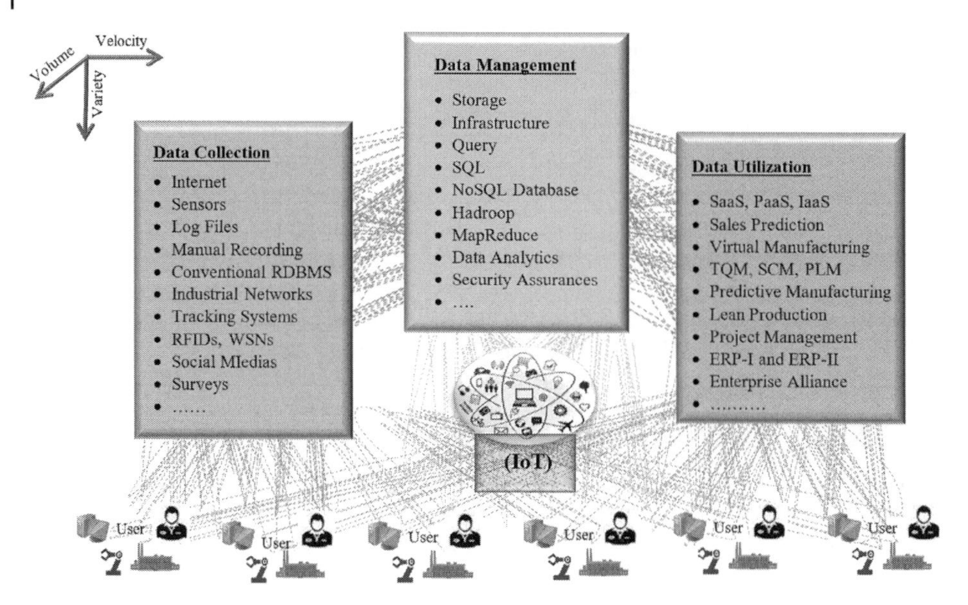

Figure 12.11 Big data in digital manufacturing.

1. For *data collection*, IoT makes it possible for the cloud to acquire all the necessary data from different sources, such as portable devices, Websites, sensors, personal assisted devices, conventional Relational Database Management Systems (RDBMSs), and tracking systems. As shown in Figure 12.12 (Baumers and Ozcan 2016), with the need to incorporate more and more manufacturing businesses over the product lifecycle in a manufacturing enterprise, the volume of data is growing consistent with Moore's law.
2. For *data management*, cloud computing may offer reliable services by deploying cloud data centres. Some platform technologies, such as MapReduce and NoSQL, are needed to tackle big data and retrieve relevant data effectively.
3. For *data utilization*, different tools have been developed to analyse the retrieved data and extract knowledge to support decision-making activities. System elements in DM get their products (knowledge, wisdom, or decision) via services. Users are able to access data and data-processing tools from a cloud anywhere and anytime when it is needed. The cloud includes internet-based data access to affordable computing applications.

For the data stored in the cloud, it is operated in *write once and read many* (WORM), which is different from *write many and read many* (WMRM) in the traditional storage media. Therefore, data in the cloud can be accumulated rapidly as big data, in particular, when there are billions of things feeding data to the cloud. The concept of *Big Data* (BD) is about the characteristics of the datasets as well as the methodologies to process data. There are many definitions of BD, but most of the definitions refer to the emerging technologies used to capture, aggregate, and process an ever-greater volume, velocity, and variety of data. BD refers to the data whose *volume*, *velocity*, and *variety* exceed the capability of an organization to process and analyse it in a timely manner. Note that big data not only includes big volume, high velocity, and variety, but also includes big dimensionality. *Big dimensionality* refers to the explosion of features or variables that bring new challenges to data analytics.

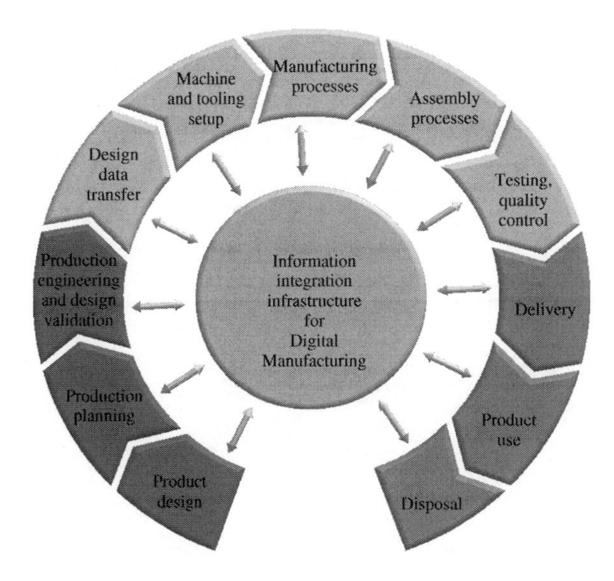

Figure 12.12 Increased data from expansion of manufacturing businesses over a product lifecycle.

BD implies fine-grained data that can describe the same event of something much more precisely to support better decision-making.

12.7.2 Big Data Analytics (BDA)

Companies are deluged with data, from *terabytes* (2^{40} bytes) to *petabytes* (2^{50} bytes) and to *exabytes* (2^{60} bytes). The organizations have gone from accumulating *structured* to *unstructured* data and from *static* to *dynamic* data in paradigm shifts. It becomes critical to process data, mine data, and convert data into information and knowledge for decision-making activities. BDA is used to analyse and mine big data to produce operational and business knowledge at an unprecedented scale and specificity (Cloud Security Alliance 2013). When BDA is applied in DM, it should be capable of utilizing big data and make faster and better decisions to respond to changes and uncertainties in the enterprise. The principles for BDA are applicable to many organizations, regardless of the size or business domains.

BDA is not just a technology; it is an integral toolset for the managements of strategies, marketing, and human resources. BDA is a set of well-established and widely used analytical methodologies and tools, such as correlations, cluster analysis, filtering, decision trees, Bayesian analysis, neural network analysis's, regression analysis, and textural analysis. The rapid development of newer technologies has cut the data processing time significantly, and makes it possible to deal with large data that would not have been feasible a few years ago.

12.7.3 Big Data Analytics (BDA) for Digital Manufacturing

Figure 12.13 shows the respective goals of information systems at six development stages of manufacturing systems: *computerization*, *connectivity*, *visibility*, *transparency*, *predictivity*, and *adaptability*. The digitization of manufacturing system begins with (i) using digital

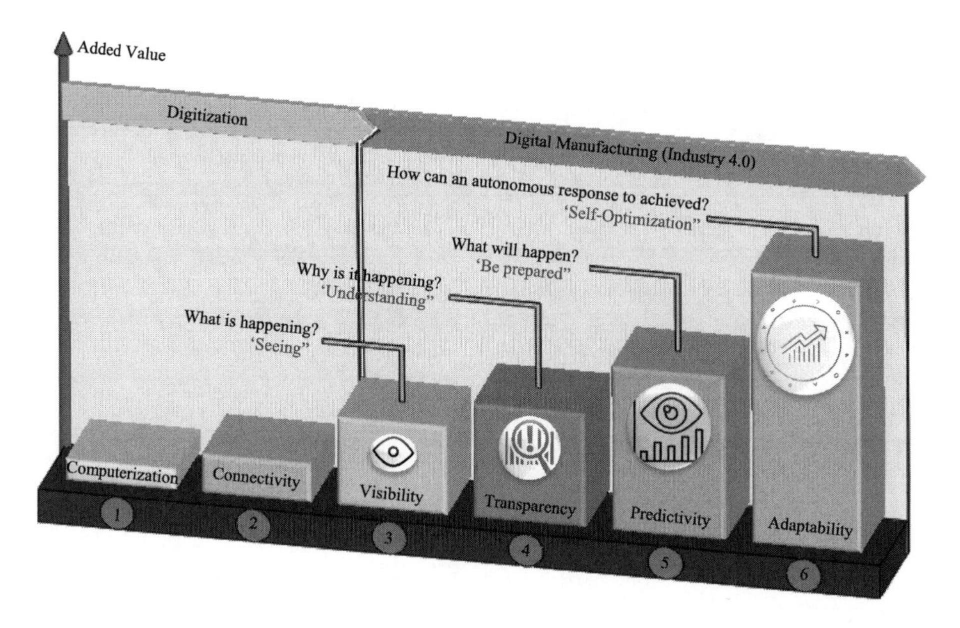

Figure 12.13 Requirements of data analysis of DM (KPMG 2018).

computers to collect data, (ii) using networking computer resources, (iii) gaining an understanding of the states of manufacturing resources and processes, (iv) anticipating and predicting whether for machine fault modes or changes in demand will affect orders and thus production levels, and finally to (v) self-optimizing and reconfiguring systems to deal with the changes and uncertainties autonomously.

BDA contributes greatly to support such an evolution and to predict what will happen on markets as well as manufacturing resources. The capabilities of BDA tools are not just to upgrade and expand legacy systems and algorithms, but require a set of more advanced tools that can be used to select relevant data and convert it into useful knowledge from big data promptly in order to support real-time decision-making processes (SAS 2014).

12.8 Computer Simulation in DM – Simio

An information unit in DM serves as a transformer to convert inputs to outputs. The relations of inputs and outputs must be modelled to determine outputs based on inputs. *Analytical*, *numerical*, or *simulation* models are used to describe the relations of input and output variables and system parameters. Due to the variety, complexity, the scale, uncertainties, and changes of manufacturing systems, *computer simulations* are the most vital and practical methods used for virtual design and optimization of manufacturing systems. In this section, a simulation tool for the performance prediction of DEDSs are discussed as an application example of computer simulation.

Simio is a multiparadigm modelling tool that combines the simplicity of objects with the flexibility of processes to provide a rapid modelling capability without requiring programming (Simio 2019a). Simio promotes a modelling paradigm shift from process orientation to an object orientation. The objects in Simio are defined by modellers. Simio is object-based but supports multiple modelling paradigms including event, process, object, and agent-based modelling.

12.8.1 Modelling Paradigms

Graphical modelling and animation are essential features in simulation models. *A graphical approach* simplifies process modelling and *a graphical animation* helps to understand, visualize, and validate simulation results dramatically. Most manufacturing systems are DEDSs whose state evolution depends on the occurrence of asynchronous discrete events over time. Early simulation tools such as Simulation Programming Language (Simscript) by Markowitz et al. (2019) and GASP (Hooper and Reilly 1983) for DEDSs are event orientated; in the simulation model, the system is viewed as a series of events that change the states of system continuously. *Events* are defined to change the *states* of a system that occur when events take place. From a process perspective, a *process flow* can be defined for the movement of system elements as a series of steps; at each step, the system state changes. An event-based or process-oriented method is efficient and flexible, but lacks details of system elements.

Integrating object-oriented modelling with event-based simulation not only preserves the efficiency and flexibility of event-based simulation but also allows defining all details of system elements in objects. *Object-oriented modelling* (OOM) is the construction of objects using a collection of objects that contain stored values of the instance where variables were found within an object. Using OOM, a system is viewed as a set of objects used to represent system elements, where system elements have their attributes and states, and events correspond to state changes of system elements. Taking an example of modelling a factory, manufacturing resources such as workers, machines, conveyors, robots, and other objects constitute the system and the events are the interactions between system elements.

Simio develops simulation models by defining *intelligent objects*. An intelligent object can be defined for any system element such as a machine, robot, aeroplane, customer, doctor, tank, bus, ship, or other comprehensive entities with certain attributes and states. An object may be animated in 3D to reflect the changing state of the object. Users can easily define intelligent objects for use in building hierarchical models. Once intelligent objects are created, they can be stored in libraries and used in any system model. The software tool also includes a set of pre-built objects in design libraries for users. As a digital twin of a physical system, a Simio model consists of objects that represent physical components of a system.

12.8.2 Object Types and Classes

To simulate a DEDS, the components of interest are defined as intelligent objects in a Simio model. Simio provides a framework for building custom objects. At the same time, Simio includes a standard object library with the most commonly used types of objects, listed in Table 12.4 (Simio 2019b).

Objects in Table 12.4 are classified into six basic object classes, shown in Figure 12.14 (Pegden 2012). In the classification, a *fixed* object is used to represent a system element that does not change its location; it is fixed in the system. Stationary equipment such as machines,

Table 12.4 Fifteen object types in Simio object library (Simio 2019b).

No.	Object	Description
1	Source:	Creates entities that arrive at the system.
2	Sink:	Destroys entities and records statistics.
3	Server:	Models a multichannel service process with input/output queues.
4	Combiner:	Combines entities in batches.
5	Separator:	Separates entities from batches.
6	Workstation:	Models a 3-phase workstation with setup, processing, and teardown.
7	Resource:	Models a resource that can be used by other objects.
8	Vehicle:	Carries entities between fixed objects.
9	Worker:	Move entities between fixed objects.
10	BasicNode:	A simple intersection of links.
11	TransferNode:	A simple intersection of links.
12	Connector:	A zero-time connection between two nodes.
13	Path:	A pathway between two nodes where entities travel based on speed.
14	TimePath:	A pathway with a specified travel time.
15	Conveyer:	An accumulating/non-accumulating conveyor device.

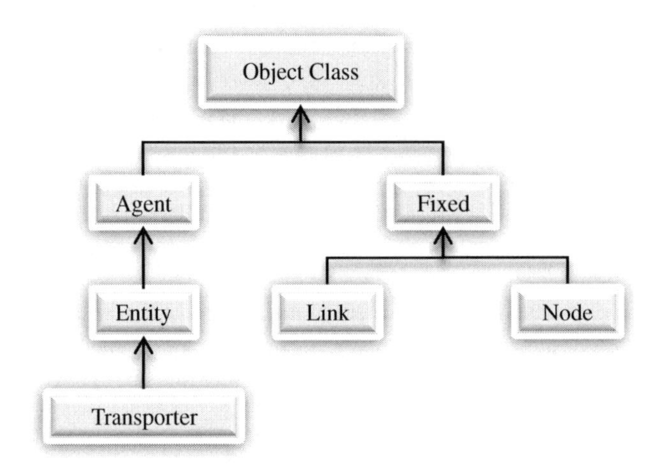

Figure 12.14 Basic classes of Simio objects (Pegden 2012).

fuelling, and stations are defined as fixed objects. Object types of *Source*, *Sink*, *Server*, *Combiner*, *Separator*, *Workstation*, and *Resource* in Table 12.4 are fixed objects. *Agents* are the objects that can move freely in space, but agents also include abstract and independently acting objects for agent-based systems. *Entities* are instantiated agents that can move from object to object in a model. Examples of entity objects are customers in a service system and work pieces in a manufacturing system. An entity has its intelligence to control its behaviour. Link and node objects are used to build networks over which entities may flow. A *link object* defines a pathway for entity movement between objects. A *node object* defines a starting or ending point for a link. Behaviours can be defined in link objects to model unconstrained or congested traffic flows, or complex material handling systems. *Connector*, *Path*, *TimePath*, and *Conveyor* in Table 12.4 are link objects. *Basic Node* and *Transfer Node* are node objects. The *Transporter* is the final class of object derived from the entity class. A transporter object is an entity that has the added capability to pick up, carry, and drop off one or more entities. A transporter object can be used to model an AGV, crane, or forklift truck. *Vehicle* and *Worker* objects fall into the transporter object class. Simio allows a wide range of object behaviours to be created from these six basic classes without coding.

Each object has the properties that control its behaviour; e.g. the Source has an interarrival time, the Path has a property to control entity passing, and the Server has a processing time. The *intelligence of an object* defines the object behaviour as a collection of processes driven by events.

12.8.3 Intelligence – Objects, Events, Logic, Processes, Process Steps, and Elements

The Simio simulation environment offers the default setting to show only commonly used properties by clicking the checkbox at the top of the properties window. The display is limited to the set of keep properties for the basic behaviour of an object; other supported features such as failures, state assignments, secondary resource allocations, financials, and custom add-on process logic are not displayed. If needed, the full set of properties of objects can be shown when this option is disabled.

A model is defined by defining the set of objects and their connections. To build a simulation model, objects are dragged from the library and placed in the *Facility View*. *Fixed objects* are defined for fixed machines, agents, entities, links, nodes, and transporters, and intelligence is added to objects as one or more processes. A *process* associated with an object is defined as the *logic*, which is executed in response to events. Each process is a sequence of *process steps* that is triggered by an event and is executed by a *token*. A token is released at the start of the process and is simply a thread of execution. A token may have the properties such as *input parameters* and *states* to control the execution of the process steps. Tokens can be customized to combine different properties and states. In Simio, six basic classes of objects come with the set of events that are automatically triggered when the process steps are available; accordingly, state changes occur in response to these events. A *step* refers to a simple process such as holding the token for a time delay, holding or releasing a resource, waiting for an event to occur, assigning a new value to a state, or deciding between alternate flow paths. Some general-purpose steps such as *Delay* are used to model objects, links, entities, transporters, agents, and groups. Other steps are specific to certain objects, such as the *Pickup* and *Dropoff* steps for adding intelligence to *transporters* and the *Engage* and *Disengage* steps for adding intelligence to *Links*. Each object class has its own set of *events*. For example, a link object has events that fire when entities enter and leave the link object, merge fully on to the link, and collide with or separate from other entities that reside on the link. By providing the model logic to respond to these events, the movements of entities across the link are completely controlled (Pegden 2012).

Process steps are used to define the underlying logic for an object. Process steps are stateless: they have input properties but no output responses or outputs. This implies that an arbitrary number of object instances can share the process steps. If the process logic is changed, this fact is reflected automatically by all object instances of the object. The states for an object instance are held in elements. *Elements* define the dynamic components of an object and may have both input and output properties. Within an object, the tokens may execute the steps that change the states of the elements that are owned by the object (Pegden 2012).

12.8.4 Case Study of Modelling and Simulation in Simio

In this section, Simio is used to develop a simulation model to evaluate the utilization rate of 3D printers at a 3D printing lab at Purdue University, Fort Wayne, Indiana, the United States.

The lab is equipped with four 3D printers, i.e. Dreamer-Flashforge, SeeMeCNC, Dreamel, and Dreamel digiLab. The lab mainly supports students' printing requests for their course projects in *solid modelling*, *CAD/CAM applications*, *senior design projects*, and *course projects* in other mechanical courses. Generally, each student is required to create and complete one printing job. The complexity of a printing job varies from one course to another.

A simplified simulation model is expected to evaluate whether available resources are sufficient to meet students' project needs in producing their parts. Since each student is required to complete one 3D printing job, the frequency of 3D printing orders from a certain class is proportional to the average enrolments in each class in a specified period of time. Therefore, the needs of printing jobs are assumed in Table 12.5.

The procedure to create and run a simulation model for the above problem is described as follows:

1. The Simio software is used to develop the simulation model. In defining objects and links for the model, note that the Simio simulation environment offers default settings to show only *commonly used properties* by clicking the checkbox at the top of the properties window. The display is limited to the set of keep properties for the basic behaviour of an object; other supported features such as failures, state assignments, secondary resource allocations, financials, and custom add-on process logic are not displayed. If needed, the

Table 12.5 Predicted needs of printing jobs for students in different courses at the lab.

Courses	Solid modelling (ME160)	CAD/CAM applications (ME546)	Senior design (ME487)	Other courses
Average enrolments in semester	24	20	15	30% of 150
Probability (%)	22.86	19.05	14.29	42.86
Printing time (hours)	(1, 3)	(5, 10)	(5, 30)	(5, 10)

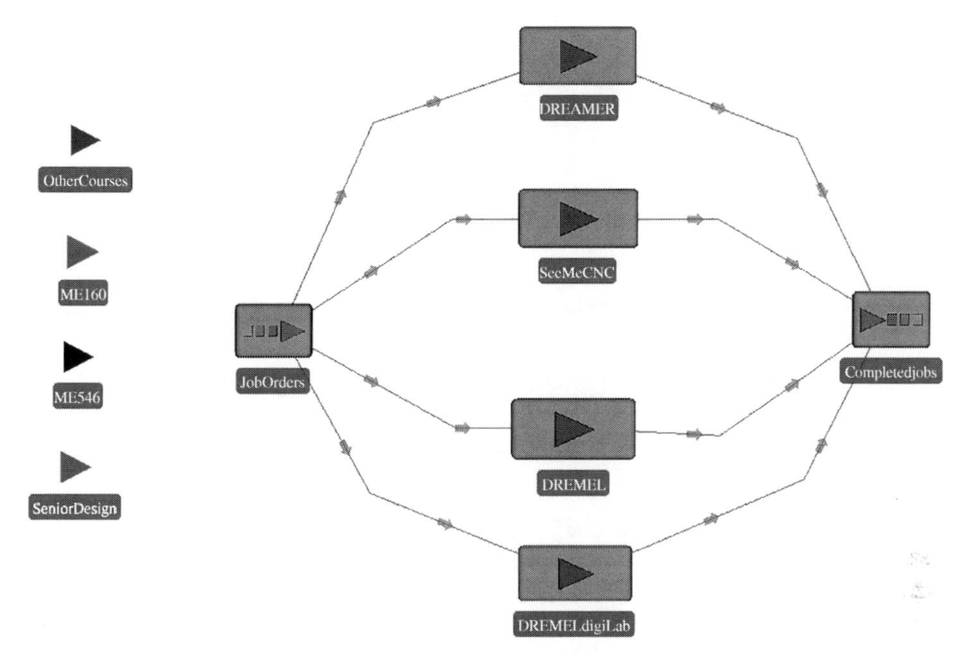

Figure 12.15 Objects in the simulation model for a 3D printing lab.

full set of properties of objects can be shown when this option is disabled. Only basic objects in the standard library are sufficient to model the scheduling problem.

2. As shown in Figure 12.15, six objects are created to represent *JobOrders* from students, four *3DPrinters*, and *JobCompletions*. Among these objects, JobOrders and JobCompletions are defined as *Source* and *Sink* objects, respectively, and 3D printers are defined as *Server* objects. Figure 12.16a shows the interface in creating a source object for JobOrders. Figure 12.16b defines a *data* file based on the information specified in Table 12.5. Figure 12.17a sets up the properties of a *Source* object where *the capacity* is set as 10 for the longest job lines and the *interarrival time* is set as random (2, 6), which means that a new order arrives randomly in two to six hours. In *add-on Process Triggers*, the *creating entities* is set as *JobCreate*. Figure 12.17b sets up the properties of a *Server* object where the *Ranking Rule* in the *Process Logic* is set as First In First Out; the *Process Time* in the *Process Logic* is set based on the date in *Processing Time* in the *CourseJobs* table; the *Capacity* of the *Input Buffer* in the *Buffer Logic* is set as 0 for the no waiting line; and the *creating entities* in *add-on Process Triggers* is still set as *JobCreate*.
3. The objects and connections as well as their process logics are now fully defined; the next step is to set up simulation parameters such as the *Starting Time* and the *Duration* of simulation. Figure 12.18 shows the interface of Simio software to enable users to specify simulation parameters. It shows the many tools available for users to review, visualize, and debug the process and data during the simulation.
4. Once the simulation parameters are set appropriately, users just need to click *Run* to activate the simulation. Figure 12.19 shows the snapshot of the simulation in three-dimensional space. After the simulation is completed, users can retrieve any data generated from the simulation. Figure 12.20 gives the result from the developed

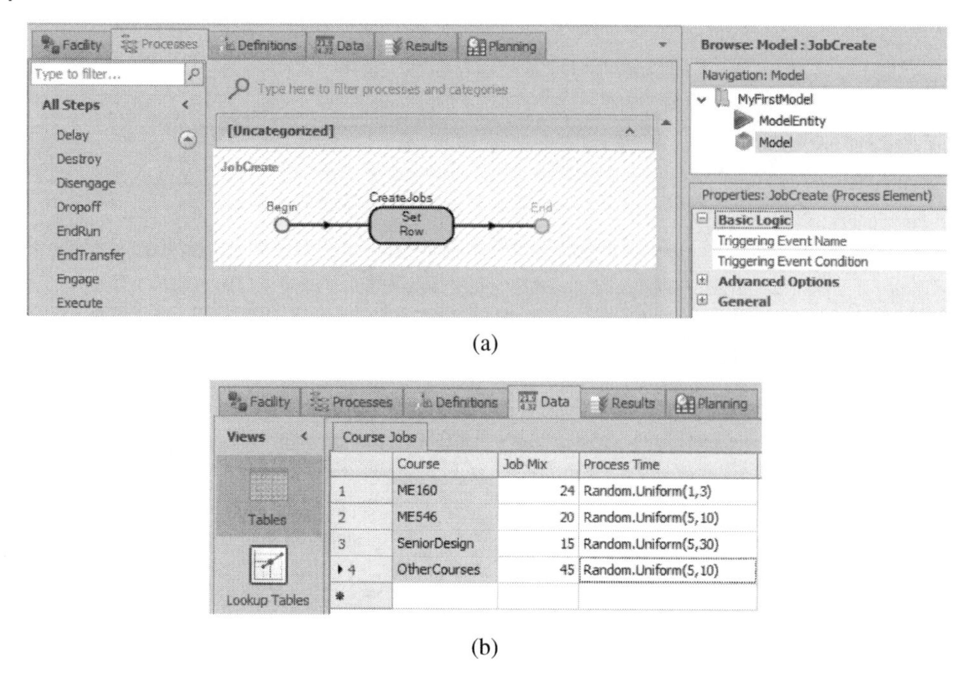

Figure 12.16 Definition of JobOrders in modelling of simulation model. (a) Creating *JobOrders* as a *Source* object. (b) Defining a date file called *CourseJobs* based on the needs of printing jobs in Table 12.5.

simulation model. It shows that four 3D printers can meet the needs of engineering students at the department satisfactorily.

12.9 Summary

Digital transformation is affecting business models, production processes, and enterprise organization. Digitizing technologies are helping enterprises to gain competitiveness in a highly turbulent business environment (Castelo-Branco et al. 2019; Kutin et al. 2018). DM is an integrated computer solution to design, manage, and operate manufacturing systems over entire product lifecycles.

In contrast to its origin concepts, such as digitized manufacturing or CIM, DM is far beyond the collection of computer-aided tools in respective domains and levels. DM is equipped with advanced decision-making tools to process, utilize, mine data for the right information, and ensure that the right data is delivered from the right source to the right place at the right time. The ultimate promise of DM is the seamless integration of all enabling tools for designing, manufacturing, assembling, using, and eventually disposing of products.

The complexity of a manufacturing system can be measured by entropy. The higher the entropy value is, the more complex the system is, the more difficult the system control is, but the higher the flexibility and adaptability of the system. Traditional static and hierarchical

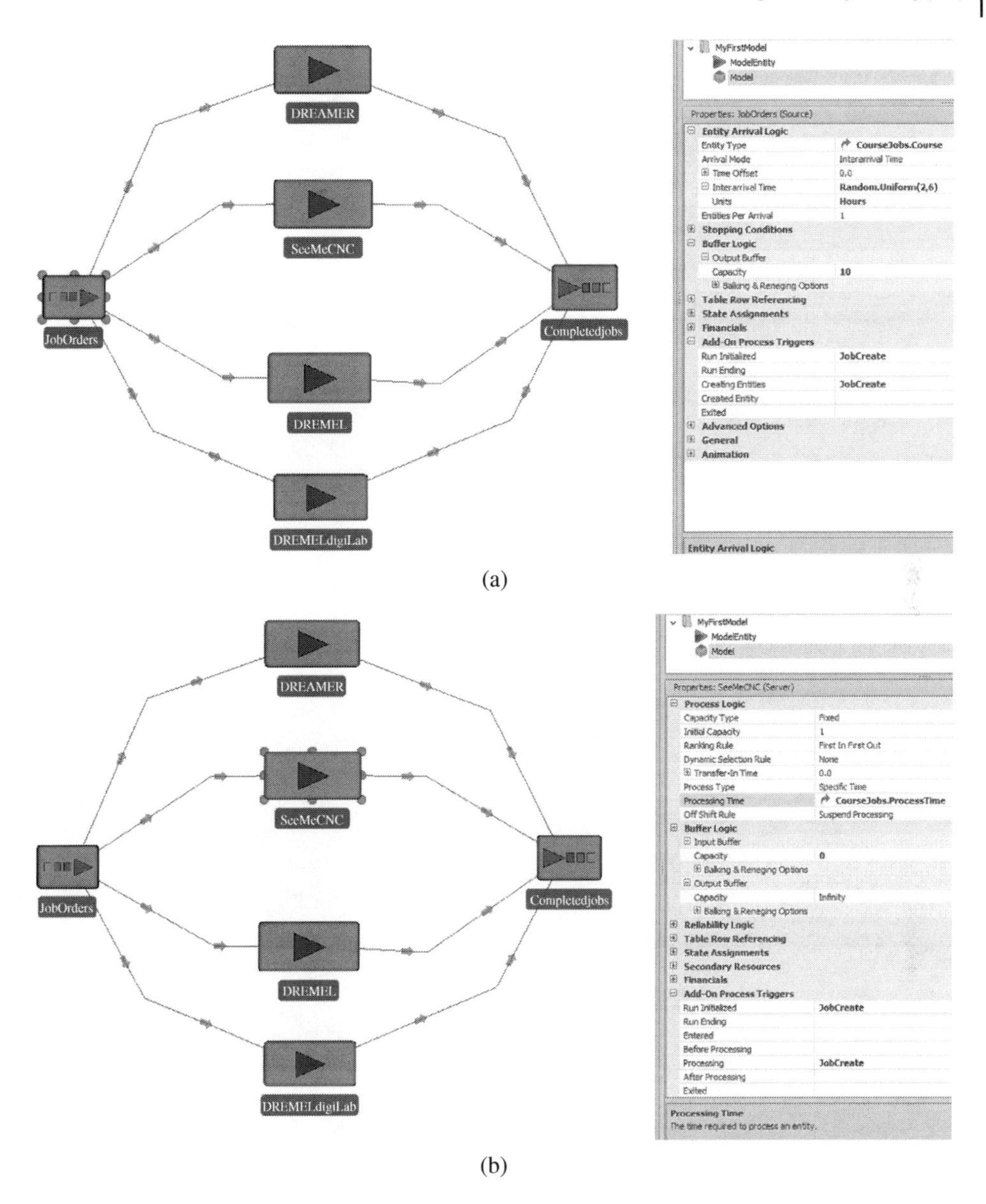

Figure 12.17 Defining properties of objects in modelling. (a) Properties of *Source* object (right). (b) Properties of *Server* object (right).

enterprise architecture (EA) reduces the system entropy and simplifies system organization and operation. However, this sacrifices system flexibility and adaptability to deal with changes and uncertainties in a prompt manner. DM should use the flatten EA to maximize the flexibility and adaptability of manufacturing systems.

Using the flatten EA for a system with a large number of system elements and interactions subjected to changes and uncertainties requires advanced computer tools to deal with ever-increasing large amounts of data to support decision-making activities at various

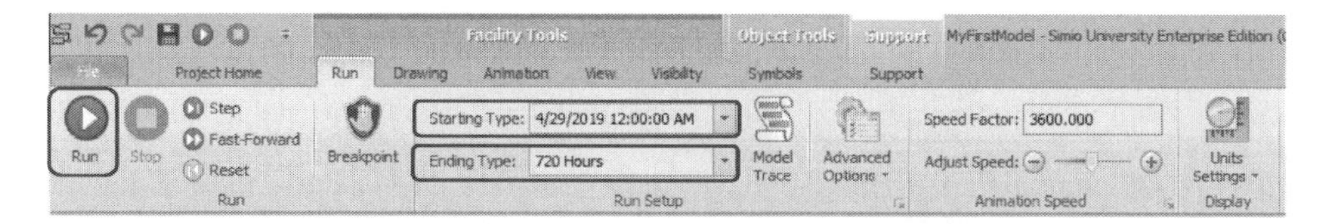

Figure 12.18 The interface to set up simulation parameters.

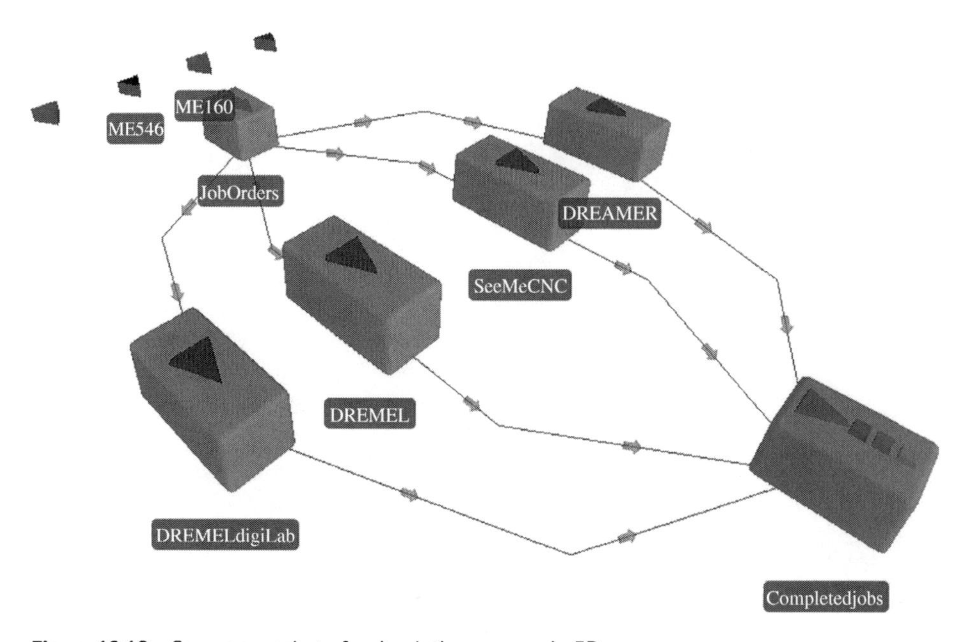

Figure 12.19 Screen snapshot of a simulation process in 3D space.

ScheduledUtilization - Percent			
Object Name	**Data Source**	**Category**	**Value**
DREAMER	[Resource]	Capacity	39.47169
DREMEL	[Resource]	Capacity	24.14239
DREMELdigiLab	[Resource]	Capacity	30.20359
SeeMeCNC	[Resource]	Capacity	82.28995

Figure 12.20 Statistic simulation results.

domains and levels. BDA becomes an essential component of the DM solution. Engineers should gain an understanding of available cutting-edge computing tools, so that they can be used effectively to achieve the visibility, transparency, predictivity, and adaptivity of manufacturing systems.

12.10 Design Projects

12.1 Figure 12.21 shows that Delmia/QUEST was used to create a simulation model to evaluate the performance of the communication protocol of Ring Extrema Determination (RED) (Bi and Wang 2010). Take the reference to learn the details of the RED algorithm, specify your criteria to evaluation criteria, and create a simulation model in Simio using RED as the communication logic, and give your conclusion about the performance of RED.

12.2 At an entertainment facility, passengers arrive at the rate of 9.6/minute and wait in a line. Every 10 minutes, up to 100 passengers move from the line to a room to watch a seven-minute training video; the actual time is randomly distributed in six to eight minutes. After the movie, the passengers line up and are loaded into the ride. Two passengers can load at a time and loading takes between 8 and 10 seconds

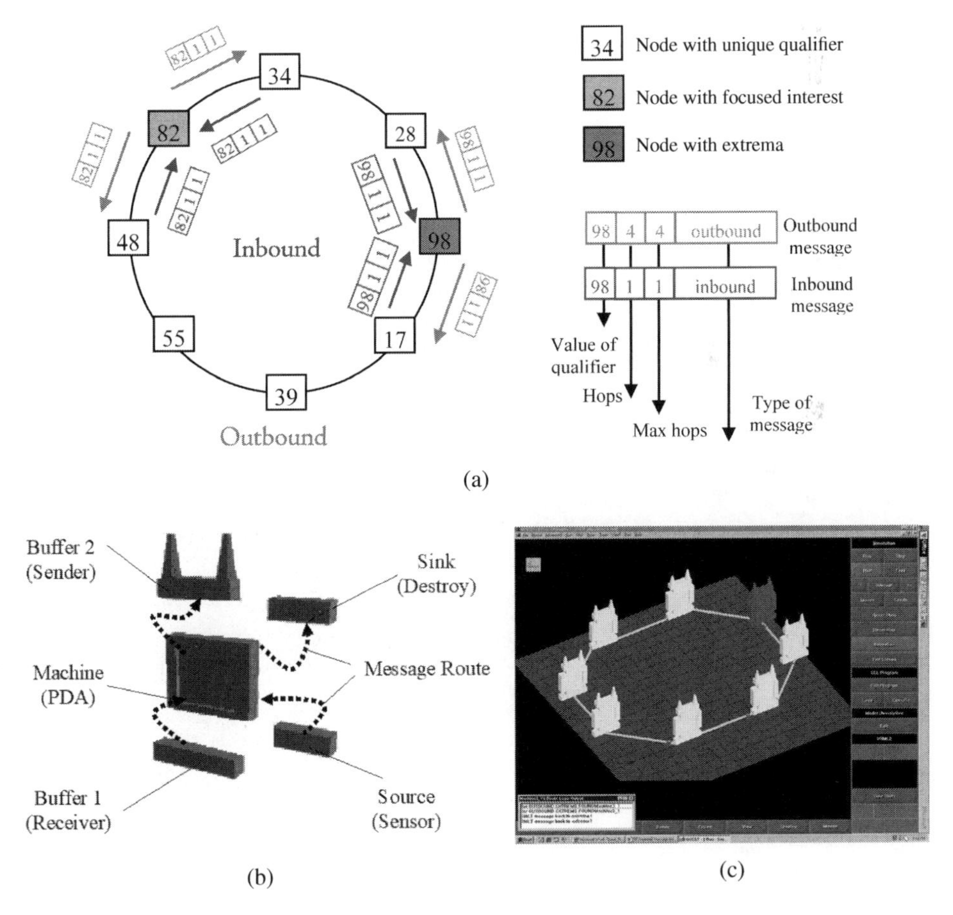

Figure 12.21 DEDS modelling example for Design Project 12.1. (a) Illustration of the Ring Extrema Determination (RED) algorithm. (b) Modelling of a wireless router. (c) Simulation for a message delivery to a designated node.

per passenger. After loading, the passengers start the ride. The ride has a capacity of 50 passengers and takes five minutes. The first video should start 15 minutes after the park opens. It is assumed that passengers start arriving immediately.

Create and run a simulation model for 10 replications for a duration of 160 hours in a numerical experiment; find out (i) the average group size, (ii) load process utilization, (iii) ride process utilization, and (iv) the waiting time from when a passenger arrives until the passenger starts the load process (Smith 2013).

References

Adeptia (2017). Islands of automation. https://adeptia.com/terms/blog/islands-automation.

Baumers, M., and Ozcan, E. (2016). Scope for machine learning in digital manufacturing. https://arxiv.org/ftp/arxiv/papers/1609/1609.05835.pdf.

Berger, R. (2019). Digital factories, the renaissance of the U.S. automotive industry. http://secure.rbj.net/collaborative-design-and-planning-for-digital-manufacturing-1st-edition.pdf.

Berman, S.J., and Bell, R. (2007). Digital transformation – creating new business models where digital meets physical. https://www-07.ibm.com/sg/manufacturing/pdf/manufacturing/Digital-transformation.pdf.

Bi, Z.M. and Cochran, D. (2014). Big data analytics with applications. *Journal of Management Analytics* 1 (4): 249–265.

Bi, Z. M., Cochran, D. (2015). Big Data Analytics with Applications, *Journal of Management Analytics*, 1 (4): 249–265.

Bi, Z.M. and Wang, L. (2010). Visualization and verification of communication protocols for networked distribute systems. In: *Enterprise Networks and Logistics for Agile Manufacturing*, 333–357. Berlin, ISBN-10: 184996243X, ISBN-13:978-1849962438: Springer.

Bi, Z.M. and Wang, L. (2012). Energy modeling of machine tool for optimization of machine setup. *IEEE Transactions on Automation Science and Engineering* 9 (3): 607–613.

Bi, Z.M., Lang, S.Y.T., Shen, W., and Wang, L. (2007). Reconfigurable manufacturing: the state of the art. *International Journal of Production Research* 46 (4): 967–992.

Bi, Z.M., Xu, L.D., and Wang, C. (2014). Internet of things for enterprise systems of modern manufacturing. *IEEE Transactions on Industrial Informatics* 10 (2): 1537–1546.

Buttner, R. and Muller, E. (2018). Changeability of manufacturing companies in the context of digitalization. *Procedia Manufacturing* 17: 539–546.

Castelo-Branco, I., Cruz-Jesus, F., and Oliveira, T. (2019). Assessing industry 4.0 readiness in manufacturing: evidence for the European Union. *Computers in Industry* 107: 22–32.

Clifford, A. (2006). Islands of automation. https://it.toolbox.com/blogs/andrewclifford/islands-of-automation-112706.

Cloud Security Alliance (2013). Big data analytics for security intelligence. https://downloads.cloudsecurityalliance.org/initiatives/bdwg/Big_Data_Analytics_for_Security_Intelligence.pdf.

Dell (2019). Dell EMC ready bundle for HPC digital manufacturing: simplify and speed design with a building-block approach to HPC. https://www.dellemc.com/en-us/collaterals/

unauth/offering-overview-documents/ready-bundle-for-hpc-digital-mftg-solution-overview .pdf.

EAdirections (2015). EA frameworks: pros and cons – inventory and insight. https://www .eadirections.com/wp-content/uploads/2015/06/EA-Frameworks-Pros-and-Cons.pdf.

FEAPO (2018). The guide to careers in enterprise architecture. https://feapo.org/wp-content/ uploads/2018/10/Guide-to-Careers-in-Enterprise-Architecture-v0.5-clean-copy-copy.pdf.

Fenner, J. (2003). Enterprise application integration technique. http://www-icm.cs.ucl.ac.uk/ staff/W.Emmerich/lectures/3C05-02-03/aswe21-essay.pdf.

Fredendall, L.D., and Gabriel, T.J. (2019). Manufacturing complexity: a quantitative measure. https://pdfs.semanticscholar.org/39c5/b53dd70414acae58532a0a8010193af221d3.pdf.

Frizelle, G. and Woodcock, E. (1995). Measuring complexity as an aid to developing operational strategy. *International Journal of Operations and Production Management* 15 (5): 26–39.

Gao (2000). A practical guide to federal enterprise architecture. https://www.gao.gov/special .pubs/eaguide.pdf.

Gregori, F., Papetti, A., Pandolfi, M. et al. (2017). Digital manufacturing systems: a framework to improve social sustainability of a production site. *Procedia CIRP* 63: 436–442.

Hooper, J.W. and Reilly, K.D. (1983). The GPSS-GASP combined (GGC) system. *International Journal of Computer and Information Science* 12 (2): 111–136.

Isik, F. (2010). An entropy-based approach for measuring complexity in supply chains. *International Journal of Production Research* 48 (12): 3681–3696.

Kawasmi, M. (2018). Digital manufacturing and operations. http://online.spellboundme.com/ sap/nowksa2018-original/assets/2-mohammad-kawasmi-digital-manufacturing-and-its-potentials.pdf.

King, W.P. (2018). Digital manufacturing. https://www.nist.gov/sites/default/files/documents/ el/msid/18_wKing.pdf.

KPMG (2018). Supply chain big data series part 1. https://advisory.kpmg.us/content/dam/ advisory/en/insights/pdfs/2018/supply-chain-big-data-part-1-shaping-tomorrow.pdf.

Kritzinger, W., Kkarner, M., Traar, G. et al. (2018). Digital twin in manufacturing: a categorical literature review and classification. *IFAC PapersOnLine* 51 (11, 2018): 1016–1022.

Kusiak, A. (2019). Fundamentals of smart manufacturing: a multi-thread perspective. *Annual Reviews in Control* https://doi.org/10.1016/j.arcontrol.2019.02.001.

Kutin, A., Dolgov, V., Sedykh, M., and Ivashin, S. (2018). Integration of different computer-aided systems in product designing and process planning on digital manufacturing. *Procedia CIRP* 67 (2018): 476–481.

Mäkiö-Marusik, E., Colomboa, A.W., Mäkiö, J., and Pechmann, A. (2019). Concept and case study for teaching and learning industrial digitalization. *Procedia Manufacturing* 31 (2019): 97–102.

Markowitz, H.M., Hausner, B., and Karr, H.W. (2019). Simscript – a simulation programming language. https://www.rand.org/pubs/research_memoranda/RM3310.html.

Pati, A. and Bandyopadhyay, P.K. (2017). Digital manufacturing: evolution and a process oriented approach to align with business strategy. *International Journal of Economics and Management Engineering* 11 (7): 1746–1751.

Pegden, C.D. (2012). Introduction to Simio. https://informs-sim.org/wsc12papers/includes/ files/vdrp102.pdf.

PWC (2017). Digital factories 2020 shaping the future of manufacturing. https://www.pwc.de/de/digitale-transformation/digital-factories-2020-shaping-the-future-of-manufacturing.pdf.

SAS (2014). Big data meets big data analytics: three key technologies for extracting real-time business value from the big data that threatens to overwhelm traditional computing architectures. http://www.sas.com/content/dam/SAS/en_us/doc/whitepaper1/big-data-meets-big-data-analytics-105777.pdf.

Siemens (2015). Industrie 4.0 – digitalization strategy. https://www.iotone.com/files/pdf/vendor/Digital_Factory_(Siemens)_Industrie_4.0_Digitalization_Strategy_2015.pdf.

Siemens (2019a). Digital manufacturing – a holistic approach to the complete product lifecycle. https://www.industry.usa.siemens.com/automation/us/en/formsdocs/Documents/AMAR_UTO1_0715_Digital_Manufacturing.pdf.

Siemens (2019b). Digital manufacturing. https://www.plm.automation.siemens.com/global/en/our-story/glossary/digital-manufacturing/13157.

Simio (2019a). Why choose Simio software? https://www.simio.com/index.php.

Simio (2019b). An introduction to Simio for beginners. https://www.simio.com/resources/white-papers/Introduction-to-Simio/introduction-to-simio-for-beginners-page-4.php.

SME (2018). Trends in digital manufacturing – implementing Industry 4.0 and building the digital factory. https://www.sme.org/globalassets/sme.org/media/white-papers-and-reports/trends-in-digital-mfg.pdf.

Smith, J.F. (2013). Simio processes and add-on processes. http://jsmith.co/node/82.

Sönmez, O.E., and Koç, V.T. (2015) On quantifying manufacturing flexibility: an entropy based approach. *Proceedings of the World Congress on Engineering 2015*, vol, II, WCE 2015, London, UK (1–3 July 2015).

Tao, F., Zhang, M., and Nee, A.Y.C. (2019). Chapter 9: Digital twin and big data. *Digital Twin Driven Smart Manufacturing* 2019: 183–202.

Vermesan, O. and Friess, P. (2016). *Digitising the Industry Internet of Things Connecting the Physical, Digital and Virtual Worlds*. River Publishers. ISBN: 978-87-93379-81-7.

Viriyasitayat, W., Xu, L.D., Bi, Z.M., and Sapsomboon, A. (2018). Blockchain-based business process management (bpm) framework for service composition in Industry 4.0. *Journal of Intelligent Manufacturing* https://link.springer.com/article/10.1007/s10845018-1422-y.

Viriyasitayat, W., Xu, L., and Bi, Z.M. (2019). The extension of semantic formalization of service workflow specification language. *IEEE Transactions on Industrial Informatics* 15 (2): 741–754, 2019.

Vrabič, R. and Butala, P. (2012). Assessing operational complexity of manufacturing systems based on statistical complexity. *International Journal of Production Research* 50 (14): 3673–3685.

Wiendahl, H.-P., ElMaraghy, H.A., Nyhuis, P. et al. (2007). Changeable manufacturing – classification, design and operation. *CIRP Annals* 56 ((2): 783–809.

Wikipedia (2019a). NIST enterprise architecture model. https://enacademic.com/dic.nsf/enwiki/11787933.

Wikipedia (2019b). Digital twin. https://en.wikipedia.org/wiki/Digital_twin.

Wikipedia (2019c). Internet of Things. https://en.wikipedia.org/wiki/Internet_of_things.

Williams, T.J. (1994). The Purdue enterprise reference architecture and methodology (PERA). http://citeseerx.ist.psu.edu/viewdoc/download?doi=10.1.1.194.6112&rep=rep1&type=pdf.

13

Direct and Additive Manufacturing

13.1 Introduction

In recent years, *direct manufacturing* has become one of the most widely explored technologies in manufacturing. With rapid improvement in the performance in terms of *process speed*, *dimensional accuracy*, *surface finish*, and *repeatability*, direct manufacturing has evolved from being a prototyping technology to a competitive technology that can substitute conventional manufacturing processes. Direct manufacturing is also referred to as *digitized manufacturing*, *additive manufacturing*, *3D printing*, or *rapid prototyping*. This chapter will focus on the commonality that, in making a product, no CAM, CAFD, or CAPP will be needed. Therefore, these concepts are used alternatively.

According to the ASTM standard F2792-10, direct manufacturing or *additive manufacturing* (AM) is the process of joining materials to make objects from 3D model data, usually layer upon layer. This is as opposed to a traditional machining process where a part geometry is generating by removing unwanted materials from the workpiece (EPMA 2019). AM has great potential benefits for users and manufacturing enterprises when low-volume and functional, highly complex, end-use products are needed. A great deal of research effort has been made to implement AM processes, where AM techniques are highly diversified in terms of applicable materials, material preparations, layer creation algorithms, phase changes, and applications. Commonly used AM techniques are *fused deposition modelling* (FDM), *additive layer manufacturing* (ALM), *laser beam melting* (LBM), *electron beam melting* (BM), *selective laser melting* (SLM), *direct energy deposition* (DED), and *direct metal deposition* (DMD). AM techniques are different from each other depending on applicable materials, patterning energy, phenomena of creating primitive geometry, nature of adding materials, and support mechanisms (Al-Ahmari 2017).

AM differs significantly from traditional subtractive manufacturing in the sense that no manufacturing relevant tools such as CAM, CAFD, and CAPP are needed in the fabrication process. This leads to a number of benefits such as creating a complex geometry, reducing assemblies for a single structure, cost savings on tooling, a shortened product design cycle, and the consistence of product quality. AM will be discussed extensively on relevant concepts, hardware and software systems, procedure of manufacturing processes, and design for additive manufacturing (DfAM).

Computer Aided Design and Manufacturing, First Edition. Zhuming Bi and Xiaoqin Wang.
© 2020 John Wiley & Sons Ltd. This Work is a co-publication between John Wiley & Sons Ltd and ASME Press.
Companion website: www.wiley.com/go/bi/computer-aided-design

13.2 Overview of Additive Manufacturing

AM is a method of building a product by adding the material layer by layer on a part continuously until the part is finished. When a layer is added, it serves as a base for more additional layers of material. Three-dimensional (3D) printing is an interchangeable umbrella term with AM. However, 3D printing typically builds parts using melted plastic or similar materials. The concept of AM goes far beyond 3D printing, as other AM processes offer more versatility in build quality and material choices (Underdahl 2015).

AM creates the desired shape by adding materials, preferably by staggering contoured layers on top of each other. Therefore, it is also called *the layer (or layered) technology*. The layer technology has been developed because any object, at least theoretically, can be sliced into *layers* and rebuilt using these layers, regardless of the complexity of its geometry. AM is an automated fabrication process based on layer technology. AM integrates two main sub-processes: (i) physically creating each single layer and (ii) joining subsequent layers in sequence to form the part. The above two processes proceed simultaneously. Therefore, AM only requires the 3D data of the part, commonly called the *virtual product model*.

AM has shown some unique features in comparison with traditional subtractive manufacturing. AM builds a part out of thousands of extremely thin layers and so it is possible to create highly complex geometries that tend to be impractical when made from other manufacturing processes. Taking as an example of a mould for injection moulding, it may include internal curved channels and holes that would be unreachable by a traditional machining process. AM may be used to substitute a complex assembly for a printed structure in one piece. AM has been widely used as an affordable tool in creating very low-volume custom pieces such as dentures and jewellery.

13.2.1 Historical Development

As a manufacturing technology, AM has been around for more than 30 years as a rapid prototyping (RP) technology in manufacturing. The process is '*rapid*' in the sense that it avoids the time for tooling design. The early layer-based technology could not fabricate anything but sticky and brittle parts that could only be used as '*prototypes*'. Therefore, AM is also referred to as *rapid prototyping*. AM is viewed as a fast and cost-effective method used to create prototypes in product development (3D Printing Industry 2015). Table 13.1 shows significant milestones in the development of AM technologies (Tamburrino et al. 2015). It is not until recently that the industry and market values of AM have become well known; AM has now become one mainstream of manufacturing technology development. AM has been widely applied from rapid prototyping to distributed manufacturing, from architecture to fashion, and from aerospace to healthcare systems (Levchenko 2015).

13.2.2 Applications

AM has a growing market capability in manufactured products in all industry sectors and, especially, is expected to see wider usage in automotive and aerospace applications and customized manufacturing. Figure 13.1 shows some examples of products from AM processes,

Table 13.1 The timeline of AM application (Tamburrino et al. 2015).

Year	Milestone	Year	Milestone
1988	Rapid prototyping	2009	Medical implants (metals)
1994	Rapid casting	2011	Aerospace (metals)
1995	Rapid tooling	2016	Nano-manufacturing
2001	Additive manufacturing for automotive	2017	architecture
2004	Aerospace (polymers)	2018	Biomedical implants
2005	Medical (polymer jigs and guides)	2022	In situ bio-manufacturing
		2030	Full body organs

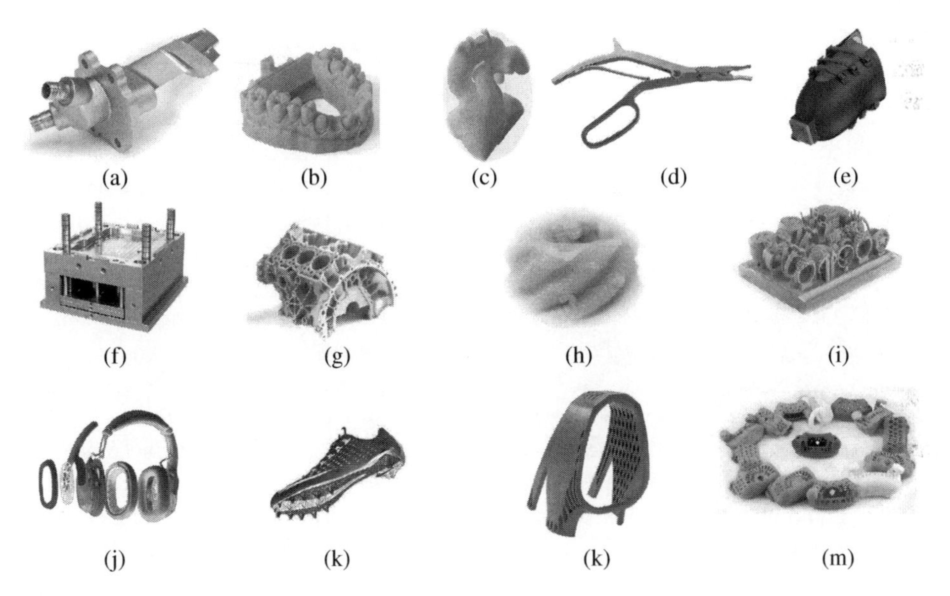

Figure 13.1 AM application examples. (a) Aerospace. (b) Denture. (c) Implants. (d) Surgical. (e) Prosthetics. (f) Molds and dies. (g) Automotive. (h) Foods. (i) Jewelry. (j) Electronics. (k) Sports. (l) Furniture. (m) Toys.

including aircraft components, dentures, implants, surgical tools, prosthetics, moulds and dies, automotive goods, foods, jewellery, electronics, sports, furniture, and toys.

AM was originally developed for *rapid prototyping* (RP) to make prototypes, models, and mock-ups; these rapid prototyped parts are usually not complete products since they only represent some isolated properties of original products for verification purposes. Therefore, RP aims to make parts as simple as possible, where time and cost are primary considerations. As shown in Figure 13.2, RP can be classified into *solid imaging concept modelling* and *functional prototyping*. The first type only produces a replica of an existing solid via virtual modelling, while the second type allows verification of certain aspect of engineering

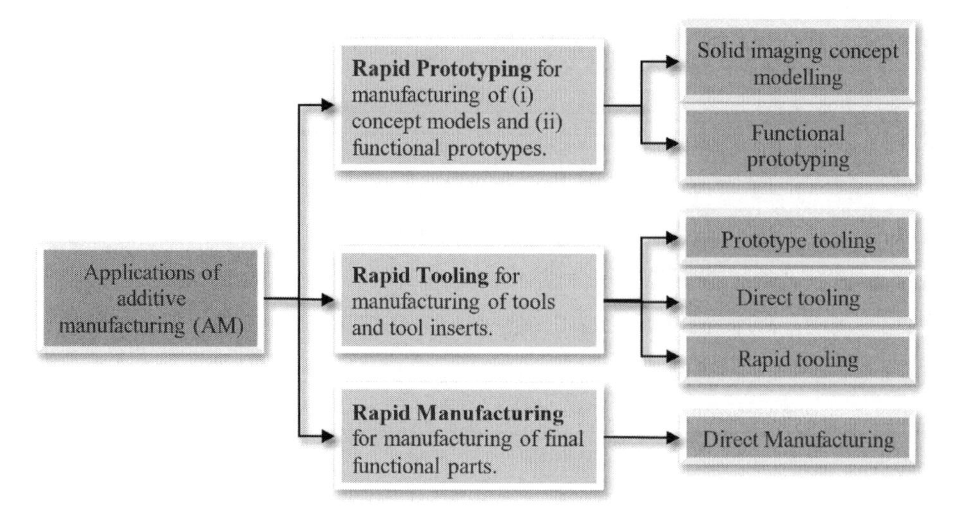

Figure 13.2 Classification of AM applications (Gebhardt and Hötter 2016).

design. Consequently, it is called a functional prototype (and the process is accordingly functional prototyping). With the technological advancement of AM, functional products can be directly generated from AM processes. *Rapid manufacturing* (RM) is used to make final parts or products directly from an AM process. If the products from AM are used as tooling, such as cutting chips or mould components, the corresponding AM processes are referred to as *rapid tooling*.

AM builds the part layer by layer and so is also called a *layer manufacturing process* (LMP). Figure 13.3 shows that raw materials in different forms can be applied and the materials can be layered and bonded in many different ways (Sinha 2016).

AM makes it possible to develop an agile manufacturing environment, which may reduce the lead-time from conception to the production stage by 70% or more, depending on the type of manufacturing desired. The impacts of AM to modern manufacturing become more and more significant. The pressure to manufacturing enterprises for cost-effective manufacturing and rapid production have led AM to favourable growth. AM has gained its popularity with the development of a heterogeneous material manufacturing capability (Frost and Sullivan 2016). Figure 13.4 shows the growth of AM technologies in North American, European, and Asian markets, and the statistics show that:

- AM has been growing at an annual rate of 15.0% globally.
- AM applications in (i) aerospace and the defence industry, (ii) medical devices, (iii) automotive industry are growing at an annual growth rate of 26%, 23%, and 34%, respectively.
- Aerospace, automotive, and medical industries are expected to account for 51% of the AM market by 2025.
- AM in the Asia-Pacific (APAC) region is set to grow at an annual growth rate of 18.6%, with China making more than 70% of the business.
- Graphene-based AM through fused filament fabrication (FFF) will be the next big innovation in the AM market.
- A very high potential of product differentiation and supporting demand for unique products will reduce commoditization of 3D printing.

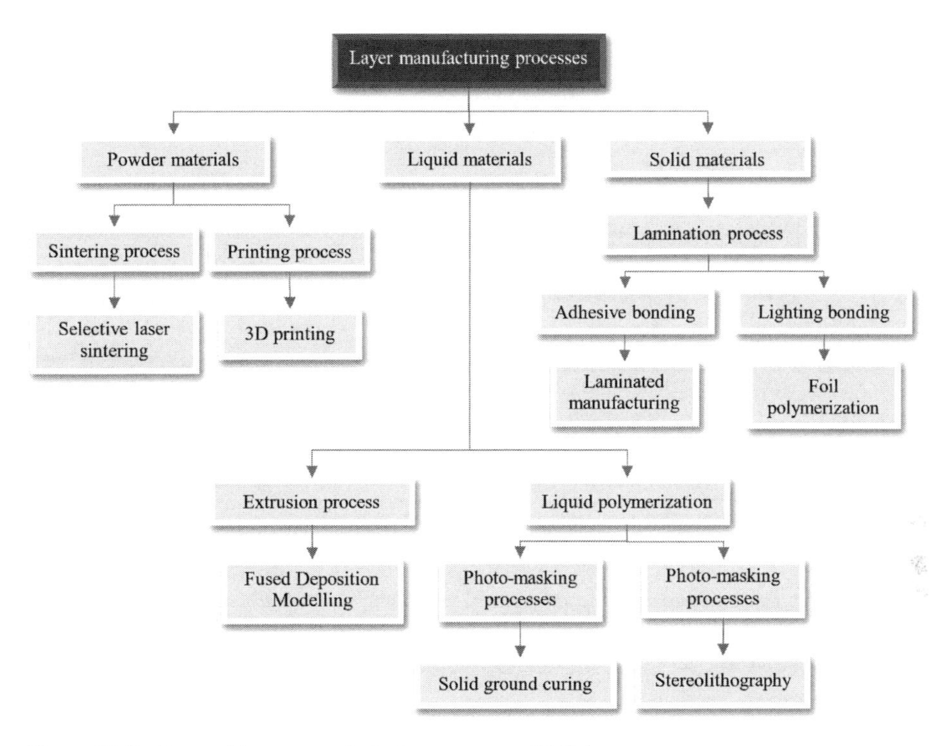

Figure 13.3 Types of layer manufacturing processes (LMPs).

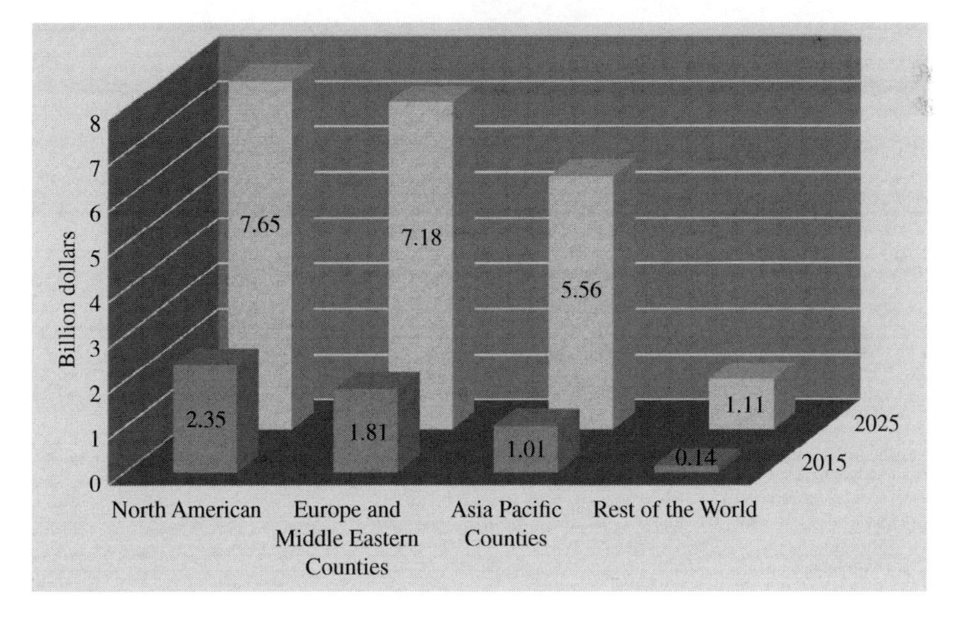

Figure 13.4 Predicted market growth of AM (billion US dollars/year) (Frost and Sullivan 2016).

13.2.3 Advantages and Disadvantages

Discrete manufactured products have the requirements of geometry, shape, and dimensions. Figure 13.5 shows three basic ways to generate the geometry of a part (Gebhardt and Hötter 2016). The formative process in Figure 13.5c needs to design a special mould assembly to match the part geometry. The material is filled into the mould cavity and is formed into the desired shape by solidification, sintering, or deformation subjected to mechanical and thermal loads. Subtractive manufacturing in Figure 13.5a starts with a block of materials and generates part geometry by removing unwanted materials from the part body. The materials can be removed by different cutting tools, where the motions of the machine and cutting tool have to be programmed to control the material removal process, so that the right geometries can be generated. Many types of manufacturing resources such as lathes, mills, fixtures, cutting tools as well as computing resources such as CAM, CAFD, and CAPP are dedicated to implementation of the manufacturing processes. An AM in Figure 13.5c starts with nothing but by adding materials layer by layer the part geometry is generated. The tooling of AM is generic to any products.

The advantages and disadvantages of AM are given in Table 13.2. One of the biggest advantages of AM over traditional manufacturing is that AM generally does not require any special tooling to make new parts. When making a prototype, this can save a lot of time, money, and effort that would normally be spent on tooling the production line or setting

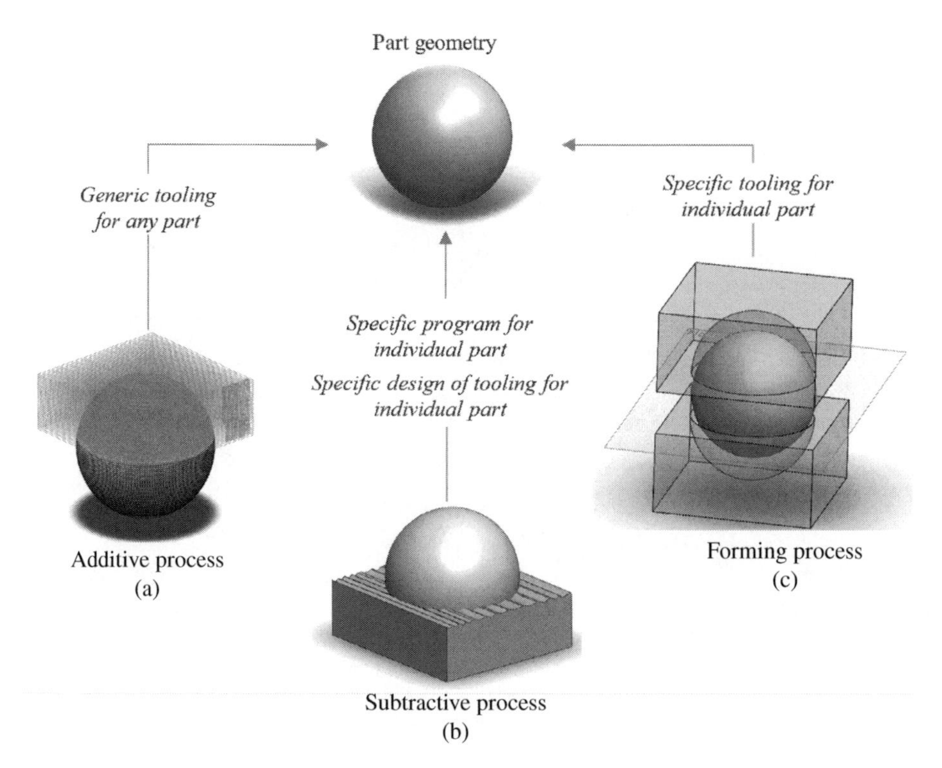

Figure 13.5 Different methods for creating part geometry. (a) Additive process. (b) Subtractive process. (c) Forming process.

Table 13.2 AM – advantages and disadvantages (Sinha 2016).

Advantages	Disadvantages
• Freedom to design and innovate without extra costs • Rapid iterations when designs are under development • Excellence for mass customization • Limited requirements of tooling • Green manufacturing • Near-net-shaping operation • Energy efficient • Enables personalized manufacturing	• Unexpected pre- and post-processing requirements • Lack of industry standards • Low speed, not suitable for mass production • Inconsistent quality • Limited number of applicable materials • High equipment cost for products with good quality

up an assembly process. Another significant advantage is that AM is a near-net-shape operation and has a high utilization of materials. Nearly all of the consumed materials are added to the object to be made. Taking an example of the 3D printing process, the material passes through the extruder of the printer and then adds on to the object. As a comparison, an injection moulding process involves wasted materials in the barrel of the injection unit and the runner system of the mould. A perforated sheet metal assembly process begins with a whole piece of sheet metal and cuts holes into it, leaving the cut material as scraps; although scraps are recycled, it adds extra time and labour costs to the manufacturing process.

Metal products with AM show a number of preferable features for product designs (EPMA 2019):

1. Design freedom for AM can be greatly increased in comparison with the product by conventional casting and machining.
2. Product weight can possibly be reduced by using a lattice structure or optimizing the material distribution subjected to other design constraints.
3. New product features such as complex internal channels can be created without an increase in the complexity of the AM process. It is possible to combine several parts in one net-shape process to reduce material waste and machining times. It has been found that AM can save up to 25 times less cost versus traditional machining processes for expensive or difficult to machine alloys.
4. The net shape capability helps to create complex parts in one step; this eliminates some assembly operations such as welding and brazing.
5. No tooling design is needed, unlike other conventional metallurgy processes, which require moulds, dies, or cutting tools.
6. The production cycle time can be reduced. A part with a complex geometry may take a few hours to make in an additive machine and post-processing may take an additional few days or weeks. However, the total time is usually significantly shortened when compared with conventional metallurgy processes, which often require a production cycle of several months.

Table 13.2 also shows some disadvantages of AM such as low productivity, inconsistent quality, and lack of industrial standards. To use AM in a broader scope of applications,

Table 13.3 Supply chain for subtractive and additive manufacturing (Frost and Sullivan 2016).

Conventional supply chain	Additive manufacturing supply chain
• Supply is a crucial part of production and a delay in any one of the tributaries of a supply chain has a ripple effect across the production value chain. • Logistics management is key to functionality. Large amounts of money and time are invested to ensure the seamless flow of logistics to and from the production sites. • Large-scale warehousing is crucial to ensure effective storage of raw material inventory, inflow of material from suppliers, and storage of finished goods before shipping. • Complex systems and processes are required to handle parts from suppliers to maximize production efficiencies. • Customization of products is almost impossible as this requires significant adjustment in the standard supply chain paths followed. • Lead time is very high compared to additive manufacturing.	• The existing conventional supply chain will see a complete overhaul to meet additive manufacturing requirements. • The number of supplier and vendors is reduced. • Logistics cost will be reduced as most of the material is supplied in powder form and the finished product is distributed locally or within the unit itself. • The need for warehouses will reduce exponentially as most products are made-to-order, eliminating the need to store finished goods. • Operations, with respect to tooling and maintenance of multiple machines, are completely ruled out. • Short lead and cycle times; logistics costs are to be reduced by more than 80%.

a number of identified challenges include (i) the high cost for acquisition, operation, maintenance, and depreciation in high-performances AM machines and materials, (ii) the need for new AM technologies with faster operating speeds, better resolution and accuracy, larger build volumes, as well as more-optimized loading and unloading procedures, (iii) the lack of consistency and maturity in quality assurance over the part, and (iv) the need for more sophisticated software tools for the full potential of AM. In summary, the entire AM process chain would benefit from faster and cheaper methods, improved industry standards, as well as superior intellectual property protection and security measures (Wu et al. 2017).

It is worth noting that benefits of using AM in a system are far beyond the manufacturing processes themselves, since manufacturing processes affect many no-value-added activities, including supply chain management (SCM). Frost and Sullivan (2016) discussed the potential benefits of using AM from the perspective of SCM in Table 13.3. SCM can be significantly simplified since the limited types of materials are available, the number of suppliers and vendors is reduced, there are no warehouse needs for workpiece in progress, and the lead time and cycle time can be greatly shortened.

13.3 Types of AM Techniques

AM is featured as a layer-by-layer fabrication of three-dimensional objects; however, different techniques are applied in the implementation. The American Society for Testing

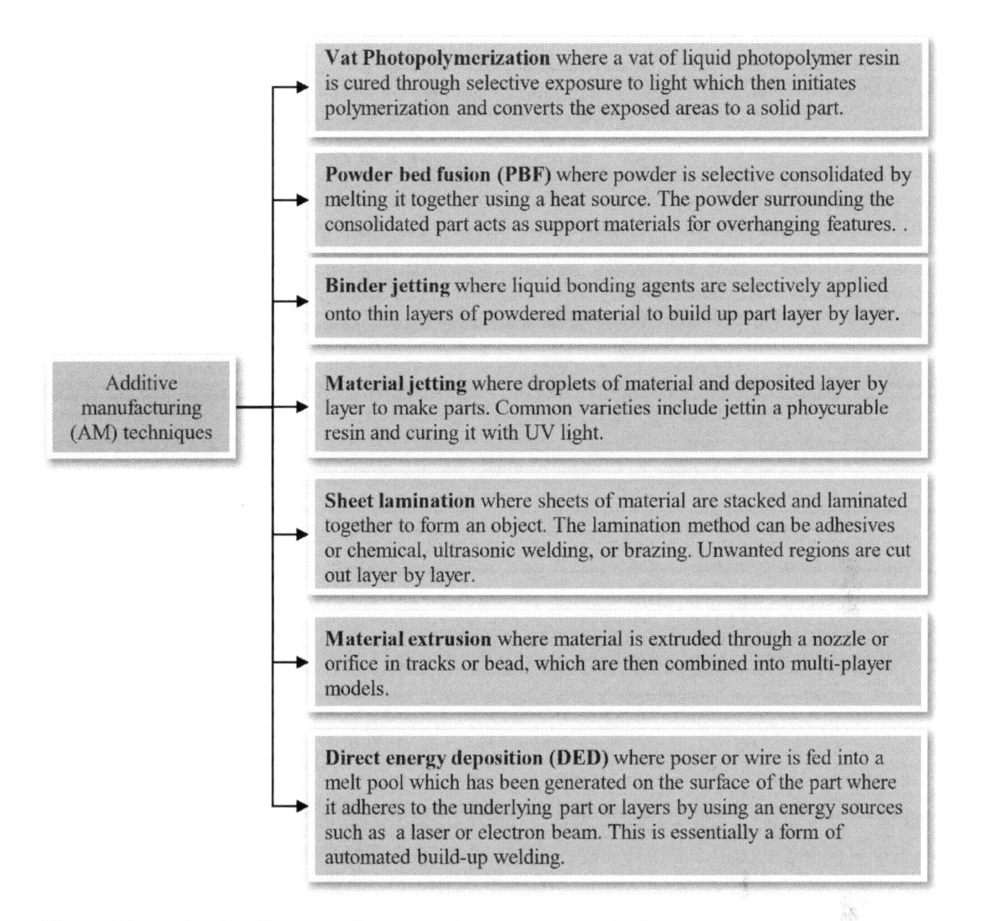

Figure 13.6 Classification of additive manufacturing technologies.

and Materials (ASTM) classified AM processes into seven categories in its standard, called *ASTM F42 – Additive Manufacturing* (GE 2019). As shown in Figure 13.6, these types of AM are (i) vat photo-polymerization, (ii) powder bed fusion, (iii) binder jetting, (iv) material jetting, (v) sheet lamination, (vi) material extrusion, and (vii) directed energy deposition (Additivemanufacturing 2019; Frost and Sullivan 2016; Redwood 2019).

13.3.1 Vat Photo-Polymerization

As shown in Figure 13.7, *vat photo-polymerization* differs from other AM processes since the material is in a liquid state rather than a powder or a filament. Photo-polymer resins are used that are often tough, transparent, and castable. One of the earliest AM processes called a *stereolithography apparatus* (SLA) is a vat photo-polymerization method. SLA makes use of a build platform in a tank of liquid polymer with ultraviolet (UV) lights, where the laser light comes from beneath the object, which is the reverse of other AM processes that feature a heat source directed from above. *Direct light processing* (DLP) is very similar to SLA. However, it creates each layer of an object by projecting laser light on to tiny mirrors, resulting in the projection of square pixels, layer-by-layer.

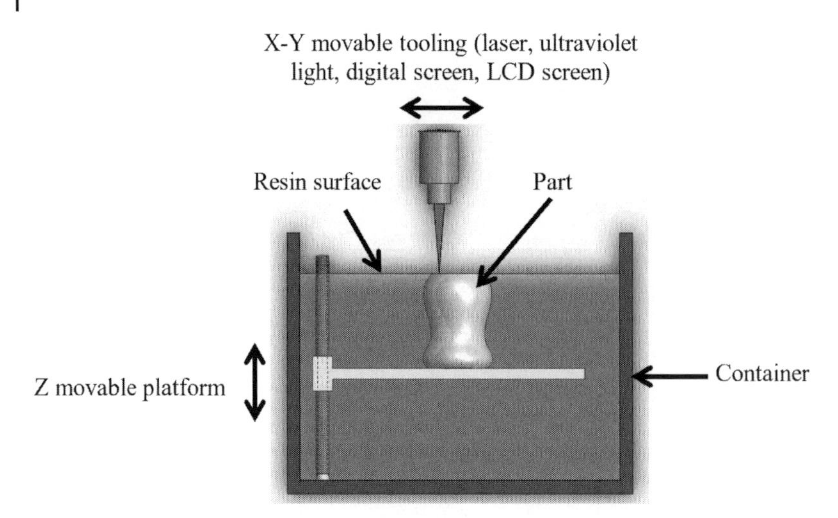

Figure 13.7 Vat photo-polymerization process.

Typical materials for the vat photo-polymerization are UV-curable photopolymer resins. The strengths are a high level of accuracy and complexity, a smooth surface finish, and it accommodates large build areas. The process is ideal for jewellery, medical applications, and low-run injection moulds.

13.3.2 Powder Bed Fusion

As shown in Figure 13.8, *powder bed fusion* (PBF) melts powder materials to a sufficiently high degree for the powder particles to fuse together. Particles are either 'sintered' (partially melted) or fully melted in various PBF processes. Thermal energy is supplied in the form

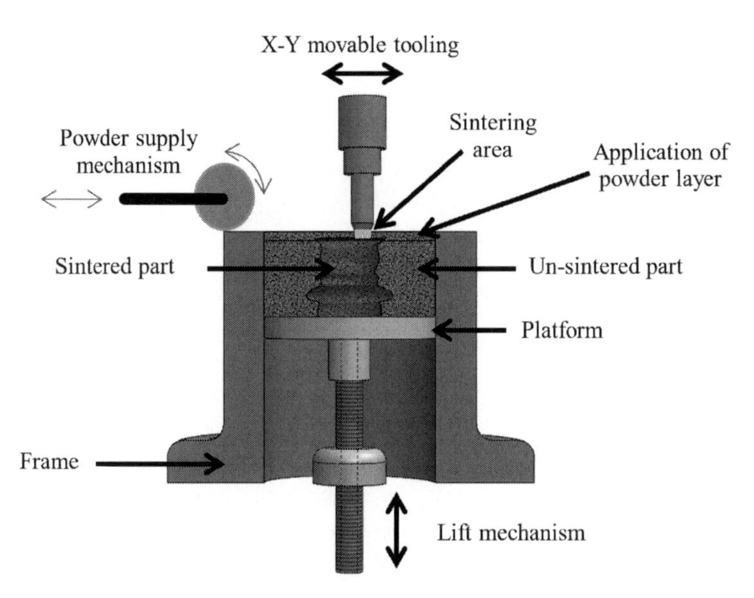

Figure 13.8 Powder bed fusion (PBF) process.

of a laser or beams of electrons. It may be built as a heated print head that partially or fully melts plastic or metal powder. An ultrathin layer of material is spread by a roller or blade over the preceding layer. The powder is fed from a reservoir beneath or next to a build platform that lowers to accommodate each successive layer of powder. After the additive process is completed, the unfused powder is blown or blasted away.

Typical materials for PBF are plastics, metal and ceramic powders, and sand. The strengths of the parts from PBF are the high level of complexity, the powder acts as a support material in the building process, and it is applicable to a wide range of materials. PBF is ideal for almost all types of end manufacturing, allowing for the easy design and build of complex geometries.

13.3.3 Binder Jetting

As shown in Figure 13.9, a *binder jetting process* uses powder material and a binding agent. The nozzle on the AM machine deposits tiny droplets of a binder on an ultrafine layer of powdered metal, ceramic, or glass. Multiple layers result from the powder bed moving downward after each layer is created. The resulting object is *in a green state*, so post-processing is required to improve its mechanical properties enough to make a functional component. A cyanoacrylate adhesive is a common infiltrant when the object is ceramic. Ceramic objects made by binder jetting are fairly brittle and are primarily used as architectural models or models for sand casting.

Typical materials of binder jetting are powdered plastics, metal, ceramics, glass, and sand. The binder jetting process has its advantages of high productivity, widely applicable materials, and support for full colour printing. They are ideal for aesthetic applications like architectural and furniture design models.

13.3.4 Material Jetting

As shown in Figure 13.10, *material jetting* uses the *drop-on-demand* (DOD) technology where tiny nozzles dispense tiny droplets of a waxy photo-polymer, layer by layer. UV

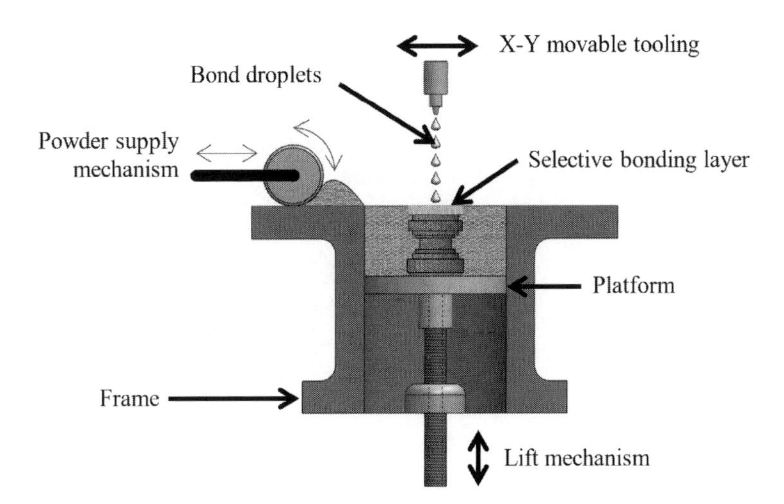

Figure 13.9 Binder jetting process.

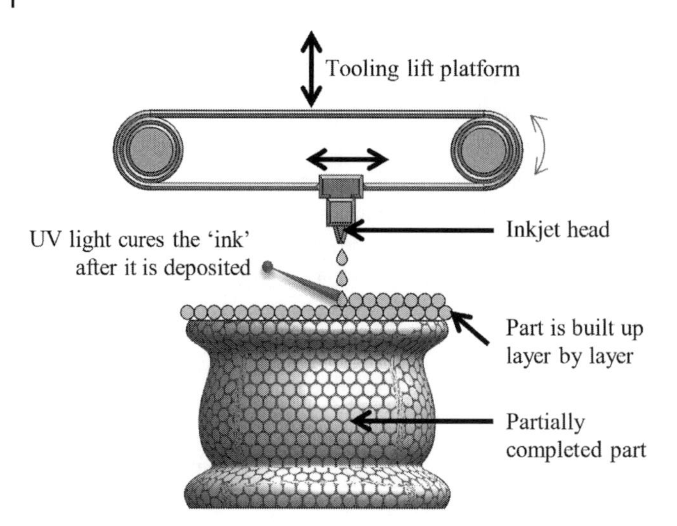

Figure 13.10 Material jetting process.

light cures and hardens the droplets before the next layer is created. Since this additive technology relies heavily on support structures, a second series of nozzles dispenses a dissolvable polymer that supports the object as it is printed. When printing is complete, the support material is dissolved away. Material jetting produces patterns used in lost-wax casting, investment casting, and mould making. Material jetting is one of the most precise AM processes; the technology can print layers as thin as 15–16 μm.

Typical materials of material jetting are photopolymers, polymers, and waxes. The material jetting process has a high level of accuracy, supports full colour parts, and enables multiple materials in a single part. It is ideal for realistic prototypes with high detail, high accuracy, and a smooth finish.

13.3.5 Material Extrusion

Material extrusion processes, such as FDM or FFF, are perhaps the most well-known AM processes. As shown in Figure 13.11, a thermoplastic filament is extruded through a heated nozzle and on to the build platform. The material solidifies as it cools. FDM is applicable to a wide variety of thermoplastic filaments, such as acrylonitrile butadiene styrene (ABS), polylactic acid or polyactide (PLA), nylon, and polycarbonate (PC). A material extrusion process is usually fast and inexpensive but the dimensional accuracy is low. A material extrusion is directionally dependent because the material is deposited along the horizontal x- and y-axes, but the strength along the vertical z-axis is weaker.

A material extrusion process is inexpensive and economical, it allows for multiple colours, and it is safe and portable and can be used in office environment. Typical materials are thermoplastic filaments, and pellets, liquids, and slurries. Material extrusion is often the go-to method to make non-functional prototypes or rapid prototyping that needs several design iterations.

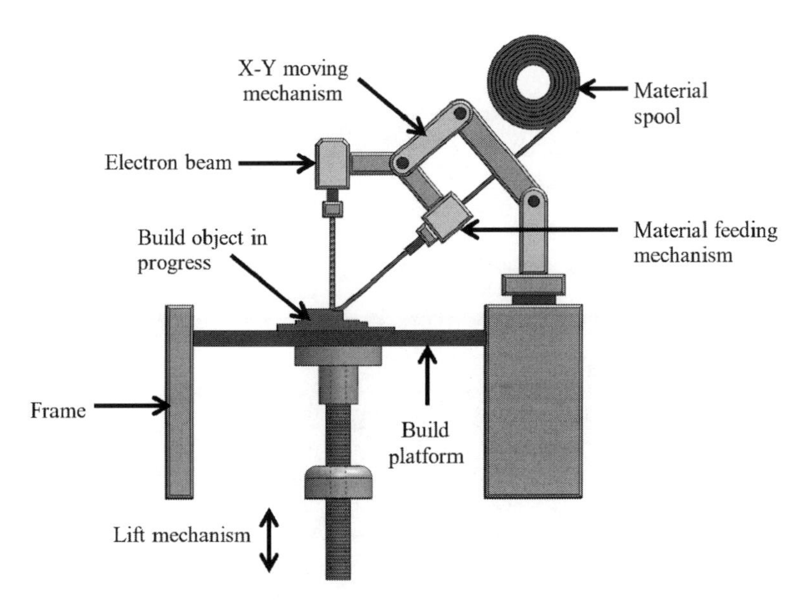

Figure 13.11 Material extrusion process.

13.3.6 Sheet Lamination

In a *sheet lamination process*, shown in Figure 13.12, ultrathin layers of solid materials are bonded by alternating layers of adhesive. It is also referred to as *laminated object manufacturing* (LOM). If the material is paper, the corresponding process is called *paper lamination technology* (PLT), where paper sheets are combined by and adhesive as a layered object resembling plywood. After the bonding process, a laser, metal knife, or tungsten carbide blade directed by a CAD file cuts away the unwanted materials to create the final 3D object.

Ultrasonic additive manufacturing (UAM) uses metal sheets, ribbons, or foils to build objects using a single layer one at a time. Applicable materials are titanium, stainless steel, copper, and aluminium. Metal layers are typically conjoined through ultrasonic welding and compression via a rolling sonotrode, a device that generates the ultrasonic vibrations. The UAM process does not require melting and uses less energy than most AM processes. Computer numerical control (CNC) machining may be used to further refine the surface of the object and remove excess, non-bonded metal. Sheet lamination is ideal for non-functional models as the parts can be made in a short time, at a low cost, and material handling is simple.

13.3.7 Directed Energy Deposition

DED is also called *direct metal deposition* or *metal deposition*. As shown in Figure 13.13, DED utilizes highly focused thermal energy delivered via laser, electron beam, or plasma arc to melt and fuse material jetted into the heated chamber from either powdered metal or wire filament. This AM process is most commonly used with metal, while some DED systems can be used with ceramic powder or polymers. The system usually features metal

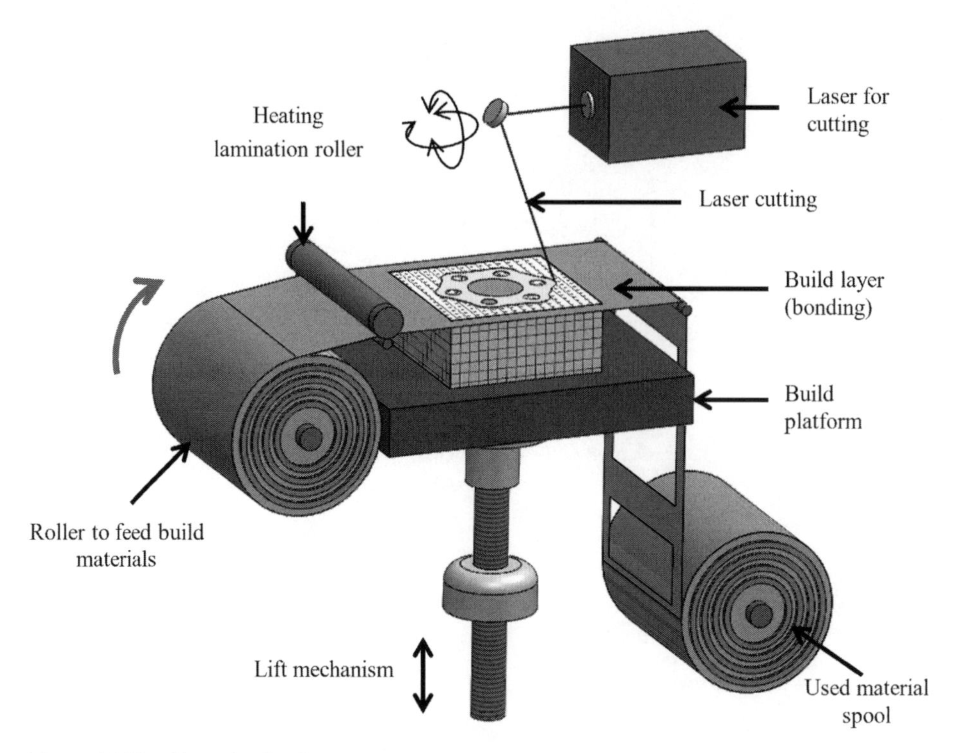

Figure 13.12 Sheet lamination process.

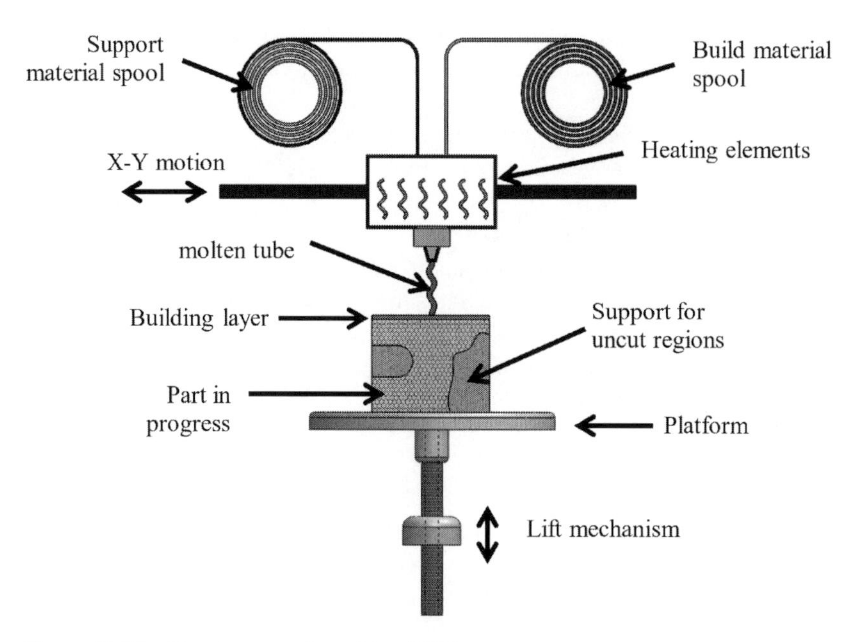

Figure 13.13 Direct energy disposition (DED) process.

deposition along four or five axes. This makes it one of the few AM processes that can be used to repair worn tools and parts in the aerospace, defence, and automotive industries.

Typical materials for DED are metal wire, metal powder, and ceramics. DED is not limited by direction or axis, but is effective for repairs and adding features, allows multiple materials in a single part, and has the highest single-point deposition rate.

13.4 AM Processes

Despite the highly diversified AM techniques, the procedure of AM processes is straightforward and generic. Figure 13.14 shows the steps in the procedure of AM along with the hardware and software resources required to fulfil the tasks at these steps.

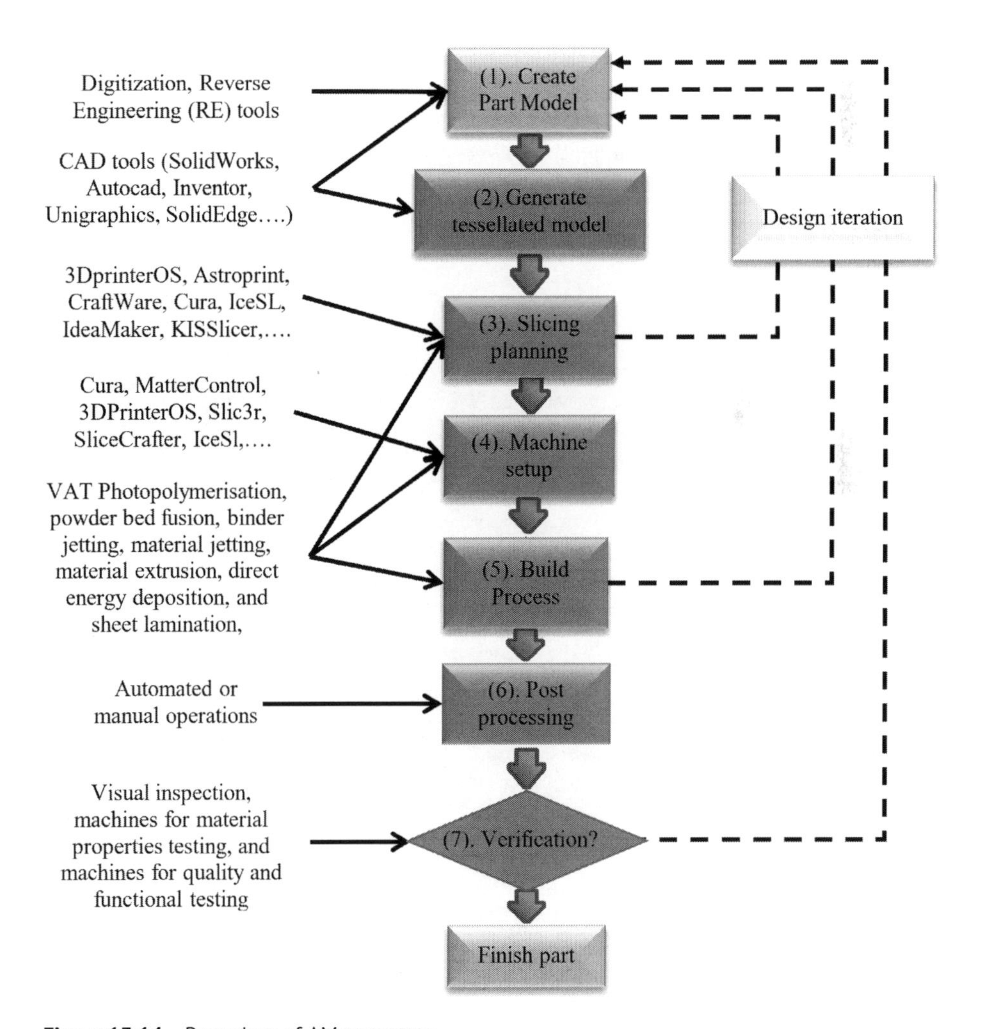

Figure 13.14 Procedure of AM processes.

13.4.1 Preparation of CAD Models

The first step is to create a virtual model of the part. If AM is for an existing physical object, reverse engineering techniques should be used to acquire surface data from the object, process data, and reconstruct the data into a valid surface or solid model of the physical object. If AM is to prototype a new object, the activities at this step may begin with determination of the design intents and then to convert the design intents into a solid model through the process of conceptual design to detailed design. The model represents the geometry as well as design rationales of the part. Moreover, appropriate materials should be selected since that determines the manufacturability of the part. If the part will be made especially from AM, some special requirements relating to AM, such as the flexibility in part features, functions, and materials, must be taken into consideration. Many commercial CAD tools such as SolidWorks, SolidEdge, Unigraphics, and free CAD tools such as FreeCAD, 3D Builder, and LibreCAD can be used to create part models.

13.4.2 Preparation of Tessellated Models

AM supports a model in *tessellated format*, more specifically, the .STL format that stands for *Stereolithograpy*, *Standard Triangle Language*, or *Standard Tessellation Language*. The STL format can be either *ASCII* or *binary* representations. Since binary STL files are more compact, they are more common than ASCII files unless the user needs to review and change data manually. A model in the .STL format only includes the surface geometry of

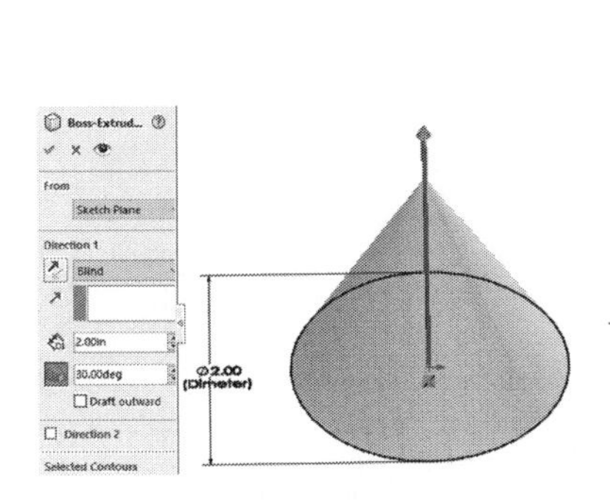

Diameter: 2.00 inch
Draft Angle: 30. 00 degree
Extrusion: Blind
Sketch Plane: Front

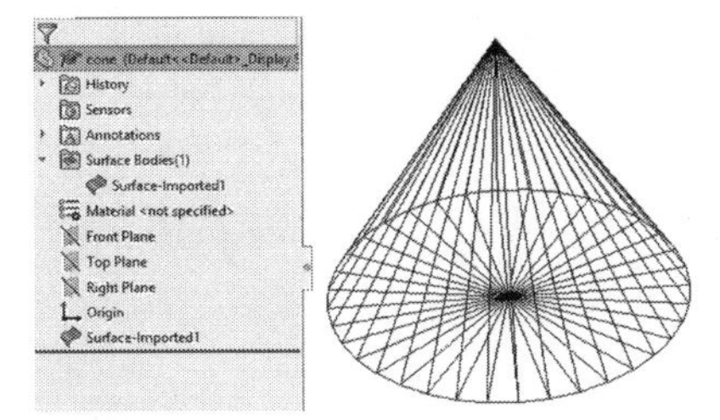


```
solid cone
  facet normal 8.419796e–001 –1.994943e–001 5.012708e–001
    outer loop
      vertex 4.926819e+001 1.671269e+001 0.000000e+000
      vertex 5.041412e+001 2.098934e+001 0.000000e+000
      vertex 2.540000e+001 2.540000e+001 4.399409e+001
    endloop
  endfacet
  facet normal 7.945461e-001 –3.426717e–001 5.012708e–001
    outer loop
      vertex 4.926819e+001 1.671269e+001 0.000000e+000
      vertex 2.540000e+001 2.540000e+001 4.399409e+001
      vertex 4.739705e+001 1.270000e+001 0.000000e+000
    endloop ......
```

(a) (b)

Figure 13.15 The same solid geometry with different information in SLDPRT and STL formats. (a) Native SW part format (.SLDPRT). (b) STL file format (ASCII).

a solid object without information of colour, texture, or other high-level design features or attributes.

Figure 13.15 shows an example of a cone model in native .SLDPRT in SolidWorks and .STL formats, respectively. In Figure 13.15a, geometry and shapes are not the only information included in a 3D model file, but in the .STL model in Figure 13.15b they are. The STL file describes a raw, unstructured triangulated surface by the unit normal and vertices (ordered by the right-hand rule) of the triangles using a three-dimensional Cartesian coordinate system. Note that even for the same geometry and shape, the information can be represented very differently in different file formats. Engineers should do their best practice in order to preserve as much information as possible before the CAD model is finally exported to the .STL model.

Due to the importance and popularity of the .STL format, nearly all CAD tools can export models in that format. As an example, Table 13.4 shows the interface, the steps, and the settings in SolidWorks where a tessellated model is generated.

Curvy features in a part model are tessellated in the .STL file; therefore, the *resolution* setting might affect the quality of AM until the resolution of the .STL file is higher than the AM machine. The resolution setting also affects the file size even though it is now a very minor issue. Figure 13.16 gives an example where the resolution setting affects the smoothness and the file size of the part.

13.4.3 Slicing Planning and Visualization

An AM process creates a part layer by layer. A slicer software is applied to cut the solid in a number of horizontal layers with specified resolution; accordingly, it produces a path

Table 13.4 The interface, steps, and settings to export an .STL file from SolidWorks.

Steps in exporting an .STL file 1. Click *File* menu choose Save As. 2. Select *file type* as STL. 3. Click the *Options* button in the Save dialog box. 4. Select *Output* as *Binary* and select the desired *units* (inches or millimetres). 5. Set *Resolution* to *Fine* (recommended for most parts). 6. If you want to preview the STL model before saving, check the 'Show STL info before saving' box. 7. Name and *Save* your STL file. **Note:** STL files can be created from a SolidWorks part or assembly models. Check the box at the bottom of the Export Options dialog box to save the assembly as one STL file or individual STL files.	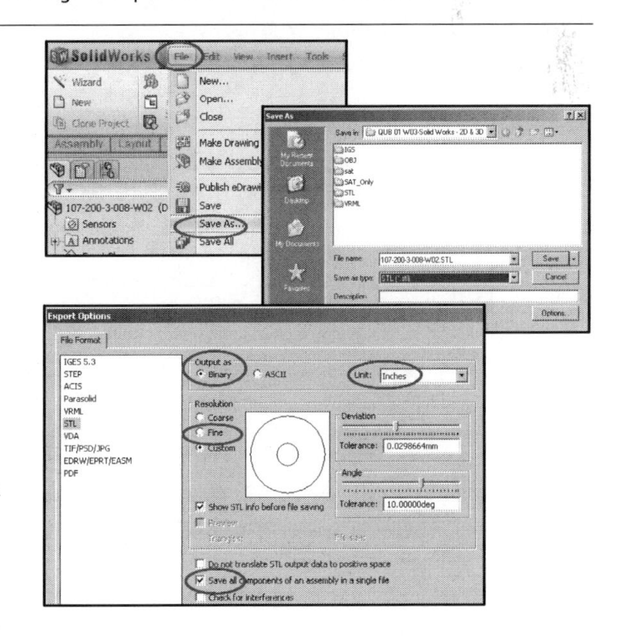

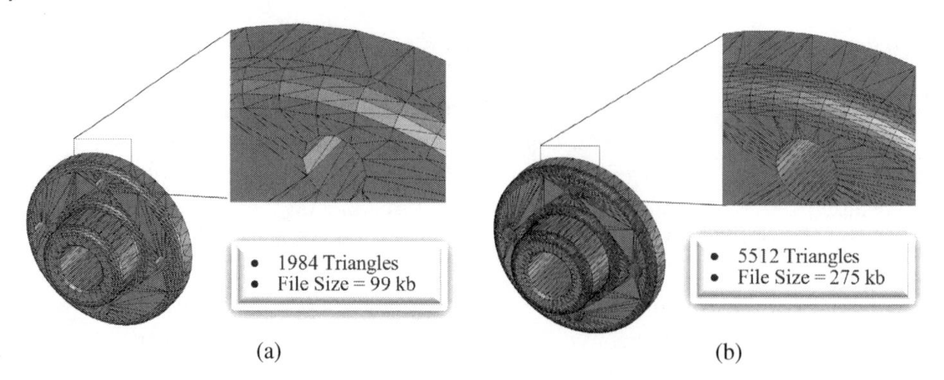

Figure 13.16 Resolution setting affects the part quality and file size. (a) Coarse resolution. (b) Fine resolution.

where the end-effector tool of the AM machine has to follow – line segment by line segment and layer by layer. A slicing plan from a 3D slicer software usually includes the information of (i) the toolpath (more or less intelligently) based on the geometry of the given STL file, (ii) the percentage of infills that determine the amount of time and materials in AM, and (iii) the construction of support materials. If a part feature cannot be self-supporting, supporting structures are needed, which need to be removed after the AM process (All3DP 2019). The generated toolpaths can usually be graphically visualized. Users should review and justify whether toolpaths can be used reasonably to produce the given part model.

13.4.4 Machine Setups

Other than the toolpath based on the part geometry, AM involves many process parameters that are associated with the type of AM machine, raw materials, and quality requirements. In generating the control program based on the tool path, these process parameters should be set up appropriately. The control program must be tailored to the AM machine one uses, which determines the quality of the finish part. Figure 13.17 gives an example of the process parameters in the FDM (Ha 2016). Even though the control software is able to generate the program for parts using default settings and computer algorithms without manual intervenes, it is desirable that designers know what types of process parameters there are and how these parameters will affect the quality of the finished parts, so that the process parameters can be refined when parts are fabricated unsatisfactorily.

13.4.5 Building Process

A control program for AM consists of the codes to initialize and terminate an AM process, move the tool along toolpaths, and operate auxiliary tools based on a given process parameter. Each layer corresponds to a set of toolpaths and parameters that are specified in the control program. However, there is no standard format among different types of AM machines for data generation and management of AM processes.

After the STL file is processed and converted into the control program and the machine is calibrated and set up appropriately, users can run the machine to produce the part. Since the

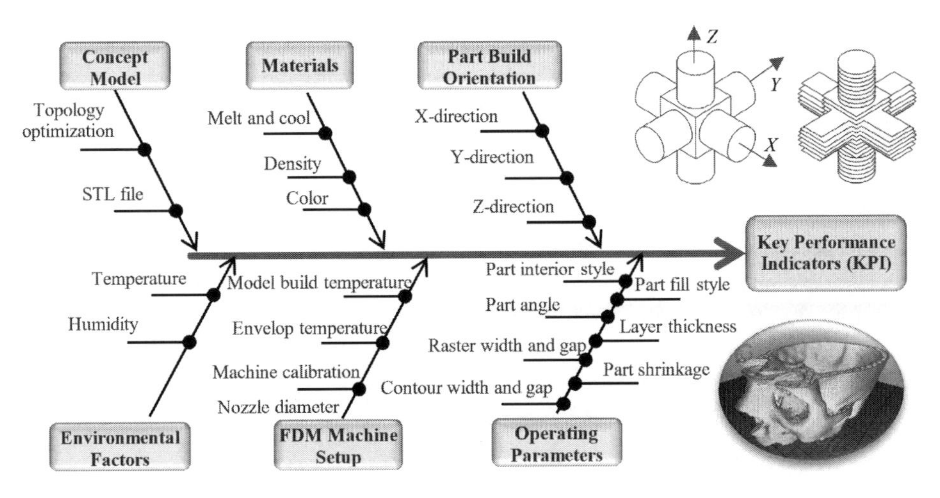

Figure 13.17 Process parameters in AM by the FDM machine (Ha 2016).

part is built layer by layer, it usually takes a long time to build the whole part. The machining time depends on a number of factors such as *part geometry, the speed of the end-effector tool, the setting of process parameters*, and *the effectiveness of the control programs.*

13.4.6 Post-Processing

After the building process is completed, the part can be removed from the AM machine. The post-processing begins with the clean-up of the debris and supportive structure over the part. In some cases, secondary machining processes are followed to create some features that are not able to be formed with the required dimensional accuracy or tolerances. Here, some post-processes related to the improvement of surface finishes are discussed as examples.

Although AM provides some significant advantages in reducing tooling cost and shortening the product development cycle in contrast to traditional machining processes, some disadvantages are staircase effects, surface quality, and dimensional accuracy. The dimensional accuracies from some popular AM processes are shown in Table 13.5 as examples of

Table 13.5 Surface roughness of AM technique (Campbell et al. 2002; Kumbhar and Mulay 2018).

No.	Process type	Minimum layer thickness (mm)	Surface roughness (Ra) in μm
1	Stereolithography (SLA)	0.100	2 ~ 40
2	Selective laser sintering (SLS)	0.125	5 ~ 35
3	Fused deposition modelling (FDM)	0.254	9 ~ 40
4	Material extrusion (3D printing)	0.175	12 ~ 27
5	Laminated object manufacturing (LOM)	0.114	6 ~ 27
6	Material jetting	0.100	3 ~ 30

Table 13.6 Practical options for post-processing for improvement of surface finish.

Conventional	Non-conventional
1. Vibratory bowl abrasion/abrasive blast (grit and ceramic)/shot peen vibratory grinding	1. Ultrasonic abrasion
2. Hot cutter machining (HCM)	2. Chemical treatment
3. Manual optical polishing	3. Electrochemical polishing
4. CNC finishing/machining	4. Electroplating
5. Micro-machining process (MMP)	5. Laser micro-machining
6. Filling the gaps by epoxy resin/part painting	

the limited surface quality of AM processes where the surface roughness could be as poor as $9 \sim 40\ \mu$m from FDM (Campbell et al. 2002; Kumbhar and Mulay 2018).

Surface finish can be improved basically in two ways: (i) optimizing part orientations, toolpaths, and lay thickness and (ii) applying secondary processes such as milling, abrasive machining, chemical machining, laser surface finishing operations, or abrasive flow machining. Note that due to the nature of the AM processes, the surface roughness caused by the *stair casing effect* can only be improved by post-processing. Table 13.6 shows some practical options of post-processing for improvement of surface finish for parts from an AM process (Kumbhar and Mulay 2018).

13.4.7 Verification and Validation

Finished parts from an AM process are mostly used as prototyped parts for verification and validation of design concepts. The testing data collected from this step will be utilized to improve the design of a part or process in an iterative process until the final version of the part is made satisfactorily. Simple verification and validation such as part defects can be completed by a manual visual inspection. If some quantified specifications such as surface finishes, material strengths, or hardness are required, sophisticated testing machines are needed. For functional parts, measurements of part properties, such as tensile strength, toughness, fatigue strength, surface roughness, and porosity, are often necessary to qualify the part and determine whether design requirements have been met. Recent research interests on verification and validation have been shifted to microstructures, residual stresses, fatigues, and mechanical properties of manufactured AM parts (Kim et al. 2015).

13.5 Design for Additive Manufacturing (DfAM)

With gradually and progressively maturing of AM technologies, AM has been increasingly applied in different products. From a design perspective of products and manufacturing processes, selecting manufacturing processes depends on many factors. These factors are called *process selection drivers* and usually include (i) quantity of product, (ii) cost for tooling, manufacturing machines, and equipment, (iii) time required for processing, (iv) level of skilled labour required, (v) process supervision, (vi) energy consumption, (vii) availability

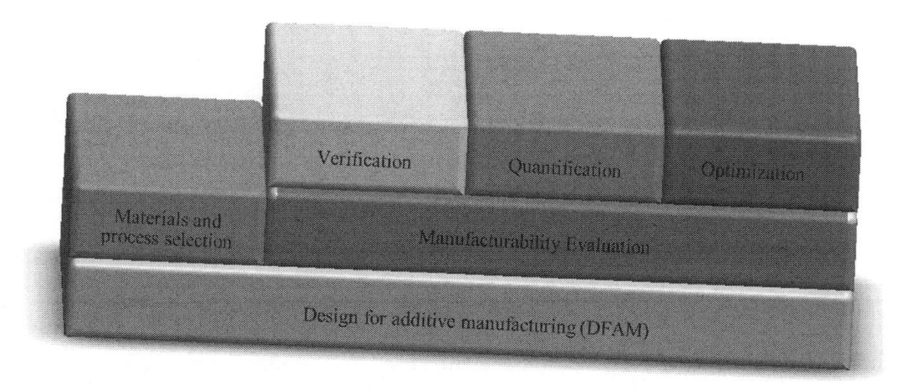

Figure 13.18 Design for additive manufacturing.

of material and cost of material, (viii) capabilities required to process material, (ix) product dimensions and size, (x) surface finish required and design tolerances, (xi) waste produced by the process, and (xii) maintenance and other costs (EngineeringClicks 2016).

DfAM is a methodology for designing products with consideration of constraints and capabilities of AM processes (Johansson and Sandberg 2016). The manufacturing process should be considered in the first phase when the product is designed. Otherwise, applicable manufacturing processes might be ineffective and inefficient. The purpose of DfAM is to maximize product performance through the synthesis of shapes, sizes, hierarchical structures, and material compositions, subject to the capabilities of AM technologies (Rosen 2014). Figure 13.18 shows some aspects where DfAM should be taken into consideration (Asadollahi-Yazdi et al. 2016).

DfAM is similar to design for manufacturing (DFM) of other manufacturing processes. Firstly, a designer must ensure that AM is the right choice for a product based on aforementioned criteria. Then designers should take into consideration the constraints from AM at the stage of product design.

13.5.1 Selective Materials and AM Processes

In comparison with the vast number of other conventional and non-conventional manufacturing processes, the variants of AM techniques are limited, and so are the number of applicable types of materials for these AM processes. Firstly, designers should ensure that the selected materials can be fabricated by AM and the corresponding AM machines are available. Table 13.7 gives examples of some common AM techniques and applicable materials together with the ranges of material strengths (Underdahl 2015).

13.5.2 Considerations of Adopting AM Technologies

AM has proven its advantage in applications; however, AM techniques are still facing a number of challenges. When AM is selected as the manufacturing process, the following aspects have to be taken into consideration at the stage of product design.

Table 13.7 Examples of commonly used AM techniques and applicable materials (Underdahl 2015).

Types	Strengths (psi)	Materials
Stereolithography (SLA), Digital Light Processing (DLP)	2 500 ~ 10 000	Thermoplastic and similar photopolymers
Fused deposition modelling (FDM)	5 200~9 800	ABS, PC, PC/ABS, PPSU, polytherimide (PEI)
Polymer jetting process	7 200~8 750	Acrylic-based photo-polymers, elastomeric photo-polymers
Selective laser sintering (SLS)	5 300~11 300	Nylon, metals
Direct metal laser sintering (DMLS)	37 700~190 000	Stainless steel, titanium, chrome, aluminium, and Inconel

1. *Shape optimization.* When the geometry and shape of a part are optimized, the constraints from an AM machine such as workspace and directional strengths are taken into consideration. This will have a significant impact on the amount of materials to be used and the time to manufacture the part.
2. *Pre- and post-processing needs.* Not all parts from an AM process are completely finished parts. Product designs have to take into accounts the needs of pre-processing and post-processing; for example, considering some post-processes such as removing supports, improving surface quality, or adding new features by conventional machining. Both pre- and post-processing bring challenges that may affect the process of AM. Dimensional accuracy is an important issue, but post-processes for high accuracy may lead to significant and expensive equipment.
3. *Options of AM techniques.* Currently, there are seven types of AM processes (Section 12.3), where each type of AM processes has specific advantages and disadvantages. The selection of an AM technique affects the mechanical or physical properties of the part.
4. *Troubleshooting and error correction.* An AM process, with no exception from other manufacturing processes, may involve numerous uncertainties and introduce errors from many sources. For AM, three main sources of error are *data preparation*, *abnormal process*, and *abnormal material supply*. The latter two are difficult to avoid and troubleshooting may be a better solution. Build failure and quality issues can lead to increased costs associated with process predictability and repeatability.
5. *Hardware and maintenance issues.* Similar to the challenges with troubleshooting and error correction, no available AM machines are flawless and they require anticipated and unanticipated repairs. This could, for instance, include issues associated with materials or specific processes.
6. *Cost effectiveness.* When the required volume increases, i.e. medium to high production volumes, the process economics become unfavourable and the cost effectiveness decreases.
7. *Part orientation.* The orientation of a part should be optimized to maximize the strengths at the most critical area and minimize the manufacturing time. Not all desired criteria can be optimized simultaneously, so designers need to determine the part orientation that meets certain requirements at the expense of other properties. Part orientation

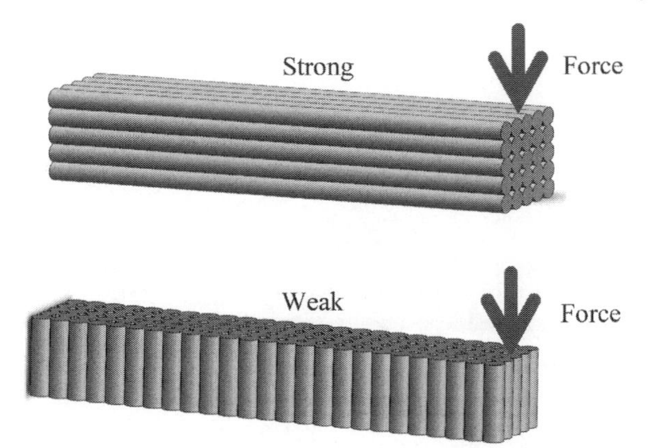

Figure 13.19 Print orientation affects bulk and local mechanical properties.

affects the strengths of a part when it is subjected to the external loads in application. Figure 13.19 shows an example of the varying capabilities in resisting loads due to the difference of part orientation in an AM process. Therefore, special attention should be paid to anisotropic parts.

13.5.3 Part Features

When AM is selected to make a part, the features on part should be favourable to an AM process. The following guides should be used to determine part features (Rosen 2014; Geometric 2013):

1. Verify whether the part size is within the workspace of an AM machine.
2. Check the minimum wall thickness to see if it is less than the allowable minimum thickness in a certain AM process. It also applies to checking the minimum distance of generic pockets (e.g. holes, cutouts, or pockets) and the minimum distance from an edge to a pocket.
3. Identify the faces requiring a support and minimize the number of such faces.
4. Check the minimum thickness of faces with supports and the minimum feature sizes.
5. Review rib parameters and check rib reinforcement and verify the feasibility based on the ratios of (i) rib-base thickness to nominal wall thickness and (ii) rib height to nominal wall thickness.
6. Check the minimum ratio of inside and outside diameters of the boss and the minimum ratio of the height and outside diameter of the boss.
7. Check the minimum hole diameter to thickness or depth ratio.
8. Recognize knife edges and eliminate as many of them as possible.
9. Determine a corner radius to see if it is larger than an allowable minimum radius.

13.5.4 Support Structures

One of the significant advantages of AM is that the manufacturing process is not subjected to significant mechanical force and no fixture is needed during the process. However, a *support structure* must be considered when a part has the geometry of undercuts in the

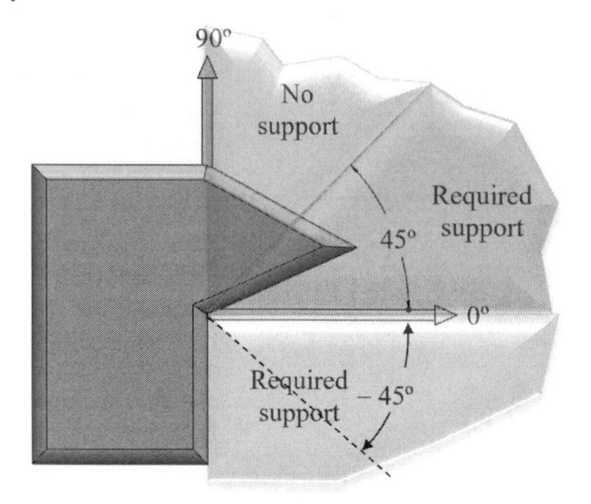

Figure 13.20 Definition of overhanging angle.

direction where the materials are added. Generally, the following principles can be applied for the design of support structures (Jiang et al. 2018):

1. The support structure must prevent a part from collapsing or warping as outer contours with undercuts would require supports. For metal processes, stress and strain should be analysed via thermal analysis to determine possible breaks.
2. The strength at a connection of a support structure and original part should be minimized for easy removal.
3. The contact area of a support structure and original part should be minimized to reduce surface deterioration after the support structure is removed.
4. The material consumption and the build time should be considered for a trade-off with part quality.

Figure 13.20 gives the definition of an overhanging angle. As *the rule of thumb* for a selective laser sintering (SLS) or FDM process, a support structure must be provided when the overhanging angle is less than 45° or negative, as indicated in the figure (Perez 2015).

Many AM software tools such as CURA and Slic3r for FDM processes and Magic for metallic processes have the capability to generate support structures automatically.

13.5.5 Process Parameters

Figure 13.17 has shown that numerous process parameters are involved in AM, the cause–effect relationships of these parameters and part quality, manufacturing time, material consumption, and cost need to be analysed. For example, the amount of materials added to the part depends on the size and the extruding speed of the extruder in FDM. Therefore, the extruding speed must be tuned to ensure that the appropriate amount of materials is supplied and added to the part. Figure 13.21 shows that either too fast or too low an extruding speed would cause the part to have a quality problem.

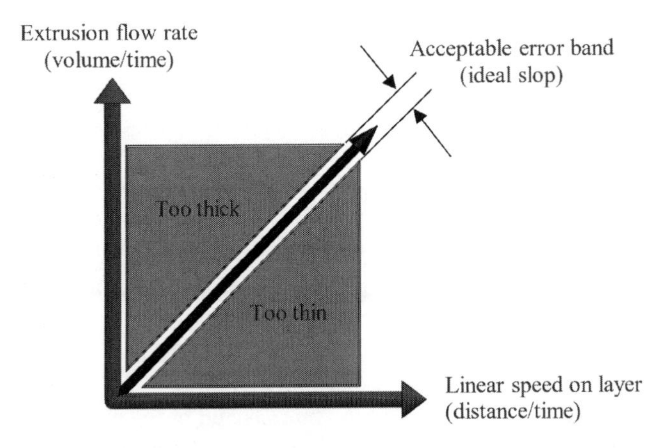

Figure 13.21 Ideal extrusion flow rate to avoid quality problems on the part.

13.6 Summary

Different types of AM processes, such as stereolithography apparatus (*SLA*), digital light processing (*DLP*), selective laser sintering (*SLS*), selective laser melting (*SLM*), multijet fusion (*MJF*), 3D printing, Plyjet, selective deposition lamination (*SDL*), ultrasonic additive manufacturing (*UAM*), fused deposition modelling (*FDM*), and direct metal deposition (*DMD*), share the commonalities that products are directly made from their CAD models. In contrast to conventional or non-conventional subtractive manufacturing, AM makes products directly from CAD models. An AM process eliminates the need to use manufacturing relevant CAD tools such as CAM, CAFD, and CAPP. Therefore, AM processes possess many advantages, such as creating complex geometry, reducing assemblies for a single structure, cost savings on tooling, a shortened product design cycle, and consistence of the product quality.

AM has been identified as the cornerstone of realizing Industry 4.0 and has placed manufacturing competitiveness and higher productivity in both national and supranational agenda worldwide. AM is anticipated as a key to secure a robust industrial base. Despite the rapid development of various AM techniques, a great deal of research is needed to address the challenges related to the two key enabling technologies, namely 'materials' and 'metrology', to achieve this functionality in a predictive and reproductive way (Tofail et al. 2018).

Future AM are expected to meet the demands of production lines based on the required quality standards. AM will not largely replace traditional manufacturing, but it will expand greatly in terms of applicable materials, features, sizes and quality of products, manufacturing speeds, and hybrid processes. Future AM will present various types of new technologies that enable direct manufacturing from digital product files without process planning requirements that are required in conventional methods (Al-Ahmari 2017).

The procedure of an AM process is straightforward and very simplified in comparison with that of conventional subtractive manufacturing since no special fixture or tooling is needed and programs are generic to any objects. However, a large number of process parameters are involved; these process parameters affect the quality of parts. An AM process has some specific requirements on part features, materials, post-processing, and operation conditions. The DfAM approach should be adopted at the phase of product design.

13.7 Design Project

13.1 Read the article by Greene et al. (2019) on the prosthetic hand design shown in Figure 13.21 or other similar designs in the literature, understand the functional requirements (FRs) of custom prosthetic hands, discuss the limitations of existing designs, and create your own products to meet the specified FRs.

Create CAD models for the parts in your design and make all of the structural parts from 3D printing processes. At manufacturing stages, use the machine software to estimate printing times and record actual printing times to evaluate the performance of the planning software of the machine (Figure 13.22).

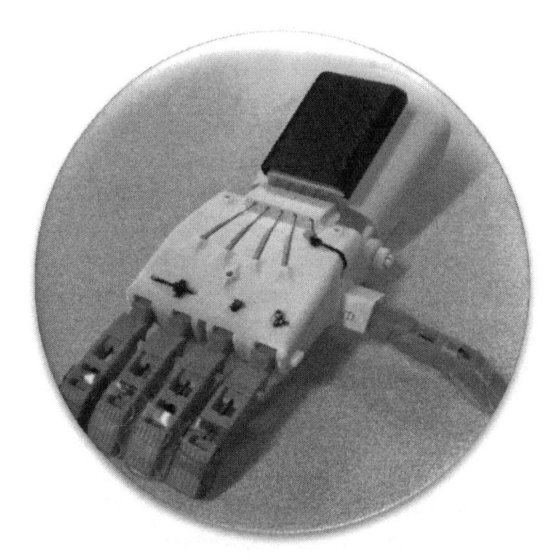

Figure 13.22 Prototyped prosthetic hand by Greene et al. (2019) as reference for Design Project 13.7.1.

References

3D Printing Industry (2015). The top 10 3D printing stocks for 2015. https://3dprintingindustry.com/news/fiscal-2014-revenue-results-3d-printings-top-10-guns-44789/.

Additivemanufacturing (2019). 7 Families of Additive Manufacturing. https://www.additivemanufacturing.media/cdn/cms/7_families_print_version.pdf.

Al-Ahmari, A.M. (2017). Editorial direct digital manufacturing. *Journal of King Saud University – Engineering Sciences* 29: 203.

All3DP (2019). Want to get the best results from your 3D printer? https://all3dp.com/1/best-3d-slicer-software-3d-printer/#what.

Asadollahi-Yazdi, E., Gardan, J., and Lafon, P. (2016). Integrated design in additive manufacturing based on design for manufacturing. *World Academy of Science, Engineering and Technology: International Journal of Industrial and Manufacturing Engineering* 10 (6): 1144–1151.

Campbell, R.I., Martorelli, M., and Lee, H.S. (2002). Surface roughness visualization for rapid prototyping models. *Computer-Aided Design* 34 (2002): 717–725.

EngineeringClicks (2016). Manufacturing process selection for your product. https://www.engineeringclicks.com/manufacturing-process-selection.

EPMA (2019). Introduction to Additive Manufacturing technology, a guide for designers and engineers. http://s550682939.onlinehome.fr/CommissionsThematiques/DocComThematiques/EPMA_Additive_Manufacturing.pdf.

Frost and Sullivan (2016). Global Additive Manufacturing Market, Forecast to 2025. http://namic.sg/wp-content/uploads/2018/04/global-additive-manufacturing-market_1.pdf.

GE (2019). Additive manufacturing processes. https://www.ge.com/additive/additive-manufacturing/information/additive-manufacturing-processes.

Gebhardt, A. and Hötter, J.-S. (2016). *Additive Manufacturing: 3D Printing for Prototyping and Manufacturing*. Carl Hanser Verlag GmbH & Co., KG. ISBN: 978-1-56990-582-1. https://www.hanserpublications.com/SampleChapters/9781569905821_9781569905821%20SAMPLE%20PAGES%20Additive%20Manufacturing%20Gebhardt%20Hotter.pdf.

Geometric (2013). Design for additive manufacturing. https://dfmpro.geometricglobal.com/files/2013/07/Whitepaper-Additive-Manufacturing.pdf.

Greene, S., Lipson, D., Mercado, A., and Soe, A.H. (2019). Design and manufacture of a scalable prosthetic hand through the utilization of additive manufacturing. https://web.wpi.edu/Pubs/E-project/Available/E-project-042816-081456/unrestricted/Final_Report.pdf.

Ha, S. (2016). 3D printing/process parameters. https://worldmaterialsforum.com/files/Presentations/WS1-1/WMF%202016%20-%20WS%201.1%20-%20Sung%20Ha%20Final.pdf.

Jiang, J., Xu, X., and Stringer, J. (2018). Support structures for additive manufacturing: a review. *Journal of Manufacturing and Materials Processing* 2: 64. https://doi.org/10.3390/jmmp2040064.

Johansson, M. and Sandberg, R. (2016). How additive manufacturing can support the assembly system design process. https://www.diva-portal.org/smash/get/diva2:943228/FULLTEXT01.pdf.

Kim, D.B., Witherell, P., Lipman, R., and Feng, S.C. (2015). Streamlining the additive manufacturing digital spectrum: a systems approach. *Additive Manufacturing* 5: 20–30.

Kumbhar, N.N. and Mulay, A.V. (2018). Post processing methods used to improve surface finish of products while are manufacturing by additive manufacturing technologies: a review. *Journal of The Institution of Engineers (India): Series C* 99 (4): 148–487.

Levchenko, A. (2015). Additive manufacturing as a mean of rapid prototyping: from words to the actual model. https://www.theseus.fi/bitstream/handle/10024/95525/Levchenko_Anastasia.pdf?sequence=1&isAllowed=y.

Perez, A.J. (2015). How to think like a 3D design: a practical guide to additive manufacturing. https://www.mitefgreece.org/wp-content/uploads/2015/11/1500_1630_AJ-Perez_How-to-think-as-a-3D.pdf.

Redwood, B. (2019). Additive manufacturing technologies: an overview. https://www.3dhubs.com/knowledge-base/additive-manufacturing-technologies-overview.

Rosen, D. (2014). Research supporting principles for design for additive manufacturing. *Virtual and Physical Prototyping* 9 (4): 225–232.

Sinha, N. (2016). Additive manufacturing. http://home.iitk.ac.in/~nsinha/Additive_Manufacturing%20I.pdf.

Tamburrino, F., Perrotta, V., Aversa, R., and Apicella, A. (2015). Additive technology and design processes: an innovative tool to drive and assist product development. In: *Heritage and Technology Mind Knowledge Experience, Aversa, Capril* (11–13 June 2015), 1742–1747. https://www.researchgate.net/publication/315661330_Additive_technology_and_design_process_an_innovative_tool_to_drive_and_assist_product_development.

Tofail, S.A.M., Koumoulos, E.P., Bandyopadhyyay, A. et al. (2018). Additive manufacturing: scientific and technological challenges, market uptake and opportunities. *Materials Today* 21 (1): 22–37.

Underdahl, B. (2015). Digital manufacturing for dummies. https://www.protolabs.pl/media/1011252/digital_manufacturing_for_dummies_uk.pdf.

Wu, B., Myant, C., and Weider, S.Z. (2017). The value of additive manufacturing: future opportunities. www.imperial.ac.uk/media/imperial-college/imse/IMSE-Briefing-paper-2-AM.pdf.

14

Design for Sustainability (D4S)

14.1 Introduction

With the continuous increase in global wealth and the improvement of human living standards, people are becoming more and more conscious of the deterioration of today's global environment. Some buzzwords, such as *global warming*, *pollution*, *shortage of oil*, and *extinction of species*, have frequently been used in the news headlines and major subjects of political disputations. How to pursue the sustainability of the economy, society, and environment has been recognized as a nation's priority for both developed and developing countries. In manufacturing, many terminologies related to sustainability, such as *Green Manufacturing*, *Lean Manufacturing*, *Environmentally Conscious Manufacturing*, and *Sustainable Manufacturing*, have been proposed.

This chapter gives an introduction to the rationales, metrics, enabling technologies, and computer aided tools for system sustainability. In Section 14.2, the concept of sustainability is discussed. In Section 14.3, the main drivers for the sustainability of manufacturing systems are presented. In Section 14.4, the relationships between manufacturing and sustainability are discussed. In Section 14.5, a set of quantified metrics are proposed to evaluate the sustainability quantifiably. From Section 14.6 to Section 14.8, reconfigurable systems, lean production (LP), product lifecycle assessment (LCA), and continuous improvement (CI) are introduced as four examples of the main enabling technologies for sustainability. In Sections 14.10 and 14.11, environmental sustainability is quantified and *SolidWorks Sustainability* is presented as an example of computer aided tools for the design for sustainability (D4S). Section 14.12 gives a summary on computer aided technologies in the field of sustainable manufacturing.

14.2 Sustainable Manufacturing

According to the US National Research Council, *sustainability* is defined as the level of human consumption and activity, which can continue into the foreseeable future, so that the systems that provide goods and services to humans persist indefinitely. Interactions within and across these levels is critical to the fundamental understanding of sustainable manufacturing systems (Rachuri 2009). Sustainable manufacturing is developing technologies to transform materials without the emission of greenhouse gases and the use of

Computer Aided Design and Manufacturing, First Edition. Zhuming Bi and Xiaoqin Wang.
© 2020 John Wiley & Sons Ltd. This Work is a co-publication between John Wiley & Sons Ltd and ASME Press.
Companion website: www.wiley.com/go/bi/computer-aided-design

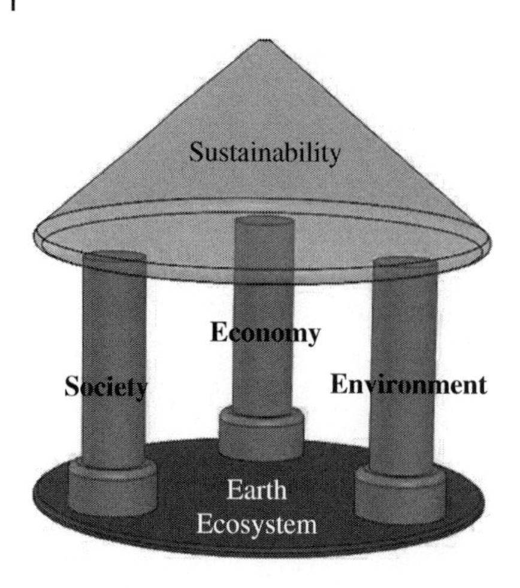

Figure 14.1 Three pillars or dimensions of sustainability.

non-renewable or toxic materials or generation of waste (Allwood et al. 2008). Sustainable systems are characterized by interlinked interactions at various levels spanning *economic*, *environmental*, and *societal* issues (Kibira et al. 2009). Therefore, three dimensions or supporting pillars of sustainability are the economy, environment, and society, as shown in Figure 14.1. These three pillars are interdependent, and in the long run none can exist without the others. They are together designed to serve as a common ground for numerous sustainability standards and certification systems (Wikipedia 2019).

The US Department of Commerce (2010) has defined *sustainable manufacturing* as the creation of manufacturing products that use materials and processes that minimize negative environmental impacts, conserve energy and natural resources, are safe for employees, communities, and consumers, and are economically sound. There is a need for the system as a whole to be sustainable.

As shown in Figure 14.2, sustainable manufacturing is rooted in *lean production* and *green manufacturing*. Lean production focuses on the cost saving and waste reduction and green manufacturing (GM) aims to meet product design requirements and minimize environmental impact simultaneously. However, minimizing environmental impacts is a necessary but not a sufficient condition for sustainability manufacturing. The three most important components of a sustainable manufacturing system are: (i) the selection and application of appropriate metrics to measure manufacturing sustainability, (ii) the completion of comprehensive, transparent, and repeatable lifecycle assessments, (iii) the adjustment and optimization over the system to minimize cost, environmental impact, and adverse social impact based on the chosen metrics and the LCA (Reich-Weiser et al. 2008). The innovative elements in sustainable manufacturing are *remanufacture*, *redesign*, *recover*, *reuse*, *recycle*, and *reduce* (6R), and all of the enabling technologies for lean production and green manufacturing can be fully utilized in sustainable manufacturing.

Traditionally, manufacturing enterprises put a higher priority on profits and the reactions to reduce environmental impacts were the last resolutions when they were needed, but the importance of sustainability has gradually been recognized by manufacturing enterprises. Figure 14.3 shows the evolution of the concepts and practices relevant to sustainability. With regards to sustainability, the practical methods have shifted from passive remedies to active preventions, and the scope of manufacturing activities have been expanded from

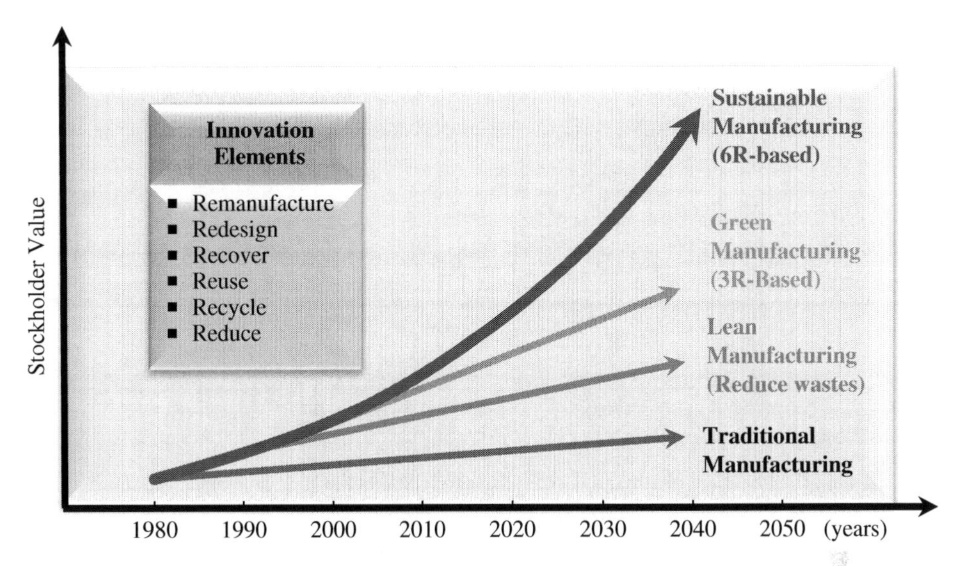

Figure 14.2 From *traditional manufacturing* to *sustainable manufacturing.*

specific manufacturing operations to all business operations over the product lifecycles (PLCs) from selecting raw materials, to manufacturing, assembling, using, and disposing of products. The ultimate goal is to develop the *industrial ecology* theory and methods, so that all of the enabling technologies are integrated and partnered with the environment and society as an eco-manufacturing society for the long-term sustainability of the economy, the environment, and society.

Industrial ecology is the study of the physical, chemical, and biological interactions and interrelationships in industry, the environment, and society. Industrial ecology aims to move industry from a linear to a cyclical or closed system where waste is used as an input rather than to be disposed of and cause damage to society. Industry ecology looks into both the material and energy flows through the system to find and eliminate inefficiency and waste (Ashrafi 2014).

14.3 Drivers for Sustainability

A manufacturing system used to be a closed system in the economy domain. The manufacturer runs its businesses to gain profits by transferring raw materials into finished products, and selling products to customers to meet their needs via product uses. The system has clear boundaries for all of the manufacturing activities. However, with the advancement of manufacturing technologies and the continuous growth of manufacturing capacities, manufacturing businesses are facing fiercer and fiercer competitions due to the saturation of manufacturing capacities for global market needs. Traditional closed and isolated manufacturing systems are no longer able to survive for very long since manufacturers have to take into consideration the changes of business environmental factors and make their manufacturing systems adapt to these changes to catch emerging business opportunities (Bi et al. 2014).

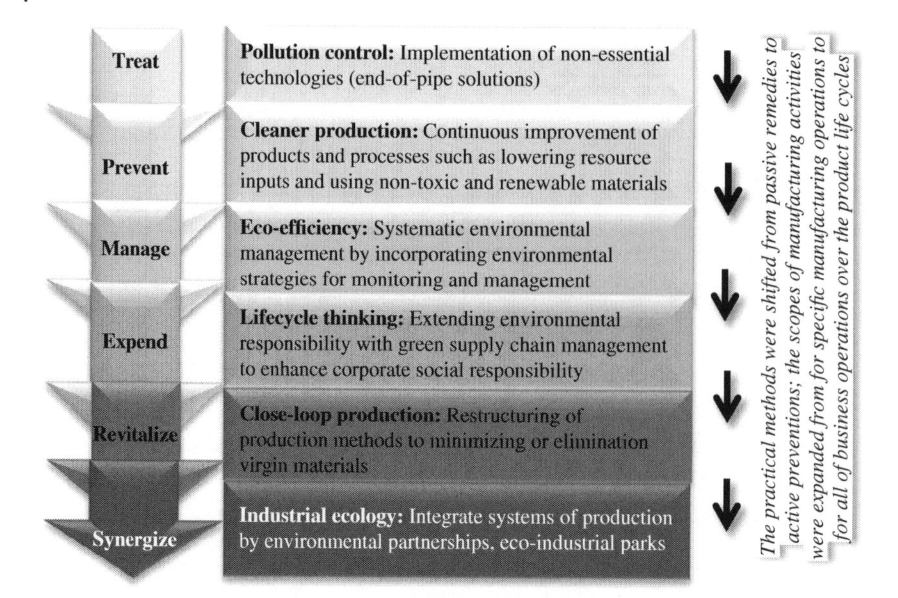

Figure 14.3 Evolution of sustainable manufacturing and concepts. (*See color plate section for color representation of this figure*).

As shown in Figure 14.4, the business environment of a manufacturing system becomes an interlinked part of our living environment and society. The boundaries between a manufacturing system and the living environment and society become invisible. In other words, the environmental and social factors become the input parameters of the system and, in return, the outputs of the system will affect not only the economic domain but also the environmental and social domains. From an environmental perspective, important factors *are shortage of natural resources*, *energy consumption*, *globalized markets environmental concerns*, and environmental regulations and laws to enforce manufacturing companies to practice 6R. From a society perspective, human beings affect manufacturing systems directly as operators, designers, managers, marketing or service staff, consumers, and living beings on the earth environment. The ways for society to interact with manufacturing systems are *services as products*, *human machine interactions*, *product or machine replacements*, *remote working or services*, *wealth distribution*, and *decentralization manufacturing*.

Today, public perception of sustainability has been shaped by news and documentations such as *global warming*, *rising cost of energy*, and the *paucity of non-renewable resources* (Seidel et al. 2006). The identified five drivers for sustainable manufacturing are: (i) the shortage of natural resources, (ii) the dramatic increase in world population, (iii) global warming, (iv) pollution, and (v) an unstoppable global economy (Westkamper et al. 2007), which are discussed briefly below.

14.3.1 Shortage of Natural Resources

The shortage of natural resources is caused by two factors: (i) the reduction of unexploited or undiscovered nature resources and (ii) the accelerated consumption of nature resources.

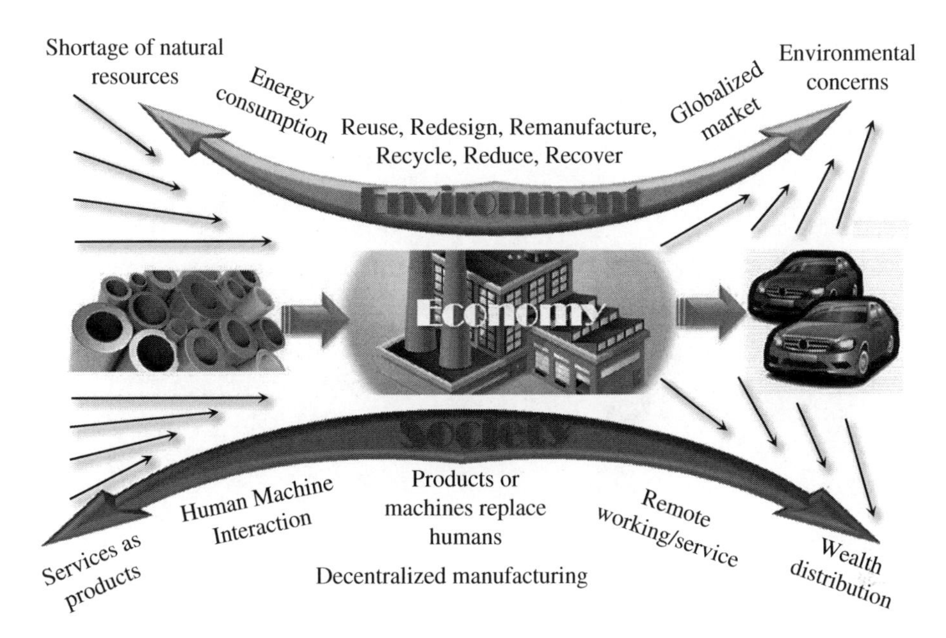

Figure 14.4 Changes of manufacturing business environment. (*See color plate section for color representation of this figure*).

Due to the expansion in developing countries, the demand of global energy is predicted to increase by 35% over the period 2015–2040. The total demand of the primary energy was forecasted to be increased by 96 *million barrels of oil equivalent per day* (*mboe/d*) from 276 mboe/d in 2015 to an estimated 372 mboe/d in 2040. On a yearly average, the growth rate would be 1.2% in the forecasted period (Opec 2017). Natural resources are mainly used to generate energy to meet the needs of human civilization. Since 1950, the rapid increase in global energy consumption can be observed in Figure 14.4 (Ritchie and Roser 2019). It shows that the global primary energy consumption is measured in terawatt-hours (tWh) per year, and the total energy consumption in 2017 reached nearly 160 000 tWh per year. The worst thing is that the majority of energy is from *non-renewable resources* such as coal, crude oil, and natural gas.

The consumed energy came from discovered and explored energy. Figure 14.5 shows the relative trends of energy exploration, discovery, and consumption (GEP project 2019). In the figure, the energy is measured by billions of barrels, and the data spans the years from 1950 to 2000. It shows that consumed energy has been greater than discovered energy since the early 1990s.

The energy from natural resources is mostly non-renewable. Therefore, accessible energy will be continually declining when the consumed energy is greater than the explored and discovered energy. It was predicted that discovered fossil fuels (*coal*, *oil*, and *natural gas*) can last for energy consumption for 114, 52.8, and 50.7 years at the present production rate, respectively (Ritchie 2017). There is an emerging need to reduce energy consumption and utilize more renewable energy sources in our living environment (Figure 14.6).

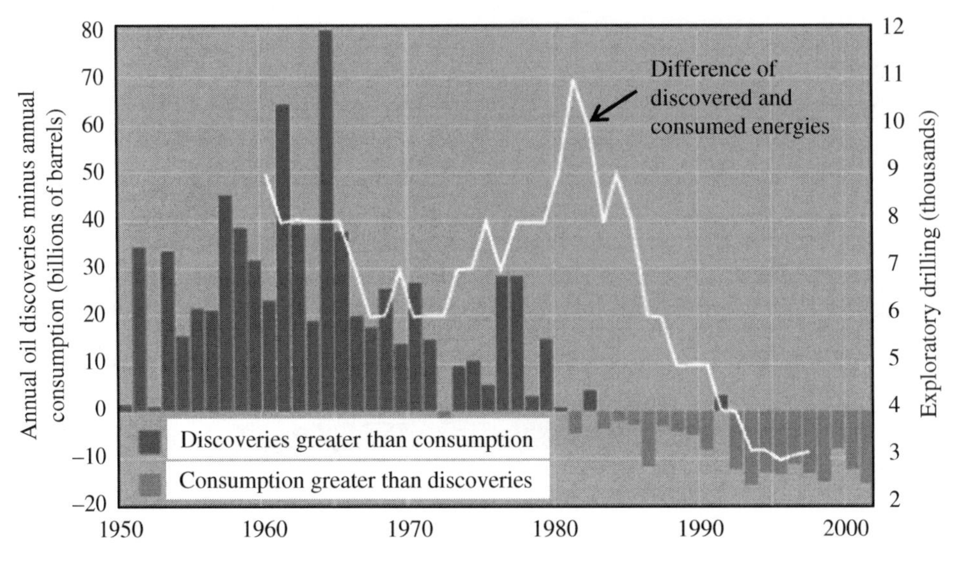

Figure 14.5 Energy exploration, discovery, and consumption (GEP Project 2019).

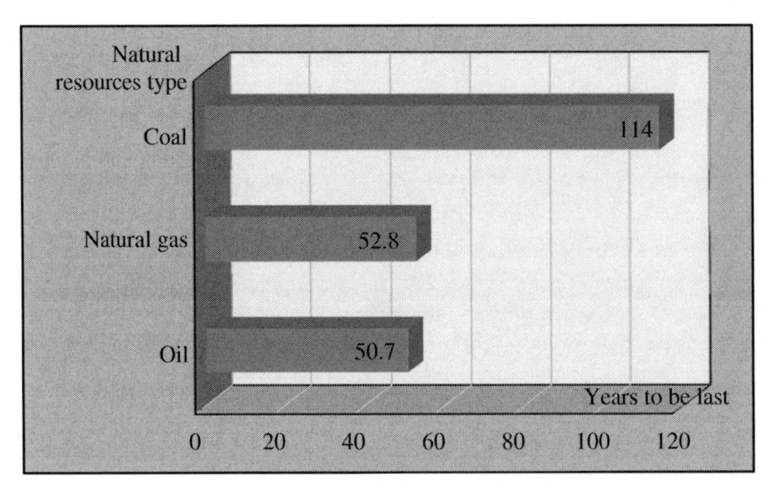

Figure 14.6 Predicted years of fossil fuel reserves left (Ritchie 2017).

14.3.2 Population Increase

The population growth affects the quality of the environment adversely. More land is used for agriculture or living purposes and the environment changes drastically. As the population of humans grows in certain cities or rural areas, more resources are needed to sustain the well-being of the population. This increases the pressure on accessible resources and, as a consequence, many habitats are being destroyed. Humans are using up more and more resources and nature cannot replenish these resources fast enough to supply our needs. The atmosphere is also negatively impacted by population growth (Blanc 2017). During the twentieth century, the human population increased from less than 2 billion to over 6 billion

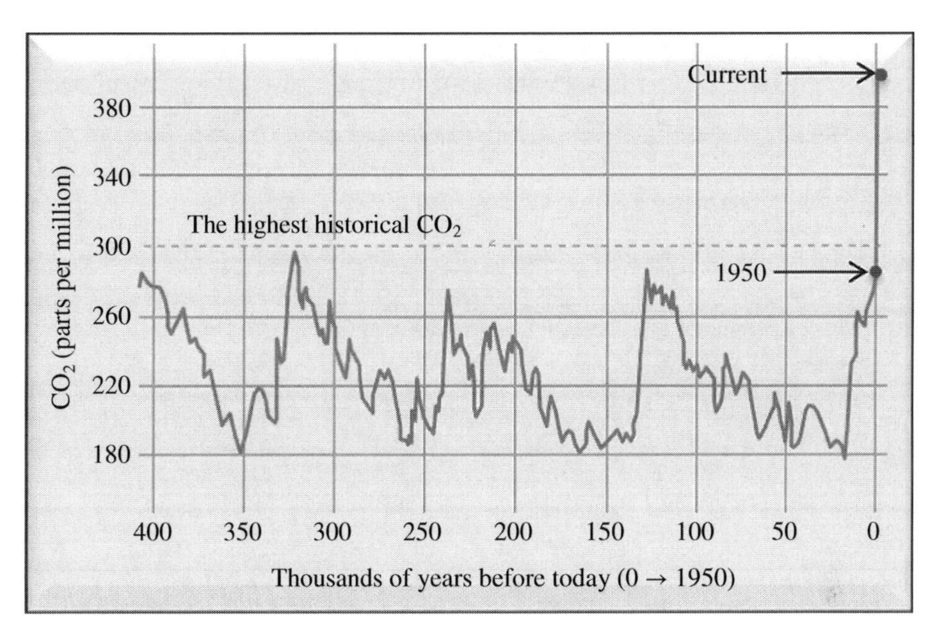

Figure 14.7 Increase of CO_2 level over time (UCSUSA 2016).

people. The number of cities with more than a million people has grown from less than 20 to more than 300 in the last 75 years (Malain and Walrond 2010). The population growth also affects economic development adversely. A 1% increase in the population could decrease the income per capita by 2% (Thuku et al. 2013).

14.3.3 Global Warming

The United Nations has declared that the evidence of a warming trend is *unequivocal* and that human activities have very likely been the driving force on this change over the past 50 years (Ramani et al. 2010). Whether or not one agrees with the prediction about global warming, it is clear that energy and resources of production are costly and the costs are likely to be increased (Dornfeld 2009). *Global warming* is a planetary crisis that threatens the survival of our civilization and the habitability of the Earth. Global warming is caused by carbon dioxide and ita equivalents, such as carbon monoxide and methane, that are released into the atmosphere. Figure 14.7 shows that the level of carbon dioxide (CO_2) has increased rapidly since 1950 (UCSUSA 2016). The level of CO_2 in the atmosphere is 100% greater today than it was 100 years ago, while the highest historical data on CO_2 in the past thousands of years never exceeded 300 parts of CO_2 per million in the atmosphere. Current levels of greenhouse gases, carbon dioxide, methane, and nitrous oxide, in our atmosphere are higher than at any point over the past 800 000 years, and their ability to trap heat is changing our climate in multiple ways (National Geographic 2019).

Global warming has caused a number of severe environmental issues. For example, Figures 14.8 and 14.9 show the corresponding changes of *annual global temperature* (UCSUSA 2016) and *sea level* (Four Peaks Technologies 2012), respectively. Figure 14.8 shows the change of the anomaly (°F) relative to the twentieth century average with respect

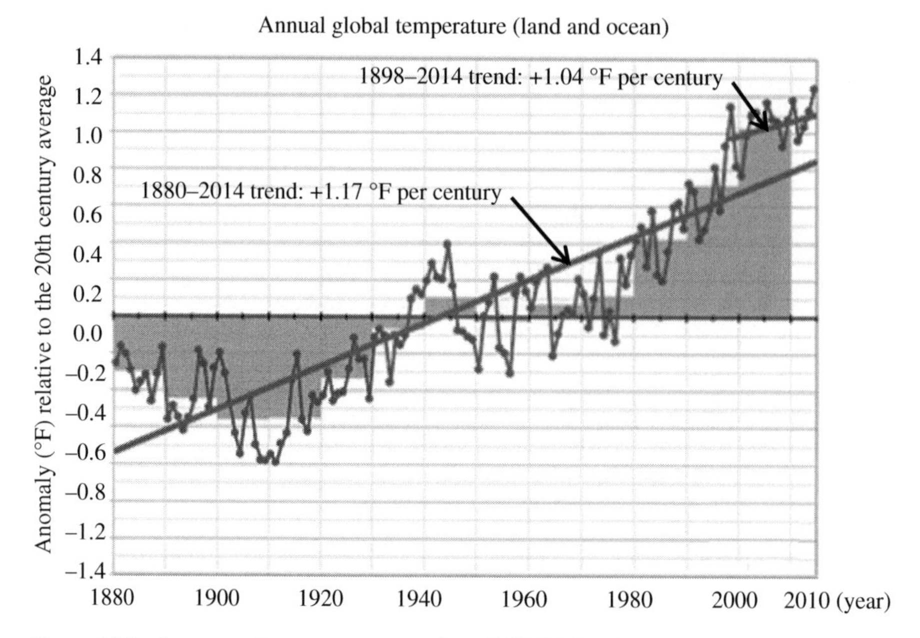

Figure 14.8 Increase of temperature over time (UCSUSA 2016).

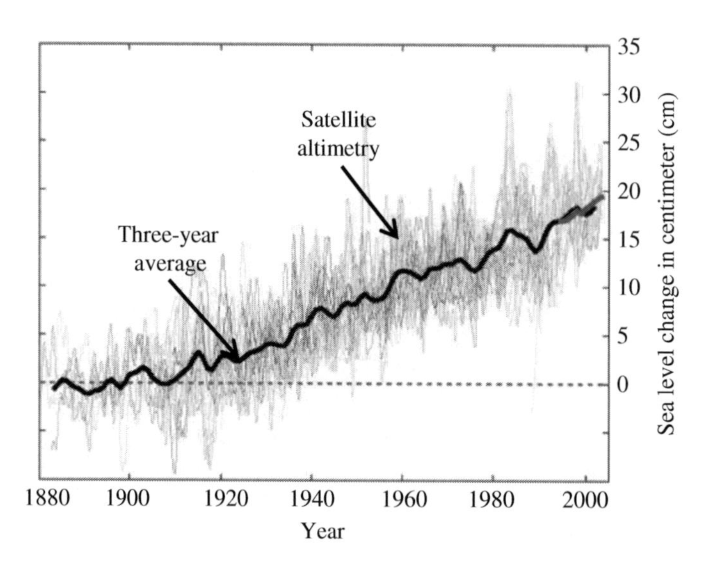

Figure 14.9 Sea level rise in recent years (Four Peaks Technologies 2012).

to time, with the increase rate in the duration of 1880 to 2014 was 1.17 °F per century. The sea level has increased as the global temperature increased. Figure 14.9 shows that the sea level has increased by around 20 cm since 1880.

It is worth mentioning that for every centimetre rise in the vertical sea level, around one to 10 m of the coastline will be inundated. Moreover, two-thirds of the major urban areas are partially in the sea level zone of zero to 10 m, 21% of the urban populations of the least developed nations are also in that zone, and 75% of people in the zone are in Asia (Neumann et al. 2015). Therefore, the consequence of a global temperature increase is that the sea level rise from global warming will quickly deteriorate our living environment.

14.3.4 Pollution

Each American used an average of 90 000 kWh of power in 1997, equivalent to 8000 l of oil. The world consumes 75 million barrels of oil a day (Lerouge and McDonald 2008). Each American produced a daily average of 2 kg of solid waste, or approximately 163 million tons of municipal solid waste that is eventually land filled. That figure could climb to 363 billion tons annually by 2030, which is enough to bury Los Angeles 100 m deep (Lee et al. 2001).

Pollution and environmental deterioration bring the risk of *food and water scarcity*. Food and water scarcity increase the stress on water supplies, desertification, and shifts the growth bands that affects water and food production. It causes *agflation* since food costs will continuously increase due to the scarcity of water, oil shortage for fertilizer, and scarce land (Purser 2019).

The percentage of humanity subject to water scarcity is predicted to increase from 10% in 2025 up to 33% in 2050. In addition, there is growing concern over food scarcity; for example, there are 19 grain export embargoes and 10 riots due to the anomalies of food distribution. It has been predicted that there will be an 80% increase in energy demand, which will cause 70% of global greenhouse gas emissions by the mid-century. To make our living environment sustainable, greenhouse gas emissions need to decrease to 60% below the present levels by 2050 if humans are to avoid catastrophic climate changes (FAO 2017). As resources are getting the harmful effect of waste and pollution, this is causing a measurable negative impact on our living environment and governments around the world are getting actively involved in the development of products that are not only profitable and add value to our societies, but also cause less damage to the environment (Sarkis 2001).

14.3.5 Globalized Economy

A globalized economy relates closely to trades, capital flows, the movement of labour, and, especially, the globalization of manufacturing businesses. With the gradual dismantling of trade barriers and capital flows becoming easier, globalization of production has flourished (Islam 2015).

The global manufacturing capacities exceeds market needs, and enterprises are facing fiercer and fiercer competition worldwide. Therefore, the globalized economy raises expectations for customers, shareholders, governments, and the public on environmental sustainability and manufacturers are highly pressured to improve the environmental performance for their products and processes.

The world is more crowded, more polluted, urban, more ecologically stressed, and warmer than ever before in recorded history. As the expectation of society, the public, and customers on products increases, the transparency of manufacturing businesses becomes more prevalent and companies are recognizing the need to act on sustainability. Pursuing sustainability becomes one of the most critical business strategies to add long-term value on products and to run businesses healthily in the ecological, social, and economic environment. Sustainability is built on the assumption that developing such strategies foster the company's longevity (Haanaes 2016).

14.4 Manufacturing and Sustainability

Section 14.3 has discussed five main drivers for sustainability. All of these drivers are closely relevant to manufacturing.

14.4.1 Natural Resources for Manufacturing

Manufacturing is the leading sector to consume natural resources. As shown in Figure 14.10 (EIA 2017), nearly half of the electricity produced in the United States was used to operate machinery (48.2%); the other share of the electricity use was for manufactured products, i.e. process heating and boiler use (14.4%), facility heating, ventilation, air conditioning, and cooling (9.5%), electrochemical processes (6.8%), process cooling and refrigeration (7.3%), lighting (6.5%), and other miscellaneous processes and facility uses (7.3%). In addition, the amount of the energy consumption has increased exponentially as shown in Figure 14.11.

Moreover, the raw materials for all manufactured products are from nature. Other than fossil fuels and water, the natural resources such as phosphorous and rare earth elements are becoming drained as well (Ruz 2011).

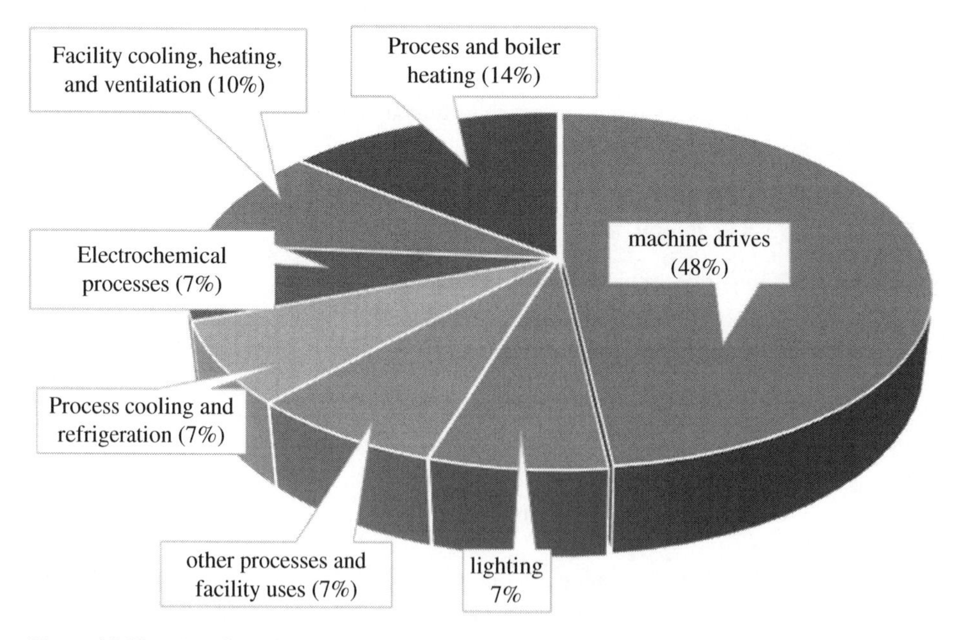

Figure 14.10 Manufacturing electricity consumption by major end users (EIA 2017).

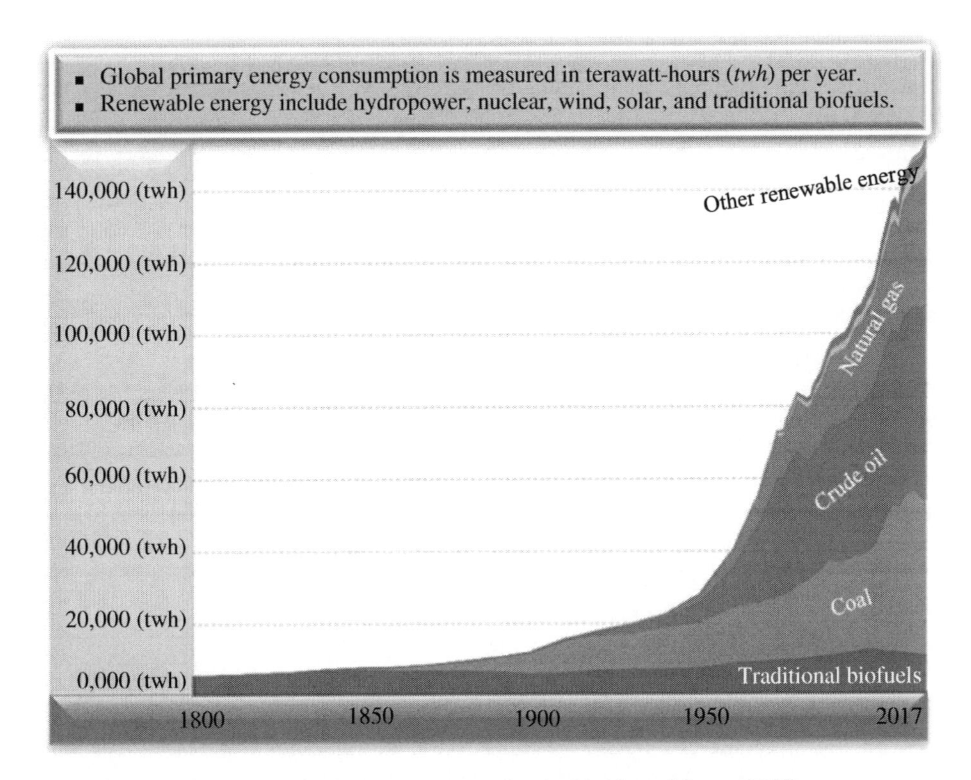

Figure 14.11 The prediction of energy consumption by Ritchie and Roser (2019).

14.4.2 Population Increase and Manufacturing

The relationships between population and manufacturing seem to be the relationships of chicken and eggs.

On the one hand, the *Industrial Revolution* (*IR*) marked a major turning point in Earth's ecology and humans' relationship with their environment. The IR dramatically changed every aspect of human life and lifestyles. Humans have been around for about 2.2 million years. By the first millennium before Christ (BC), the estimated human population ranged from 150 to 200 million, and the total population was 300 million in the year 1000. The world human population growth rate was ~0.1% for the next seven to eight centuries. Since the dawn of IR in the mid-1700s, the world's human population grew by about 57% to 700 million. In only 100 years after the onset of IR, the world population grew by 100% to two billion people in 1927 (about 1.6 billion by 1900). During the twentieth century, the world population would take on exponential proportions, growing to six billion people just before the start of the twenty-first century. That is a 400% population increase in a single century (McLamb 2011).

On the other hand, the growing population demanded more subsistence to meet its living, cultural, and material needs. Subsistence is either from nature and agricultural production or from manufacturing. The larger the population of the world, the higher the living standards is, the greater is its demand for manufactured goods.

14.4.3 Global Warming and Manufacturing

The manufacturing section is responsible for around 19% of direct emissions and 11% of indirect emissions through electricity use. In addition, over 90% of emissions from powered products such as appliances, electronics, and autos are manufactured products. Taking into consideration the greenhouse gas footprint, it is clear that the manufacturing sector is mainly responsible; meanwhile, the manufacturing sector will be greatly affected by global warming and any future climate change regulatory regime. Manufacturing enterprises are confronting the risks and opportunities that global warming presents (C2ES 2017).

14.4.4 Pollution and Manufacturing

A manufacturing system turns raw materials into finished products to meet customers' needs via a series of manufacturing processes. However, the system outcomes are not just finished products but waste and the environmental pollution caused by consumed energy and natural resources. The by-products of manufacturing including waste materials or substances produced by the manufacturing process itself are usually harmful to the environment.

Manufacturing generates over 60% of annual non-hazardous waste. Increasingly severe legislation demands a reduction in the environmental impact of products and manufacturing processes (Ijomah et al. 2007). Kaebernick and Kara's survey (Kaebernick and Kara 2006) concluded that most companies acknowledge the importance of environmental issues. Between 80% and 90% of companies ranked environmental issues at fairly important to very important. A recent survey of 1000 US manufacturers have found that 90% have environmental strategies and 80% of them made environmental-friendly operation mechanisms (Sarkis 2001).

14.4.5 Manufacturing in a Globalized Economy

Firstly, it is because of advanced technologies to make manufacturing capabilities exceed market needs thatthe globalization economy was formed. *Secondly*, the globalized economy affects the manufacturing industry in multiple ways. Among these impacts, *free trade*, *environmental issues*, and the *demands of responsible sourcing* are the three most critical ones that affect the supply chains and end-users of products for any enterprise (Styles 2017).

From the perspective of free trade, these agreements help to create equal and open access to markets for consumers; however, free trade agreements are not always the case and trade partnerships can fall apart quickly and dramatically. They might bring more challenges to manufacturers, especially when competing against international counterparts who might face better business conditions. *From the perspective of environmental impacts*, international treaties have been drawn up to unite the world to address pollution, resource depletion, and climate change; leading manufacturers will see environmental protection hindering business opportunities rather than regulatory burdens. *From the perspective of material sources*, raw materials are frequently sourced from, and shipped to, many places around the world. As a supply chain extends worldwide, it poses challenges to track, plan, and schedule material flows and ensure that it meets all international standards for compliance with environmental regulations and honest trade practices.

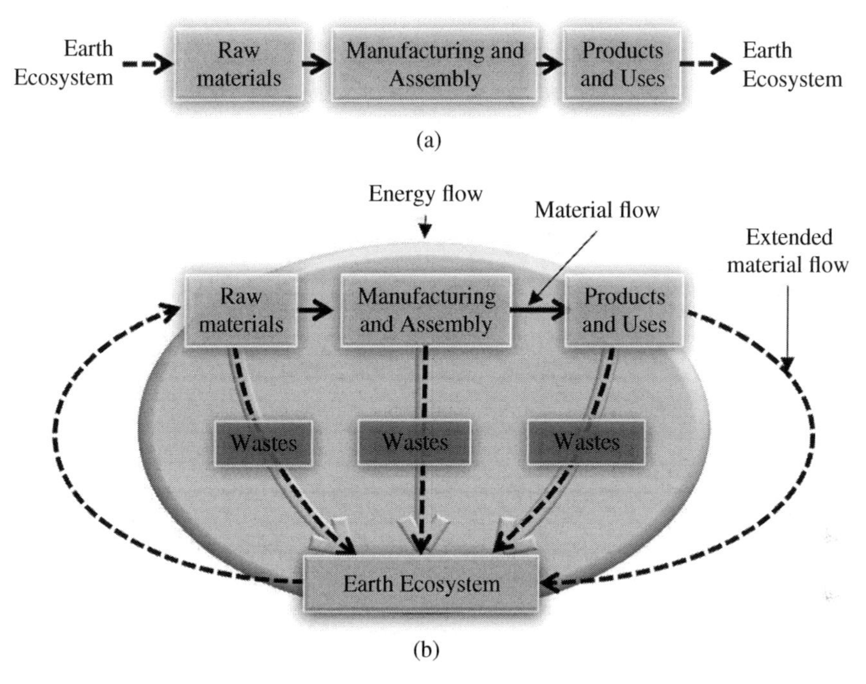

Figure 14.12 Inclusion of sustainability in manufacturing. (a) Traditional open and linear system. (b) Closed-loop system.

The globalized economy to promote worldwide trade share and exchange advanced manufacturing technologies, better products, and design and process innovations. However, unique challenges come with these opportunities. The manufacturers should purse sustainable manufacturing to take full advantage of the globalized economy (Styles 2017).

Figure 14.12 shows the shifts from traditional manufacturing to sustainable manufacturing. The traditional manufacturing system in Figure 14.12a is generally a linear and closed system in the economic domain to transfer raw materials to finished products. In contrast, the sustainable manufacturing in Figure 14.12b is a highly nonlinear, dynamic, and open system whose businesses and operations have expanded greatly to society and the environment.

14.5 Metrics for Sustainable Manufacturing

The role of metrics in engineering design and analysis cannot be overstated. Metrics serve as *enabling technologies* in the designing processes (Dornfeld 2009). Sustainable manufacturing requires metrics for decision making at all levels of an enterprise. It is suggested that it follows the framework of goal and the scope definition. The distinction is made between environmental cost metrics and sustainability metrics (Reich-Weiser et al. 2008). Various standards, such as ISO 14000 and ISO 14064, have been developed in the last two decades. Kibira et al. classified environmental policy procedures, which are used to determine the incentives to achieve compliance with environmental safety

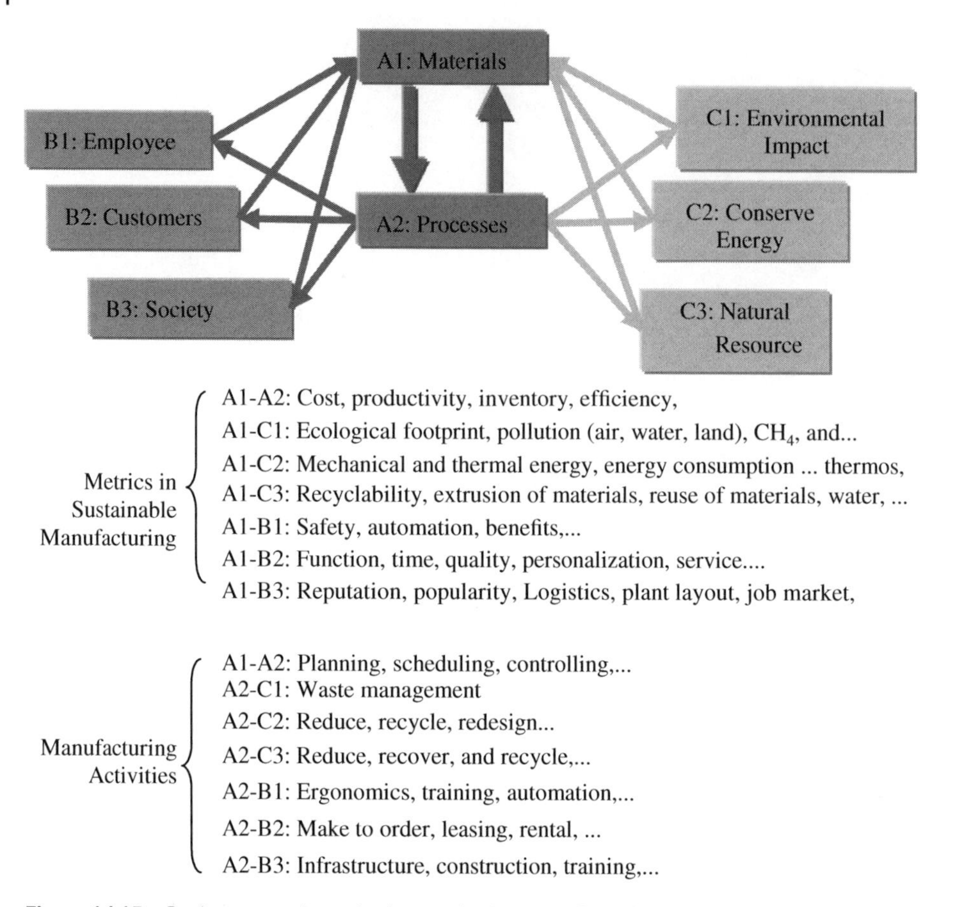

Figure 14.13 Businesses and metrics in sustainable manufacturing.

requirements (Kibira et al. 2009). Dornfeld has suggested ways to measure sustainability in terms of global warming gas emissions (CO_2, methane CH_4, and N_2O), per capital, per gross domestic product (GDP), per area/nation, recyclability, reuse of materials, energy consumption, pollution (air, water, land), and the ecological footprint – 'fair share', exergy (available energy), or other thermodynamic measures (Dornfeld 2009).

According to the definition of sustainable manufacturing (The US Department of Commerce 2010), the activities in a manufacturing system can be classified into two types, i.e. activities on materials and activities on processes. As shown in Figure 14.13, these activities make an impact on the environment, economics, and society. The manufacturing environment has aspects of 'environmental impact', 'conserve energy', and 'natural resource'. The society has aspects of 'customers', 'employee', and 'community'. Various criteria have been identified to evaluate the performances of a sustainable system paradigm.

Manufacturers have begun to realize the need for the responsible use and management of resources in the life cycle of a product (Ozel and Yildiz 2009). The role of manufacturing systems to sustainability really relies on how the boundaries of a manufacturing system are defined. As shown in Figure 14.14, traditional manufacturing systems did not take into

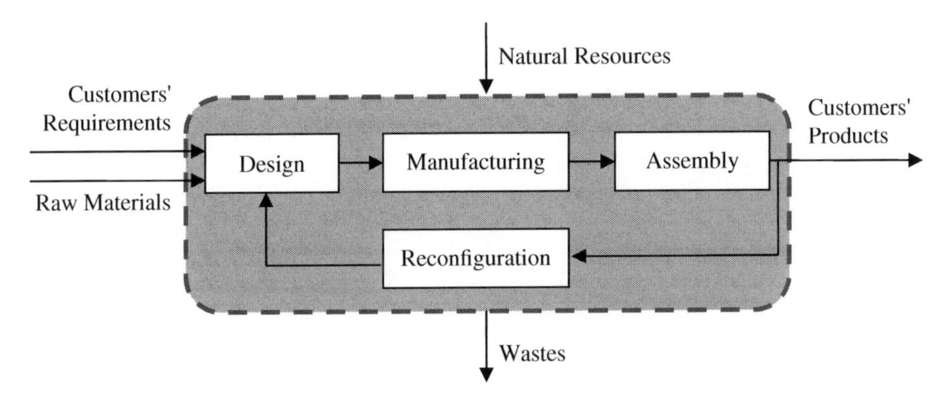

Figure 14.14 Traditional manufacturing system.

consideration many aspects such as waste management, pollution, recovery, and reuse of used products. As shown in Figure 14.15, with the increasing consciousness of environmental problems, many activities have been introduced in the lifecycle of products. As a result, manufacturers have a choice of either including these activities in the operations and optimization of the overall structure based on the required tasks or they would pay for the cost to solve the problems of waste management and disposal, etc. Therefore, the roles of manufacturing systems are dependent on the definition of a manufacturing system, which varies from one company to another. Within the scope of manufacturing, we will later discuss further the paradigm of reconfigurable manufacturing that fits the goal of system sustainability seamlessly. Through the reconfiguration of system components, the manufacturing sources can be reused so that the wastes from manufacturing can be reduced.

Jayal et al. (2010) provided a case study to measure the contribution from manufacturing to overall sustainable manufacturing, which showed that the contribution from manufacturing has been estimated to be 25.7%. Figure 14.15 shows the main constituents of a sustainable manufacturing system. In this figure, the sustainability consists of three pillars, i.e. environment, society, and economy. The life cycle of a product is divided into four phases: 'pre-manufacture', 'manufacture', 'use', and 'post-use'. The impact on sustainability from the four phases has been shown individually and the overall impact has accumulated. Obviously, the contributions to sustainability from the manufacturing phase are limited since considerable portions of the contributions are from the activities at the phases of 'pre-manufacture', 'use', and 'post-use'. In Figure 14.16 one can conclude that while sustainable manufacturing is extremely important to sustainability, it is unrealistic to expect that a next-generation manufacturing paradigm can meet the requirements of sustainability completely.

Another issue is the relation between sustainability and sustainable manufacturing. Although there is some overlapping between *design of sustainability* and *design of sustainable manufacturing*, it can be completely different in terms of the scope of the manufacturing. On the one hand, a manufacturing system can be confined to one product with a limited consideration of its product life; the main difference from a traditional manufacturing system is that some new metrics on waste management and environmental impact have to be taken into consideration in its design. On the other hand, an extreme

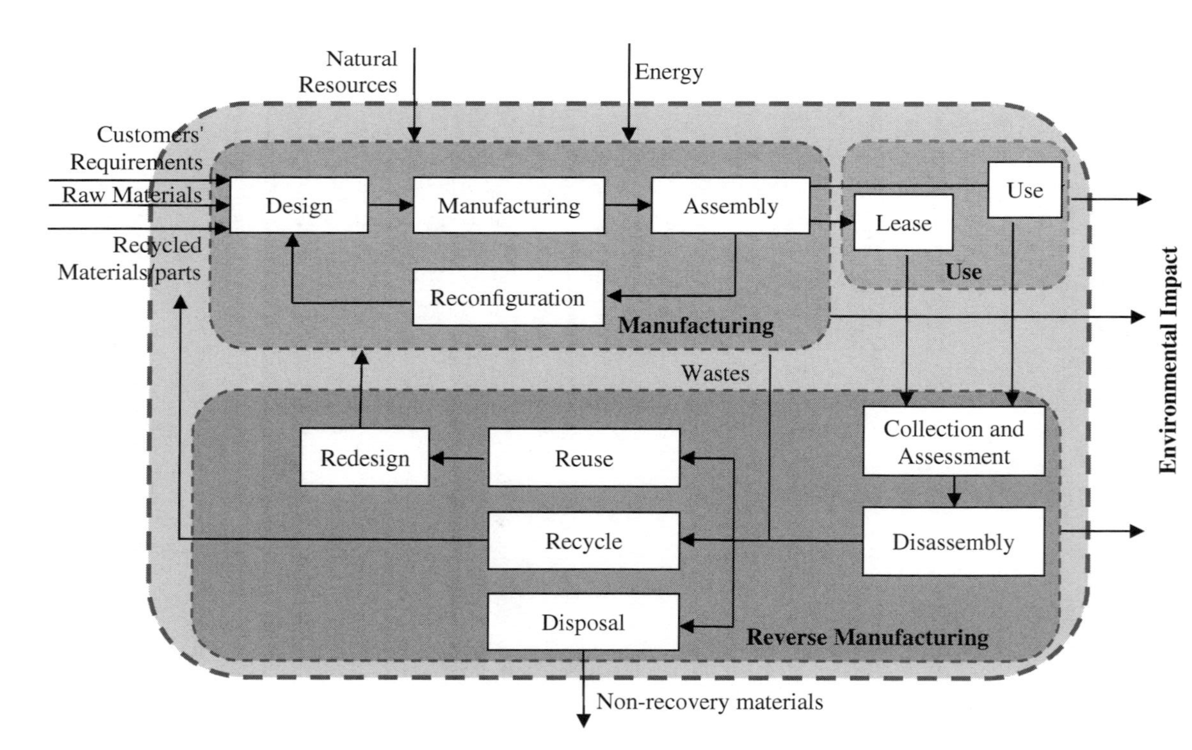

Figure 14.15 Constitutions of a sustainable manufacturing system.

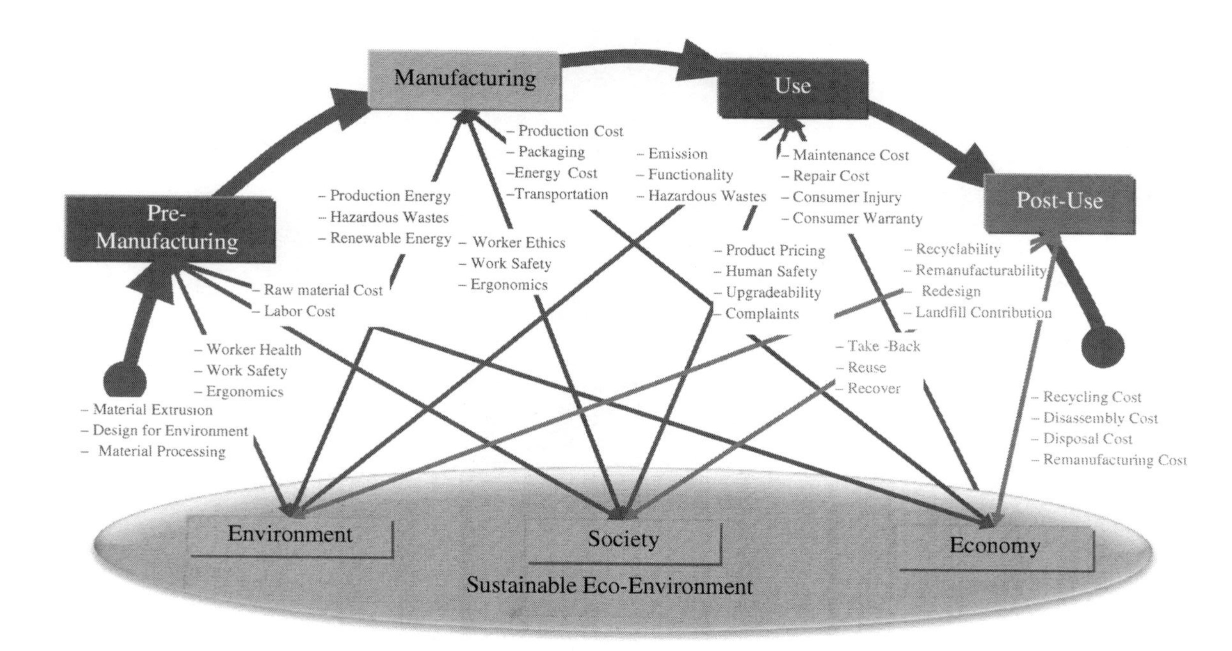

Figure 14.16 Evaluation of sustainability in a manufacturing system.

case is to include all of the activities that occurred in its product lifetime in one company so that the design of a sustainable manufacturing system becomes a D4S. Note that when the boundaries of a system are considered, the inputs and outputs beyond system control have to be valued or de-valued for the sake of system optimization, and this raises more concerns on the uncertainties and fidelity of a design result. In the following sections, a few examples are discussed of enabling technologies used to enhance the sustainability of manufacturing systems.

14.6 Reconfigurability for Sustainability

A clear definition of system boundaries is essential to determine a correct system paradigm. As illustrated in Figure 14.15, existing system paradigms have been developed to meet design metrics of cost, quality, personalization, and time. Since sustainability has become a new dimension to evaluate a manufacturing system paradigm, it is worth examining the gaps in the desired performances of existing system paradigms and the requirements of sustainability. From our perspective, a reconfigurable manufacturing system (RMS) paradigm is one of the most advanced paradigms to deal with all of today's manufacturing requirements. In this section, RMSs are used as examples to examine the gap with a sustainable manufacturing system.

In an RMS, system reconfigurability can be classified in terms of the levels where the reconfigurable actions are taken. Reconfigurability at lower levels is mainly achieved by changing hardware resources, while reconfigurability at higher levels is mainly achieved by changing software resources and/or by choosing alternative methods or organizations. The resources at different levels must work together so that system reconfigurability can be maximized cost-effectively. As shown in Figure 14.17, an RMS consists of a reconfigurable hardware system and a reconfigurable software system. The hardware system includes *reconfigurable machining systems*, *reconfigurable fixturing systems*, *reconfigurable assembly systems*, and *reconfigurable material-handling systems* (Bi et al. 2007). RMS characteristics include 'modularity', 'scalability', 'integrability', 'convertibility', and 'diagnosability' (Mehrabi et al. 2000). From the perspective of sustainability, the relevant objectives of RMSs are (i) to reduce the wastes through the reuse of manufacturing resources and (ii) to reduce energy cost through the optimization of manufacturing processes and system reconfiguration. Numerous researches have been published on design and control of reconfigurable systems to achieve these two objectives.

To examine the contributions of an RMS to sustainability, the activities in a sustainable manufacturing process are classified into 6R, i.e. 'reuse', 'recover', 'redesign', 'remanufacture', 'reduce', and 'redesign'; these activities are applied on either 'materials' or 'tools'. Based on the above discussions on system components and design metrics of an RMS, the involved activities for the purposes of sustainability have been highlighted in Figure 14.18. Currently, the focus of an RMS is on reconfigurability and sustainability of machines and tools. It seems that all of the sustainable activities have not been involved in an RMS; this observation is consistent with the conclusion in Figure 14.15 that an RMS only makes a fractal portion of contribution to the overall sustainability of a product in its lifecycle.

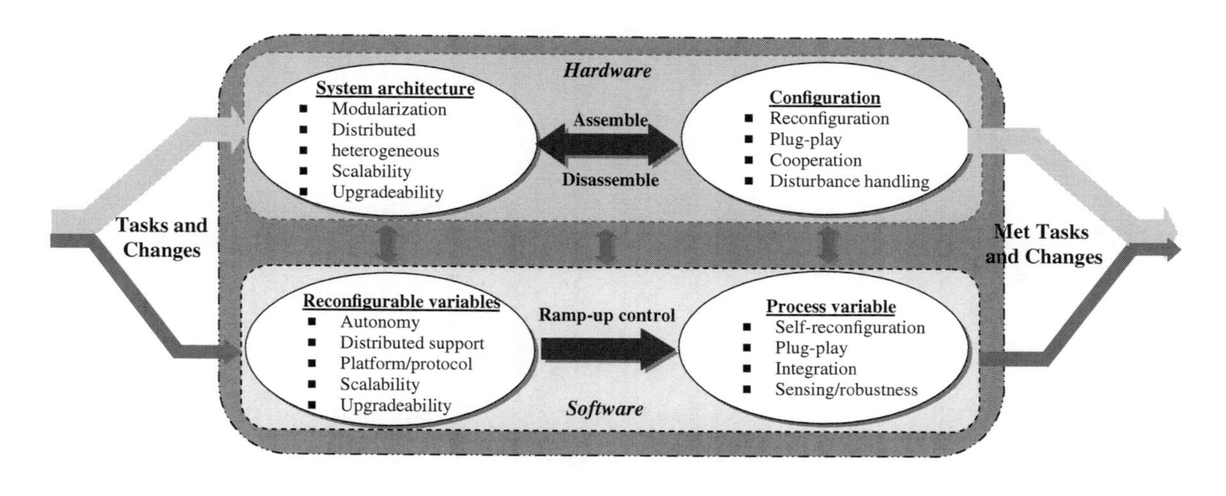

Figure 14.17 Hardware and software in an RMS.

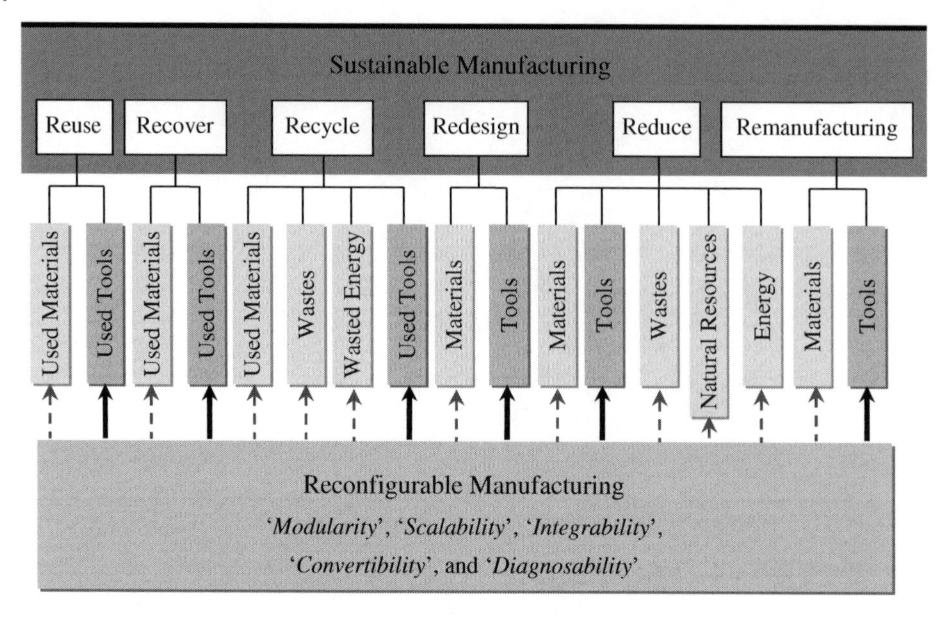

Figure 14.18 Reconfigurability for sustainability of a manufacturing system.

Besides, the traditional manufacturing systems have been optimized to fulfil the customers' requirements in functions, quality, cost, and delivery time, so there is a long way to take an additional requirement of sustainability and evolve an RMS paradigm in order to produce a sustainable manufacturing paradigm. Two other reasons worth mentioning are (i) an RMS is still a closed system, which needs a clear definition of system boundaries, and the system paradigm is still a result of sub-optimization. (ii) The reconfiguration and optimization of manufacturing resources, in particular machines and tools, are focused, but limited progress has been made in planning and scheduling for materials and waste.

System paradigms such as those of an RMS, which are based on local sub-optimization of a system, make a fractal portion of the contribution to sustainability and the value of a product in its lifecycle. The role of a manufacturing system in the value-added chain of a product has to be re-examined. As far as sustainability is concerned, the boundaries of a manufacturing system in a product life cycle can no longer be defined clearly, but at least the boundaries of a manufacturing system have to be defined in different ways based on the perspectives of designers, for example, by taking system dynamics into consideration. Finally, some research efforts towards the evolution of today's manufacturing system paradigms to sustainable manufacturing are discussed in the next section.

14.7 Lean Production for Sustainability

Although the concept of *lean production* did not traditionally focus on environmental issues, the ideas of eliminating waste are aligned seamlessly with sustainable manufacturing. Therefore, the practice of sustainable manufacturing can begin with lean production. If one company already uses lean manufacturing, it will be easier to integrate

Table 14.1 DOWNTIME and environmental impact (EPA 2017).

DOWNTIME	Environmental impact
Defective parts, components, or products	Energy and materials are consumed to make defective products and defective products need to be disposed of.
Overproduction of parts, components, or products	Materials and energy are consumed to make unnecessary products. Products may spoil or become obsolete. Products may require hazardous materials.
Waiting	Materials may spoil or become damaged. Downtime wastes energy for heating, cooling, and lighting.
Non-utilized resources	Resources are devalued with respect to time, the maintenance on idle resources involves in additional wastes
Transport	Energy is used for transport and produces emissions. Transport can cause damage or spills. More space is needed for additional motion, requiring heating and cooling.
Inventory	More packaging and space are needed for excess inventory. Storage could cause deterioration of products and waste. Requires energy to heat, cool, and light inventory space.
Motion	Energy is needed to operate machines and tools in motion. More packaging is required to protect components during movement.
Extra processing	If processing is unnecessary, increases waste and energy use. Consumes more parts and raw materials.

environmental principles into the lean processes. The Green Suppliers Network (GSN) found that companies could save up to 30% more by integrating *lean* and *green* than if they focused on lean production alone (EPA 2015).

Sustainable manufacturing aims to reduce environmental waste, where environmental waste refers to any unnecessary use of resources or a substance released into the air, water, or land that could harm human health or the environment. Environmental waste mainly includes (i) any energy, water, or other materials used that are more than what is really needed to meet the customer's needs, (ii) hazardous materials and substances, (iii) pollutants, residuals, and other material waste released into the environment, such as air emissions, wastewater discharges, hazardous waste, and solid waste (Muralikrishna and Manicham 2017).

As mentioned previously, adding ***green*** to *lean* can lead to significantly greater returns for manufacturers. Table 14.1 outlines the relationship between lean waste and environmental impacts (EPA 2017). Note that traditional waste addressed by lean production is 'DOWNTIME', which stands for ***D****efects*, ***O****verproduction*, ***W****aiting*, ***N****on-utilized resources*, ***T****ransportation*, ***I****nventory*, ***M****otion*, and ***E****xtra processing*.

Numerous performance criteria have been proposed for the evaluation of manufacturing systems; some criteria conflict with each other in certain circumstances. For example, the mass production for a high productivity may affect the system flexibility. It may not be the case for a company to pursue sustainability. Other than the good agreement of sustainability and lean manufacturing, sustainability can also be easily aligned with reduced cost and increased profit.

Sustainable manufacturing practices increase production efficiency, primarily through increased resource efficiency. Resource efficiency includes things like energy, water, and material efficiency. Increasing the resource efficiency will lower the material uses and costs. Sustainable manufacturing can also lower the cost of waste removals, as the company produces less waste and by-products, and reduces transportation costs through lower product weight and more efficient transportation.

14.8 Lifecycle Assessment (LCA) and Design for Sustainability (D4S)

Figures 14.4 and 14.16 show that the operations and businesses in sustainable manufacturing have expanded to entire product lifecycles. As shown in Figure 14.19, when beginning to think about your efforts, it is tempting to start with the impacts within your operations or just within your facility. However, the effect of your product or service does not stop at the factory gate. Think of the entire product life cycle and take a cradle to grave approach from the inputs used to make the product to the impact when it is disposed of (The US Department of Commerce 2011).

The D4S method has been developed based on the product LCA, where the product journey starts with the extraction, processing, and supply of raw materials and energy needed for the product. It then covers the production of the product, its distribution, use (and possibly reuse and recycling), and its ultimate disposal. Environmental impacts of all kinds occur in different phases of product life cycle and should be taken into consideration in an integrated way (Figure 14.15). Key factors of the assessment models are material and energy consumptions (water, non-renewable resources, energy in each of the lifecycle stages), and processing of output materials (waste, water, heat, emissions, and waste) and other pollution factors like noise, vibration, radiation, and electromagnetic fields (Crul and Diehl 2019). Figure 14.20 provides some commonly used strategies in the D4S practices over product life cycle (The US Department of Commerce 2011).

It is worth noting that manufacturing will always lead to some adverse environmental impact. It is unrealistic to make a company with no environmental impact as there is no sustainability destination. The D4S practice is continuous improvement (CI), i.e. making constant advances in the company's overall sustainability performance.

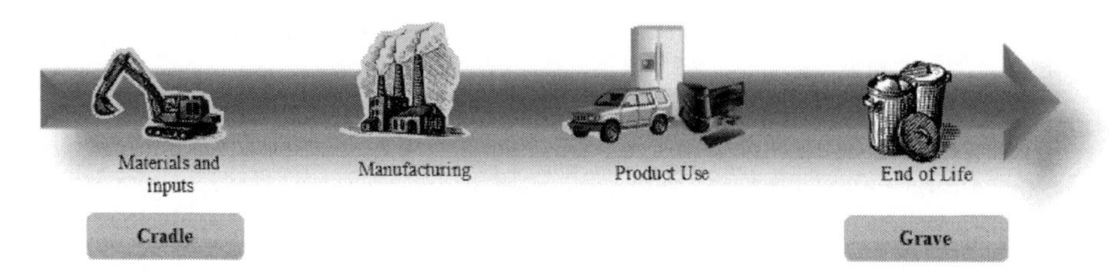

Figure 14.19 Assessment of sustainability in lifecycles of products from cradle to grave (The US Department of Commerce 2011).

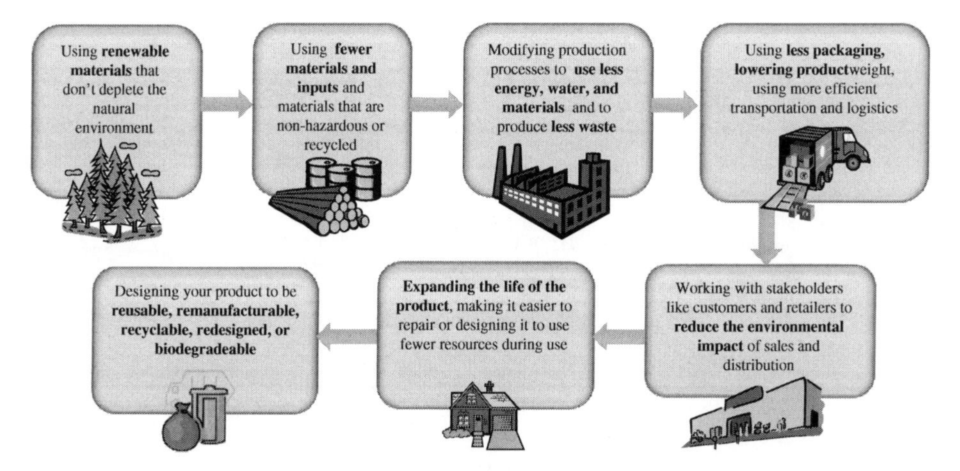

Figure 14.20 Practices for design for sustainability (The US Department of Commerce 2011).

14.9 Continuous Improvement for Sustainability

Crul and Diehl (2019) suggested 10 steps for CI practice in D4S, shown in Figure 12.21. The figure shows that each step corresponds to a process of product development. *Step 1* is to create a CI team that is responsible for introducing and implementing D4S procedures at the organizational and technical levels. *Step 2* is to connect the D4S process with normal product development and business processes in a company. Strengths, weaknesses, opportunities, and threats (SWOTs) are analysed in order to outline the strengths and weaknesses and determine the goal of CI. *Step 3* is to select a product that affects the identified D4S drivers, so that CI can be aligned with the project goals from step 2. *Step 4* is to validate the selected product to see if it is the best starting point to maximize the improvement of sustainability. *Step 5* is to evaluate the sustainability impacts of the target product during its lifecycle based on three sustainability pillars, i.e. environment, society, and economy. It helps to identify lifecycle stages that may be overlooked. *Step 6* is to develop the D4S strategies by creating conceptual designs for each process. *Step 7* is to generate and prioritize ideas for improving the sustainability of products. From *steps 8 to 10*, the detailed design of technical solutions is developed, evaluated, and finally implemented for the cycle of CI. Interested readers for the aforementioned steps can learn the details in the literature (Crul and Diehl 2019).

14.10 Main Environmental Impact Factors

In this section and the next one, the main environment impacts are discussed. D4S seeks to reduce negative impacts on the environment by reducing consumption of non-renewable resources, minimize waste, and create healthy, productive environments (GSA 2019). From the environment perspective, the following four criteria are commonly used in D4S.

14.10.1 Carbon Footprint

Carbon dioxide (CO_2) and other gases are produced by burning fossil fuels, which accumulate in the atmosphere; CO_2 and equivalent gases in turn increase the earth's average temperature and cause global warming. Global warming is responsible for the loss of glaciers, extinction of species, more extreme weather, and other environmental problems. The carbon footprint acts as a proxy for the larger impact factor, referred to as *global warming potential* (GWP). The carbon footprint is measured by equivalent tons of carbon dioxide (CO_2).

14.10.2 Total Energy

It is easy to understand that the total energy is included as an indicator since it quantifies the amount of energy consumption directly. The energy consumed in the lifecycle of a product is measured in *megajoules* (MJ) and the total energy includes (i) upstream energy required to obtain and process these fuels, (ii) embodied energy of materials that would be released if burned, (iii) electricity of fuels used during the product's lifecycle, and (iv) transportation. In addition, efficiencies in energy conversion (e.g. power, heat, steam) are taken into account.

14.10.3 Air Acidification

Air acidification measures the impacts relevant to acid rain such as sulfur dioxide (SO_2), nitrous oxide (N_2O), and other acidic emissions to air. Acid rain makes the land and water toxic for plants and aquatic life. Air acidification also damages constructions since it slowly dissolves man-made building materials such as concrete. Air acidification is measured in units of *kilograms of sulfur dioxide equivalent* (SO_2-e).

14.10.4 Water Eutrophication

Water eutrophication is used to measure the environmental impacts on the water ecosystem caused by an overabundance of nutrients. Nitrogen (N) and phosphorous (P) from water and agriculture fertilizers cause an overabundance of algae to bloom, which depletes the water of oxygen and results in the death of plant and animal life. Water eutrophication is measured in units of *kilograms of phosphate equivalent* (PO_4-e).

14.11 Computer Aided Tools – SolidWorks Sustainability

Sustainable manufacturing is the integration of social, environmental, and economic conditions into a product or process. Very soon all designs will be sustainable. SolidWorks Sustainability allows designers to be environmentally conscious about their designs. Successful products are developed by integrating LCA directly into engineering design processes.

In SolidWorks, LCA is a method to quantitatively assess the environmental impact of a product throughout its entire lifecycle, from the procurement of the raw materials, through the production, distribution, use, disposal, and recycling of that product. LCA is the detailed analysis that gives the companies the information they need to make

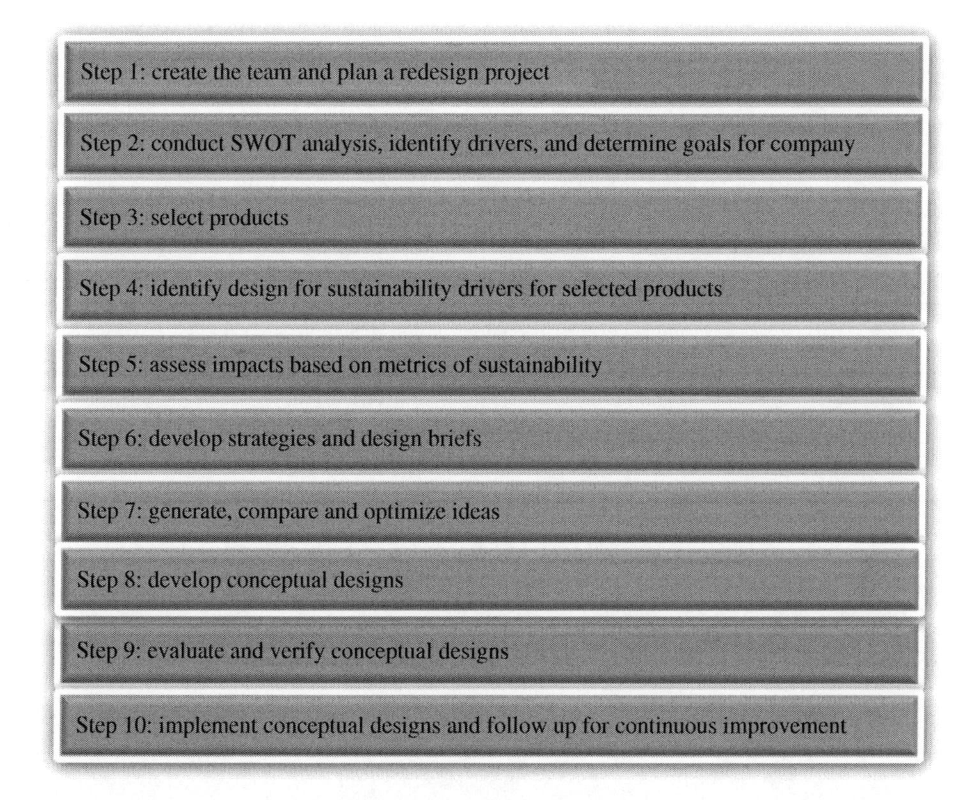

Figure 14.21 Continuous improvement (CI) project for D4S (Crul and Diehl 2019).

the most environmentally friendly decisions throughout product design. The analysis looks at a product's entire life, which encompasses raw material extraction, material production, manufacturing, product use, end-of-life disposal, and all of the transportation that occurs between these stages. LCA is conducted on parts or assemblies directly within the design window. It is capable of searching comparable materials and see in real-time how they affect environmental impacts. The key elements in LCA are to (i) identify and quantify environmental loads such as (i) the energy and raw materials consumed and (ii) the emissions and wastes generated (iii) evaluate potential environmental impacts of these loads; (iv) assess the options available for reducing these environmental impacts. Figure 14.21 and Table 14.2 show the six stages of LCA in SolidWorks Sustainability.

Figure 14.22 shows the architecture of SolidWorks Sustainability. It qualifies the environmental impacts in terms of carbon footprints, total energy, air acidification, and water eutrophication. The tool is equipped with a comprehensive library of materials with the rich data of their environmental factors and allows users to select and compare materials and support the simulation-based optimization based on the product assessment (PLA).

14.11.1 Material Library

Figure 14.23 shows the extraction and processing of raw materials that make environmental impacts in manufacturing processes. The SolidWorks Sustainability includes the materials

Table 14.2 Six stages of LCA in SolidWorks Sustainability (Planchard 2019).

Stage	Assessment
Raw material extraction	Planting, growing, harvesting of trees, mining of raw ore (example: bauxite), drilling and pumping of oil, etc.
Material processing	The processing of raw materials converted into engineering materials, such as oil into plastic, iron into steel, and bauxite into aluminium.
Manufacturing process	The processing of forming or shaping materials into desired part geometries and properties by milling and turning, casting, and stamping. etc.
Assembly	Assemble all of the finished parts into a final product.
Product use	End consumer uses the product for intended lifespan of the product.
End of life	Once the product reaches the end of its useful life, how it is disposed of, such as landfill, recycled, or incinerated.

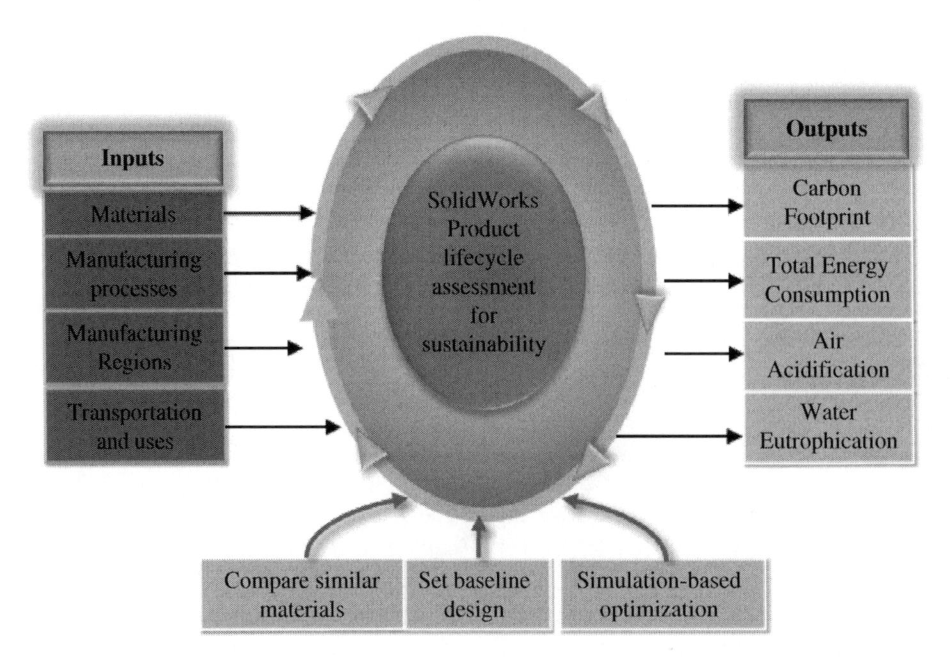

Figure 14.22 Constitutions and supportive tools in SolidWorks Sustainability.

library with all of the information of environmental impacts for the majority of engineering materials. As shown in Figure 14.24, the material library is organized as a two-level hierarchical structure; the first level is for *material classes* much as *steel*, *iron*, *aluminium alloys*, *rubber*, and *woods*. Each entity at the first level corresponds to one material class, which includes a list of material sub-types at the type level.

14.11.2 Manufacturing Processes and Regions

The inputs for *manufacturing processes* begin with the region where the products are made, as shown in Figure 14.25, since the environmental impacts depend on geographic locations.

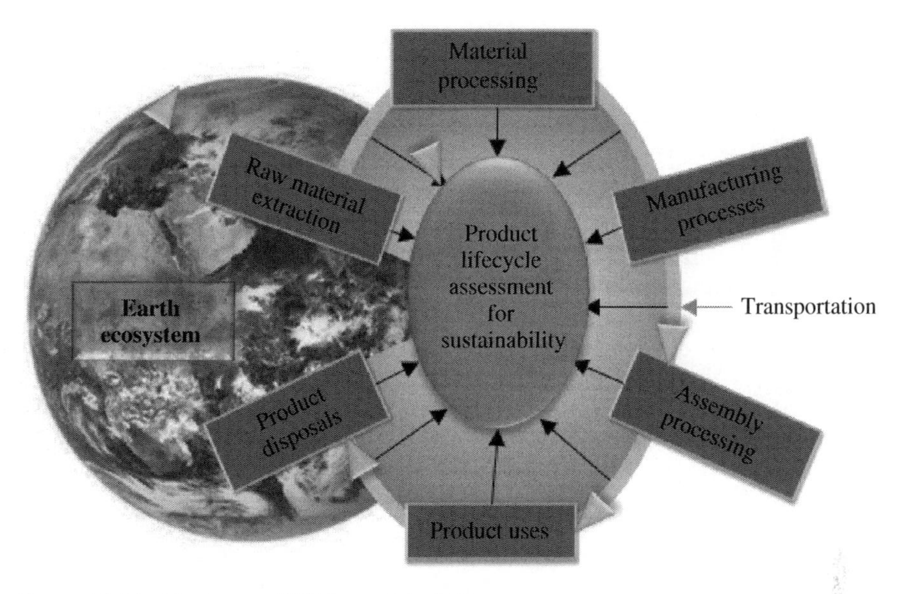

Figure 14.23 Six stages of LCA for sustainability.

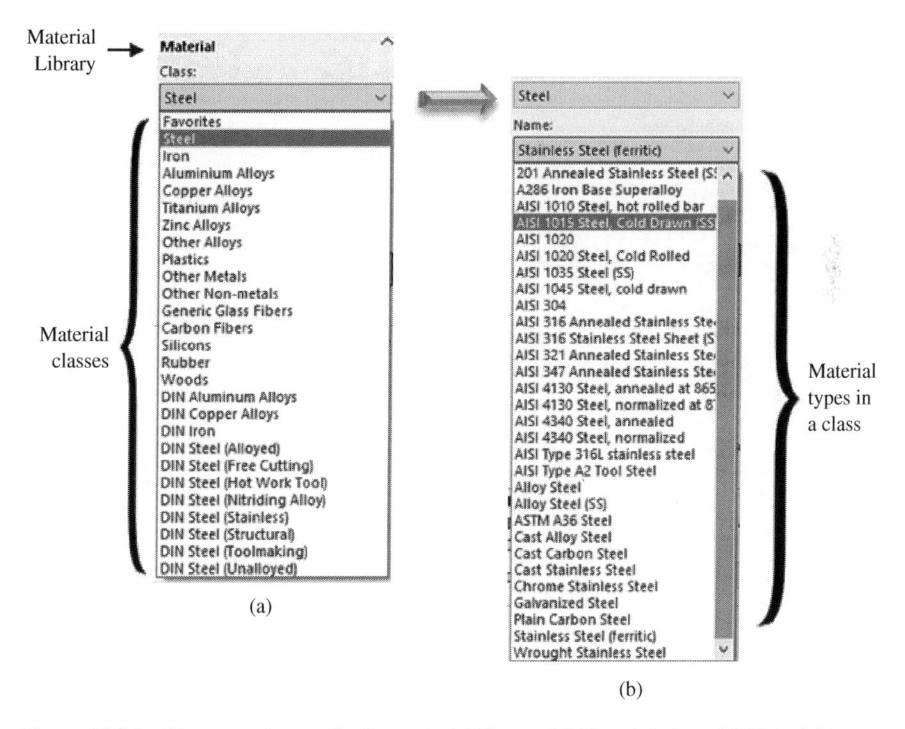

Figure 14.24 Classes and types in the material library. (a) Material class. (b) Material type.

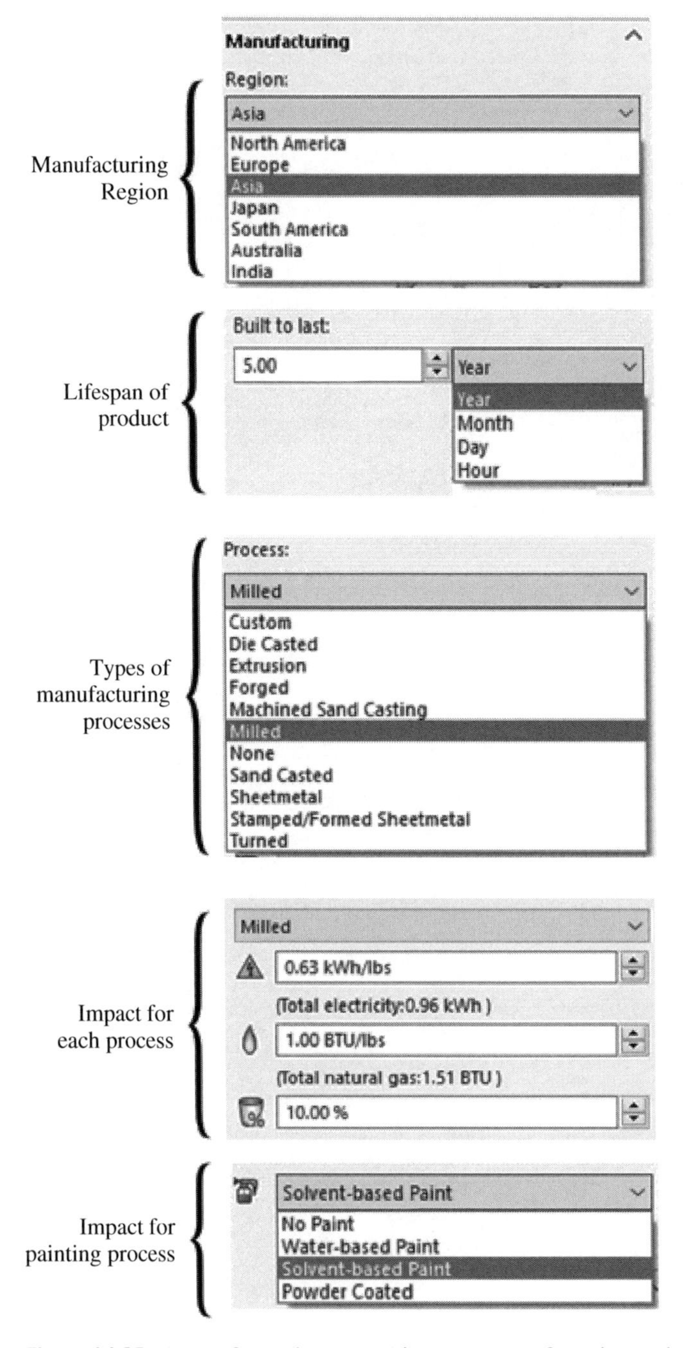

Figure 14.25 Inputs for environmental impact – manufacturing region and process.

Each region produces energy by a different method of combinations. The impact of a kWh is different for each region. The region input determines the resources consumed by manufacturing processes in that region and the choices for the region include North America, Europe, Asia, Japan, Australia, and India. The inputs for transportation and use region determine the energy sources consumed during the product's use phase (if applicable) and the destination for the product at its end-of-life. It is then used to estimate the environmental impacts associated with transporting the product from its manufacturing location to its use location.

14.11.3 Transportation and Use

The SolidWorks Sustainability takes into consideration the environmental impacts by non-value-added activities such as *transportation* and some activities outside the manufacturing system such as *product uses*. Both of them are affected by the geographic regions where these activities take place. Figure 14.26 shows in the interface in SolidWorks for users to input the regions for transportation and product use.

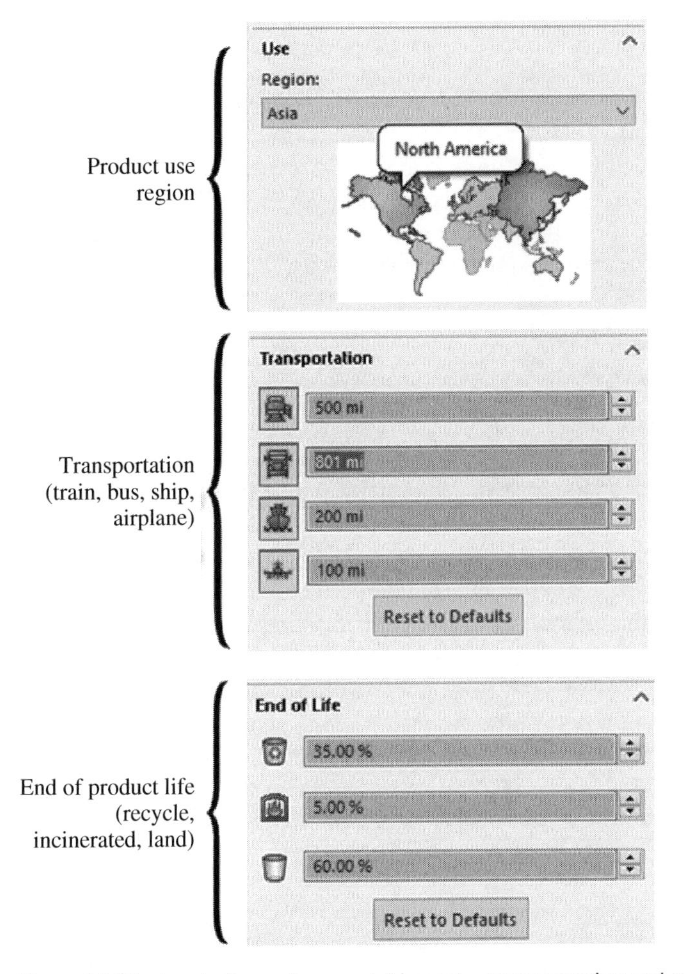

Figure 14.26 Inputs for environmental impact – transportation and use.

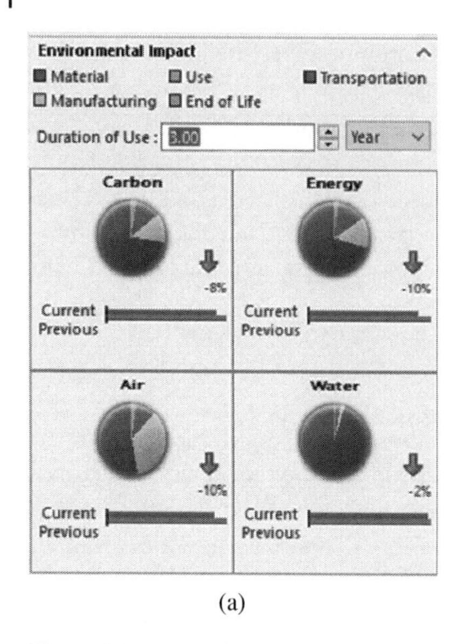

(a)

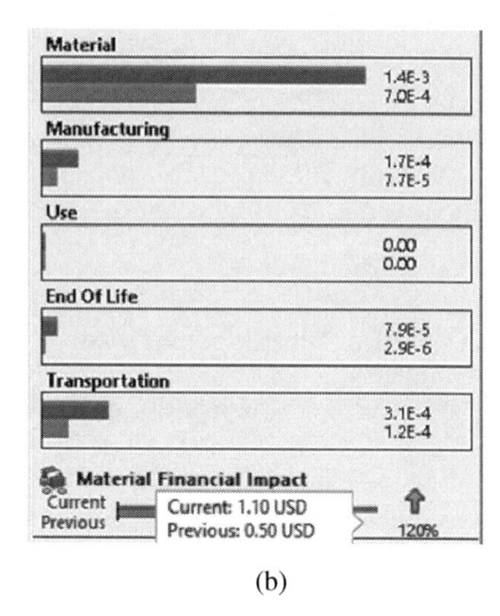

(b)

Figure 14.27 The example of assessment results. (a) Environmental impact. (b) Material financial impact.

Figure 14.27 shows examples of assessment results. The outputs of the environmental impacts consist of (i) four environmental indicators and (ii) material financial reports. Four environmental indicators are *carbon footprint*, *air acidification*, *water eutrophication*, and *total energy consumption*. It gives the factor percentages of environmental impact from *material*, *manufacturing*, *use regions*, and *product disposal*.

14.11.4 Material Comparison Tool

The SolidWorks Sustainability allows environmental impacts to be compared and minimized by simulation-based optimization. However, the comparison has to be made between a baseline design and a new design. Figure 14.28 shows the interface for users to save, set, and import a baseline design as a reference. In addition, the online tool helps users to convert the environmental impacts into human scale parameters, for example by converting the carbon footprint into miles driven for a car.

As shown in Figure 14.29, in comparing different materials, similar materials can be identified based on different criteria relevant to the functionalities of products, such as yield strength, thermal conductivities, or density.

SolidWorks Sustainability can assess the environmental impacts of an assembled product directly. In such a case, the inputs of the part can be defined. Figure 14.30 shows an

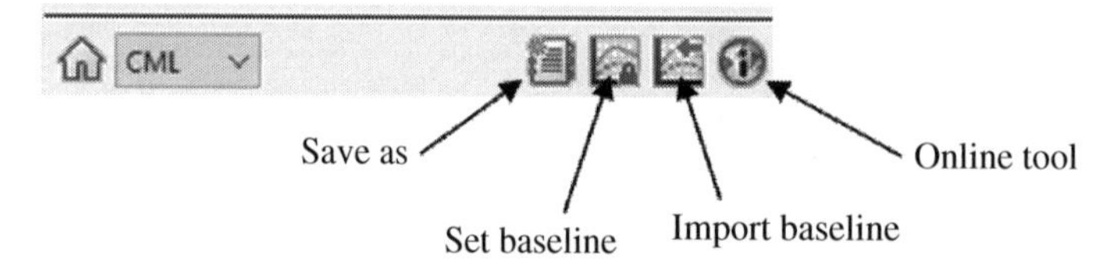

Figure 14.28 Interface to save, set, and import a baseline design.

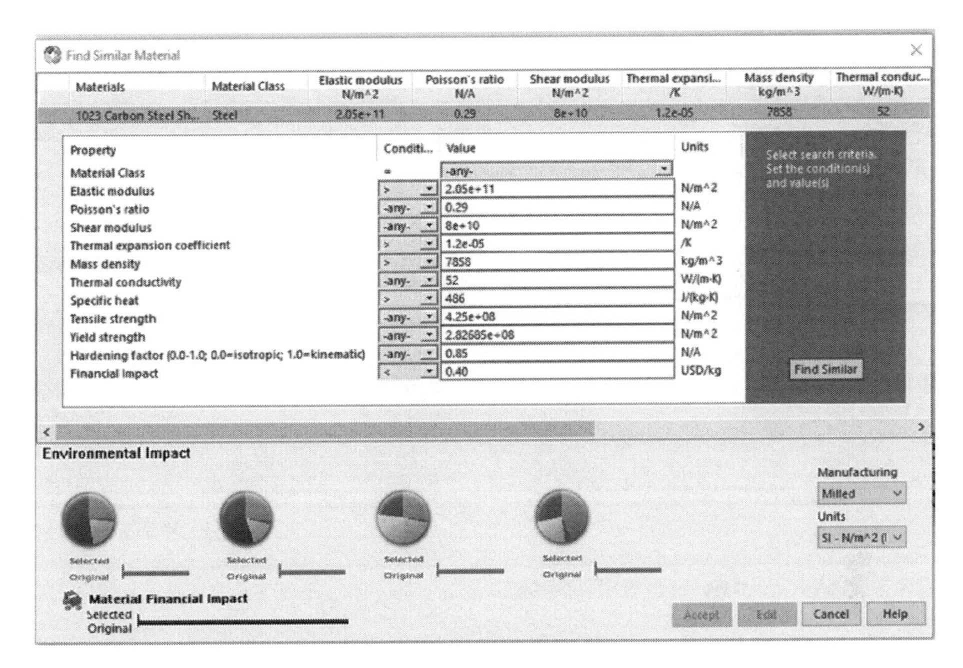

Figure 14.29 Materials selection for environmental impacts.

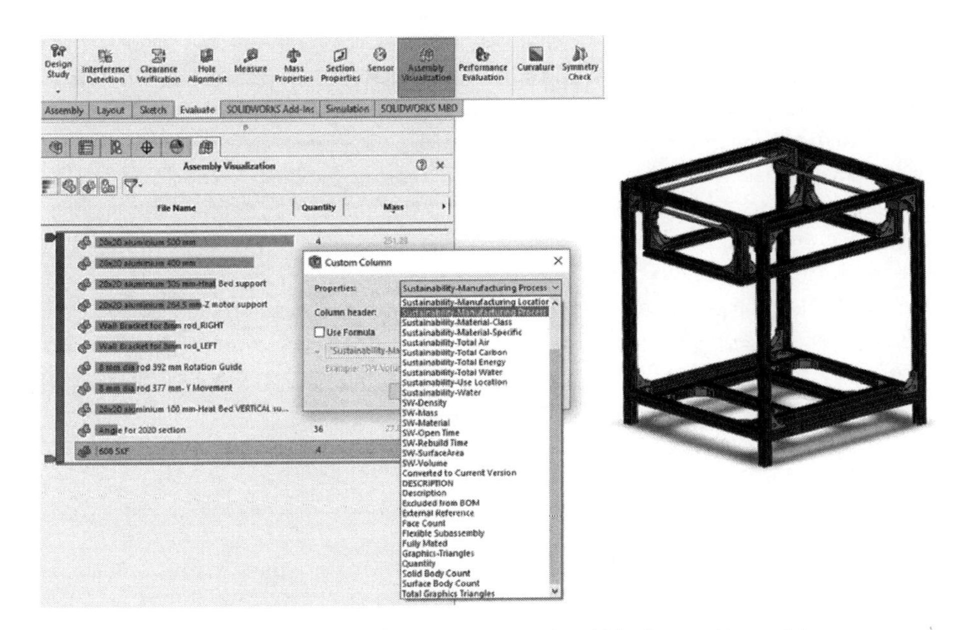

Figure 14.30 SolidWorks Sustainability supports product LCA of assembly model.

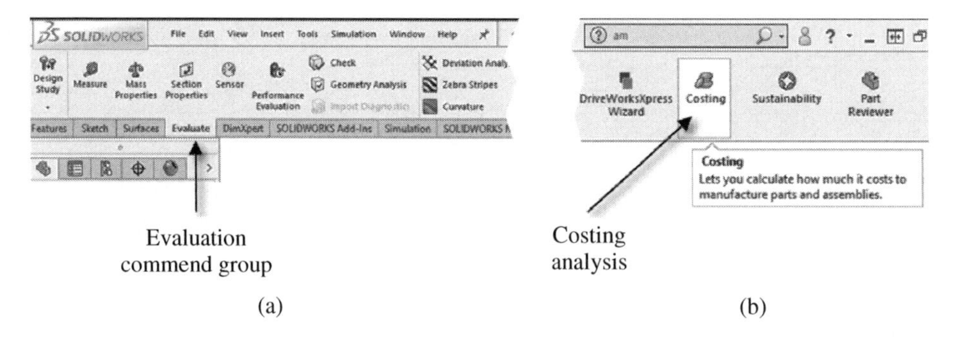

Figure 14.31 Costing analysis tool in SolidWorks. (a) Evaluation commend group. (b) Costing analysis.

example where an assembled structure is assessed for its environmental factors. Note that the SolidWorks has a concise version of Sustainability called SustainabilityXpress; it does not support the assessment of an assembly model.

14.11.5 Costing Analysis in SolidWorks

When the environmental factors are taken into consideration when designing products or processes, these environmental factors must be converted into economic factors. Therefore, a traditional costing analysis must be expanded to include visible direct cost, hidden indirect cost, cost for contingent liability, less tangible cost, and initial capital cost of environmental factors. Typically, overhead environmental costs include (i) *monitoring and reporting*, (ii) *waste management and disposal*, (iii) *capital depreciation*, (iv) *employee training*, (v) *utilities (electricity and water)*, (vi) *permits and fees*, (vii) *equipment*, (viii) *fines and penalties*, (ix) *equipment cleaning*, (x) *legal support*, and (xi) *sampling and testing* (Barr 2012).

The SolidWorks has the costing analysis tool called *costing* to take into consideration the costs from environmental factors. Figure 14.31 shows that the costing analysis tool is under the *evaluation* CommendGroup.

As shown in Figure 14.32, costing analysis begins with the selection of *processing types* and *material*. Once the processing types are selected, more details about the processes can be defined as shown in Figure 14.33. Finally, the costing analysis provides the assessment result of the product cost and its breakdowns, including materials, manufacturing, and markup costs, which are shown in Figures 14.33 and 14.34.

14.12 Summary

Sustainable manufacturing is the creation of manufacturing products that use materials and processes that minimize negative environmental impacts, conserve energy and natural resources, are safe for employees, communities, and consumers, and are economically sound. Sustainable manufacturing is driven by five main factors: shortage of natural resources, population increase, global warming, pollution, and globalization. Due to the saturation of global manufacturing capabilities to meet market needs, enterprises are facing pressure to take sustainability in manufacturing systems into consideration.

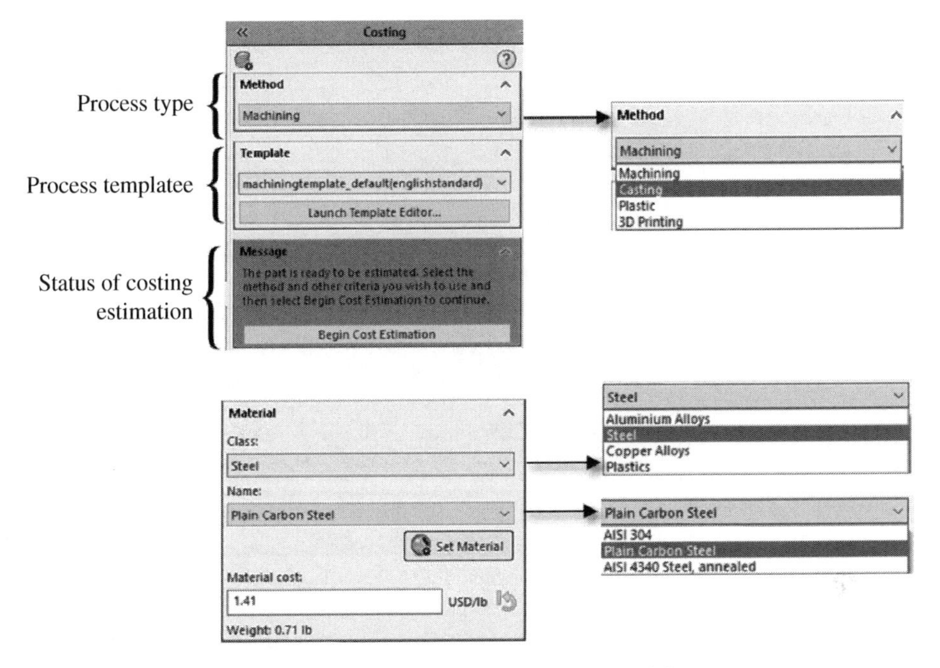

Figure 14.32 Inputs of costing analysis – process type and material.

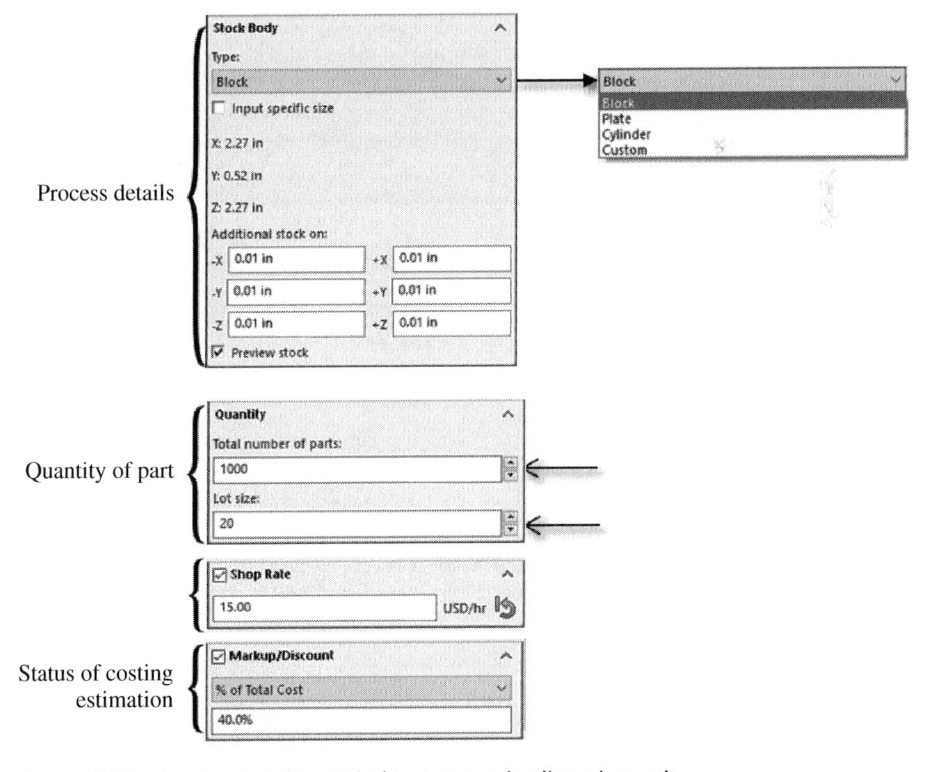

Figure 14.33 Inputs of costing analysis – process details and quantity.

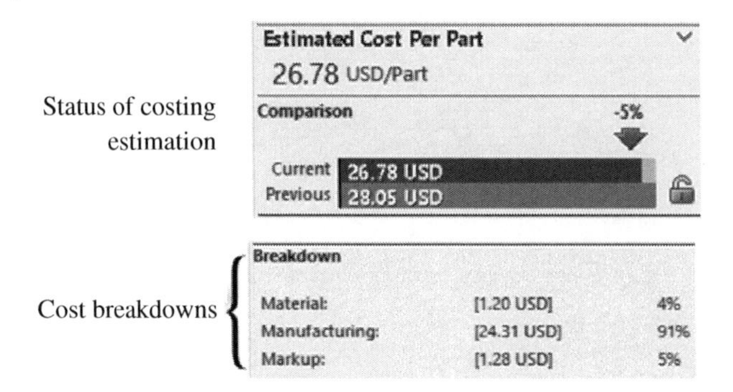

Figure 14.34 Example of the results from costing analysis.

Sustainable manufacturing involves manufacturing businesses in LCA from 'the cradle to the grave'. Manufacturing businesses have expanded greatly, and especially should perform reduce, reuse, redesign, recycle, remanufacture, and recover (6R) in order to minimize adverse impacts on the environment and society. A few of the critical enabling technologies for sustainable manufacturing are reconfigurable systems, lean production, LCA, and continuous improvement.

In designing products or processes, environmental impacts are measured by carbon footprints, total energy, air acidification, and water eutrophication. The practice of LCA requires a large amount of data on the environmental impacts by different materials, processes, transportations, and regions of manufacturing and product uses. Computer aided tools for sustainability assessment include available data on environmental impacts for designers to evaluate and optimize product design based on sustainability and cost.

14.13 Design Project

14.1 Use one of the products you designed in previous chapters. Use the SolidWorks Sustainability and Costing Analysis tool to evaluate its environmental impacts and cost. Modify the materials, processes, materials, and business regions to reduce environmental impacts and cost.

References

Allwood, J.M., Laursen, S.E., Russell, S.N. et al. (2008). An approach to scenario analysis of the sustainability of an industrial sector applied to clothing and textiles in the UK. *Journal of Cleaner Production* 2008, 16 (12): 1234–1246.

Ashrafi, N. (2014). A review of current trend in design for sustainable manufacturing. *IOSR Journal of Mechanical and Civil Engineering* 11 (4 Ver. II (Jul–Aug. 2014)): 53–58.

Barr, E. (2012). OECD sustainable manufacturing metrics toolkit and sustainable manufacturing 101 module. https://www.nist.gov/document-6868.

Bi, Z.M., Wang, L., and Lang, S.Y.T. (2007). Current state of reconfigurable assembly systems. *International Journal of Manufacturing Research* 2 (3): 303–328.

Bi, Z.M., Xu, L.D., and Wang, C. (2014). Internet of Things for enterprise systems of modern manufacturing. *IEEE Transactions on Industrial Informatics* 10 (2): 1537–1546.

Blanc, R. (2017). How does population growth affect the environment sustainability? https://wordpress.clarku.edu/id125-envsus/2017/04/07/how-does-population-growth-affects-the-environment-sustainability.

C2ES (2017). A climate of change: manufacturing must rise to the risks and opportunities of climate change. https://www.c2es.org/document/a-climate-of-change-manufacturing-must-rise-to-the-risks-and-opportunities-of-climate-change-2.

Crul, M.R.M., and Diehl, J.C. (2019). Design for sustainability – a practical approach for developing economies. http://www.d4s-de.org/manual/d4stotalmanual.pdf.

Dornfeld, D.A. (2009). Opportunities and challenges to sustainable manufacturing and CMP. *Materials Research Society Symposium Proceedings*: 1157. https://doi.org/10.1557/PROC-1157-E03-08.

EIA (2017). Electricity consumption in the United States was about 3.82 trillion kilowatthours (kWh) in 2017. https://www.eia.gov/energyexplained/index.php?page=electricity_use.

EPA (2015). The green suppliers network offers the lean and clean advantage. https://www.epa.gov/sites/production/files/2015-03/documents/backgrounder.pdf.

EPA (2017). Lean and environment toolkit: chapter 2 identifying environmental wastes. https://www.epa.gov/lean/lean-environment-toolkit-chapter-2.

FAO (2017). The state of food security and nutrition in the world. http://www.fao.org/3/a-i7695e.pdf.

Four Peaks Technologies (2012). Oceans overview. http://www.climatewarmingcentral.com/ocean_page.html.

GEP Project (2019). World energy supply. http://www.theglobaleducationproject.org/earth/energy-supply.php.

GSA (2019). Sustainable design. https://www.gsa.gov/real-estate/design-construction/design-excellence/sustainability/sustainable-design.

Haanaes, K. (2016). Why all businesses should embrace sustainability. https://www.imd.org/research-knowledge/articles/why-all-businesses-should-embrace-sustainability.

Ijomah, W.L., McMahon, C.A., Hammond, G.P., and Newman, S.T. (2007). Development of design for remanufacturing guidelines to support sustainable manufacturing. *Robotics and Computer-Integrated Manufacturing* 23: 712–719.

Islam, R. (2015). Globalization of production, work and human development: is a race to the bottom inevitable? http://hdr.undp.org/sites/default/files/islam_hdr_2015_final.pdf.

Jayal, A.D., Badurdeen, F., Dillon, O.W., and Jawahir, I.S. (2010). Sustainable manufacturing: modeling and optimization challenges at the product, process and system levels. *CIRP Journal of Manufacturing Science and Technology* 2: 144–152.

Kaebernick, H., and Kara, S. (2006). Environmentally sustainable manufacturing: a survey on industry practice. www.mech.kuleuven.be/lce2006/key5.pdf.

Kibira, D., Jain, S., and Mclean, C.R. (2009). A system dynamics modeling framework for sustainable manufacturing. http://www.systemdynamics.org/conferences/2009/proceed/papers/P1285.pdf.

Lee, S.G., Lye, S.W., and Khoo, M.K. (2001). A multi-objective methodology for evaluating product end-of-life options and disassembly. *International Journal of Advanced Manufacturing Technology* 18: 148–156.

Lerouge, C., and McDonald V. (2008). Manufacturing in 2020: new study reveals future vision of the global manufacturing industry. www.hu.capgemini.com/m/hu/tl/Manufacturing_in_2020.pdf.

Malain, L., and Walrond, W. (2010). The path to sustainability. http://www.risiinfo.com/technologyarchives/environment/The-path-to-sustainability.html.

McLamb, E. (2011). The ecological impact of the industrial revolution. https://www.ecology.com/2011/09/18/ecological-impact-industrial-revolution.

Mehrabi, M.G., Ulsoy, A.G., and Koren, Y. (2000). Reconfigurable manufacturing systems: key to future manufacturing. *Journal of Intelligent Manufacturing* 11: 403–419.

Muralikrishna, I.V. and Manicham, V. (2017). *Learn more about environmental pollution, Environmental Management*, ISBN 978-0-12-811989-1. Elsevier Inc.

National Geographic (2019). Causes and effects of climate change. https://www.nationalgeographic.com/environment/global-warming/global-warming-causes.

Neumann, B., Vafeidis, A.T., Zimmermann, J., and Nicholls, R.J. (2015). Future coastal population growth and exposure to sea-level rise and coastal flooding – a global assessment. *PLoS One* 10 (3): e0118571. https://doi.org/10.1371/journal.pone.0118571.

Opec (2017). World oil outlook 2040. https://www.opec.org/opec_web/flipbook/WOO2017/WOO2017/assets/common/downloads/WOO%202017.pdf.

Ozel, T., and Yildiz, S. (2009). A framework for establishing energy efficiency and ecological footprint metrics for sustainable manufacturing of products. *ASME International Manufacturing Science and Engineering Conference*, MSEC2009, West Lafayette, Indiana, USA (4–7 October 2009), MSEC2009–84365. ASME.

Planchard, M. (2019). SolidWorks Sustainability. http://blogs.solidworks.com/teacher/wp-content/uploads/sites/3/solidworkssustainabilitywpi.pdf.

Purser, J. (2019). Joint operating environment towards 2035. http://openvce.net/resources/expo/joint-futures/Joint_Futures_JOE_Current%20_Briefing.ppt.

Rachuri, S. (2009). NIST workshop on sustainable manufacturing: metrics, standards, and infrastructure. http://www.nist.gov/el/msid/sustainable_workshop.cfm.

Ramani, K., Ramanujan, D., Zhao, F., et al. (2010). Integrated sustainable lifecycle design: a review. https://netfiles.uiuc.edu/hmkim/www/pdf/SustainabilityReview.pdf.

Reich-Weiser, C., Vijayaraghavan, A., and Dornfeld, D.A. (2008). Metrics for sustainable manufacturing. *Proceedings of the International Manufacturing Science and Engineering Conference*, Paper No. MSEC2008, USA (7–10 October).

Ritchie, H (2017). How long before we run out of fossil fuels? https://ourworldindata.org/how-long-before-we-run-out-of-fossil-fuels.

Ritchie, H., and Roser, M. (2019). Energy production and changing energy sources. https://ourworldindata.org/energy-production-and-changing-energy-sources.

Ruz, C. (2011). The six natural resources most drained by our 7 billion people. https://www.theguardian.com/environment/blog/2011/oct/31/six-natural-resources-population.

Sarkis, J. (2001). Manufacturing's role in corporate environmental sustainability: concerns for the new millennium. *International Journal of Operations and Production Management* 21 (5/6): 666–686.

Seidel, R., Shahbazpour, M., and Oudshoorn, M. (2006). Implementation of sustainable manufacturing practices in SMEs – case study of a New Zealand furniture manufacturer. www.mech.kuleuven.be/lce2006/165.pdf.

Styles, G. (2017). Three ways globalization is influencing manufacturing. https://www.manufacturing.net/article/2017/03/three-ways-globalization-influencing-manufacturing.

The US Department of Commerce (2010). Sustainable manufacturing initiative (SMI) and public–private dialogue. https://trade.gov/competitiveness/sustainablemanufacturing/docs/2010_Next_Steps.pdf.

The US Department of Commerce (2011). Getting started: understand your impacts and set priorities. https://www.eea.europa.eu/publications/GH-07-97-595-EN-C/Issue-report-No-6.pdf.

Thuku, G.K., Paul, G., and Almadi, O. (2013). The impact of population change on economic growth in Kenya. *International Journal of Economics and Management Sciences* 2 (6): 43–60.

UCSUSA (2016). The planet's temperature is rising. https://www.ucsusa.org/global-warming/science-and-impacts/science/temperature-is-rising.

Westkamper, E., Alting, L., and Arndt, G. (2007). Life cycle management and assessment: approaches and visions towards sustainable manufacturing. *Proceedings of the Institution of Mechanical Engineers, Part B: Journal of Engineering Manufacture* 215 (B): 599–625.

Wikipedia (2019). Sustainability. https://en.wikipedia.org/wiki/Sustainability.

Index

Computer Aided Design and Manufacturing, First Edition. Zhuming Bi and Xiaoqin Wang.
© 2020 John Wiley & Sons Ltd. This Work is a co-publication between John Wiley & Sons Ltd and ASME Press.
Companion website: www.wiley.com/go/bi/computer-aided-design

d

e

f

i

j

k

l

q

r